Interventional Radiology Essentials

Interventional Radiology Essentials

Editor

Jeanne M. LaBerge, M.D.
Professor of Radiology
University of California, San Francisco
San Francisco, California

Principal Authors

Roy L. Gordon, M.D.
Professor of Radiology
Chief, Interventional Radiology
University of California, San Francisco
San Francisco, California

Jeanne M. LaBerge, M.D.
Professor of Radiology
University of California, San Francisco
San Francisco, California

Robert K. Kerlan, Jr., M.D.
Clinical Professor of Radiology
University of California, San Francisco
San Francisco, California

Mark W. Wilson, M.D.
Assistant Professor of Radiology
University of California, San Francisco
San Francisco, California

LIPPINCOTT WILLIAMS & WILKINS
A **Wolters Kluwer** Company
Philadelphia • Baltimore • New York • London
Buenos Aires • Hong Kong • Sydney • Tokyo

Acquisitions Editor: James Ryan
Developmental Editor: Carol Field
Production Editor: Robert Pancotti
Manufacturing Manager: Tim Reynolds
Cover Designer: Jeane Norton
Printer: Maple Press

530 Walnut Street
Philadelphia, PA 19106 USA
LWW.com

Printed in the USA

Library of Congress Cataloging-in-Publication Data

Interventional radiology essentials / editor, Jeanne M. LaBerge
p.; cm.
Includes bibliographical references and index.
ISBN 0-7817-2010-9
1. Interventional radiology. 2. Blood-vessels—Interventional radiology. 3. Angiography. I. LaBerge, Jeanne M.
[DNLM: 1. Radiography, Interventional—methods. 2. Digestive System Diseases—radiography. 3. Vascular Diseases—radiography. WN 160 I6137 2000]
RD598.67.I58 2000
617′.05—dc21

00-030961

Care has been taken to confirm the accuracy of the information presented and to describe generally accepted practices. However, the authors, editor, and publisher are not responsible for errors or omissions or for any consequences from application of the information in this book and make no warranty, expressed or implied, with respect to the currency, completeness, or accuracy of the contents of the publication. Application of this information in a particular situation remains the professional responsibility of the practitioner.

The authors, editor, and publisher have exerted every effort to ensure that drug selection and dosage set forth in this text are in accordance with current recommendations and practice at the time of publication. However, in view of ongoing research, changes in government regulations, and the constant flow of information relating to drug therapy and drug reactions, the reader is urged to check the package insert for each drug for any change in indications and dosage and for added warnings and precautions. This is particularly important when the recommended agent is a new or infrequently employed drug.

Some drugs and medical devices presented in this publication have Food and Drug Administration (FDA) clearance for limited use in restricted research settings. It is the responsibility of the health care provider to ascertain the FDA status of each drug or device planned for use in their clinical practice.

10 9 8 7 6 5 4 3 2 1

To our residents and fellows, past and present, who have taught us what they needed to know. They are a remarkable group of physicians who have enriched our lives and our profession.

Contents

Contributing Authors

Timothy A. Chuter, M.D.
Assistant Professor of Surgery, University of California, San Francisco, 505 Parnassus Avenue, Room M-488, San Francisco, California 94143-0222

Roy L. Gordon, M.D.
Professor of Radiology and Chief, Interventional Radiology, University of California, San Francisco, 505 Parnassus Avenue, San Francisco, California 94143-0628

Robert K. Kerlan, Jr., M.D.
Clinical Professor of Radiology, University of California, San Francisco, 505 Parnassus Avenue, San Francisco, California 94143-0628

Jeanne M. LaBerge, M.D.
Professor of Radiology, University of California, San Francisco, 505 Parnassus Avenue, San Francisco, California 94143-0628

Shelley R. Marder, M.D.
Assistant Professor of Radiology, University of California, San Francisco, 505 Parnassus Avenue, San Francisco, California 94143-0628

Rajiv Sawhney, M.D.
Assistant Professor of Radiology, San Francisco Veterans Administration Hospital, 4150 Clement Street, San Francisco, California 94121

Susan D. Wall, M.D.
Professor of Radiology and Chief, Interventional Radiology, Veterans Affairs Medical Center, San Francisco, 4150 Clement Street, San Francisco, California 94121

Mark W. Wilson, M.D.
Assistant Professor of Radiology, University of California, San Francisco, 505 Parnassus Avenue, San Francisco, California 94143-0628

Preface

"For those who like this kind of a book, this is the kind of a book they will like."
—*Abraham Lincoln* (1)

This is a review book. It is intended for those who are looking for a concise, illustrated review of interventional radiology.

In 1994, the American Board of Medical Specialties recognized interventional radiology as a unique subspecialty of radiology with an established body of knowledge pertinent to the field. Through this action, the board recognized both the important role that IR plays in modern medical practice and the diverse range of skills and knowledge that the specialty entails.

In this book, we have attempted to consolidate information about this new subspecialty into a single volume covering the three main areas of IR—vascular diagnosis, vascular intervention, and nonvascular interventions. We have tried to make the information clear and accessible through the liberal use of tables and outlines that function as guideposts for the reader.

The images that illustrate this book have been adapted from the University of California at San Francisco (UCSF) IR teaching file. I am indebted to the many UCSF residents and fellows who over the years have helped us to accumulate cases. In addition, I am grateful to Dr. Ernest J. Ring for opening up his extensive personal teaching file to this project.

As the editor of this project, I thank my colleagues in the Interventional section at UCSF for their time and effort in writing this book. The format of our book is different from that of an ordinary textbook and thus its production has placed extra demands on its authors. As well, I appreciate the support of Jim Ryan at Lippincott Williams & Wilkins, who gave us the freedom to design a book that would uniquely address the needs of our readers.

In conclusion, this is a review book and not a definitive textbook. For those who are looking for a book like this, here it is.

1. Lincoln Reviews a Book. In: Woods RL, ed. *Treasury of the Familiar.* New York: Macmillan Press, 1980: 180.

VASCULAR DIAGNOSIS

CHAPTER 1 PREVIEW

Case 1

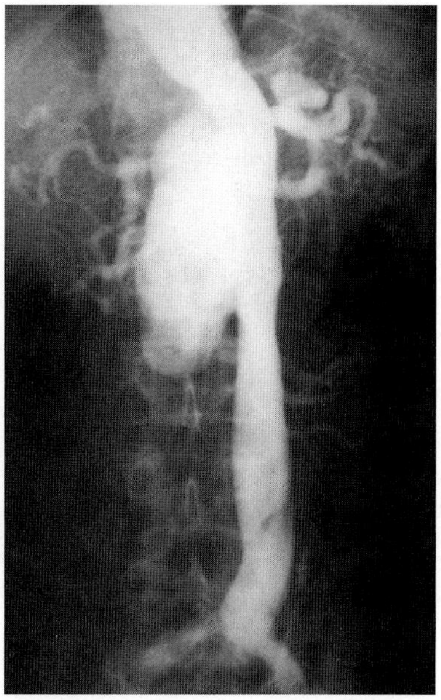

History: A 65-year-old man presented with fever and back pain. Abdominal aortography was performed.
Question: What is your diagnosis?

Case 2

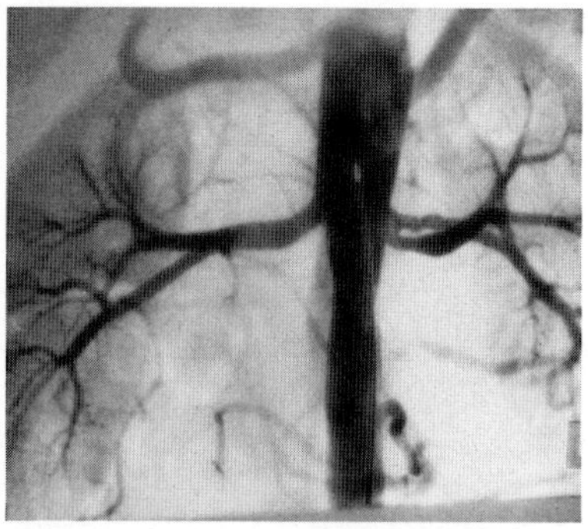

History: A 30-year-old woman presented with hypertension. Abdominal aortography was performed.
Question: What is your diagnosis?

Case 3

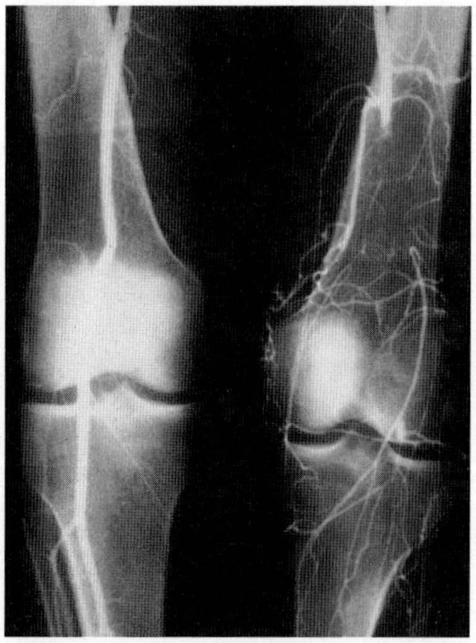

History: A 30-year-old man complained of left leg pain. A runoff arteriogram was performed.
Question: What is your diagnosis?

Case 4

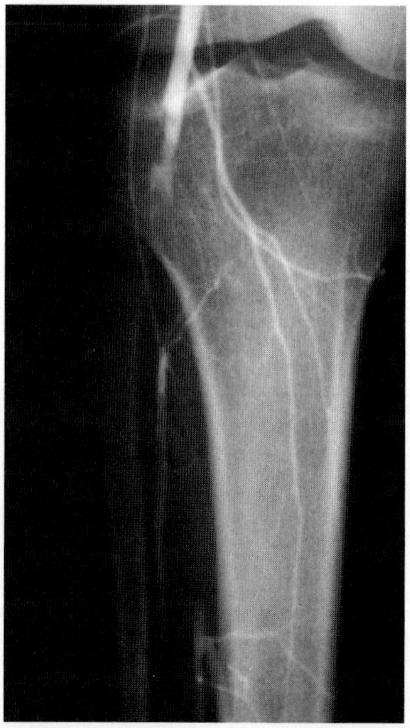

History: A 45-year-old man presented to the ER with an acutely cold right foot. A runoff arteriogram was performed.
Question: What is your diagnosis?

Chapter 1
Vascular Disease Classification

Jeanne M. LaBerge, MD

OUTLINE OF VASCULAR PATHOLOGY

- Atherosclerosis
  - occlusive
  - aneurysmal
- Intimal hyperplasia (restenosis)
- Inflammatory disease
  - infection (mycotic aneurysm)
  - radiation
  - idiopathic (polyarteritis, Takayasu's, Buerger's, etc.)
- Fibrodysplasia
- Abnormal arteriovenous communications
  - arteriovenous and venous malformations
- Neoplasia
  - benign (hemangioma)
  - malignant (sarcoma, pericytoma)
- Traumatic injury
  - accidental, penetrating or blunt trauma
  - iatrogenic
- Embolic disease
  - thromboembolism
  - atheroembolism
- Physiologic disturbances
  - extrinsic compression syndromes (thoracic outlet and popliteal entrapment)
  - pharmacologic abnormalities

INTRODUCTION

This book is designed to provide the reader with a concise, illustrated review of interventional radiology. Since much of interventional radiology in current practice involves the diagnosis and therapy of vascular disease, it is appropriate to begin this book with a review of vascular disease and its angiographic manifestations.

The main emphasis of this chapter is on diseases of the blood vessels themselves. In times past, angiography was commonly used to assess the vascular features of neoplastic masses in the liver, pancreas, kidney and brain. However, more recently computed tomography and magnetic resonance imaging combined with percutaneous biopsy techniques have obviated the need for this type of angiography in most cases. Today, diagnostic angiography is used primarily to evaluate diseases of blood vessels, and in rare circumstances to differentiate the etiology of a mass lesion.

The topics discussed in this chapter are listed in the outline above. A brief discussion of the essential anatomic, pathophysiologic and angiographic features of each entity is presented.

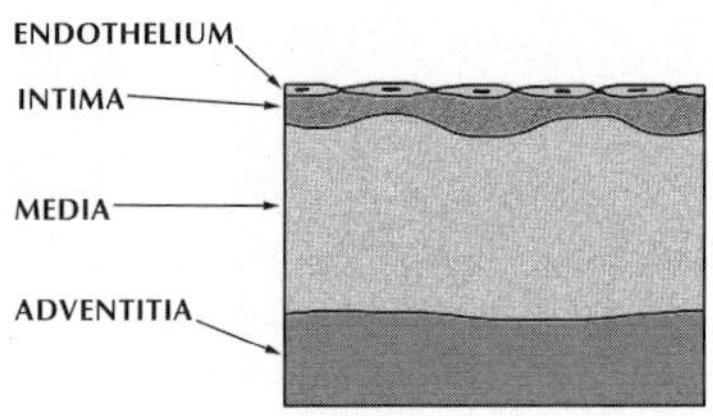

Figure 1. *Cross-sectional anatomy of a blood vessel wall.*

VASCULAR DISEASE

- Primary
  - atherosclerosis
  - intimal hyperplasia
  - inflammatory arteritis
  - fibrodysplasia
  - vascular neoplasm
- Secondary
  - extrinsic compression
  - drug-induced vasospasm
  - emboli
  - trauma

THE NORMAL BLOOD VESSEL

Anatomy: Blood vessels are composed of three tissue layers: intima, media and adventitia **(Fig. 1)**. The intima is composed of a single cell layer of endothelium resting on a layer of connective tissue. While only a single cell layer thick, the intima is vital to the normal function of the vessel. Its endothelial cells secrete a variety of factors in response to injury such as prostacyclin, tissue plasminogen activator and heparin sulfate. The media is composed of smooth muscle cells mixed with collagen on a connective tissue substrate. The amount of smooth muscle in the media is determined by the vessel function; muscular arteries (eg. coronary and visceral) have more smooth muscle than elastic arteries (eg. aorta), which in turn have more than arterioles. The adventitia is a poorly defined layer of investing connective tissue through which run nerve fibers and small nutrient vessels known as the vaso vasorum.

Pathophysiology: Vascular disease may be categorized as primary or secondary. Primary vascular diseases result from a pathologic process that develops within the wall of the vessel itself. Atherosclerosis, intimal hyperplasia, inflammatory arteritis and vascular neoplasia are all primary vascular disorders. Secondary vascular diseases result from processes extrinsic to the vessel wall such as extrinsic muscle compression or acute embolic occlusion. This distinction is important because, with secondary disease, vessels remote from the affected site may appear normal; whereas, with primary vascular disease, vessels remote from the affected site are frequently abnormal.

Angiography: Important descriptive features concerning the appearance of a blood vessel on angiography include: the luminal contour (smooth or irregular), the luminal diameter (aneurysmal or stenotic), the pattern of tapering of the vessel along its course, the branching pattern and appearance of enlarged branching vessels (collaterals) and the presence or absence of calcification. A normal blood vessel is smooth-walled without calcification and demonstrates smooth tapering as it progresses distally to give off branches of equal or smaller caliber.

ATHEROSCLEROSIS

- Degenerative disease of large and medium-sized vessels
- The hallmark is a fibrofatty plaque composed of a lipid core covered by a fibrous cap
- Consequences are stenoses, aneurysms and distal emboli

ATHEROSCLEROSIS

Anatomy: Atherosclerosis is a degenerative disease of large and medium-sized blood vessels. The hallmark of atherosclerosis is the atheroma or fibrofatty plaque that develops within the intima. A simple plaque consists of a core of lipid covered by a fibrous cap. Over time, plaques may become more complex as they develop calcification, surface erosion or ulceration, internal hemorrhage or surface thrombus.

Pathophysiology: Initially, a small simple plaque does not produce adverse consequences. However, as it grows and becomes more complex, atherosclerotic plaque may cause three different problems: stenosis due to encroachment of the plaque on the vessel lumen, aneurysm formation due to weakening of the vessel wall, and atheroemboli due to the formation and discharge of thrombus or platelet aggregates on the surface of the plaque.

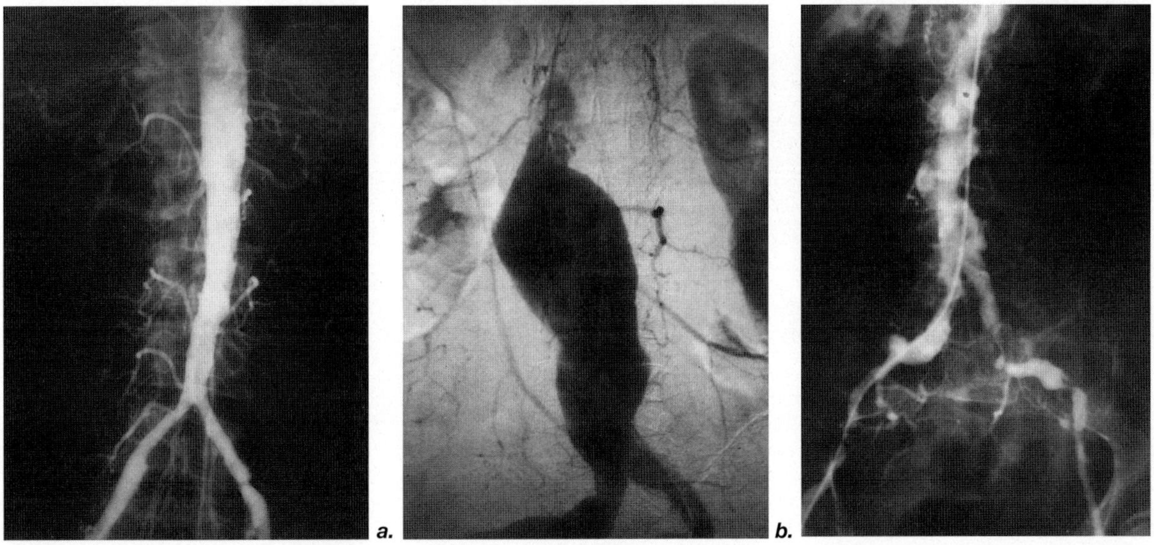

a. b. c.

Figure 2. Three manifestations of atherosclerosis. a) Occlusive disease with narrowing of the aortoiliac vessels. b) Aneurysmal disease with dilatation of the abdominal aorta, and c) embolic disease—in this patient, the aorta was the source of small atheromatous emboli to the distal digits.

Angiography: Atherosclerosis becomes apparent angiographically in the form of aneurysms and stenoses **(Fig. 2)**. The angiographic sign of an aneurysm is increased luminal diameter compared to adjacent vessels or compared to an absolute standard. There is also loss of the normal vascular tapering. Descriptive features should include the luminal diameter and location. A limitation of angiography in the depiction of aneurysms is that the lumen of the aneurysm may be partially filled in with thrombus making the aneurysm itself hard to detect.

The angiographic sign of stenosis is focal narrowing of the lumen of a vessel. Descriptive features should include the length, severity, morphology of the stenosis (eccentric vs. concentric, calcified or not, ulcerated or not), and the extent and appearance of collateral vessels.

Other Related Issues: The more general term arteriosclerosis includes three disorders: atherosclerosis, Monckeberg's medial calcification and arteriolosclerosis. Of these, atherosclerosis is the dominant disease of greatest relevance to angiographers. Monckeberg's is characterized by calcification in the media. It is of no clinical significance but may be seen in association with atherosclerosis. Arteriolosclerosis causes proliferative hyaline thickening in the wall of small vessels leading to diffuse small vessel stenoses. Arteriolosclerotic changes are seen most prominently in the kidney.

INTIMAL HYPERPLASIA

- The healing process that occurs following vascular trauma: PTA, stenting, surgery
- Vascular injury stimulates platelet and macrophage migration which stimulate SMC migration and proliferation, resulting in endothelialization
- Normally stabilizes in 8-12 weeks

INTIMAL HYPERPLASIA

Anatomy: "Neointimal hyperplasia" is a term commonly used to describe the healing process that occurs after vascular injury. The histologic sequence of events involved in neointimal hyperplasia is as follows. Vascular trauma, usually from percutaneous transluminal angioplasty or surgery, results in denudation of the vascular endothelium. Platelets and macrophages migrate to the site of injury and release growth factors which stimulate medial smooth muscle cell (SMC) migration **(Fig. 3)**. SMC proliferation and differentiation occurs and a new endothelial lining is created. SMC proliferation also results in the elaboration of connective tissue elements, collagen and fibrin, which contribute to the thickness of the neointima. Generally, after a contiguous endothelial layer is reestablished (8 to 12 weeks), SMC proliferation ceases and, thereafter, the neointima remains constant or regresses.

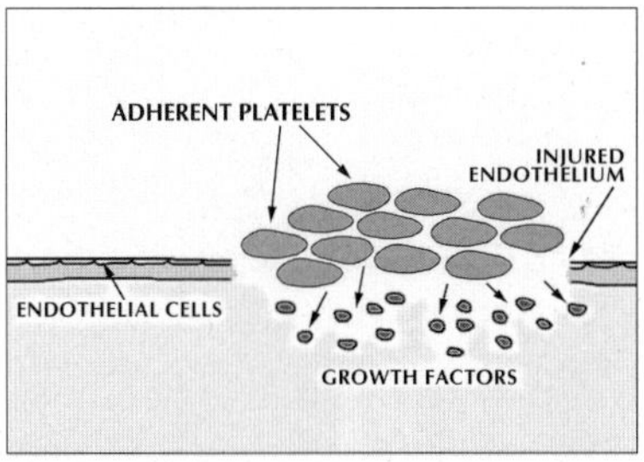

Figure 3. The process of intimal hyperplasia. Platelet deposition results in the elaboration of growth factors that stimulate smooth muscle cell migration.

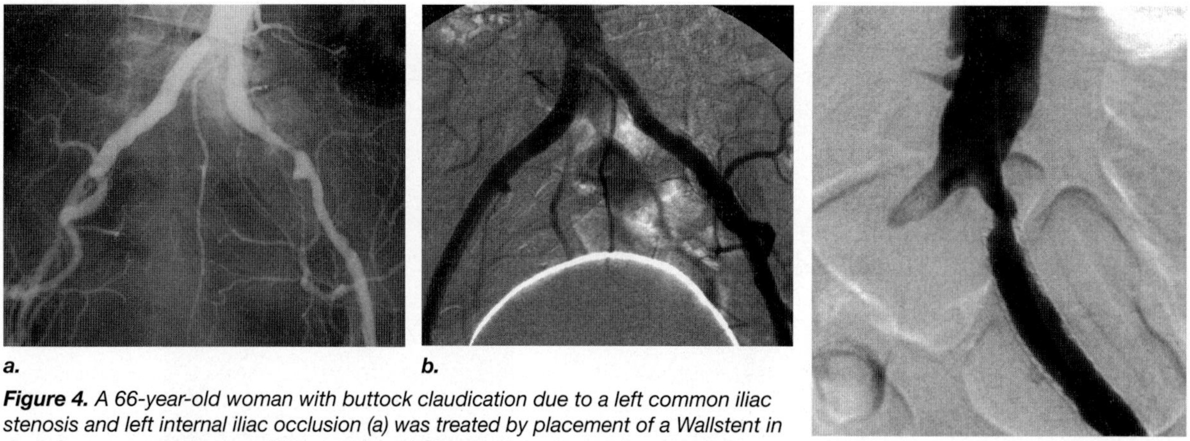

a. b. c.

Figure 4. A 66-year-old woman with buttock claudication due to a left common iliac stenosis and left internal iliac occlusion (a) was treated by placement of a Wallstent in the left common and external iliac arteries. b) She developed recurrent symptoms one year later. Repeat angiogram (c) showed minimal intimal hyperplasia within the stent and a stenosis in the unstented common iliac artery that could be due to progressive atherosclerosis or intimal hyperplasia. The lesion responded well to restenting.

Pathophysiology: In some patients, the healing response to vascular injury is exuberant and excessive **(Fig. 4)**. Rather than stabilizing at 8 to 12 weeks, the process of intimal hyperplasia continues unabated, resulting in "restenosis." Factors that may influence the development of "restenosis" include tissue growth factors, hemodynamic characteristics of the vessel and the extent of vessel injury.

Angiography: "Restenosis" is an angiographic term used to indicate recurrence of a stenosis at the site of percutaneous or surgical intervention. A stenosis with either a >50% luminal diameter or a loss of more than 50% of the gain in lumen achieved by PTA is required to meet the definition of restenosis.

INFLAMMATORY VASCULITIS

- Infiltration of the media by histocytes
- The angiographic consequences are varied: stenosis, thrombosis, vessel rupture, pseudoaneurysm, aneurysm
- Etiologies include infection, radiation, idiopathic

INFLAMMATORY DISEASE—VASCULITIS

Anatomy: The term vasculitis is commonly applied to a diverse group of vascular diseases in which inflammatory infiltration in the wall of the vessel is the predominant histologic abnormality.

Pathophysiology: Invasion of the wall by histiocytes can lead to a variety of pathologic outcomes. Inflammation can result in edema, fibrin precipitation and narrowing of the vessel lumen, thrombosis or vessel rupture with pseudoaneurysm formation **(Table 1)**.

Angiography: The category of inflammatory vascular disease includes a variety of processes with a variety of angiographic appearances. Bacterial infection is the most common cause of infectious vasculitis and it can result in vessel disruption and the formation of a "mycotic" pseudoaneurysm **(Fig. 5)**. Radiation vasculitis is suspected by an appropriate history and focal vascular involvement. Vascular changes generally consist of stenosis, occlusion, or beaded areas of narrowing. These vascular abnormalities may lead to ischemic symptoms 5-10 years after radiation therapy. Non-specific or idiopathic arteritis can have a variety of appearances **(Figs. 6, 7)**. The salient features of several notable vasculitides are listed in **Table 1**.

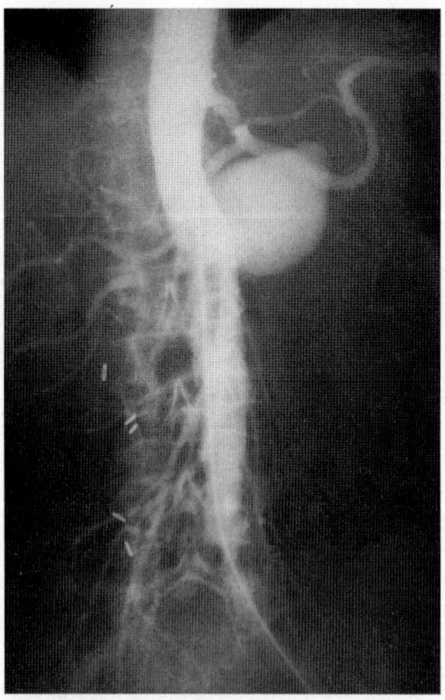

Figure 5. *Mycotic pseudoaneurysm of the abdominal aorta.*

TABLE 1: FEATURES OF VASCULITIS

Syndrome	Organ Affected	Angiographic Findings
Polyarteritis nodosa	kidney, liver, and pancreas	microaneurysms of small and medium-sized vessels
Takayasu's arteritis (Pulseless disease)	aorta and pulmonary artery	stenosis or occlusion of the aorta or proximal arch vessels
Buerger's disease (Thromboangitis obliterans)	upper and lower extremities	segmental occlusion of small to medium-sized vessels
Temporal arteritis (Giant cell arteritis)	subclavian artery, proximal arch vessels, profunda	stenoses or occlusions
Kawasaki's disease (Mucocutaneous lymph node syndrome)	coronary arteries in children, occasionally aorta and viscerals	aneurysms, similar to polyarteritis
Behçet's disease	aorta and major branch vessels	arterial aneurysms venous thrombosis

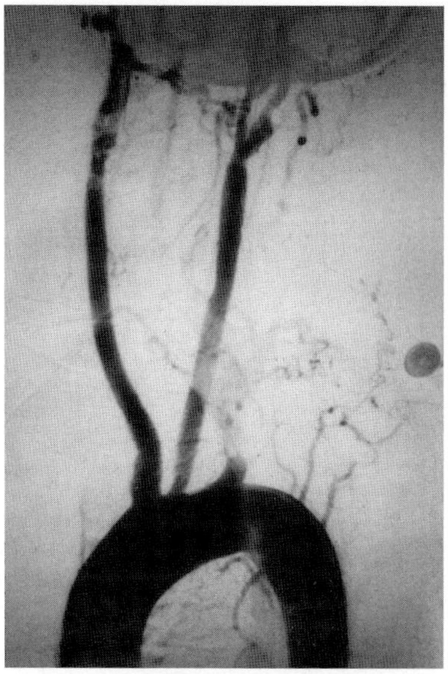

Figure 6. Bilateral subclavian artery occlusions due to Takayasu's arteritis.

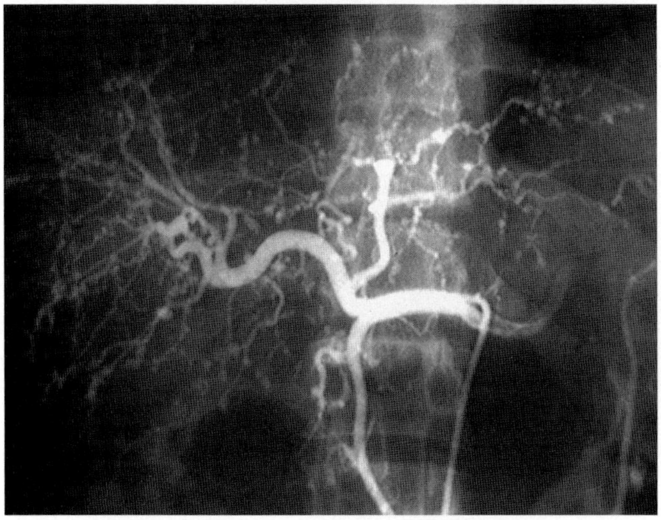

Figure 7. Selective hepatic arteriography in a patient with polyarteritis demonstrates multiple small peripheral saccular aneurysms.

FIBRODYSPLASIA

Anatomy: Arterial fibroplasia is a mixed group of nonatherosclerotic, noninflammatory diseases of the vessel wall. This disease can affect the intima, media or adventitia of the vessel wall. The underlying abnormality is replacement of the normal components of the vessel wall by fibrous connective tissue.

FIBRODYSPLASIA

- Nonatherosclerotic, noninflammatory disease due to replacement of the intima, media or adventitia by fibrous connective tissue
- Medial fibroplasia (85%), perimedial dysplasia (10%), intimal fibroplasia (5%)
- "String of beads", multiple stenoses, focal stenosis

Pathophysiology: Fibrous replacement of the vessel wall leads to stenosis and/or aneurysm formation. The renal arteries are most commonly affected. Other vessels that can be affected include common carotid, SMA, and the external iliacs. The most common forms of fibrodysplasia are intimal fibroplasia (5%), medial fibroplasia (85%), and perimedial dysplasia (10%).

Angiography: A "string of beads" appearance with stenoses alternating with areas of mild dilatation is typical of medial fibroplasia **(Fig. 8)**. Intimal fibroplasia usually results in a focal stenosis of the main renal artery or a web-like stenosis of a segmental vessel **(Fig. 9)**. Multiple stenoses without mural aneurysm is characteristic of perimedial dysplasia.

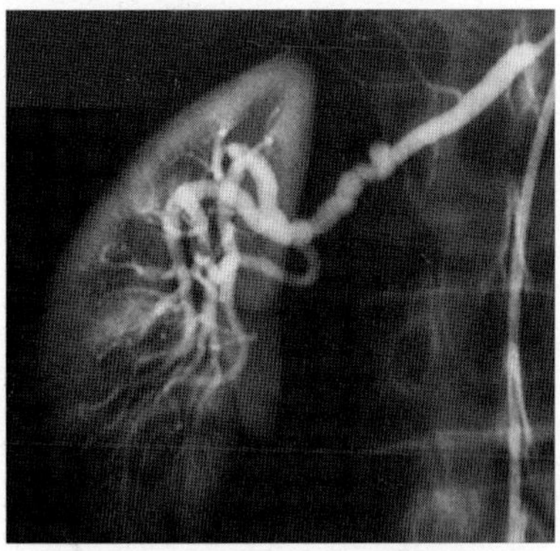

a. *Selective right renal arteriogram.*

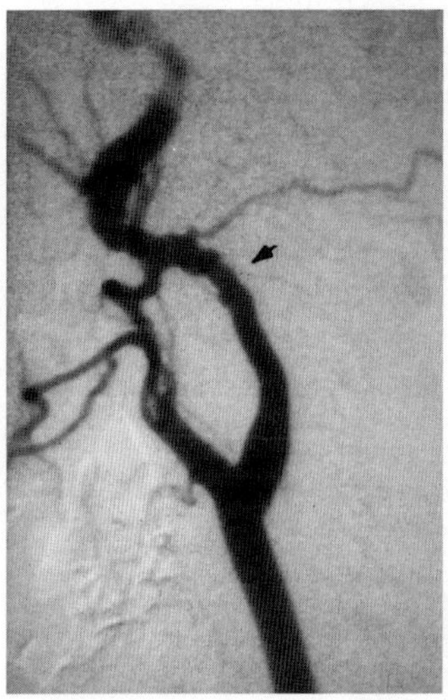

b. *Selective common carotid arteriogram.*

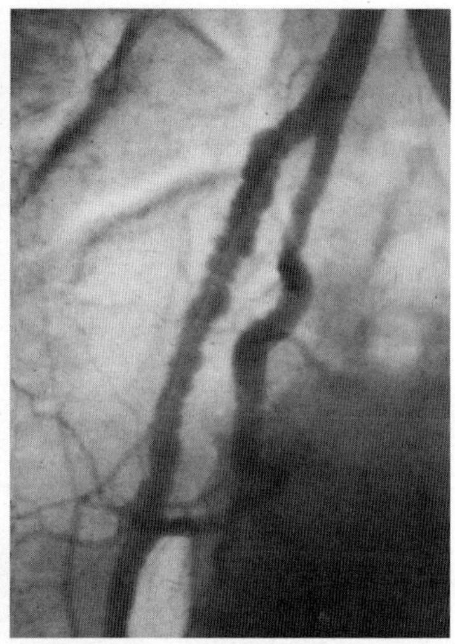

c. *Right iliac arteriogram.*

Figure 8. *Medial fibroplasia. String of beads appearance of the (a) renal, (b) internal carotid and (c) external iliac arteries.*

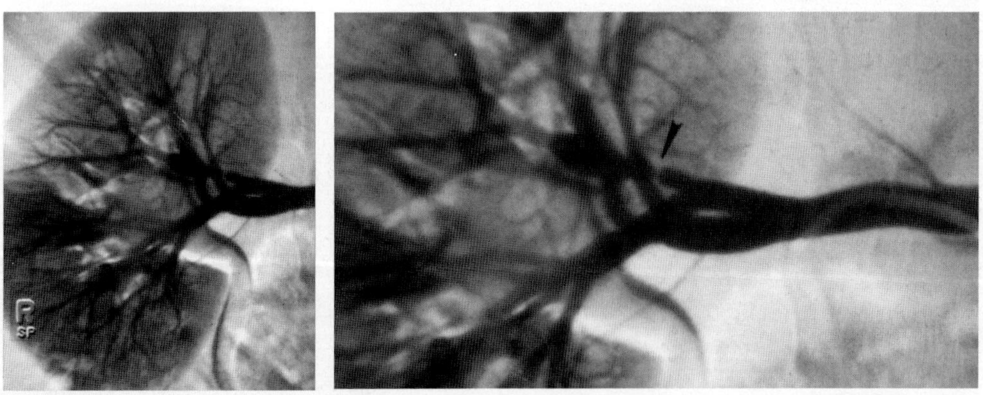

a. ***b.***

Figure 9. *Intimal fibroplasia. Tight web-like stenosis of a segmental renal artery with post-stenotic dilatation. a) Selective right renal arteriogram. b) Magnification view showing web-like stenosis (arrowhead) and post-stenotic dilatation.*

ABNORMAL ARTERIOVENOUS COMMUNICATIONS

Arteriovenous communications can be classified in a variety of ways. For the purpose of brevity, these lesions will be divided into four groups: true congenital arteriovenous malformations (AVM) **(Fig. 10)**, venous malformations, hemangiomas and arteriovenous fistulas (AVF) **(Fig. 11)**. AVF is usually an acquired traumatic lesion and will be mentioned in the trauma section. Hemangioma is a benign vascular tumor and will be discussed in the neoplasia section.

ARTERIOVENOUS MALFORMATION (AVM)

- Congenital aberration of embryonic development at 4-10 weeks
- Persistent communications between arteries and veins at any level (capillaries to large arteries)
- Angiographic findings: multiple hypertrophied feeding vessels, hypervascular nidus, rapid AV shunting

Anatomy: Congenital AVMs result from a focal aberration in embryonic development occurring between the 4th and 10th weeks causing persistence of primitive vascular communications. These abnormal persistent communications can occur at any level from capillaries to large vessels. With the exception of Osler-Weber-Rendu syndrome, most AVMs are sporadic and not inherited.

Pathophysiology: Congenital AVMs usually grow with the patient and remain stable in size over time. In some patients though, growth continues unchecked and symptoms develop during adolescence or adulthood. Symptoms include: cosmetic deformity, limb length discrepancy, high output cardiac failure and superficial complications such as bleeding and ulceration.

Angiography: Three findings are characteristic: 1) multiple hypertophied, tortuous feeding vessels, 2) a hypervascular nidus with numerous small central vessels, and 3) rapid arteriovenous shunting with early opacification of large draining veins.

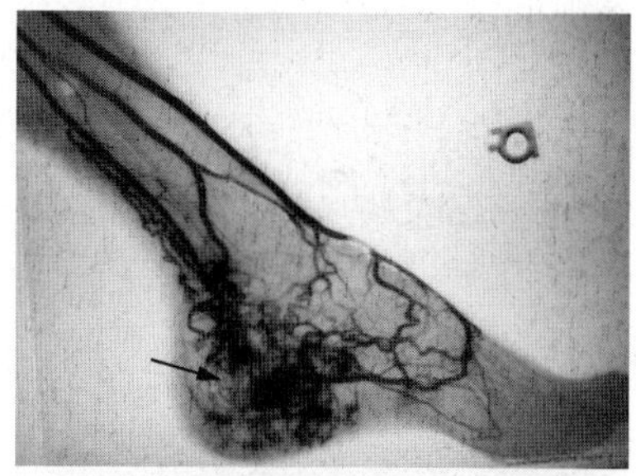

Figure 10. Peripheral AVM of the foot. Note large dilated feeding arteries and hypervascular nidus (arrow).

Venous Malformations

Anatomy: This developmental anomaly is limited to the venous side of vascular development. Aberrations in development lead to cavernous malformations or abnormal patterns of venous return.

Pathophysiology: Venous malformations may result in cosmetic deformity and limb length discrepancy. In the Klippel-Trenaunay syndrome, patients present with a triad of limb hypertrophy, cutaneous birthmark and extensive varicosities. The underlying lesion is hypoplasia of the deep venous system with the resultant development of superficial venous varicosities. When an arteriovenous malformation is present, this syndrome may be termed "Parkes-Weber" or Klippel-Trenaunay-Weber syndrome.

Angiography: The arterial phase is normal without enlarged feeding vessels or arteriovenous shunting. Late phase films show puddling of contrast in venous spaces.

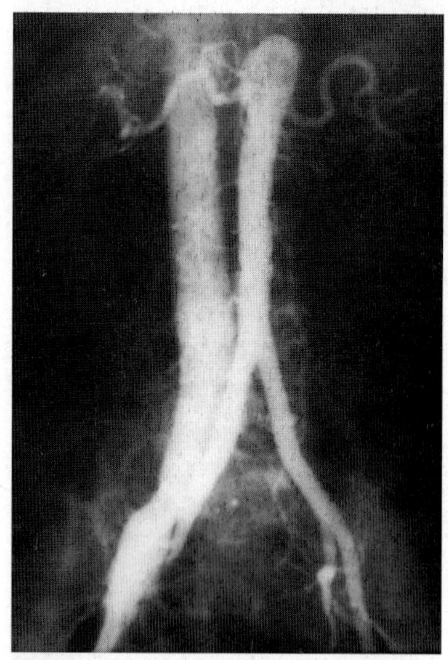

Figure 11. Iatrogenic AV fistula. This communication between the right iliac artery and vein was the result of back surgery.

NEOPLASIA

- Benign: Hemangioma
  - Skin, liver, spleen, pancreas
  - Large feeding vessels and densely staining mass
- Malignant: Hemangiosarcoma, pericytoma
  - Arsenic, thorotrast, or PVC induced
  - Neovascularity, mass effect

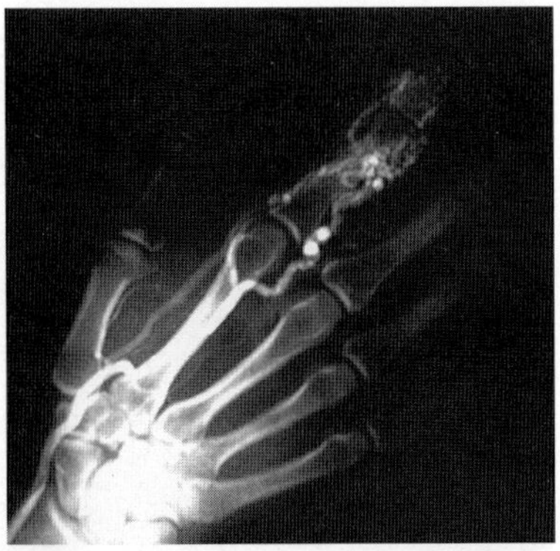

Figure 12. Hemangioma of the index finger.

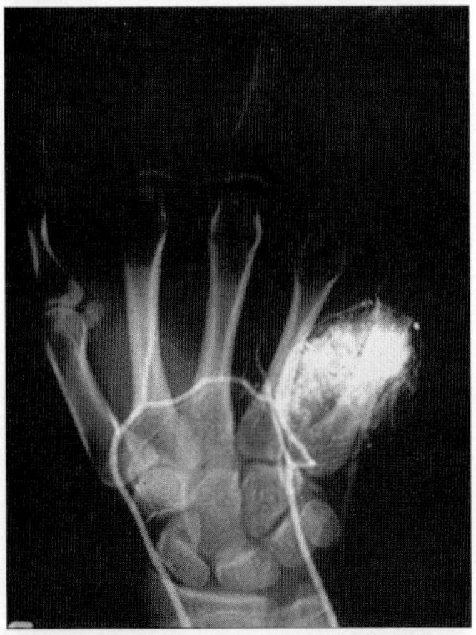

Figure 13. Dense capillary hemangioma of the palm.

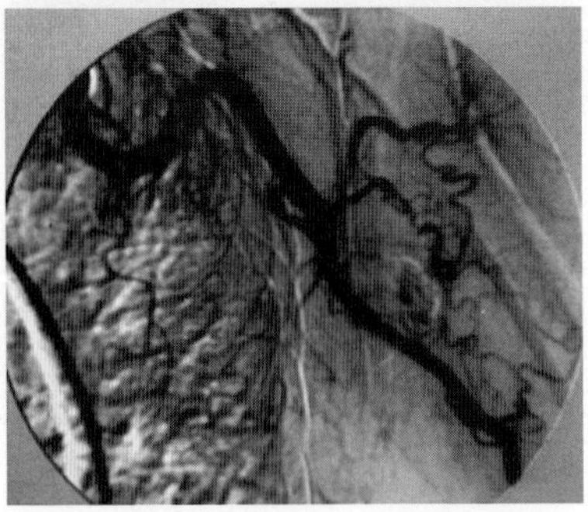

Figure 14. Hemangiosarcoma of the axilla.

NEOPLASIA

Primary Blood Vessel Neoplasms: Benign Hemangioma **(Figs. 12, 13)**

Anatomy: Hemangioma is a benign neoplasm or hamartoma that consists chiefly of well-formed vascular channels. Two basic forms, capillary and cavernous, are distinguished by the size of the vascular channels. Capillary hemangiomas ("strawberry hemangiomas") are common in children. While some are present at birth, most appear in the first year. Ninety-five percent regress spontaneously by age 7. Lesions are most commonly cutaneous but may occur in the liver, spleen, pancreas and brain.

Pathophysiology: Hemangiomas are usually asymptomatic. Large cutaneous lesions may bleed. When liver and cutaneous lesions are present in combination, high output congestive heart failure may ensue. Neurologic symptoms can result from cerebral involvement.

Angiography: Large feeding vessels lead to a densely staining mass.

Malignant Hemangiosarcoma or Hemangiopericytoma **(Fig. 14)**

These are rare malignant tumors that are associated with arsenic, thorotrast and polyvinyl chloride (PVC) exposure. Angiography usually demonstrates hypertrophy of the feeding vessel, mass effect and neovascularity.

Vascular Features of Neoplasia in General

Anatomy and pathophysiology: Neoplasms have a wide spectrum of vascularity based upon the location and type of tumor. In general, benign neoplasms have a more orderly vascular organization than malignant neoplasms. But, malignant tumors may also be hypovascular. Often, tumor vessels are not lined with endothelium and will also have bizarre morphology with areas of dilatation and abrupt termination.

Angiography: Cross-sectional imaging together with fine needle aspiration biopsy have replaced diagnostic angiography as a means of detecting and identifying tumors. However, the vascular patterns of common tumors should still be familiar to modern angiographers.

The following page provides a quick reference for angiographic patterns of common neoplasms **(Table 2)**.

TABLE 2: ANGIOGRAPHIC PATTERN OF COMMON NEOPLASMS

Location	Tumor	Angiographic Pattern
Liver	Hepatoma	Hypervascular, bizarre neovascularity, arterioportal shunting, portal vein occlusion.
	Focal nodular hyperplasia	Hypervascular, central scar, spoke-wheel appearance, not bizarre like hepatoma.
	Adenoma	Hypervascular, diffuse inhomogeneous staining, not as bizarre as hepatoma.
	Cavernous hemangioma	Focal (2 mm-5 mm) hyperdense, sharply marginated punctate staining appearing in mid-capillary phase and persisting through late venous phase. Angiography is pathognomonic.
	Metastases	Multiple round lesions of different sizes. Hypervascular: carcinoid, islet cell adenocarcinoma, thyroid, renal, leiomyosarcoma, melanoma. Hypovascular: GI adenocarcinoma, pancreatic adenocarcinoma, lung (usually), breast.
Kidney	Renal cell carcinoma	Usually extremely hypervascular with multiple abnormal vascular channels. Frequent arteriovenous shunting. Papillary subtype is hypovascular.
	Oncocytoma	Hypervascular, occasionally with spoke-wheel appearance. Cannot be reliably differentiated from renal cell carcinoma.
	Angiomyolipoma	Hypervascular, often with microaneurysms. Cannot be reliably differentiated from renal cell carcinoma angiographically. CT shows fat.
Pancreas	Adenocarcinoma	Hypovascular, encasing both arteries and veins.
	Islet cell tumors	Often hypervascular, may be occult.
	Cystadenocarcinoma	Hypervascular with dense blush.

TRAUMATIC VASCULAR INJURY

- Blunt trauma
  - Shear stress results in lacerations at points of fixation
- Penetrating trauma
  - Local injury in the path of a projectile
- Joint or fracture dislocations

TRAUMATIC INJURY

Anatomy: Vascular injury may be caused by external penetrating trauma (e.g. gunshot, knife), internal penetrating trauma (e.g. long bone fracture dislocation), blunt trauma (e.g. automobile collision) or iatrogenic injury (e.g. arteriography). The mechanism of injury will determine the site and type of vascular injury. Blunt trauma can cause spasm, intimal disruption, dissection, or thrombotic occlusion. Penetrating trauma usually causes disruption of all three vessel wall layers and may result in pseudoaneurysm formation, arteriovenous fistula or thrombotic occlusion.

Pathophysiology: The adverse clinical consequences of vascular trauma are dependent on the nature of the injury. Spasm, dissection and thrombosis may all lead to absent pulses and end-organ ischemia.

Angiography: The spectrum of angiographic findings seen after trauma includes: stenosis, spasm, dissection, acute occlusion, AV fistula, pseudoaneurysm and frank extravasation **(Figs. 15-17)**.

ANGIOGRAPHIC FINDINGS OF VASCULAR TRAUMA

- Occlusion
- Pseudoaneurysm
- Extravasation
- Arteriovenous fistula
- Dissection
- Stenosis

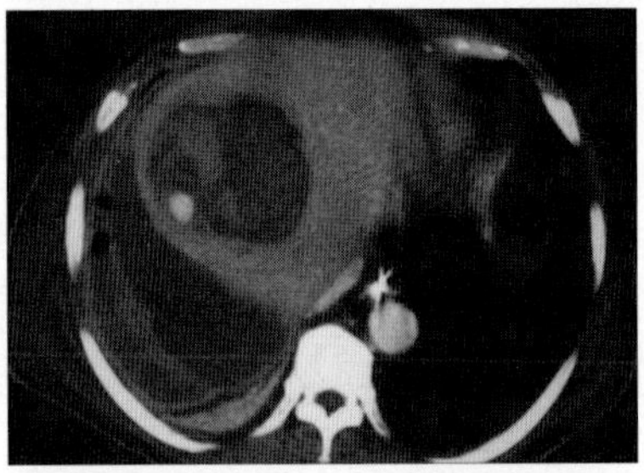

Figure 15a. *Traumatic pseudoaneurysm in the liver. CT demonstrates a large intra-parenchymal hematoma with a densely opacified central pseudoaneurysm.*

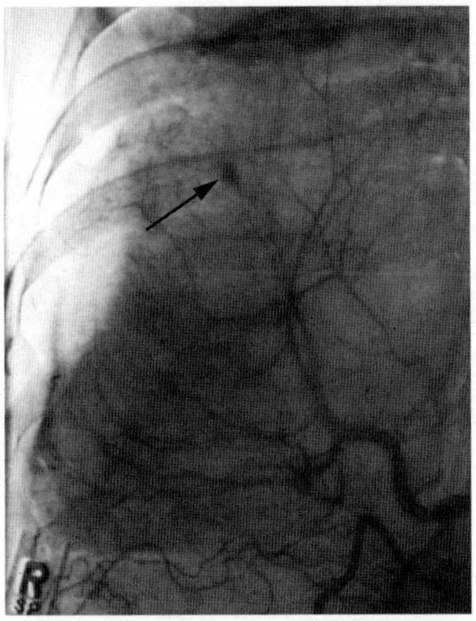

Figure 15b. *Selective common hepatic arteriogram reveals a pseudoaneurysm of a peripheral right hepatic artery branch (arrow).*

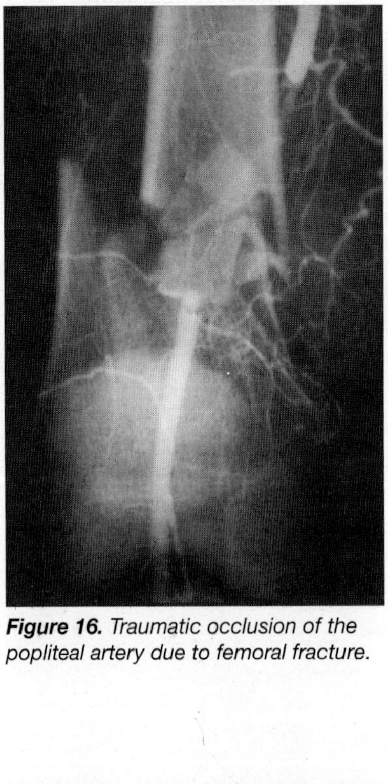

Figure 16. *Traumatic occlusion of the popliteal artery due to femoral fracture.*

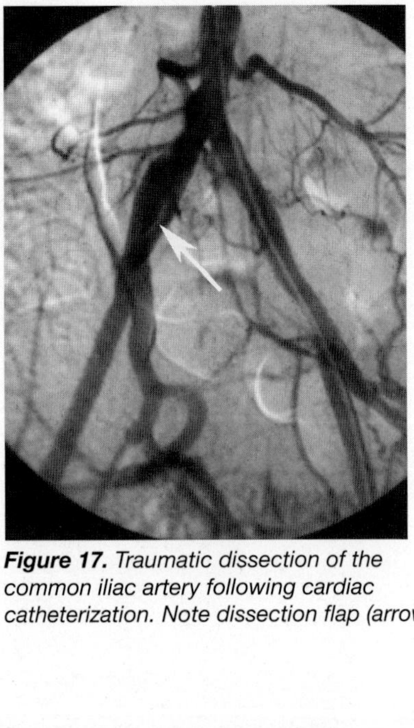

Figure 17. *Traumatic dissection of the common iliac artery following cardiac catheterization. Note dissection flap (arrow).*

EMBOLIC DISEASE

An embolus is by definition a core of material that flows downstream to lodge at a distal location. The embolic material may be thrombus, atheroma or, uncommonly, a foreign body. The majority of clinically apparent arterial emboli are thrombotic. Atheromatous emboli are less common and cause a distinct clinical syndrome that will be discussed separately.

EMBOLIC DISEASE
• Thromboembolism — Source 85% cardiac (afib or MI) 10% aneurysms (aortoiliac, fem pop) — Abrupt occlusion, no collaterals, meniscus • Atheroembolism — Small microemboli that lodge distally — Blue toe syndrome

Thromboembolism

Anatomy: Over 85% of thrombotic arterial emboli originate in the heart in patients suffering from atrial fibrillation or myocardial infarction. In 10%, the thrombotic source is a peripheral aneurysm of the aortoiliac or femoropopliteal vessels. Arterial emboli typically lodge at bifurcation points causing occlusion or stenosis. The common femoral artery bifurcation and the popliteal artery trifurcation are common sites of obstruction **(Fig. 18)**.

Pathophysiology: Peripheral emboli cause acute obstruction. Patients without underlying vascular disease present with pain, pallor and pulselessness. Patients with underlying vascular disease may be less symptomatic because of the presence of well-developed collaterals.

Angiography: The abrupt occlusion of a vessel at its bifurcation is suggestive of embolic disease. A filling defect or meniscus may be present at the level of the obstruction **(Fig. 19)**. Collaterals are typically absent. Smaller emboli may be detected downstream. For example, a large clot may lodge at the femoral bifurcation and a smaller clot may be detected in a tibial vessel.

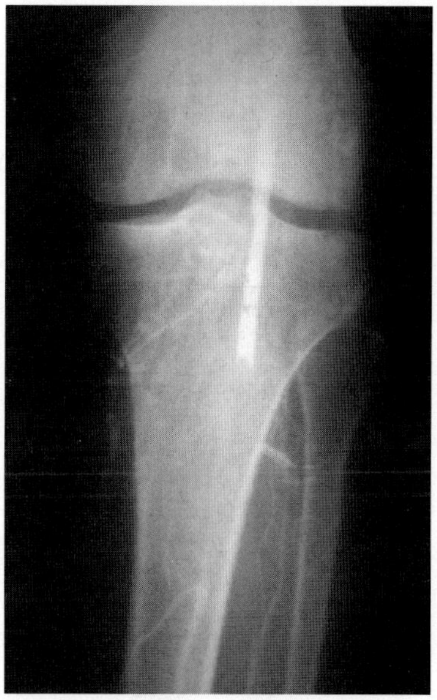

Figure 18. *Acute embolic occlusion of the popliteal artery. Note abrupt occlusion, lack of collaterals and characteristic location at the distal popliteal artery.*

Atheroembolism

Anatomy and pathophysiology: Degenerative fragmentation of a complex atheroma may result in small (micro) emboli that break free, and lodge in very small peripheral vessels. The syndrome caused by such small atheromatous emboli is termed the "blue toe syndrome." By definition, blue toe syndrome is marked by the sudden onset of cool, painful, cyanotic toes in the presence of palpable distal pulses. Distal pulses are intact because the embolic material—platelet aggregates, cholesterol debris or small clots—is so small that it lodges in peripheral arterioles. The embolic source is typically a complex plaque in the infrarenal aortoiliac or femoropopliteal vessels.

Angiography: Angiography is performed to uncover the source of the microemboli. The vasculature distal to the offending atheroma is usually normal and in-line flow to the foot is preserved.

NOTE DIFFERENCE

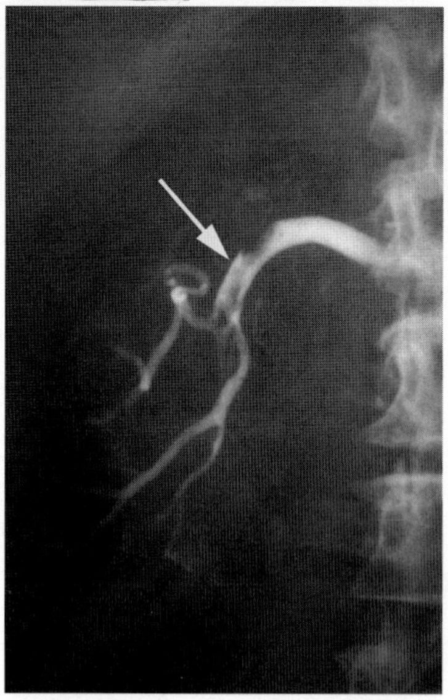

Figure 19. *Embolism to the right renal artery in patient with atrial fibrillation. Note filling defect in the main renal artery (arrow).*

PHYSIOLOGIC DISTURBANCES

- Muscle compression
  - Popliteal entrapment
  - Thoracic outlet syndrome
- Drug-induced vasospasm
  - Ergotamine, cocaine, vasopressin

PHYSIOLOGIC DISTURBANCES

Some vascular disorders are activity dependent with symptoms appearing only under prescribed conditions. External compression from muscle contraction and drug-induced stensosis are examples of such physiologic vascular disorders.

Popliteal Artery Entrapment Syndrome

Anatomy: This syndrome is due to an anomalous course of the popliteal artery that passes medial to and beneath the medial head of the gastrocnemius muscle with consequent compromise of arterial flow.

Pathophysiology: Symptoms are most commonly observed in young athletic men who present with unilateral foot and calf claudication. While the anomaly is frequently bilateral (25%-50%), symptoms are usually unilateral.

Angiography: The pathognonomic finding is medial deviation of the medial portion of the popliteal artery **(Fig. 20)**. The vessel may appear normal in the nonstressed situation; however, medial deviation may become apparent when the vessel is imaged with the foot in active plantar flexion. Longstanding disease may lead to popliteal occlusion.

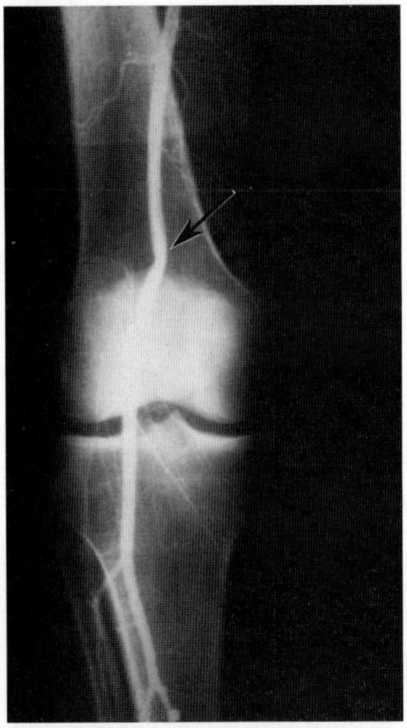

Figure 20. *Medial deviation of the popliteal artery (arrow) due to popliteal artery entrapment syndrome.*

Thoracic Outlet Syndrome

Anatomy: The subclavian artery and vein and the nerves of the brachial plexus may be compressed by the bony or muscular structures of the thoracic outlet (scalene muscles, pectoralis minor muscle, cervical or first ribs, and clavicle) leading to unilateral upper extremity pain or swelling. Symptoms are most commonly due to nerve compression (95%) or venous thrombosis (4%) and are less commonly due to arterial compression (1%).

Pathophysiology: Longstanding compression of the subclavian artery at the thoracic outlet may lead to a focal stenosis and/or the development of poststenotic aneurysm. Poststenotic aneurysm may result in peripheral embolization and profound distal upper extremity ischemia.

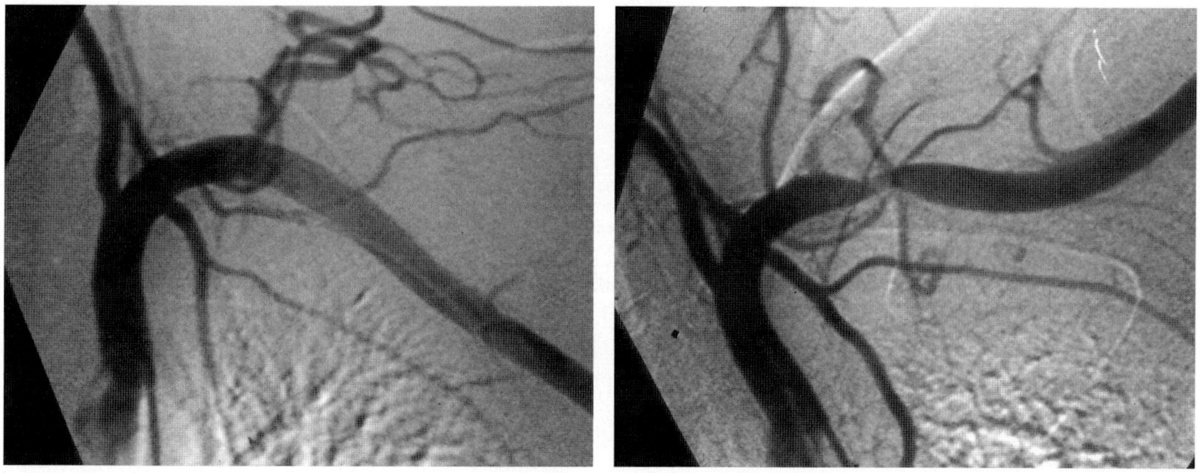

a. b.

Figure 21. *A 30-year-old weightlifter complained of numbness and tingling in his left arm and occasional episodes of pain in his hand. a) Angiography reveals a normal subclavian artery in the neutral position. b) A stenosis is evident when the arm is hyperabducted.*

Angiography: Angiography performed in the neutral position may reveal a stenosis or fusiform aneurysm of the subclavian artery. When angiography in the neutral position is normal, stress view should be obtained with the arm maximally abducted **(Fig. 21)**. However, it should be noted that slight compression of the subclavian artery is normal when the arm is maximally hyperabducted. Arteriography of the arm and hand is performed to detect distal emboli.

Pharmacologic Abnormalities

Anatomy and pathophysiology: A variety of prescribed and illicit drugs can lead to vascular smooth muscle cell contraction and vasospasm. A notable feature of drug-induced vasospasm is that it can be reversed by discontinuation of the offending drug. Ergotamine is perhaps the best known agent causing focal peripheral vasoconstriction. Other drugs that can cause reversible lower extremity ischemia include cocaine, methysergide and vasopressin.

Angiography: Characteristic findings include bilateral, symmetric vasoconstriction that may be focal or diffuse.

REFERENCES

General

Kumar V, Coltran RS, Robbins SL. Blood vessels. In: Robins S, ed. Basic pathology. 6th edition. Philadelphia: W.B. Saunders, 1997; 281-307.

Atherosclerosis

Haimocici H, DePalma RG. Atherosclerosis: biologic and surgical considerations. In: Haimovici H, ed. Vascular surgery: principles and techniques. 3rd edition. San Mateo: Appelton & Lange, 1989; 161-187.

Zairns CK, Glagov S. Artery wall pathology in atherosclerosis. In: Rutherford RB, ed. Vascular surgery. 5th edition. Philadelphia: W.B. Saunders, 2000; 313-333.

Intimal Hyperplasia

Clowes AW. Pathologic intimal hyperplasia as a response to vascular injury and reconstruction. In: Rutherford RB, ed. Vascular surgery. 5th edition. Philadelphia: W.B. Saunders, 2000; 408-418.

Inflammatory Disease (Vasculitis)

Rooke TW, Joyce JW. Uncommon arteriopathies. In: Rutherford RB, ed. Vascular surgery. 5th edition. Philadelphia: W.B. Saunders, 2000; 418-434.

Fibroplasia

LaBerge JM, Kerlan RK Jr. SCVIR Annual Meeting Film Panel Session: Case 3. Segmental arterial mediolysis (SAM) resulting in spontaneous dissections of the middle colic and left renal arteries and occlusion of the SMA. JVIR 1999, 10:509-13.

Stanley JC, Wakefield TW. Arterial fibroplasia. In: Rutherford RB, ed. Vascular surgery. 5th edition. Philadelphia: W.B. Saunders, 2000; 387-408.

AVM

Riles TS, Rosen RJ. Arteriovenous fistulae and vascular malformations. In: van Breda A, Strandness DE, eds. Vascular diseases: surgical and interventional therapy. New York: Churchill Livingstone, 1994; 1107-1147.

Neoplasia

Kumar V, Coltran RS, Robbins SL. Blood vessels. In: Robins S, ed. Basic pathology. 6th edition. Philadelphia: W.B. Saunders, 1997; 281-307.

Trauma

Espinosa GA, Chiu JC, Samett EJ. Clinical assessment and arteriography for patients withpenetrating extremity injuries: a review of 500 cases with the Veterans Affairs West Side Medical Center. Military Medicine, 1997; 162:19-23.

Modrall JG, Weaver FA, Yellin AE. Diagnosis and management of penetrating vascular trauma and the injured extremity. Emergency Medicine Clinics of North America, 1998; 16:129-44.

Embolic Disease

Greenberg RK, Ouriel K. Arterial thromboembolism. In: Rutherford RB, ed. Vascular surgery. 5th edition. Philadelphia: W.B. Saunders, 2000; 822-835.

Physiologic Disturbances

Fowl RJ, Kempczinski RF. Popliteal artery entrapment. In: Rutherford RB, ed. Vascular surgery. 5th edition. Philadelphia: W.B. Saunders, 2000; 1087-1101.

Sanders RJ, Cooper MA, Hammond SL, Weinstein ES. Neurogenic thoracic outlet syndrome. In: Rutherford RB, ed. Vascular surgery. 5th edition. Philadelphia: W.B. Saunders, 2000; 1184-1200.

CHAPTER 1 PREVIEW ANSWERS

Case 1

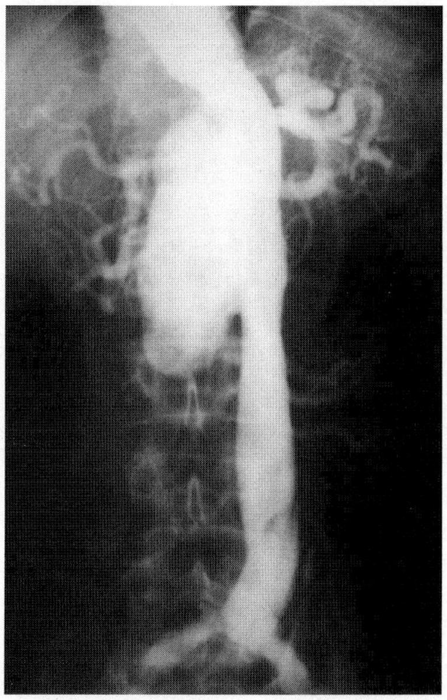

Findings: Large focal eccentric aneurysm of the abdominal aorta at the level of the visceral vessels.
Diagnosis: Mycotic pseudoaneurysm.

Case 2

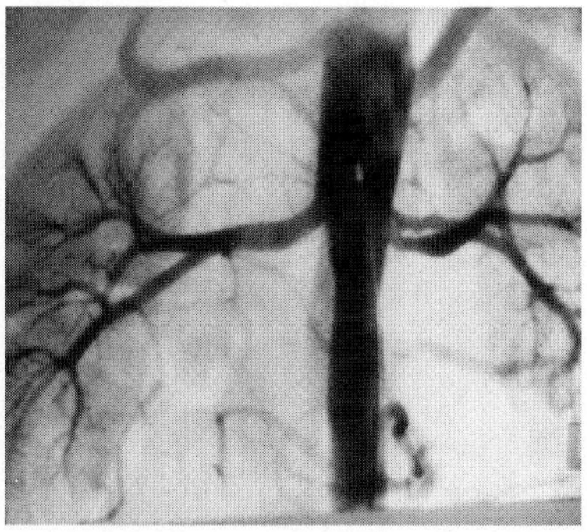

Findings: Narrowing of the proximal left renal artery and infrarenal abdominal aorta.
Diagnosis: Midaortic syndrome due to Takayasu's arteritis.

Case 3

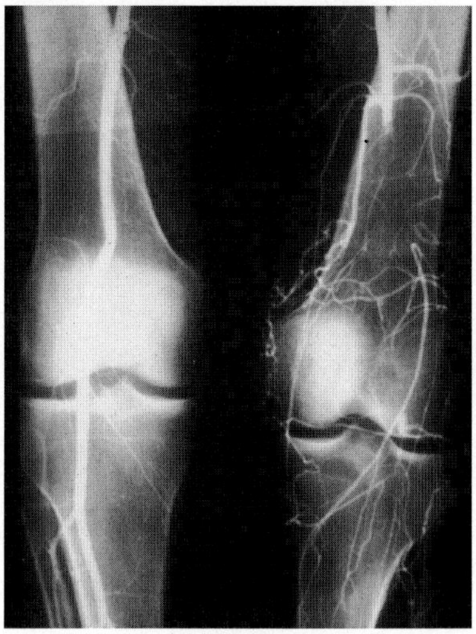

Findings: Bilateral popliteal artery disease with occlusion on the left and medial deviation of the popliteal artery on the right.
Diagnosis: Popliteal artery entrapment syndrome.

Case 4

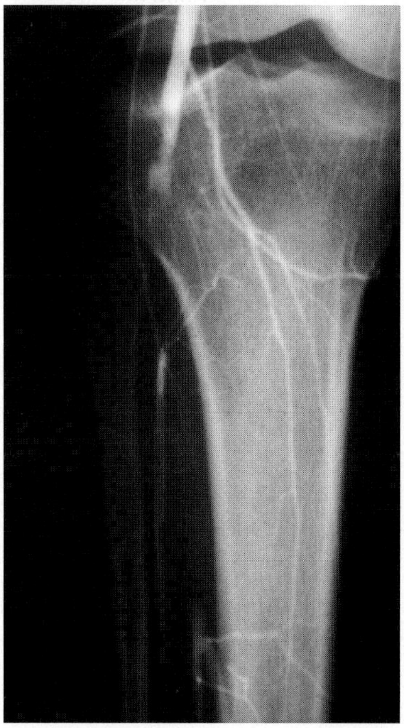

Findings: Filling defect in the distal popliteal artery with distal occlusion and small geniculate collaterals.
Diagnosis: Embolic occlusion of the popliteal artery.

CHAPTER 2 PREVIEW

Case 1

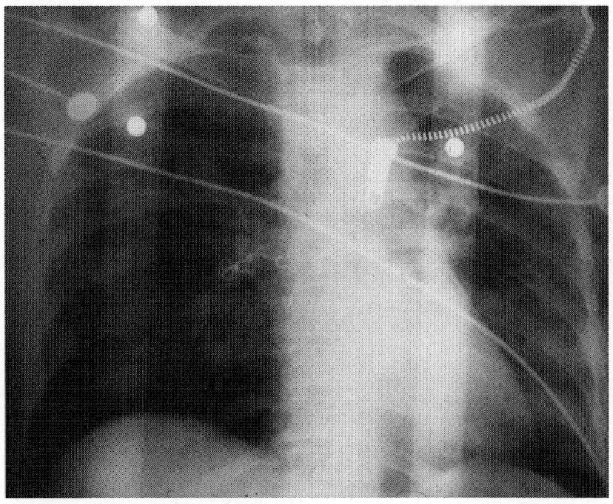

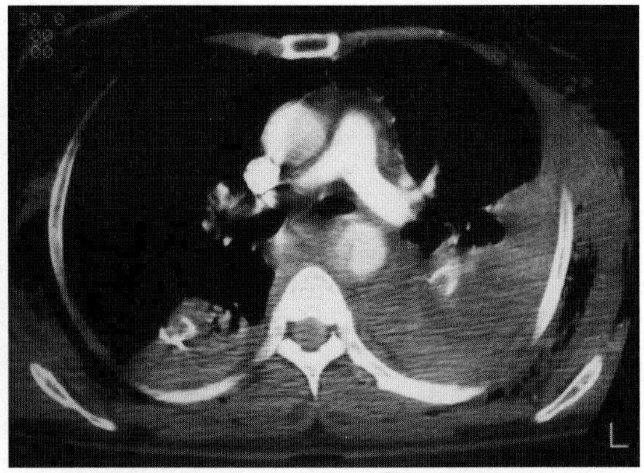

History: A 70-year-old unrestrained passenger in a motor vehicle accident sustained head and chest trauma. On exam, her lower extremity pulses were absent.
Questions: What is your diagnosis? What study would confirm your diagnosis?

Case 2

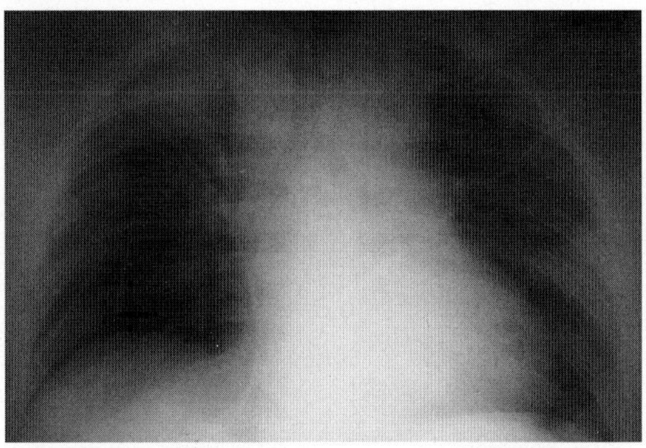

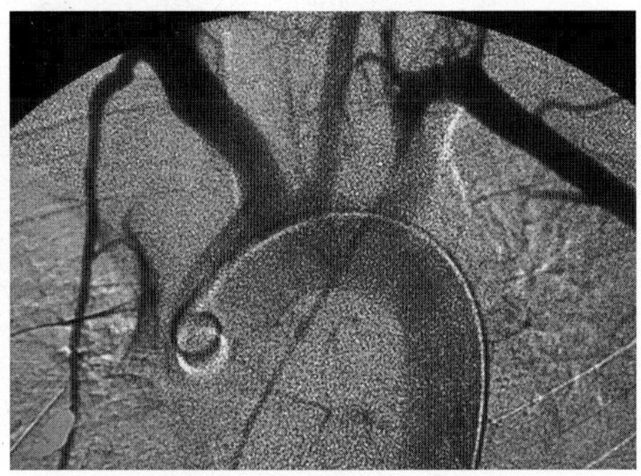

History: A 23-year-old man was thrown from his motorcycle and sustained fractures of the right first rib.
Questions: What is your diagnosis? What would you do next?

Case 3

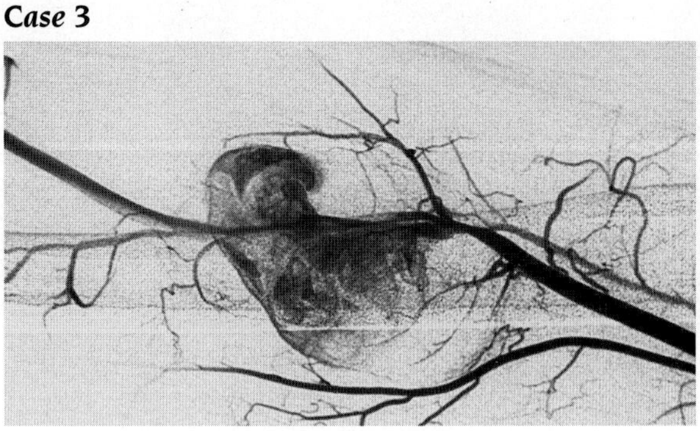

History: A 21-year-old man was stabbed in the arm. On exam, his radial and ulnar pulses were intact. 6 weeks later, he returned with a claw hand deformity, motor and sensory loss, flexed elbow and a pulsatile mass over the midhumerus.
Question: What is your diagnosis?

Chapter 2
Thoracic Aortography and Upper Extremity Arteriography

Shelley R. Marder, MD

OUTLINE OF THORACIC AORTOGRAPHY

- Anatomy of the thoracic aorta
  - — normal and variant anatomy
  - — developmental and congenital conditions
    - aberrant right subclavian artery
    - right-sided aortic arch
    - coarctation of the aorta
- Pathologic conditions of the thoracic aorta
  - — aortic dissection
  - — traumatic aortic injury
  - — aortic aneurysms
  - — aortitis

INTRODUCTION

For two decades, catheter arteriography has been the "gold standard" method of evaluating vascular pathology of the chest and upper extremities. However, over the past few years the role of arteriography has faded in favor of improved noninvasive imaging techniques such as magnetic resonance angiography (MRA) and spiral CT. Today, noninvasive imaging is the preferred technique for evaluating many vascular diseases. In this chapter, we review vascular pathology of the chest and upper extremities and comment on the extent to which new imaging modalities are complementing or supplanting catheter arteriography.

THORACIC AORTA: NORMAL ANATOMY

Normal configuration of three great vessels, 70%

- Right brachiocephalic
  - — Right subclavian
  - — Right common carotid
- Left common carotid
- Left subclavian

NORMAL AND VARIANT ANATOMY OF THE THORACIC AORTA

The thoracic aorta arises immediately distal to the aortic valve and ascends obliquely toward the right to the level of the first intercostal space (ascending aorta). It continues horizontally from right to left anterior to the trachea and gives off the great vessels (transverse aorta or aortic arch) **(Fig. 1)**. Once it reaches the left side of the mediastinum, it forms the aortic isthmus which is the portion of the aorta between the origin of the last great vessel and the ligamentum arteriosum (obliterated ductus arteriosum). The aorta then descends through the posterior mediastinum along the left aspect of the trachea, anterolateral to the spine, until it passes through the diaphragmatic aortic hiatus at the level of the 12th thoracic vertebral body.

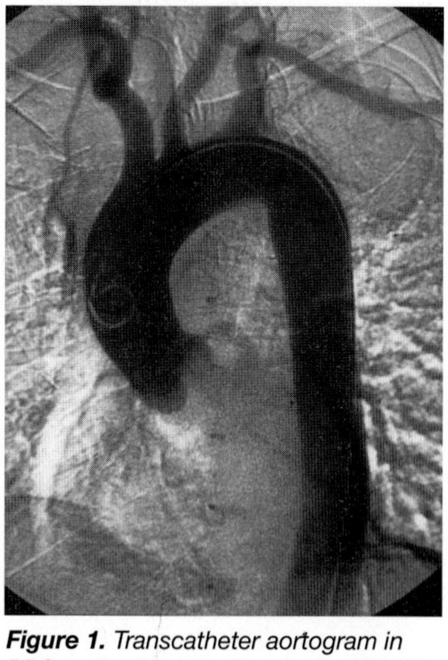

Figure 1. *Transcatheter aortogram in RPO projection revealing typical (70%) arrangement of great vessel origins: right innominate, left common carotid, left subclavian arteries.*

Normally, three great vessels originate from the thoracic aorta in the following order: right brachiocephalic (or innominate artery which divides into the right subclavian and right common carotid arteries), left common carotid artery, and left subclavian artery. This configuration of great vessels is found in 70% of individuals. Because the embryologic development of the thoracic aorta is complex (involving paired primitive aortic arches and subsequent involution of the primitive right aortic arch), numerous normal variants can occur.

THORACIC AORTA: NORMAL ANATOMIC VARIANTS

- Bovine arch, 22%
- Left vertebral from aorta, 5%
- Common origin of carotids, 1%
- Bilateral innominate arteries, 1%
- All four great vessels originate separately, 0.1%

Normal Anatomical Variants

In 22% of individuals, the left common carotid artery arises from the right innominate forming a variant called a "bovine arch" **(Fig. 2)**. In 5% of individuals, the left vertebral artery arises directly from the aorta instead of originating from the left subclavian artery. In 1% of individuals, there is a common origin of the carotid arteries and in 1% there is common origin of the left carotid and left subclavian arteries (i.e. bilateral innominate arteries). In 0.1% of individuals, all four great vessels originate separately.

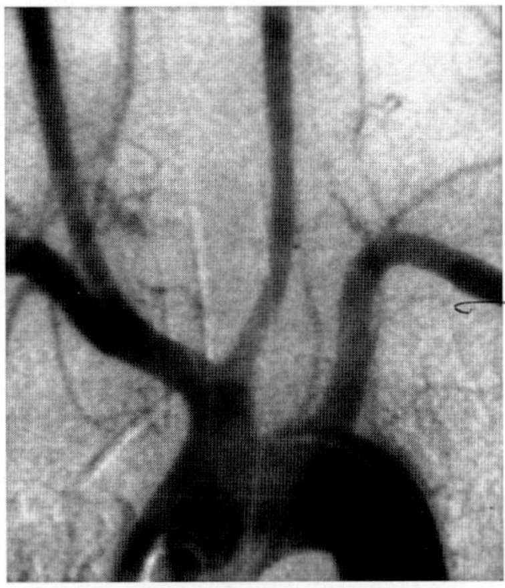

Figure 2. Transcatheter aortogram in RPO projection revealing "bovine" arch with a common trunk for the right innominate and left common carotid arteries.

The bronchial arteries arise between the T4 and T7 vertebral bodies with the majority arising at the T5-T6 levels. The anatomy of the bronchial arteries is highly variable **(Table 1)**. When performing bronchial or intercostal arteriography, one should be mindful that the arterial supply to the spinal cord can have a common origin with the bronchial trunk. The main spinal artery (arteria radicularis magna or the artery of Adamkiewicz) typically originates from a left-sided intercostal or lumbar artery at the T8 to L4 level, and ascends cranially until it reaches the midportion of the spinal canal, where it makes a sharp hairpin turn to descend caudally. Spinal arteries can originate from the bronchial, intercostal, thyrocervical, costocervical or thyroidal arteries. Angiography of any of these vessels, especially prior to embolization, mandates a thorough evaluation to exclude the presence of the spinal artery.

Developmental Anatomic Variants and Congenital Conditions **(Table 2)**

Aberrant Right Subclavian Artery with a Left-Sided Aortic Arch: One percent of individuals have a left-sided aortic arch with an aberrant right subclavian artery **(Fig. 3)**. The aberrant right subclavian artery arises as the last great vessel and ascends obliquely toward the right side of the body. In 80%, it travels behind the esophagus; in 15%, it lies between the esophagus and trachea; while in 5%, it courses anterior to the trachea. Congenital heart disease is present in 10%-15% of individuals with this anomaly. Persistence of the primitive right aortic arch at the origin of an aberrant right subclavian artery can give rise to the diverticulum of Kommerell.

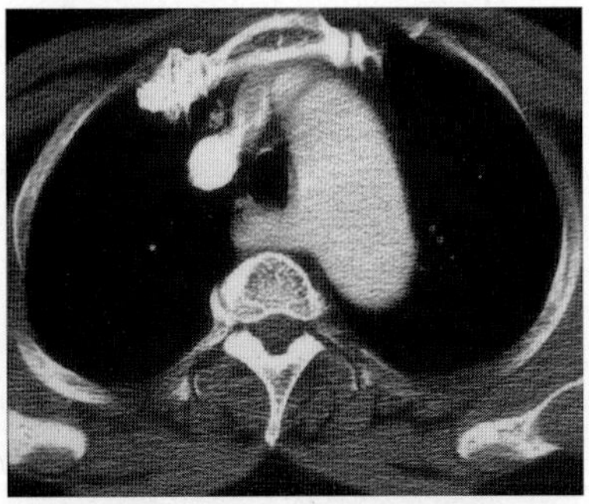

Figure 3. Contrast-enhanced CT scan revealing a left aortic arch with a retroesophageal aberrant right subclavian artery.

TABLE 1: BRONCHIAL ARTERY ANATOMY

- 2 left, 1 right common intercostal bronchial trunk (40%)
- 1 left, 1 right common intercostal bronchial trunk (30%)
- 2 left, 2 right bronchial arteries (20%)
- 1 left, 2 right (1 right common trunk) (10%)

TABLE 2: DEVELOPMENTAL ANATOMIC VARIATIONS

Left arch with aberrant right subclavian artery, 1%
- Arises as last great vessel
- Associated with congenital heart disease (CHD), 10%-15%
- Diverticulum of Kommerell, persistence of primitive right aortic arch

Right aortic arch, 1%
- 2/3 mirror image, 99% associated with cyanotic CHD
- 1/3 aberrant left subclavian, 10% CHD, vascular rings

Coarctation, narrowing usually immediately adjacent to ligamentum arteriosum
- 50% isolated finding
- 50% associated with CHD (bicuspid aortic valve, patent ductus arteriosus, ventricular septal defect)

Right-Sided Aortic Arch: A right aortic arch is present in 1% of the population **(Fig. 4)**. Two-thirds have mirror image branching while one-third have an aberrant left subclavian artery (nonmirror image branching). With mirror image branching, the order of the great vessels is as follows: left innominate artery, the right common carotid, and right subclavian. Patients with right aortic arch and mirror image branching almost always (99%) have cyanotic congenital heart disease.

When an aberrant left subclavian artery is present (i.e. nonmirror image branching), the order of vessel origins is as follows: left common carotid, right common carotid, right subclavian, aberrant left subclavian. This anomaly is associated with vascular rings (left ligamentum arteriosum and retroesophageal aberrant left subclavian) and an increased incidence of congenital heart disease (10%) compared with the normal population.

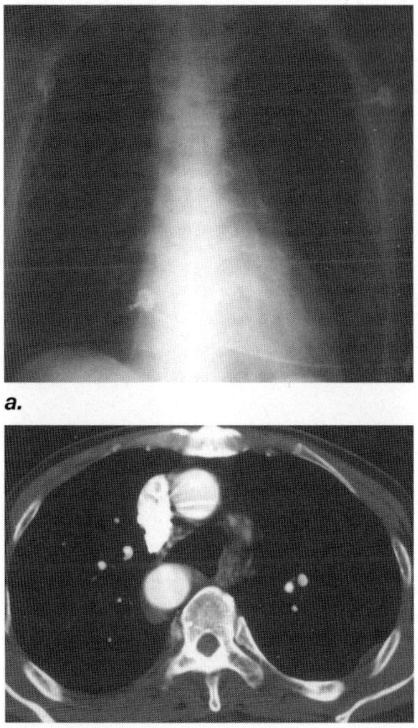

a.

b.

Figure 4. a) Chest radiograph of a right aortic arch. Note higher than usual location with no deviation of the trachea or double density. b) Contrast-enhanced CT of right aortic arch.

Vascular Rings: Anomalous development of the primitive aortic arch can result in symptomatic vascular rings. The most common types of vascular rings are as follows: left arch with aberrant right subclavian artery, right arch with aberrant left subclavian artery, complete double aortic arch, and incomplete double arch with an atretic portion which is usually the left arch. Symptoms develop from compression of the airway or esophagus. These symptoms include stridor, wheezing and dysphagia. T1-weighted spin echo MR imaging in coronal, axial, and sagittal planes is useful for determining the anatomy of the vascular ring and the extent of tracheal or esophageal compression.

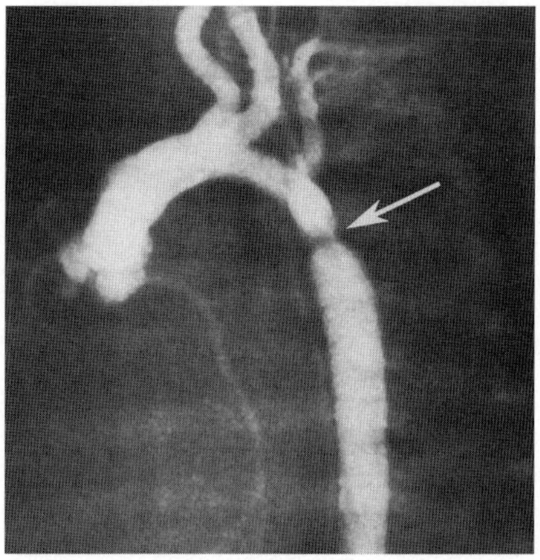

a.

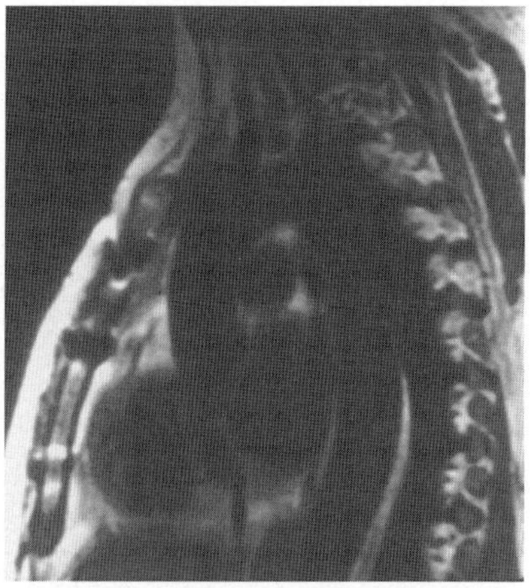

b.

Figure 5. *a) Transcatheter aortogram depicting typical juxtaductal coarctation (arrow). b) T1- weighted spin echo MR image of coarctation in sagittal projection.*

Coarctation: Coarctation of the aorta is due to a developmental defect in the media that results in narrowing of the aorta immediately adjacent to the ligamentum arteriosum **(Fig. 5)**. In 50% of patients, coarctation is an isolated finding and in 50% it is associated with congenital heart disease [bicuspid aortic valve, patent ductus arteriosus (PDA) or ventricular septal defect (VSD)]. Diagnostic evaluation prior to surgical repair should include measurement of the length and severity of the aortic stenosis as well as documentation of collateral supply. T1-weighted spin echo MR imaging in an LAO projection is suitable for measuring the caliber of the aorta before and after surgical repair. Cine gradient recall echo (GRE) imaging is of value in demonstrating a flow jet and establishing the hemodynamic significance of the coarctation. Three-dimensional gadolinium-enhanced MRA may provide physiologic evidence of significance by demonstrating dilated collaterals. Transcatheter aortography is currently performed only when the significance of the stenosis cannot be determined by MRA or when therapeutic intervention, such as balloon angioplasty and/or endoluminal stenting, is indicated.

PATHOLOGIC CONDITIONS OF THE THORACIC AORTA

Aortic Dissection

Aortic dissection occurs when a tear in the intima allows blood to travel in the wall of the aorta between the middle and outer thirds of the media. The underlying pathologic cause for dissection is weakening of the inner layer of the aortic wall. Intimal weakening occurs with age and is accelerated by hypertension (a condition found in 78% of patients with dissection). In younger patients, pregnancy and connective tissue diseases such as Marfan syndrome, or Ehlers-Danlos syndrome may result in a decreased structural integrity of the intima that can lead to dissection. Patients who have bicuspid aortic valve or coarctation of the aorta are also at increased risk for aortic dissection.

AORTIC DISSECTION: CLASSIFICATION

- DeBakey
  - Type I ascending and descending (30%)
  - Type II ascending only (20%)
  - Type III descending only (50%)
- Stanford
  - Type A ascending involved
  - Type B ascending not involved

Classification: There are two commonly used classification systems for aortic dissection. The DeBakey system classifies dissection into three categories. Type I (30%) involves both the ascending and descending thoracic aorta. Type II (20%) involves only the ascending aorta. Type III (50%) involves only the aorta distal to the left subclavian artery (that is, only the descending thoracic aorta). The Stanford classification has two categories, Type A and Type B, which are differentiated by the status of the ascending aorta. If the ascending aorta or the great vessels are involved, regardless of the site or origin of the dissection, it is a Type A dissection.

If only the descending aorta is involved, it is a Type B dissection. Stanford A therefore encompasses DeBakey I and II. Determining the status of the ascending aorta is a high priority because immediate surgery is the standard treatment for patients with Type A dissection.

AORTIC DISSECTION: ASSOCIATIONS

- Normal aging, aggravated by hypertension
- Marfan's
- Ehlers-Danlos
- Bicuspid aortic valve, coarctation

Clinical Presentation: Patients with dissection often present acutely. The classic symptom of acute dissection is sudden onset of severe "tearing," chest and/or back pain. Acute dissection progresses to hemodynamic shock in 25% of cases. Less commonly, patients present with pain due to end-organ ischemia (to the extremities, bowel, kidneys, or brain). If left untreated, 36%-72% of patients with dissection will die within the first 2 days. Thus, making a quick and accurate diagnosis of acute aortic dissection is critical. Patients with chronic dissection are often asymptomatic and their condition is discovered on imaging obtained for another reason. In these patients, follow-up surveillance is necessary to monitor progression of disease.

The most common cause of immediate death (in the first 24 hours) in patients with acute dissection is aortic rupture—into the pleural space or the pericardium. Patients with Stanford Type A dissections should undergo surgical repair. Survival rates for patients with Stanford Type A dissection treated operatively are 60%. If treated nonoperatively, survival is less than 10%. For Stanford Type B dissections either medical or surgical management can be employed with a survival rate of 40%-70%. Abdominal aortic extension with involvement of the visceral branches or lower extremity vessels may necessitate surgical intervention. The role of endovascular stenting and percutaneous fenestration is under investigation.

AORTIC DISSECTION: TREATMENT

- Involvement of ascending aorta: surgical
- Descending only: medical
- Involvement of visceral branches/lower extremity vessels: surgical
- For nonoperative patients, endovascular stenting or percutaneous fenestration may be performed

Imaging of Dissection: **(Table 3)**. Plain film findings of dissection are nonspecific.

TABLE 3: IMAGING OF AORTIC DISSECTION

Plain Film Findings
- Mediastinal widening
- Double aortic contour
- Enlargement of aortic silhouette
- Displacement of the calcified intima by 1 cm

Aortography
- Intimal tear or intimal stripe, delayed filling of false lumen
- Compression of true lumen by false lumen
- Obstruction of branch vessels
- Abnormal catheter position
- Aortic valve regurgitation

Computed Tomography
- Intimal flap
- Delayed opacification false lumen
- Displaced intimal calcification on non-contrast CT
- Compression true lumen
- Branch vessel involvement

Magnetic Resonance
- T1-weighted spin echo for aortic diameter; axial, sagittal visualize intimal flap, compression of true lumen; slow flow in false lumen
- Cine GRE: functional assessment, sagittal, axial through aortic valve for regurgitant flow
- 3D gadolinium-enhanced MRA: measure aortic diameter, depict thrombus, intraluminal detail, assess branch vessels, and projection angiography for surgical planning

The most common finding on chest radiography is widening of the mediastinum (seen in 81% of patients). Other findings suggestive of aortic dissection include double aortic contour, irregular or indistinct aortic outline, displacement of the tracheal air shadow, enlarged cardiac silhouette, pleural or pericardial effusion, displacement of the calcified intima by more than 1 cm, and enlargement of the aortic silhouette on serial chest films.

If dissection is suspected clinically, further evaluation should be performed with CT **(Fig. 6)**, MR **(Fig. 7)** or aortography **(Fig. 8)**. These modalities are required to directly visualize the true and false lumens. Regardless of the modality used, the great vessels and abdominal aorta must also be evaluated to exclude extension of the dissection beyond the thoracic aorta.

Specific findings of dissection on aortography include visualization of the intimal tear as evidenced by an intimal stripe and opacification of the false channel. These are the two most specific and common findings of dissection. Other abnormalities include compression of the true lumen by the false lumen, aortic wall thickening (>1 cm), obstruction of branch vessels, abnormal catheter position, and aortic valve regurgitation.

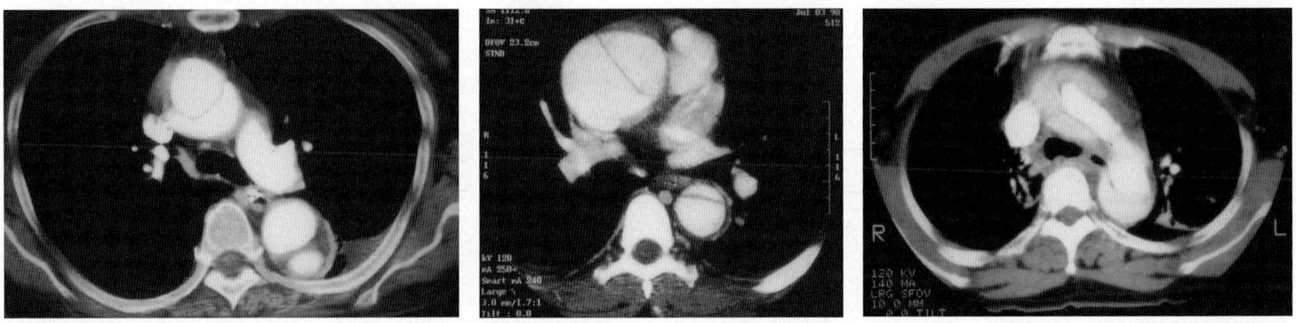

a. b. c.

Figure 6. *CTA of three Type A dissections. a, b) Dissection flaps in the ascending and descending aorta, and c) circumferential intimal detachment in the transverse aortic arch. (Note denser contrast in true lumen.)*

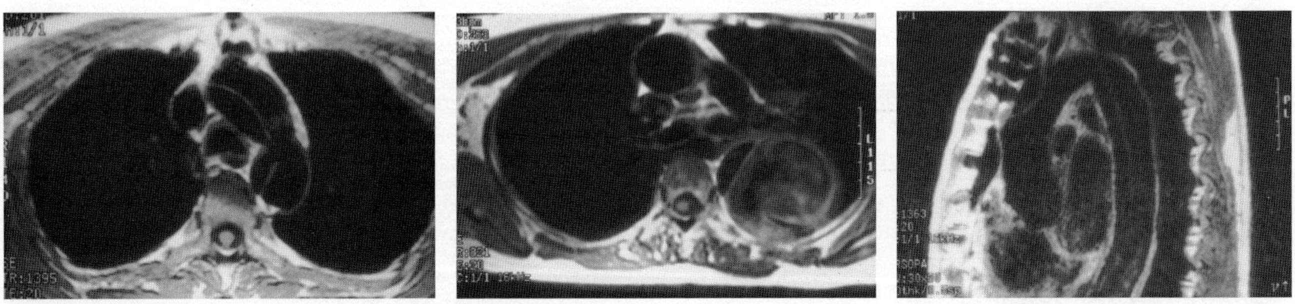

a. b. c.

Figure 7. *MR image of Type A dissections with a) transverse image of arch, b) massively dilated descending aorta. Chronic dissection with slow flow in false lumen compromising true lumen. c) Sagittal MR image showing compression of true lumen by false lumen.*

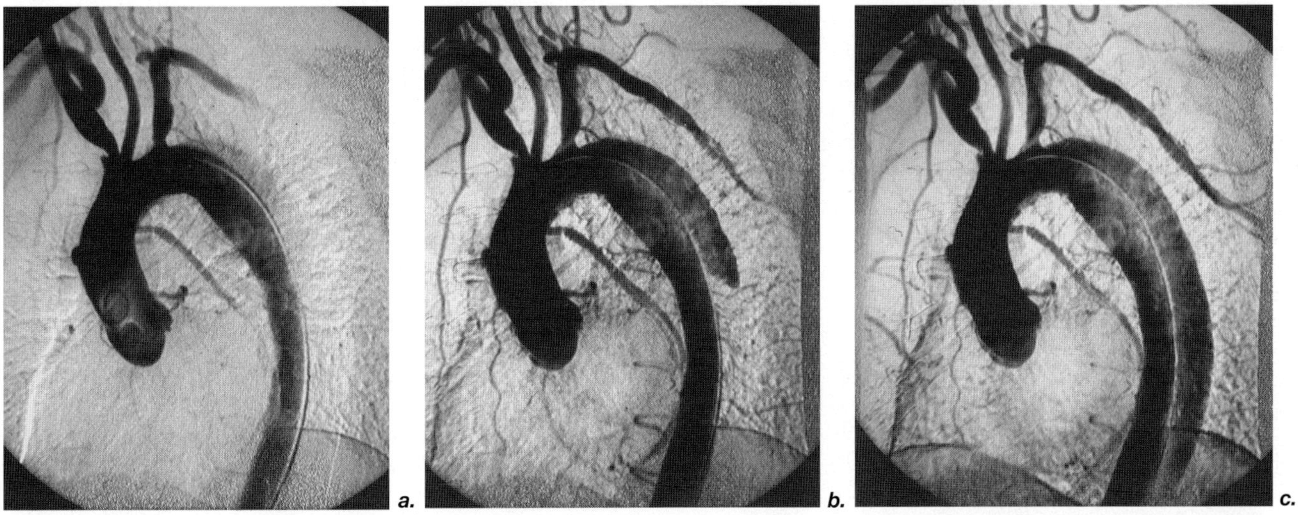

Figure 8. Aortogram of Type B dissection with a) early filling of the true lumen, b, c) intimal flap and delayed filling of the false lumen. Subsequent filming revealed delayed washout of the false lumen.

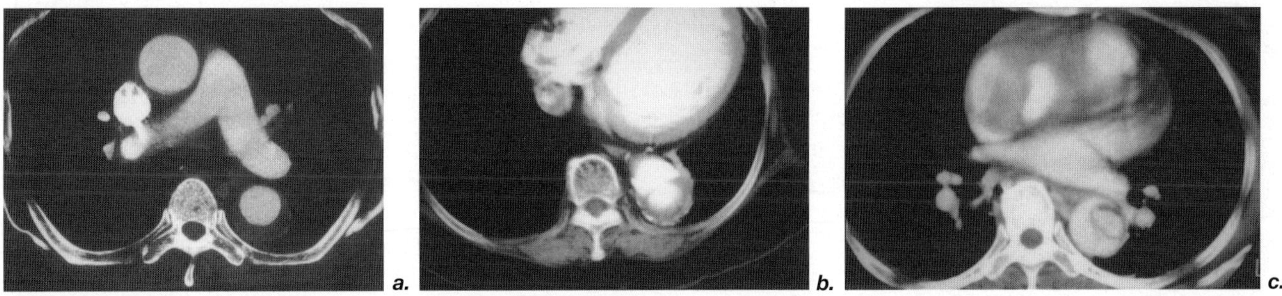

Figure 9. Use of CTA in distinguishing three distinct entities. a) Intramural hematoma with nonopacified blood and no mass effect. b) Penetrating atherosclerotic ulcer into the aortic wall. c) Atypical circumferential dissection with compromise of the true lumen.

To a large extent, MR imaging is supplanting aortography in the diagnosis of aortic dissection. Common MR imaging techniques include axial T1-weighted spin echo sequences, cine GRE imaging and 3-D gadolinium-enhanced MRA. Using these techniques, the patency of the false lumen, the extent of branch vessel involvement, and the status of the abdominal aorta can be determined. Furthermore, MRA and/or CTA can be of value in distinguishing two conditions that should be included in the differential diagnosis of dissection: intramural hematoma and penetrating atherosclerotic ulcer **(Fig. 9)**. With intramural hematoma, there is a focal dissection of blood in the wall of the aorta due to subintimal rupture of the vasa vasorum. Intramural hematoma is best demonstrated on 3-D gadolinium-enhanced MRA. Characteristic findings include a normal intima with crescentic wall thickening. Penetrating atherosclerotic ulcer will be discussed later in conjunction with traumatic aortic injury.

AORTIC INJURY: PLAIN FILM FINDINGS

- Widened mediastinum
- Obscuration of aortic contour
- Displacement of NG tube or trachea
- Depression of left mainstem bronchus
- Apical cap
- Indirect signs of significant thoracic trauma

Traumatic Aortic Injury (TAI)

Evaluation of the thoracic aorta to exclude vascular injury is one of the most common indications for thoracic aortography. Aortic injury may result from sudden deceleration and tearing of the aorta at points of fixation such as the aortic isthmus. The most common location for traumatic aortic injury (among patients surviving long enough to be studied) is the aortic isthmus (85%-90%) just distal to the origin of the left subclavian artery. Less commonly, injury occurs at the aortic root, or in the descending aorta at the diaphragmatic hiatus. It is estimated that only 10%-20% of patients with traumatic aortic injury from blunt chest trauma survive to be evaluated in the emergency room. Of these, 25% will exsanguinate within 24 hours. Only 1%-2% of patients left untreated could be expected to become long-term survivors. Fortunately, immediate surgical repair can result in a salvage rate of 80%.

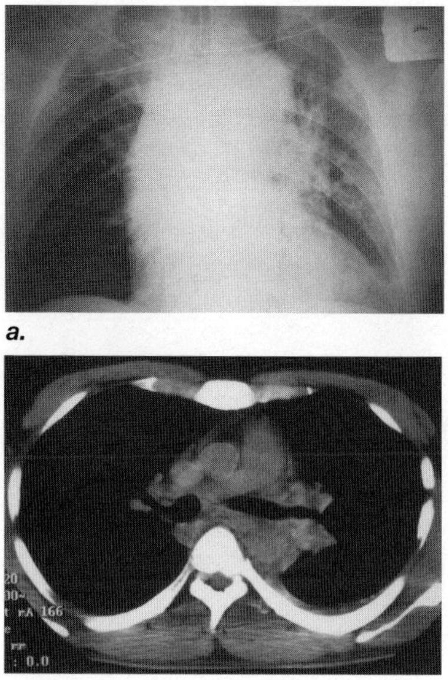

a.

b.

Figure 10. *a) Chest radiograph of TAI with wide mediastinum, loss of distinct aortic contour and apical cap. b) Noncontrast CT confirms mediastinal hematoma without direct visualization of the tear.*

Pathophysiology: The most common traumatic injury to the aorta is a transverse tear. The depth of the tear varies from a superficial intimal defect to a complete transection. In 60% of the cases, the intima and media are involved and the adventitia is intact. If the adventitia expands, a pseudoaneurysm is formed. When the adventitia is also torn, surrounding clot may temporarily contain the bleeding long enough to allow imaging evaluation and treatment. A chronic false aneurysm may be observed in untreated injuries who are studied at a later date. The specific location of the aortic injury has been tabulated among survivors who make it to the hospital. Among patients who undergo aortography, 90% of TAI occurs at the ligamentum arteriosum, just distal to the left subclavian artery; 9% occurs at the aortic root. Only 1%-3% occurs at the diaphragmatic hiatus.

Various theories for the mechanism of TAI have been proposed. These include: (1) horizontal deceleration with differential shearing force on the aorta when the body is suddenly decelerated; (2) vertical deceleration with sudden flexion of the arch over the fulcrum of the hila with transection or avulsion of the ascending aorta or arch; (3) osseous pinch in which the arch and great vessels are compressed between the anterior osseous structures and the spine; (4) hydrostatic forces with rupture from sternal compression, arch flexion and increased hydrostatic pressure as high as 2,500 mm Hg.

TRAUMATIC AORTIC INJURY: CT FINDINGS

- Intimal flap
- Luminal clot
- Irregular wall or contour
- Intramural hematoma, dissection
- Pseudocoarctation
- Pseudoaneurysm
- Extravasation

Chest Radiography: Traumatic aortic injury should be suspected in patients with a history of ejection from a car, deformation of the steering wheel by the driver in an motor vehicle accident (MVA), or passenger compartment crush. Chest radiography plays a central role in the evaluation of aortic injury as clinical findings may be minimal, confusing or obscured by other injuries **(Fig. 10)**. Plain film findings of aortic injury include mediastinal widening (>25% the width of the chest at the level of the aortic knob on an upright film), obscuration of the aortic contour, displacement of the trachea or nasogastric tube to the right, depression of the left mainstem bronchus, or apical cap. Indirect signs such as 1st and 2nd rib fractures, scapular fractures and vertebral body fractures are signs of significant thoracic trauma. Radiologic or clinical suspicion of aortic injury, should be investigated by direct imaging of the thoracic aorta. CTA **(Fig.11)** is performed in the stable patient, while transcatheter aortography **(Fig.12)** is performed in the unstable patient.

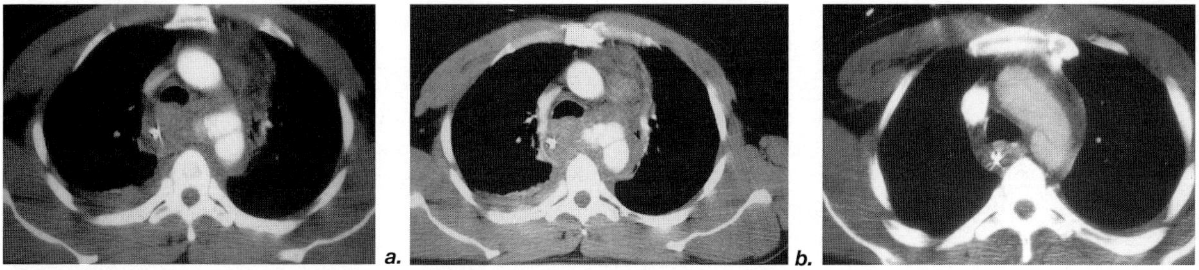

a. b. c.

Figure 11. *a) CTA with direct visualization of pseudoaneurysmal collection of contrast and extensive mediastinal hematoma. b) Same patient with extravasation. c) Another patient with traumatic pseudoaneurysm at the isthmus.*

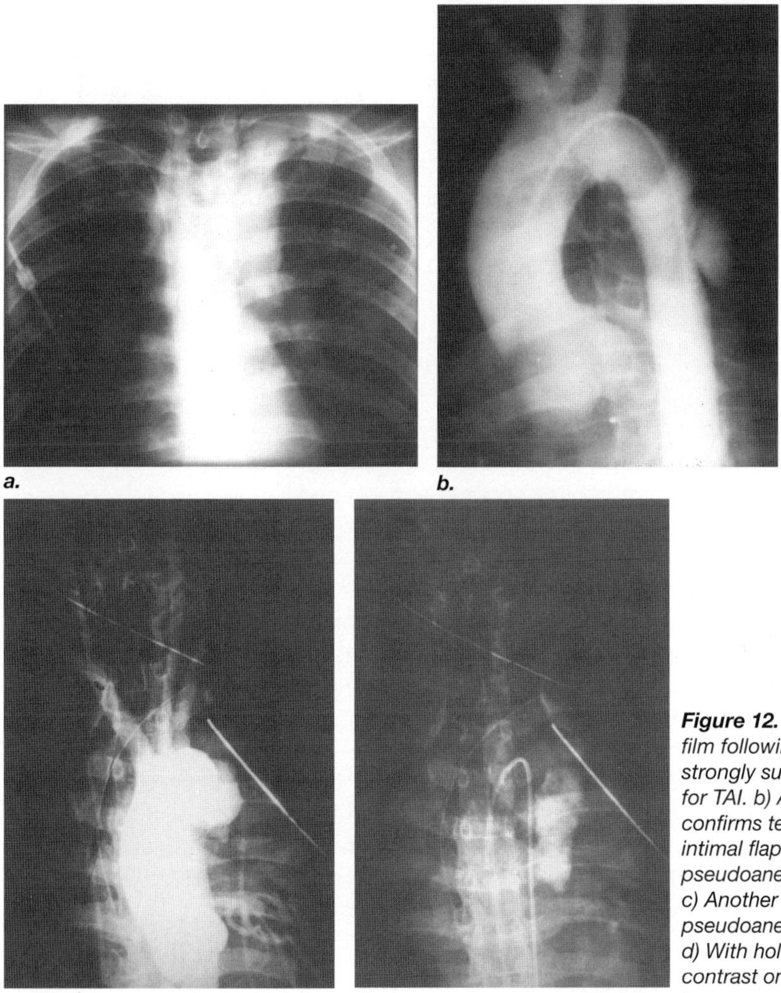

a. b. c. d.

Figure 12. *a) Chest film following MVA strongly suspicious for TAI. b) Aortogram confirms tear with intimal flap and pseudoaneurysm. c) Another traumatic pseudoaneurysm. d) With hold-up of contrast on delayed images.*

TRAUMATIC AORTIC INJURY: ANGIOGRAPHIC FINDINGS

- Asymmetric contour
- One or both acute margins
- Narrowed adjacent lumen
- Intimal flap
- Extravasation
- Overlap density
- May retain contrast

TRAUMATIC AORTIC INJURY: DIAGNOSTIC ALGORITHM

- Normal chest film: no further evaluation
- Wide mediastinum or mechanism in a stable patient: CTA
- Wide mediastinum, unstable patient: transcatheter aortogram

CT Evaluation: Initial studies of CT for aortic trauma showed that CT of the mediastinum with or without contrast is useful in triaging patients and deciding who needs an aortogram. In those patients with equivocal plain film findings or low clinical suspicion, a normal mediastinal CT obviates the need for thoracic aortography. More recent studies of helical CT have shown that CT aortography is accurate enough to replace transcatheter aortography in evaluating patients with severe chest trauma. Our current approach is to perform thoracic helical CT angiography on any patient with a suspicious chest film or mechanism who is at low to moderate risk of thoracic injury, is stable clinically, or is already on the CT scanner to have an abdominal or head CT.

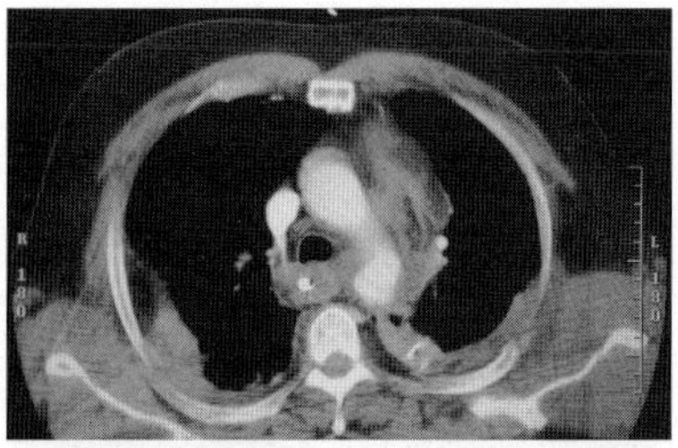

a.

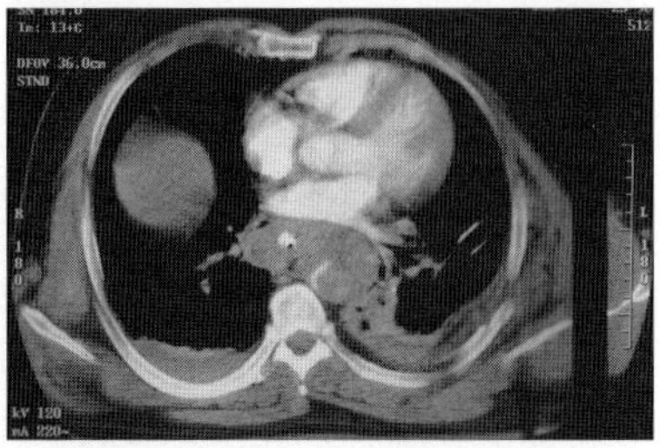

b.

Figure 13. *a) Pseudoaneurysm following MVA. b) Traumatic pseudocoarctation; narrowing of the aortic lumen by surrounding mediastinal hematoma such that groin pulses are compromised.*

Optimal CT technique requires dynamic scanning during rapid bolus infusion (2 ml/sec of 60% iodinated contrast medium). Special care must be taken to avoid artifacts from ECG leads, nasogastric tubes, arms in the gantry and motion. CT findings of aortic injury include pseudoaneurysm, irregular aortic contour, a linear lucency of aortic wall extending into the lumen, marginal luminal irregularity, periaortic hematoma or intramural clot, dissection or pseudocoarctation (abrupt decrease in caliber of the descending thoracic aorta), or extravasation. A CT scan that demonstrates an aortic wall abnormality, pseudoaneurysm formation, or pseudocoarctation **(Fig. 13)** is diagnostically sufficient to permit the cardiothoracic surgeon to take the patient directly to surgery.

Only in the setting of an equivocal study, a periaortic hematoma, or the presence of a superior or middle mediastinal hematoma that cannot be explained by vertebral or sternal fractures, is a transcatheter aortogram required. For the hemodynamically unstable patient in whom aortic injury is suspected, direct triage to aortography remains the standard of care.

Transesophageal Echo: Transesophageal echocardiography (TEE) is under evaluation as a portable diagnostic modality for rapid evaluation of accessible portions of the thoracic aorta. There are reports of detection of small intimal tears in the descending aorta that are not detectable at aortography. The significance of these small tears is unknown. Moreover, the limitations of TEE have so far prevented its widespread use. These include the need to get the Cardiology Service emergently involved, the failure to image the great vessels **(Fig. 14)**, and a 3-cm to 5-cm "blindspot" along a portion of the upper ascending aorta. Thus, the place of TEE in the work-up of TAI has not yet been defined.

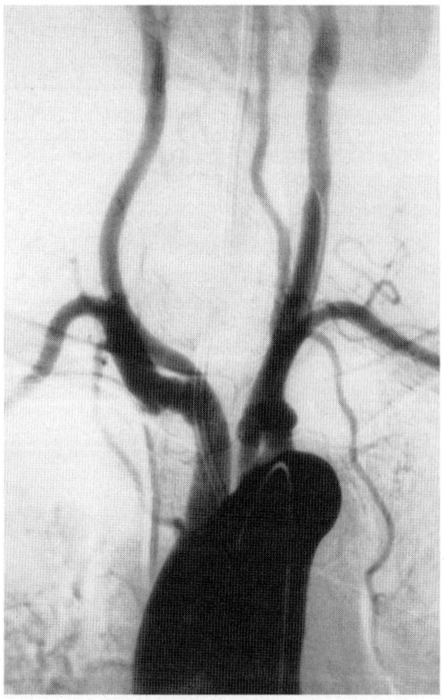

Figure 14. *Multiple great vessel injuries in a shoulder harness-wearing MVA victim not seen on TEE. These included thrombosis of right vertebral artery, pseudoaneurysms of right subclavian, left common carotid and left subclavian arteries, and dissection of left common carotid artery.*

Catheter Arteriography: Aortography is generally a safe procedure with only a 2% incidence of minor complications. Its drawbacks are cost, invasiveness, and time. Of these, time is the most significant factor as it delays other interventions in these patients for whom laparotomy, neurosurgical and orthopedic procedures are commonly required. Arteriography can be performed using either conventional cut-film or digital subtraction arteriographic (DSA) techniques. Intraarterial DSA is faster than conventional cut-film aortography, but DSA has the limitations of poorer spacial resolution and a smaller, more restricted field of view. Recent studies have documented a higher false positive rate and a greater difficulty in distinguishing anatomic variants such as ductus diverticulum **(Fig. 15)** by DSA than by conventional filming.

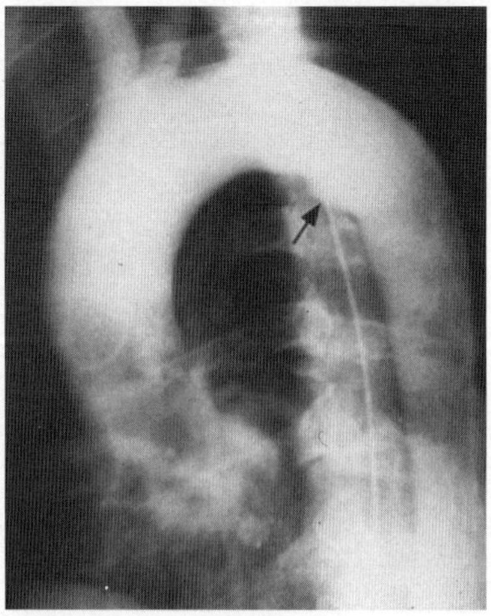

Figure 15. *Aortogram of normal variant ductus diverticulum with smooth, obtuse margins with the aortic lumen.*

Technique for Trauma Aortography: The common femoral artery is punctured and a 15-mm J-guidewire is passed up into the thoracic aorta. If difficulty is encountered, a test injection of contrast is performed by hand to determine whether the aortic lumen is intact. In most cases, a soft-tipped guidewire will pass easily into the ascending aorta. When the arch cannot be crossed, a right-sided axillary puncture is undertaken. A 5-F or 6-F pigtail catheter capable of delivering 35 ml/sec for 2 seconds is placed in the ascending aorta. A forceful hand injection is performed to document position by visualizing the coronary arteries and aortic valve cusps; the catheter is repositioned to avoid recoil into the coronary arteries or left ventricle approximately midway between the brachiocephalic artery and the aortic valve. The contrast injection rate for a cut-film study is generally 30-35 ml of nonionic contrast per second for a total of 60-70 ml (concentrated contrast 370 mg Iodine per ml). This dose may be adjusted up or down based on aortic caliber and subjective flow rate. We start with the 45-degree RPO projection which best visualizes the aortic isthmus, the most likely site of injury. Filming is performed at 3 films per second for 4 seconds and 1 film per second for 8 seconds. Only if no clear injury is present, do we proceed with the AP projection. In an equivocal study, filming in an LPO or true lateral projection is also performed. In general, if the exam is positive for an aortic disruption, only one view is necessary, and the patient is transferred directly to the operating room. It is important to visualize the entire thoracic aorta and intrathoracic great vessels on the "trauma arch" study.

Arteriographic findings indicative of aortic injury include: irregular aortic margin, intimal flap, hangup of contrast on delayed films, abrupt change in aortic contour (posttraumatic coarctation), extraluminal collection with an acute angle with the aorta, dissection, and rarely, frank extravasation. Findings that can simulate TAI and account for a 1% false positive rate include ductus diverticulum (9% of the population), ulcerated aortic plaque **(Fig. 16)**, aortic aneurysm, syphilitic aortitis, aortic dissection, and artifacts related to streaming, mixing, and motion on DSA. Distinguishing features between ductus diverticulum (9%-26% of normals) and TAI are listed in **Table 4**, and between penetrating aortic ulcer and TAI are listed in **Table 5**.

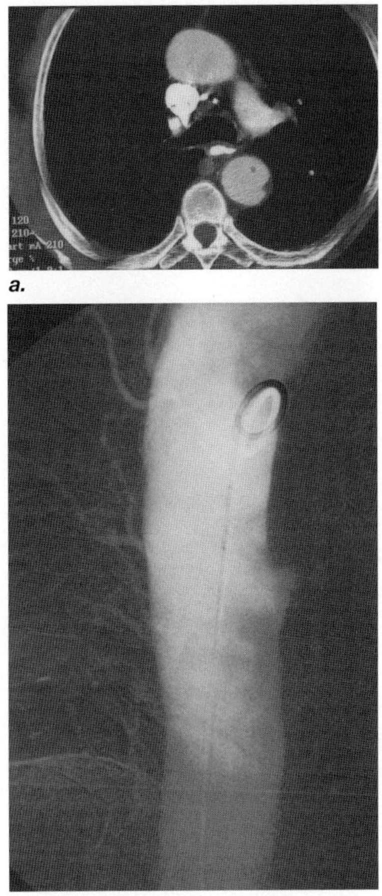

Figure 16. *Penetrating atherosclerotic ulcer with a) CT showing no surrounding hematoma and multifocal atherosclerosis, and b) sharp margins on aortogram in descending aorta, an atypical location for TAI, with subsequent brisk washout of contrast.*

TABLE 4: DISTINGUISHING BETWEEN DUCTUS DIVERTICULUM AND TAI

Ductus Diverticulum	Traumatic Pseudoaneurysm
Smooth symmetric contour	Irregular or asymmetric contour
Obtuse margins with aortic lumen	One or both acute margins with lumen
Normal aortic lumen adjacent	Narrowed aortic lumen adjacent to pseudoaneurysm
No intraluminal intimal flap	Intraluminal intimal flap
No double density or overlap	Overlap density between lumen and pseudoaneurysm
No retention of contrast after washout	May retain contrast after washout

TABLE 5: DISTINGUISHING BETWEEN PENETRATING AORTIC ULCER AND TAI

Penetrating Ulceration	Traumatic Pseudoaneurysm
Associated atherosclerotic changes	Limited or no atherosclerosis
Mid descending aorta location	Proximal medial descending aorta
Lack of compression (or ectasia)	Frequent luminal compression adjacent
Intramural hemorrhage by CT/MR	Infrequent intramural hemorrhage
Enhancement by CT	No enhancement of aortic wall
Rare mediastinal hemorrhage	Common mediastinal hemorrhage

Thus, even aortography, the "gold standard," is not perfect. Occasionally, false positive aortograms result in unnecessary thoracotomy. In one study, this occurred in 1% of all the trauma arches and in 14% of all the "not clearly normal" aortograms. It is possible that with the increasing application of spiral CT, TEE, and with long-term nonsurgical observation of some cases of minor intimal injury, the incidence of both transcatheter aortography and unnecessary thoracotomy may significantly decrease over the next decade. Also, on the horizon, is the possibility of commercially available endovascular stent-grafts for the nonoperative treatment of acute aortic injury **(Fig. 17)**.

THORACIC ANEURYSM: SHAPE

- Fusiform: circumferential involvement
- Saccular: asymmetric, involves only a portion of the wall

Aneurysms

Aneurysms of the thoracic aorta are common and are estimated to occur in 10% of the elderly population. The normal thoracic aorta measures 3 cm to 4 cm. When asymptomatic aneurysms progress to twice the normal diameter, operative or endoluminal intervention should be considered. Once a diameter of 6 cm is reached, there is a 76% two-year mortality with further conservative management (worse depending on absolute size, age > 50 and the presence of diastolic hypertension). Factors that should be considered in the surgical assessment are the size, type, location, relation to branch vessels, presence of mural thrombus and periaortic hematoma – all of which are well documented on MR imaging.

THORACIC ANEURYSM: TYPE

Layers involved
- True: all three layers intact, dilated, usually fusiform
- False: all three layers disrupted, contained extravasation
- Pseudoaneurysm: one or two layers intact, contained, usually saccular

Pathophysiology: Atherosclerotic aneurysms are produced by gradual weakening of the aortic wall due to loss of muscle fibers resulting in dilatation of the aorta. The descending aorta is frequently involved. Men are affected more than women (4:1). This condition is also associated with high blood pressure, coronary artery disease and abdominal aneurysms. Aneurysms may be either fusiform or saccular. True aneurysms involve all three layers of the aortic wall, while false aneurysms are the result of a disruption in one or more layers of the vessel wall. Saccular aneurysms are more likely to be pseudoaneurysms and have a greater likelihood of rupture. For psuedoaneurysm, etiologies such as infection or post-traumatic and post-surgical complication must be considered.

TABLE 6: THORACIC AORTA, ANEURYSMS

Atherosclerotic
- Gradual weakening of wall
- Exacerbated by hypertension
- Descending aorta
- Typically fusiform
- Mural calcification

Posttraumatic
- Isthmus location
- 1%-2% unrecognized at time of injury
- Saccular

Syphilitic
- Decreasing in frequency
- Destruction of muscle fibers of aortic wall
- Majority involve ascending aorta
- Associated aortic valvular insufficiency
- Commonly saccular, 15% pencil thin

Mycotic
- Vasa vasorum seeded by systemic bacteremia
- Atherosclerotic plaque may become infected
- Contiguous infection, ie., postoperative
- Typically saccular

Congenital
- Defects in wall due to underlying connective tissue disorders
- Marfan's syndrome
  - Ascending aorta
  - Aortic sinuses
  - Predisposed to dissection
- Poorly developed media
- Aortic rupture from fragile tissue in Elhers-Danlos syndrome

Arteritis
- Takayasu's
- Giant cell arteritis
- Relapsing polychondritis, rheumatoid arthritis, ankylosing spondylitis

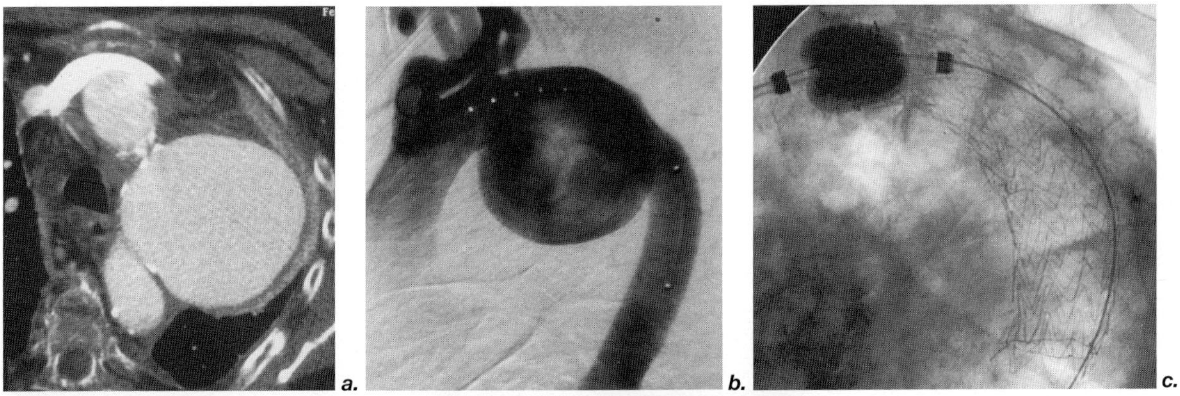

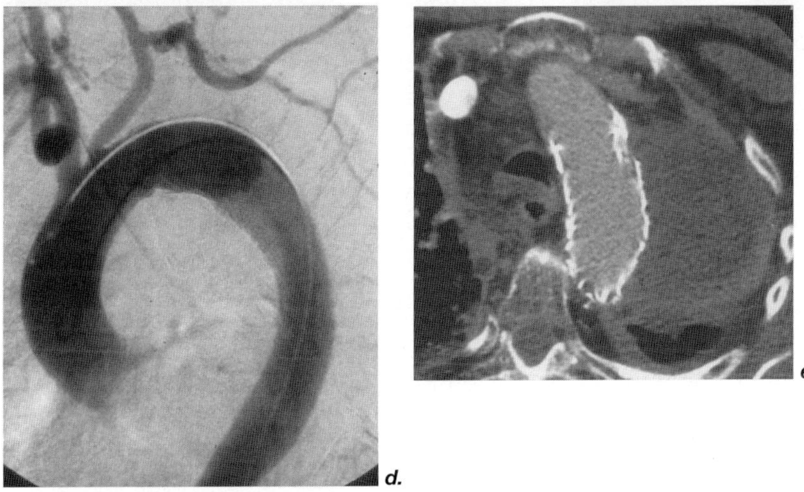

Figure 17. *a) Traumatic pseudoaneurysm in a 30-year-old, 20 years after MVA. Prestent graft CT used for measuring involved aorta. b) Angiogram with measuring catheter (Cook, Bloomington, Ind.) confirmed length to be stented and need for surgical transposition of left subclavian artery to carotid artery to provide an adequate neck for stent-graft placement without traversing vessel origin. c) Thoracic excluder stent-graft (WL Gore, Flagstaff, Arizona; Stanford Medical Center Phase II FDA trial) dilated using a 30-mm diameter balloon. d) Completion aortogram without evidence of endoleak. e) Six-month follow-up CT scan demonstrating stent-graft and thrombosed pseudoaneurysm. (Images courtesy of Michael Dake, MD, and Stephen Kee, MD, Stanford University School of Medicine.)*

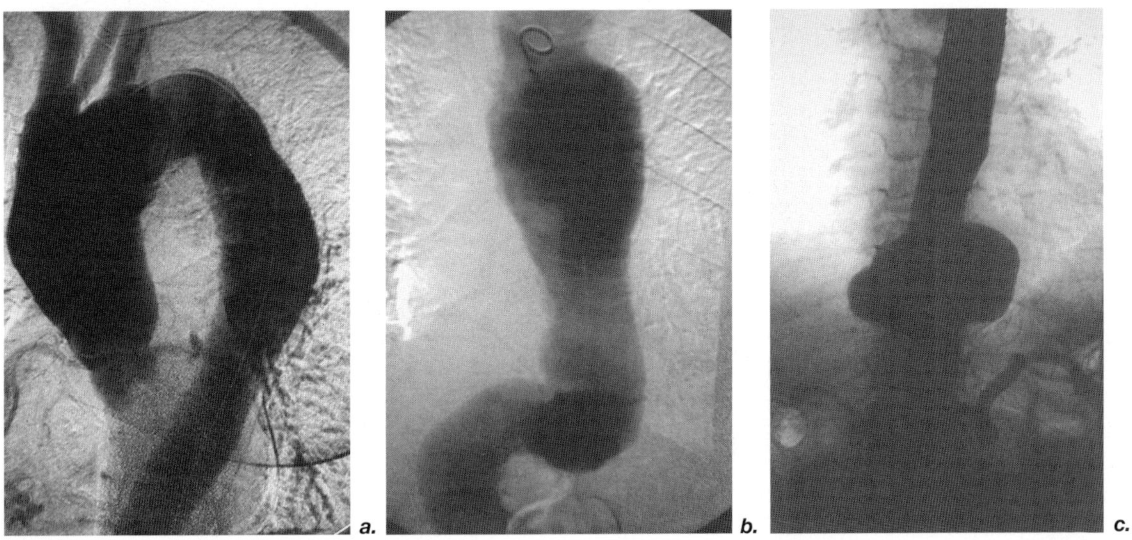

Figure 18. *a) Ascending and b) descending aortic dilatation and irregularity due to atherosclerotic aneurysms. c) Isolated descending aortic mycotic aneurysm.*

Etiology: Thoracic aortic aneurysms results from a variety of causes including atherosclerosis, bacterial infection, trauma, congenital defects, syphilis, and inflammatory aortitis **(Table 6)**. The majority (75%-80%) of thoracic aneurysms are atherosclerotic in origin **(Fig. 18)**. Typically, these aneurysms are fusiform and involve the arch, the descending aorta and often the abdominal aorta. Isolated ascending aortic aneurysms are not usually due to atherosclerosis and are more likely due to cystic medial necrosis or syphilis.

Mycotic aneurysms develop via several different mechanisms. The vasa vasorum may be seeded from systemic bacteremia or an atherosclerotic plaque may become infected. Contiguous infection (i.e., after coronary bypass) may also involve the wall of the aorta and result in aneurysm formation. Whatever the mechanism, infection results in weakening of the wall with aneurysm formation that is typically saccular in nature.

Congenital aneurysms are frequently secondary to defects in the wall (i.e., absent or deficient media) due to underlying connective tissue disorders as found in patients with Marfan's syndrome. Marfan's patients also have dilatation of the aortic sinuses and are predisposed to dissection and aortic valvular insufficiency. The abdominal aorta and iliac arteries are less frequently involved. Only 2% of thoracic aortic aneurysms are of congenital origin. However, aortic complications account for 55% of deaths in Marfan's patients.

Syphilitic aneurysms have decreased in frequency since the advent of antibiotics. Syphilis has a predilection for the aortic wall that results in destruction of the muscle fibers and consequent aneurysm formation. The majority of syphilitic aneurysms are located in the ascending aorta (36%) or arch (24%) and there may be associated aortic valvular insufficiency.

Imaging: For all types of aneurysms, MR is the imaging technique that is likely to provide the most information. T1-weighted imaging depicts aortic morphology, caliber, mural thrombus, and periaortic structures. MRA with sophisticated postprocessing techniques can demonstrate branch vessel involvement and the extent of mural thrombus. Sagittal or coronal cine MR imaging through the aortic valve can assess functional involvement and impairment.

Aortitis

Takayasu's aortitis **(Fig. 19)**, also known as "pulseless disease," is an inflammatory condition of the aortic wall of unknown etiology. It occurs predominately in young women and is characterized by intimal thickening and destruction of the muscular and elastic fibers of the vessel wall. Smooth segmental stenoses are present in the thoracic aorta and branch vessels. Atypical coarctation of the abdominal aorta (middle aortic syndrome) may also be seen, and occasionally, the pulmonary arteries are involved. Arteriography demonstrates irregular areas of narrowing or occlusion of large and medium-sized arteries including the subclavians, carotids, or renal arteries. Ascending thoracic aneurysm is a rare manifestation of Takayasu's aortitis.

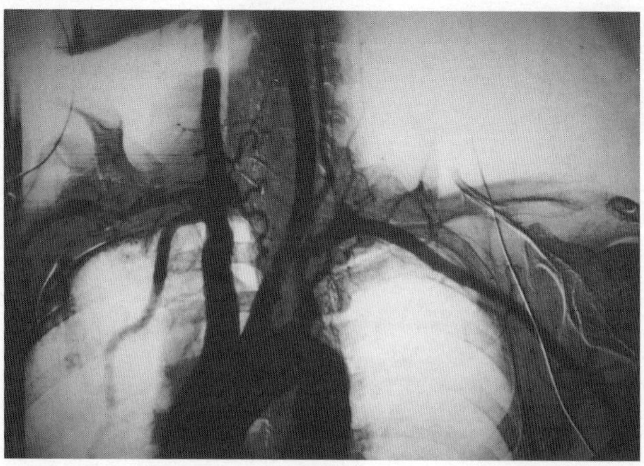

Figure 19. *Takayasu's arteritis involving multiple great vessels, including the subclavian arteries bilaterally and the right common carotid artery.*

Giant cell arteritis (GCA) occurs equally in males and females over age 50 and is an unusual cause of aortic aneurysm. More commonly, isolated segments of large and medium-sized arteries (aorta, coronary, carotid, vertebral, subclavian, visceral, iliac, renal, opthalmic, or retinal) manifest alternating areas of smooth narrowing and mild dilatation or occlusion. Axial T1-weighted spin echo MR imaging may demonstrate both aneurysmal dilatation and diffuse wall thickening. These findings are nonspecific, but potentially helpful in determining the extent of involvement and response to therapy. The diagnosis of GCA is confirmed by temporal artery biopsy.

UPPER EXTREMITY ARTERIOGRAPHY

Arteriography of the upper extremities is a relatively uncommon procedure. In contrast to the lower extremity arteries, the upper extremity arteries are often affected by nonatherosclerotic diseases and are more prone to traumatic injury. In this section we review selected disorders affecting the upper extremity vessels.

OUTLINE OF UPPER EXTREMITY ARTERIOGRAPHY

- Anatomy of the upper extremity
- Pathologic conditions of the upper extremity
  - atherosclerotic disease and subclavian steal
  - thoracic outlet syndrome
  - upper extremity trauma
  - vasospastic and inflammatory disorders
  - thromboangiitis obliterans
  - arteriovenous malformations

ARTERIOGRAPHIC TECHNIQUE

Diagnostic arteriography of the upper extremity is generally performed following transfemoral catheterization and placement of a 4-F or 5-F Berenstein-type catheter in the subclavian artery. Selective catheterization beyond the axillary artery must be performed with care, as there is a high risk of inducing vasospasm. When more selective catheterization is required, attempts should be made to reduce pain by using isosmolar contrast (ioxaglate or iodixanol) and to promote vasodilation with intraarterial injections of nitroglycerine (100-200 μg) or tolazoline (25-50 mg) and/or local warming of the hand. Injection rates range from 3 ml to 5 ml per second for up to 30 ml of contrast.

NORMAL AND VARIANT ANATOMY

The right subclavian artery **(Fig. 20)** is a branch of the brachiocephalic artery. Its main intrathoracic branches—the vertebral, thyrocervical and internal mammary arteries—take off proximal to the anterior scalene muscle while the costocervical trunk arises beneath the anterior scalene muscle. On the left, the subclavian artery originates directly from the aortic arch. In 4%-6% of people, the left vertebral artery originates directly from the aorta.

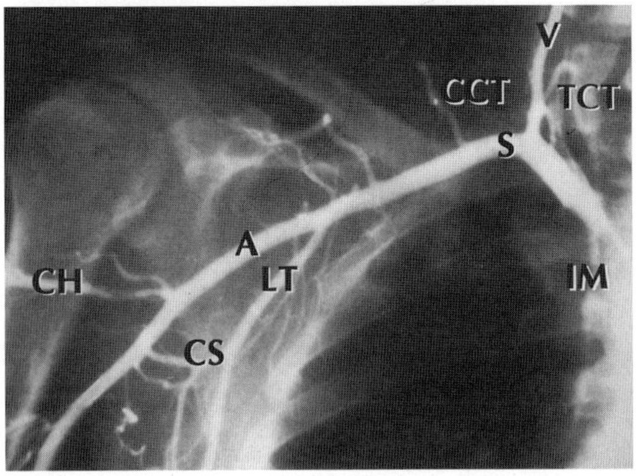

a.

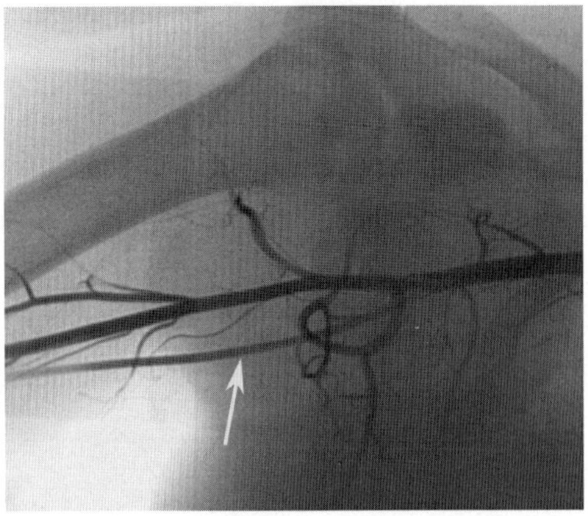

b.

Figure 20. a) Normal anatomy of the subclavian (S) and axillary (A) arteries. IM= internal mammary, TCT= thyrocervical trunk, V= vertebral, CCT= costocervical trunk, LT= lateral thoracic, CH= circumflex humeral, CS= circumflex scapular. b) High take-off of the radial artery (arrow) from the axillary artery.

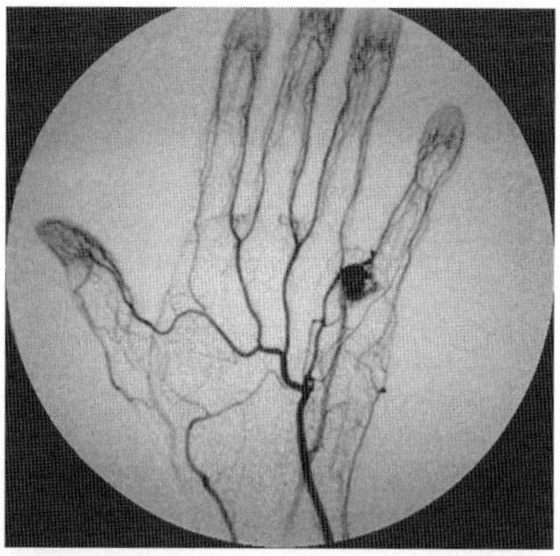

a.

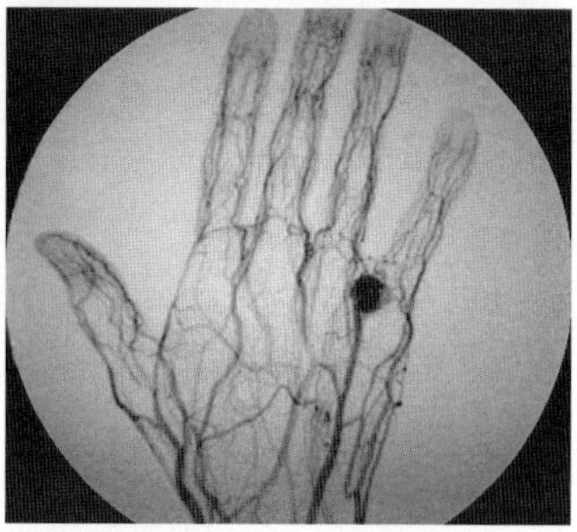

b.

Figure 21. *a) Arterial phase of a hand angiogram with incomplete deep and superficial palmar arches in a patient who also had the rare occurrence of arteriovenous fistula of the common digital artery and vein caused by blunt trauma. Note aneurysm and early venous filling on arterial phase. b) Venous phase.*

The subclavian artery becomes the axillary artery at the lateral margin of the first rib. Its main branches are the superior thoracic (to the upper intercostal spaces), the lateral thoracic (to the lateral chest wall), and the subscapular and humeral circumflex arteries (to the shoulder). At the lateral margin of the teres major muscle, the axillary artery becomes the brachial artery. One anatomic variant of significance is a high take-off of the radial artery from the axillary or brachial artery (14%-17% of the population). High origin of the ulnar artery is a rare variant with a 2%-3% occurrence. From its anteromedial course, the brachial artery typically divides in the region of the antecubital fossa (a few centimeters below the elbow) into the radial and ulnar arteries. The interosseous artery is a branch of the ulnar artery (and is analogous to the peroneal artery in the leg).

The arterial supply to the hand is quite variable. The radial and ulnar arteries supply two arches in the palm: one deep and one superficial. The deep arch is the larger of the two and is supplied predominately by the radial artery. The superficial arch is smaller, more distal and is supplied predominately by the ulnar artery. Most people (up to 95%) have a complete deep palmar arch; 5% have an incomplete deep palmar arch **(Fig. 21)**. The superficial palmar arch is anatomically complete in 78% of people. The supply to the fingers originates from the superficial arch as four common palmar digital arteries and/or from the metacarpal arteries of the deep palmar arch.

PATHOLOGIC CONDITIONS OF THE UPPER EXTREMITY VESSELS

Atherosclerosis **(Table 7)**

Atherosclerosis is far and away the most common disease process affecting the lower extremity arteries, but it affects the upper extremity arteries much less frequently. Disease severe enough to cause symptoms is uncommon. Symptomatic upper extremity atherosclerosis usually involves the proximal vessels, most commonly the subclavian artery. Exceptions to this pattern of involvement are diabetic patients and patients with renal failure who typically have distal disease. Symptoms of upper extremity arterial insufficiency include weakness, pain, numbness and coolness. When the internal mammary has been used for cardiac bypass, a stenosis at the origin of the subclavian artery can result in symptoms of cardiac ischemia.

TABLE 7: UPPER EXTREMITY ATHEROSCLEROSIS

- Prevalence: less common than in the lower extremity
- Proximal: subclavian, vertebral
    - Subclavian steal
    - Internal mammary artery bypass graft: angina, ischemia
    - Treatment: PTA, stent
- Distal: diabetics, end-stage renal disease
- Angiographic findings
    - Tapered narrowing: stenosis, occlusion with collaterals
    - Convex meniscus: embolism, central cardiac source or ulcerated plaque or proximal aneurysm, associated with thoracic outlet syndrome or prior trauma

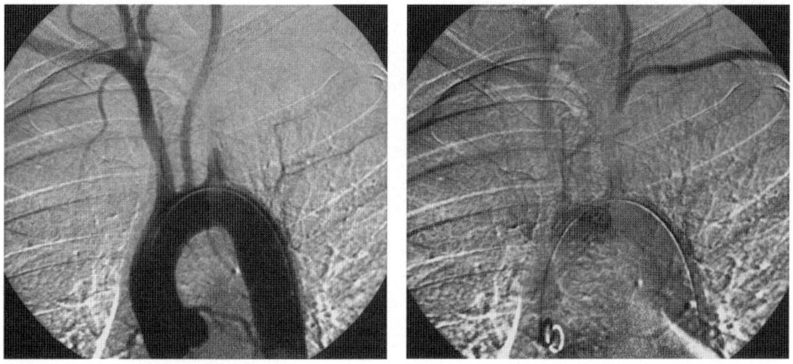

a. b.

Figure 22. *a) Subclavian steal syndrome with proximal occlusion of the left subclavian artery. b) Delayed filling of the left subclavian artery distal to the occlusion from retrograde flow in the left vertebral artery.*

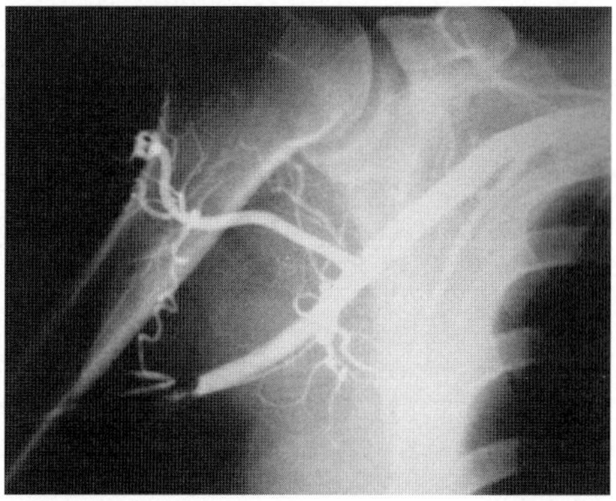

Figure 23. *Occlusion of the proximal brachial artery from an embolus. Findings include abrupt occlusion with filling defect and poor collateral flow.*

Severe stenosis of the subclavian artery proximal to the origin of the vertebral artery can also result in subclavian steal syndrome **(Fig. 22)**. In this condition, flow in the vertebral artery on the affected side is reversed in an attempt to supply blood to the arm. Patients develop symptoms of basilar insufficiency, including vertigo, syncope, and dysarthria. Symptoms of upper extremity weakness, paresthesias, and pain are infrequent. The left arm is affected more frequently (75%) than the right. Angiographic findings of subclavian steal include: stenosis or occlusion at the origin or in the proximal subclavian artery, absent antegrade filling of the vertebral artery, and delayed retrograde filling of the vertebral artery after opacification of the contralateral vertebral and basilar arteries.

Another type of upper extremity "steal" syndrome can occur in dialysis patients with upper extremity dialysis access. In these patients, arterial flow to the upper extremity is compromised by disproportionate flow through the surgical arterial-to-venous fistula or graft. Treatment may necessitate diminishing the caliber of the graft or occluding the shunt completely in order to perfuse the hand.

Distal upper extremity ischemia can also be the result of embolization **(Fig. 23)**. The source of the distal emboli may be the heart (90% cardiac chambers or valves) or the great vessels in patients with thoracic outlet syndrome or a proximal ulcerated plaque or aneurysm. (In contrast to the lower extremities, aneurysms of the upper limbs are more commonly posttraumatic in etiology than atherosclerotic.) Angiographic findings of embolization include a sharp cut-off or a meniscus in a patient with poorly developed collaterals. Importantly, angiography in patients with distal emboli should include a thorough evaluation of the great vessels to locate the embolic source.

Thoracic Outlet Syndrome

The term "thoracic outlet" refers to the space through which the subclavian vessels and the brachial plexus must pass in exiting the chest. More specifically, this area includes the scalene triangle, the costoclavicular space and the pectoralis minor space. Thoracic outlet syndrome (TOS) is a condition that results from compression of the neurovascular bundle in the region of the thoracic outlet. Symptoms may be due to compression of the nerves (brachial plexus), the subclavian vein or the subclavian artery. Symptoms are most commonly due to nerve impingement (>95%) or venous compression (2%-3%). Symptoms due to arterial involvement comprise only 1% of all cases of TOS.

THORACIC OUTLET SYNDROME

- Costoclavicular space
  - 1st rib, clavicle, scalenus medius muscle, costoclavicular ligament
  - Narrowed by cervical ribs, muscles, fascial slips, traumatic changes
- Angio: neutral position and stressed (hyperabduction)
  - Focal stenosis, poststenotic dilatation, occlusion
- Rx: resection of 1st rib

The most common site of compression of the subclavian artery is the costoclavicular space that is bounded inferiorly by the first rib, superiorly by the clavicle, posteriorly by the scalenus medius muscle and anteriorly by the costoclavicular ligament. Cervical ribs, anomalous muscles, fascial slips, or posttraumatic distortions in the rib or clavicle may narrow this space leading to impingement on the brachial plexus, subclavian artery or vein.

When TOS is suspected, arteriography should be performed in both the neutral position and following a provocative maneuver (either hyperabduction or Adson's maneuver with hyperextended neck and head turned towards symptomatic side) **(Fig. 24)**. Abnormal findings indicative of TOS include focal stenosis with poststenotic dilatation, aneurysm formation, and mural thrombus. Occasionally, an artery that appears normal on nonstressed views will become occluded on stress views. Because some degree of narrowing can be produced in up to 30% of normal people with stress maneuvers, the angiographic abnormalities brought on by stress maneuvers should be corroborated by symptoms and signs of ischemia such as muscle cramping, pain and hand pallor.

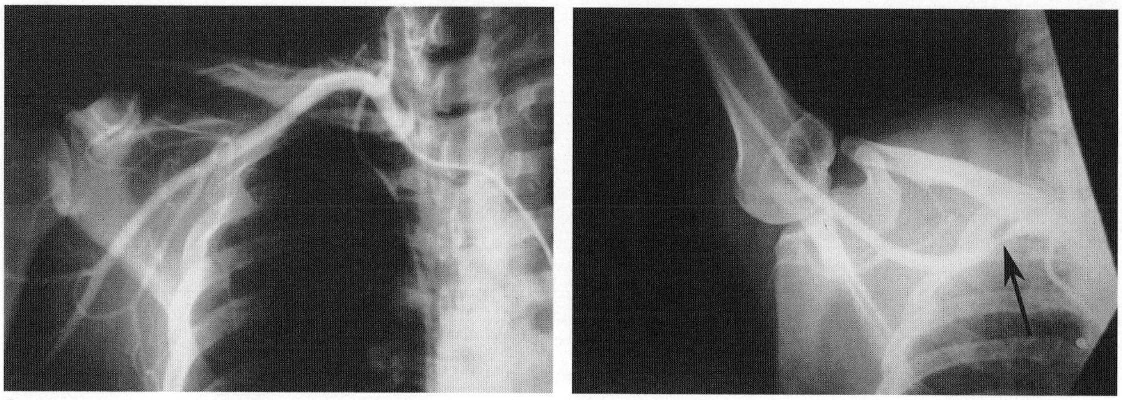

a. **b.**

Figure 24. *Thoracic outlet syndrome. a) Patent subclavian in neutral position. b) With the arm hyperabducted, there is narrowing (arrow) and reproduction of symptoms.*

The most severe complication of arterial TOS is distal embolization. Microembolization results from platelet aggregation at the point of subclavian artery compression. Larger thrombotic emboli can result from thrombus within a poststenotic aneurysm or can be released when the subclavian artery becomes totally occluded. Emboli can result in severe distal ischemia that is difficult to treat.

The treatment of choice for TOS is surgical decompression of the thoracic outlet. Resection of the first rib and scalene muscle is usually required. Unfortunately, there is a high rate of recurrence reported (15%-20%) following first rib resection and scalenectomy. Stenting of the affected artery is usually not advisable because stents placed in this region are subject to repetitive external compressive forces that may cause the stent to collapse.

Upper Extremity Trauma

Penetrating injury to the upper extremity is a common indication for upper extremity arteriography. Because the arteries to the upper limbs reside in an exposed location, traumatic arterial damage is relatively frequent. Common mechanisms of injury include gunshot and knife wounds, motor vehicle accidents, orthopedic fractures, industrial accidents, and self-inflicted injuries.

Diagnostic Workup: The evaluation of patients with extremity trauma depends on the likelihood of vascular injury as assessed by history and physical exam. Stable patients with a relatively low likelihood of injury (i.e., significant bleeding which has stopped, large but stable hematoma, possible nerve compression, or proximity less than 3 cm) undergo noninvasive ultrasound evaluation. Stable patients with a high likelihood of injury (i.e., distal pulse abnormalities or ischemia, expanding hematoma, pulsatile bleeding, or bruit at the site of injury) usually undergo arteriography.

UPPER EXTREMITY TRAUMA

- Angiographic findings
  - Spasm or extrinsic compression from hematoma
  - Dissection or intimal flap
  - Occlusion, thrombosis
  - Active extravasation
  - Pseudoaneurysm
  - Arteriovenous fistula
- Treatment includes surgery, embolization, stent-grafts

Unstable patients are evaluated in the operating room where operative arteriography may be performed before and after surgical repair. A brief emergency room evaluation includes plain films marking the site of entry for a knife or the entry and exit for a bullet. These films will show projectile fragments and may help predict the location of the injury (although the path of a projectile is not usually very predictable).

Noninvasive Evaluation: This includes bilateral arm pressure measurements, calculation of arm pressure index, and color Doppler imaging. Normally the ratio of the two arm pressures (termed arterial pressure index [API]) is 1.0. Occult vascular injury is suspected if the API is less than 0.9 (affected/unaffected limbs). The API is a useful screening test for upper extremity vascular injury that has been shown to have 95% sensitivity, 97% specificity, and 99% negative predictive value. Thus, if the API is greater than 0.9 and the color Doppler imaging is normal, "occult" vascular injury can be excluded.

Sonographically detectable lesions include intimal tears or flaps, pseudoaneurysm with characteristic flow patterns, stenosis, hematoma or arteriovenous fistulae. The jet of flow into the neck of a saccular aneurysm may be visualized. AV fistulae present as loss of normal arterial triphasicity with continuous high diastolic flow in the artery and arterialized flow in the vein. The communication may be seen as turbulent flow. Although sonography may miss small injuries, these are not usually clinically significant. False negative evaluation is most common at vessel bifurcations in patients with aberrant anatomy and in regions that are difficult to scan such as the axilla. Thus, ultrasound provides an inexpensive, noninvasive means of deriving information that the emergency room physician can use to triage patients to surgery, observation, or immediate or delayed angiography.

Arteriography: Angiographic evaluation of the upper extremities is usually performed through a catheter positioned in the subclavian artery beyond the thyrocervical trunk. If the catheter is positioned in the axillary artery, test injections must be performed to exclude a high take-off of the radial artery. Imaging is performed proximal to the level of the injury and distal to the injury to the level of the palmar arch. Demonstrating the supply to the hand prior to surgical repair is critical as this information will help determine the type of repair that is performed (ligation versus primary repair or grafting). Occasionally, angiography is performed after surgical exploration has failed to uncover the source of bleeding. In this situation the patient may be transported from the operating room with the wound packed. If the initial diagnostic angiogram fails to identify a bleeding site, it is imperative that the surgeons remove the packing prior to repeating the upper extremity angiogram.

Arteriographic manifestations of acute upper extremity vascular injury include extrinsic compression or displacement by hematoma, dissection or intimal flap **(Fig. 25)**, laceration (partial or complete transection), occlusion **(Fig. 26)**, thrombosis, active extravasation of contrast, pseudoaneurysm **(Fig. 27)**, spasm, and arteriovenous fistula formation **(Fig. 28)**. A chronic manifestation of trauma is pseudoaneurysm (that may be mycotic following injection drug usage) **(Fig. 29)**.

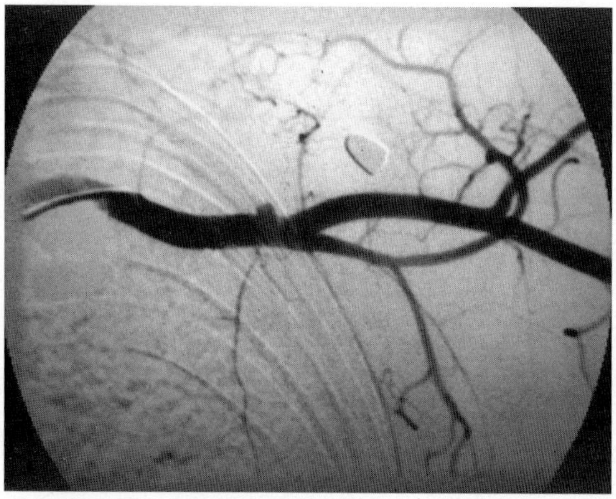

Figure 25. *Gunshot wound resulting in intimal injury with contrast extending into the vessel wall. Although this angiographic appearance suggests a pseudoaneurysm, there was no holdup of contrast on delayed images and at surgical repair, the injured region was not dilated.*

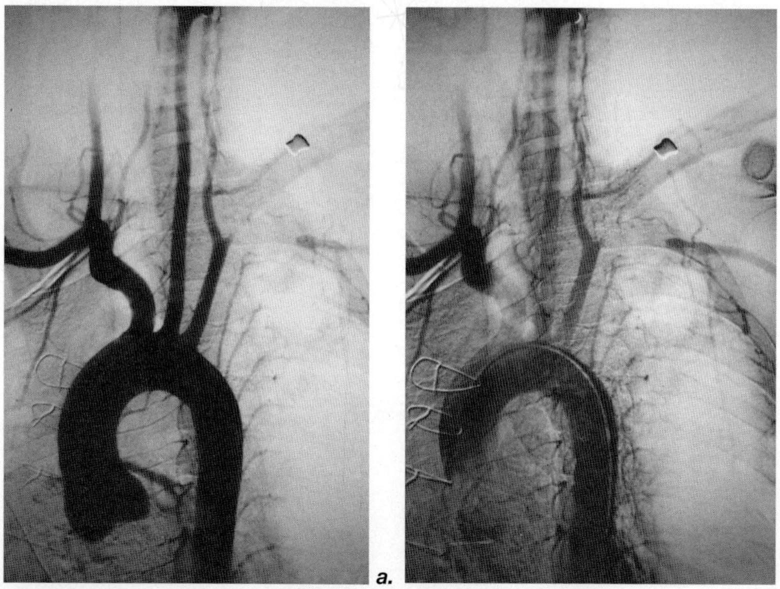

a. **b.**

Figure 26. *a) Traumatic occlusion of the left subclavian artery with b) distal reconstitution.*

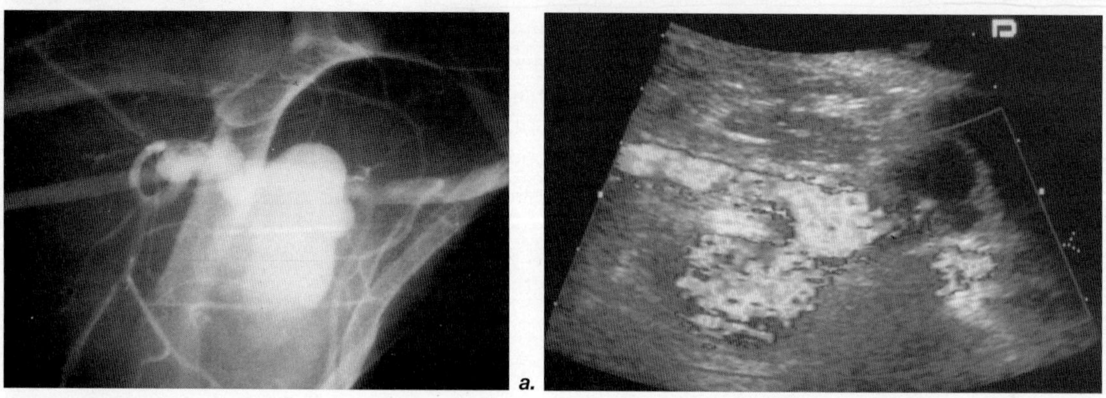

a. **b.**

Figure 27. *a) Stab wound to the axilla with delayed presentation of a false aneursym. b) Color ultrasound demonstrating jet and turbulent flow.*

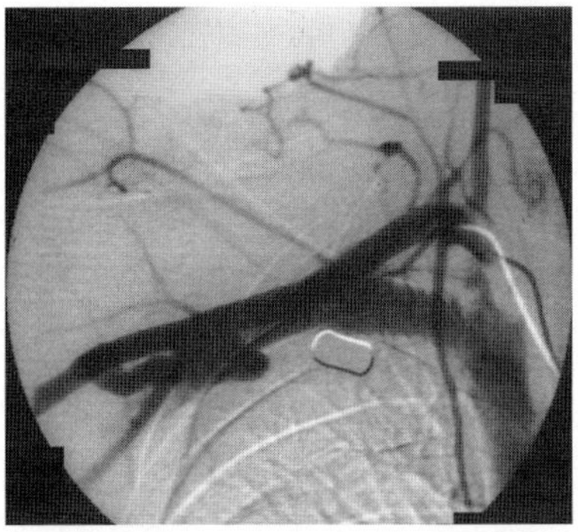

Figure 28. This patient presented with an audible bruit and diminished distal pulse. Arteriography demonstrates an axillary arteriovenous communication; note drainage into the innominate vein on the arterial phase.

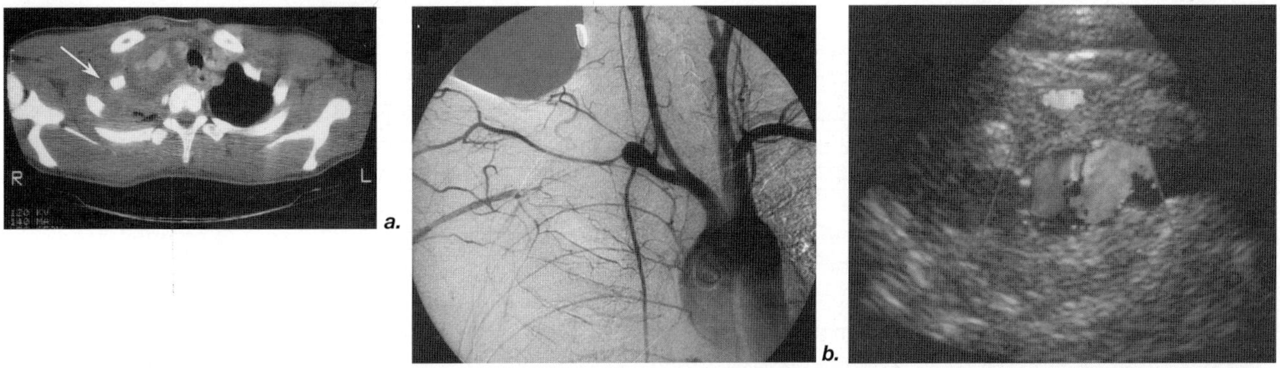

Figure 29. a) CT findings of high-density contrast (arrow) within apical abscess in an IV drug user who injected into her right neck. These findings are suspicious for a mycotic aneurysm. b) Angiography reveals a subclavian artery aneurysm with distal occlusion (note that the relationship of the occlusion to the vertebral and internal mammary arteries is important in operative planning). c) Ultrasound appearance of the pseudoaneurysm with "yin yang" flow.

A special subset of chronic arterial injury is the hypothenar hammer syndrome **(Fig. 30)**, which is characterized by occlusion of the ulnar or digital arteries secondary to repetitive trauma to the palm of the hand in the region of the hook of the hamate. This is often an occupational disorder occurring in jackhammer operators, painters, mechanics and other individuals who use their palms for pushing or pounding. Repetitive trauma leads to arterial spasm, thrombosis, or pseudoaneurysm formation. Emboli may cause digital ischemia especially in those patients with an incomplete palmar arch. Raynaud's phenomenon may be present and hypothenar hammer syndrome should be suspected in a male with Raynaud's phenomena who has an appropriate occupational history.

HYPOTHENAR HAMMER SYNDROME

- Repetitive occupational trauma
- Occlusion of ulnar or digital arteries
- Arterial spasm
- Pseudoaneurysm
- Emboli
- Raynaud's phenomenon

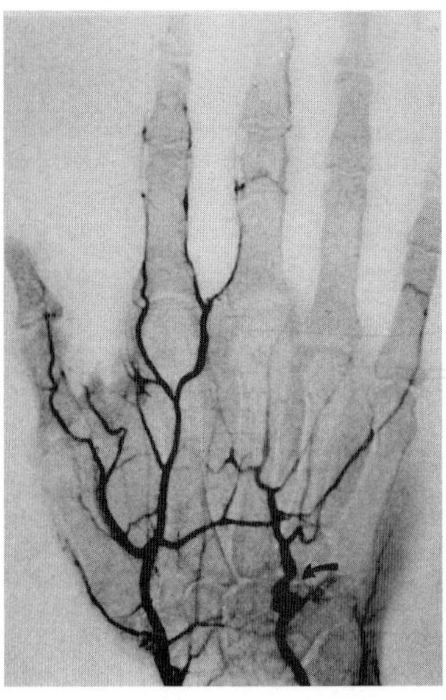

Figure 30. Hypothenar hammer syndrome with aneurysm and irregularity of the distal ulnar artery (arrow). The poor flow to the distal digital arteries may be secondary to emboli, spasm or occlusion.

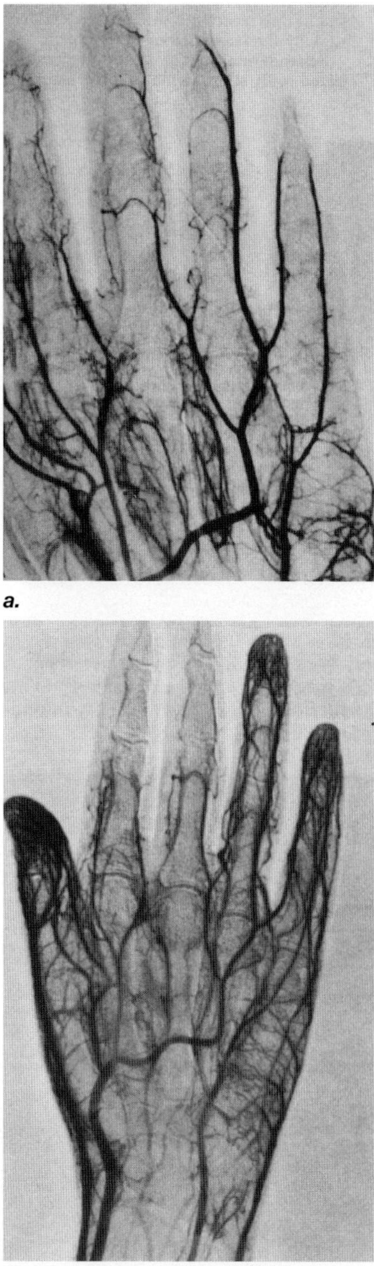

a.

b.

Figure 31. *a) Raynaud's phenomenon affecting the proper palmar digital arteries (2nd through 4th fingers). b) Following intra-arterial vasodilator administration, impoved flow to the 4th and 5th digits is noted.*

Treatment: Most patients with acute arterial injury to the upper extremity are treated surgically. If surgery is unsuccessful or undesirable, transcatheter embolization may be performed. (*Appropriate embolization techniques are discussed in Chapter* 13.) Occasionally with planned subclavian repair, we are asked to position an occlusion balloon in the proximal subclavian artery that can be used intraoperatively for proximal control. The role of stents in the setting of acute vascular injury is currently under investigation. Case reports suggest an emerging role for endovascular stent-grafts in the treatment of traumatic arteriovenous fistulae and pseudoaneurysms of the subclavian arteries. The long-term outcomes of endoluminal, nonoperative approaches are unknown, but it is predicted that the advent of commercially available stent-grafts will greatly expand the capabilities of the interventional radiologist in treating vascular trauma.

Vasospastic and Inflammatory Disorders

Raynaud's Disease **(Fig. 31)** consists of intermittent vasoconstriction of the digital arteries often induced by cold temperatures or emotional stimuli with resultant pallor, paresthesias, cyanosis and rubor with reperfusion **(Table 8)**. Young women are characteristically affected; both upper extremities are involved; ischemic changes are rare; and underlying disease is absent. If an underlying condition (such as rheumatoid arthritis, systemic lupus erythematosus, progressive systemic sclerosis or polyarteritis nodosa) is present, the disorder is known as Raynaud's phenomenon. The purpose of angiography is threefold: 1) to exclude another etiology such as an aneurysm causing emboli or thoracic outlet compression; 2) to assess the response of the arteries to vasodilators; and 3) to evaluate the extent of arterial occlusion. Typical arteriographic findings include distal small-vessel spasm, nonopacification and/or slow flow to the hand with improvement after intraarterial administration of tolazoline (15-25 mg) or reserpine (0.5-1.0 mg).

TABLE 8: VASOSPASTIC DISEASE

- Raynaud's disease
  - Young women, stimulated by cold or emotions
  - Intermittent vasoconstriction of the digital arteries bilaterally
  - Partially responsive to vasodilators and rare ischemic changes
- Raynaud's phenomenon
  - Underlying condition, collagen vascular disease, thoracic outlet syndrome, endocrine disorder, malignancy, drug-related
  - Ischemic changes and ulcers more frequent
- Role of angiography
  - Evaluate extent of stenoses and occlusion, exclude underlying condition like aneurysms in PAN, hypothenar hammer syndrome
  - Assess response to vasodilator

Thromboangiitis Obliterans (Buerger's Disease): Raynaud's phenomenon can also be associated with thromboangiitis obliterans. Patients with Buerger's disease are typically young (30-40), male, and cigarette smokers. Symptoms include intermittent claudication, as the lower extremity vessels are frequently involved, and peripheral upper extremity ischemic changes including ischemic finger ulcers. Rest pain, decreased skin temperature and sensitivity to cold can be present in the hands. The angiographic appearance includes stenoses and occlusions in small and medium-sized arteries with tortuous, "corrugated" or "corkscrew" appearing collaterals, lack of calcification, segmental occlusions of the palmar arches and digital arteries, and occlusion of the radial and ulnar arteries.

Arteriovenous Malformations

True congenital arteriovenous malformations (AVMs) result from embryologic failure of development of normal capillaries with preservation of primitive vascular connections. In this condition, innumerable collaterals form a network or "nidus" which is fed by enlarged arteries and drained by enlarged, early-filling veins. AVMs are generally isolated, with no sexual predilection, usually not inherited and, if stable, will likely require no specific treatment. Patients with symptoms of enlargement may require treatment. These include overgrowth of a limb or digit, cosmetic deformities, ulceration and bleeding, or rarely, high output cardiac failure. Thorough imaging evaluation is performed prior to treatment. Cross-sectional techniques such as CT or MR are used to determine the extent of adjacent soft tissue, and muscle and bone involvement. MR imaging and MRA can be employed to identify pretreatment flow characteristics and provide a baseline study for follow-up surveillance. Angiography **(Fig. 32)** is undertaken to distinguish the malformation from a vascular tumor, and to determine the arterial supply to the malformation and the venous drainage. Typical angiographic findings include: 1) multiple hypertophied, tortuous feeding vessels, 2) a hypervascular nidus with numerous small central vessels, and 3) rapid arteriovenous shunting with early opacification of large draining veins.

THROMBOANGIITIS OBLITERANS

- Male, cigarette smokers
- Claudication, ischemic changes, Raynaud's phenomenon
- Angiographic findings
  - Stenoses, occlusions in small to medium-sized arteries
  - Tortuous, corkscrew or corrugated-appearing collaterals

Hemorrhage, secondary ischemia and high output heart failure are indications for attempting treatment. The specific objectives of treatment include pain reduction, cosmesis, improved function, and limb preservation. Transcatheter embolization appears to be the best treatment option. Surgical ligation of large feeding vessels is not an attractive option because, at best, surgery provides only temporary relief. If the nidus is left intact, it will recruit collateral feeders from any adjacent vascular supply. Only embolization of the nidus or complete surgical resection of the lesion will have a lasting result.

Superselective catheterization and transcatheter embolization for obliteration of the nidus is performed at limited centers. In performing transcatheter embolization, attempts are made to avoid proximal occlusion. Numerous embolic agents have been used for this application including particulate polyvinyl alcohol, absolute ethanol and rapidly polymerizing acrylic adhesives. Complications of embolization include tissue necrosis, inadvertent nontarget embolization, and passage to the venous circulation. The most encouraging results have been reported by the experienced interventionalists at New York University Medical Center who achieved success in 25% of 215 patients with no residual disease at 3 years and partial success in 30% who are asymptomatic, but have imaging evidence of a residual or recurrent nidus.

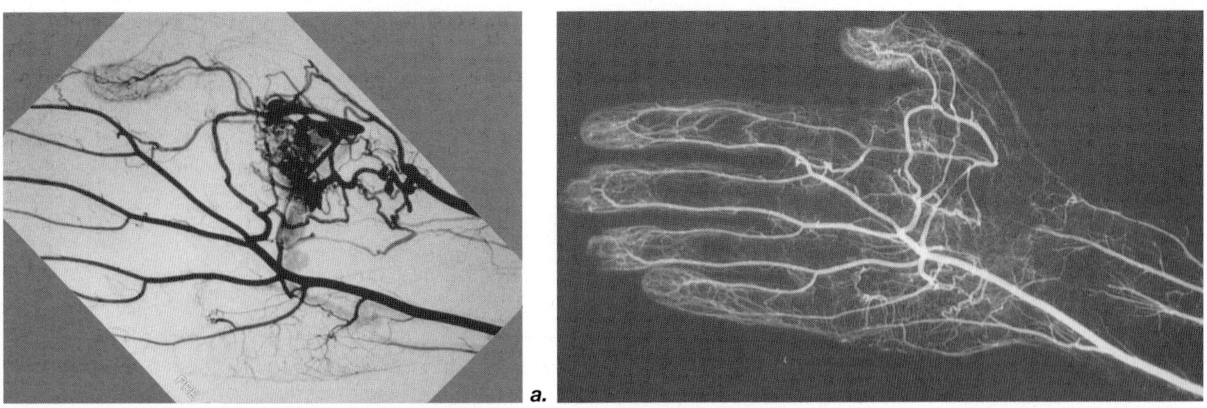

a. b.

Figure 32. *a) Arteriovenous malformation over the thenar eminence fed by radial artery branches. b) Obliteration of the nidus following embolization with experimental rapidly polymerizing N-butyl-cyanoacrylate (NBCA) adhesive (Image courtesy of Robert J. Rosen, MD, New York University Medical Center).*

REFERENCES

Thoracic Aortography

Chung JW, Park JH, IM JG, et al. Spiral CT angiography of the thoracic aorta. Radiographics 1996; 16:811-824.

Dyer DS, Moore EE, Mestek MF, et al. Can chest CT be used to exclude aortic injury? Radiology 1999; 213:195-202.

Fisher RG, Chasen MH, Lamki N. Diagnosis of injuries of the aorta and brachiocephalic arteries caused by blunt chest trauma: CT vs. aortography. AJR 1994; 162:1047-1052.

Fischer RG, Sanchez-Torres M, Thomas JW, Whigham CJ. Subtle or atypical injuries of the thoracic aorta and brachiocephalic vessels in blunt thoracic trauma. Radiographics 1997; 17: 835-849.

Fischer RG, Sanchez-Torres M, Whigham CJ, Thomas JW. "Lumps" and "bumps" that mimic acute aortic and brachiocephalic vessel injury. Radiographics 1997; 17:825-834.

Fishman JE, Nunez D Jr., Kane A, et al. Direct versus indirect signs of traumatic aortic injury revealed by helical CT: performance characteristics and interobserver agreement. AJR 1999; 172;1027-1031.

Gavant ML, Flick P, Menke P, Gold RE. CT aortography of thoracic aortic rupture. AJR 1996; 166:955-961.

Gavant ML, Menke PG, Fabian T, Flick PA, Graney MJ, Gold RE. Blunt traumatic aortic rupture: detection with helical CT of the chest. Radiology 1995; 197:125-133.

Hartnell GG, Finn JP, Zenni M, et al. MR imaging of the thoracic aorta: comparison of spin-echo, angiographic, and breath-hold techniques. Radiology 1994; 191:697-704.

Kadir S. Arteriography of the thoracic aorta in diagnostic angiography. Philadelphia: W.B. Saunders 1986; 124-171.

Kadir S. Regional anatomy of the thoracic aorta in atlas of normal and variant angiographic anatomy. Philadelphia: W.B. Saunders 1991; 19-54.

Kearney PA, Smith DW, Johnson SB, et al. Use of transesophageal echocardiography in the evaluation of traumatc aortic injury. J Trauma 1993; 34:696-703.

Kato N, Dake MD, Miller DC, Semba CP, Mitchell RS, Razavi MK, Kee ST. Traumatic thoracic aortic aneurysm: treatment with endovascular stent grafts. Radiology 1997; 205:657-662.

Levy JR, Heiken JP, Gutierrez FR. Pictorial essay: imaging of penetrating atherosclerotic ulcers of the aorta. AJR 1999; 173:151-154.

Link KM, Lesko NM. The role of MR imaging in the evaluation of acquired diseases of the thoracic aorta. AJR 1992; 158:1115-1125.

Nunez D Jr., Rivas L, McKenney K, et al. Pictorial essay: helical CT of traumatic arterial injuries. AJR 1998; 170:1621-1626.

Rose SC, Moore EE. Angiographic artifacts that simulate arterial pathology in acute trauma. Am J Emerg Med 1990; 8:109-117.

Yee J, Hung RK. Spiral CT angiography of aortic dissection. Emergency Radiology 1999; 6:24-31.

Upper Extremity Angiography

Chait A. Arteriography of the upper extremity in Abrams' angiography. Baum S, ed. 3rd edition. Boston: Little, Brown and Company, 1997; 1755-1766.

Dymarkowski S, Bosmans H, Marchal G, Bogaert. Three-dimensional MR angiography in the evaluation of thoracic outlet syndrome. AJR 1999; 173:1005-1008.

Johansen K, Lynch K, Paun M, Copass M. Noninvasive vascular tests reliably exclude occult arterial trauma in injured extremities. J Trauma 1991; 31:515-519.

Johnson SP, Durham JD, Subber SW, et al. Acute arterial occlusions of the small vessels of the hand and forearm: treatment with regional urokinase therapy. JVIR 1999; 10:869-876.

Kadir S. Arterial Anatomy of the upper extremities in atlas of normal and variant angiographic anatomy. Philadelphia: W.B. Saunders, 1991; 55-96.

Kadir S. Arteriography of the upper extremities in diagnostic angiography. Philadelphia: W.B. Saunders, 1986; 172-206.

Knudson MM, Lewis FR, Atkinson K, Neuhaus A. The role of duplex ultrasound arterial imaging in patients with penetrating extremity trauma. Arch Surg 1993; 128:1033-1037.

Ordog GJ, Balasubramanium S, Wasserberger J, Kram H, Bishop M, Shoemaker W. Extremity gunshot wounds: Part one- Identification and treatment of patients at high risk of vascular injury. J Trauma 1994; 36:358-368.

Patel AV, Marin ML, Veith FJ, Kerr A, Sanchez LA. Endovascular graft repair of penetrating subclavian artery injuries. J Endovasc Surg 1996; 3:382-388.

CHAPTER 2 PREVIEW ANSWERS

Case 1

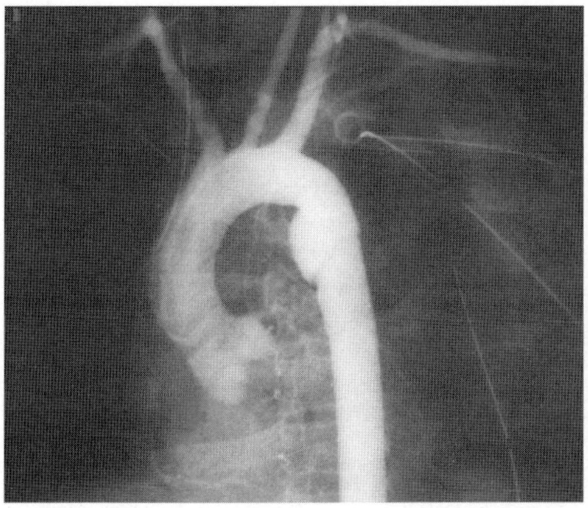

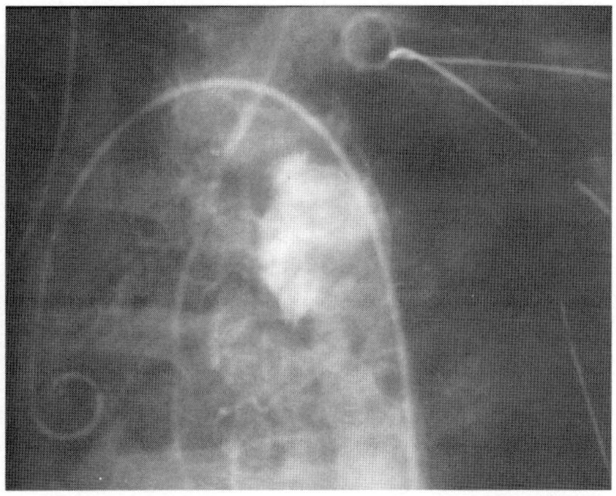

Findings: Chest film showed loss of definition of the descending aorta and a left apical cap. CT showed periaortic mediastinal hematoma, luminal contour irregularity, an intimal flap and bilateral hemothoraces. Angiograph demonstrates a pseudoaneurysm at the aortic isthmus and holdup of contrast on delayed images.
Diagnosis: Traumatic aortic injury.

Case 2

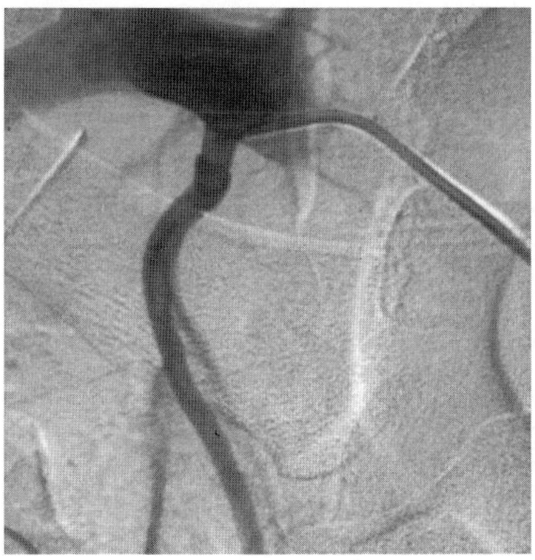

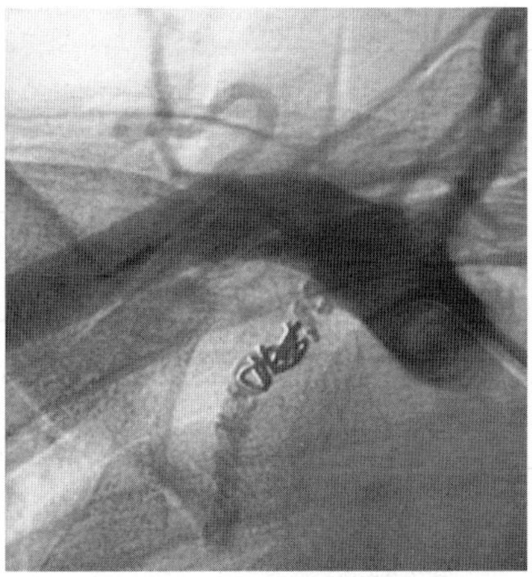

Findings: Chest film showed widened mediastinum and an irregular right paratracheal contour consistent with mediastinal hematoma. On aortography the arch appears normal but there is an irregular angulated contour of the proximal right internal mammary artery. Aneurysmal dilatation and intimal flap are also identified on selective catheterization.
Diagnosis: Internal mammary artery injury successfully embolized with 3-mm and 4-mm coils.

Case 3

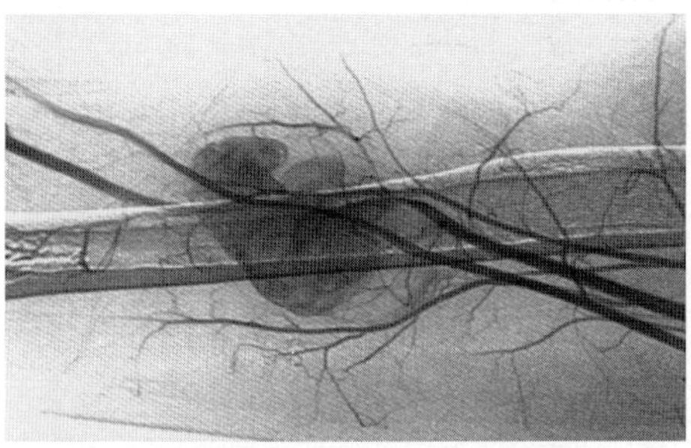

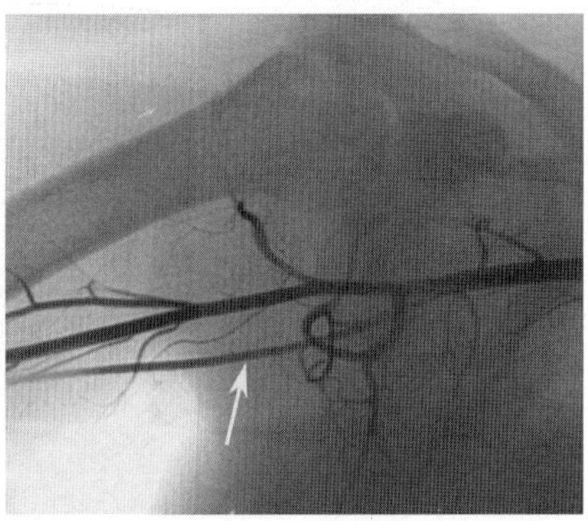

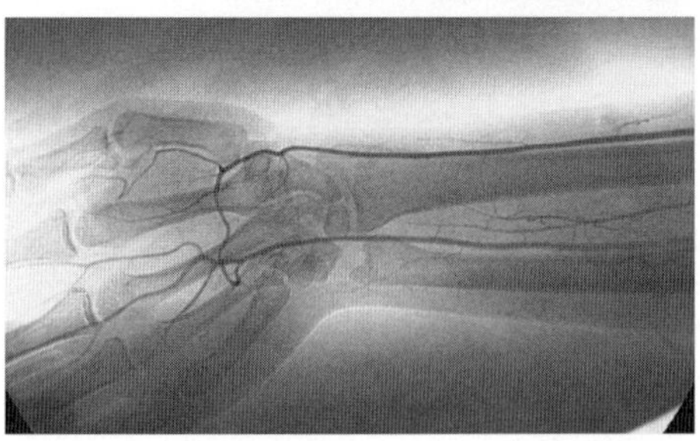

Findings: There is a false aneurysm (contained extravasation) of the brachial artery causing compressive ulnar and median neuropathy. The high take-off of the radial artery in the axilla (arrow) points out the importance of proximal catheter position and explains the strong radial pulse (potentially confounding the initial physical exam).
Diagnosis: Traumatic injury (large false aneurysm) of the brachial artery in a patient with variant anatomy (high radial artery take-off).

CHAPTER 3 PREVIEW

Case 1

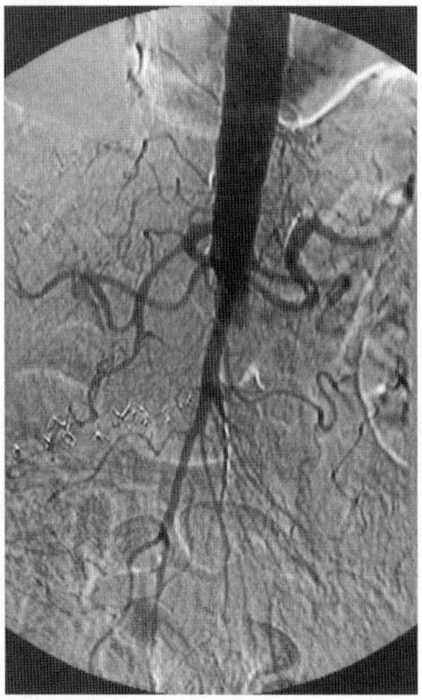

Figure 1. *Abdominal aortogram.*

History: A 70-year-old woman presented with acute, bilateral lower extremity ischemia. On exam, her femoral pulses were absent. A catheter was placed in the ascending thoracic aorta from an axillary artery approach and an abdominal aortogram was obtained **(Fig. 1)**.

Question: What is your diagnosis?

Case 2

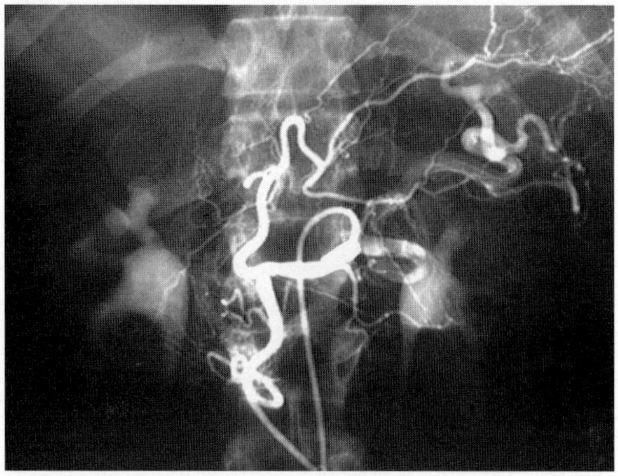

Figure 2. *Selective common hepatic arteriogram.*

History: A 49-year-old woman complained of right mid-abdominal pain. Abdominal CT revealed a mass in the right lobe of the liver. Preoperative visceral arteriography was performed **(Fig. 2)**.

Question: Where is the right hepatic artery?

Case 3

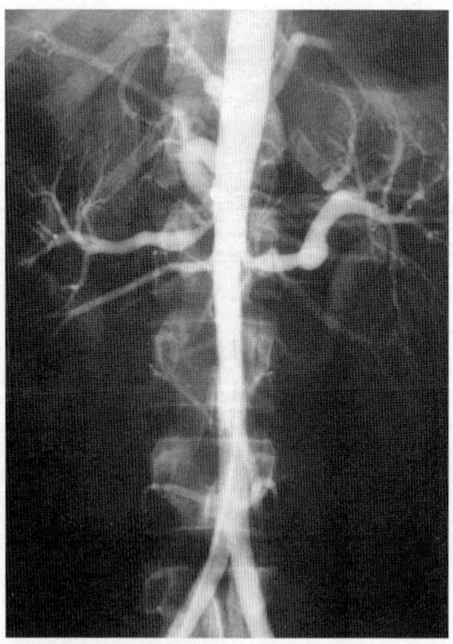

Figure 3. *Abdominal aortogram.*

History: A 24-year-old woman was evaluated for hypertension. Additional history is withheld. An abdominal aortogram was performed **(Fig. 3)**.

Question: What is your diagnosis?

Chapter 3
Abdominal Aortography, Visceral and Renal Arteriography

Jeanne M. LaBerge, MD,

and Susan D. Wall, MD

PATHOLOGIC CONDITIONS AFFECTING THE ABDOMINAL AORTA AND ITS BRANCHES

Abdominal Aortic Pathology
- Occlusive atherosclerosis
- Abdominal aortic coarctation
- Abdominal aortic aneurysm
- Aortic occlusion

Visceral Pathology (Celiac, SMA, IMA)
- Mesenteric ischemia
  - Atherosclerosis, median arcuate ligament compression syndrome
  - Embolic occlusion
  - Nonocclusive mesenteric ischemia
- Hepatic tumors
  - Cavernous hemangioma
  - Hepatic adenoma
  - Focal nodular hyperplasia
  - Regenerating nodules
  - Hepatocellular carcinoma (hepatoma)
  - Fibrolamellar hepatocellular carcinoma
  - Hypervascular and hypovascular metastases
  - Peliosis hepatis
- Splenic artery aneurysm

Renal Pathology
- Renovascular hypertension
  - Fibromuscular disease, atherosclerosis, other
- Renal artery aneurysms
  - Atherosclerosis, polyarteritis nodosa
- Renal tumors

INTRODUCTION

In this chapter, we review the normal and pathologic anatomy of the aorta, visceral and renal vessels. A list of the pathologic conditions that will be presented is shown above.

ABDOMINAL AORTA: NORMAL ANATOMY

Course and Caliber

The aorta courses to the left of the spine from the aortic hiatus to its bifurcation at the L4 level **(Fig. 1)**. It is normally 1.5 cm - 2.0 cm in diameter. Three centimeters is the upper limit of normal diameter. The aorta gradually tapers from its entry into the abdomen to its bifurcation **(Fig. 2)**. High bifurcations are rare **(Fig. 3)**. The average bifurcation angle is 60°.

(* *Anatomic diagrams in this chapter are reproduced courtesy of* Eastman Kodak Inc.)

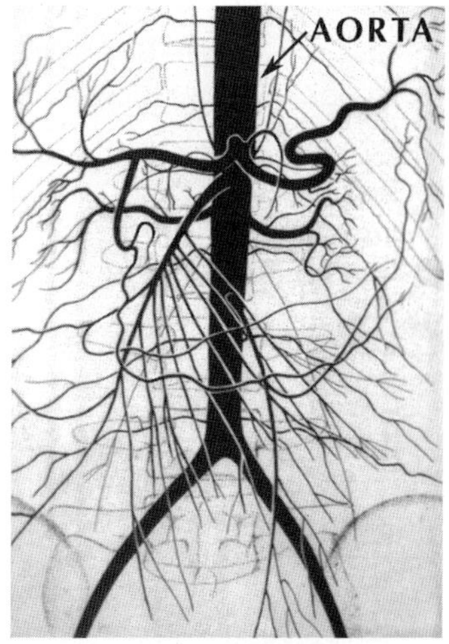

Figure 1. The abdominal aorta and its branches. Note the gradual taper of the distal aorta.*

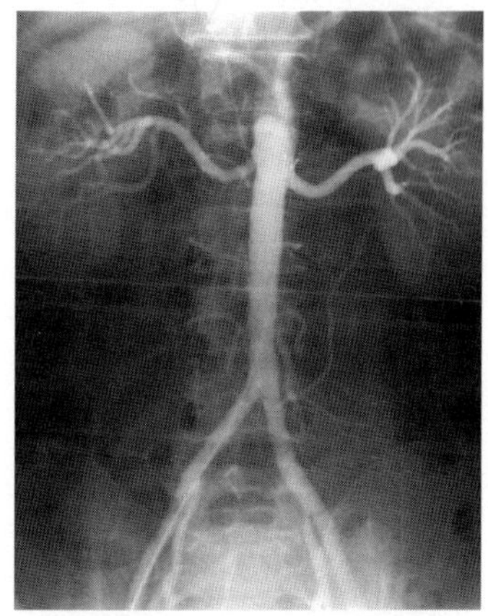

***Figure 2.** Mid-stream aortogram showing the normal course and caliber of the aorta. Note the aortic bifurcation at the L-4 level.*

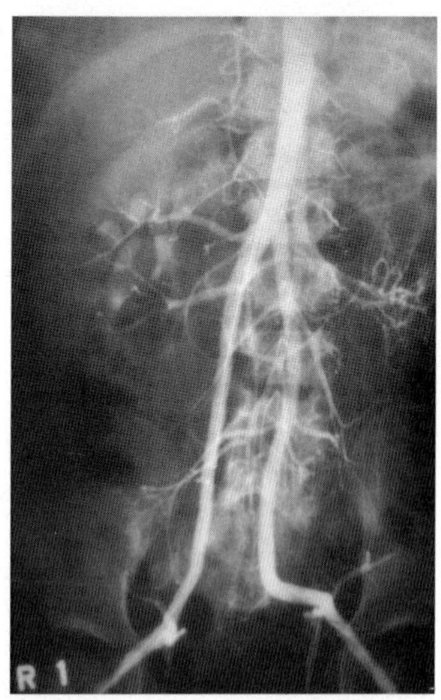

***Figure 3.** Mid-stream aortogram in a patient with an unusually high bifurcation. This is a rare vascular anomaly.*

TABLE 1: BRANCHES OF THE ABDOMINAL AORTA

Dorsal branches	Ventral branches	Terminal branches
Lumbar arteries	Celiac	Common iliac arteries
Inferior phrenic artery	— Splenic artery	Median sacral artery
Middle adrenal artery	— Common hepatic artery	
Gonadal arteries	— Left gastric artery	
Renal arteries	Superior mesenteric artery	
Median sacral artery	Inferior mesenteric artery	

Branches

The major branches of the abdominal aorta are listed in **Table 1**. Salient anatomic features of these branches are mentioned below.

The four paired lumbar arteries provide important collaterals that anastomose with branches of the internal iliac arteries in patients with aortoiliac occlusive disease. The lumbars typically anastomose with the posterior division branches of the internal iliac artery **(Fig. 4)**. The median sacral artery is a small terminal branch of the aorta that can also serve as a collateral in occlusive disease. The inferior phrenic arteries typically originate as a common trunk from the aorta above the level of the celiac artery. These vessels can serve as collaterals in occlusive disease or as a source of parasitic arterial supply to liver tumors.

The middle adrenal arteries take origin at the level of the SMA. They can provide collaterals to the kidneys in patients with renal occlusive disease. The gonadal arteries arise from the aorta at the L2-3 level in most patients and otherwise take origin from the proximal renal artery.

Anatomic features of the renal arteries and the ventral visceral branches—the celiac, superior and inferior mesenteric arteries—will be described in greater detail later.

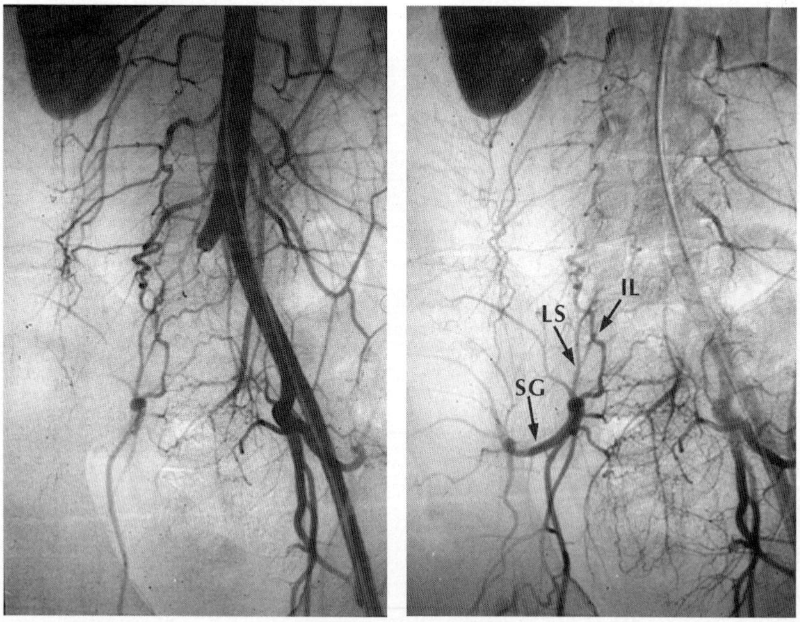

a. b.

Figure 4. *Pelvic arteriography in a patient with a right iliac graft occlusion demonstrates collateral flow from the lumbar arteries to the posterior division of the internal iliac artery. a) Early arterial phase image shows enlarged lumbar arteries and initial filling of the right internal iliac artery via the iliolumbar artery. b) Later image shows complete opacification of the posterior division branches of the internal iliac artery – lateral sacral (LS), iliolumbar (IL), and superior gluteal (SG) arteries.*

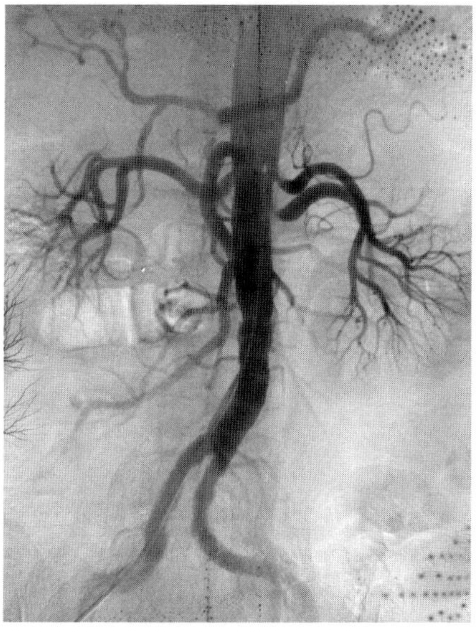

Figure 5. *Midstream aortogram shows atherosclerotic occlusive changes. The infrarenal aorta is tortuous, the lateral walls are irregular, and there are stenoses of the left renal artery origins.*

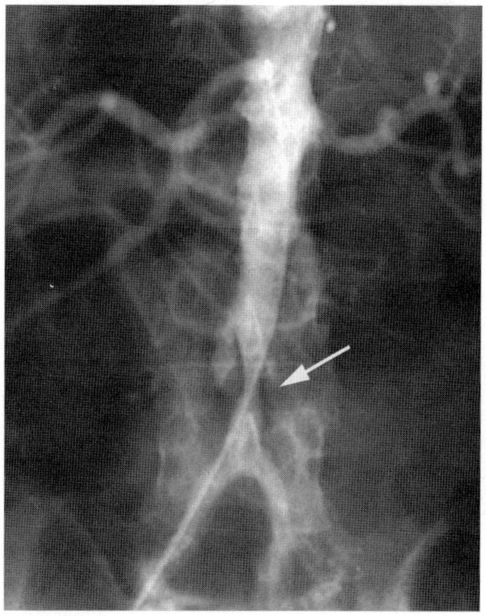

Figure 6. *Infrarenal aortogram in a patient with buttock claudication reveals a flow limiting stenosis of the infrarenal aorta at the bifurcation (arrow).*

ABDOMINAL AORTA: PATHOLOGY

Occlusive Atherosclerosis

Atherosclerosis of the aorta can be manifest as either occlusive or aneurysmal disease. Angiographic findings with occlusive disease include tortuosity, mural irregularity, and luminal narrowing **(Fig. 5)**. Because of the large diameter of the aorta, occlusive disease must be severe before it becomes hemodynamically significant. However, when present, a flow limiting stenosis of the infrarenal aorta can result in buttock claudication or impotence **(Fig. 6)**.

Occlusive disease tends to be most severe along the posterior wall of the aorta. It usually begins below the origin of the renal arteries and is more common distally. The end result of distal occlusive disease is aortic occlusion. If aortic side branches are hypertrophied, the remaining infrarenal aorta will remain patent **(Fig. 7)**. More typically, side-branch collateral flow is minimal and slow flow in the infrarenal aorta eventually leads to thrombosis of the entire infrarenal aorta **(Fig. 8)**. An important subset of young male patients with infrarenal aortic occlusion have Leriche syndrome, which is a syndrome characterized by aortic occlusion, impotence, and common iliac artery occlusive disease.

LERICHE SYNDROME

The clinical triad includes:

- Claudication
- Impotence
- Diminished femoral pulses

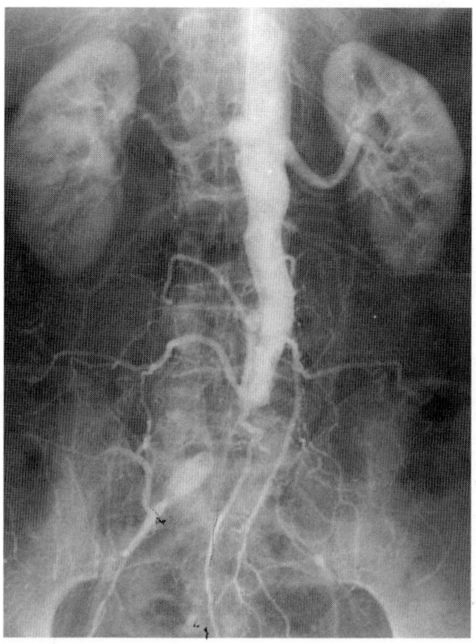

Figure 7. *Midstream aortogram. The aorta is occluded at the bifurcation but the infrarenal aorta remains patent. Flow through the side-branches (lumbars, IMA and middle sacral artery) helps to maintain infrarenal aortic patency.*

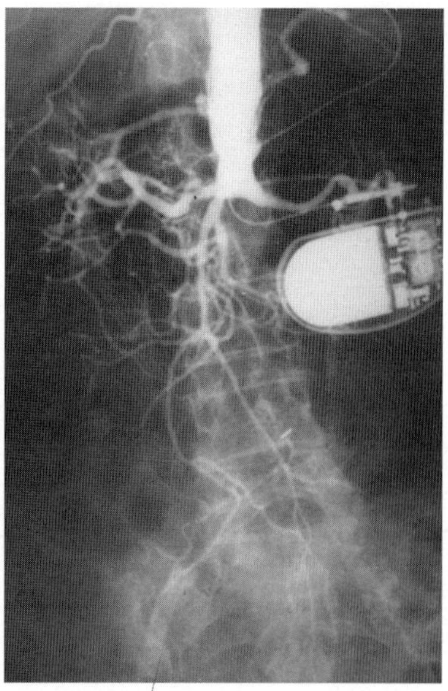

Figure 8. *Infrarenal aortic occlusion in a patient with long-standing claudication. Note occlusion of the aorta up to the level of the renal arteries.*

ABDOMINAL AORTIC COARCTATION

- Pan-aortitis leading to stenosis or occlusion
- Segmental, diffuse smooth narrowing
- Claudication, hypertension, abdominal angina
- Atherosclerosis, radiation, Takayasu's, neurofibromatosis, FMD, rubella, congenital

Abdominal Aortic Coarctation

While infrarenal aortic stenosis is usually due to atherosclerosis, the differential diagnosis of focal aortic stenosis (or coarctation) includes a variety of uncommon nonatherosclerotic processes. Abdominal coarctation may be congenital (2% of aortic coarctations occur in the abdomen) or acquired. Diseases such as Takayasu's, rubella, neurofibromatosis, idiopathic hypercalcemia or radiation fibrosis can all result in narrowing of the abdominal aorta. Patients usually present with symptoms of claudication, hypertension or abdominal angina.

The angiographic appearance of abdominal coarctation is quite striking. Characteristically, there is a severe stenosis of the perirenal aorta. The visceral vessels are massively enlarged and provide collateral flow to the lower extremity. Of note, Takayasu's arteritis is an inflammatory vasculitis most frequently detected in young women who develop long, smooth, concentric stenosis of the thoracic or abdominal aorta.

ABDOMINAL AORTIC ANEURYSM

- 90% are infrarenal
- Two-thirds extend into iliacs
- 10% have perianeurysmal fibrosis
  - inflammatory process of unknown etiology
  - can obstruct ureters, simulates retroperitoneal fibrosis
- Rupture occurs in body of aneurysm, not at juncture with nondilated aorta

Abdominal Aortic Aneurysm (AAA)

By definition, an abdominal aortic aneurysm is any widening of the aorta greater than 3 cm. Most aneurysms of the abdominal aorta are true aneurysms due to atherosclerosis (90%) **(Fig. 9)**. The remainder (10%) are pseudoaneurysms due to bacterial infection (mycotic) **(Fig. 10)**, trauma, or breakdown at a surgical anastomosis.

Atherosclerotic true aneurysms are usually located in the infrarenal aorta (90%), and two-thirds extend into the iliac arteries. Most patients are asymptomatic or present with a palpable mass. A minority of patients present with abdominal pain. Ten percent of AAA's are "inflammatory" and exhibit perianeurysmal fibrosis that can lead to entrapment, deviation or obstruction of the ureters.

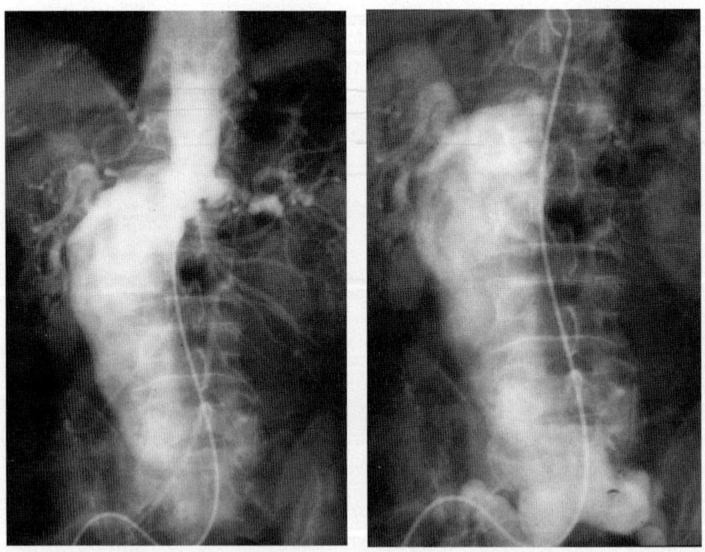

a. b.

Figure 9. *True aneurysm of the abdominal aorta demonstrated by catheter angiography. a) Early arterial phase image shows that the aneurysm begins at the level of the renal arteries. b) Later image demonstrates the extent of the aneurysm which involves the entire infrarenal aorta and proximal common iliac arteries.*

The chief clinical concern in patients with abdominal aortic aneurysm is rupture. The risk of rupture is related to size. That is, for aneurysms larger than 4 cm in diameter, the annual rate of rupture increases with diameter. Although the exact risk for a given diameter is not clear, most experts agree that the annual risk of rupture for aneurysms 6 cm or greater is at least 10% per year. For aneurysms greater than 7.5 cm, the risk may be as high as 30% per year.

Surgical repair of AAA is usually undertaken for symptomatic aneurysms > 6 cm in diameter or if the aneurysm increases in size at a rate of > 0.5 cm per year. Morbidity for elective repair is less than 2% but is as high as 60% for emergent procedures.

Pre-operative imaging studies are performed to evaluate: 1) the renal artery locations and presence of concomitant disease; 2) the extent of disease in the iliacs along with an exact measurement of the aneurysm dimension, and 3) side branch patency, particularly the status of IMA and internal iliac arteries. In the past catheter angiography has been performed to evaluate patients with AAA, but it has now been superseded by CTA at UCSF **(Fig. 11)**.

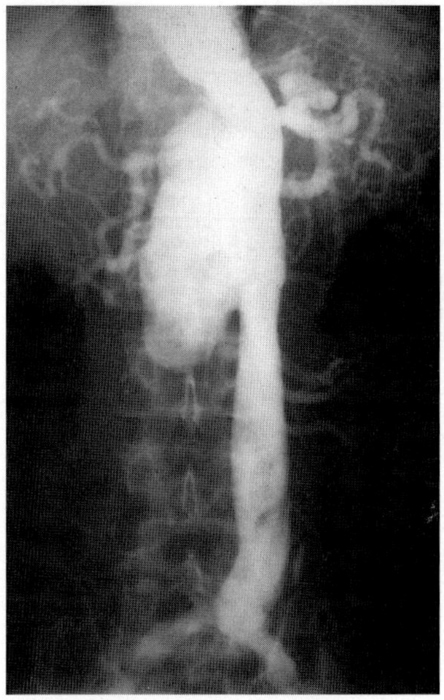

Figure 10. *Midstream aortogram demonstrates a large eccentric saccular aneurym of the perirenal aorta. This proved to be a mycotic pseudoaneurysm.*

Abdominal Aortic Occlusion

The differential diagnosis of aortic occlusion includes progressive occlusive disease **(Fig. 12)** and massive thromboembolism. Atherosclerotic occlusion of the aorta typically extends up to the level of the renal arteries. The presence of collaterals provides evidence of chronic longstanding vascular disease. Embolic occlusion is usually acute and is not associated with significant collaterals. In addition, with embolic occlusion there may be a stump of aorta below the renal arteries that has not yet thrombosed (*see Chapter Preview- Case 1*).

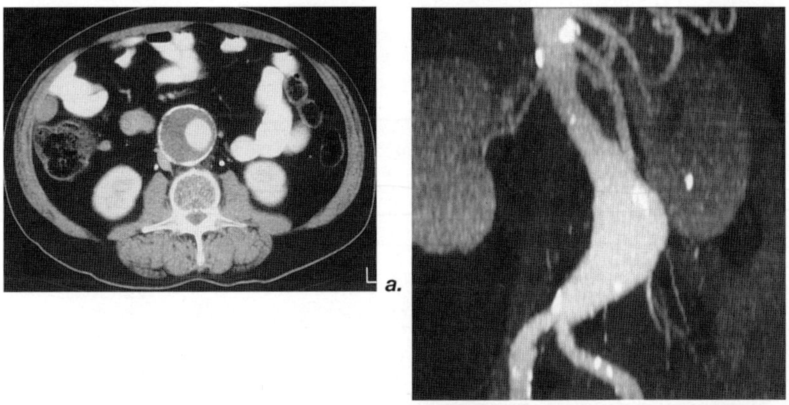

a. b.

Figure 11. *Currently, preoperative imaging of AAA includes axial CT with CT reconstruction. a) Axial imaging provides an accurate measure of the dimensions of the patent lumen and aneurysm sac. b) Coronal maximal intensity projection image (MIP) reconstruction demonstrates the relationship of the aneurysm to the renal and visceral vessels.*

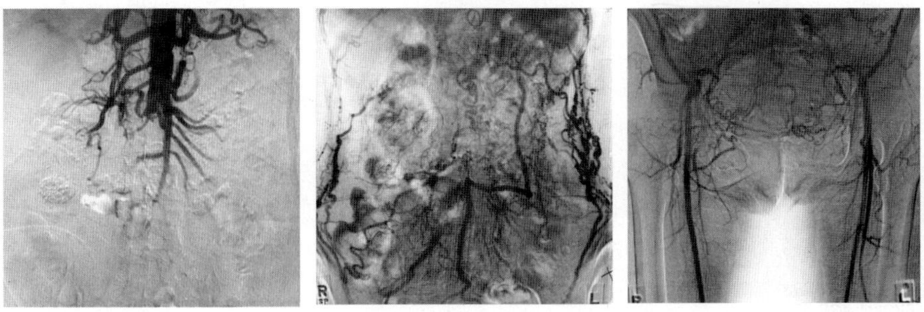

a. b. c.

Figure 12. *Infrarenal aortic occlusion in a patient with longstanding aortic occlusive disease. a) Early arterial image demonstrates occlusion of the aorta at the level of the renal arteries. b) Later image shows collateral flow to the pelvis and lower extremities via the SMA-IMA-superior hemorrhoidal-internal iliac pathway and via the lumbar, intercostal to circumflex iliac artery pathway. c) Reconstituted flow to the lower extremity.*

VISCERAL ARTERIES
• Celiac — Left gastric, splenic, hepatic • SMA — IPD, jejunal, MCA, RCA, ileocolic • IMA — Left colic, superior hemorrhoidal

Visceral Vasculature: Anatomy

Within the context of this chapter, the visceral vessels are defined as the celiac, SMA and IMA. The celiac artery typically takes origin at the T12-L1 level **(Fig. 13)**. It has three branches: the left gastric, hepatic, and splenic arteries.

The left gastric artery is the first branch of the celiac and supplies the distal esophagus and stomach. Its branches anastomose to the gastroepiploic and short gastric arteries. As well, the left gastric artery joins with the right gastric artery to form an arcade along the lesser curvature of the stomach **(Fig. 14)**. The right gastric artery usually arises from the proper hepatic artery (40%) or the left hepatic artery (40%).

The splenic artery supplies the spleen and pancreas. Branches to the pancreas include the dorsal pancreatic (DP) and the pancreatica magna. Alternatively, the DP may arise from the celiac or SMA. The other splenic artery branches of note are the short gastric branches to the stomach and the left gastroepiploic to the stomach and omentum.

The common hepatic artery supplies the liver, gallbladder and duodenum **(Fig. 15)**. The first branch is the gastroduodenal (GDA) which in turn gives off pancreaticoduodenal branches and then continues as the gastroepiploic **(Fig. 16)**. The cystic artery usually arises from the right hepatic artery.

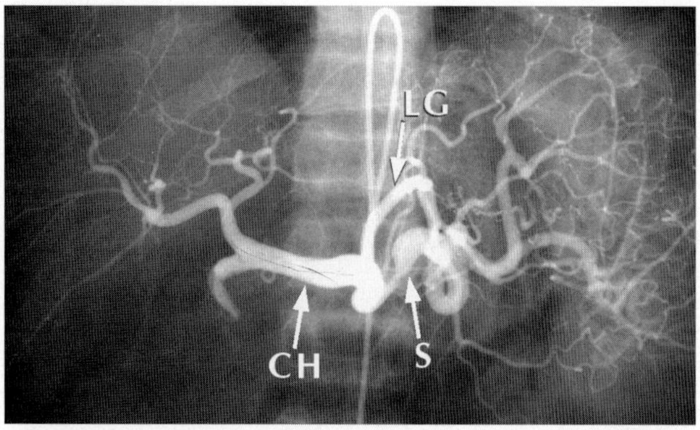

Figure 13. *Celiac arteriogram demonstrating the three major branches: left gastric (LG), splenic (S), and common hepatic (CH) arteries.*

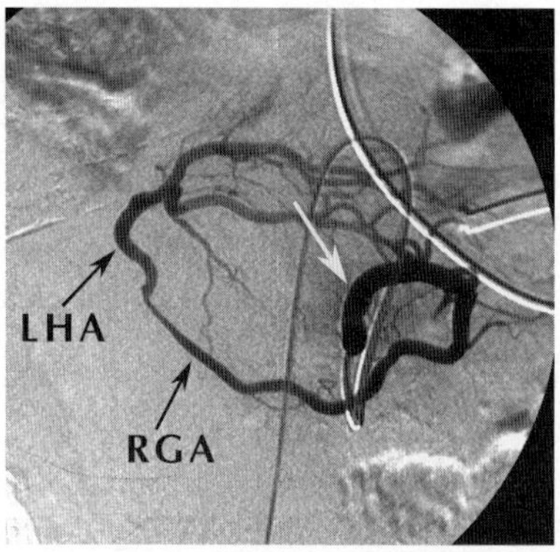

Figure 14. *Selective left gastric arteriogram. Note enlargement of the left gastric artery (arrow) and filling of the arcade along the lesser curvature from the left gastric to the right gastric (RGA) to fill the left hepatic artery (LHA).*

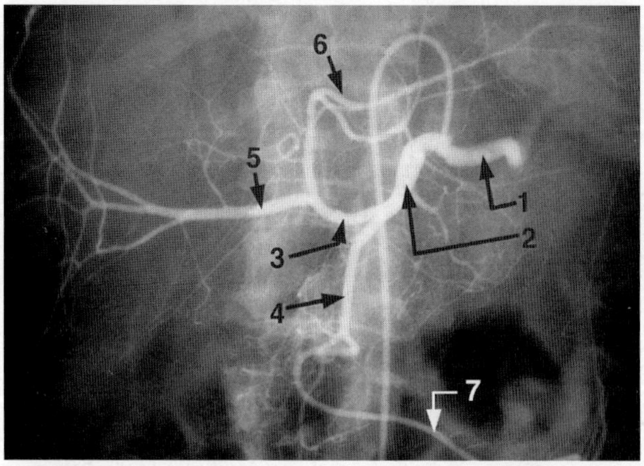

Figure 15. *Selective common hepatic arteriogram. There is reflux into the splenic artery (1). The common hepatic artery (2) bifurcates into the proper hepatic (3) and gastroduodenal (4) arteries. The PHA in turn bifurcates into right (5) and left (6) hepatic arteries. The terminal branch of the GDA (7) is the gastroepiploic artery.*

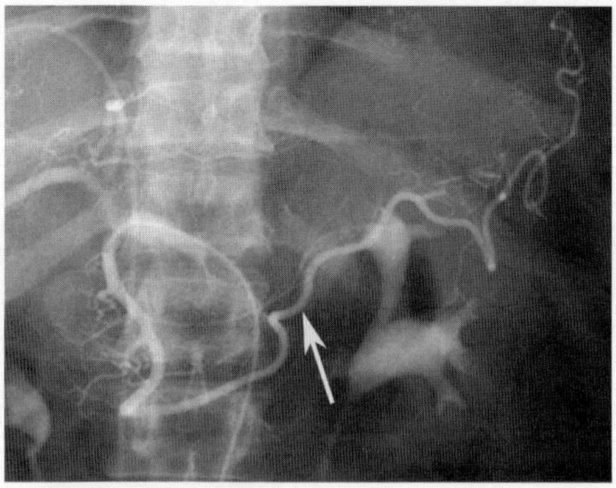

Figure 16. *Selective gastroduodenal arteriogram. Note the duodenal branches and the gastroepiploic artery (arrow).*

The branching pattern of the CHA is quite variable. Common variations are described in **Tables 2-4**. Examples are shown in **Figures 17-19**. (CH=common hepatic, PH=proper hepatic, LH=left hepatic, RH=right hepatic, GDA=gastroduodenal artery, SMA=superior mesenteric artery, LGA=left gastric artery.)

TABLE 2: OVERALL HEPATIC ARTERY BRANCHING

Hepatic Artery Branching Pattern	Prevalence (approximate)
Conventional: all branches off CHA	75%
Replaced or aberrant • One vessel, 30% • Two or more, 10%	40%

TABLE 3: ABERRANT RIGHT HEPATIC ARTERY (Overall Prevalence, 25%)

- Replaced - 18%
  - — from SMA, 12%
  - — from Aorta, < 2%
- Accessory - 7-8%
  - — from GDA, 2%
  - — from SMA, 5%
- Overall: SMA gives rise to RH in 20%

TABLE 4: ABERRANT LEFT HEPATIC ARTERY (Overall Prevalence, 25%)

- Replaced - 18%
  - — from LGA, 12%
  - — from SMA (via CH), 2%
- Accessory - 8%
  - — from LGA, 12%
- LH/LGA trunk from Aorta, 2%
- Overall: LG gives rise to LH in 25%

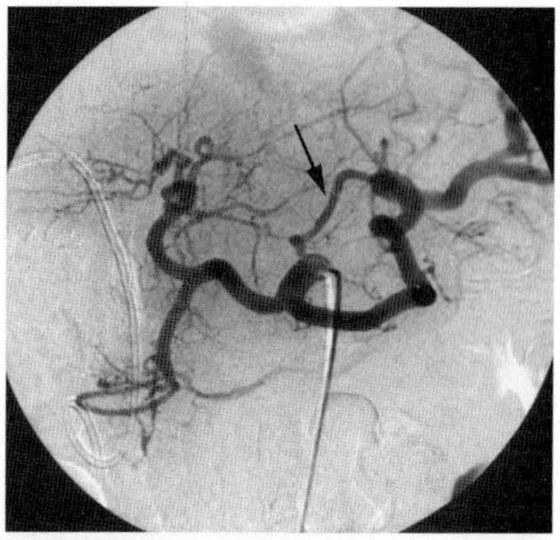

a.

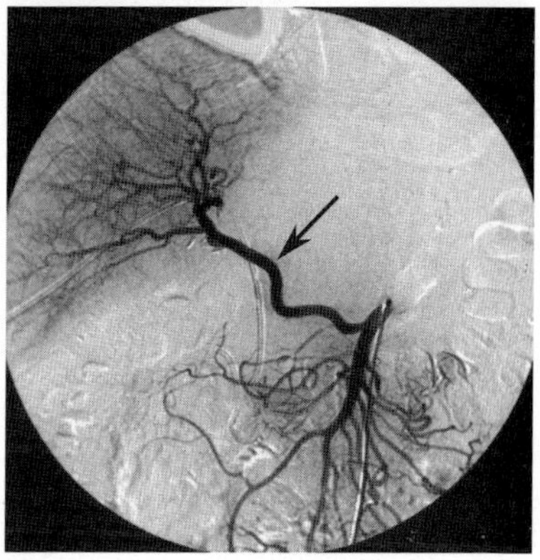

b.

Figure 17. A preoperative visceral angiogram was performed on a patient with metastatic colon cancer. a) Selective celiac arteriogram opacifies the left gastric (arrow), splenic and common hepatic arteries. Note that the proper hepatic artery supplies only the left lobe of the liver. b) Selective SMA injection reveals a replaced right hepatic artery (arrow).

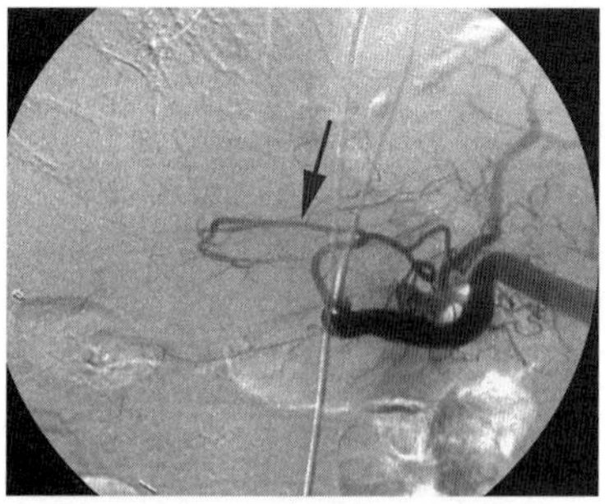

***Figure 18.** Preoperative arteriogram showing replaced left hepatic artery (arrow) from left gastric artery.*

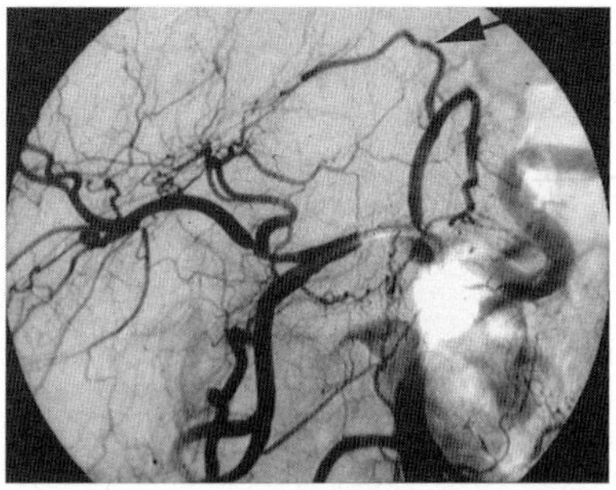

***Figure 19.** Accessory left hepatic artery (arrow) from left gastric artery.*

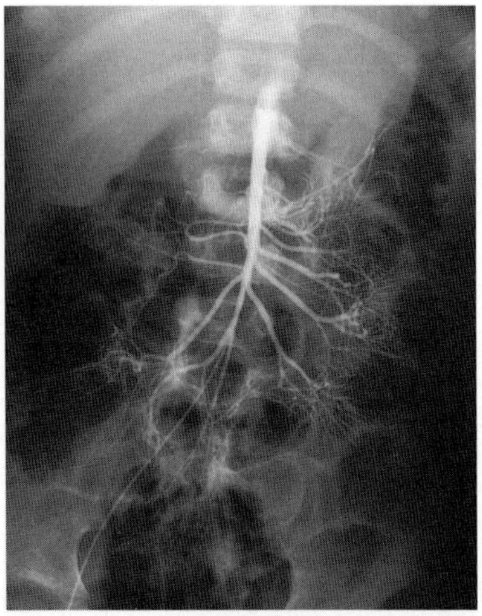

Figure 20. *Selective visceral arteriogram displays the normal branches of the SMA (the jejunal, ileal, middle colic, right colic and iliocolic arteries).*

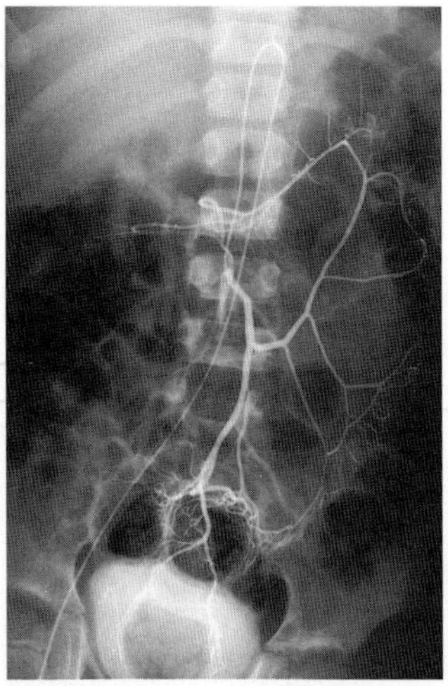

Figure 21. *Selective IMA angiogram in a child demonstrates normal anatomy. The IMA bifurcates into superior hemorrhoidal and left colic arteries. The left colic anastomoses with the middle colic at the splenic flexure.*

The SMA arises at the level of L-1 **(Fig. 20)** approximately 1 cm below the celiac. Its branches include:

- Inferior pancreaticoduodenal. This vessel divides into anterior and posterior branches which anastomose with corresponding branches of the superior pancreaticoduodenal arteries from the GDA to form the duodenal arcades.
- Jejunal and ileal arteries (10 -14 branches).
- Middle colic artery. This vessel divides into left and right branches. (Uncommonly, it may originate from the celiac, splenic or dorsal pancreatic artery.)
- Right colic artery. This vessel divides into ascending and descending branches that supply the ascending colon and hepatic flexure.
- Ileocolic artery. This vessel supplies the cecum, terminal ilium, and appendix.

The inferior mesenteric artery arises at the L3-L4 level just to the left of midline **(Fig. 21)**. Its branches include the left colic artery, 2-3 sigmoid arteries and the superior hemorrhoidal artery.

VISCERAL COLLATERALS (EPONYMS)

- Marginal artery of Drummond
- Arc of Riolan
- Arc of Buehler
- Arc of Barkow

Visceral Collaterals

The pathways for flow between the visceral vessels are numerous. Several named collaterals are listed below:

- Marginal artery of Drummond: An anastomotic, marginal colonic artery that runs along the mesenteric border of left colon providing vasa recta to the bowel. It is formed by the anastomosis of the ascending branch of left colic artery from the IMA with the transverse branch of middle colic artery from SMA **(Fig. 22)**.
- Arc of Riolan: This is an inconstant, more direct, medial anastomotic connection between middle colic and left colic arteries. It serves as an important collateral in occlusive mesenteric disease.
- Arc of Buehler: This is a persistent embryologic anastomosis between celiac and SMA. The more common pathway joining the celiac to the SMA is via the GDA and its branches **(Fig. 23)**.
- Arc of Barkow: (also known as the greater omental arcade)—this eponym applies to arteries and veins that run parallel to and anastomose with the right and left gastroepiploic arteries. This is the collateral portal venous pathway that is seen in splenic vein occlusion **(Fig. 24)**.

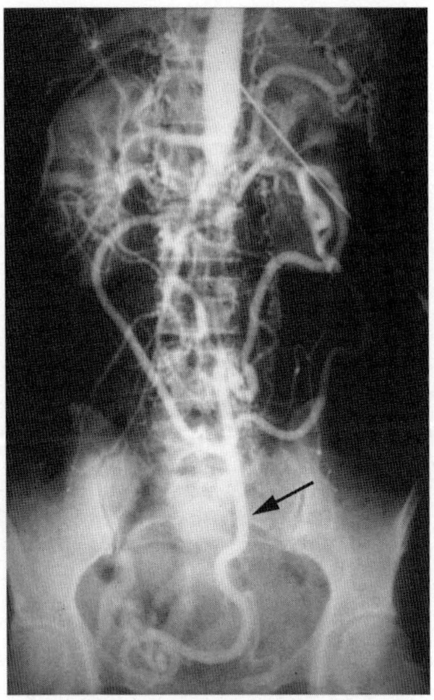

Figure 22. *Markedly enlarged collateral between the SMA and IMA in a patient with aortic occlusion. This collateral is providing vascular supply to the lower extremity through the superior hemorrhoidal artery (arrow).*

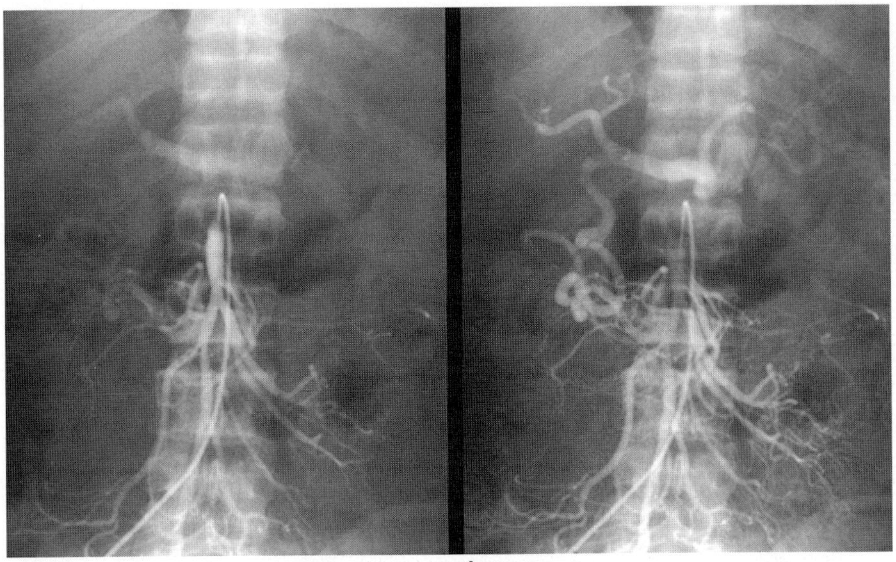

a. b.

Figure 23. *Selective SMA arteriogram in a patient with arcuate ligament compression of the celiac artery. a) Early arterial phase image shows opacification of the SMA. b) Later there is opacification of the celiac artery. Contrast flows from the SMA to the celiac via the inferior pancreaticoduodenal artery (IPD) to the GDA and common hepatic artery.*

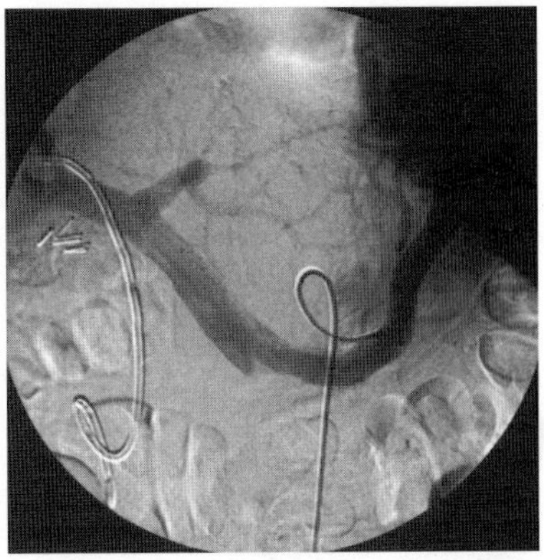

a.

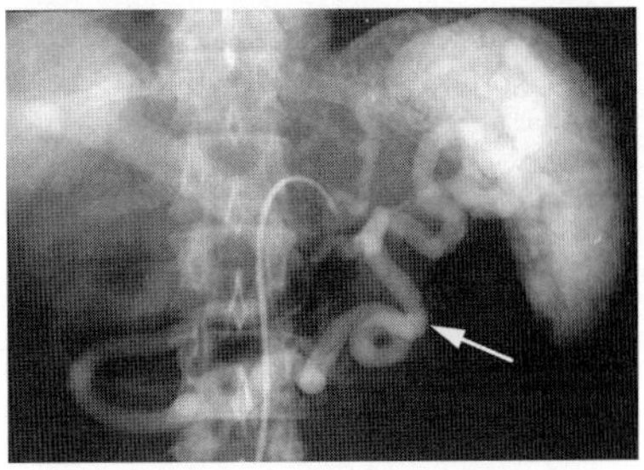

b.

Figure 24. *Angiographic appearance of the splenic vein and its collaterals. a) Normal splenic vein. This is a late phase image from a splenic arteriogram demonstrating the normal straight course of the splenic vein as it exits the spleen and joins with the unopacified blood from the SMV to form the portal vein. b) Splenic vein occlusion. Above the splenic vein, the left gastric vein drains into the portal vein. Below the spleen, a prominent gastroepiploic vein drains into the portal vein.*

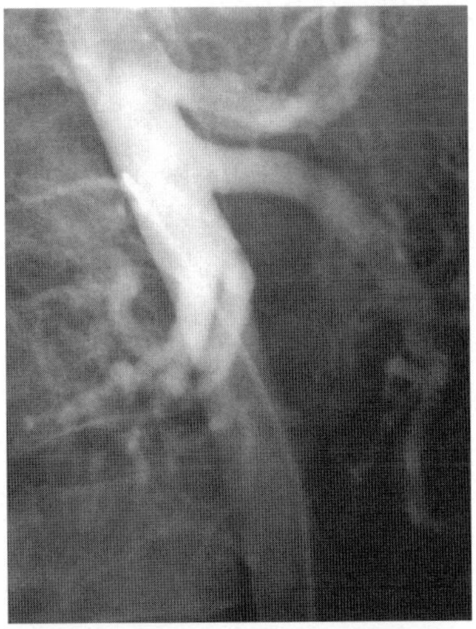

Figure 25. *Normal lateral aortogram. Note the course and caliber of the SMA and celiac artery origins.*

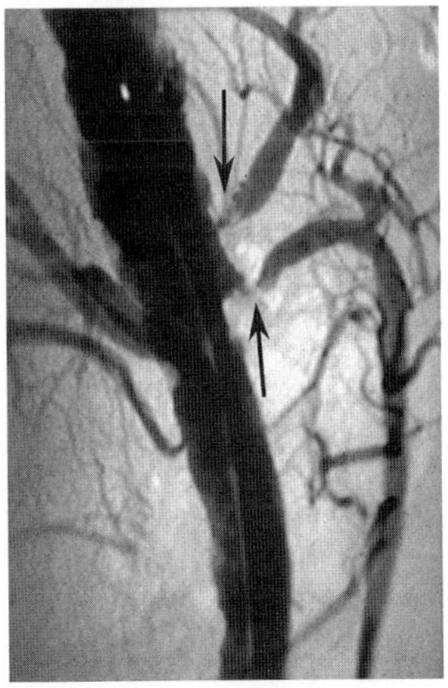

Figure 26. *Lateral aortogram in a patient with postprandial pain. Note severe stenoses at the origins of the celiac and SMA (arrows).*

VISCERAL VASCULATURE: PATHOLOGY

Mesenteric ischemia: Visceral ischemia is usually due to longstanding disease at the origin of the visceral vessels. Lateral aortography **(Fig. 25)** is the best examination to detect this disease. The two most common causes of stenosis at the origin of the visceral vessels are atherosclerosis and median arcuate ligament compression.

Vascular compromise severe enough to cause postprandial pain is uncommon because of the generous collaterals between the three main visceral vessels. Symptoms of visceral ischemia do not occur unless there is severe stenosis or occlusion of at least two of the three visceral vessels **(Fig. 26)**.

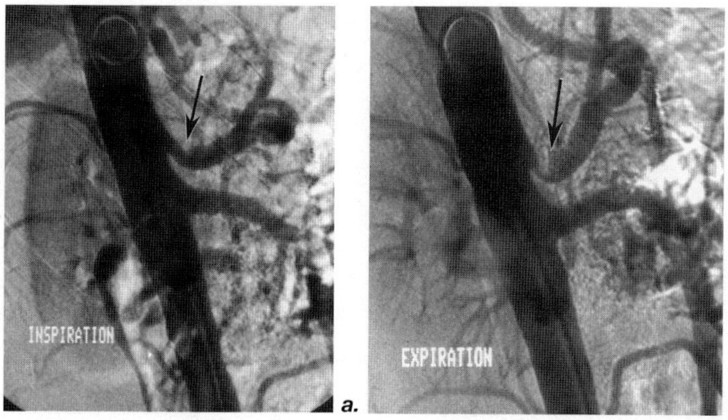

Figure 27. *Lateral aortograms in a patient with arcuate ligament compression syndrome. a) In inspiration, only a slight contour defect is seen along the upper surface of the celiac (arrow). b) This defect is accentuated in expiration.*

Median arcuate ligament syndrome (MALS) is a less common and perhaps controversial cause of visceral ischemia. The median arcuate ligament of the diaphragm may cross over the origin of the celiac artery (and less commonly, the SMA). A characteristic concave impression is formed on the cranial surface of these vessels just beyond their origin. The impression is accentuated in expiration **(Fig. 27)**. MALS is seen in young women with abdominal pain and an epigastric bruit.

TABLE 5: ARTERIOGRAPHIC CHARACTERISTICS OF BENIGN HEPATIC MASSES

Cavernous Hemangioma
- Hepatic artery branches are normal, not dilated
- No neovascularity or AV shunting
- Amorphous, well-defined vascular spaces fill in mid-late arterial phase with very delayed washout (> veins), sometimes amorphous calcifications
- Early peripheral ring or C-shaped accumulation of contrast
- Women > men

Hepatic Adenoma
- Hepatic artery branches:
  - May be dilated
  - Stretch around the mass
  - Feed tumor from periphery
- Highly vascular in 50%; no pooling
- Venous phase: homogeneous, dense, lumpy
- Displaces, but does not invade portal vein
- Not premalignant according to most pathologists
- Strong association with birth control pills, may bleed spontaneously

Focal Nodular Hyperplasia
- Hepatic artery:
  - Penetrates toward center of mass
  - Branches, "spoke-wheel" pattern from central fibrous core
  - Irregular distribution
- No neovascularity
- Venous phase: homogeneous, granular
- Sometimes a lucent ring around periphery

Regenerating Nodules
- Cirrhosis; following partial hepatectomy
- Isodense or hypovascular, no neovascularity
- Hepatic artery, normal size
- Portal vein displaced; hepatic vein distorted

Peliosis Hepatis
- Endothelial-lined, blood-filled vascular spaces of uncertain significance
- Multiple, round collections of contrast on parenchymal phase angiography
- Associated with infections (cat scratch fever), steroids, HIV disease, renal failure

Mesenteric ischemia may also be attributed to conditions that cause diffuse narrowing of the visceral vessels (particularly the SMA). Causes of nonocclusive mesenteric ischemia include hypovolemic shock, digitalis toxicity and ergotism **(Fig. 28)**.

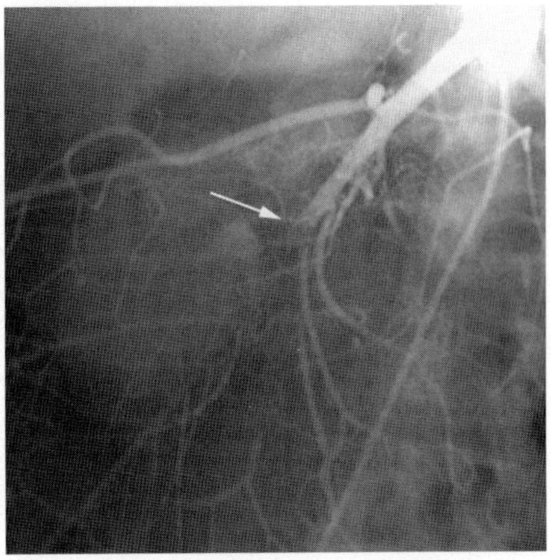

Figure 28. *Selective SMA arteriogram in a patient with severe acute abdominal pain. Note the absent opacification of the distal SMA (arrow) due to embolic occlusion.*

Hepatic Tumors

In the past, arteriography was commonly used to detect and characterize masses in the liver. Now these tasks are accomplished more precisely with cross-sectional imaging modalities – CT and/or MR. However, it is still useful to be aware of the arteriographic appearance of liver masses. Benign lesions include cavernous hemangioma, adenoma, focal nodular hyperplasia, regenerating nodules and peliosis hepatis. Malignant lesions include: hepatocellular carcinoma (hepatoma), fibrolamellar hepatocellular carcinoma, hypervascular metastasis, and hypovascular metastasis. Several important features of each of these entities are listed in **Tables 5 and 6**.

TABLE 6: ARTERIOGRAPHIC CHARACTERISTICS OF MALIGNANT HEPATIC MASSES

Hepatocellular Carcinoma (Hepatoma)
- Hepatic artery, dilated
- Very vascular; prominent neovascularity
- Abnormal vascular spaces, stellate pattern of dilated sinusoids
- Arteriovenous shunting
- Portal vein invasion
- Cirrhosis
- Very elevated alpha-fetoprotein

Fibrolamellar Hepatocellular Carcinoma
- Distinct clinical variant of hepatoma; much better prognosis
- Younger patients, mean age of presentation = 20 yrs.
- Large, often calcified
- Hypervascular, fibrous septa, areas of necrosis
- Difficult to differentiate from adenoma or focal nodular hyperplasia
- FNA often not helpful because of pleomorphism of tissue within the tumor

Hypervascular Metastases
- Usually multiple, round, peripheral
- Hyperdense compared with normal liver
- Central necrosis (lucent) in large lesions
- Most carcinoid, choriocarcinoma, hypernephroma, endocrine, leiomyosarcoma

Hypovascular Metastases
- Hepatic artery, branches may be increased in number
- Generally no neovascularity
- "Swiss cheese" appearance due to filling defect seen on capillary phase
- Lung, pancreas, gastrointestinal

The angiograms obtained in four patients with liver masses are shown below in **Figures 29-32**. Can you make the correct diagnosis in each case? (*The diagnoses are given at the bottom of this page.*)

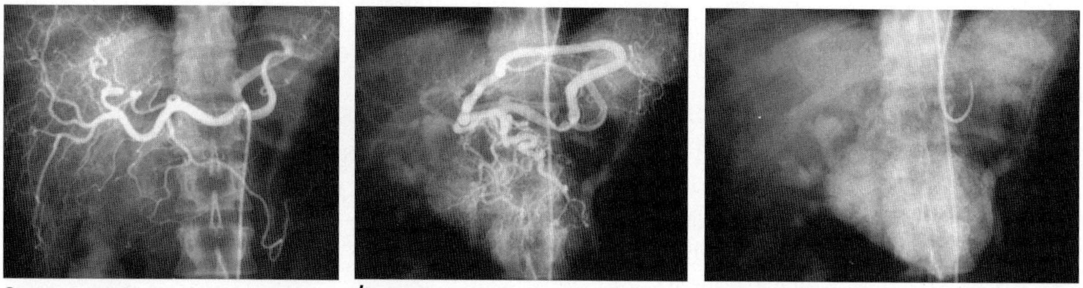

a. b. c.

Figure 29. *Selective visceral angiography was performed to evaluate a left lobe mass. a) Celiac injection does not opacify the left hepatic artery. b) Left gastric artery injection reveals a large dilated replaced left hepatic artery. c) Late phase image demonstrates diffuse granular pattern.*

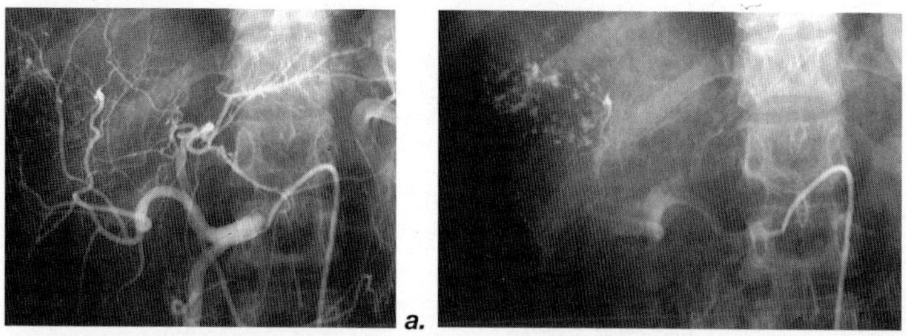

a. b.

Figure 30. *A 60-year-old woman complained of right upper quadrant pain. a) Early arterial phase image of common hepatic arteriogram shows normal size vessel and punctate areas of contrast pooling in the right lobe. b) The late phase image shows continued pooling of contrast in the right lobe.*

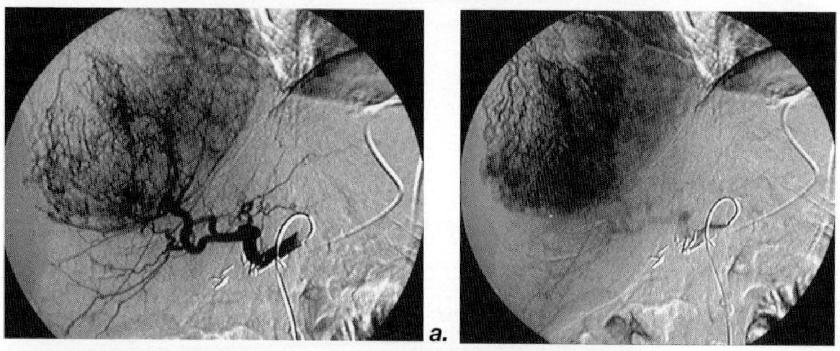

a. b.

Figure 31. *Preoperative arteriogram in a 50-year-old woman with metastatic disease. a) Selective hepatic arteriogram reveals a large hypervascular mass in the right lobe of the liver. b) Late arterial image reveals diffusely hypervascular mass.*

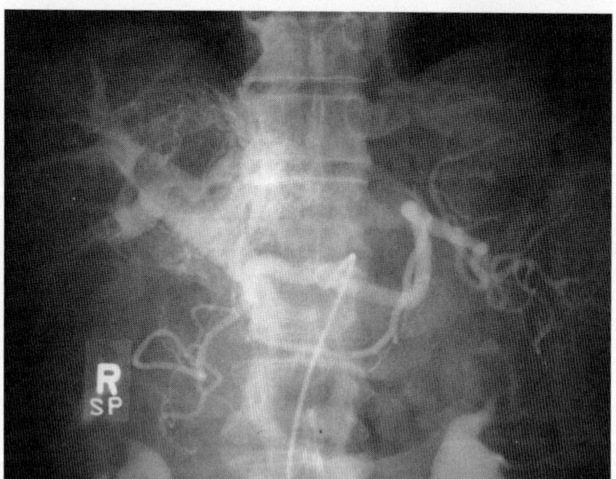

Figure 32. *A 55-year-old man with hepatitis B complained of right upper quadrant pain. CT was obtained and then a pre-operative visceral arteriogram was performed. An early arterial phase image from the celiac arteriogram is shown.*

Diagnoses

Figure 29: Focal nodular hyperplasia
Figure 30: Cavernous hemangioma
Figure 31: Metastatic leiomyosarcoma
Figure 32: Hepatocellular carcinoma with arterial portal shunting

Splenic Artery Aneurysm

Visceral artery aneurysms are uncommon, but when they occur they most frequently involve the splenic artery **(Figs. 33, 34)**. Unlike aortic and peripheral arterial aneurysms, splenic artery aneurysms are seen more often in women than in men. This unusual gender association may be due to underlying fibromuscular dysplasia or the vascular effects of repeated pregnancy. Main splenic artery aneurysms can also be due to atherosclerosis. Branch aneurysms may be seen with splenomegaly, portal hypertension, or polyarteritis nodosa. False aneurysms may be seen with pancreatitis, surgery or trauma.

Splenic artery aneurysm is usually asymptomatic and may be discovered by the presence of signet-ring left upper quadrant calcification—a finding that is seen in up to 70% of patients with splenic artery aneurysm. Rupture with resultant intraperitoneal hemorrhage is the chief complication. Aneurysms larger than 2 cm in diameter are treated by surgical excision or transcatheter embolization.

SPLENIC ARTERY ANEURYSM

- Women > men
- Atherosclerosis, FMD, pregnancy, pancreatitis
- Branch aneurysms
  - Splenomegaly
  - Portal hypertension
  - Polyarteritis nodosa
  - Atherosclerosis

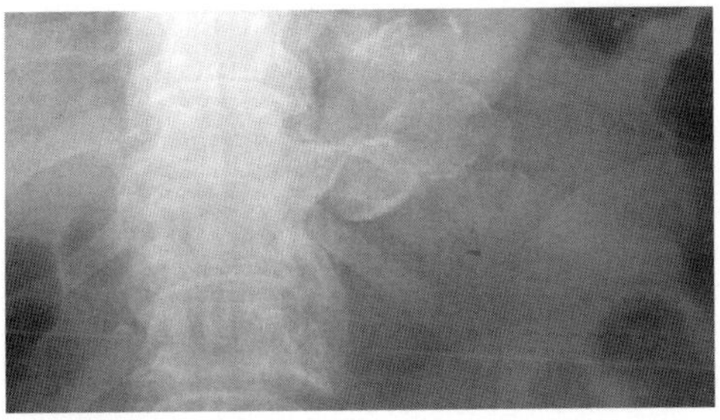

Figure 33. *An elderly woman presented with abdominal pain. Plain film shows a rounded vascular calcification in the left upper abdomen.*

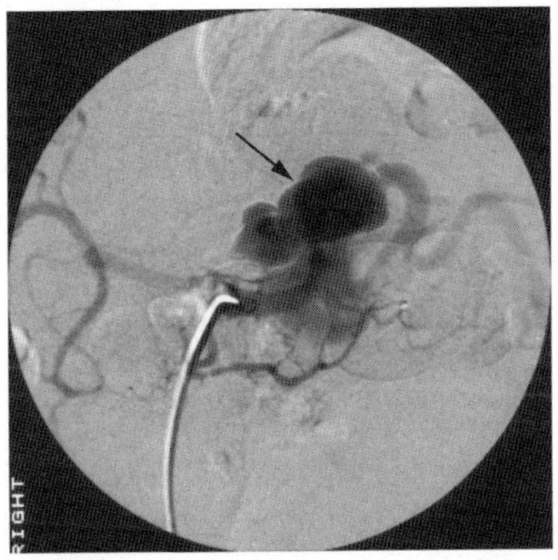

Figure 34. *Selective splenic arteriogram discloses a splenic artery aneurysm (arrow). This was treated by coil embolization.*

RENAL ARTERIES: ANATOMY

The renal arteries arise from the lateral wall of the aorta at the L1-L2 level **(Fig. 35)**. Typically, the origin of the right renal artery is slightly anterolateral and the left slightly posterolateral. In 70% of people, the renal arteries are single. Multiple renal arteries (2-4) occur with equal frequency on both sides, are unilateral in 32% and bilateral in 12% **(Fig. 36)**. An accessory superior pole renal artery occurs in 5%-10%.

The main renal artery divides into anterior and posterior divisions. The bloodless "line of Brodel" is the watershed of the anterior and posterior renal artery branches. It is a relatively avascular plane and, consequently, is the safest area for needle puncture during percutaneous nephrostomy.

Extraparenchymal branches of the renal artery include:

- Inferior adrenal arteries
- Capsular arteries
- Inferior phrenic artery
- Branches to renal pelvis and proximal ureter
- Gonadal arteries (in about 20%).

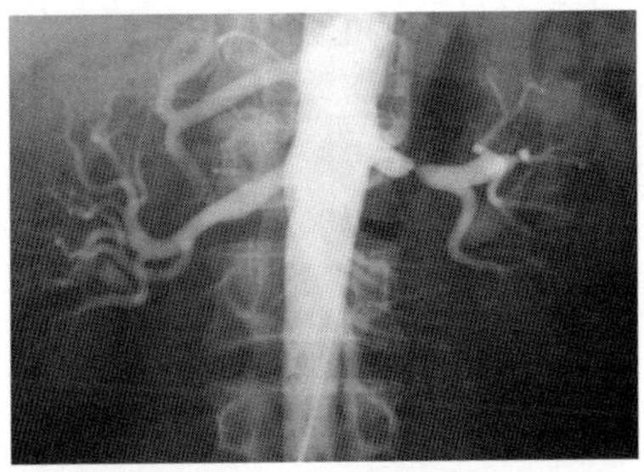

Figure 35. *Midstream aortogram demonstrates single renal arteries (and a left renal artery stenosis) in this middle-aged man with hypertension.*

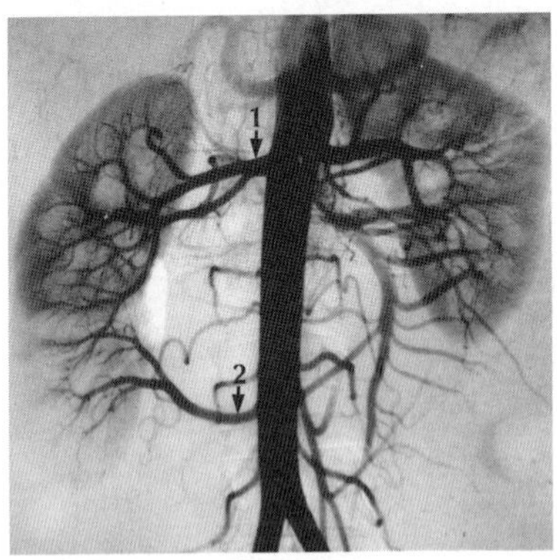

Figure 36. *Subtraction film from midstream aortogram performed in a potential renal donor. There are two right renal arteries. The inferior right renal artery (2) takes origin low in the infrarenal aorta.*

RENOVASCULAR HYPERTENSION

- 5% of patients with hypertension
- Predictors:
  - Early or late onset (<30 or >50)
  - Rapid deterioration of renal function
  - Worsening on ACE inhibitor
- Etiology:
  - Atherosclerosis
  - Fibromuscular dysplasia
  - Other: pheo, neurofibromatosis, arteritis, hematoma

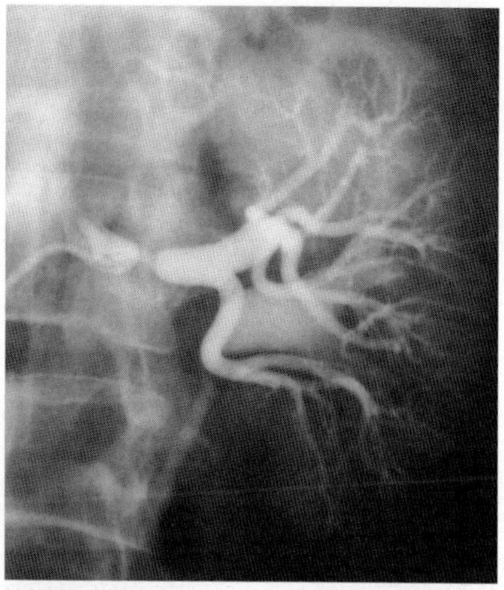

Figure 37. *Selective left renal arteriogram in a patient with renovascular hypertension. The short focal stenosis of the main renal artery is likely due to atherosclerosis.*

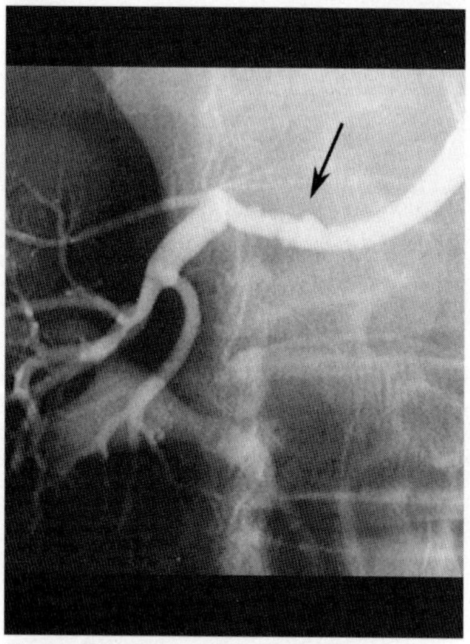

Figure 38. *Selective right renal arteriogram in a young woman with hypertension. The string of beads appearance in the distal main renal artery (arrow) is consistent with medial fibroplasia.*

RENAL ARTERIES: PATHOLOGY

Renovascular Hypertension

Nearly 25% of the adult US population have hypertension—approximately 5% - 10% of whom have renovascular hypertension due to an underlying anatomic cause. Renovascular hypertension should be suspected in young women (< 30 yrs), those with late onset hypertension (> 50 yrs), those with rapid deterioration of renal function and worsening hypertension, or those with renal deterioration after the addition of an angiotensin converting enzyme (ACE) inhibitor.

Almost all nonessential hypertension is due to atherosclerosis or fibromuscular dysplasia (FMD) in 95%. About 5% is attributable to other causes: neurofibromatosis, pheochromocytoma, fibrous bands, arteritis, emboli, thrombosis, aneurysm, vascular malformations, radiation, cysts, neoplasms, or perirenal hematoma.

Pathophysiology

The ischemic kidney produces excessive renin which stimulates the production of angiotensin II and aldosterone, promoting sodium and fluid retention. The contralateral kidney is thus stimulated to produce less renin. A renal vein renin ratio of 1.5 : 1 or greater is predictive of renovascular hypertension that will respond to revascularization. However, there are many problems with this test such as the presence of accessory renal veins, contralateral ischemic renal artery disease, and 20% - 50% of patients without lateralization of renin levels also improve with treatment.

Treatment

Nonoperative intervention is generally preferred as a first therapy for renovascular disease. However, surgical revascularization is preferred over angioplasty in the following situations:

- Aortic reconstructive surgery is to be performed
- Extensive aortic plaque is present—increasing risk of renal, visceral, or peripheral emboli
- Restenosis follows technically successful second dilatation
- No safe percutaneous vascular access route is available

Atherosclerosis

Atherosclerosis is the most common cause of renovascular hypertension in middle-aged and older patients. Disease may be located at the origin of the vessel (ostial) or in the main renal artery (nonostial) **(Fig. 37)**. While atherosclerosis usually leads to stenosis, it may also result in dissection or aneurysm.

Fibromuscular Disease

This process is responsible for renovascular hypertension in about one-third of patients. Disease typically is present in the mid-to-distal portion of the main renal artery (80%) but it may extend into the branch vessels in 20%. Disease is bilateral in two-thirds of patients. When it is unilateral, it is more frequent on the right. There are various forms of fibroplasia or hyperplasia involving the media, intima, perimedia, adventitia. The most common are:

- Medial fibroplasia (60% - 70%) produces a "string of beads" appearance. Multiple fibromuscular ridges produce alternating areas of stenosis and mural thinning with aneurysms **(Fig. 38)**. This process involves the mid-to-distal renal arteries and sometimes branch vessels.

- Perimedial fibroplasia (15% - 25%); produces irregular stenoses and beading (without aneurysm formation), and rarely true aneurysm. It is caused by fibroplasia of outer media.
- Medial hyperplasia (5% - 15%); produces a long, smooth stenosis.
- Intimal fibroplasia (1% - 2%); the characteristic finding is a focal stenosis **(Fig. 39)**.
- Medial dissection (~5%) with false channels and aneurysms.
- Adventitial fibroplasia (< 1%) with long, segmental stenoses.

FIBROMUSCULAR DYSPLASIA

- One-third of renovascular hypertension
- Mid–distal renal artery (17% branches)
- Medial fibroplasia (60%-70%), "string of beads"
- Perimedial fibroplasia (15%-17%), irregular stenoses
- Two-thirds bilateral (unilateral, R>L)

Page Kidney

By definition this is a type of renin-angiotensin-mediated hypertension caused by renal compression due to a mass within perinephric or subcapsular space. It is usually due to hematoma from blunt trauma but may be due to large cyst or tumor. Findings on angiography include:

- stretching and splaying of the intrarenal vessels
- slow arterial washout because of mass impression
- distortion of renal contour
- enlargement and displacement of the capsular artery.

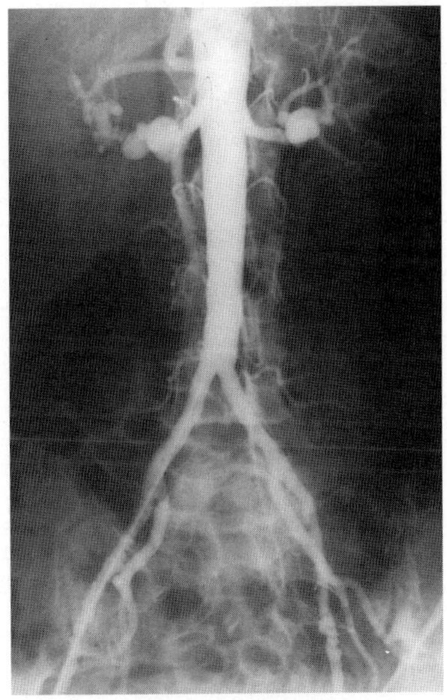

Figure 39. Midstream aortogram in a middle-aged woman. Bilateral renal artery aneurysms are consistent with fibromuscular disease. There is also involvement of the external iliac arteries bilaterally.

Polyarteritis Nodosa (PAN)

This is a systemic vasculitis associated with small artery aneurysms in multiple viscera. The kidneys are involved in 85% of cases. The characteristic pathologic feature of this disease is fibrinoid medial necrosis of medium-sized and small arteries. Angiographically, the findings of multiple small aneurysms in multiple organs is pathognomonic, but only a minority of patients with PAN have them. A similar angiographic appearance can be seen with drug-induced angiitis. Other angiographic findings of vasculitis include irregular stenosis and occlusion of intrarenal branches. The nephrogram may have a spotted or striated pattern. The temporal course of this disease is variable. Aneurysms may regress with clinical improvement. (*See Chapter 1 for angiographic illustration of* PAN.)

Neurofibromatosis

This disease may result in stenosis of the main renal artery due to intimal hyperplasia or due to the presence of a neurofibroma. The renal artery stenosis may be isolated or seen in association with midabdominal coarctation. Clinically, patients present with hypertension during childhood. Angiography demonstrates smooth or nodular-appearing stenosis and saccular aneurysms. Typically, the proximal renal artery or ostia are involved (compared with mid- and distal lesions seen in FMD). (*See Chapter 1 for angiographic illustration of neurofibromatosis of abdominal aorta.*)

Renal Artery Aneurysms

Renal artery aneurysms are relatively uncommon. They are usually asymptomatic and may be seen in 1% of patients undergoing renal arteriography. Less commonly, renal artery aneurysms may cause hematuria or pain. True aneurysms of the renal artery may be due to atherosclerosis, FMD **(Fig. 40)** or arteritis (polyarteritis nodosa, Wegener's granulomatosis, necrotizing angiitis). False aneurysms can be due to trauma or bacterial infection (mycotic).

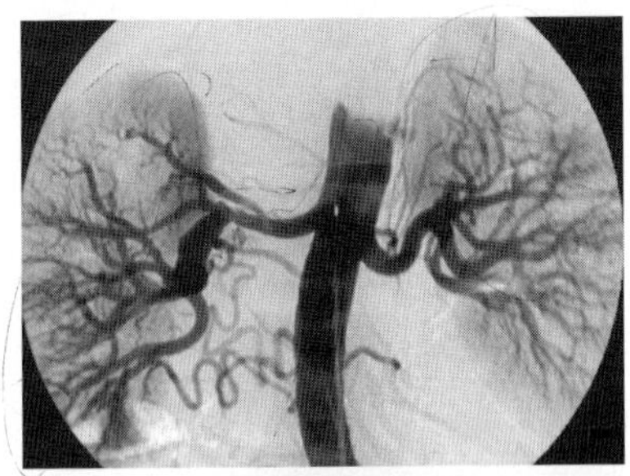

a.

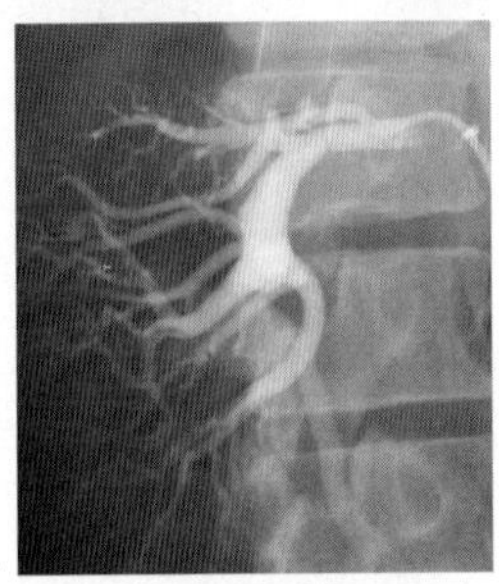

b.

***Figure 40.** This arteriogram was obtained in a young man with hypertension due to the intimal variety of fibromuscular dysplasia. a) Digital subtraction aortogram reveals an aneurysm of the right renal artery beyond the bifurcation. Note the tortuous collateral from the aorta to the lower pole of the right kidney. b) Selective right renal arteriogram shows a tight weblike stenosis of the right renal artery with an aneurysm just distal to the web extending into several branch vessels.*

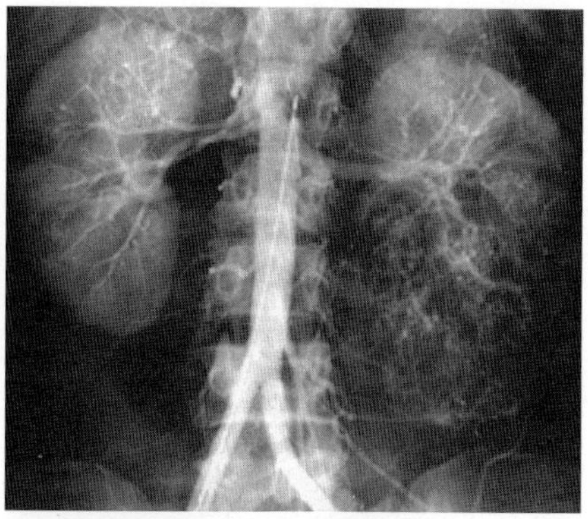

a.

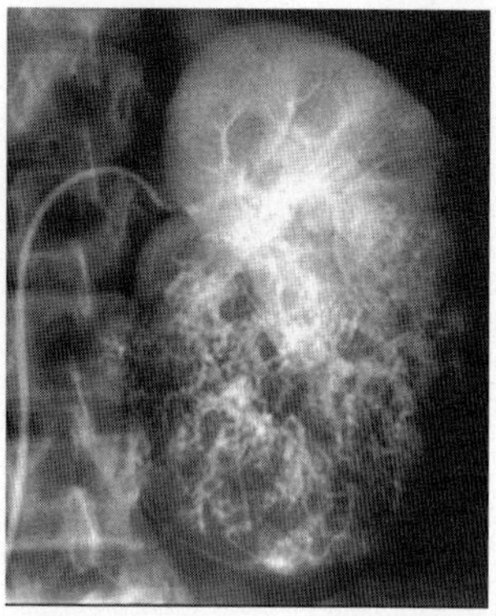

b.

***Figure 41.** A preoperative arteriogram was obtained in this elderly man with hematuria. a) Late arterial phase image from midstream aortogram reveals a vascular mass in the lower pole of the left kidney. b) Selective left renal arteriogram demonstrates a hypervascular mass with marked neovascularity. This proved to be a renal cell carcinoma.*

TABLE 7: RENAL TUMORS

- Benign
  - Adenoma
  - Oncocytoma (oxyphilic adenoma)
  - Angiomyolipoma
- Malignant
  - Adenocarcinoma
  - Mesenchymal, sarcoma
  - Uroepithelial
  - Wilms' tumor
  - Metastatic disease

Aneurysm rupture is rare, but occurs more often when caused by an inflammatory process (such as polyarteritis nodosa, Wegener's granulomatosis, necrotizing angiitis) and in pregnancy. Other complications include peripheral embolization and thrombosis.

Renal Tumors

Angiography is rarely used in the current work-up of a renal mass. However, tumors may be detected as an incidental finding during angiography performed for another purpose. As well, embolization of renal tumors may be performed preoperatively **(Fig. 41)**. A general classification of renal tumors is shown in **Table 7**.

Adenomas are often small and clinically insignificant. They may be hypervascular or avascular depending on the cell type. Oncocytoma is a subgroup of benign adenoma referred to as oxyphilic adenoma. The tumor is usually exophytic and may have a characteristic spoke-wheel pattern of hypervascularity. Angiomyolipoma is a renal tumor that is characteristically diagnosed by CT. The findings on angiography are variable but usually there is hypervascularity, and differentiation from adenocarcinoma is difficult. Angiomyolipoma may result in pain, hematuria, or capsular rupture with retroperitoneal hemorrhage. Multiple angiomyolipomas are seen with tuberous sclerosis.

Renal cell carcinoma is the most common renal tumor. It is seen in older patients who present with pain, hematuria, and weight loss. Tumors may be bilateral in 5%. Multiple lesions are seen in patients with von Hipple-Lindau disease. Most tumors are hypervascular and may recruit parasitic supply from the lumbar or adrenal arteries. Invasion of the renal vein and IVC is characteristic but is seen in only one-third of patients.

REFERENCES

Anatomy

Bergman RA, Thompson SA, Afifi AK, Saadeh FA. The cardiovascular system. In: Bergman RA, ed. Compendium of human anatomic variation. Baltimore: Urban & Schwarzenberg, 1988; 60-87.

Kadir S. Kidneys. In: Kadir S, ed. Atlas of normal and variant angiographic anatomy. Philadelphia: W.B. Saunders, 1991; 387-427.

Kadir S, Lundell C, Saeed M. Celiac, superior and inferior mesenteric arteries. In: Kadir S, ed. Atlas of normal and variant angiographic anatomy. Philadelphia: W.B. Saunders, 1991; 297-364.

Abdominal Aorta

Cronenwett JL, Krupski WC, Rutherford RB. Aortic and iliac artery aneurysms. In: Rutherford RB, ed. Vascular Sugery. 5th edition. Philadelphia: W.B. Saunders, 2000; 1246-1280.

Visceral Abdominal Vasculature

Cho KJ. Vascular diseases. In: Reuter SR, Redman HC, Cho KJ, eds. Gastrointestinal angiography. 3rd edition. Philadelphia: W.B. Saunders, 1986; 78-127.

Kadir S. Angiography of the abdominal aorta and pelvis. In: Kadir S, ed. Diagnostic angiography. Philadelphia: W.B. Saunders, 1991; 207-253.

CHAPTER 3 PREVIEW ANSWERS

Case 1

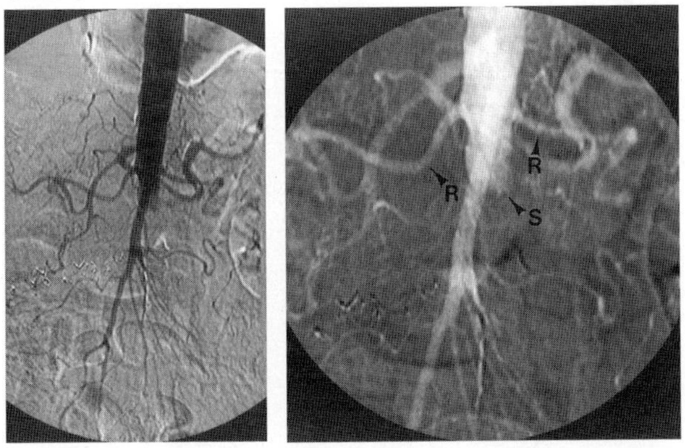

Figure 1. *Abdominal aortogram.*

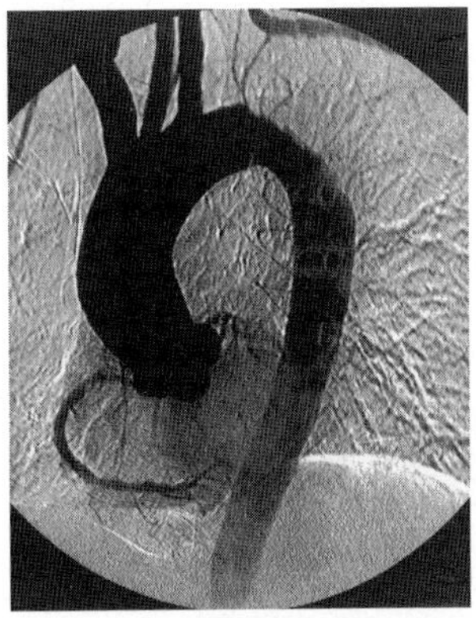

Figure 2. *Thoracic aortogram, LAO projection.*

Findings: The abdominal aorta is occluded below the level of the renal arteries (R) **(Fig. 1)**. There is a short stump of infrarenal aorta (S) and no significant collaterals. Both of these findings suggest an acute occlusion. Likely embolic sources are the heart or the thoracic aortic arch. A thoracic aortogram was performed **(Fig. 2)** that shows a large filling defect along the superior surface of the arch.
Diagnosis: Acute aortic occlusion due to massive thromboembolism from the thoracic aorta.

Case 2

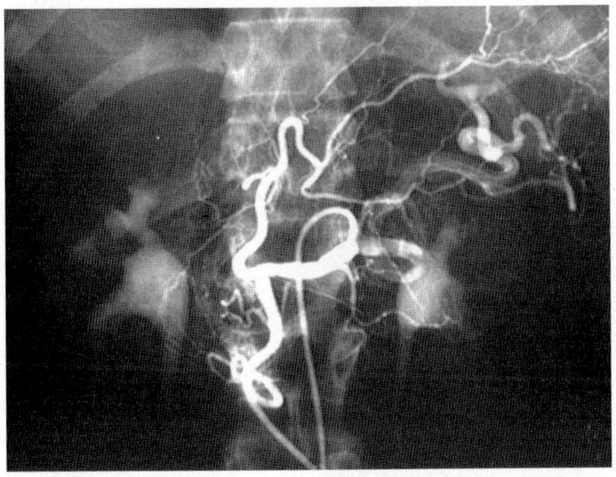

Figure 3. *Selective common hepatic arteriogram.*

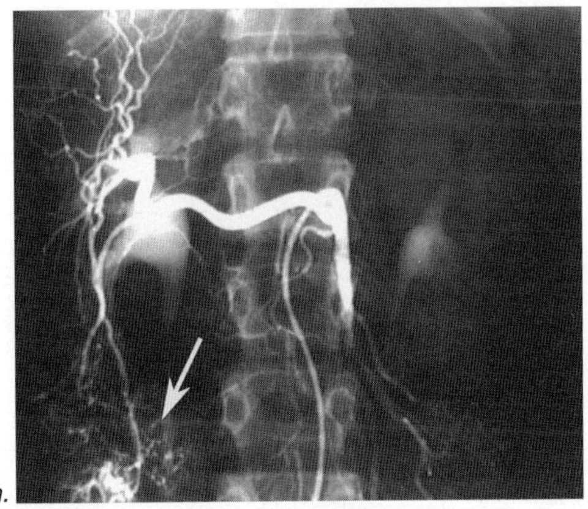

Figure 4. *Selective right hepatic arteriogram.*

Findings: Selective injection of the common hepatic artery opacifies the left hepatic and gastroduodenal arteries **(Fig. 3)**. The right hepatic artery (RHA) is not opacified. Injection of the SMA revealed a replaced RHA. Selective injection of the RHA **(Fig. 4)** revealed a pedunculated hypervascular mass (arrow) that proved to be a hepatocellular carcinoma.
Diagnosis: Aberrant origin of the right hepatic artery from the SMA.

Case 3

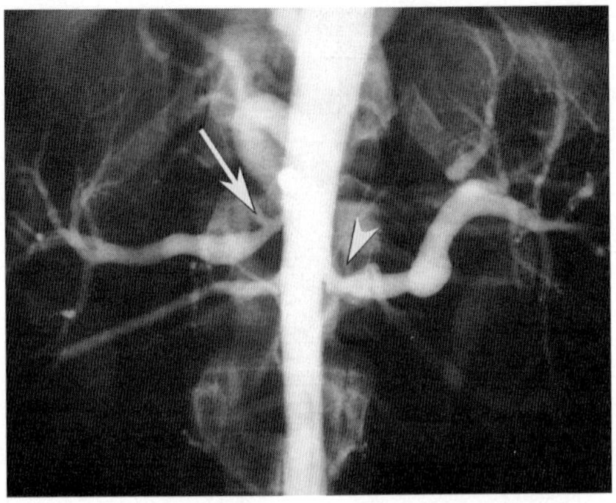

Figure 5. *Midstream aortogram.*

Findings: There are two renal arteries on the right and one on the left **(Fig. 5)**. The origin of the superior right renal artery is severely stenotic (arrow). The proximal left main renal artery is also stenotic (arrowhead) and the perirenal aorta is mildly irregular. This patient with café-au-lait spots has neurofibromatosis.
Diagnosis: Bilateral renal artery stenoses due to neurofibromatosis.

CHAPTER 4 PREVIEW

Case 1

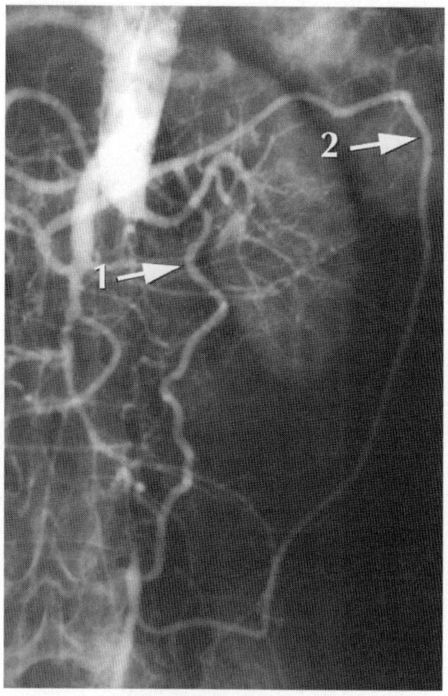

Figure 1. *Abdominal aortogram.*

History: A 60-year-old woman with absent femoral pulses complained of claudication.
Questions: What is your diagnosis? Can you name the vessels labeled 1 and 2?

Case 2

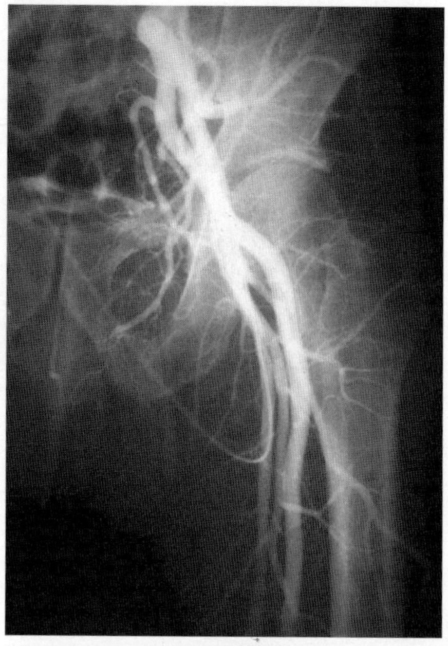

Figure 2. *Left femoral arteriogram.*

History: A 40-year-old man underwent arteriography to evaluate a gunshot injury to the left calf.
Question: What is your diagnosis?

Case 3

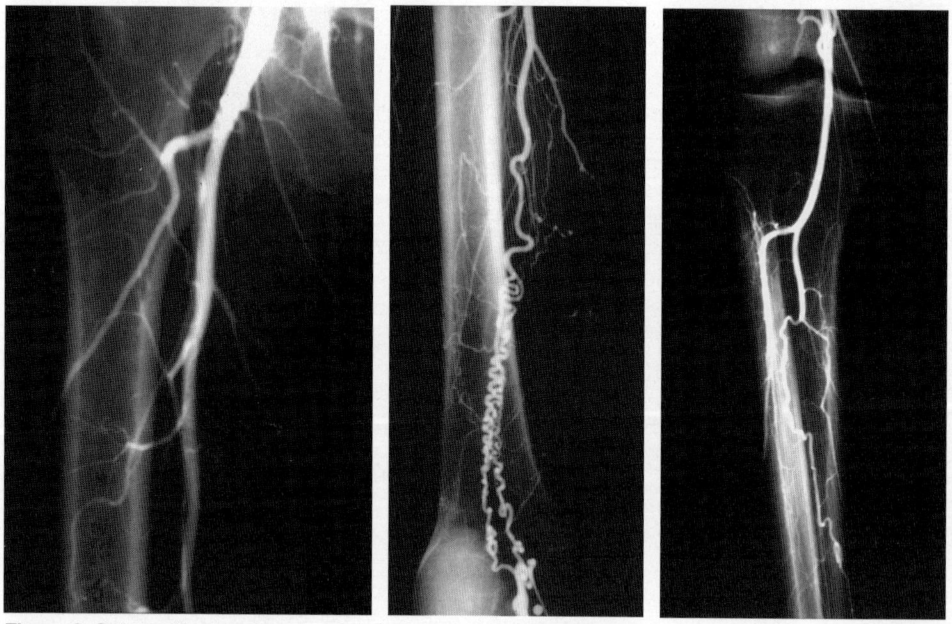

Figure 3. *Selected images from a right lower extremity arteriogram.*

History: A 43-year-old Russian immigrant complained of rest pain in the right leg.
Question: What is your diagnosis?

Chapter 4
Lower Extremity Arteriography

Jeanne M. LaBerge, MD

CHAPTER OUTLINE

- General principles of peripheral arteriography
- Anatomic considerations
- Technique
- Review of selected disease processes causing lower extremity ischemia

INTRODUCTION

Peripheral vascular angiography is probably the most commonly performed interventional procedure in practice today. Although noninvasive imaging techniques have made great strides in recent years, catheter arteriography remains the "gold standard" method for definitively evaluating patients with peripheral vascular disease. In this chapter, we review the principles and practice of catheter-based peripheral arteriography.

GENERAL PRINCIPLES

The three basic objectives of peripheral arteriography are to: 1) identify and characterize those lesions directly responsible for the patient's current symptoms; 2) demonstrate all of the vascular anatomy necessary to determine and plan therapy; and 3) identify unstable lesions not directly responsible for symptoms. The angiographer's primary goal is to identify all lesions responsible for a patient's symptoms. This is a dynamic process. After any initial abnormalities are detected, the angiographer must determine whether or not these findings explain all of the patient's symptoms. If the findings do not explain the symptoms, additional views should be obtained to detect occult lesions.

OBJECTIVES OF PERIPHERAL ARTERIOGRAPHY

- Identify and characterize lesions responsible for a patient's symptoms
- Demonstrate anatomy necessary to plan therapy
- Identify other unstable lesions

A full pulse exam should be performed prior to the angiographic study. The immediate preprocedure pulse exam serves as a reference if complications should occur during the procedure and also helps the angiographer determine whether or not the observed angiographic findings are compatible with the physical exam findings. The ankle brachial index (ABI) should also be reviewed. The ABI is normally about 1.1. An ABI of less than 0.96 is abnormal. A rough correlation can be made between the symptoms, ABI and the expected angiographic findings as shown in **Table 1**.

TABLE 1: CORRELATION OF SYMPTOMS, ABI AND EXTENT OF DISEASE

Symptom	ABI	Expected Angiographic Findings
Buttock claudication	0.5 - 0.8	Aorto-iliac occlusive disease
Calf claudication	0.5 - 0.8	Superficial femoral occlusive disease
Rest pain	< 0.3	Occlusive disease in at least two levels
Gangrene	< 0.2	Tibial and pedal occlusive disease

> **CHARACTERIZATION OF VASCULAR ABNORMALITIES**
>
> - Location: both segmental and within a segment
> - Length: focal or diffuse
> - Morphology: eccentricity, calcification
> - Severity: degree of luminal compromise

Full characterization of vascular lesions should include a description of the following items:

1) The morphology of the lesion:
   - — eccentric vs. concentric;
   - — calcified or not;
   - — % luminal compromise; and
   - — degree of collateral flow.

2) The location of disease within the vessel, that is, at the origin of the vessel or in the main artery.

3) The extent of disease within the vessel: focal or diffuse.

4) The presence of similar disease in other vessels, with particular attention to the same vessel in the opposite, unaffected leg.

Such characterization will help determine the best treatment options for the patient. The SCVIR Standards of Practice Guidelines on Angioplasty include recommendations and expected results of angioplasty based on these anatomic and morphologic descriptors. A variety of peripheral vascular abnormalities are shown below **(Figs. 1-4)** with corresponding anatomic descriptions.

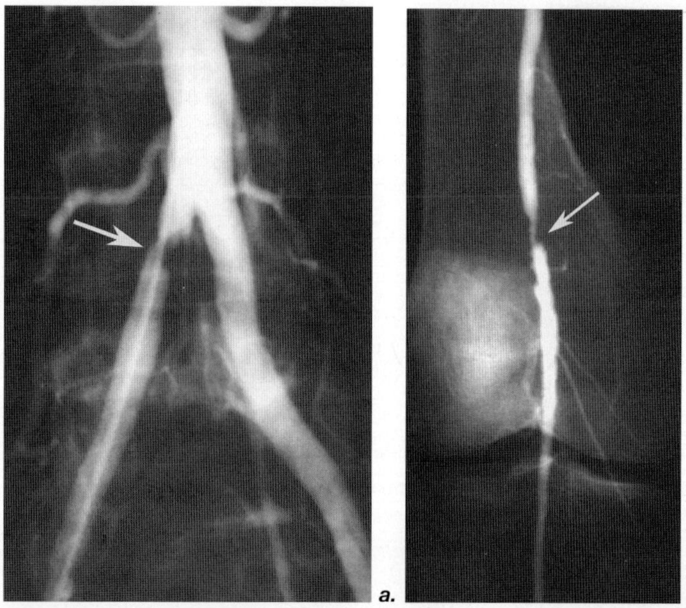

Figure 1. Short, focal, noncalcified concentric stenoses (arrows). a) Right common iliac artery. b) Right popliteal artery.

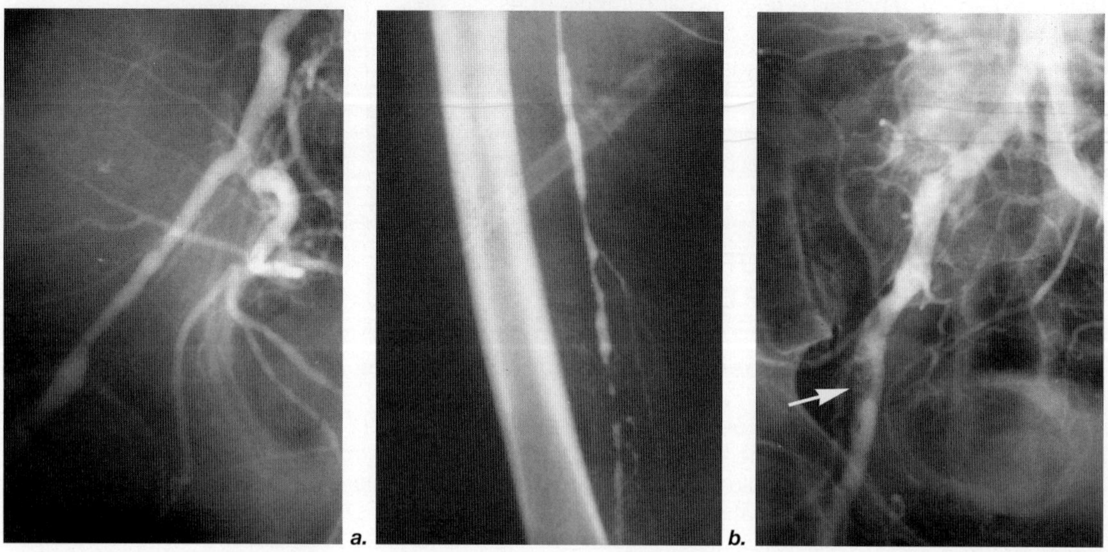

Figure 2. Long segment stenoses. a) Long, calcified external iliac stenosis. b) Diffuse stenosis of the mid- to distal SFA. c) Diffuse right iliac stenoses with a calcified external iliac lesion (arrow).

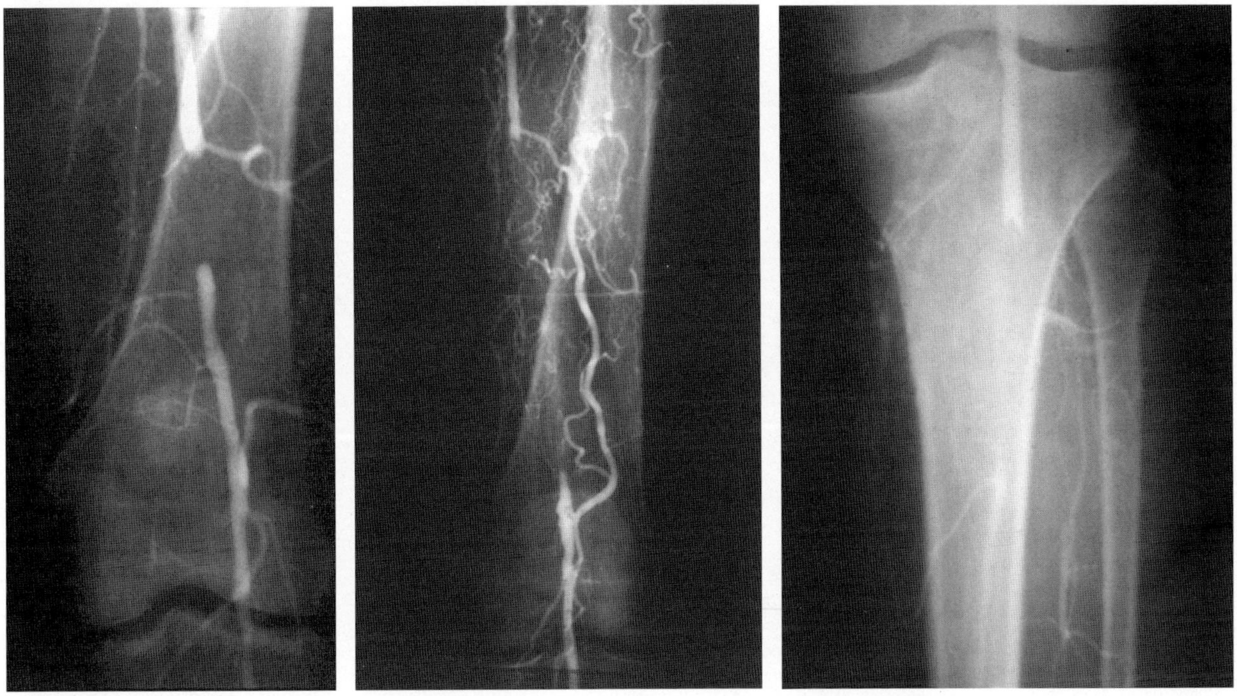

a. Subacute occlusion. *b. Chronic occlusion.* *c. Acute occlusion.*

Figure 3. *Peripheral vascular occlusions. a) Short-segment occlusion (< 7 cm) of the popliteal artery with small to moderate geniculate collaterals. (b) Long-segment occlusion (> 7 cm) of the popliteal artery with large geniculate collaterals reflecting longstanding occlusion. (c) Occlusion of the distal popliteal artery without collaterals compatible with acute popliteal embolus.*

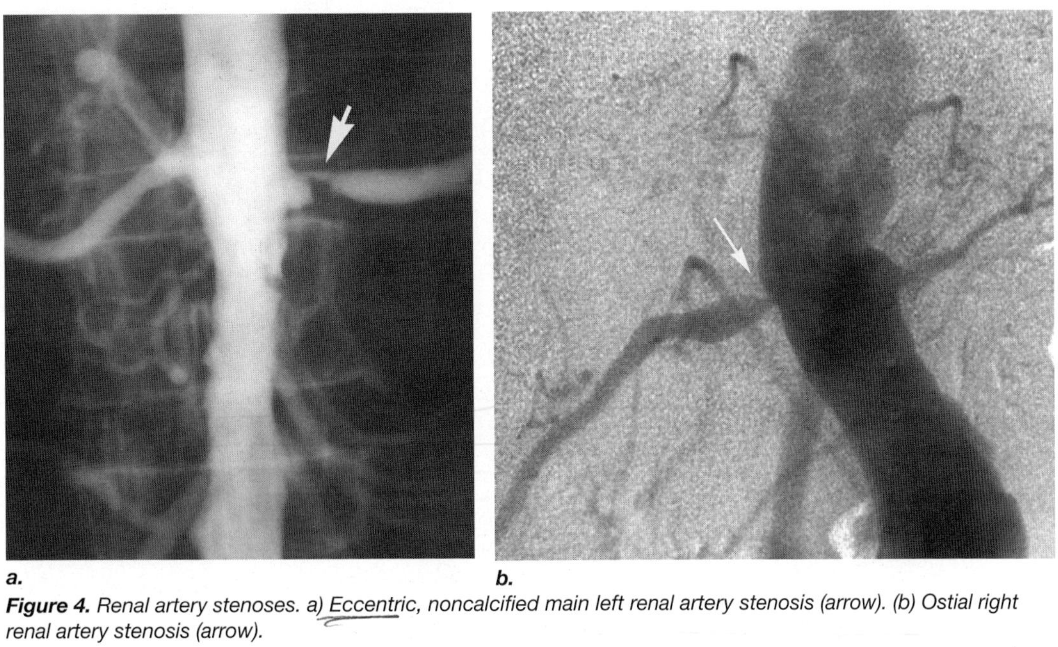

a. *b.*

Figure 4. *Renal artery stenoses. a) Eccentric, noncalcified main left renal artery stenosis (arrow). (b) Ostial right renal artery stenosis (arrow).*

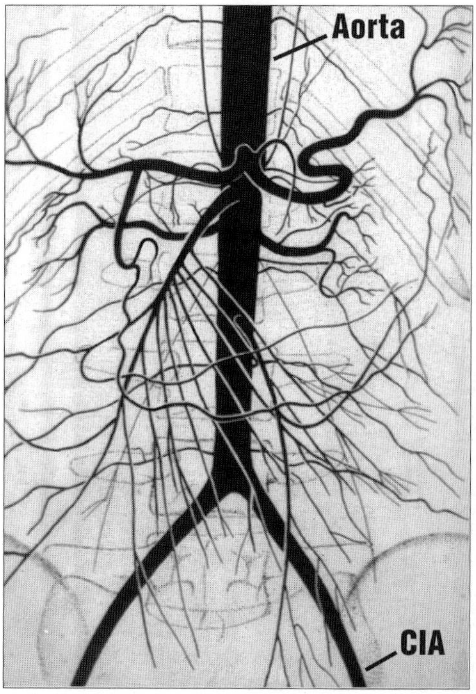

a.

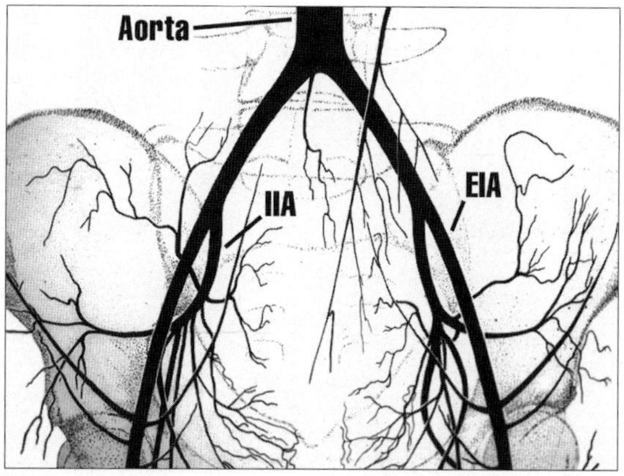

b.

Figure 5. The inflow vessels. Inflow operations such as an AFBG would involve these vessels. a) Aorta and common iliac arteries. b) The pelvic vessels, the common internal and external iliacs.*

KEY FOR FIGURES 5, 6
Common iliac artery (CIA)
Internal iliac artery (IIA)
External iliac artery (EIA)
Common femoral artery (CFA)
Superficial femoral artery (SFA)
Anterior tibial artery (AT)
Posterior tibial artery (PT)

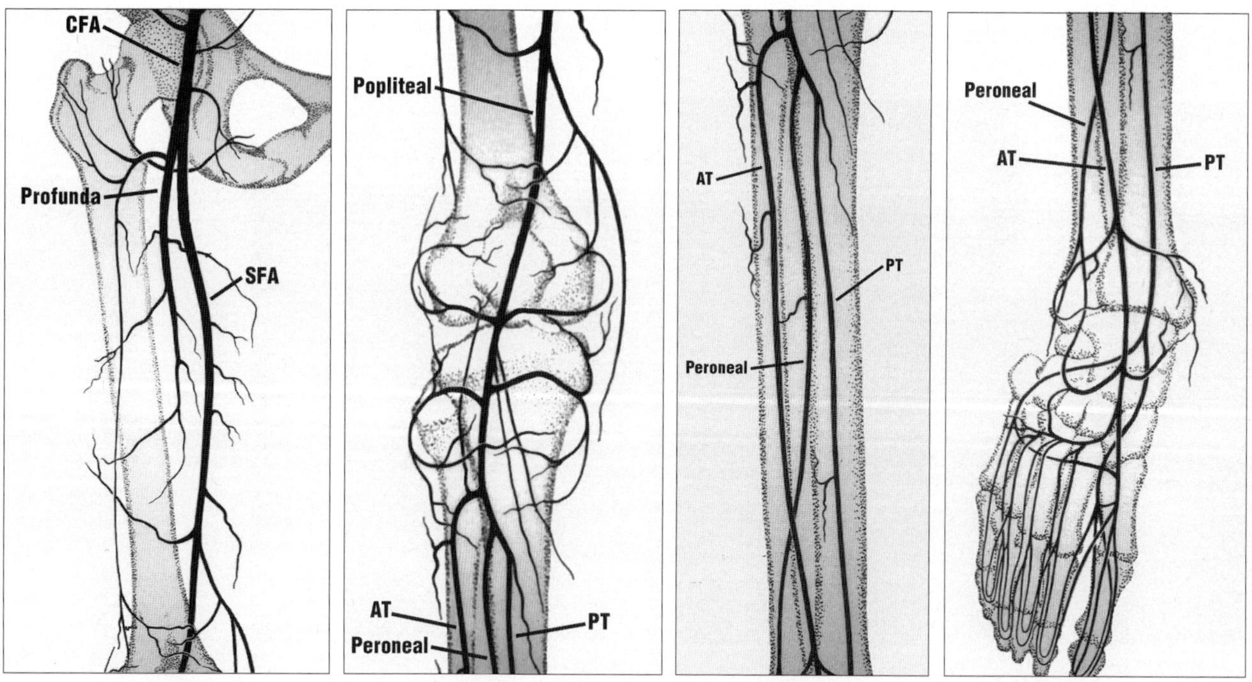

a. b. c. d.

Figure 6. The outflow vessels. Outflow operations such as femoral-popliteal bypass would involve these vessels. a) The femoral vessels: common femoral (CFA), profunda and SFA. b) The popliteal artery begins at the adductor canal; it is often divided into "above the knee" and "below the knee" segments corresponding to the operative approaches. c) The popliteal "trifurcation." This is really a bifurcation into the anterior tibial and tibial peroneal trunk which then bifurcates into the posterior tibial and peroneal vessels. Variant branching patterns are common however. d) The peroneal artery terminates at the ankle, while the anterior tibial artery continues into the foot as the dorsalis pedis and the posterior tibial artery bifurcates below the ankle into medial and lateral plantar branches.

(Anatomic diagrams in this chapter are reproduced courtesy of Eastman Kodak Inc.)*

ANATOMIC CONSIDERATIONS

The vasculature affected by peripheral vascular disease can be broadly divided into three distinct groups: the inflow vessels—the aortoiliac vasculature **(Fig. 5)**; the outflow vessels—the femoropopliteal vasculature; and the "distal" vasculature of the calf and foot—the tibial and pedal vessels **(Fig. 6)**. Such categorization is relevant because the therapies applied to treat peripheral vascular disease can also be broadly grouped as "inflow" procedures" (e.g. aortofemoral bypass grafting [AFBG]), "outflow" procedures (e.g. femoropopliteal bypass grafting [fempop grafts]), or "distal" procedures (e.g. femoral-dorsalis pedis bypass grafting [fem-distal grafts]).

> **ANATOMIC CATEGORIZATION OF PERIPHERAL VASCULAR DISEASE**
>
> - Inflow disease
>   - Aortoiliac vessels
> - Outflow disease
>   - Femoropopliteal
>   - Tibial
>   - Pedal

The nature of a patient's disease is frequently referred to by the predominant distribution of vascular abnormality. That is, when disease predominantly affects the aortoiliac vessels, it is referred to as "inflow disease." Similarly, disease chiefly affecting the femoropopliteal vessels is referred to as "outflow disease" and disease confined to the calf or foot may be referred to as "distal disease." Patients with severe disease in more than one area are said to have "multilevel disease." An important therapeutic rule of practice is that when patients have multilevel disease, initial therapy should be directed at the more proximal disease. That is, inflow procedures are generally performed before outflow or distal procedures.

The vasculature responsible for peripheral vascular disease can also be categorized as "axial" or "nonaxial." The axial or in-line vessels are the most important because they provide direct pulsatile flow to the lower extremity. The axial vasculature includes the aorta, common iliac, external iliac, superficial femoral (SFA), popliteal and tibial arteries. The nonaxial vessels play a supportive role providing collateral flow to the leg when one of the axial vessels is occluded. The nonaxial vessels include the aortic branch vessels, and the internal iliac, profunda femoris, and geniculate arteries.

The Renal Arteries

Evaluation of the renal arteries is included on the standard aortogram performed to evaluate peripheral vascular disease for two reasons: 1) clinically significant renal stenoses can be corrected at the time of an in-flow operation or nonoperative intervention, and 2) identifying the number and location of the renal arteries can help the surgeon avoid inadvertent ligation or injury of these vessels at the time of aortic surgery.

The renal arteries are usually single and arise at the L1-L2 level. Variant anatomy is common, however, and accessory vessels can be seen in up to 30% of patients. Accessory vessels usually arise from the infrarenal aorta or common iliac artery. The renal vessels take origin from the lateral aspect of the aorta; however, the right renal artery frequently originates along the anterior lateral wall and the left renal artery frequently originates along the posterior lateral wall. For this reason, imaging of the proximal portion of the main renal arteries is best achieved by filming in the RPO projection.

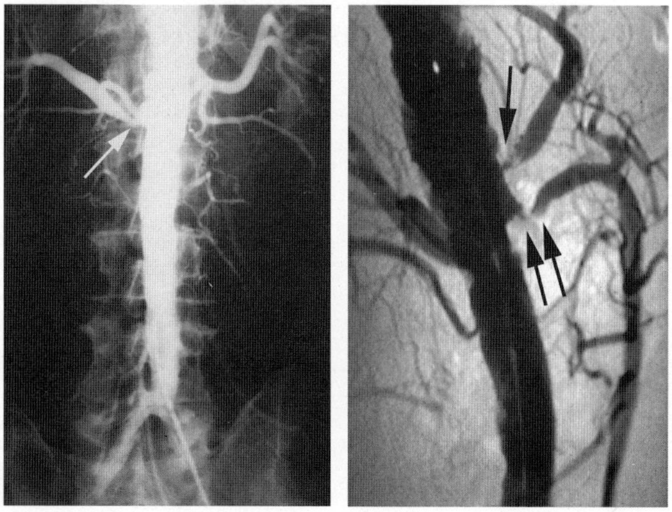

a. b.

Figure 7. Midstream aortogram performed in a patient with claudication and postprandial abdominal pain. a) AP view demonstrates narrowing of the infrarenal aorta and right renal artery stenosis (arrow). b) Lateral view shows high-grade stenoses of the celiac (arrow) and SMA (double arrow).

The Visceral Vessels

Since abdominal atherosclerosis usually begins in the infrarenal aorta, the emphasis of the peripheral vascular arteriogram is on opacification of the infrarenal aorta. A thorough evaluation of the suprarenal aorta and its branches is not usually included. Evaluation of the visceral vessels is appropriate in the following circumstances: when the patient complains of symptoms of visceral ischemia **(Fig. 7)**, when the atherosclerotic process is extensive and extends above the renal vessels to involve the visceral aorta, or when a prominent visceral collateral is observed on the initial AP aortogram.

Some important visceral collaterals are described in **Table 2**. These collaterals may serve to supply flow to the gastrointestinal tract or to the lower extremities.

TABLE 2: IMPORTANT VISCERAL ARTERIAL COLLATERALS

Collateral Connection	Vessels Involved
Celiac to SMA	Gastroduodenal (GDA) to inferior pancreaticoduodenal (IPD) arteries Arc of Buehler (persistent embryologic connection)
SMA to IMA	Arc of Riolan Marginal artery of Drummond
IMA to iliac arteries	Superior hemorrhoidal artery to inferior hemorrhoidal artery

While visceral collaterals are best imaged in the AP projection **(Fig. 8)**, the origin of the visceral vessels, which is where atherosclerosis usually occurs, is best imaged in the lateral projection.

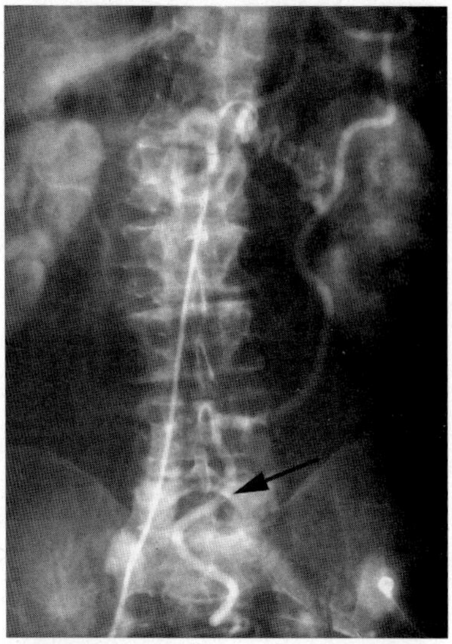

Figure 8. *Late arterial phase image from midstream aortogram in a patient with claudication. The marginal artery supplies blood to the lower extremities via the superior hemorrhoidal branch of the IMA (arrow).*

The Infrarenal Aorta

Normally, the infrarenal aorta is straight and smooth **(Fig. 9)**. The average diameter is 2 cm and the upper limit of normal diameter is 3 cm. The aorta bifurcates at the L4 level into the common iliac arteries. The branches of the infrarenal aorta of note in patients with peripheral vascular disease are the IMA, the middle sacral artery and the lumbar arteries. The IMA originates along the left lateral aspect of the aorta at the L3-L4 level. The four paired lumbar vessels arise from the posterior lateral wall of the aorta. The middle sacral artery is the terminal branch of the distal aorta. The IMA and middle sacral arteries are frequently occluded in patients with severe underlying atherosclerosis. The lumbar vessels often remain patent and provide collateral supply to the lower extremity **(Fig. 10)**. The lumbar vessels anastomose to the superior gluteal, iliolumbar and deep circumflex iliac arteries to reconstitute flow in the internal iliac or common femoral vessels.

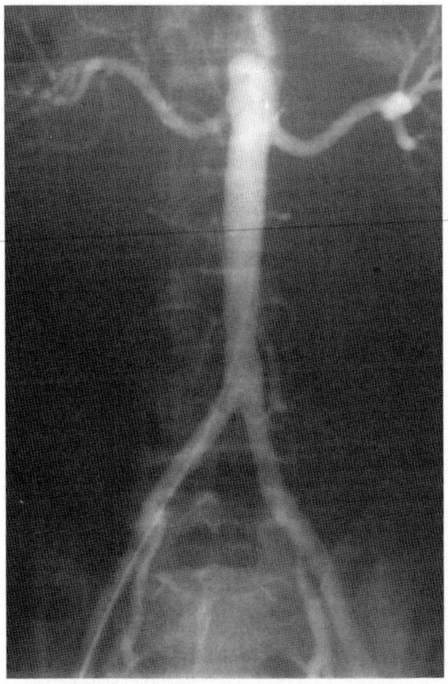

Figure 9. *An infrarenal aortogram is normally obtained before the runoff arteriogram is performed. In this case, the renal arteries and infrarenal aorta are normal.*

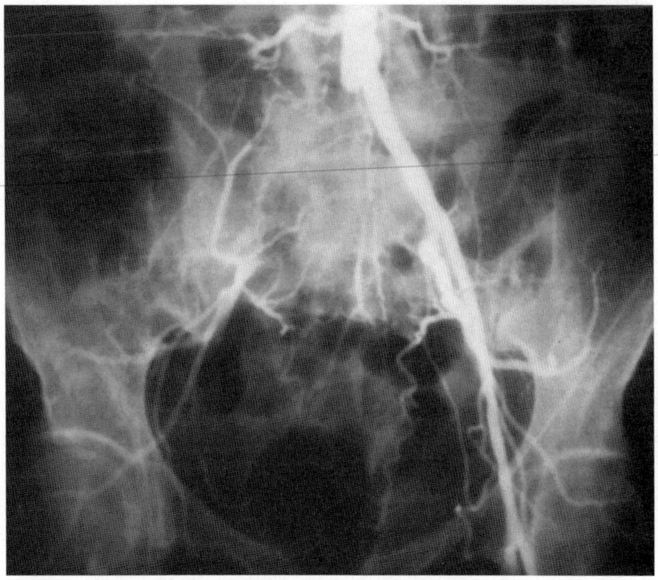

Figure 10. *Iliac occlusion with collateral reconstitution of the internal iliac artery. The right common and external iliac arteries are occluded. There is reconstitution of the right internal iliac artery via the lumbar-iliolumbar collateral. Anterior division branches of the internal iliac reconstitute flow to the lower extremity via collaterals with the profunda.*

The Iliac Vessels

The common iliac artery is the in-line continuation of the aorta. The average diameter is 1 centimeter. It has no branches and, therefore, when the common iliac artery occludes, the entire vessel fills with thrombus back to the aortic bifurcation and forward to the origin of the internal iliac artery. The common iliac artery branches into the internal and external iliac arteries. This bifurcation can be portrayed in profile in the ipsilateral posterior oblique projection **(Fig. 11)**.

The branches of the internal iliac artery provide important collateral flow to the lower extremity in patients with aortoiliac disease. The posterior division branches—the iliolumbar, lateral sacral and superior gluteal arteries—can provide collateral flow from the lumbars to reconstitute the internal and external iliac vessels while the anterior division branches—the internal pudendal, obturator and inferior gluteal arteries—can provide collateral flow to reconstitute the femoral vessels.

The external iliac artery **(Fig. 12)**, is the in-line continuation of the common iliac artery, and is, on average, 8 mm in diameter. It has only two branches, the deep circumflex iliac and the inferior epigastric arteries. These two terminal vessels can accept collateral flow from the superior epigastric artery or the internal iliac vessels to reconstitute flow in the common femoral artery (CFA).

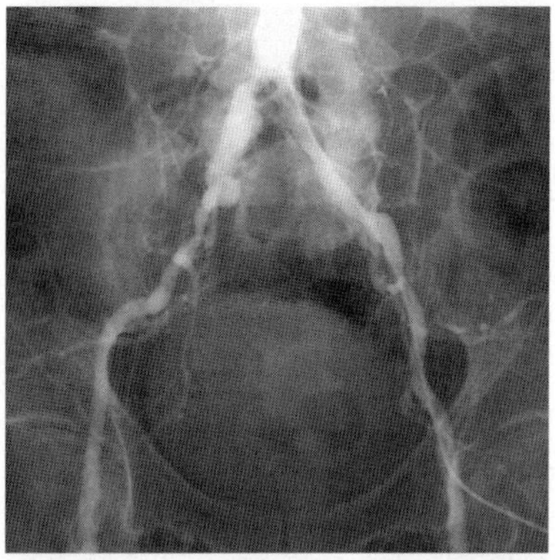

a.

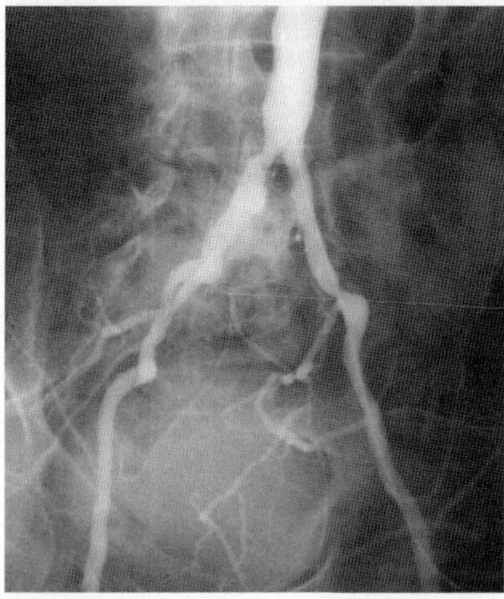

b.

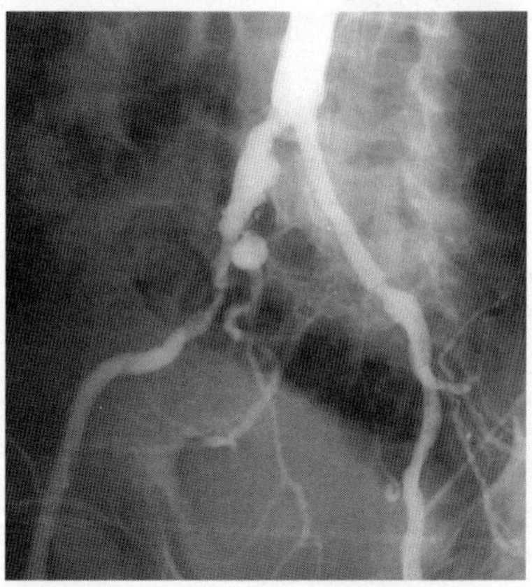

c.

Figure 11. *Pelvic arteriogram in a patient with both aneurysmal and occlusive atherosclerosis. a) AP view shows an aneurysm of right common iliac artery. b) LPO projection demonstrates the origin of the left internal iliac artery. c) RPO projection demonstrates the origin of the right internal iliac artery and an internal iliac artery aneurysm.*

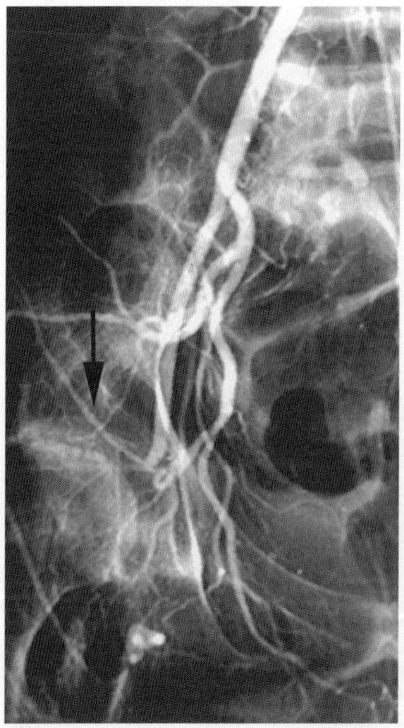

a.

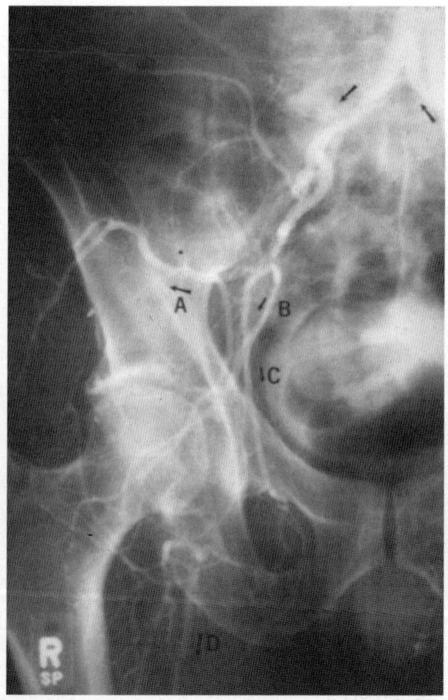

b.

Figure 12. *Consider these two pelvic arteriograms. Is the right external iliac artery patent? a) In this case the external iliac artery is patent and the CFA is occluded. Note visualization of one of the terminal branches of the external iliac artery, the circumflex iliac artery (arrow). b) In this case the external iliac artery is occluded and the internal iliac artery is supplying collateral flow to reconstitute the profunda. Can you name the collaterals a-c? (a= superior gluteal, b= inferior gluteal, c= obturator.)*

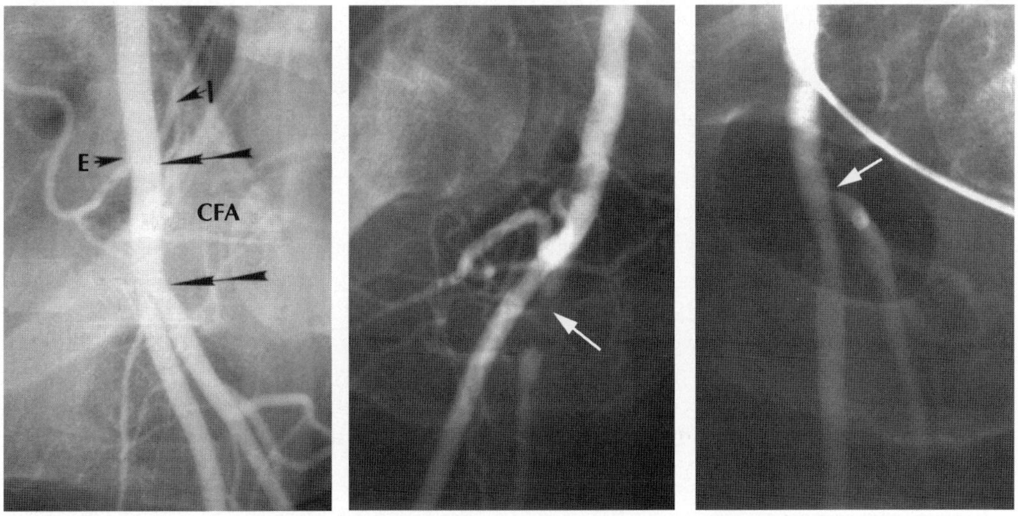

a. b. c.

Figure 13. The common femoral artery bifurcation is best displayed in the ipsilateral anterior oblique projection. a) In this case the CFA (arrows) is normal. The terminal branches of the external iliac artery are shown—the circumflex iliac (I) and the inferior epigastric (E) arteries. b) In this case the proximal SFA is occluded (arrow). c) In this case, there is a tight stenosis at the origin of the profunda (arrow).

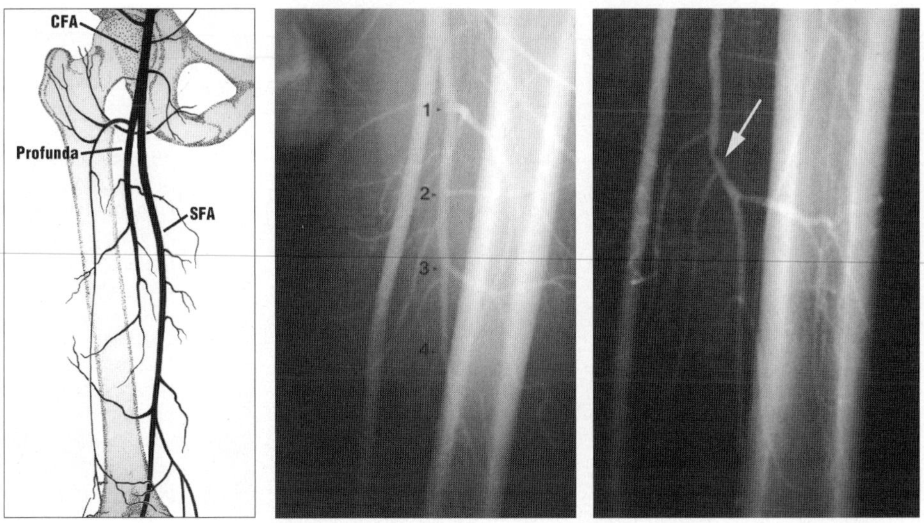

a. b. c.

Figure 14. The femoral vessels. a) The CFA birfucates into the SFA and profunda. Note that the profunda has four perforating muscular branches. b) Perforators of normal caliber are demonstrated in this patient with a patent SFA. c) A peripheral stenosis of the profunda (arrow) is evident in this patient with diabetic peripheral vascular disease.

The Femoral Vessels

The CFA is the in-line continuation of the external iliac artery below the inguinal ligament. On average, the CFA is 4 cm long and 6-8 mm wide. In some patients the CFA may be short (less than 2.5 cm long) due to a high origin of the profunda femoris artery (profunda). The CFA bifurcates into the profunda and SFA. The anatomy of the CFA bifurcation is best portrayed in the ipsilateral anterior oblique projection **(Fig. 13)**.

The two proximal branches of the profunda, the lateral and medial femoral circumflex, are of paramount importance in patients with iliac occlusive disease. These arteries form collaterals with branches of the internal iliac artery to reconstitute flow to the lower extremity. The distal branches of the profunda, the perforating arteries **(Fig. 14)**, serve as collateral pathways in patients with SFA occlusion. Hypertrophied perforators anastomose with the popliteal artery through geniculate collaterals.

The SFA extends from the femoral bifurcation to the adductor canal. It gives off multiple small muscular branches that can act as collaterals to reconstitute SFA flow in patients with short segment SFA occlusions. Absence or hypoplasia of the SFA and profunda is seen in patients with a persistent sciatic artery.

The Popliteal Artery

The popliteal artery is the continuation of the SFA below the adductor hiatus **(Fig. 15)**. Its branches—the superior, middle and inferior geniculates—provide collateral flow to reconstitute the popliteal artery or tibial vessels. The popliteal artery is usually 5-7 mm in diameter and has a gradual medial to lateral course as it transits through the popliteal fossa. A variety of disease processes can affect the popliteal artery. While these processes can occur anywhere within the vessel, several typically affect a focal segment as shown in **Table 3**. Examples of each of these pathologic conditions are shown at the end of this chapter.

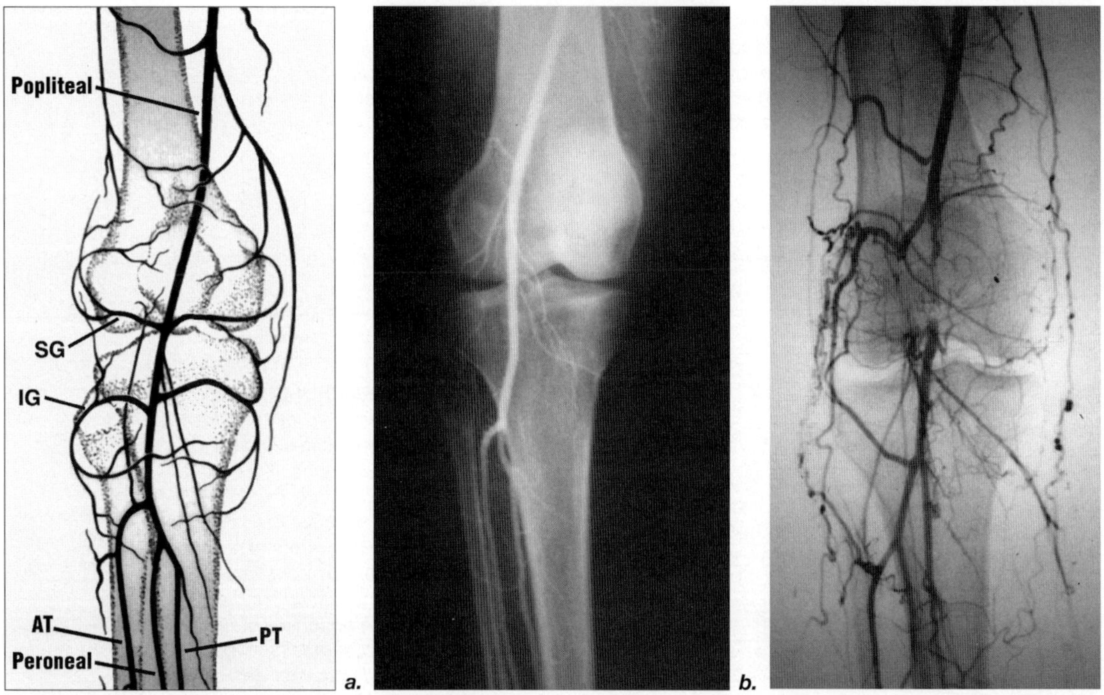

Figure 15. *The popliteal artery. a) The geniculate vessels (superior geniculate [SG] and inferior geniculate [IG]) provide collateral flow around a popliteal stenosis or occlusion. b) Normal popliteal arteriogram. c) Mid-popliteal artery stenosis with enlarged geniculate collaterals. Note variant branching of the tibial vessels. There is a common anterior tibial-peroneal trunk.*

TABLE 3: DISEASE AFFECTING THE POPLITEAL ARTERY

Disease	Involved Segment
Popliteal artery entrapment	Middle
Popliteal artery aneurysm	Proximal to middle
Emboli to the popliteal artery	Distal
Atherosclerosis	Anywhere
Cystic adventitial disease	Anywhere

The Tibial Vessels

The popliteal artery usually bifurcates into the anterior tibial artery and the tibioperoneal trunk about 5 cm below the tibial plateau. The tibioperoneal trunk courses inferiorly for 3-4 cm and then bifurcates into the posterior tibial and peroneal arteries. This normal branching pattern is seen in 90% of patients. In 10% of patients, variant branching is observed. The most common variants are trifurcation of the popliteal artery, origin of the peroneal artery from the anterior tibial artery, and high origin of the anterior tibial artery.

On a runoff angiogram, the bony cortex of the tibia and fibula may obscure the tibial arteries. Visualization of these vessels, particularly their origins, can be aided by digital subtraction techniques or by imaging the vessels in a lateral projection **(Fig. 16)**.

Pedal Vessels

The anterior and posterior tibial arteries continue into the foot as the dorsalis pedis and posterior tibial arteries, respectively. The peroneal artery terminates at the ankle but may play a pivotal role by supplying collateral flow to reconstitute either the dorsalis pedis or posterior tibial arteries **(Fig. 17)**. The primary pedal arch is a potential pathway linking the ventral and dorsal circulation in the foot. It is formed by the union of the lateral plantar branch of the posterior tibial artery and the deep plantar branch of the dorsalis pedis artery.

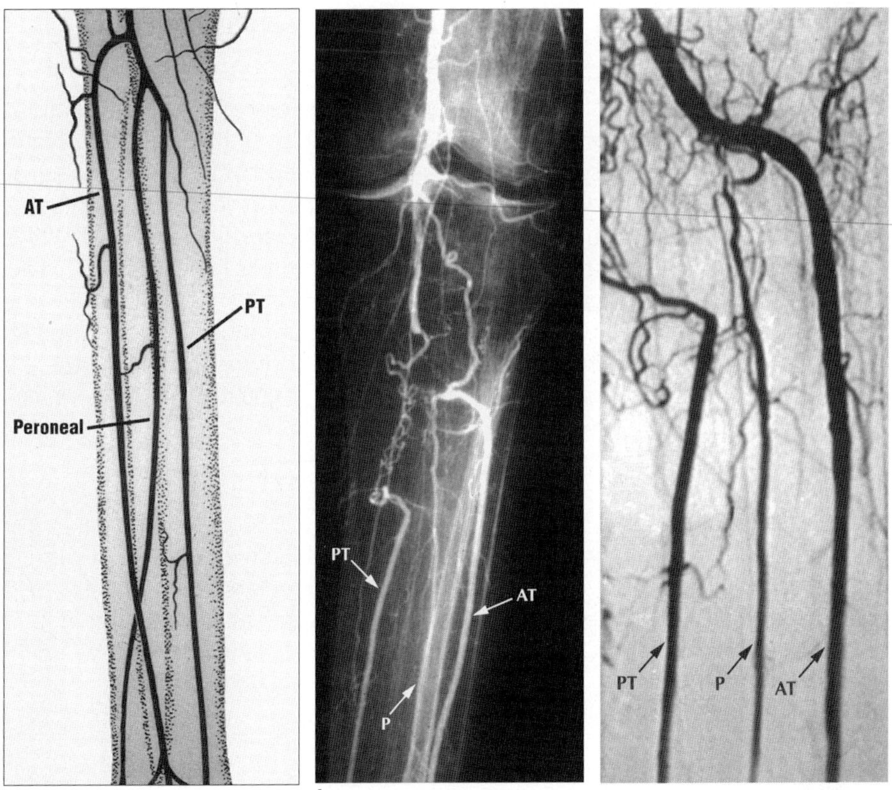

a. **b.** **c.**

Figure 16. *The tibial vessels. a) The normal trifurcation pattern. b) This patient with rest pain has occlusion of the distal popliteal artery with distal reconstitution of the three tibial arteries. c) The anatomy of the popliteal trifurcation is well displayed on the lateral projection. In this case, the tibioperoneal trunk is occluded and there is reconstitution of the posterior tibial and peroneal arteries.*

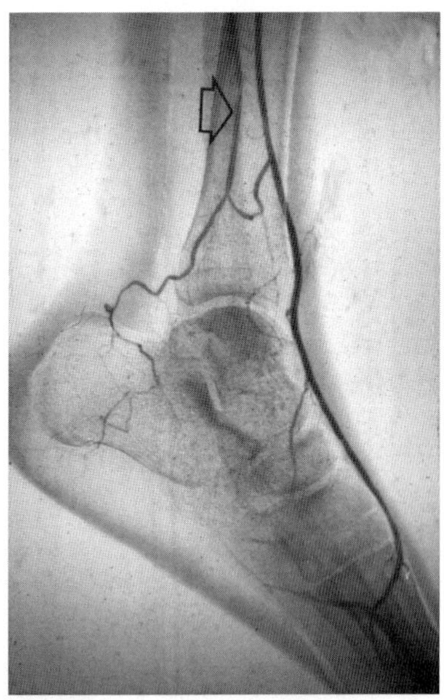

Figure 17. *Pedal arteries. Can you identify the vessels in this image? (For answers see Figure 18 on next page.)*

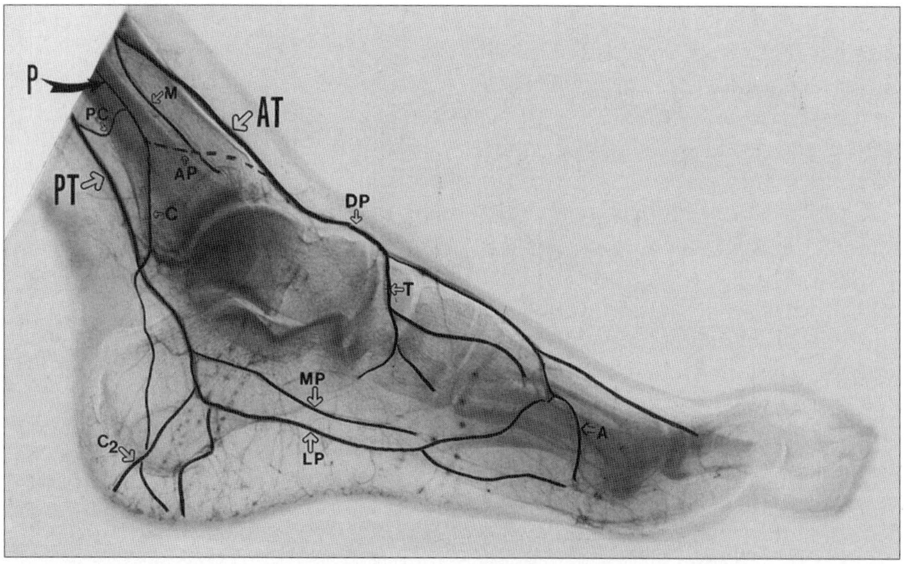

KEY FOR FIGURE 18
Posterior tibial artery (PT)
 Lateral plantar (LP)
 Medial plantar (MP)
 Calcaneal (C2)
Anterior tibial artery (AT)
 Malleolar (M)
 Dorsalis pedis (DP)
  Tarsal (T)
  Arcuate (A)
Peroneal (P)
 Calcaneal (C)

Figure 18. *The vascular anatomy of the foot is best depicted in the lateral projection. Note that the primary pedal arch is formed by the union of lateral plantar and dorsalis pedis vessels.*

The vascular anatomy of the foot is best portrayed in the lateral projection **(Fig. 18)**.

TECHNIQUE

Vascular access: Peripheral arteriography can be performed via a transfemoral, translumbar, or transaxillary approach **(Fig. 19)**. The overall complication rates according to type of access are 1%, 2%, and 3%, respectively. The approach to patients with patent aortobifemoral grafts (AFBG) varies; at UCSF, AFBG limbs are routinely punctured and this route is favored over a transaxillary approach. When a transaxillary approach is required, the left axillary route is favored as it is easier and is associated with fewer complications. The approach to anticoagulation during arteriography also varies; at UCSF, patients are not routinely heparinized for runoff studies. However, after catheterization of a stenotic vessel or when there is diminishment of the pulse following catheterization, 4,000 to 5,000 units of heparin are administered intravenously.

VASCULAR ACCESS FOR PERIPHERAL ANGIOGRAPHY

Route:	Complications:
Femoral	1%
Translumbar	2%
Axillary	3%

Equipment

Peripheral arteriography can be performed with either conventional (cut-film) or digital filming. In the past, cut-film imaging was standard and many of the images in this chapter were obtained using that technique. Today, most new angiographic rooms offer only digital imaging.

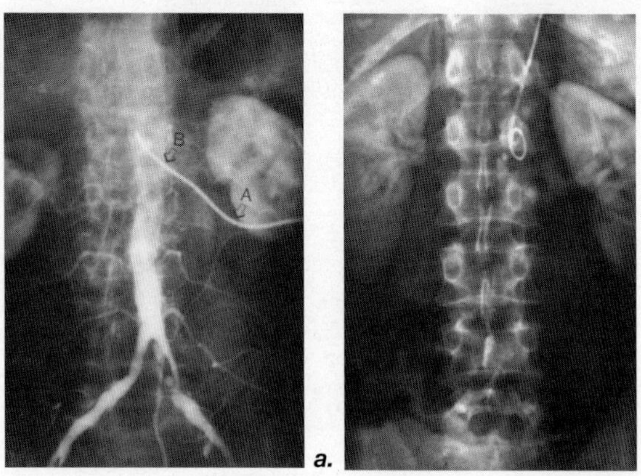

a. **b.**

Figure 19. *Alternative routes of vascular access for peripheral arteriography. a) Direct puncture of the aorta using a translumbar approach as in this case. b) Axillary or brachial artery catheterization is another alternative. In this case, the pigtail catheter was advanced into the infrarenal aorta from a left axillary approach.*

The advantages of digital filming over conventional filming include lower contrast requirements, greater flexibility in image acquisition and lower film costs. The technique described below applies to conventional filming as this is the more complex of the two approaches. The same general considerations apply to digital imaging; however, digital imaging requires less contrast (half to one-third) and imaging is not constrained to prescribed table locations.

Conventional films are obtained using a stepping table or a long-leg film changer. The stepping table is the most commonly employed imaging system. This well-known device consists of a stationary standard high-speed film changer, a standard angiographic x-ray tube, and a motorized translating angiographic table. The table is designed to move in a cephalic direction at predetermined intervals. The table motion can be preprogrammed to stop at four or five different positions **(Fig. 20)**. The table translation time from one station to the next is 1 second. Films are exposed during the motionless periods between movements. There is usually a 1-inch overlap in coverage of the 14-inch by 14-inch high-speed films. The focal film distance is 40 inches. The anatomic coverage achieved with a stepping table depends somewhat on patient size but typically is limited to the vasculature from the aortic bifurcation to the lower calf. For this reason, a separate aortogram is usually obtained prior to the runoff itself.

The preliminary aortogram is obtained by placing a pigtail catheter at the level of the renal arteries and injecting contrast at a rate of 20 ml/sec for a total of 40 ml. The catheter is then pulled down to the aortic bifurcation and contrast material is injected at a rate of 7-10 ml/sec for a total injection time of 10-12 seconds (i.e. 70-100 ml). A film delay of 4-7 seconds allows the lower extremity vasculature to fill with contrast before filming is begun **(Fig. 21)**. The transit of contrast material through each station is dependent upon several patient-specific factors such as cardiac output and the severity of underlying vascular occlusive disease. The angiographer must predict a patient's peripheral flow rate based on an estimate of both cardiac output and the severity of underlying occlusive disease and then select an injection and filming rate to match these patient specific factors. In the average patient with peripheral vascular disease, the film and injection rates that I use routinely are as follows: 4 second film delay and then 3 films at position 1 at 1 film per second, 2 films at position 2 at 1 film per second, five films at position 3 at 1 film per second, and then 1 film every other second at position 4 for a total of 10 films. (Injection rates are for conventional filming.)

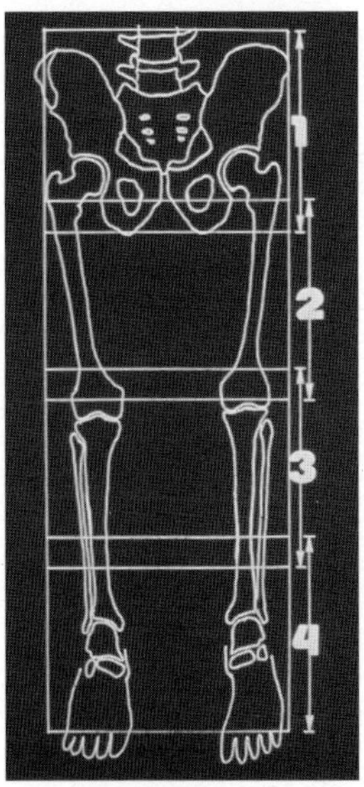

***Figure 20.** The four stations of a bilateral runoff study are shown in this graphic.*

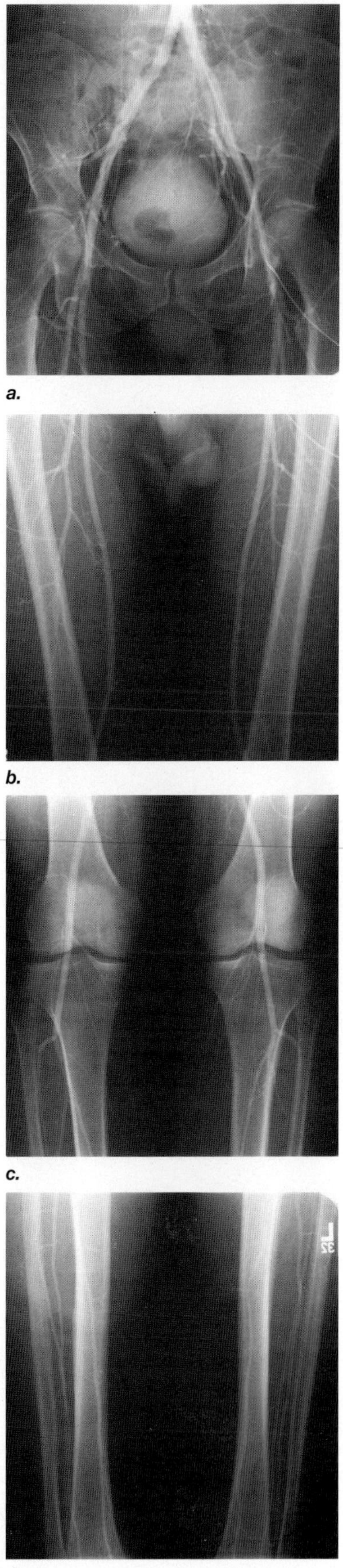

***Figure 21.** Typical bilateral runoff films using conventional filming and stepping table, a-d. Pelvis, thigh, knee and calf films.*

Single Versus Bilateral Leg Imaging

In general, a bilateral study is preferred over a unilateral study because the bilateral study provides a valuable overall assessment of the patient's vascular disease. Identification of vascular pathology in the aorta and the asymptomatic leg may help determine the etiology of a patient's vascular disease and influence the type of therapy selected. Single leg runoff **(Fig. 22)** studies are usually reserved for patients who are at risk of contrast-induced renal failure. The chief advantage of the single leg study is that it requires less contrast material than the standard bilateral study. Additionally, since contrast is injected into the ipsilateral iliac artery, timing uncertainties are lessened. The typical injection and filming rates that I use for a single leg study are as follows: 4 ml/second is injected for 11 seconds for a total of 44 ml. Filming is begun after a 2-second delay at a rate of 1 film/second in a sequence of 3 films at position 1, 2 films at position 2, 5 films at position 3, and 10 films at position 4.

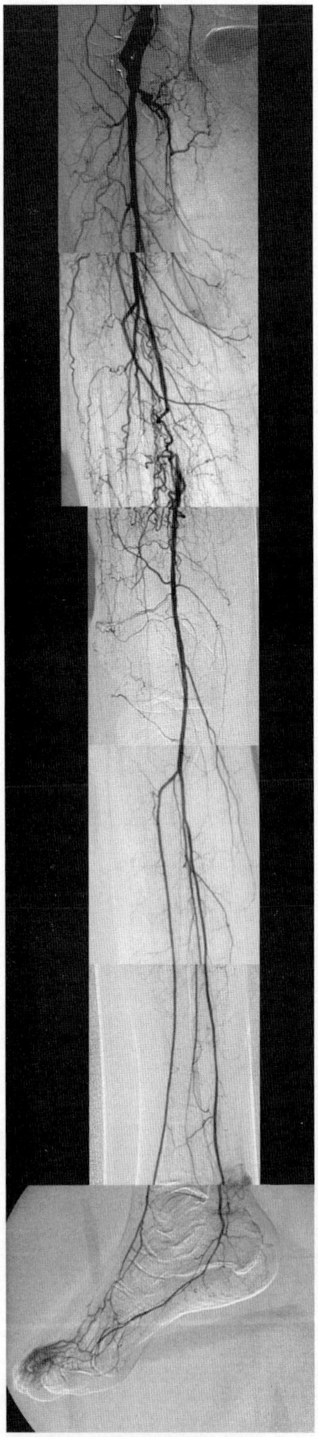

Figure 22. *Composite view of digital single leg runoff arteriogram.*

Additional Views

Supplemental views of specific areas are frequently required to clarify a patient's vascular anatomy, particularly to clarify the anatomy at branch points. The supplemental views that are commonly obtained and the reasons for obtaining them are listed in **Table 4**.

TABLE 4: SUPPLEMENTAL VIEWS FOR PERIPHERAL ARTERIOGRAPHY

Supplemental View	Indication
Lateral aorta	Evaluation of the celiac and SMA origins Evaluation of the posterior aortic wall
Oblique pelvis	Portray the common iliac bifurcation (use ipsilateral posterior oblique projection)
Oblique groin	Portray the common femoral artery bifurcation (use ipsilateral anterior oblique projection)
Trifurcation	Portray proximal tibial disease (use lateral or internal oblique projection)
Foot	Evaluate suitability of the pedal vessels for distal bypass (use lateral projection, conventional or digital filming)

Contrast

Contrast administration in the doses required for optimal conventional runoff imaging is painful. Consequently, the use of nonionic or low osmolality contrast agents is advised. Patients at risk for contrast-induced renal failure may benefit from the use of digital imaging with its lower contrast dose requirements and perhaps the use of carbon dioxide as an adjunctive or primary contrast agent. Recently, gadolinium has been described as an alternative agent in patients with contrast allergies or those at risk for contrast-induced renal failure.

CONTRAST MATERIAL FOR PERIPHERAL ARTERIOGRAPHY

- Nonionic or low-osmolar iodinated contrast
- Carbon dioxide
- Gadolinium

REVIEW OF SPECIFIC DISEASES

A systematic review of vascular disease is presented in Chapter 1. Several entities of specific note that may cause peripheral ischemia are listed below.

Variant Anatomy

1) Persistent sciatic artery: In this uncommon condition, the femoral arteries are hypoplastic and the lower extremity blood supply is derived from the internal iliac artery via a persistent sciatic artery. This vessel courses posteriorly around the hip and down the thigh to become the popliteal artery at the knee. The persistent sciatic artery may be at increased risk for aneurysm formation.

Pathologic Conditions

1) Leriche syndrome: Patients with this condition present with a clinical triad of lower extremity claudication, impotence and absent femoral pulses. Symptoms are due to atherosclerotic occlusion of the infrarenal aorta. This condition was originally described in younger men (< 50 years old) **(Fig. 23)**.

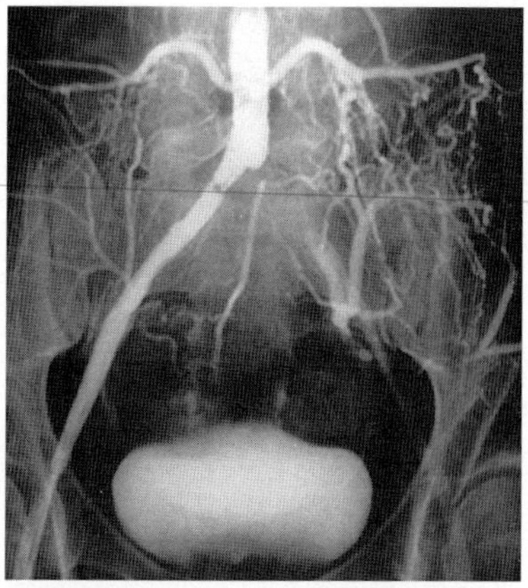

Figure 23. *Pelvic arteriogram obtained in a 40-year-old man who complained of impotence and left leg claudication. Note that atherosclerosis is centered at the aortic bifurcation. The left common iliac artery is occluded and the right common iliac artery is stenotic. This is Leriche-type anatomy.*

2) Acute vascular occlusion: Acute arterial occlusion may be due to an embolus or in-situ thrombosis. Peripheral emboli lodge most commonly at either the CFA bifurcation or the popliteal trifurcation. Symptoms include the 5 P's—pain, pallor, pulselessness, paresthesias and paralysis.

3) Thromboangitis obliterans (Buerger's disease): This is an inflammatory disease of unknown etiology involving the small and medium-sized arteries of the extremities leading to segmental occlusions. It is typically seen in young men (< 40) who smoke heavily.

POPLITEAL ARTERY PATHOLOGY

- Cystic adventitial disease
- Popliteal artery entrapment syndrome
- Aneurysm
- Occlusive atherosclerosis
- Emboli
- Trauma

4) Popliteal artery pathology **(Fig. 24)**: The differential diagnosis of popliteal artery disease includes: cystic adventitial disease, muscular entrapment, aneurysm, atherosclerosis, and emboli. Traumatic dissection could also be included if there is a history of trauma or posterior knee dislocation. These entities are described in greater detail below.

a. Cystic adventitial disease of the popliteal artery: A mucin-containing multilocular cyst of the adventitia of the popliteal artery results in a focal popliteal artery stenosis with resultant ischemia. The etiology of this entity is unknown. The angiographic finding is a segmental stenosis or occlusion in an artery with a normal course.

b. Popliteal artery entrapment syndrome: The popliteal artery is compressed by the medial head of the gastrocnemius muscle resulting in ischemia. This syndrome is typically observed in young active men. The abnormality is bilateral in 20%. On physical exam, peripheral pulses may diminish with passive dorsiflexion. The characteristic angiographic finding is medial deviation of the midportion of the popliteal artery. This finding may only be present when stress views are obtained with the foot in active plantar flexion. Occlusion of the popliteal artery is the end-stage result of this disease. The diagnosis of popliteal artery entrapment syndrome should be suspected in a patient with popliteal artery occlusion when medial deviation is observed in the contralateral popliteal artery.

c. Popliteal artery aneurysms: The normal popliteal artery is less than 0.7 cm in diameter. By definition, a diffuse or focal enlargement of the popliteal artery greater than 0.7 cm is an aneurysm. Aneurysms are most commonly atherosclerotic and are located just below the adductor hiatus in the proximal popliteal artery. Over half of popliteal aneurysms are bilateral. Abdominal aortic aneurysms are seen in one-third of patients with popliteal aneurysms.

d. Occlusive atherosclerosis: Stenosis or occlusion due to atherosclerosis may occur anywhere in the vessel. Collateral vessels are usually hypertrophied due to longstanding disease.

e. Emboli: Popliteal embolus usually results in occlusion of the distal popliteal artery and proximal trifurcation vessels. Large collaterals are not present because this is an acute injury.

5) Peripheral mycotic aneurysms: These focal aneurysms are due to bacterial infection of the vessel wall. Salmonella sepsis may result in aortic or femoral artery mycotic aneurysms. Parenteral drug abuse may result in femoral artery mycotic aneurysms due to staphylococcus.

6) Diabetic peripheral vascular disease **(Fig. 25)**: Diabetic patients are at risk for developing advanced atherosclerosis in the lower extremity that has a distinctive distribution. Severe tibial and pedal disease predominates with relative sparing of the aortoiliac and femoral vessels. Peripheral profunda stenosis is a characteristic finding that is not usually seen in patients with nondiabetic atherosclerotic disease.

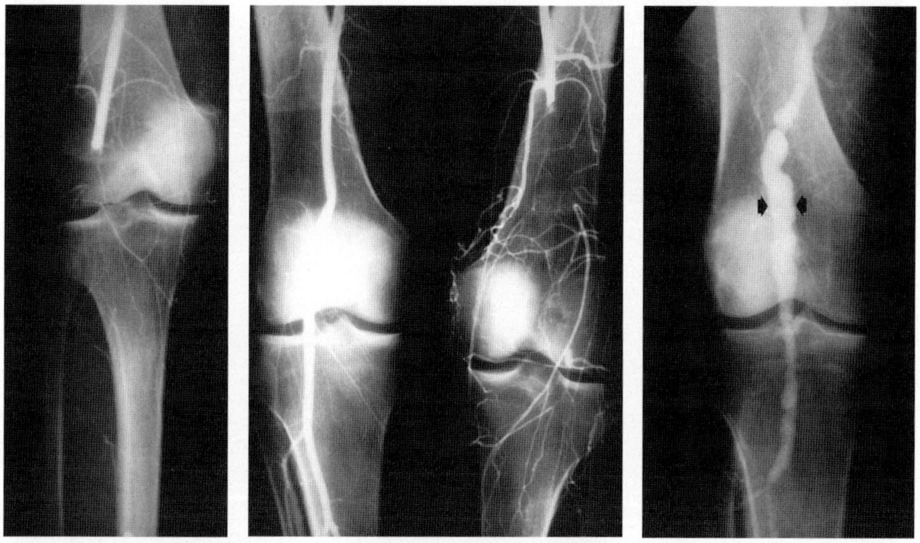

a. *b.* *c.*

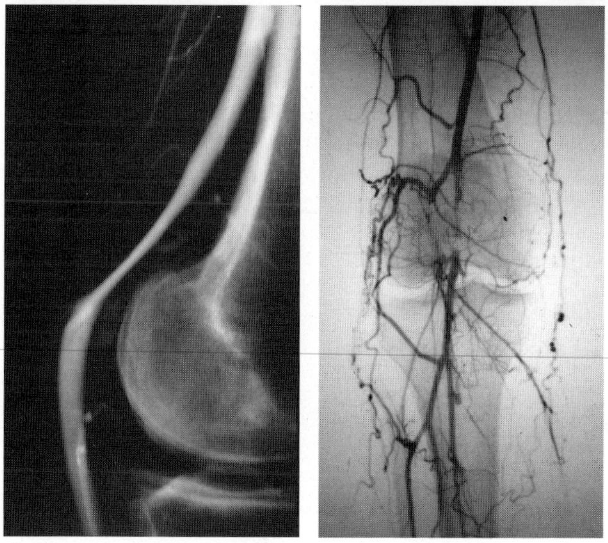

d. *e.*

Figure 24. *Examples of 5 different types of popliteal pathology are shown above. Can you make the diagnosis in each case? (See below for answers).*

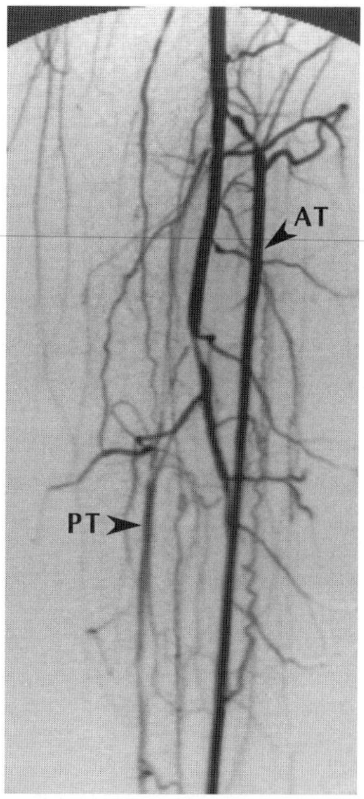

Figure 25. *Digital arteriogram of the proximal left calf demonstrates the typical findings of diabetic peripheral vascular disease. The runoff vessels above this level were normal. There is extensive narrowing of the tibial vessels involving the anterior tibial (AT), tibioperoneal trunk and posterior tibial (PT) vessels.*

Answers

a) Embolic occlusion.
b) Politeal artery entrapment syndrome.
c) Popliteal artery aneurysm.
d) Cystic adventitial disease with smooth narrowing of the upper popliteal artery.
e) Atherosclerosis with mid popliteal stenosis.

7) Klippel-Trenaunay-Weber syndrome: There is a reason that this entity is listed last. It is distinctly uncommon. The classic triad of Klippel-Trenaunay is unilateral lower extremity hypertrophy, port-wine hemangiomas and varicose veins. When, in addition, there is an arteriovenous malformation, the entity is known as Klippel-Trenaunay-Weber syndrome. The hallmark feature of this syndrome on venography is absence of the deep femoral vein.

REFERENCES

Anatomy

Bergman RA, Thompson SA, Afifi AK, Saadeh FA. The cardiovascular system. In: Bergman RA, ed. Compendium of human anatomic variation. Baltimore: Urban & Schwarzenberg, 1988; 60-87.

Kadir S. Arterial anatomy of the lower extremities. In: Kadir S, ed. Atlas of normal and variant angiographic anatomy. Philadelphia: W.B. Saunders, 1991; 123-162.

Technique

Alson MD, Lang EV, Kaufman JA. State of the art: pedal arterial imaging. JVIR 1997; 8:9-18.

Darcy M. Lower extremity arteriography: current approach and techniques. Radiology 1991; 178:615-621.

Gates J, Hartnell GG. Optimized diagnostic angiography in high-risk patients with severe peripheral vascular disease. Radiographics 2000; 20:121-133.

Peripheral Vascular Disease

Fowl RJ, Kempczinski RF. Popliteal artery entrapment. In: Rutherford RB, ed. Vascular Sugery. 5th edition. Philadelphia: W.B. Saunders, 2000; 1087-1101.

Levien LJ, Bergan JJ. Adventitial cystic disease of the popliteal artery. In: Rutherford RB, ed. Vascular Sugery. 5th edition. Philadelphia: W.B. Saunders, 2000; 1079-1087.

Olin JW. Thromboangiitis obliterans (Buerger's disease). In: Rutherford RB, ed. Vascular Sugery. 5th edition. Philadelphia: W.B. Saunders, 2000; 350-363.

Zairns CK, Glagov S. Artery wall pathology in atherosclerosis. In: Rutherford RB, ed. Vascular Sugery. 5th edition. Philadelphia: W.B. Saunders, 2000; 313-333.

CHAPTER 4 PREVIEW ANSWERS

Case 1

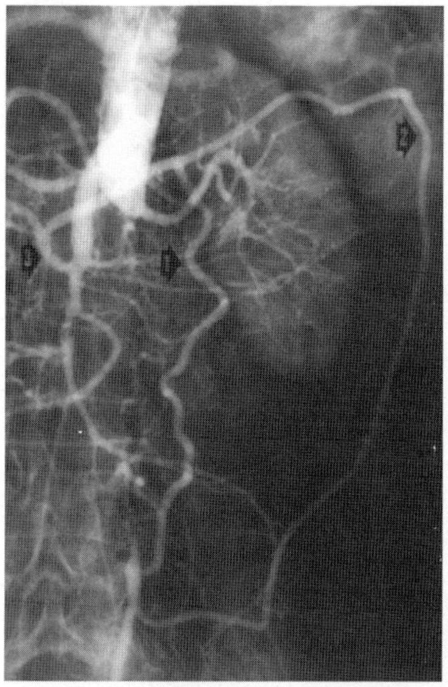

Figure 1. *Abdominal aortogram.*

Findings: The aorta is occluded at the level of the renal arteries. Collateral flow to the lower extremities is supplied via SMA-IMA-Internal iliac artery collaterals. The two vessels depicted by the arrows are the Arc of Riolan (#1), and the marginal artery of Drummond (#2). Both are branches of the middle colic artery (#3).

Diagnosis: Infrarenal abdominal aortic occlusion with collateral reconstitution of the IMA via the marginal artery and arc of Riolan.

Case 2

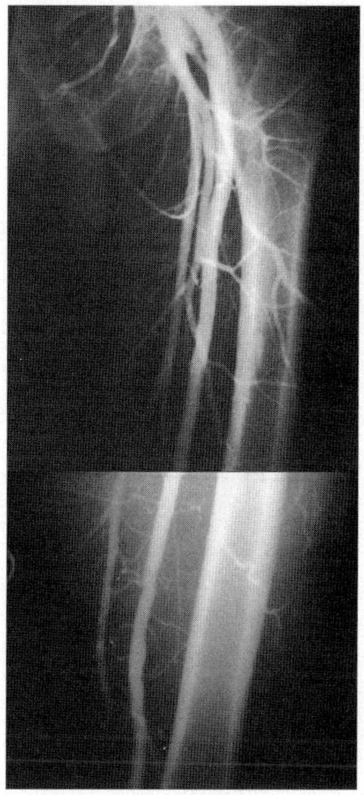

Figure 2. *Left leg runoff, composite of groin and thigh stations.*

Findings: There is aberrant origin of the main vascular supply to the left leg. The common femoral artery, profunda and SFA vessels are small. The main supply to the leg takes origin from the internal iliac artery.

Diagnosis: Variant anatomy of the left lower extremity, persistent sciatic artery.

Case 3

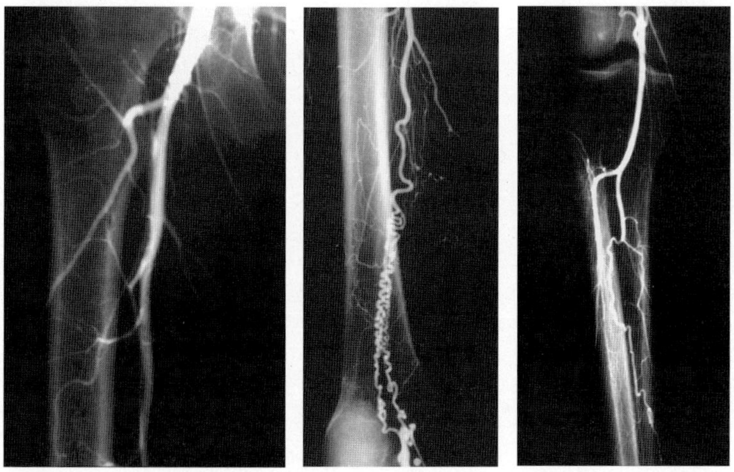

Figure 3. *Three stations from a right leg runoff.*

Findings: The SFA is occluded. Hypertrophied profunda collaterals terminate in a prominent corkscrew pattern that reconstitutes flow in the midpopliteal artery. There is abrupt occlusion of the midtibial vessels although the proximal tibial vessels are normal.

Diagnosis: Thromboangitis obliterans (Buerger's disease).

CHAPTER 5 PREVIEW

Case 1

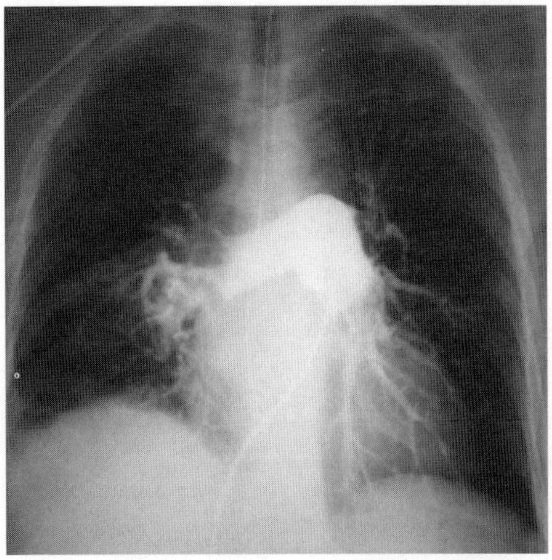

History: This is a pulmonary angiogram performed by injecting contrast into the main pulmonary artery.
Question: What is your diagnosis?

Case 2

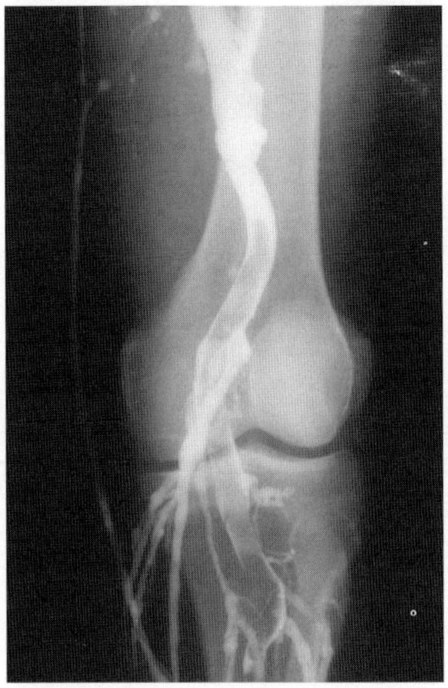

Questions: What is the study illustrating? What are the findings and diagnosis?

Case 3

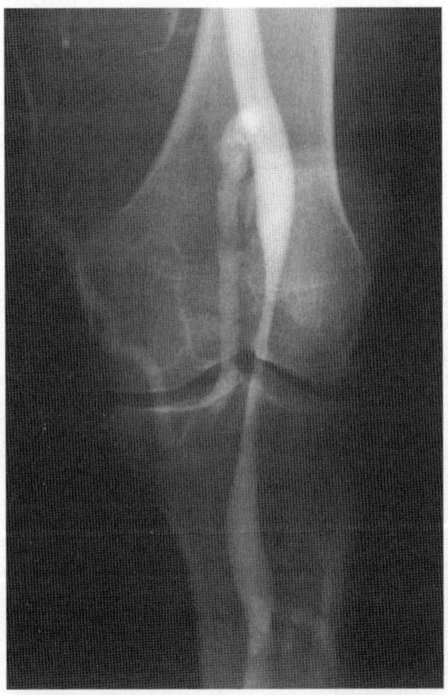

Questions: What is the study illustrating? What are the findings and diagnosis?

Case 4

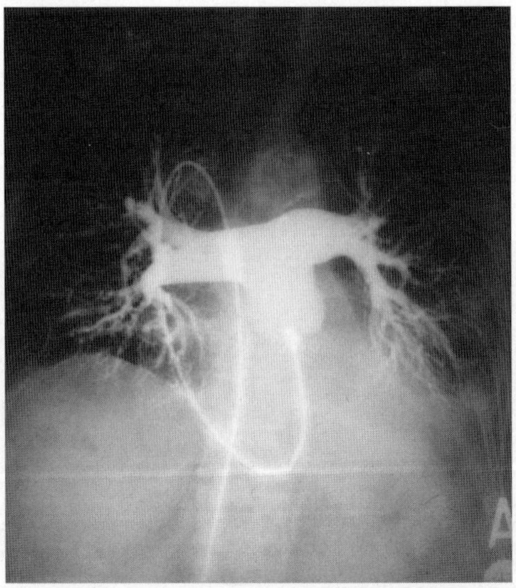

Question: This is difficult, for extra credit. There is something interesting about this pulmonary angiogram. What is it?

Chapter 5
Pulmonary Angiography and Systemic Venography

Roy L. Gordon, MD

PULMONARY ANGIOGRAPHY

- Thromboembolic disease
  - Indications
  - Diagnostic evaluation
  - Angiography, technique, findings
- Arteriovenous malformations

INTRODUCTION

The role of angiography in the evaluation of patients with pulmonary vascular disease is evolving. Previously, pulmonary angiography was often performed to evaluate patients with suspected pulmonary embolism. Now, with recent advances in contrast-enhanced spiral CT and 3-dimensional reformatting, angiography is less frequently performed for this indication. Current indications for pulmonary angiography are shown in **Table 1**. Because of rapid advances in CT and MR, even these indications may soon be eliminated. Interestingly, the most common use of pulmonary angiography overall in our hospital is in the evaluation of children prior to and after cardiopulmonary surgery.

In this chapter, we review the current practice of pulmonary angiography. The second half of this chapter is devoted to the related subject of systemic venography.

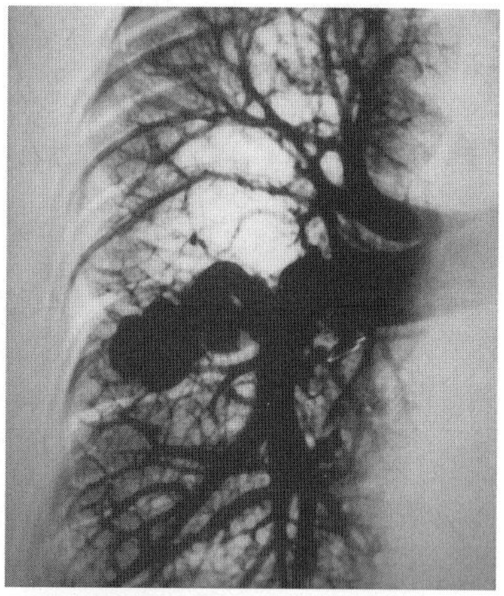

Figure 1. *Selective right pulmonary angiography shows a large arteriovenous malformation. The angiogram was performed in the course of a therapeutic embolization.*

TABLE 1: CURRENT INDICATIONS FOR PULMONARY ANGIOGRAPHY

- Suspected PE with equivocal CT
- Pulmonary AVM prior to intervention **(Fig. 1)**
- Prior to foreign body removal
- Evaluation of uncommon diseases such as hepatopulmonary syndrome, pulmonary vasculitis, stenosis
- In children, evaluation of cardiopulmonary disorders prior to or following surgery

PULMONARY EMBOLISM

It is estimated that between 120,000 to 150,000 people die annually from acute pulmonary thromboembolism. Another 600,000 to 750,000 individuals have significant pulmonary emboli each year. The diagnosis of acute pulmonary thromboembolism is difficult and complex, often with a confusing clinical presentation. The importance of establishing a prompt and accurate diagnosis is highlighted by comparing the mortality rates of treated and untreated pulmonary emboli. Untreated, the mortality rate for patients with acute pulmonary emboli is between 20% and 35% while for patients with properly treated pulmonary emboli, the rate is reduced to 8%. The diagnostic algorithm and therapeutic alternatives with pulmonary thromboembolism continue to evolve. The introduction of rapid helical CT scans into clinical use represents a significant advance in the diagnosis of PE and will probably become the diagnostic method of choice for evaluating patients suspected of having PE. The clinical presentation of patients with suspected pulmonary emboli is associated with presenting signs, symptoms and associated laboratory findings that are nonspecific. These should not be utilized to exclude the diagnosis of pulmonary embolism but rather should function to heighten or decrease one's suspicion for the diagnosis and serve as an entry to a diagnostic pathway based on definitive imaging studies.

Clinical Presentation

Patients with symptoms of pulmonary embolism commonly present with pleuritic chest pain, dyspnea, apprehension, cough, tachypnea, rales, tachycardia and fever. Hemoptysis is considered a hallmark of pulmonary embolism (PE) but this sign will be present in only 30% of patients with PE. Despite the many signs and symptoms associated with PE, only 50% of patients suspected of having a PE on clinical grounds will actually present with one of these signs. Other tests utilized to improve the specificity of the diagnosis of PE include electrocardiography and arterial blood gas analysis. Several characteristic findings may be present on the ECG but these occur in the minority of patients with PE. Hypoxemia and hypocarbia on blood gas evaluation are suggestive of the diagnosis, but up to 14% of patients with documented emboli will have normal values. The clinical signs, symptoms and laboratory test results associated with pulmonary embolism are nonspecific and overlap with many other clinical entities including congestive heart failure, pneumonia, septic shock or acute myocardial infarction.

SYMPTOMS OF PE

- Pleuritic chest pain
- Dyspnea, tachycardia
- Cough
- Fever
- Hemoptysis

Radiographic Findings

A chest x-ray (CXR) is often utilized to evaluate patients presenting with the signs and symptoms described above. The chest x-ray can provide valuable information and suggest diagnoses other than PE, such as pneumonia or congestive heart failure as an explanation for the observed symptoms. A few findings may provide support for the diagnosis of pulmonary embolism. A zone of hyperlucency and diminished vascular markings (Westermark's sign) may be identified. An enlarged pulmonary artery may be a sign of embolus with occlusion of a large vessel. Pleural effusions may be associated with an embolus but this finding is very nonspecific. Similarly, the presence of atelectasis is nonspecific. A wedge-shaped infiltrate whose margins are convex and point toward the hilum (Hampton's hump) may be present but this may be mimicked by a pneumonia. In the majority of cases, the CXR will remain normal. Thus, although the CXR can provide further support for the diagnosis of PE or may provide alternative explanations for the patient's symptoms, the CXR findings are not diagnostic and further workup must be instituted to establish the diagnosis of acute pulmonary embolism.

CHEST RADIOGRAPH FINDINGS OF PE

- Westermark's sign
- Enlarged pulmonary artery
- Hampton's hump
- Pleural effusion and atelectasis (nonspecific)

Pulmonary Scintigraphy

The Prospective Investigation of Pulmonary Embolism Diagnosis (PIOPED) study established the scan interpretation criteria and categorization of probability of PE. Correlation with pulmonary angiography has shown that patients with a high probability scan have a 90% chance of harboring a PE while a normal scan virtually eliminates a PE. Scans in the low to intermediate categories have a 10%-70% chance of representing a PE. Since a significant percentage of the ventilation perfusion (V/Q) scans will fall into the latter categories, it is important to systematically evaluate these patients. Historically, patients with low to intermediate probability lung scans went straight to pulmonary angiography to confirm or exclude the diagnosis of PE. However, the advent of new diagnostic tests such as lower extremity duplex sonography and spiral CT have changed the diagnostic algorithm.

Adjuvant Tests in the Diagnosis of Pulmonary Embolism

D-dimer is a fibrin degradation product that is elevated in the presence of intravascular thrombosis. Several studies have indicated that evaluation of D-dimer levels in patients with low to intermediate probability scans can effectively exclude the presence of pulmonary embolism. Values of less than 500 g/L have 97% negative predictive value for excluding pulmonary embolism. Elevation of the D-dimer level is nonspecific and pulmonary embolism will be present in only 50% of patients with an elevated level. Obtaining a D-dimer level can further stratify patients and guide diagnostic workup.

DIAGNOSTIC TESTS TO EVALUATE PULMONARY THROMBOEMBOLIC DISEASE

- Chest radiograph
- V/Q scan
- D-dimer level
- Lower extremity sonography
- Pulmonary angiography

Doppler Sonography

Based on clinical observation that over 90% of pulmonary emboli originate from the lower extremities, several studies have been performed to examine the role of bilateral lower extremity Doppler sonography in evaluating patients with suspected PE. Since the treatment of deep venous thrombosis and pulmonary embolism is similar, the demonstration of deep venous thrombosis may obviate the need for additional diagnostic imaging for PE. Although Doppler sonography can reliably and accurately detect the presence of deep venous thrombosis, thrombus will be demonstrated in the deep veins of the legs in only 75% of patients with PE. In the face of a negative lower extremity sonographic evaluation, one should proceed with further workup as a significant percentage of pulmonary emboli will be missed if one assumes that a PE cannot be present if there are no demonstrable thrombi in the legs.

The lack of clinical symptoms suggestive of DVT in the lower extremity should not dissuade one from imaging the lower extremity venous system, as 30% of patients with pulmonary emboli can have clinically silent deep venous thrombosis. The thrombus may be incomplete or only cause partial occlusion. There is a high incidence of duplicated superficial femoral veins (20%) and popliteal veins (5%) and thrombosis involving such a limb may only produce minimal clinical symptoms. In addition to the high incidence of clinically silent DVT's, thrombi can originate in other veins. With the increasing use of central venous catheterization, there are an increasing number of upper extremity venous thrombi which possibly could embolize to the lung. Thrombus involving the gonadal or deep pelvic veins will not be detectable with lower extremity Doppler sonography.

New approaches combine contrast-enhanced CT of the pulmonary vessels followed by rapid CT scanning of the leg veins for thrombi, using the same contrast injection. These techniques may greatly influence the future workup of this group of patients.

PRE-ANGIO EVALUATION

- ECG
  - R/O acute MI and left bundle branch block (transvenous pacer)
- PA pressure above 70 mm Hg
  - Selective catheter
- See CT or V/Q for side

ANGIOGRAM

- R or L, 15–25ml/sec x 2
- Main 35 ml/sec, max 50 ml
- Sub-selective 5–15 ml/sec x 2
- Nonionic contrast
- Inspiration
- Fast filming

Pulmonary Angiography Technique

As previously mentioned, the indications for performing angiography in adult patients today are limited. In our institution they fall mainly into three groups:

a) patients in whom the clinician is convinced the patient has a PE despite a negative CT or when the CT is technically unsatisfactory
b) patients with pulmonary AVM's, as a prelude to embolization
c) as part of a procedure to extract a foreign body from the pulmonary arterial tree.

In addition to the routine evaluations performed prior to angiography, the patient's ECG should be evaluated and a left bundle branch block excluded. If a left bundle branch block is present, there is a danger of inducing total heart block and a temporary prophylactic pacemaker should be inserted before the catheter is introduced through the right atrium and ventricle. Catheterization can be performed from the brachial, jugular or preferably, the femoral approach. The guidewire introduced via Seldinger technique is advanced only a short distance into the iliac vein to minimize any chance of breaking off clot that may be present in the iliac vein or inferior vena cava (IVC). A dilator is inserted a few centimeters over the wire and a preliminary venogram performed to ensure that the iliac vein and the IVC are free of thrombus. A variety of catheters is available with which to perform pulmonary angiography and the choice of catheter is at the discretion of the angiographer. Most commonly, a 6-F or 7-F straight pigtail with tip deflecting wire or a Grollman catheter is utilized. Continuous ECG monitoring should be performed while transversing the right atrium and right ventricle to carefully monitor for any induced cardiac arrhythmias. If ventricular tachycardia or a run of premature ventricular contractions are noted, the catheter should be immediately withdrawn from the right atrium or right ventricle into the inferior vena cava. Instructing the patient to cough usually will break a short run of PVC's. If necessary, Lidocaine or other antiarrhythmic drugs can be administered. The catheter is selectively advanced into either the left or right pulmonary artery and appropriate runs are obtained with multiple projections as needed. Nonionic contrast is utilized to minimize cough reflex as well as to minimize any chances of acute fatality secondary to cardiac overload. Prior to injecting any contrast, the pulmonary arterial pressures should be measured. The patient has severe pulmonary hypertension if the pulmonary arterial systolic pressure is greater than 70 mm of mercury. Pressures of 30 mm to 40 mm of mercury are considered above normal.

Patients with significantly elevated pulmonary arterial pressures are at increased risk of severe complications with performance of pulmonary arteriography. In these patients, only selective runs with nonionic contrast should be performed. Pulmonary angiography was usually performed using cut-film, but now is very often performed using digital angiographic equipment. Recent comparative studies have found that digital subtraction angiography (DSA) is as good as or better than cut-film angiography for the detection of pulmonary embolus. The key to getting satisfactory subtracted views despite respiratory and cardiac motion is to acquire films using short exposure times at a rate of 6 films per second or faster and to use some injector delay to generate a variety of masks. This usually provides a reasonable subtracted view although it may take some effort to find a suitable match between mask and image. Sufficient contrast should also be used so that good unsubtracted digital images are available (15 to 30 ml/sec for a total of 30 to 50 mls of nondilute contrast). The angiogram should be performed in full inspiration, which is not always easy to accomplish in ill, uncooperative patients with dyspnea. Often patients are on a ventilator. Usually angiographic runs in more than one projection will be needed **(Fig. 2)**. It may be helpful to view continuous, slowly cycling loops of DSA angiograms on the television monitor. Viewing at the workstation also facilitates optimizing subtracted and unsubtracted images.

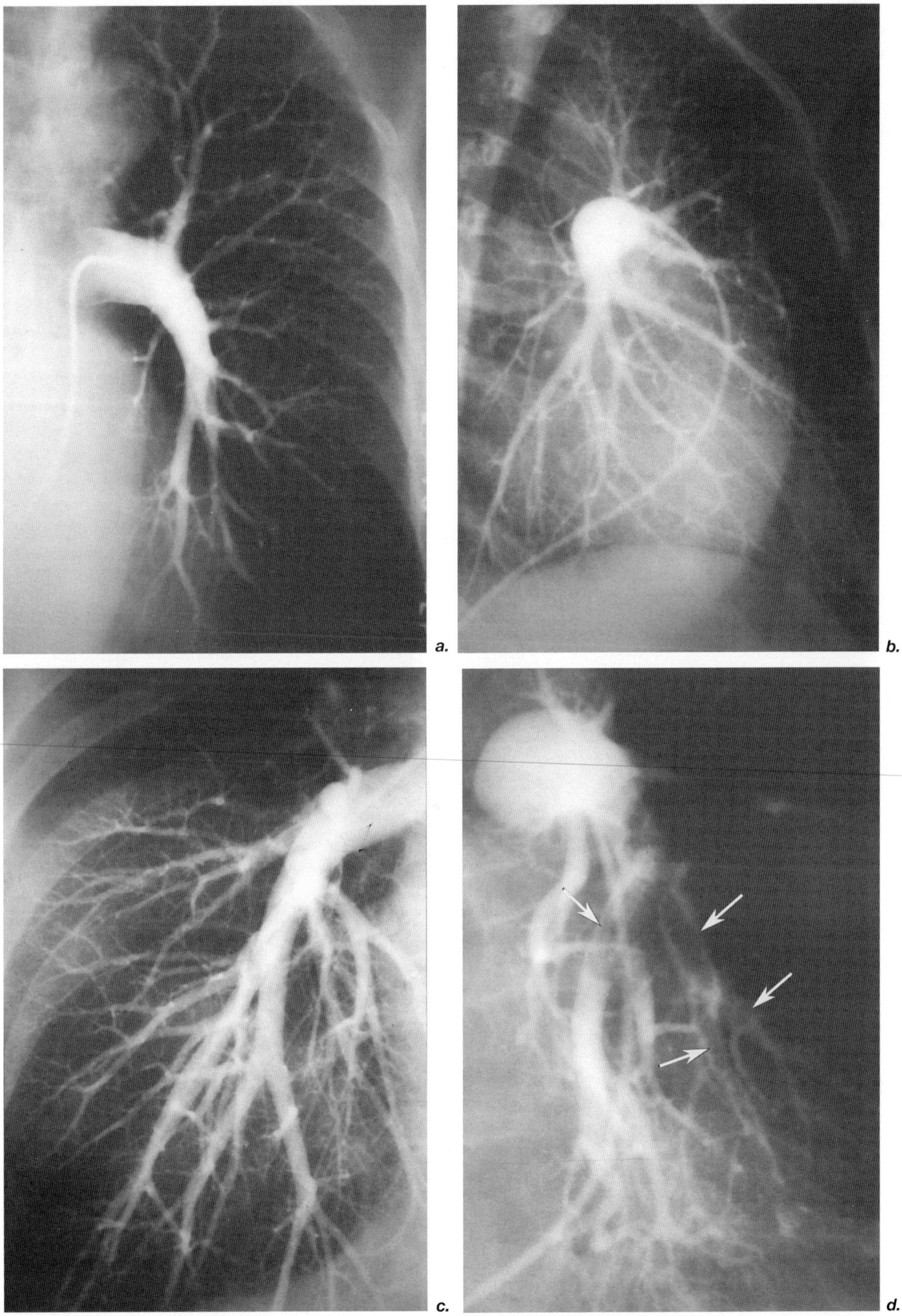

Figure 2. *a) Normal selective left pulmonary angiogram, frontal projection. b) Normal selective left pulmonary angiogram. The oblique projection separates out all the vessels, which are demonstrated clearly with little overlapping. c) Normal subselective right lower lobe pulmonary angiogram. Even though there is excellent demonstration of the vessels, an oblique projection should also be performed to eliminate any areas of vessel overlap. d) This left lower lobe pulmonary angiogram in a different patient shows large filling defects in a number of vessels (arrows). This is highly typical for pulmonary emboli. There is some crowding of vessels that can be seen with atelectasis.*

ANGIOGRAPHIC FINDINGS OF PE

- Acute
  - intraluminal filling defect
  - abrupt vascular cutoff
  - diminished capillary stain (nonspecific)
- Chronic
  - arterial webs, stenoses, occlusions

Once the pulmonary arteriogram is obtained, the diagnosis of PE is made angiographically by demonstrating intraluminal filling defects or abrupt vascular cut-offs **(Fig. 3)**. Indirect signs such as diminished capillary stain or delayed opacification are very nonspecific.

Chronic PE

Chronic PE presents with a different clinical and angiographic picture. Rather than abrupt onset of chest pain as noted in acute PE, patients with a chronic PE will present with exertional dyspnea, right-sided heart failure, and cyanosis. Pulmonary angiography will demonstrate arterial webs, stenoses, or irregular occlusions. An acute PE will lyse within one to two weeks. If the patient has an acute clinical event and has not been anticoagulated, useful clinical information can still be obtained with a pulmonary angiogram up to one week after the acute event. If patients are anticoagulated, there will not be a significant error in diagnosis if a 12- to 24-hour period of time elapses before the pulmonary arteriogram is obtained.

Complications

In a recent review from Duke University (Smith, 1996) of 1,434 patients undergoing pulmonary angiography using nonionic contrast agents, there were no procedure-related deaths. There were four major complications (0.3%) including: two cases of respiratory arrest, one from contrast allergy and one from fluid overload; and two patients with recurrent ventricular arrhythmias who had pulmonary artery hypertension (77 and 80 mm Hg, systolic.) They had 11 minor complications (0.8%) including: six reversible catheter-induced arrhythmias, two vasovagal episodes, two minor contrast reactions and one temporary, unexplained episode of chest pain.

Interventional Therapeutic Approaches to Massive Acute Pulmonary Embolism

A variety of different approaches has been described including: a) surgical embolectomy, b) percutaneous transvenous catheter clot aspiration, c) percutaneous transvenous mechanical clot fragmentation, d) percutaneous transvenous selective catheter-directed thrombolysis and e) combinations of the preceding techniques. The data is of a preliminary nature and there is no consensus regarding optimum techniques or indications for intervention as opposed to purely medical therapy. The ability to perform pulmonary angiography is an obvious prerequisite for the utilization of these invasive techniques.

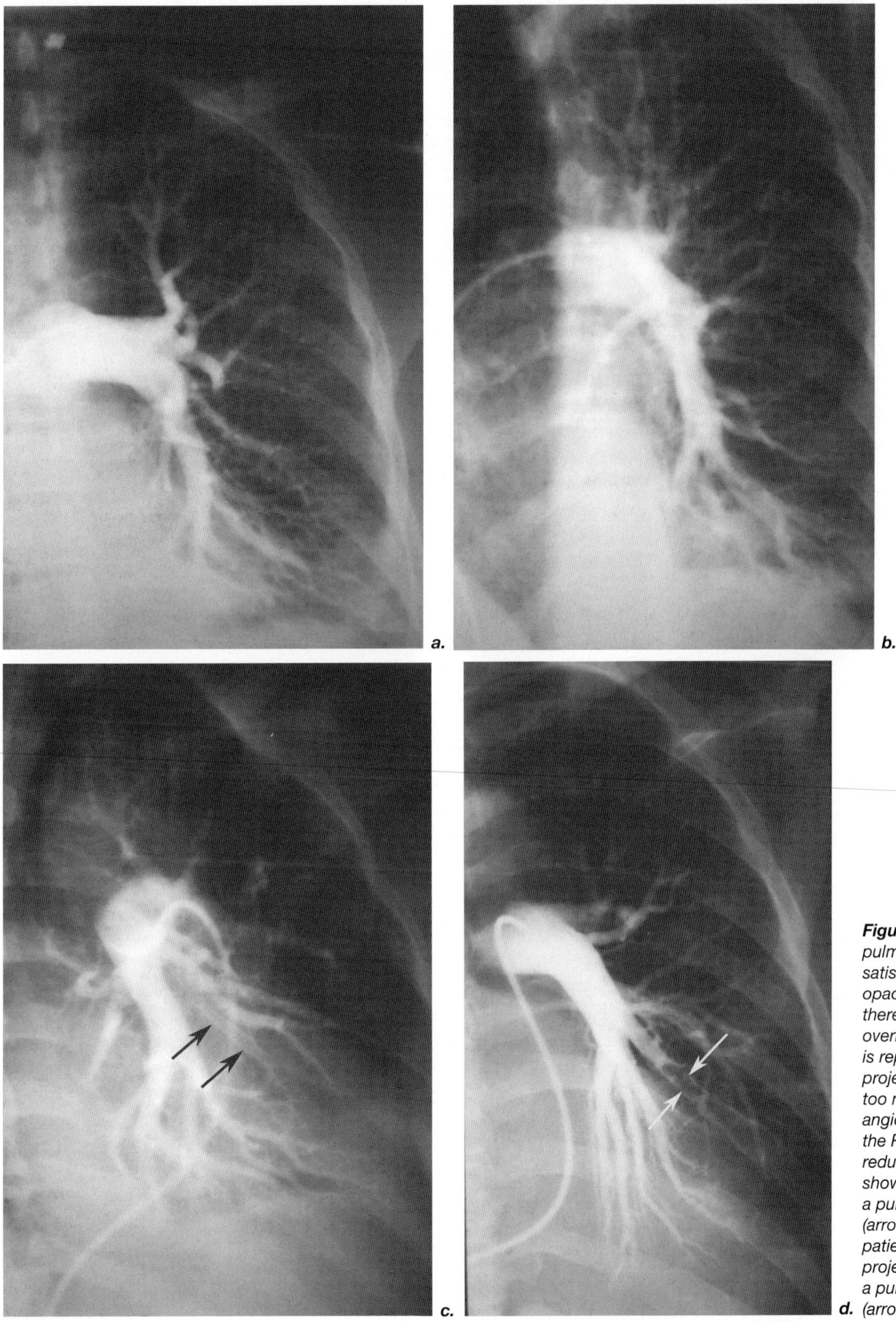

Figure 3. *a) Left frontal pulmonary angiogram with satisfactory vessel opacification; however there is significant vessel overlap. b) The angiogram is repeated in the LPO projection, but there is still too much overlap. c) The angiogram is repeated in the RPO projection, which reduces the overlap and shows the filling defect of a pulmonary embolus (arrows). d) A different patient with a similar projection clearly showing a pulmonary embolus (arrows).*

PULMONARY ARTERIAL VENOUS MALFORMATIONS

Pulmonary AVM's are an important indication for performing pulmonary arteriography. Pulmonary arterial venous malformations can occur as an isolated abnormality or as part of the hereditary hemorrhagic telangiectasia syndrome (Osler-Weber-Rendu syndrome). In Osler-Weber-Rendu, the pulmonary AVM's are multiple, whereas those patients with isolated pulmonary AVM's usually do not have Osler-Weber-Rendu. Patients usually present with dyspnea, cyanosis, digital clubbing, and hemoptysis. Paradoxical PE may produce a CVA or brain abscess. A simple pulmonary AVM will be characterized by a direct artery-to-vein communication with a dilatation of the venous limb giving rise to a venous aneurysm. Complex pulmonary AVM's (20%) will have multiple arterial feeders and multiple draining veins. In the simple form, embolization with detachable balloons or coils is a very efficient way of occluding the pulmonary AV malformation and eliminating most of the clinical symptoms associated with pulmonary AVM.

SYSTEMIC VENOGRAPHY

- Lower extremity
- Upper extremity
- Vena cava

SYSTEMIC VENOGRAPHY

The widespread use of sonography, CT and MR has resulted in an ever decreasing need to perform contrast venography. However, there are areas as outlined above where venography may be helpful. Leg venography, once the mainstay for diagnosis of deep venous thrombosis (DVT), is now used only in rare cases of equivocal sonography. It is, however, important in mapping uncommon venous anomalies of the leg, particularly in children prior to surgery **(Fig. 4)**. Surgery on the superficial veins in the absence of normal deep veins should be avoided. Such conditions have been described by Klippel and Trenaunay and Parkes Weber. Venography is of increasing use as part of interventional procedures to treat venous thrombosis in a variety of venous beds, for foreign body retrieval and for vascular device placement. The rapid progress in CT and MR with 3-dimensional reformatting may reduce the use of venography even for those applications where it is today the method of choice. In our institution, we currently perform venography in regions listed above, either directly or in the venous phase following arterial injection. Our most frequent use of venography is associated with dialysis access fistulas in complex central venous access patients, and in patients with thoracic outlet or SVC occlusions.

SONOGRAPHIC EVALUATION OF DVT

With the introduction of duplex sonography with color Doppler imaging, there are very few indications for lower extremity venography. The principles of sonographic evaluation of the lower extremities have been well described. Real-time high-resolution sonographic equipment is utilized to perform compression ultrasound of the deep venous system. The normal vein shows complete coaptation of the walls with application of compression. Patients with acute deep venous thrombosis will show expansion of the vein with noncompressibility of the vein when pressure is applied. Chronic DVT may show thickened walls, noncompressibility and small channels within the thrombus. Acute thrombus is usually echogenic whereas chronic thrombus may be hypoechoic. In addition to compression, the normal vein will increase in diameter with the Valsalva maneuver and this may be helpful in equivocal cases. The venous flow is always evaluated for phasicity (alteration of flow with respiration), as well as augmentation (increase in flow velocity with compression of the distal leg). No intraluminal material should be identified within the normal vein.

These same principles can be applied to veins of the upper extremity as well as the subclavian vein. In the subclavian vein and brachiocephalic vein, there should be transmitted pulsations from the heart. Dampened pulsations may suggest a more central occlusion. In addition, with the sniff test, there should be a decrease in diameter of the vein by at least 50% in a normally patent subclavian vein.

Color Doppler imaging may show partial nonocclusive thrombus along the wall. These same principles can be applied to veins of the upper extremity as well as the subclavian vein. With the increased utilization of central venous catheterization, there is a corresponding increased incidence of thrombosis of upper extremity deep veins and of all central venous pathways. These can be reliably evaluated with ultrasound. MR venography is being investigated and is shown to be as sensitive as sonography and contrast venography. Although it is very sensitive and highly specific, the expense as well as cumbersome nature of the procedure may limit its clinical utility.

VENOGRAPHY OF THE LOWER LIMB

Anatomy

The veins of the leg are divided into the superficial and the deep systems. The superficial veins are located in the subcutaneous tissues superficial to the fascia that envelopes the muscles of the leg, while the deep veins are all located deep to the fascia. Short perforating veins cross the fascia and link the superficial to the deep system. Valves in both systems direct flow from the superficial to the deep veins. The superficial system includes the lesser and the greater saphenous veins of the calf. The greater saphenous vein continues along the medial aspect of the thigh and drains into the common iliac vein. The deep system is more complex than the superficial system. Three paired veins drain the calf: the anterior tibial, the peroneal, and the posterior tibial veins. These veins run parallel to the tibial arteries of the same name. The tibial veins drain into the popliteal vein at knee level. The popliteal forms the superficial femoral vein at the adductor canal. The deep femoral vein drains the thigh muscles and joins the superficial femoral vein to form the common femoral vein.

Contrast venography is indicated for diagnostic purposes in patients with suspected venous abnormalities and in cases of suspected DVT with equivocal findings on sonography. A certain percentage of sonographic studies will be technically suboptimal due to a patient's large body habitus, edema of the subcutaneous tissues, or inability to evaluate the adductor canal. Although calf vein thrombosis is not of great clinical significance, some patients with symptoms will need evaluation of their calf veins.

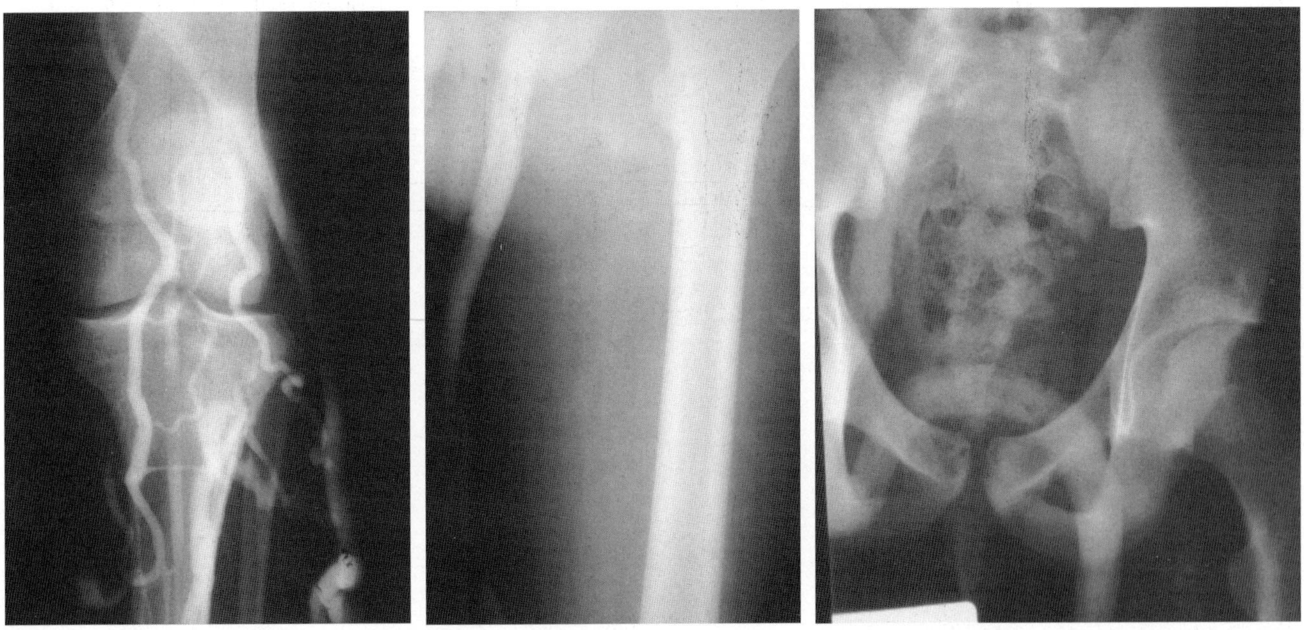

a. **b.** **c.**

Figure 4. *(a, b, c) Films from a venogram in a child with congenital aplasia of the deep veins. All drainage from the left leg goes through the large saphenous vein which crosses over to the right side at the femoral level.*

Although this can be done with Doppler sonography, sonographic evaluation of the calf veins is time consuming and technically challenging. Because of the multitude of veins (two veins for each artery), it may be simpler to perform a contrast venogram. There are various methods for performing contrast venography of the lower extremity **(Fig. 5)**.

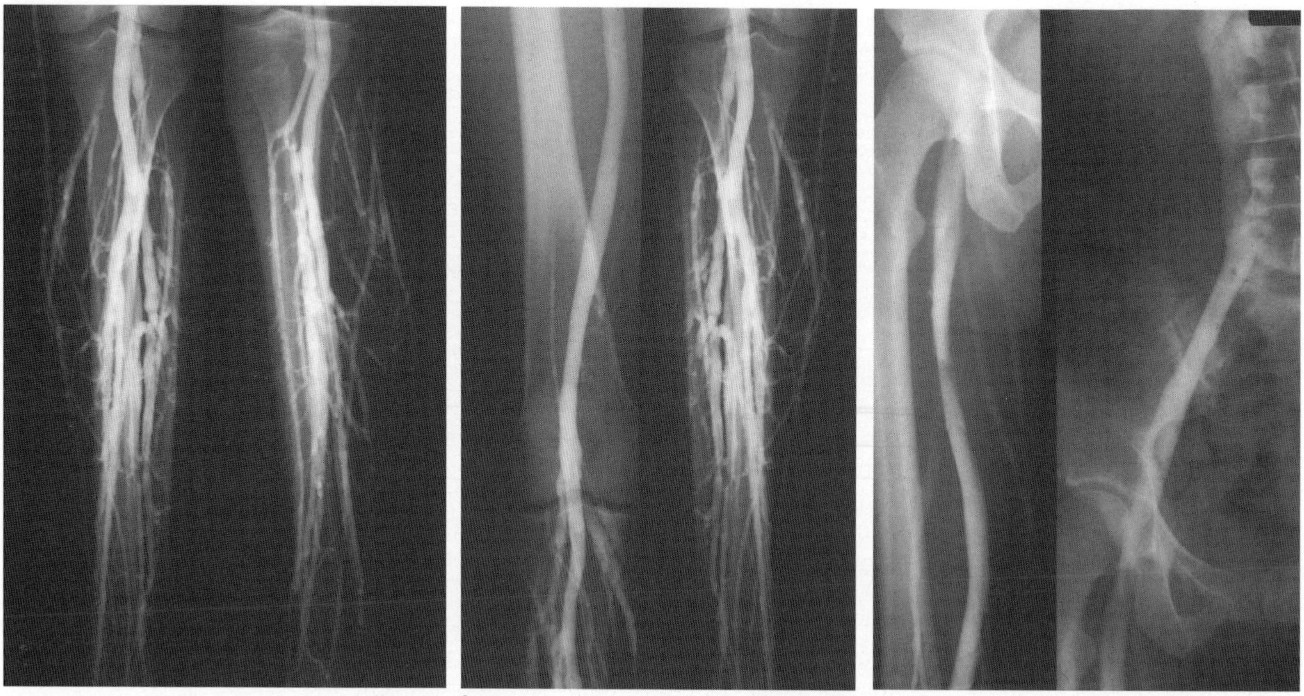

a. b. c.

Figure 5. *(a, b, c) A normal lower limb contrast venogram filmed on large conventional x-ray films.*

With the demise of cut-film capability at many institutions, older proven methods based on full-sized films have had to be adapted to the capabilities of the new digital suites. In the profoundly swollen foot, there are helpful methods to unmask a suitable foot vein such as: prolonged elevation of the leg; wrapping the foot with Ace bandages or a blood pressure cuff to displace the edema; and wrapping the foot in a warm towel which increases blood flow to the foot, makes the veins more prominent and helps to clean the foot. If all else fails, a surgical cutdown may be needed. At our institution, we use a semi-upright non-weight bearing venographic technique. Half-strength contrast medium is injected through a butterfly needle or other I.V. access route on the dorsum of the foot. Contrast is injected by hand at a brisk rate, while carefully observing the injection site to check that there is no extravasation of contrast agent into the soft tissues. The opacification of the veins is monitored fluoroscopically and spot films taken to provide complete coverage of all the veins.

An ankle tourniquet is only used when contrast flows predominantly via the superficial veins and there is no opacification of the deep system. By blocking the superficial flow, contrast is forced to the deep system and can demonstrate filling defects in the deep veins or verify their complete obliteration. For demonstration of incompetence of the perforating veins linking deep and superficial venous networks, tourniquets are required at both the ankle and knee levels. Once the contrast column reaches the area of the groin, the fluoroscopy table is returned to the horizontal position while continuing the injection of contrast. This provides better filling of the large capacity pelvic veins. This filling can be further enhanced by having the patient perform the Valsalva maneuver or by temporary compression of the femoral vein, which is only released when the patient is horizontal. In centers with specific interests in chronic venous insufficiency of the lower limb, techniques such as descending venography are used.

Findings of acute thrombosis include filling defect with smooth surface **(Fig. 6)**. As the clot becomes older, there is retraction of the clot away from the wall and slight irregularity of the surface.

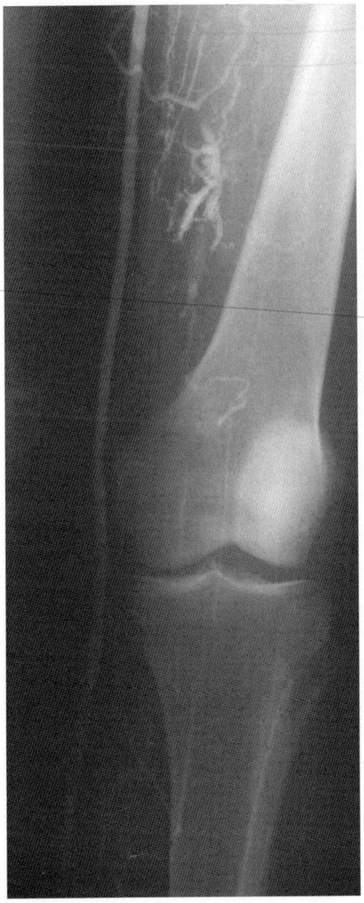

a.

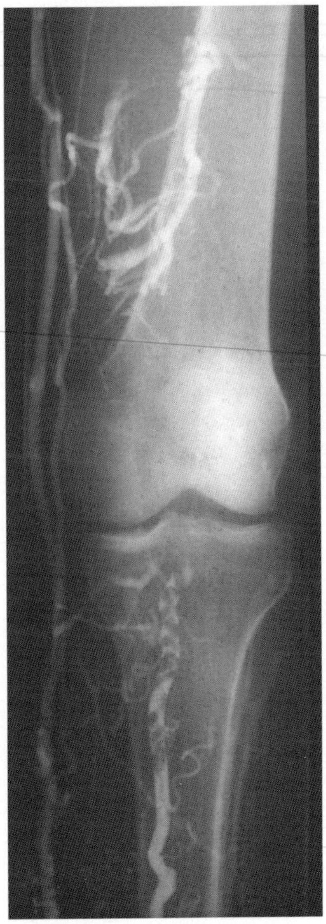

b.

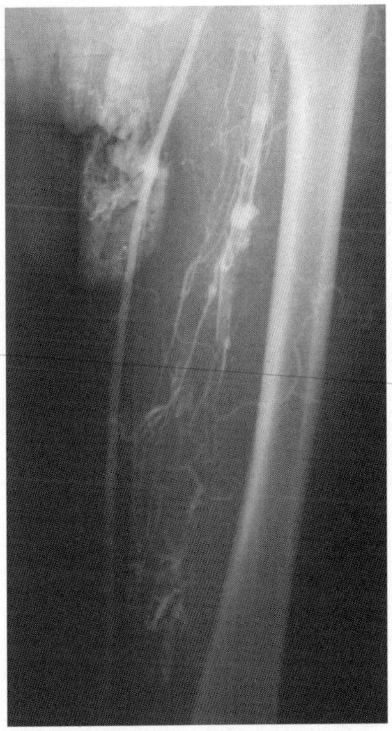

c.

Figure 6. *a) The venogram shows filling of the saphenous vein, but little filling of the deep venous system. b) A later film in the course of the same venogram demonstrates serpentine-like filling defects in the deep vein below the knee, typical for DVT. c) Filling defects are now seen in the partially filled superficial femoral vein, which is the main deep venous channel.*

Over time, a chronic thrombus may show focal areas of recanalization with small irregular channels and no visualization of adjacent valves **(Fig. 7)**. Enlarged collateral veins also may be identified bypassing the occlusion. Pitfalls that should be watched for include inflow of unopacified blood and streaming artifacts. Incomplete filling may also produce a pseudothrombosis due to persistence of unopacified blood, especially behind valve leaflets.

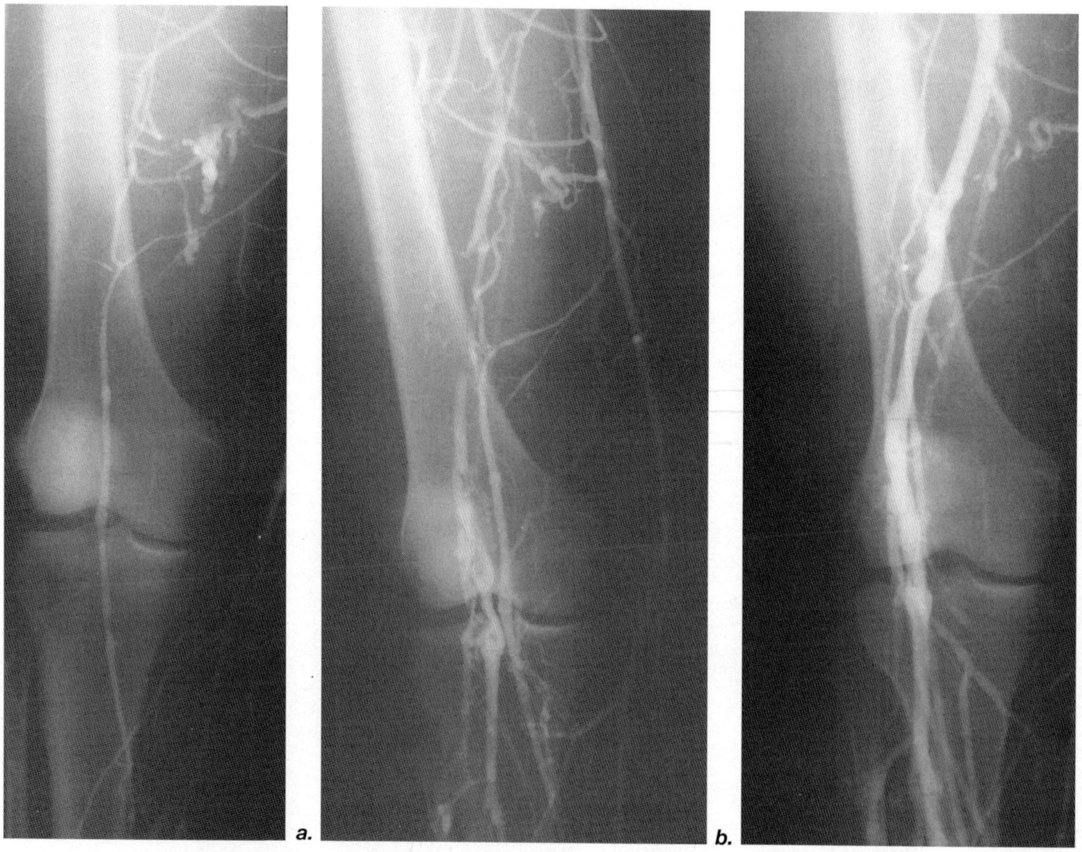

a. b. c.

Figure 7. *a) There is no filling of the deep veins in this patient with an acute leg DVT. b) A venogram 3 weeks later shows some venous filling as partial recanalization is taking place. c) After waiting a further 5 weeks, the new venogram shows almost complete recanalization of the deep system, with just some hint of a shaggy or stringy appearance typical of a recanalized vein.*

May-Thurner Syndrome

In this syndrome, obstruction of the left common iliac vein causes reduction of blood flow and frequently consequent deep vein thrombosis of the left iliofemoral system with swelling, edema and pain. It is an unusual condition, occurring typically in female patients of about 40 years of age. The obstruction occurs at the level at which the left common iliac vein passes in front of the spine and behind the right common iliac artery. There appears to be compression of the vein with impedance to blood flow. In addition, there is intraluminal web or spur formation within the iliac vein. These spurs have been described as lateral spurs, central spurs or membranous-type spurs. Successful interventional treatment by thrombolysis, followed by dilatation and stent placement at the stenosed site, has been described. Contrast venography is an important element in the management of these patients.

MAY-THURNER SYNDROME

- Also known as "iliac compression syndrome" or "pelvic venous spur"
- Left common iliac vein obstruction
- Intraluminal spurs
- 70% to 87% female
- Approximately 40 years of age

VENOGRAPHY OF THE UPPER LIMB

Indications

Venography is indicated to evaluate the cause for swelling of the arm and to map the venous anatomy prior to vascular procedures such as fistula placement or central venous access. A common site for obstruction is the axillosubclavian segment secondary to indwelling central venous lines or as a result of thoracic outlet compression **(Fig. 8)**.

Technique

Upper extremity venography is similar to lower extremity venography with venous access being established in the hand or medial antecubital fossa. If the subclavian and axillary veins are to be evaluated, one should try to start an I.V. in the medial antecubital vein. As in the lower limb, great care should be exercised to avoid contrast extravasation into the soft tissues at the venipuncture site. It may be necessary to insert a catheter up the brachial vein and perform the contrast injection at the level of the axillary vein or higher to obtain optimal opacification of any pathology in the subclavian vein **(Fig. 9)**.

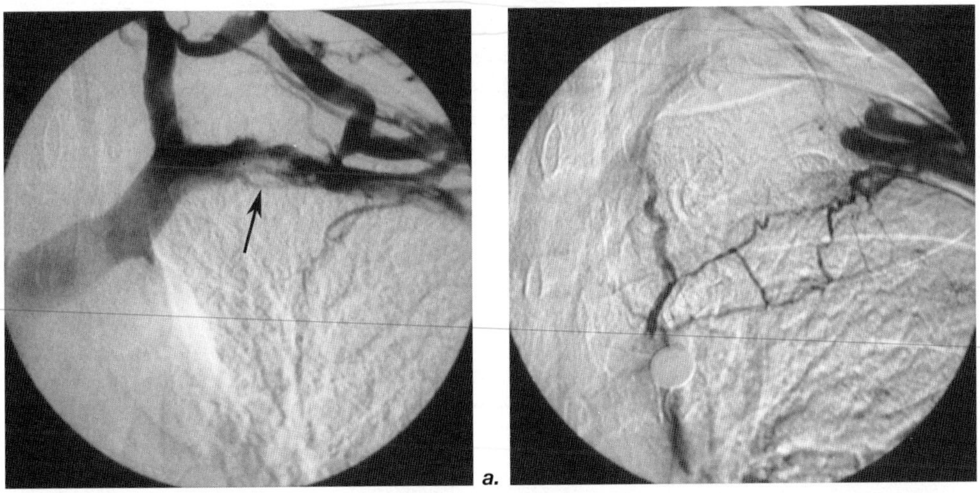

a. b.

Figure 8. *a) Venogram of left upper extremity showing some narrowing (arrow) of the subclavian vein at the thoracic outlet. There is also filling of collateral veins. The arm is in the neutral adducted position. b) With the arm in the hyperabducted position with the patient's hand behind her head, the venogram shows complete occlusion of the subclavian vein and all collateral veins.*

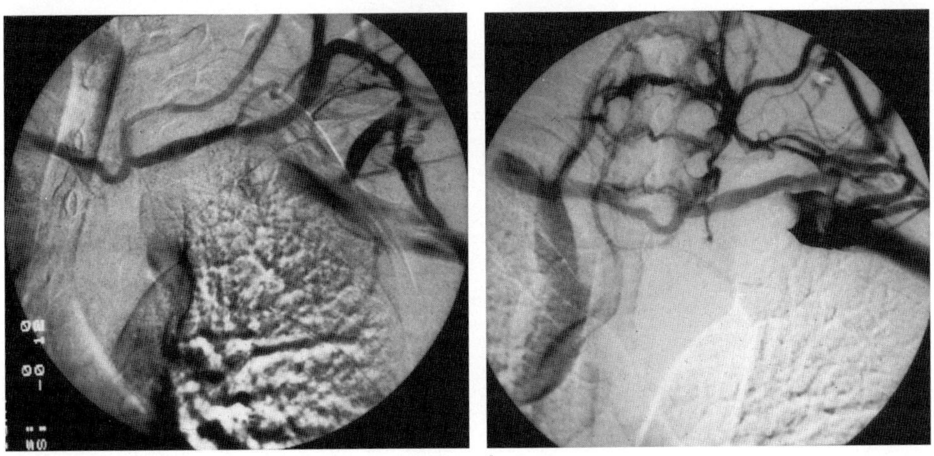

a. b.

Figure 9. *a) A patient with a swollen left arm had this venogram performed by injection of contrast material into an antecubital vein. It appears that the left subclavian vein does not fill and that there are collateral veins filling. b) A long catheter was threaded up the vein and the venogram was repeated with the tip of the catheter tip in the axiliary vein. A much better venogram is obtained, which demonstrates the complete occlusion of the subclavian vein with a convex filling defect and multiple collateral veins.*

Veins are prone to spasm, so that a vascular sheath should be used if it is anticipated that there will be a number of catheter exchanges needed. Acute thrombus will show as a filling defect within the vein. Totally occluded veins will not be opacified and so will be absent from their expected location. Thoracic outlet syndrome will be demonstrated by extrinsic narrowing and/or occlusion which is exaggerated with the arm in a hyperabducted position. When evaluating the thoracic outlet, the venogram should be performed in the neutral position with the arm adducted to the patient's side, and also in the extreme hyperabducted position with the hand behind the patient's neck. Venograms in the intermediate positions may sometimes be helpful. The filling of collateral veins should be carefully observed and documented in the various positions.

PAGET-SCHROETTER SYNDROME

- Also known as "effort thrombosis"
- Accounts for 1%–9% arm thrombosis cases
- Young, M>F
- Following exercise
- Right>left
- Axillosubclavian compression
- First rib, clavicle, scalenus anticus, subclavius and pectoralis minor muscles

Paget-Schroetter Syndrome

Also known as idiopathic, traumatic, spontaneous or effort thrombosis, Paget-Schroetter syndrome is an uncommon condition in which there is occlusion and thrombosis of the axillosubclavian vein as it passes over the first rib and under the clavicle. It accounts for only some 1% to 9% of all cases of arm extremity thrombosis. It occurs most often in young males after vigorous exercise or activity involving abduction and/or hyperabduction of the arm, more commonly on the right than the left arm. Weight lifting, kayaking and similar activities are typical activities associated with the syndrome, which is manifest by the fairly acute onset of discomfort, a feeling of heaviness, swelling and occasionally cyanosis of the involved extremity. Venography as described above may be the first diagnostic step as well as the start of thrombolytic therapy. The management of this condition usually includes initial thrombolysis with subsequent surgical relief of the thoracic outlet narrowing and may also utilize balloon dilatation and rarely stenting.

COMPLICATIONS OF VENOGRAPHY

- Contrast allergy
- Contrast extravasation
- Thrombophlebitis
- Pulmonary embolus

COMPLICATIONS OF VENOGRAPHY

Complications following contrast venography are uncommon provided that after all venographic procedures heparinized saline is infused through the I.V. to flush out all contrast from the venous system under study. This will limit the possibility of aggravating or causing venous thrombosis. Patients sometimes have pain which is self-limited once the contrast agent has been washed out. During the period of contrast injection, great care should be taken to avoid extravasation of contrast agent from the access vein into the soft tissues. Contrast extavasation can cause severe pain and may lead to tissue damage. Careful technique is key to avoiding this important complication. Rare complications from contrast allergy or nephrotoxicity in patients with impaired renal function are similar in venography to other procedures involving the use of intravascular contrast agents. A feared but very rare complication is pulmonary embolism as a result of squeezing thrombus-containing veins during venography.

SUPERIOR VENA CAVA SYNDROME

The superior vena cava (SVC) can be stenosed or occluded, resulting in severe venous congestion of the head, neck and upper extremities. There is typically edema, swelling and cyanosis of the face, neck and arms, often with symptoms of headache, blurred vision and dyspnea. In some cases, there may be marked swelling of the female breast. The SVC obstruction may involve some or all of the following elements: direct tumor ingrowth, external tumor encasement, external pressure from tumors, metastatic deposits, or enlarged lymph nodes, radiation fibrosis and venous thrombosis. Neoplastic conditions account for over 90% of cases of SVC syndrome with thrombosis secondary to central venous catheters accounting for the remainder of cases. Benign causes include tuberculosis and fibrosing mediastinitis. Sonography or other noninvasive imaging can easily confirm the diagnosis. Stenting of the SVC with or without preliminary thrombolysis often provides good palliative relief. Venography is the first step in any therapeutic approach and will usually require bilateral upper extremity venograms with simultaneous venous injection of both arms.

INFERIOR VENA CAVA (IVC)

Studies of the IVC using catheter venography are infrequently used in a diagnostic role today, because of the widespread use of noninvasive techniques such as US, CT and MR. The main use of IVC venography is as an adjunct to a therapeutic intervention such as IVC filter placement or thrombolysis. Less common applications are in patients with venous shunts between the IVC and portal system, in patients after liver transplantation and in patients after TIPS. Direct venous catheter access has the advantage of allowing the measurement of intravenous pressure across venographically demonstrated stenoses.

REFERENCES

Pulmonary Angiography

Bergin CJ, Sirlin CB, Hauschildt JP, et al. Chronic thromboembolism: diagnosis with helical CT and MR imaging with angiographic and surgical correlation [see comments]. Radiology 1997; 204:695-702.

Bookstein JJ, Silver TM. The angiographic differential diagnosis of acute pulmonary embolism. Radiology 1974; 110:25-33.

Diffin DC, Leyendecker JR, Johnson SP, Zucker RJ, Grebe PJ. Effect of anatomic distribution of pulmonary emboli on interobserver agreement in the interpretation of pulmonary angiography. AJR 1998; 171:1085-1089.

Drucker EA, Rivitz SM, Shepard JA, et al. Acute pulmonary embolism: assessment of helical CT for diagnosis [see comments]. Radiology 1998; 209:235-241.

Fava M, Loyola S, Flores P, Huete I. Mechanical fragmentation and pharmacologic thrombolysis in massive pulmonary embolism. JVIR 1997; 8:261-266.

Ferris EJ. Deep venous thrombosis and pulmonary embolism: correlative evaluation and therapeutic implications. AJR 1992; 159:1149-1155.

Garg K, Sieler H, Welsh CH, Johnston RJ, Russ PD. Clinical validity of helical CT being interpreted as negative for pulmonary embolism: implications for patient treatment [see comments]. AJR 1999; 172:1627-1631.

Garg K, Welsh CH, Feyerabend AJ, et al. Pulmonary embolism: diagnosis with spiral CT and ventilation-perfusion scanning—correlation with pulmonary angiographic results or clinical outcome [see comments]. Radiology 1998; 208:201-208.

Goldhaber SZ. Contemporary pulmonary embolism thrombolysis. Chest 1995; 107:45s-51s.

Goodman LR, Curtin JJ, Mewissen MW, et al. Detection of pulmonary embolism in patients with unresolved clinical and scintigraphic diagnosis: helical CT versus angiography. AJR 1995; 164:1369-1374.

Hagspiel KD, Polak JF, Grassi CJ, Faitelson BB, Kandarpa K, Meyerovitz MF. Pulmonary embolism: comparison of cut-film and digital pulmonary angiography. Radiology 1998; 207(1):139-145.

Harrison KA, Haire WD, Pappas AA, et al. Plasma D-dimer: a useful tool for evaluating suspected PE. J Nuclear Medicine 1993; 34:896-898.

Hudson ER, Smith TP, McDermott VG, et al. Pulmonary angiography performed with Iopamidol: complications in 1,434 patients. Radiology 1996; 198:61-65.

Hull RD, Raskob GE, Ginsberg JS, et al. A noninvasive strategy for the treatment of patients with suspected pulmonary embolism. Arch Intern Med 1994; 154:289-297.

Johnson MS, Stine SB, Shah H, Harris VJ, Ambrosius WT, Trerotola SO. Possible pulmonary embolus: evaluation with digital subtraction versus cut-film angiography—prospective study in 80 patients. Radiology 1998; 207:131-138.

Killewich LA, Nunnelee JD, Auer AI. Value of lower extremity venous duplex examination in the diagnosis of pulmonary embolism. J Vasc Surg 1993; 17:934-939.

Kim KI, Muller NL, Mayo JR. Clinically suspected pulmonary embolism: utility of spiral CT. Radiology 1999; 210:693-697.

King MA, Ysrael M, Bergin CJ. Chronic thromboembolic pulmonary hypertension: CT findings. AJR 1998; 170:955-960.

Lang EV, Barnhart WH, Walton DL, Raab SS. Percutaneous pulmonary thrombectomy. JVIR 1997;8:427-432.

Loud PA, Grossman ZD, Klippenstein DL, Ray CE. Combined CT venography and pulmonary angiography: a new diagnostic technique for suspected thromboembolic disease. AJR 1998; 170:951-954.

Malloy PC. Tutorial 22: Diagnosis and treatment of pulmonary embolism. In SCVIR Syllabus on Venous Interventions. 1995; 252-275.

Mayo JR, Remy-Jardin M, Muller NL, et al. Pulmonary embolism: prospective comparison of spiral CT with ventilation-perfusion scintigraphy. Radiology 1997; 205:447-452.

Oser RF, Zuckerman DA, Gutierrez FR, Brink JA. Anatomic distribution of pulmonary emboli at pulmonary angiography: implications for cross-sectional imaging [see comments]. Radiology 1996; 199:31-35.

Perrier A, Bounameaux H, Morabia A, et al. Contribution of D-dimer plasma measurement and lower-limb venous ultrasound to the diagnosis of pulmonary embolism: a decision analysis model. Am Heart J 1994; 126:624-635.

Remy-Jardin M, Remy J, Cauvain O, Petyt L, Wannebroucq J, Beregi JP. Diagnosis of central pulmonary embolism with helical CT: role of two-dimensional multiplanar reformations. AJR 1995; 165:1131-1138.

Remy-Jardin M, Remy J, Deschildre F, et al. Diagnosis of pulmonary embolism with spiral CT: comparison with pulmonary angiography and scintigraphy. Radiology 1996; 200:699-706.

Rozenshtein A, Saffran L, Snider A, Ost D. Spiral computed tomography has a high negative predictive value for pulmonary embolism. American Roentgen Ray Society Annual Meeting 1999; 172(3): 17.

Schlueter FJ, Zuckerman DA, Horesh L, Gutierrez FR, Hicks ME, Brink JA. Digital subtraction versus film-screen angiography for detecting acute pulmonary emboli: evaluation in a porcine model. JVIR 1997; 8(6):1015-1024.

Schmitz-Rode T, Gunther RW, Pfeffer JG, et al. Acute massive pulmonary embolism: use of a rotatable pigtail catheter for diagnosis and fragmentation therapy. Radiology 1995; 197:157-162.

Smith LL, Iber C, Sirr S. Pulmonary embolism: confirmation with venous duplex US as adjunct to lung scanning. Radiology 1994; 191:143-147.

Smith TP, Lee VS, Hudson ER, et al. Prospective evaluation of pulmonary artery pressures during pulmonary angiography performed with low-osmolar nonionic contrast media. JVIR 1996; 7:207-212.

Stein PD, Henry JW. Prevalence of acute pulmonary embolism in central and subsegmental pulmonary arteries and relation to probability interpretation of ventilation/perfusion lung scans [see comments]. Chest 1997; 111:1246-1248.

The PIOPED investigators: value of the ventilation/perfusion scan in acute pulmonary embolism. JAMA 1990; 263:2753.

Uflacker R, Strange C, Vujic I. Massive pulmonary embolism: Preliminary results of treatment with the Amplatz thrombectomy device. JVIR 1996; 7:519-528.

Urokinase pulmonary embolism trial: a national cooperative study. Circulation 1973; 47:1-108.

van Rossum AB, Pattynama PM, Ton ER, et al. Pulmonary embolism: validation of spiral CT angiography in 149 patients. Radiology 1996; 201:467-470.

van Rossum AB, Treurniet FE, Kieft GJ, Smith SJ, Schepers-Bok R. Role of spiral volumetric computed tomographic scanning in the assessment of patients with clinical suspicion of pulmonary embolism and an abnormal ventilation/perfusion lung scan [see comments]. Thorax 1996; 51:23-28.

Zuckerman DA, Sterling KM, Oser RF. Safety of pulmonary angiography in the 1990s. JVIR 1996; 7:199-205.

Systemic Venography

Binkert CA, Schoch E, Stuckmann G, et al. Treatment of pelvic venous spur (May-Thurner Syndrome) with self-expanding metallic endoprostheses. Cardiovasc Intervent Radiol 1998; 21:22-26.

Cronan JJ. Ultrasound evaluation of deep venous thrombosis. Seminars in Roentgenology 1992; 27:39-52.

Gorenstein A, Gordon RL, Shiffrin E, et al. Phlebography of the deep venous system of the leg in children. Radiology 1982; 144:519-524.

LaBerge JM, Callen PW. Diagnosis of deep venous thrombosis. In: Trerotola SO, Savader SJ, Durham JD, eds. SCVIR Syllabus: Venous Interventions. San Francisco: Society of Cardiovascular & Interventional Radiology, 1995; 190-201.

Longley DG, Finlay DE, Letourneau JG. Sonography of the upper extremity and jugular veins. AJR 1993; 160:957-962.

Malloy PC. Diagnosis and treatment of pulmonary embolism. In: Trerotola SO, Savader SJ, Durham JD, eds. SCVIR Syllabus: Venous Interventions. Eds. San Francisco: Society of Cardiovascular & Interventional Radiology, 1995; 252-275.

Raghavendra BN, Rosen RJ, Lan S, Riles T, Horii SC. Deep venous thrombosis: detection by high-resolution real-time ultrasonography. Radiology 1984; 152:789-793.

Rose SC, Zwiebel WJ, Nelson BD et al. Symptomatic lower extremity deep venous thrombosis: accuracy, limitation, and role of color duplex flow imaging in diagnosis. Radiology 1990; 175:639-644.

Sheeran SR, Hallisey MJ, Murphy TP, Faberman RS, Sherman S. Local thrombolytic therapy as part of a multidisciplinary approach to acute axillosubclavian vein thrombosis (Paget-Schroetter Syndrome). JVIR 1997; 8:242-260.

Vaccaro JP, Cronan JJ, Dorfman GS. Outcome analysis of patients with normal compression US examinations. Radiology 1990; 175:645-649.

Winick AB, Venbrux AC. Descending venography. In: Trerotola SO, Savader SJ, Durham JD, eds. SCVIR Syllabus: Venous Interventions. San Francisco: Society of Cardiovascular & Interventional Radiology, 1995; 239-251.

CHAPTER 5 PREVIEW ANSWERS

Case 1

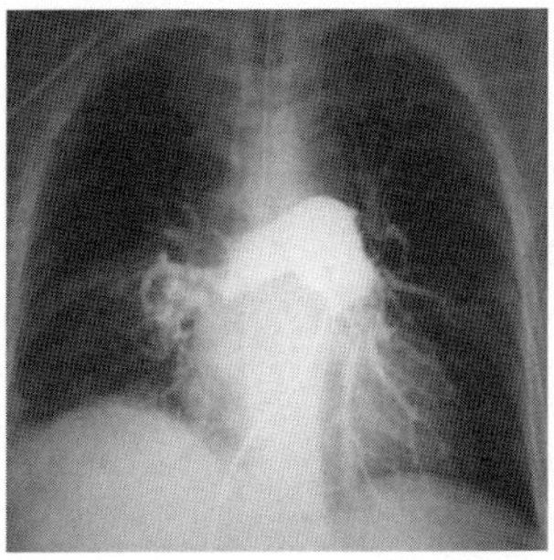

a.

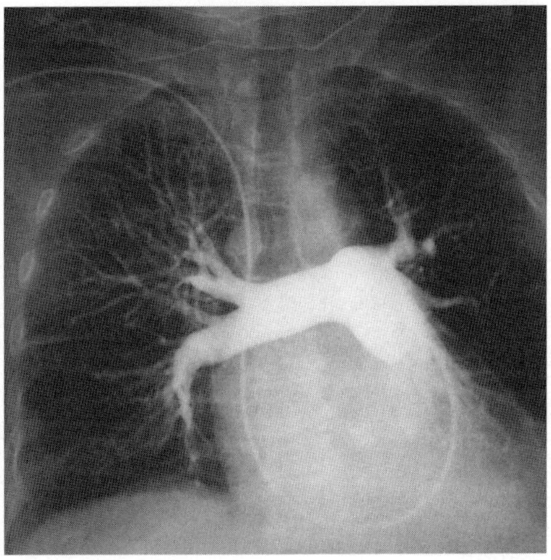

b.

Findings: a) There is a large filling defect in the main right pulmonary artery with very few peripheral vessels filling with contrast. This is a large pulmonary embolus. Compare this to the normal pulmonary angiogram b) which shows no filling defects. There is good filling of the peripheral vessels. Did you notice that the catheter was inserted via the right arm?

Case 2

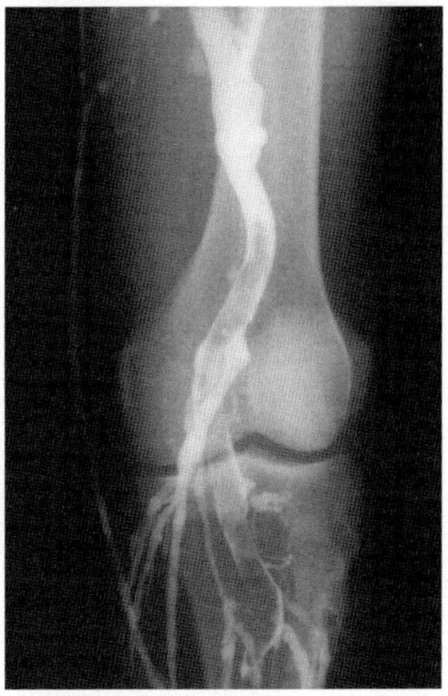

Findings: This is a spot film from a contrast venogram taken at the level of the knee. There is a filling defect in the popliteal vein typical of a deep vein thrombus.

Case 3

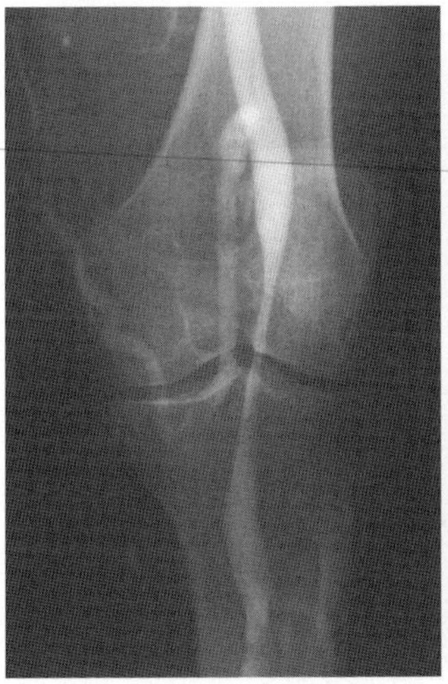

a.

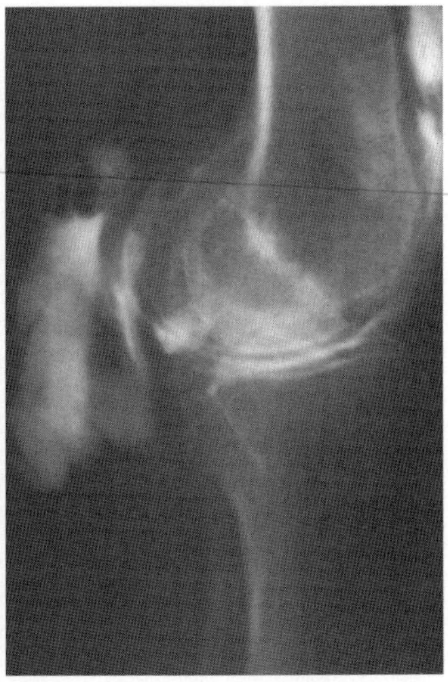

b.

Findings: a) This is a spot film from a contrast venogram at the level of the knee. The popliteal vein shows a smooth segment of narrowing. This may be caused by external pressure in this region from a popliteal artery aneurysm or a Bakers cyst from the knee joint. Sonographic examination should be helpful. b) This shows a contrast arthrogram with filling of a Bakers cyst.

Case 4

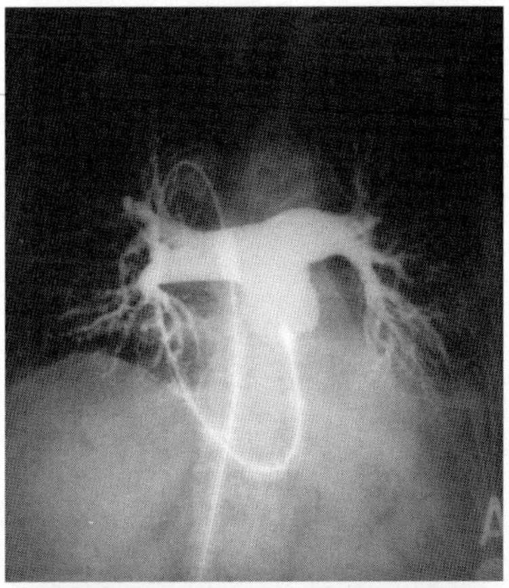

a.

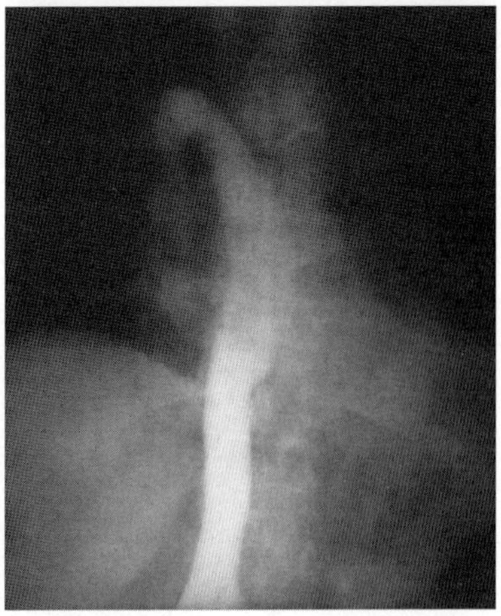

b.

Findings: a) The vascular catheter has a very abnormal course. Although it comes up from below in the inferior vena cava it does not go straight onto the right atrium but makes an extra loop! b) The venogram shows the anomaly of an azygous continuation of the IVC, which explains the odd course of the catheter. The pulmonary angiogram is otherwise normal.

CHAPTER 6 PREVIEW

Case 1

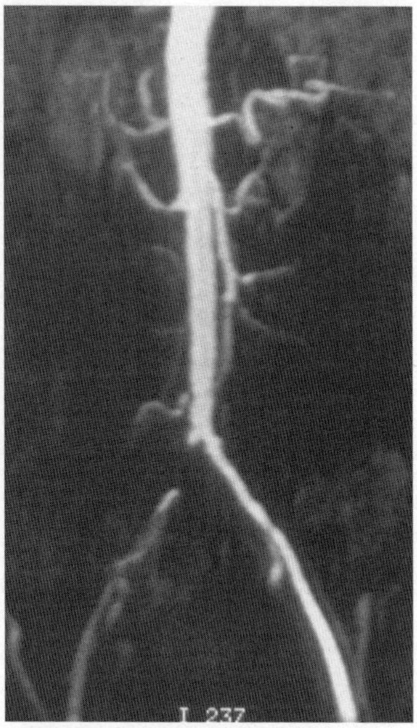

History: Gd-MRA scan in a patient with right thigh claudication.
Question: What is your diagnosis and suggested treatment?

Case 2

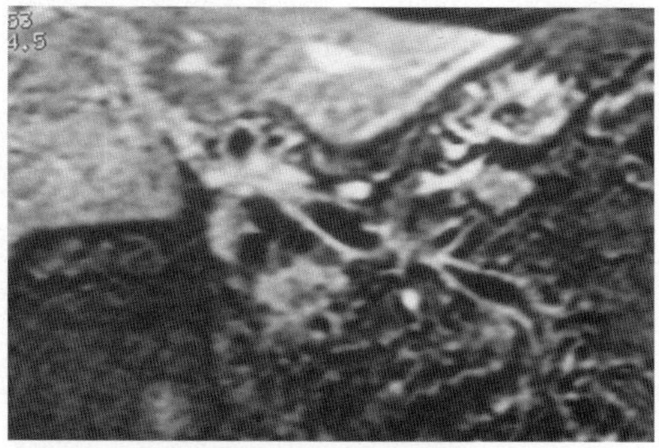

History: Gd-MRA scan in a patient with cirrhosis.
Question: What is your diagnosis and suggested treatment?

Case 3

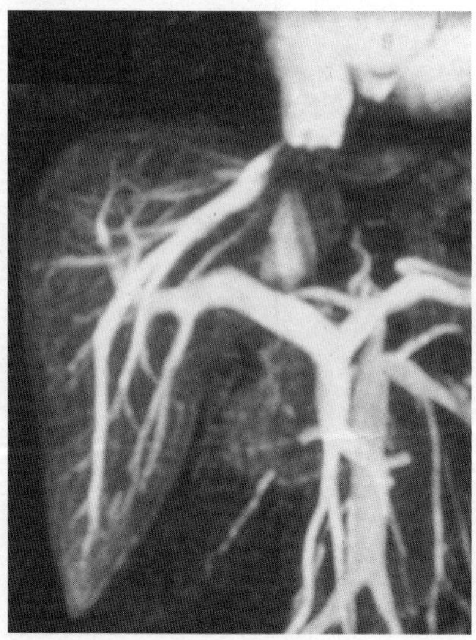

History: Gd-MRA scan in a patient with recurrent ascites.
Question: What is your diagnosis and suggested treatment?

Chapter 6
Magnetic Resonance Angiography

Mark W. Wilson, MD

CHAPTER OUTLINE

- Magnetic resonance angiography (MRA) techniques
  - Flow-void imaging
  - Time-of-flight MRA
  - Phase contrast MRA
  - Gadolinium-enhanced MRA
- Arterial applications of MRA
  - Aortic dissection
  - Abdominal aortic aneurysm
  - Aortic and peripheral vascular atherosclerotic disease
  - Renal artery stenosis
- Venous applications of MRA
  - Portal vein abnormalities
  - Budd-Chiari syndrome
  - Inferior vena cava abnormalities
- Miscellaneous applications of MRA
  - Pseudoaneurysms
  - Vascular malformations
  - Interventional MR imaging

INTRODUCTION

Over the past decade there has been a virtual explosion of magnetic resonance angiography (MRA) techniques and applications. Today MR is an integral part of the evaluation of several arterial and venous vascular disorders. This chapter will provide a basic overview of the commonly used MRA techniques and a review of its application in the diagnosis of specific vascular abnormalities. Special attention will be paid to the use of MRA in planning interventional therapies.

MAGNETIC RESONANCE ANGIOGRAPHY TECHNIQUES

To begin, it should be noted that MRA has several advantages and disadvantages. Among the advantages are the noninvasive nature of MRA, the elimination of iodinated contrast, the elimination of ionizing radiation, flow velocity measurement, and the ability to provide a full three-dimensional representation of the vasculature and surrounding structures. Among the limitations of MRA are the demonstration of very small vascular structures with adequate resolution, the need to overcome motion artifacts, and the need to perform the examination within a comfortable setting and time period for the patient. Several MRA techniques have been developed to exploit these advantages and to address these limitations.

ADVANTAGES OF MRA

- No iodinated contrast
- No radiation
- Flow velocity measurement
- 3D representations

LIMITATIONS OF MRA

- Poor spatial resolution
- Sensitive to motion artifacts
- Long study time

MRA TECHNIQUES

- Flow-void imaging
- Time-of-flight
- Phase contrast
- Gadolinium enhancement

FEATURES OF FLOW-VOID IMAGING

- Absence of MR signal in flowing blood is imaged
- Low signal-to-noise ratio (SNR)
- Suited for evaluating large vessels
- Simplest MRA technique

An in-depth discussion of MR physics and imaging protocols is beyond the scope of this review. The purpose of this section is to give a basic description of the commonly used MRA techniques and the information they provide, so that the reader will be able to identify the different methods and understand their appropriate applications.

Flow-Void Imaging

Flow-void imaging, also known as black-blood imaging, takes advantage of the fact that fluid in motion produces little or no signal on spin echo MR images. This results in a signal void in arteries and veins with flowing blood. Given that it is the absence of MR signal that is being imaged, this technique has a relatively low signal-to-noise ratio (SNR). For that reason, flow-void imaging is better suited for evaluating large vascular structures such as the aorta, main pulmonary arteries, and inferior vena cava **(Fig. 1)**. Flow-void imaging allows for good discrimination of the vessel wall and vessel diameter. Flow-void imaging is also the easiest and fastest method of MR vascular imaging.

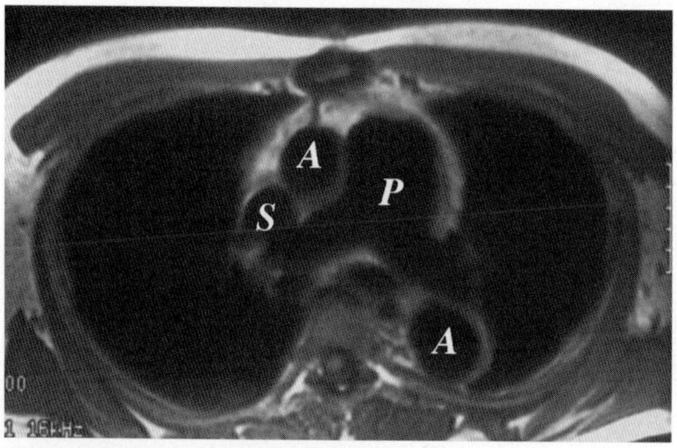

Figure 1. *Spin echo T1-weighted MR image through the thorax. Flow-voids enable identification of the aorta (A), the pulmonary arteries (P), and the superior vena cava (S).*

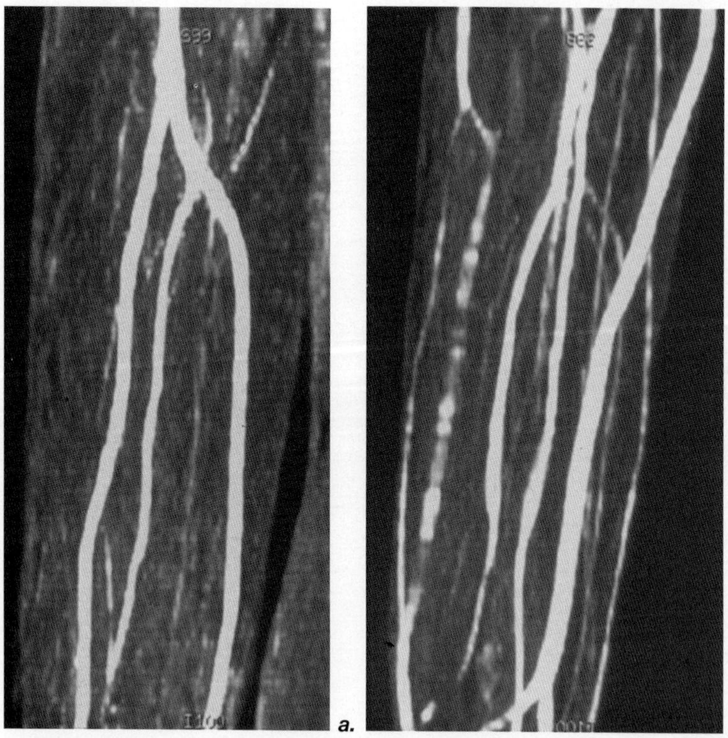

a. **b.** ***Figure 2.*** *Two-dimensional time-of-flight MRA of the right forearm. a) A caudally positioned saturation band eliminates the venous signal, resulting in visualization of arteries only. b) Conversely, a cranially positioned saturation band results in visualization of veins only.*

Time-of-Flight MRA

Contrary to flow-void imaging, time-of-flight MRA (TOF-MRA) imparts signal to flowing blood. More precisely, TOF-MRA utilizes a special gradient echo imaging sequence which acquires high signal from the moving protons in flowing blood while simultaneously suppressing signal from stationary protons in surrounding tissues. This results in a stronger SNR than flow-void imaging. TOF-MRA also allows for the placement of a saturation band above or below a vascular region of interest which removes signal from blood flowing into that region. For example, if a saturation band is placed caudal to a region of interest in the thigh, then femoral arterial flow will have signal, while femoral venous flow will not. This feature is quite useful for separating out arteries from veins in a given region **(Fig. 2)** and for discerning the arterial versus venous contribution to a vascular malformation.

FEATURES OF TIME-OF-FLIGHT MRA

- Direction of flow can be discriminated
- May be applied in 2D or 3D
- Improved SNR
- Better suited for imaging smaller vessels

TOF-MRA can be applied to a single axial image slice, known as two-dimensional (2D) TOF, or to an entire imaging volume, known as three-dimensional (3D) TOF. While 3D TOF inherently produces a 3D representation of a vessel within a volume, several 2D images can be stacked together in postprocessing to effectively produce a 3D image. However, as 3D TOF may suffer from loss of signal as the protons move through larger imaging volumes, 2D TOF may suffer from misregistration between the stacked images. 2D TOF has a shorter imaging time, while 3D TOF has improved resolution for smaller structures.

Phase Contrast MRA

Phase contrast MRA (PC-MRA) takes advantage of the phase shift produced by moving protons in flowing fluid with respect to stationary protons. The magnitude of the phase shift is proportional to the velocity of flow. The data obtained can be displayed as an angiogram and may be used for velocity determination **(Fig. 3)**. Another feature of PC-MRA is the loss of signal that occurs with turbulent flow, as with flow across a stenosis. This is useful in assessing the hemodynamic significance of a stenosis. Cine MRA is also possible with phase contrast.

FEATURES OF PHASE CONTRAST MRA

- Turbulence causes loss of signal
- May be applied in 2D or 3D
- Velocity determination is possible
- Cine MRA is possible

As with TOF-MRA, phase contrast MRA can be performed in two or three dimensions and direction of flow can also be determined. Contrary to TOF-MRA, phase contrast is less susceptible to saturation effects. PC-MRA also provides improved vessel-to-background contrast. Cine MRA is also possible with phase contrast. However, PC-MRA is more time consuming than flow-void imaging and TOF-MRA.

Gadolinium-Enhanced MRA

Gadolinium is the most commonly used MR contrast agent. As with iodinated contrast media, gadolinium is administered as a chelate with other molecules, such as dimeglumine (Gd DTPA). In contrast to iodinated agents, gadolinium is not nephrotoxic in recommended doses in patients with impaired renal function. There is also a very low incidence of idiosyncratic reactions to gadolinium. Gadolinuium-enhanced MRA (Gd-MRA) is performed by administering an intravenous bolus injection of gadolinium and then acquiring the images during the peak of contrast concentration in the phase and region of interest. An initial low-dose test bolus is administered to determine the precise time to begin imaging after the injection. Recent software modifications now allow this timing run to be automated. Most studies are performed using a gadolinium dose of 0.2 to 0.4 mmol/kg, which is between 20 ml to 40 ml in the standard formulations.

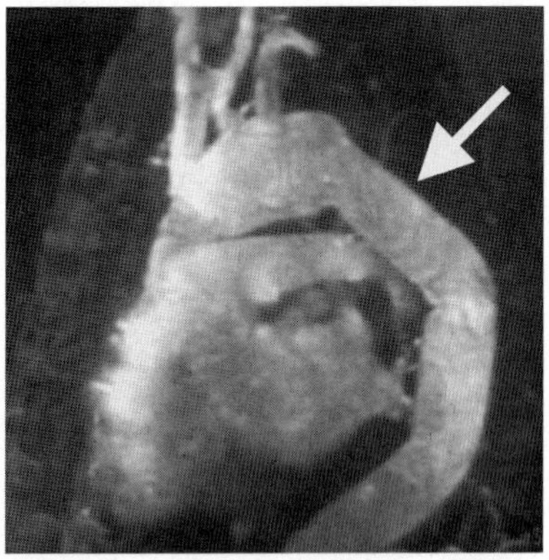

Figure 3. *A parasagittal phase contrast MRA of the thorax demonstrating an ectatic and tortuous thoracic aorta (arrow). The cardiac structures are indistinct owing to motion artifact.*

GADOLINIUM-BASED CONTRAST MEDIA

- Most commonly used contrast agent for MR
- Not nephrotoxic in recommended doses
- Little or no cross-reactivity with iodinated contrast agents
- Very low incidence of idiosyncratic reactions

GADOLINIUM-ENHANCED MRA

- Allows coverage of an extensive volume in a single breath-hold
- Excellent SNR
- Accurate timing of gadolinium bolus and image acquisition are essential
- Minimizes effects of turbulence and slow flow on image quality

Gadolinium produces high signal on T1 weighted MR images. Three-dimensional gradient echo sequences with gadolinium enhancement allow imaging over an extensive volume of tissue. The entire abdomen can usually be covered with a single breath-hold **(Fig. 4)**. Gd-MRA has an excellent SNR, which can be improved even more by digital subtraction of an unenhanced image obtained prior to the bolus reaching the region of interest. Gd-MRA minimizes the effects of turbulence and slow flow on image quality. However, the images can be degraded by poor timing of the contrast bolus, which may result in too little contrast in the phase of interest or concurrent enhancement in both the arterial and venous phases in the same image. Furthermore, Gd-MRA does not inherently provide information on the direction and velocity of flow. Gadolinium enhancement does add expense to the examination, but the price is decreasing.

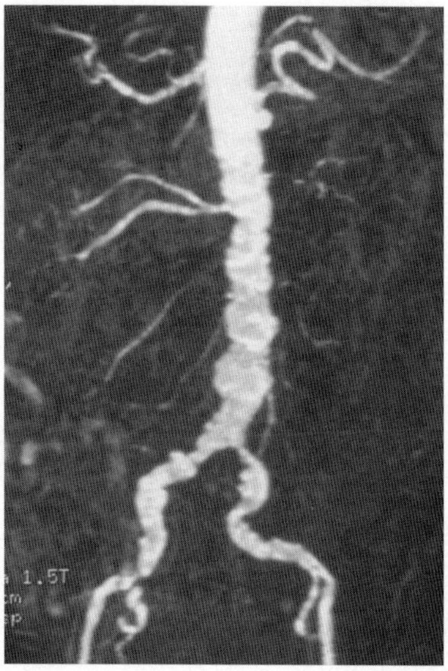

Figure 4. *Gadolinium-enhanced MRA demonstrating extensively diseased abdominal aorta, iliac arteries, and renal arteries.*

ARTERIAL APPLICATIONS OF MRA

Aortic Dissection

MRA OF AORTIC DISSECTION

- Characterizes dissection type
- Identifies major branch vessel involvement
- Defines intimal flap and thrombus
- Eliminates risk of iodinated contrast

Aortic dissection can be described by two classification schemes, the Stanford and the DeBakey schemes **(Table 1)**. The imaging workup of aortic dissection may include plain film radiography, computed tomography, MRI, and transesophageal echocardiography. Currently, catheter arteriography is usually reserved for more precise definition of branch vessel involvement or for the interventional treatment of aortic dissection.

MRA has become the method of choice at several centers for the evaluation of aortic dissection. MRA accurately classifies the dissection **(Fig. 5)**. It also identifies major branch vessel involvement, defines the intimal flap, identifies thrombosis, and eliminates the risk of iodinated contrast media, especially in patients with renal artery involvement. MRA is also useful in the postoperative evaluation of aortic dissection. Aortic dissection may be evaluated by several techniques, including flow void imaging, Gd-MRA, and PC-MRA. All methods are accurate with reported sensitivities of 98%-100% and specificities of 94%-100%.

TABLE 1: CLASSIFICATION OF AORTIC DISSECTION

Stanford type A
Ascending aortic involvement with or without descending aortic involvement
Stanford type B
Descending aortic involvement only
DeBakey type I
Ascending and descending aortic involvement
DeBakey type II
Ascending aortic involvement only
DeBakey type III
Descending aortic involvement only

(Stanford type A [DeBakey types I and II] are treated surgically, while Stanford type B [Debakey type III] are treated medically unless there is clinically significant branch vessel involvement.)

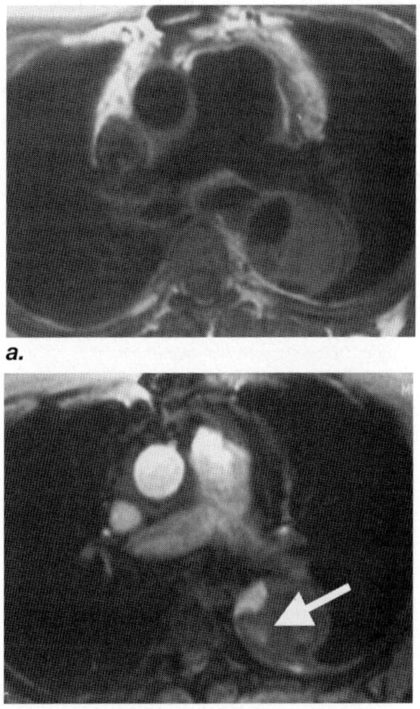

a.

b.

Figure 5. *a) Flow-void image of a type B aortic dissection demonstrates apparent thrombosis of the false lumen. b) Gd-MRA confirms the dissection type and reveals a small degree of flow in the false lumen (arrow).*

Aortic Aneurysm

Aortic aneurysm may be classified as thoracic, thoracoabdominal, and abdominal **(Fig. 6)**. Abdominal aortic aneurysm (AAA) can be further classified as infrarenal, juxtarenal, and pararenal **(Fig. 7)**. Infrarenal aneurysms account for 95% of AAA's. The imaging workup of aortic aneursym may utilize transabdominal ultrasound, CT, and/or MR **(Fig. 8)**. With the recent development of aortic endograft placement for treatment of AAA's, there is now a strong interest in catheter angiography, CT angiography, or MRA in the preprocedure evaluation of these patients.

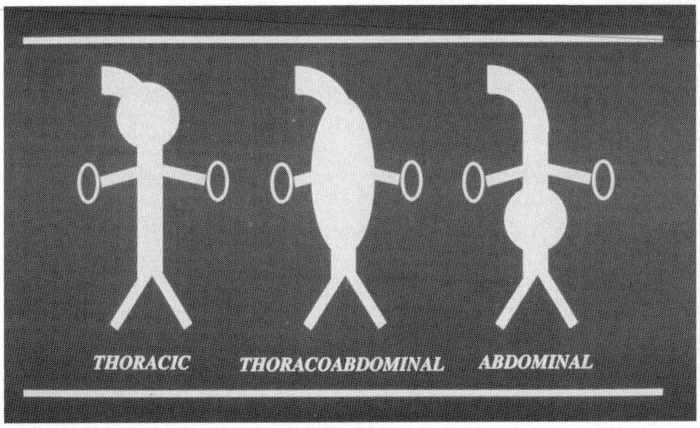

Figure 6. *Aortic aneurysm classification.*

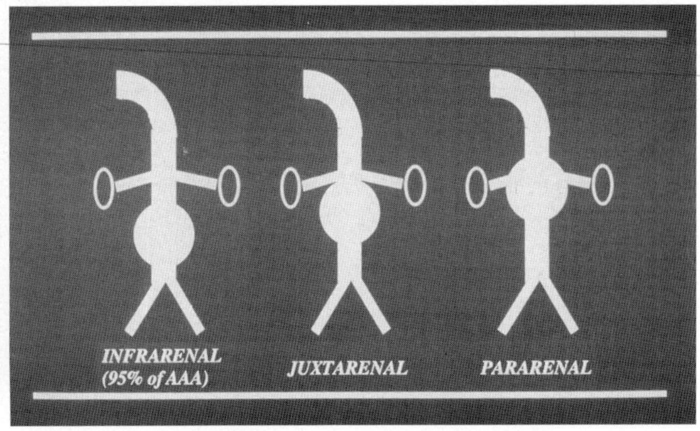

Figure 7. *Abdominal aortic aneurysm (AAA) classification.*

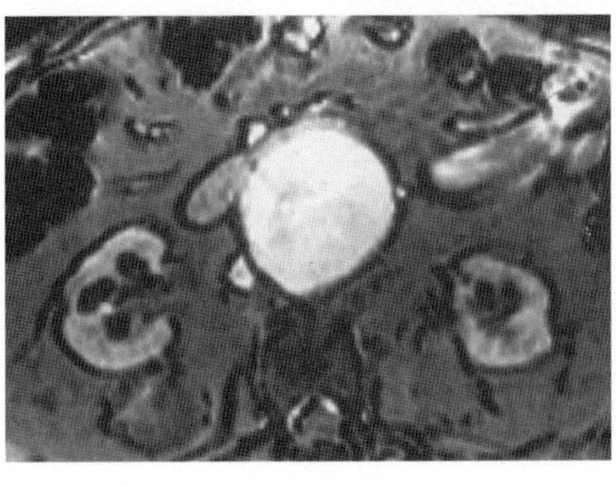

Figure 8. *Axial gradient echo MR scan obtained for aneurysm measurement.*

MRA OF AORTIC ANEURYSM

- Characterizes aneurysm type
- Allows aneurysm measurement
- Allows assessment of complications
- Suitable for pre-endograft evaluation

MRA allows aneurysm classification, accurate diameter measurement, accurate aneurysm neck measurement, assessment of branch vessel involvement, and assessment of complications **(Fig. 9)**. As with aortic dissection, AAA may be evaluated by several techniques, including flow void imaging, Gd-MRA, and PC-MRA.

MRA and CT angiography both can effectively work up AAA's preoperatively. A major difference is the need for iodinated contrast media for CTA, which is eliminated with MRA. CTA, on the other hand, is better suited for the postoperative evaluation of endograft patients due to the presence of the metallic stents, which would cause artifacts on MR **(Fig. 10)**.

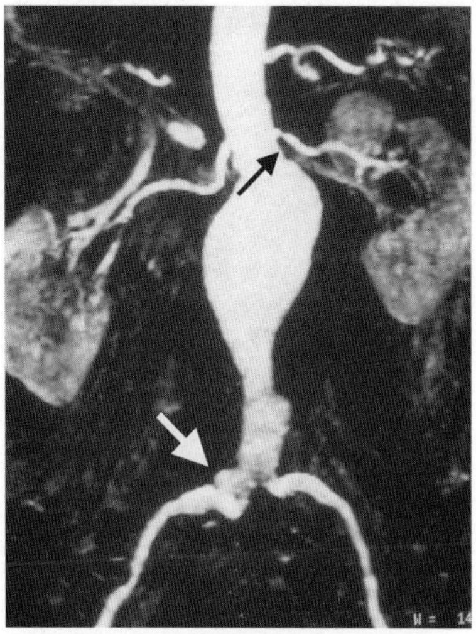

Figure 9. Gd-MRA demonstrating an infrarenal AAA with a short neck (black arrow) and involvement of the right common iliac artery (white arrow). The patent aneurysm lumen is clearly identified; however, the remaining thrombosed portion of the lumen is not well defined. A left proximal renal artery stenosis is also depicted.

Aortic and Peripheral Vascular Atherosclerotic Disease

MRA can be used as the sole method of evaluation of atherosclerotic occlusive disease or as a supplement for an incomplete catheter angiogram, while eliminating the risk of iodinated contrast. A complete evaluation includes the abdominal aorta and pelvic vessels, which are effectively imaged by Gd-MRA **(Fig. 11)**. The remainder of the lower extremity arterial tree can be imaged by 2D TOF **(Fig. 12)** or Gd-MRA with stationary imaging **(Fig. 13)**, or more recently, with moving table protocols to completely cover the legs.

The sensitivity of peripheral vascular MRA ranges from 81%-98% and the specificity ranges from 81%-98%. Peripheral vascular MRA is accurate enough to guide both iliac and infrainguinal peripheral vascular interventions **(Fig. 14)**.

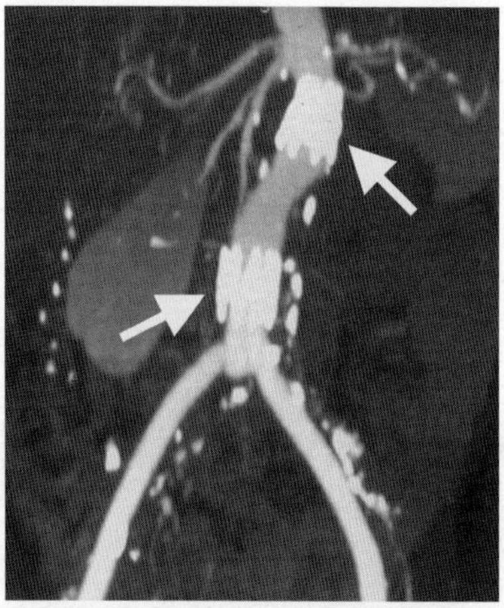

Figure 10. CT angiogram after placement of a tubular aortic stent-graft. The metallic stents at either end (arrows) would cause significant artifacts on MR.

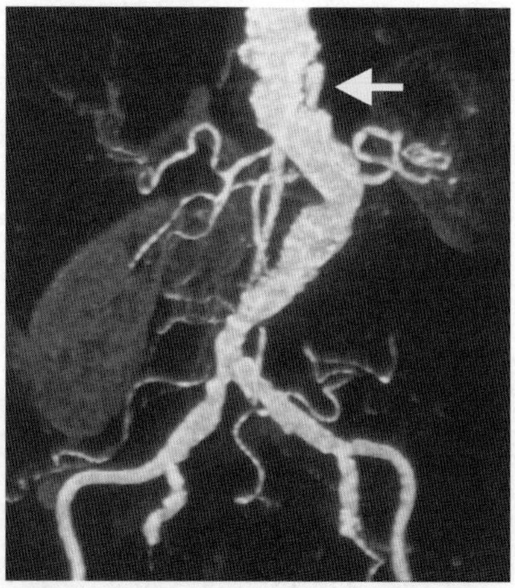

Figure 11. Gd-MRA demonstrating very tortuous, degenerated, atheromatous disease affecting the abdominal aorta. An ulcerated plaque is identified in the suprarenal aorta (arrow). The iliac abnormalities are less severe.

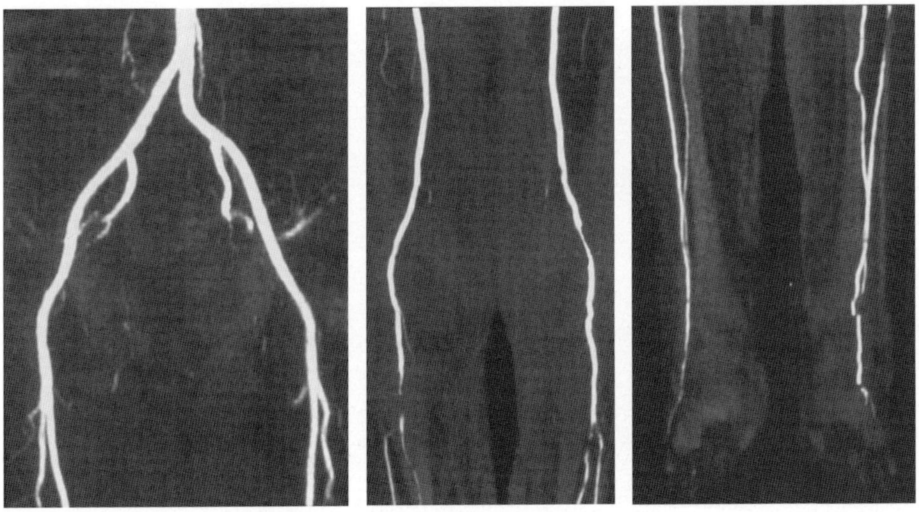

a. b. c.

Figure 12. *Stacked 2D TOF MRA of the a) iliofemoral, b) femoropopliteal, and c) runoff vessels. Although the patient has left foot ischemia, multifocal infrainguinal disease can be identified in both legs.*

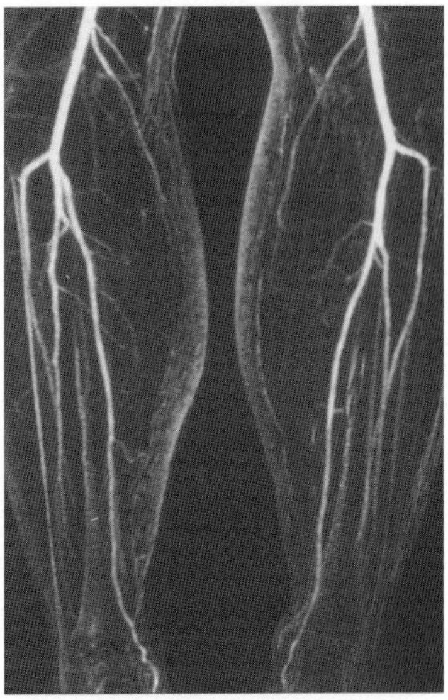

Figure 13. *Stationary Gd-MRA study of the runoff vessels in a normal patient. The resolution here is superior to that of Figures 12b and 12c; however the coverage is less.*

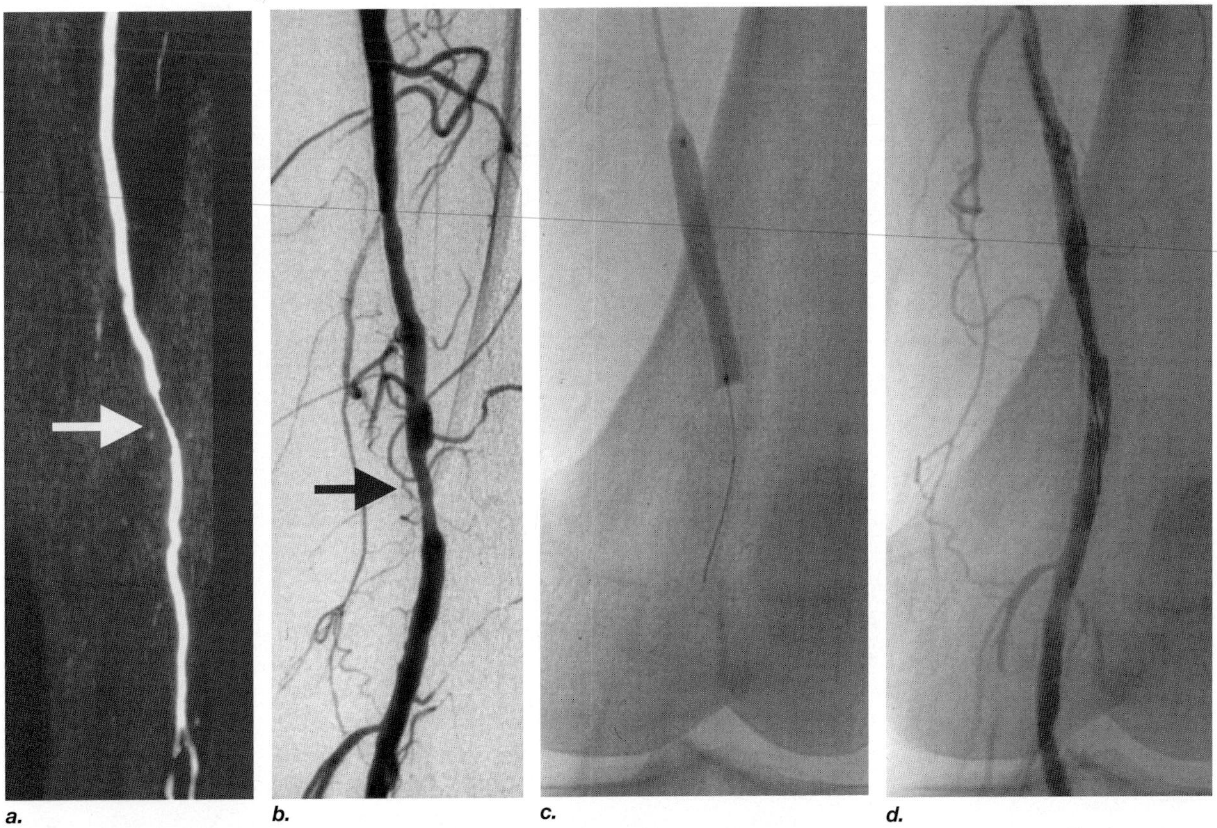

a. b. c. d.

Figure 14. *MRA planning of percutaneous interventional therapy. a) 2D TOF MRA of the left leg demonstrates a popliteal artery stenosis (arrow). b) This MRA finding was confirmed on limited arteriography prior to percutaneous intervention (arrow). c, d) The popliteal lesion was subsequently successfully balloon dilated.*

MRA OF RENAL ARTERY STENOSIS

- Gaining importance as a screening modality
- Demonstrates associated parenchymal abnormalities
- Eliminates risk of iodinated contrast
- Very useful in guiding interventions

Renal Artery Stenosis

MRA is currently enjoying an expanding role in the evaluation of renovascular disease. In a patient with impaired renal function or with clinical evidence of renovascular hypertension **(Table 2)**, MRA is fast-becoming the initial screening examination of choice. MRA offers several advantages: it is noninvasive; it eliminates the risk of iodinated contrast; it can also demonstrate associated abnormalities such as atheromatous disease elsewhere, renal atrophy, and renal masses.

Some authors have advocated a three-component MRA exam for the workup of renovascular disease:

1) Axial T2 weighted images with fat suppression to assess for parenchymal abnormalities;
2) Coronal 3D Gd-MRA of the abdominal aorta to assess for vascular abnormalities;
3) Axial 3D PC-MRA at level of the renal arteries to check for turbulent dephasing accompanying a stenosis.

The sensitivity of renal MRA ranges from 91%-100%, with a specificity from 93%-95%.

Renal MRA is very useful in the preprocedure evaluation and guidance of renal artery interventions, as illustrated in **Figure 15**. Conversely, renal MRA can also indicate when revascularization would not be helpful **(Fig. 16)**.

TABLE 2: CLINICAL FEATURES OF RENOVASCULAR HYPERTENSION

- Abdominal or flank bruit
- Accelerated worsening of hypertension
- Severe hypertension as a child or arising after the age of 50 years
- Impaired renal function, especially with administration of an ACE inhibitor agent
- Hypertension refractory to 3-drug therapy

PARTIAL VENOUS MRA DEFECTS

- Occlusion
- Tumor invasion
- Varices
- Spontaneous shunts

VENOUS APPLICATIONS OF MRA

Portal Vein Abnormalities

Delayed coronal 3D Gd-MRA images are very effective in demonstrating normal **(Fig. 17)** as well as abnormal portal venous anatomy. Portal venous abnormalities which can be evaluated by MRA include thrombosis **(Fig. 18)**, varices and shunts **(Fig. 19)**, and tumor invasion.

MRA is particularly useful in guiding portal venous interventions such as transjugular intrahepatic portosystemic shunt procedures. Its most important role here probably lies in establishing the status of the portal vein (*refer to Chapter Preview case* 2).

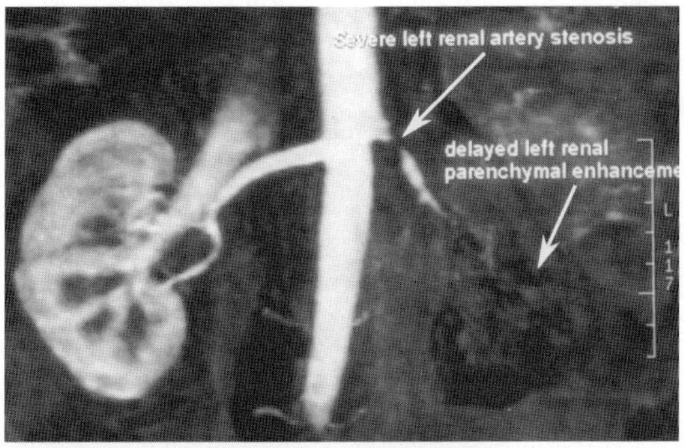

a.

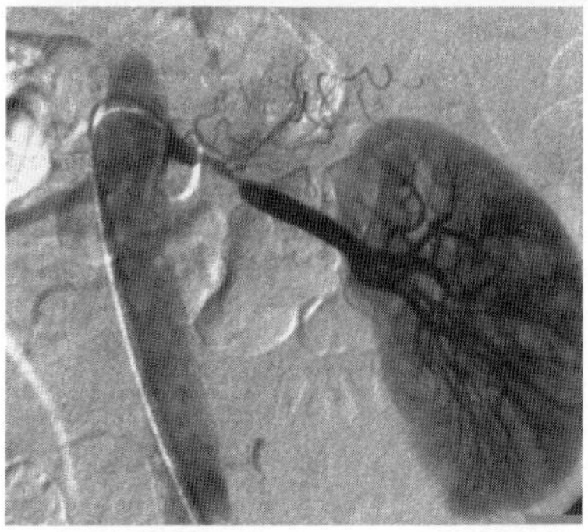

b.

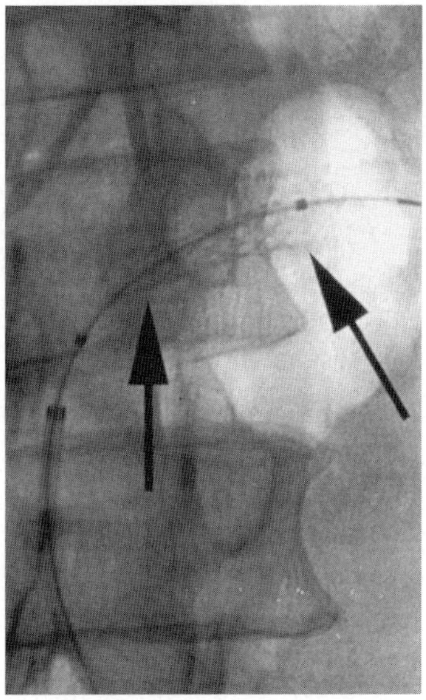

c.

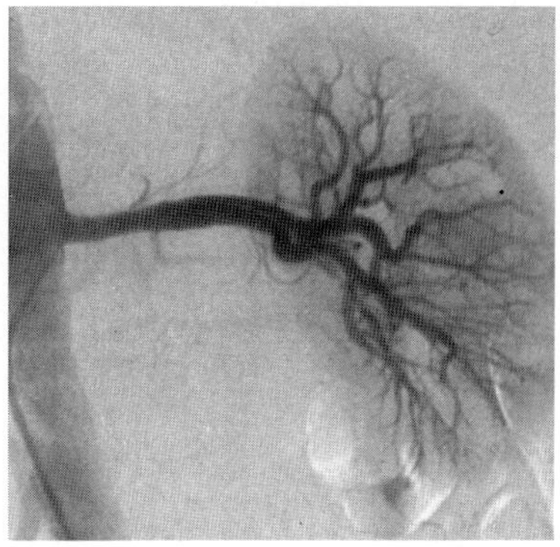

d.

Figure 15. *a) Gd-MRA depicting left renal artery stenosis and delayed parenchymal enhancement in a patient with renovascular hypertension. b) This was confirmed by catheter angiography prior to c, d) renal stent placement (arrows).*

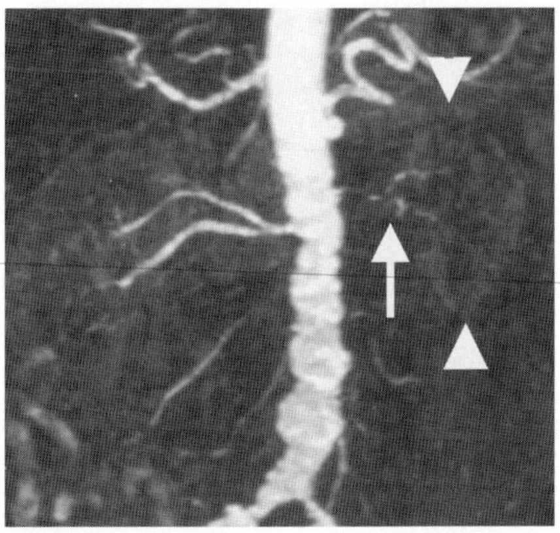

Figure 16. *Gd-MRA of a patient with declining renal function. MRA demonstrates left-sided renal atrophy (demarcated by arrowheads) with a very diseased, diminutive left renal artery (arrow). The right-sided intrarenal arterial branches are also diseased and diminutive. The MRA data is enough to indicate that this patient would likely not benefit for renal revascularization.*

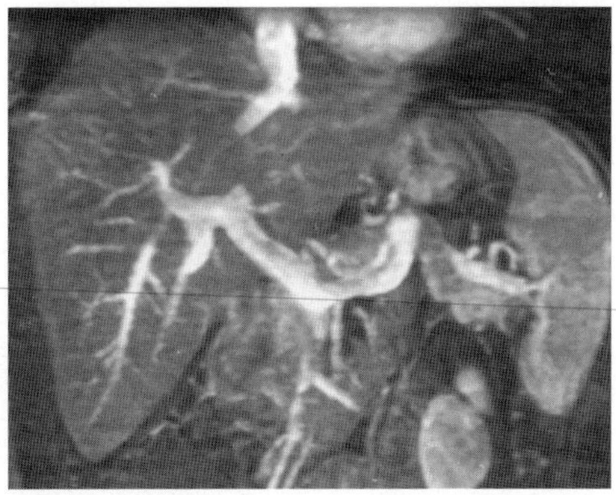

Figure 17. *Delayed abdominal Gd-MRA examination demonstrating the intra- and extrahepatic portal veins and surrounding organs.*

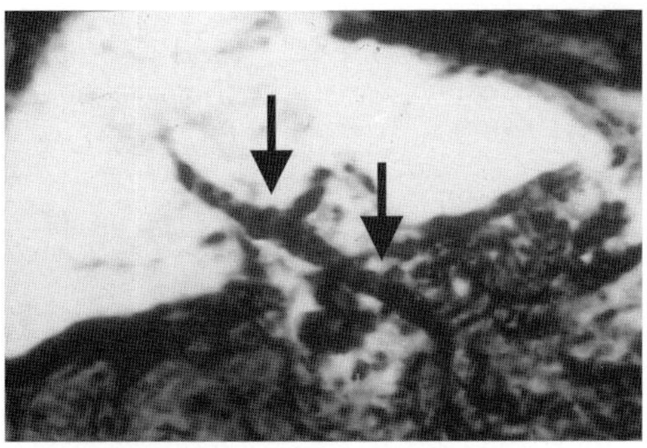

Figure 18. *Gd-MR scan demonstrated hepatic parenchymal enhancement, however, there is absence of flow in the intra- and extrahepatic portal veins (arrows).*

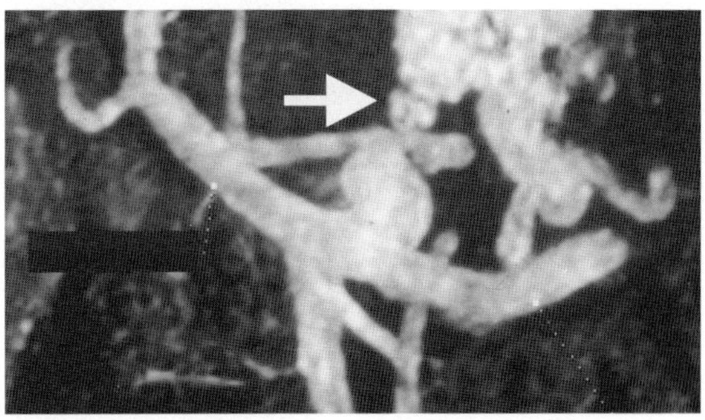

Figure 19. *Portal venous Gd-MRA depicting patent portal and splenic veins, with large varices (arrow) fed by the coronary vein.*

Budd-Chiari Syndrome

Budd-Chiari Syndrome (BCS) is defined as hepatic vein outflow obstruction **(Fig. 20)**. It may occur at the hepatic venular, hepatic venous, or inferior vena caval level. BCS has several etiologies. MRI and MRA can demonstrate several of the findings seen in the Budd-Chiari Syndrome, which are summarized in **Table 3**. MRA is particularly useful in guiding the interventional radiologic treatment of BCS (*refer to Chapter Preview case* 3).

Inferior Vena Cava Abnormalities

MRA is effective in demonstrating IVC anomalies and IVC tumor involvement. In the case of renal cell carcinoma, adrenal carcinoma, lymphoma, testicular carcinoma, and other neoplasms, Gd-enhanced MRA and MRI can show the extent of venous involvement as well as the extravascular extent to aid in tumor staging **(Fig. 21)**.

TABLE 3: IMAGING FEATURES OF BUDD-CHIARI SYNDROME DEMONSTRABLE BY MRI AND MRA

- Hepatic vein webs
- Hepatic vein stumps
- Complete absence of hepatic veins
- Caval webs
- Compression or narrowing of the hepatic inferior vena cava
- Hepatomegaly
- Heterogeneous hepatic parenchymal enhancement
- Ascites
- Caudate lobe hypertrophy
- Portal venous abnormalities resulting from postsinusoidal portal hypertension

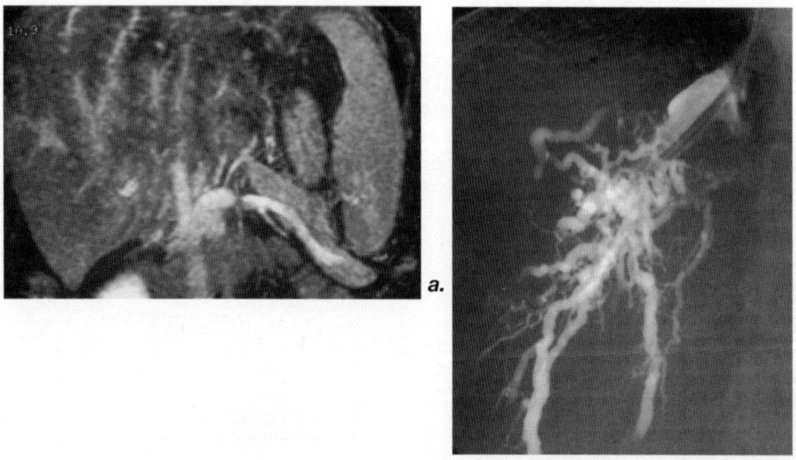

a. b.

Figure 20. *a) Gd-enhanced MR scan in a patient with Budd-Chiari syndrome (BCS). The hepatomegaly, heterogeneous liver enhancement, and absence of discernable normal hepatic veins are all features of BCS. b) The diagnosis was confirmed by catheter hepatic venography, which depicts the classic "spider-web" appearance of intrahepatic venous collaterals.*

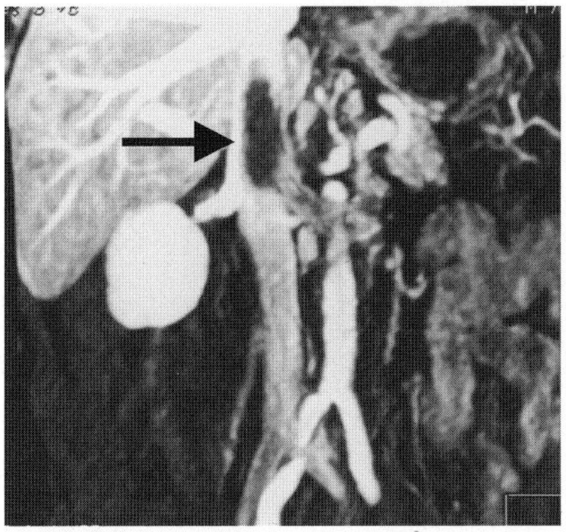

Figure 21. Gd-MRA of a patient with left adrenal adenocarcinoma following left adrenalectomy and nephrectomy. Extension of recurrent tumor into the left renal vein and inferior vena cava is clearly depicted (arrow).

MISCELLANEOUS APPLICATIONS OF MRA

Pseudoaneurysms

The presence of moderately and large-sized traumatic and mycotic peripheral vascular pseudoaneuryms can often be established clinically and confirmed unequivocally with duplex Doppler imaging. MRA gives the added advantage of distinguishing pseudoaneuryms from other vascular masses, demonstrating the extent of the vessel(s) affected, and demonstrating the relationship of the pseudoaneurysm to adjacent musculoskelatal and neurovascular structures **(Fig. 22)**.

MRA OF PSEUDOANEURYSMS

- Distinguishes them from other masses
- Demonstrates extent of vessel affected
- Demonstrates relationship to neighboring structures

Vascular Malformations

Peripheral vascular malformations may be classified as pediatric hemangiomas, venous malformations, and arteriovenous malformations. Pediatric hemangiomas are usually manifested by the age of one month and most commonly regress by the age of 7-10 years making treatment unnecessary. Venous malformations are present at birth and continue into adulthood. They may be associated with other syndromes such as Maffucci Syndrome, Klippel-Trenaunay-Weber, and Gorham Syndrome. Peripheral arteriovenous malformations are shunt lesions which may require treatment if they become enlarged or symptomatic.

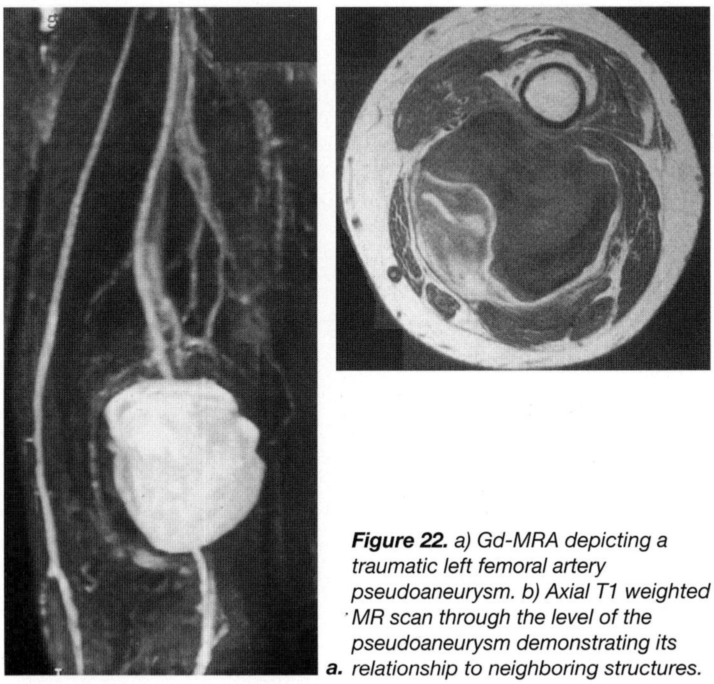

a. b.

Figure 22. a) Gd-MRA depicting a traumatic left femoral artery pseudoaneurysm. b) Axial T1 weighted MR scan through the level of the pseudoaneurysm demonstrating its relationship to neighboring structures.

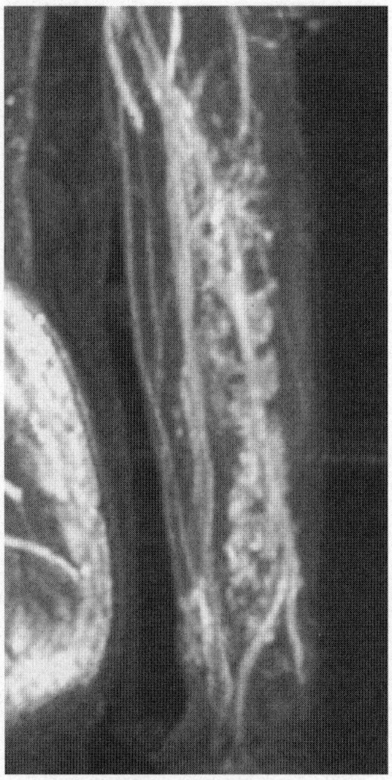

Figure 23. *Gd-MRA of the forearm demonstrating enhancement of a complex network of serpiginous vascular structures compatible with a venous malformation.*

MR is now a mainstay in the evaluation of vascular malformations. MR can characterize the vascular malformation **(Fig. 23)**, demonstrate the full extent of the abnormality, establish the relationship of the malformation to adjacent musculoskeletal and neurovascular structures, and help distinguish high-flow from low-flow lesions.

Interventional Magnetic Resonance Imaging

Interventional MR (i-MR) is an emerging modality which presently has many forms and demonstrates significant potential for MR-guided therapies. Several of the advantages and applications of i-MR are listed in **Table 4** and **Table 5**.

INTERVENTIONAL MR UNIT CONFIGURATIONS

- Horizontal open magnet
- Vertical open magnet
- Hybrid MR/C-arm fluoroscopy system

The most common i-MR units are open magnet configurations, which allow the operator to have direct access to the patient. The most common open magnet configurations currently in operation are the horizontal "double doughnut" and the vertical "double doughnut." They range in field strength between 0.2 and 0.5 Tesla. These systems are ideal for nonvascular interventions; however vascular interventions can be very successfully performed on these units as well.

A more recent development is the hybrid interventional MR system, combining a C-arm fluoroscopy unit with a short-bore, high field strength (1.5 T) MR unit. The patient is seamlessly transferred between the two components on a sliding table. This unit will be used primarily for guiding endovascular therapies.

TABLE 4: ADVANTAGES OF INTERVENTIONAL MR
• No ionizing radiation • Excellent soft-tissue discrimination • Multiplanar imaging capabilities • Ability to measure physiologic parameters (e.g., flow, diffusion)

TABLE 5: EXAMPLES OF MR-GUIDED INTERVENTIONS
• Percutaneous drainages and biopsies • Angioplasty • TIPS • Hepatic tumor ablation (e.g., laser ablation, radiofrequency ablation, cryotherapy)

REFERENCES

Blasbalg R, Mitchell DG, Outwater EK, Ito K, Gabata T, Chiowanich P. MRA of the abdomen: postprocessing dynamic gadolinium-enhanced 3D axial MR images. Abdom Imaging 2000; 25(1):62-66.

Ho VB, Choyke PL, Foo TK, et al. Automated bolus chase peripheral MR angiography: initial practical experiences and future directions of this work-in-progress. J Magn Reson Imaging 1999; 10(3):376-388.

Kelekis NL, Semelka RC, Worawattanakul S, Molina PL, Mauro MA. Magnetic resonance imaging of the abdominal aorta and iliac vessels using combined 3-D gadolinium-enhanced MRA and gadolinium-enhanced fat-suppressed spoiled gradient echo sequences. Magn Reson Imaging 1999; 17(5):641-651.

Kilgour DA. Contrast-enhanced MR angiography. Radiol Technol 1999; 71(2):133-138.

Krinsky G, Rofsky N, Flyer M, et al. Gadolinium-enhanced three-dimensional MR angiography of acquired arch vessel disease. AJR 1996; 167:981-987.

Omary RA, Frayne R, Unal O, Grist TM, Strother CM. Intraarterial gadolinium-enhanced 2D and 3D MR angiography: a preliminary study. J Vasc Interv Radiol 1999; 10(10):1315-1321.

Prince MR. Gadolinium-enhanced MR aortography. Radiology 1994; 191:155-164.

Prince MR, Narasimham DL, Jacoby WT, et al. Three-dimensional gadolinium-enhanced MR angiography of the thoracic aorta. AJR 1996; 166:1387-1397.

Rak KM, Yakes WF, Ray RL, et al. MR imaging of symptomatic peripheral vascular malformations. AJR Am J Roentgenol 1992; 159(1):107-112.

Schoenberg SO, Wunsch C, Knopp MV, et al. Abdominal aortic aneurysm. Detection of multilevel vascular pathology by time-resolved multiphase 3D gadolinium MR angiography: initial report. Invest Radiol 1999; 34(10):648-659.

Wacker FK, Reither K, Branding G, Wendt M, Wolf KJ. Magnetic resonance-guided vascular catheterization: feasibility using a passive tracking technique at 0.2 Telsa in a pig model. J Magn Reson Imaging 1999; 10(5):841-844.

Westenberg JJ, van der Geest RJ, Wasser MN, et al. Vessel diameter measurements in gadolinium contrast-enhanced three-dimensional MRA of peripheral arteries. Magn Reson Imaging 2000; 18(1):13-22.

Wilson MW, Hamilton BH, Dong Q, et al. Gadolinium-enhanced magnetic resonance venography of the portal venous system prior to transjugular intrahepatic portosystemic shunts and liver transplantation. Investigative Radiology 1998; 33:644-652.

Yakes WF, Haas DK, Parker SH, et al. Symptomatic vascular malformations: ethanol embolotherapy. Radiology 1989; 170(3 Pt 2):1059-1066.

CHAPTER 6 PREVIEW ANSWERS

Case 1

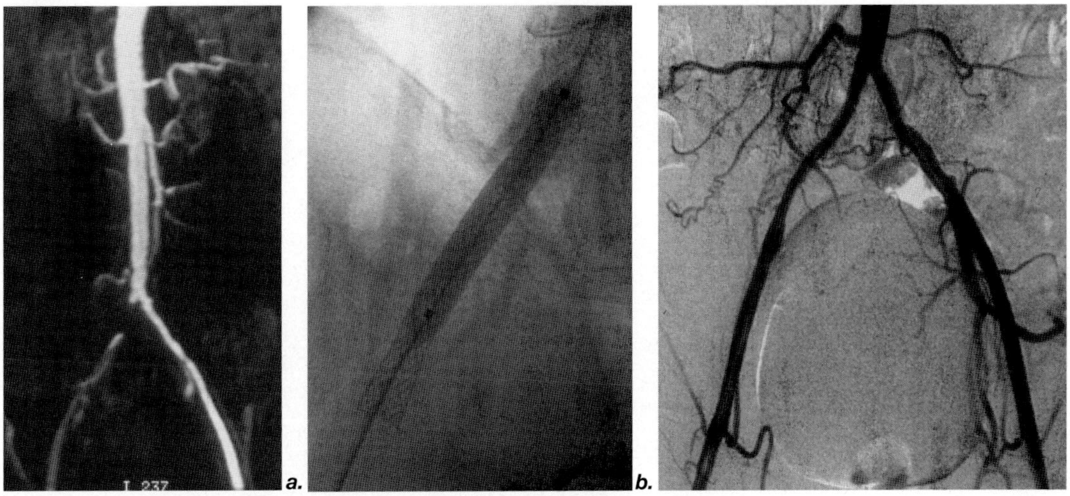

a. b. c.

Findings: a) The aortoiliac Gd-MRA scan depicts right common iliac artery occlusion. b) This was treated successfully by percutaneous iliac stent placement and dilatation. c) The arteriogram demonstrates restored patency of the right common iliac artery.

Case 2

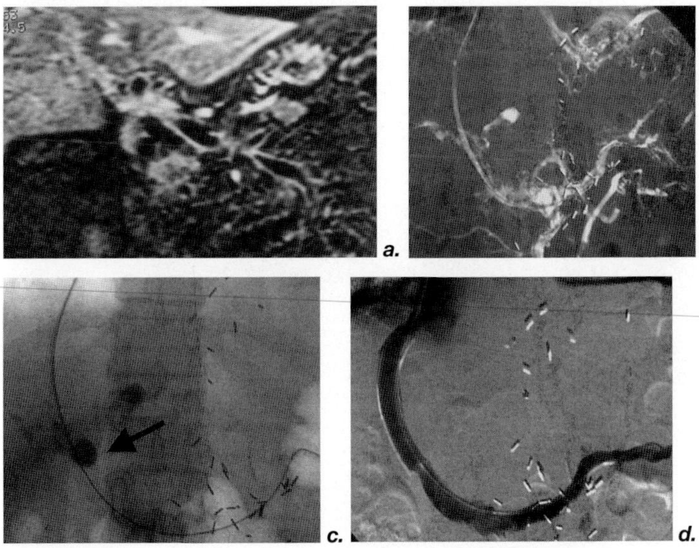

a. b. c. d.

Findings: a) The abdominal Gd-MRA scan demonstrates lack of blood flow in the portal vein, with enhancement of the portal vein wall and the liver parenchyma, compatible with acute portal vein thrombosis. b) The patient subsequently underwent TIPS for esophageal bleeding, confirming portal vein thrombosis along with marked esophageal varices. c) The thrombus was swept out with a balloon catheter (arrow). d) The TIPS was completed with stent placement in the parenchymal tract, resulting in decompression of the varices.

Case 3

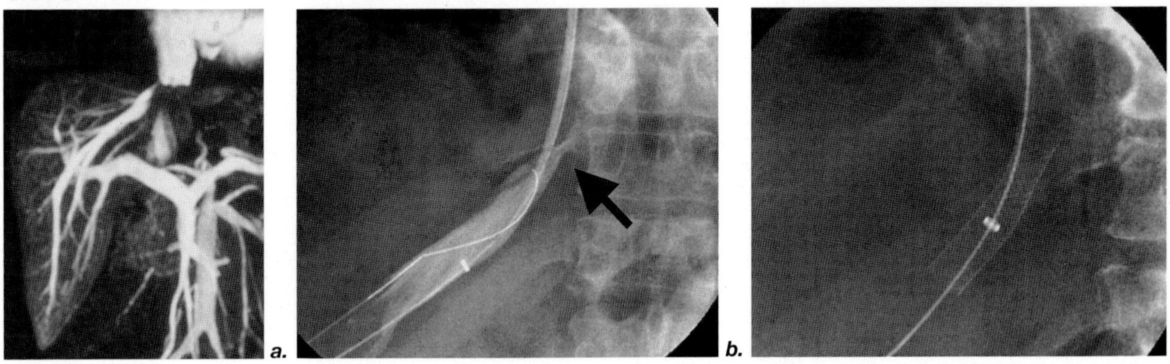

a. b. c.

Findings: a) The abdominal Gd-MRA scan depicts a stenosis at the junction of the right hepatic vein (RHV) with the inferior vena cava just proximal to the right atrium. b) This MRA finding along with the history suggests the diagnosis of Budd-Chiari Syndrome. The lesion was confirmed on catheter venography (arrow). c) It was successfully treated with balloon dilatation and stent placement.

CHAPTER 7 PREVIEW

Case 1

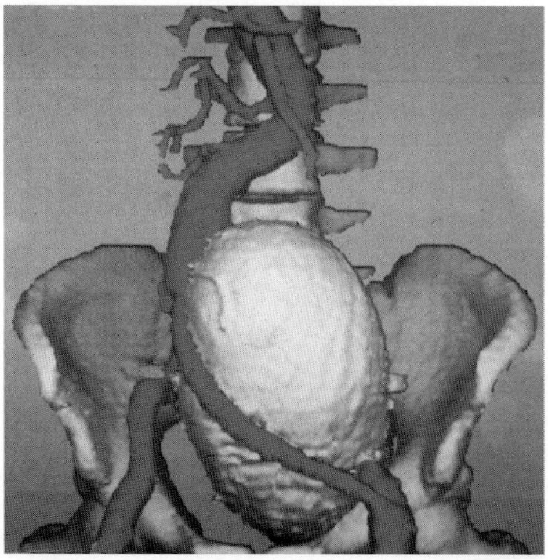

History: Computed tomographic angiogram in a 66-year-old man who previously underwent a vascular surgical procedure.
Question: What reconstruction algorithm has been used to create this image?

Case 2

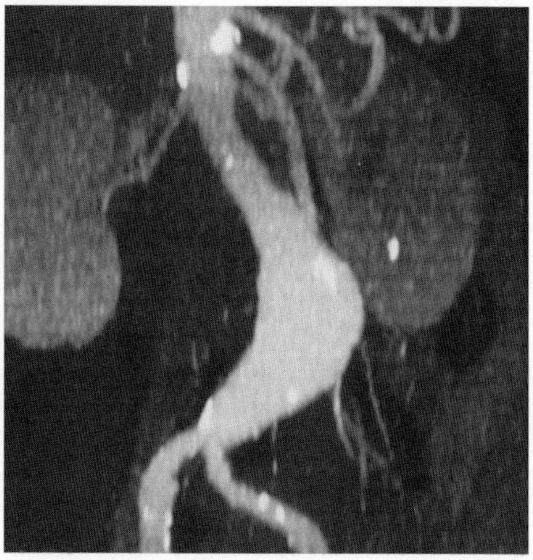

History: 56-year-old man undergoing evaluation of a pulsatile abdominal mass.
Question: What reconstruction algorithm has been used to create this image?

Case 3

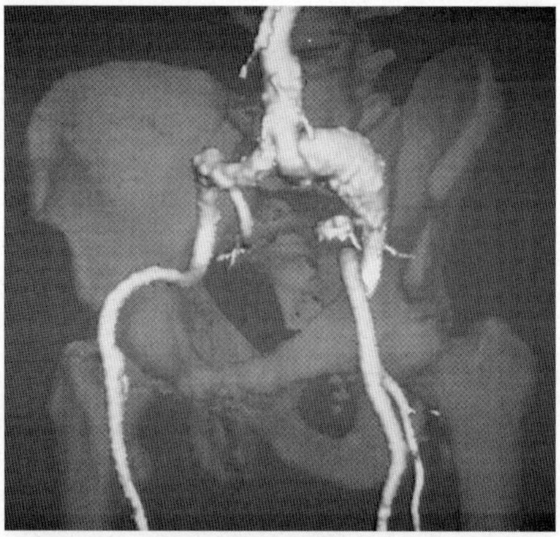

History: 62-year-old man with ischemic toes on the left foot.
Question: What is your diagnosis?

Case 4

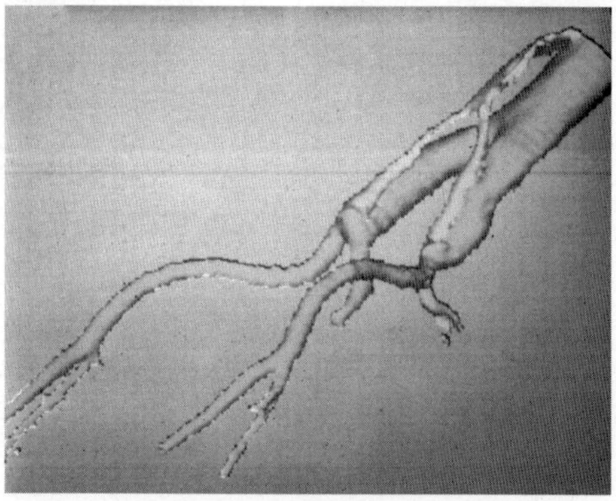

History: 47-year-old-woman with back pain.
Question: What is your diagnosis?

Chapter 7
Computed Tomographic Angiography

Robert K. Kerlan, Jr., MD

CHAPTER OUTLINE

- History
- Technique
- Image display
  - Source image
  - Multiplanar reconstructions
  - Maximum intensity projection
  - Shaded surface display
  - Volume rendering
- Clinical applications
  - Abdominal aortic aneurysms
  - Renal artery evaluation
  - Atherosclerotic peripheral vascular disease
  - Thoracic aortic aneurysms
  - Pulmonary thomboembolism

INTRODUCTION

Over the past ten years the options for imaging aortic and peripheral vascular disease have increased substantially. These options include ultrasound coupled with physiologic testing, magnetic resonance angiography (MRA), computed tomographic angiography (CTA), digital subtraction angiography (DSA), and conventional film screen angiography.

VASCULAR IMAGING

- Ultrasound (US)
- Magnetic resonance angiography (MRA)
- Computed tomographic angiography (CTA)
- Digital subtraction angiography (DSA)
- Conventional angiography

CTA can demonstrate blood vessels with a higher degree of resolution than magnetic resonance imaging. However, this technique requires the administration of intravenous contrast media as well as considerable patient cooperation to diminish respiratory motion artifacts. The precise role of this technology is still being defined. This discussion will focus on clinical applications of CTA and review in what situations this modality is most appropriate.

HISTORY

Computed tomography with intravenous contrast material has always been a valuable method to evaluate large blood vessels. The original applications dealt mainly with the assessment of aortic dissection and delineating the size of aortic aneurysms. Sagittal and coronal reformations could be performed, but extensive artifacts limited the usefulness of CTA as a clinical tool.

COMPUTED TOMOGRAPHIC ANGIOGRAPHY: HISTORY

Two major advances allowed the development of spiral CT:

- Slip ring technology
- Improved tube cooling technology

Clinically useful CTA did not develop until helical (aka. spiral) scanning was created. There were two technological advances which allowed for the development of helical scanning: 1) slip ring technology and 2) tube cooling. The development of slip rings allows for the x-ray tube to rotate continuously in one direction. Improvements in x-ray tube cooling allow the tube to be on for a prolonged (up to a minute) period of time, creating an opportunity to scan reasonably extensive regions of the body. By applying different reconstruction algorithms to a continuous data set, including maximum intensity projection and shaded surface display, an angiographic-like depiction of selected blood vessels could be achieved.

TECHNIQUE

The specific technique used for CTA depends upon the region being evaluated. However, generalizations can be made which apply to most situations:

1) A digital scout image is obtained to define the region to be scanned.

2) A secure intravenous access is established in a reasonably sized vein.

3) A series of stationary images are performed at the most proximal level to establish the optimal delay from contrast injection to initiation of scanning. The scans are obtained at five-second intervals after a 20 ml bolus of intravenous contrast injected at a rate of 4 ml/sec. These preliminary scans should be performed at 100 mA to minimize heating of the x-ray tube.

4) The images are reviewed and the delay is set at the time of maximal opacification, usually between 15 and 25 seconds.

5) The scan volume is selected from the scout view. The volume should be the minimal amount to cover the region of interest.

6) The slice thickness and pitch are selected to cover the volume in 30 to 50 seconds. The slice thickness is the width of collimation of the x-ray beam. Pitch is defined as the table speed in mm/sec divided by the slice thickness:

> **CTA: PITCH DEFINITION**
>
> Pitch is defined as:
>
> $$\frac{\text{table speed (mm/sec)}}{\text{slice thickness (mm)}}$$

$$\textbf{Pitch} = \frac{\textit{Table Speed (mm/sec)}}{\textit{Slice Thickness (mm)}}$$

For example, with a table speed of 6 mm/sec and a slice thickness of 3 mm, the pitch would be 2.

The selection of slice thickness and pitch is crucial to the final appearance of the image. To increase the field of coverage for a given time, either the pitch or the slice thickness may be increased. All other parameters being equal, increasing the pitch increases coverage at the cost of increased quantum mottle. Increasing the slice thickness increases coverage at the cost of edge resolution due to partial volume averaging.

However, to a certain degree, the optimal pitch may be influenced by the specific scanner used. Newer multiple detector (twin or quad) scanners may actually maintain acceptable scanning resolution up to a pitch of 6, allowing increased coverage in comparison to other types of scanners.

7) The helical scan is then performed with injection of 125 ml to 175 ml of intravenous contrast at a rate of 3 ml to 5 ml/sec. For most patients, 3.5 ml to 4.0 ml/sec is sufficient.

8) The images are reconstructed with at least 1 mm of overlap and transmitted to the workstation for further processing.

Scanning parameters that influence the visualization of a stenosis include collimation, pitch, voltage, current, reconstruction increment, and linear interpolation algorithm (LIA). Luboldt et al analyzed each of these parameters and concluded that a collimation of 3 mm, a pitch of 1, and a > 250 Hounsfield units contrast density differential were optimal to allow detection of stenoses down to .5 mm in diameter. Smaller widths of collimation lead to increased noise and decreased visualization of stenoses. A collimation of 3 mm with a pitch of 2 was superior to a collimation of 5 mm and a pitch of 1. Moreover, increasing the voltage and current beyond standard values was not helpful, nor was increasing the reconstruction interval to less than one-third of the collimated width.

CTA: RECONSTRUCTION FORMATS

- Multiplanar reformation (MPR)
- Maximum intensity projection (MIP)
- Shaded surface display (SSD)
- Volume rendering

IMAGE DISPLAY

Several methods of image display are available following CTA. The type of display used is most dependent upon the type of information desired from the study. Display formats are described below.

Computed Tomographic Axial Source Images

In reality, all of the information necessary can be extracted from the axial source images **(Fig. 1)**. When reconstructions are performed, this pure information can only be degraded, therefore, most radiologists rely most heavily on the original images to synthesize the diagnostic impression. These images are less satisfactory to referring surgeons who are more familiar with the traditional angiographic examination.

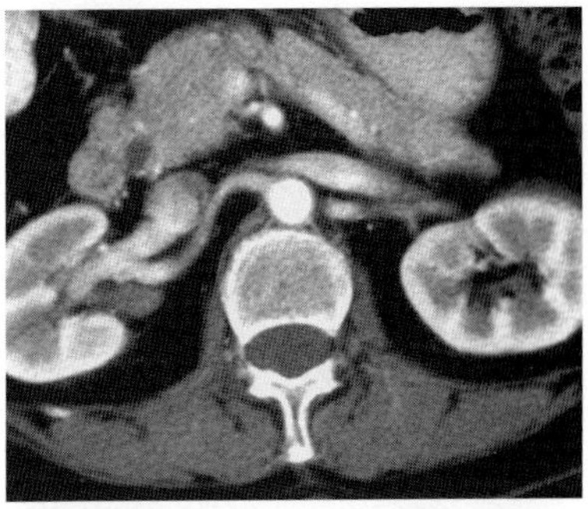

Figure 1. *Computed tomographic axial source image shows a nonstenotic right renal artery and a segment of left renal artery.*

Reformatted Computed Tomographic Images

The earliest form of CTA relied upon reformatting axial images into the coronal and sagittal plane. Oblique and curved reformatted images can also be created **(Fig. 2)**. These images have a superficial resemblance to an angiogram, but contain an extensive amount of nonangiographic information which can be both elucidating and distracting.

MULTIPLANAR REFORMATION

- Two-dimensional depiction
- May combine orthogonal and oblique or curved views
- Excellent for showing weblike abnormalities and calcium, but appearance is not anatomic

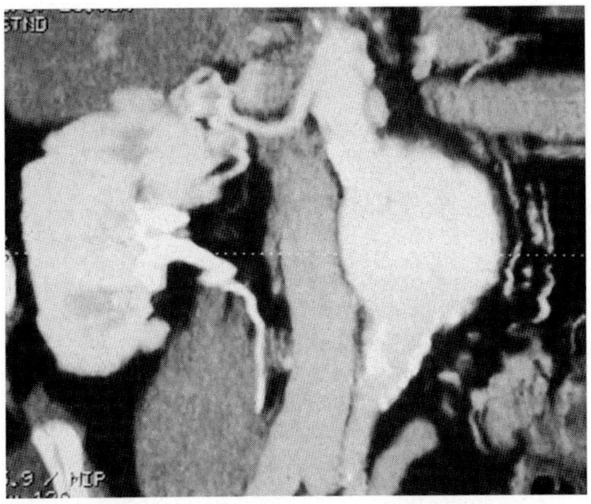

Figure 2. *Oblique computed tomographic reconstruction (multiplanar reformation) shows the nonstenotic right renal artery. An abdominal aortic aneurysm is also present.*

MAXIMUM INTENSITY PROJECTION

- Projects the single brightest pixel along any line of sight
- Excellent for showing calcium
- More reliable than SSD for showing degree of stenosis

Maximum Intensity Projection

Maximum intensity projection (MIP) is a reprojection technique that depicts the most intense pixel along any given line of sight **(Fig. 3)**. This is the most common technique employed with MRA and, though requiring a significant amount of exclusionary data editing, is considerably less time intensive than shaded surface display. MIP is the most valuable technique for demonstrating stenoses.

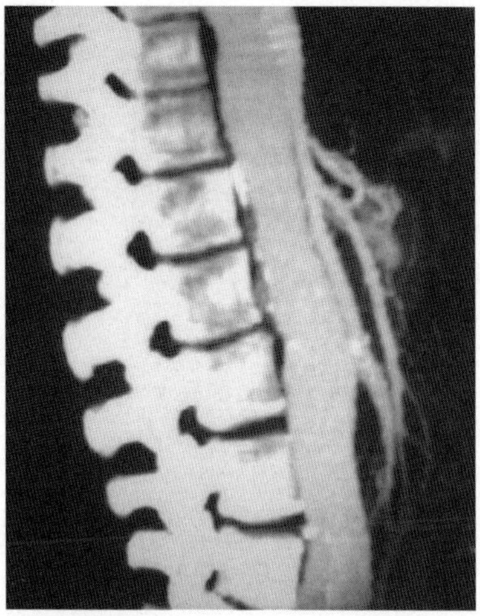

Figure 3. *Maximum intensity projection (MIP) of aortic dissection with true lumen supplying the celiac and superior mesenteric arteries.*

Shaded Surface Display

Shaded surface display (SSD) is the algorithm by which the surface features of the vessel are depicted in a three-dimensional format **(Fig. 4)**. Creation of these images can require a considerable physician effort to exclude nonvascular features. However, the resulting images can be readily interpreted by the vascular surgeon and give an appearance unlike any other technique. The principal disadvantage of SSD is obscuration of internal vascular architecture. Moreover, stenoses can be underestimated with this type of reformation.

SSD images are often reconstructed with color depictions to increase the clarity of the image **(Fig. 5)**.

Figure 4. *SSD of midabdominal aorta shows a patent aorta to left renal bypass graft.*

SHADED SURFACE DISPLAY

- Three-dimensional depiction of the external aspect of the contrast column
- Excellent to demonstrate relationship of aorta to branch vessels

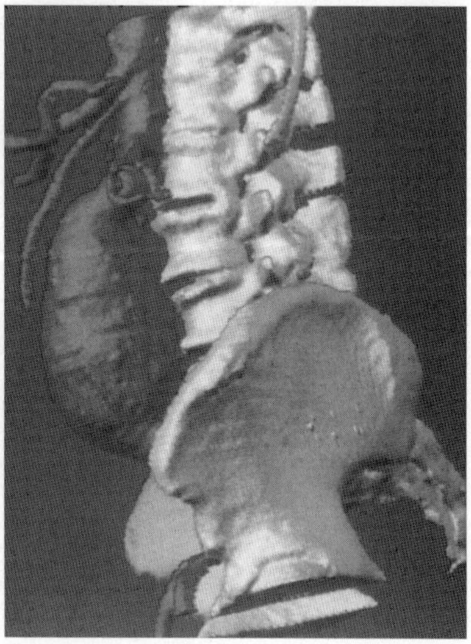

Figure 5. *SSD of infrarenal abdominal aortic aneurysm in the lateral view.*

Volume Rendering

Volume rendering (VR) involves applying shading to all voxels in the spiral data set. A trapezoidal transfer function is applied to the data set with the x-axis representing Hounsfield units and the y-axis opacity. Sliding the trapezoid higher along the x-axis excludes lower-attenuation voxels from the display, yielding images of vessels and other high-attenuation structures. Because of the rapidity of image generation, and the decreased editing required, volume rendering has become a very popular display mode for spiral CTA **(Fig. 6)**.

VOLUME RENDERING: TECHNIQUE

- Shading is applied to all voxels in the spiral data set
- Trapezoidal transfer function is applied with x-axis representing Hounsfield units and the y-axis opacity
- Lower attenuation voxels can be excluded

VOLUME RENDERING: BENEFITS

- Less physician and technologist time is involved
- Information allows virtual endoscopic applications
- Image resembles shaded surface display but contains more information than either SSD or MIP

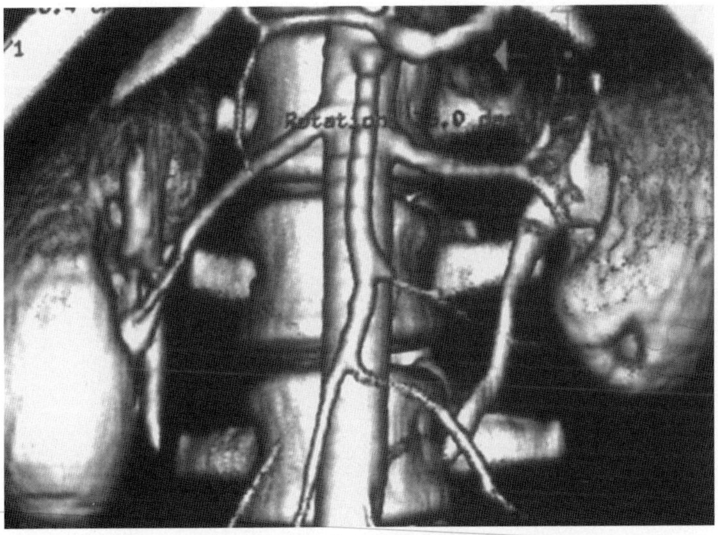

Figure 6. *Volume-rendered image shows nonstenotic single renal arteries bilaterally. (Case courtesy of Edward L. Baker M.D., California Pacific Medical Center.)*

> **CTA: ABDOMINAL AORTIC ANEURYSM**
>
> Preoperative assessment
> - Relationship to renal arteries
> - Accessory renal arteries
> - Caudal extension
> - Associated aneurysms
> - Associated occlusive disease
> - Assessment of left renal vein anatomy

CLINICAL APPLICATIONS

Abdominal Aortic Aneurysm

The most common indication for spiral CTA at our institution is elective, preoperative evaluation of an abdominal aortic aneurysm. The goals of the study are to demonstrate the extent of the aneurysm, detect significant renal or visceral artery stenoses, identify accessory renal arteries, determine the patency of the inferior mesenteric artery, and show the position of the left renal vein or veins. These goals can be achieved in over 90% of patients **(Figs. 7-10)**.

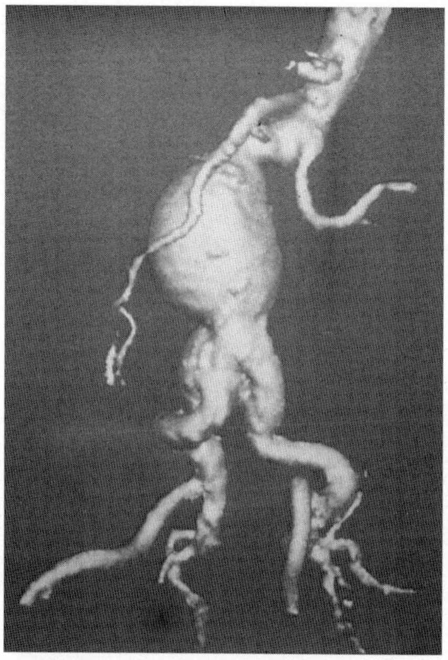

Figure 7. *SSD shows infrarenal abdominal aortic aneurysm. An aneurysm neck is present but angled with respect to the long axis of the aorta.*

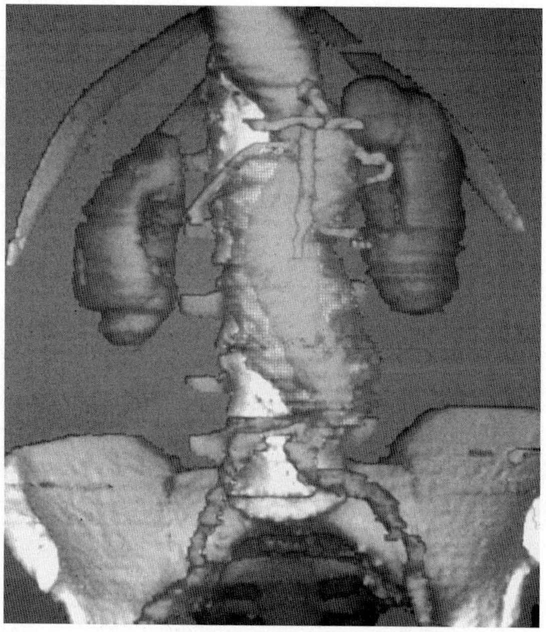

Figure 8. *SSD shows the aortic aneurysm extending to the renal arteries (juxtarenal aneurysm).*

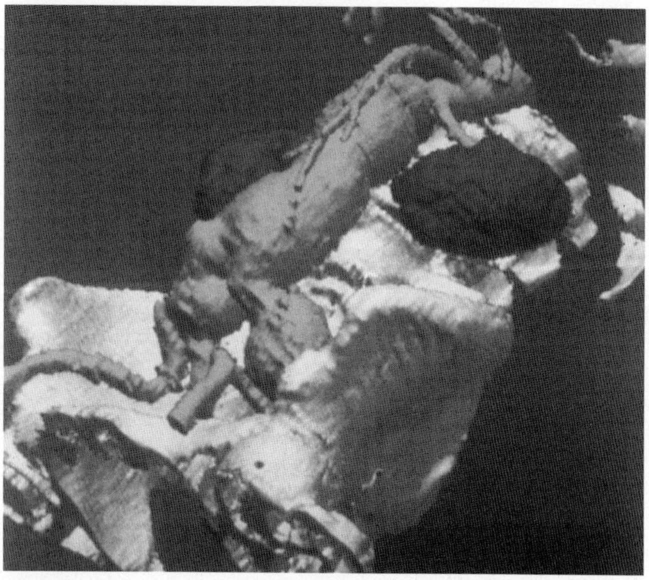

Figure 9. *SSD depicts an infrarenal abdominal aortic aneurysm extending into both right and left common iliac arteries.*

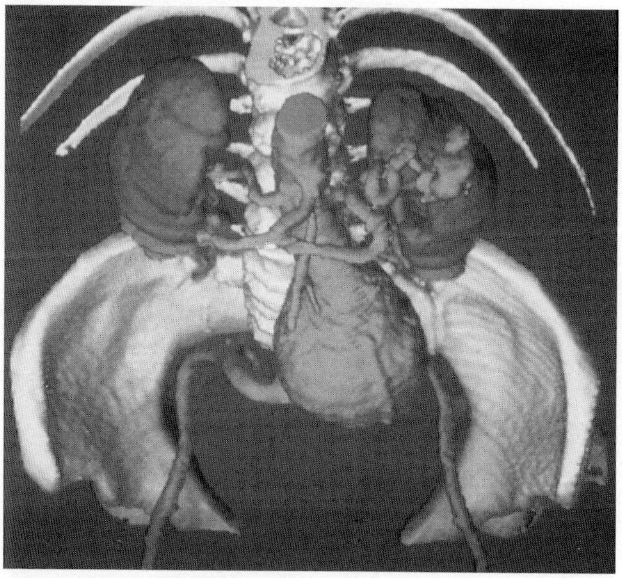

Figure 10. *SSD shows the neck of the abdominal aortic aneurysm to be angled ventrally.*

Costello and Gaa recently reported their experience with spiral CT in the evaluation of abdominal aortic aneurysms. The examination was successful in 26 of 28 patients. In 17 of the 26 patients, there was surgical confirmation of the CT findings. The remaining nine were not operated upon. Twelve accessory renal arteries were identified in the surgical group.

The importance of CTA in the evaluation of abdominal aortic aneurysms is likely to increase as the management of this disease changes. Endoluminal stent grafts can be placed avoiding the necessity of an open surgical procedure. Accurate morphologic assessment has become critical to select the appropriate form of therapy. Specifically, the size, irregularity, calcification and length of the infrarenal neck have become crucial to determine if the patient is a candidate for an endoluminal procedure. Very precise measurements are required so that the appropriate endoluminal device can be selected. An accurate assessment of the iliac arteries is also very useful in preprocedural planning.

Armon et al compared measurements of the abdominal aorta obtained with CTA to measurements obtained with aortography. These investigators found that 90% of the correlative measurements did not deviate by more than 2 cm. Using an experimental model that was validated in two patients, Rubin et al (1998) found an even higher level of accuracy of aortic measurement using volumetric data. Though some investigators would argue that gadolinium-enhanced MRA provides similar information, in most institutions CTA is considered the modality of choice for preprocedural assessment of abdominal aortic aneurysms prior to endovascular repair.

Following endovascular repair of abdominal aortic aneurysms, CTA also plays a crucial role. Accurate assessment of stent position as well as the adequacy of aneurysmal sac exclusion can be reliably made with this technique. Moreover, when flow into the aneurysmal sac persists, it is possible to identify the source of the endoleak. By accurately identifying the source, the most appropriate therapeutic revision can be selected.

Renal Artery Evaluation

The most efficient and reliable technique to evaluate patients for potential renal artery stenosis is not established. Options include radionuclide flow evaluation, ultrasound (with Doppler), CTA, MRA, as well as angiography. Each of these techniques has its own advantages and disadvantages. Radionuclide flow studies are noninvasive and virtually free of risk, but lack morphologic delineation and may be less sensitive in the presence of bilateral disease. Ultrasound is emerging as a useful modality, but is difficult in large patients and requires a high level of expertise for reliable assessment. CTA and MRA are somewhat competitive. Both require patient cooperation and may be prone to artifacts induced by respiratory motion and pulsation. CTA also requires injection of iodinated contrast media, which is a disadvantage in patients with impaired renal function. Catheter angiography offers the best morphologic depiction and may be coupled with a therapeutic intervention. However, angiography is expensive and invasive.

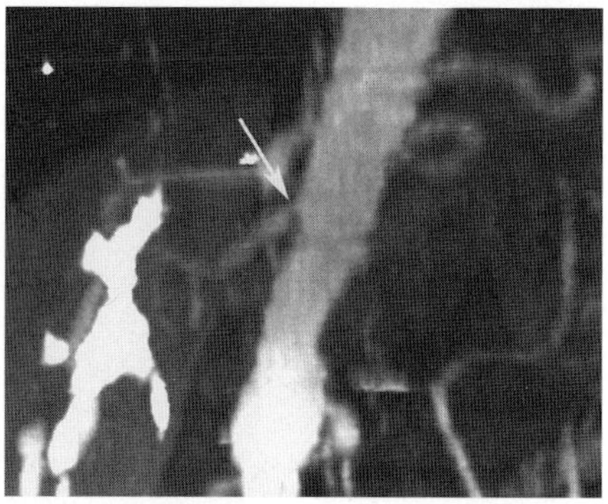

Figure 11. Renal arterial evaluation in MIP format shows a tight right renal artery stenosis (arrow) and occluded left renal artery in a patient with known Takayasu's aortitis.

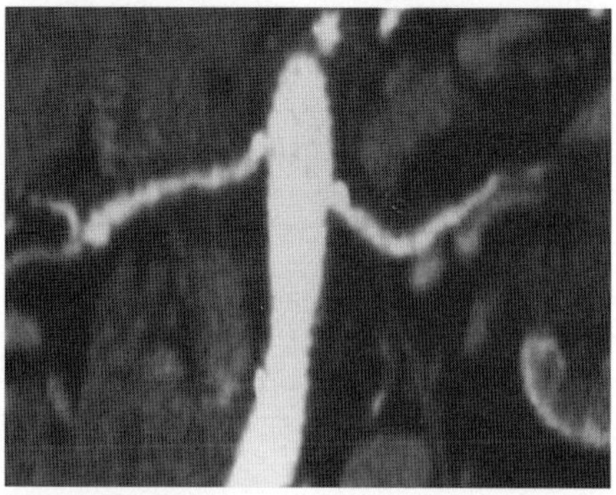

a.

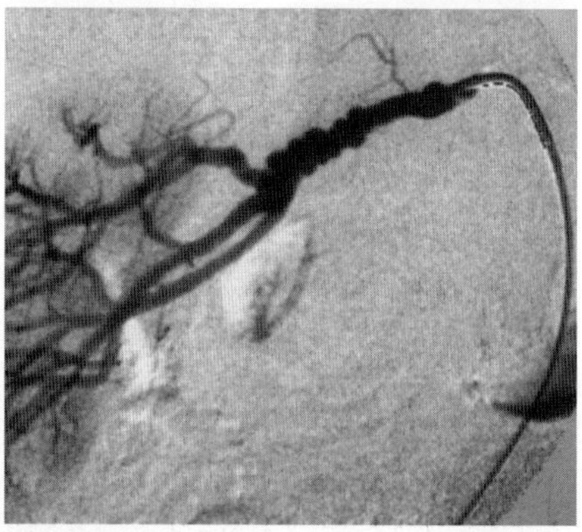

b.

Figure 12. 44-year-old woman with hypertension. a) VR image suggests medial fibroplasia bilaterally. b) Selective right renal arteriography confirms the presence of fibromuscular dysplasia.

For these reasons, CTA has been used as a screening evaluation for selected patients with suspected renovascular hypertension **(Figs. 11-13)**.

Technically, renal artery evaluations are usually performed with 3-mm collimation with a pitch of 1 followed by reconstruction at 1-mm intervals. Though early work suggested that MIP reconstructions were most reliable in the depiction of stenoses, more recent work by Johnson et al suggests that volume rendering may be more accurate.

Early reports by Galanski et al and Rubin et al indicate that CTA is a reliable method to detect renal artery stenosis. Galanski et al reported a 95% sensitivity when the axial images were reviewed on a serial paging display. Rubin et al reported a 92% sensitivity and 83% specificity in the detection of renal artery stenoses in excess of 70%. In a more recent series by Kim et al, 50 consecutive patients underwent both CTA and intraarterial DSA. In this study, for the detection of main renal artery stenosis greater than 50%, CTA had a sensitivity of 100% and a specificity of 97%.

In addition to the detection of renal artery stenosis, CTA has been used in the preoperative evaluation of living related renal donors **(Fig. 14)**. Rubin et al reported a series of 12 potential living related renal donors who were evaluated preoperatively with CTA. The study was 100% sensitive in the identification of accessory renal arteries and 93% sensitive for the identification of perihilar renal artery branches. These authors conclude that CTA could potentially serve as the single imaging test for patients undergoing renal donor evaluation. Lionel et al reported similar results in a more recent series of 16 consecutive patients and arrived at the same conclusion.

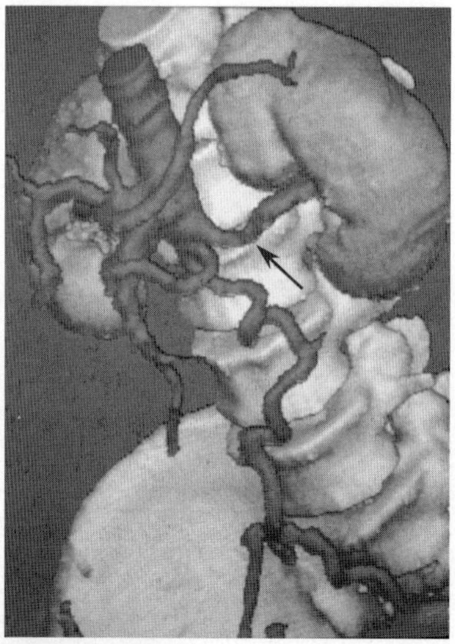

Figure 13. 67-year-old woman with absent femoral pulses and hypertension. CTA shown in SSD reveals a mild to moderate stenosis of the left renal artery (arrow). Assessment of renal artery stenosis with SSD may be less accurate than analysis of MIP images.

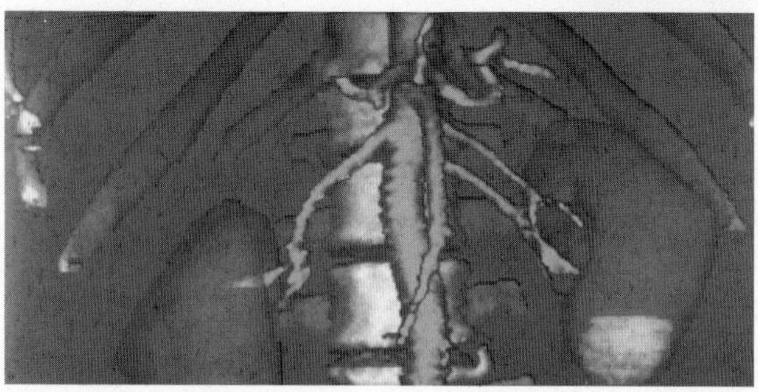

Figure 14. 34-year-old woman who is a potential living related renal donor. CTA SSD shows two left and a single right renal artery.

Aortoiliac Occlusive Disease

Since the advent of spiral CT, computed tomographic angiography was able to precisely define the extent of aortic occlusion in order to plan an aortofemoral bypass **(Figs. 15, 16)**. However, due to the distance between the proximal abdominal aorta and the distal aspect of the calf runoff vessels, complete evaluation of patients with occlusive peripheral vascular disease was difficult. The advent of multidetector scanning may make this evaluation feasible.

For reasons inherent in the system design and software, some multidetector scanners can operate at a pitch of 6 without unacceptable loss of image resolution. Using a pitch of 6 and a collimation of 3 mm or 5 mm, it is possible to visualize the lower extremity outflow in its entirety. Clinical evaluation of these new applications of CTA is currently underway.

At the time of this writing, several articles have documented the accuracy of CTA in the evaluation of peripheral vascular occlusive disease. In a pilot study of six patients, Lawrence et al compared CTA to catheter angiography and found that CTA correctly identified stenoses > 50% or occlusions in the femoral and popliteal vessels with a sensitivity of 92.9% and an accuracy of 95.5%. Subsequently Ricker et al evaluated the femoral, popliteal and tibial vessels in 50 patients. These investigators found sensitivities of 100%, 100% and 94% for detection of femoral, popliteal and tibial arterial occlusions respectively. For detection of stenoses > 75%, the sensitivities were 88% and 73% for the femoral and popliteal arteries.

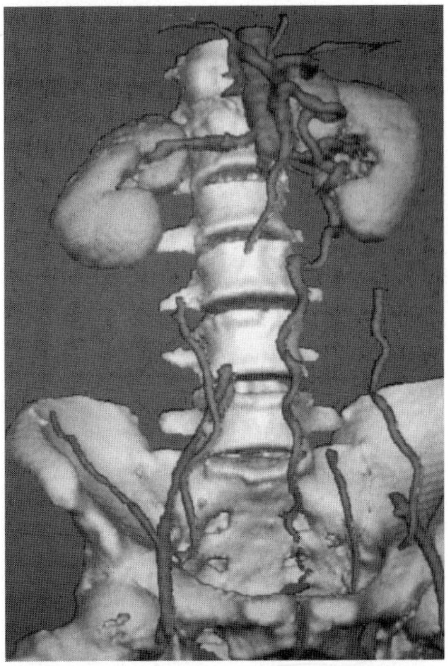

***Figure 15.** 64-year-old woman with bilateral claudication. CTA with SSD shows aortic occlusion extending to the common femoral artery on the left and to the common iliac bifurcation on the right. Surgical revascularization would be the best option if the patient was a suitable candidate.*

Thoracic Aortic Aneurysms

Both dissecting and nondissecting thoracic aortic aneurysms can be evaluated with CTA. However, there are occasionally problems encountered due to artifacts induced by simultaneously opacified structures including the brachiocephalic vein, superior vena cava, heart and pulmonary vasculature. Moreover, cardiac motion can be somewhat problematic. Despite these limitations, CTA remains an excellent diagnostic modality to evaluate thoracic aortic pathology. Even if the reconstructions are degraded by motion artifact, the axial images provide crucial anatomic information that is critical in the selection of the best therapeutic option. This is particularly true if an endovascular intervention therapy is being considered.

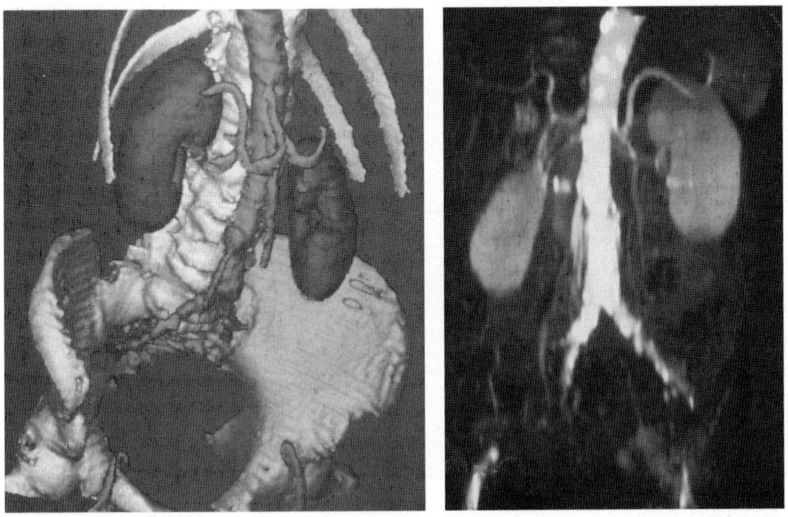

a. **b.**

***Figure 16.** 70-year-old man with bilateral claudication. a) SSD reconstruction of CTA shows bilateral external iliac artery occlusion. b) MIP also shows the occlusion but also reveals the extensive calcification. The MIP is the best modality to show calcification.*

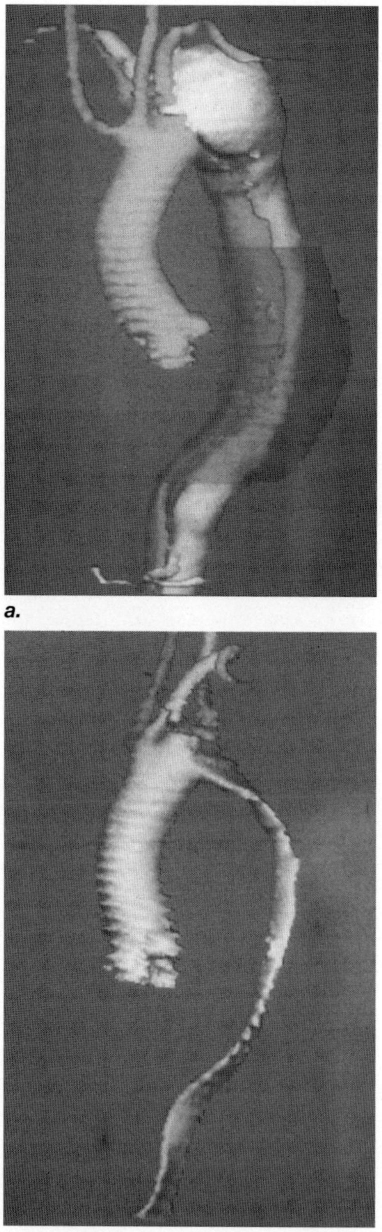

a.

b.

Figure 17. 44-year-old man with back pain. a) CTA with SSD shows a dissection of the thoracic aorta with the entry just beyond the left subclavian artery origin. The dissection extends into an aberrant right subclavian artery, which takes origin distal to the left subclavian artery. b) SSD of the true lumen shows that the remainder of the great vessel origins are not involved.

For thoracic aortic dissections, CTA is ideal to demonstrate the entry site as well as the relationship of the intimal flap to the great vessels. If reconstructions are performed by manual seeding, the true and false lumens can be depicted separately **(Fig. 17)**.

For atherosclerotic and mycotic aneurysms, CTA is very useful to reveal the relationship of the aneurysm to branch vessels and the diaphragm, which are critical features in planning an open surgical or endovascular repair.

Pulmonary Thromboembolism

Spiral CT has emerged as an important diagnostic test in the evaluation of patients suspected of having pulmonary thromboembolism. The evaluation is usually performed with 3-mm collimation at a pitch of 1.3 to 2 (depending upon the size of the patient) during intravenous contrast infusion. Some investigators recommend using diluted contrast media to diminish streak artifacts from the superior vena cava.

In most cases, the axial images provide the critical diagnostic information and three-dimensional reconstructions are usually not performed.

Pulmonary emboli appear as a filling defect **(Fig. 18)** within the opacified pulmonary arterial branch. The embolus may be surrounded by a thin layer of contrast giving rise to a "doughnut" (if seen axially) or "railroad-track" (if seen in plane) sign. If the embolus is completely occluding, absence of opacification may be the only finding. Care must be taken not to misinterpret lymph nodes or pulmonary veins as unopacified pulmonary arteries. Moreover, partial volume averaging is occasionally problematic.

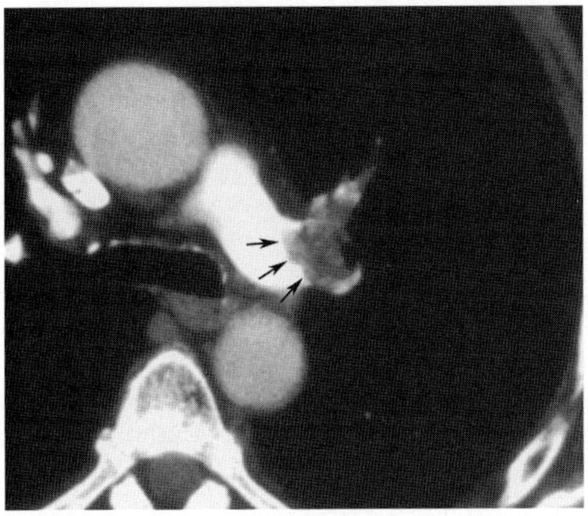

Figure 18. *A large pulmonary embolus (arrows) is present in the left main pulmonary artery. (Case courtesy of W. Richard Webb MD, UCSF.)*

Despite these limitations, spiral CT appears to have replaced both scintigraphy and pulmonary arteriography in many institutions in the evaluation of potential pulmonary emboli. A recent review by Remy-Jardin and Remy summarizes the current status of this technique. In a literature survey, these authors report a sensitivity of this technique ranging from 53% to 100% with specificities ranging from 78% to 100%. However, in the largest reported series by Herold et al reporting the experience of the European multicenter trial of 401 patients, the sensitivity was 88% and the specificity was 94%. Several papers have appeared documenting short- and intermediate-term follow-up in patients with negative spiral CT scans for pulmonary embolism who underwent no additional diagnostic testing and were not placed on anticoagulant therapy. One hundred such patients were recently reported by Lomis et al and no patient experienced morbidity or mortality attributable to pulmonary emboli over a 6-month follow-up period.

REFERENCES

Armon MP, Whitaker SC, Gregson RH, Wenham PW, Hopkinson BR. Spiral CT angiography versus aortography in the assessment of aortoiliac length in patients undergoing endovascular abdominal aortic aneurysm repair. J Endovasc Surg 1998; 5:222-227.

Brink JA. Technical aspects of helical (spiral) CT. Radiologic Clinics of North America 1995; 33(5):825-841.

Costello P, Gaa J. Spiral CT angiography of abdominal aortic aneurysms. Radiographics 1995; 15(2):397-406.

Galanski M, Prokop M, Chavan A, Schaefer CM, Jandeleit K, Nischelsky JE. Renal artery stenoses: spiral CT angiography. Radiology 1994; 193:555-560.

Herold C, Remy-Jardin M, Grenier PH, et al. Prospective evaluation of pulmonary embolism: initial results of the European Multicenter Trial (ESTIPEP) (abstr). Radiology 1998; 209:299.

Johnson PT, Halpern EJ, Kuszyk BS, et al. Renal artery stenosis: CT angiography– comparison of real-time volume rendering and maximum intensity projection algorithms. Radiology 1999; 211:337-343.

Kim TS, Chung JW, Park JH, Kim SH, Yeon KM, Han MC. Renal artery evaluation: comparison of spiral CT angiography to DSA. J Vasc Interv Radiol 1998; 9:553-559.

Lawrence JA, Kim D, Kent KC, Stehling MK, Rosen MP, Raptpoulos V. Lower extremity spiral CT angiography versus catheter angiography. Radiology 1995; 194:903-908.

Lionel G, Sebben RA, Costello P, Rao MM. The use of spiral computed tomographic angiography for the assessment of living kidney donors. Australian and New Zealand Journal of Surgery 1999; 69:217-219.

Lomis NN, Yoon HC, Moran AG, Miller FJ. Clinical outcomes of patients after a negative spiral CT pulmonary arteriogram in the evaluation of acute pulmonary embolism. J Vasc Interv Radiol 1999; 10:707-712.

Luboldt W, Weber R, Seeman M, Desantis M, Reiser M. Influence of helical CT parameters on spatial resolution in CT angiography performed with a subsecond scanner. Invest Radiol 1999; 34:421-426.

Marty B, Sanchez LA, Ohki T, et al. Endoleak after endovascular graft repair of experimental aortic aneurysms: does coil embolization with angiographic "seal" lower intraaneurysmal pressure? J Vasc Surg 1998; 27:454-461.

Remy-Jardin M; Remy J. Spiral CT angiography of the pulmonary circulation. Radiology 1999; 12:615-636.

Rieker O, Duber C, Schmiedt W, von Zitzewitz H, Schweden F, Thelen M. Prospective comparison of CT angiography of the legs with intraarterial digital subtraction angiography. AJR 1996; 166:269-276.

Rubin GD, Alfrey EJ, Dake MD, et al. Assessment of living renal donors with spiral CT. Radiology 1995; 195(2):457-462.

Rubin GD, Dake MD, Napel S, et al. Spiral CT of renal artery stenosis: comparison of three dimensional rendering techniques. Radiology 1994; 190:181-189.

Rubin GD, Paik DS, Johnston PC, Napel S. Measurement of the aorta and its branches with helical CT. Radiology 1998; 206:823-829.

Rubin GD, Zarins CK. MR and spiral/helical CT imaging of lower extremity occlusive disease. Surg Clin North Am 1995; 75(4):607-619.

CHAPTER 7 PREVIEW ANSWERS

Case 1

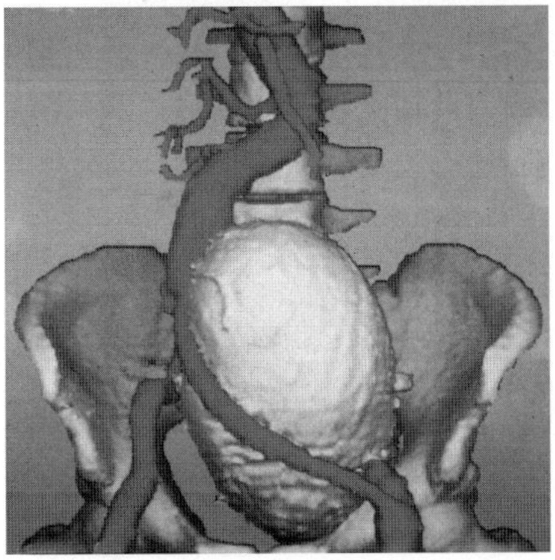

Diagnosis: Massive internal iliac artery aneurysm. This image has been created using the shaded surface display (SSD) algorithm.

Case 2

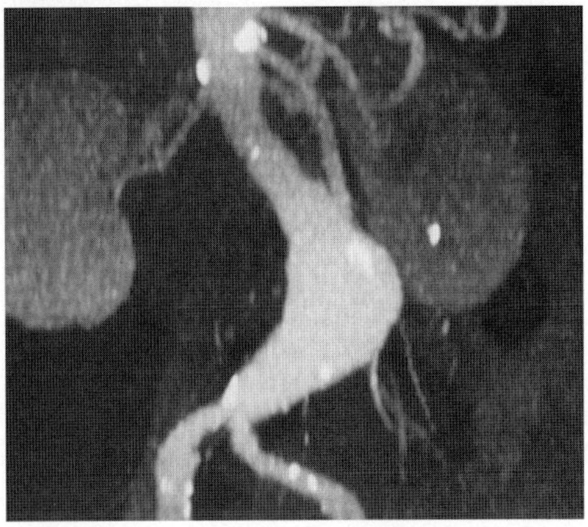

Diagnosis: Infrarenal abdominal aortic aneurysm. This image has been reconstructed using the maximum intensity projection (MIP) algorithm.

Case 3

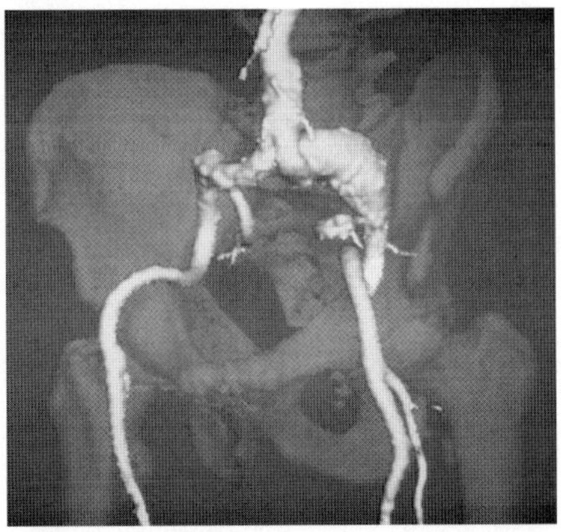

Diagnosis: Left common iliac artery aneurysm. CTA of the pelvis shown in the SSD algorithm reveals fusiform aneurysmal dilatation of the left common iliac artery. This may be a source of distal embolization to the left foot.

Case 4

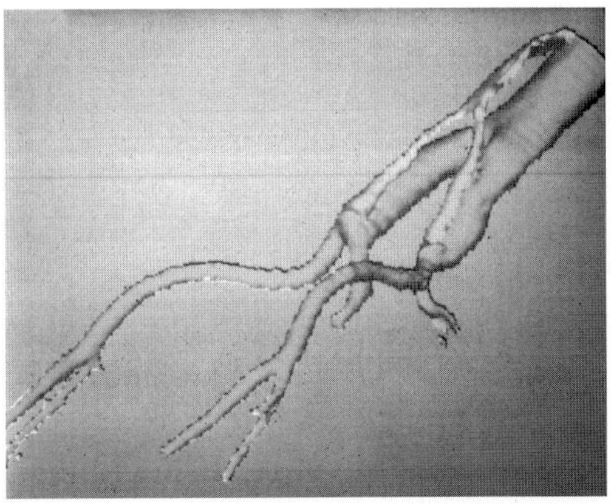

Diagnosis: Dissecting aortic aneurysm. A thin lumen is present ventral to the abdominal aorta with extensions into both common iliac arteries. The dissection appears to terminate bilaterally at the bifurcations of the common iliac arteries.

CHAPTER 8 PREVIEW

Case 1

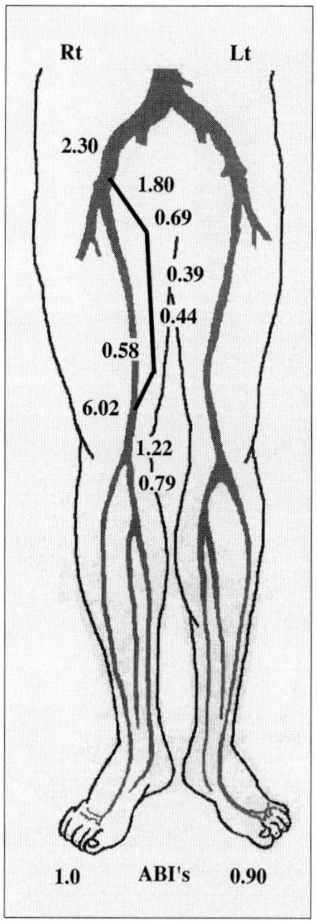

a.

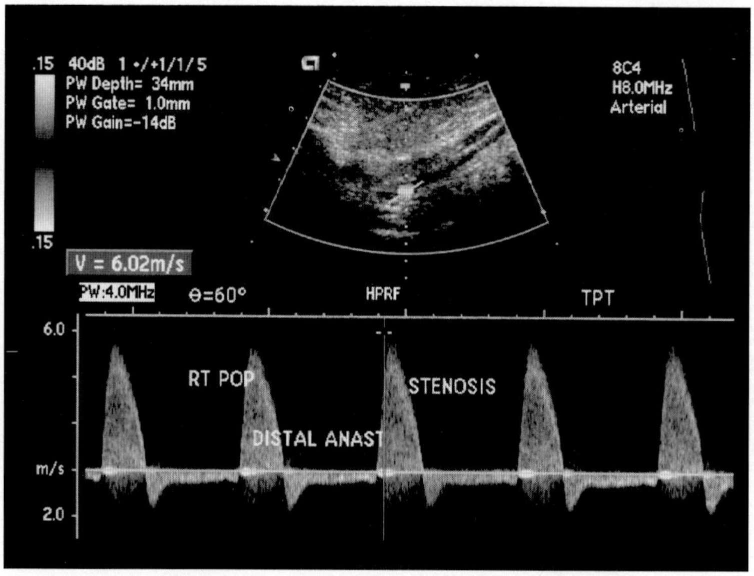

b.

History: Routine surveillance of an asymptomatic patient with a right femoral to popliteal bypass graft. a) Velocity map and b) duplex Doppler image.
Question: What is your diagnosis?

Case 2

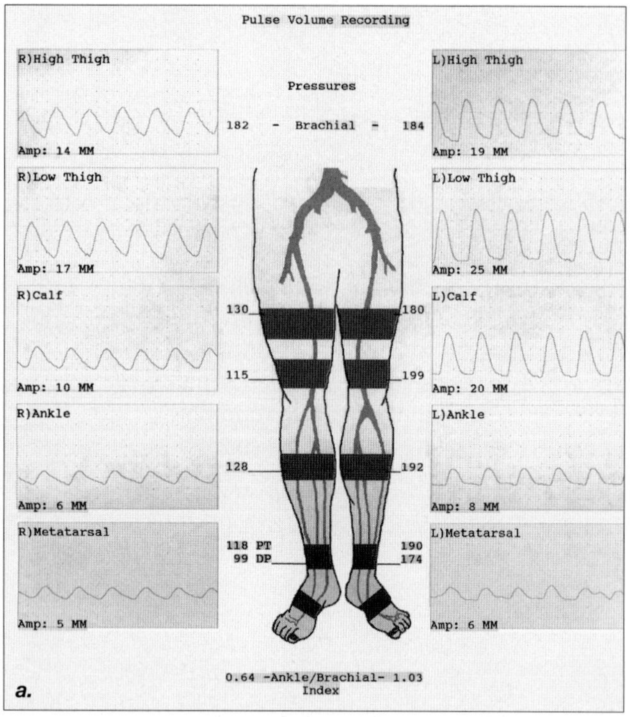

a.

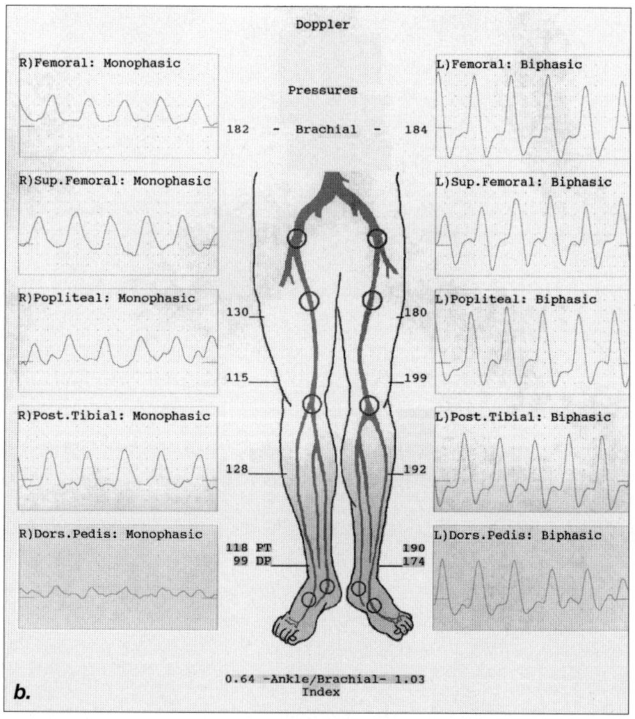

b.

History: Patient with bilateral thigh claudication. a) Plethysmography waveforms and b) Doppler waveforms.
Question: What is your diagnosis?

Chapter 8
Noninvasive Peripheral Vascular Examination

Mark W. Wilson, MD,
and Susan D. Wall, MD

CHAPTER OUTLINE

- Indications for noninvasive peripheral vascular examination
- Components of the noninvasive peripheral vascular examination
  - Segmental limb pressures
  - Ankle-brachial index
  - Treadmill stress test
  - Plethysmography
  - Doppler waveforms and duplex Doppler imaging
  - Transcutaneous oxygen tension measurements

INTRODUCTION

As minimally invasive treatment of peripheral vascular disease has gained wider acceptance, the noninvasive peripheral vascular examination has assumed a more important role in triaging patients to appropriate therapy. The strength of the noninvasive vascular evaluation lies in its ease of implementation, its ability to assess physiology, and its widespread use throughout the United States. In this chapter, we review the basic components of the noninvasive lab with an emphasis on both performance and interpretation of the basic exam.

VASCULAR LABORATORY

A well-equipped vascular laboratory should have a hemodynamic measurement station capable of segmental cuff pressure measurements, Doppler waveform generation, plethysmographic recording, an exercise treadmill, an ultrasound scanner with duplex capability, and a transcutaneous oxygen tension measurement device. Most vascular laboratories are directed by a physician with training in either vascular radiology or vascular surgery. The examinations are usually performed by a skilled ultrasound technologist with special competence in noninvasive vascular testing or by a dedicated vascular technologist. The organization which oversees vascular labs in the U.S. is the Intersocietal Commission for the Accreditation of Vascular Laboratories (ICAVL). This commission assures conformity to set standards of practice and mandates that regular quality assurance activities occur in each accredited laboratory. ICAVL accreditation must be renewed every 3 years.

INDICATIONS FOR NONINVASIVE PERIPHERAL VASCULAR EXAMINATION

The signs and symptoms of peripheral vascular occlusive disease may be as specific as an ulcerated, pulseless foot or as vague as an occasional cramping in the calf. The noninvasive examination is a useful starting point in the evaluation of both types of patients, as well as the entire spectrum in between. The objectives of the noninvasive exam include diagnosing significant peripheral vascular occlusive disease (PVOD), localizing the site(s) of disease, and assessing disease severity. In a patient with obvious tissue loss resulting from PVOD, the data obtained from the noninvasive exam can be used to predict the chance of healing, with or without intervention.

OBJECTIVES OF NONINVASIVE EXAMINATION

- Diagnosing PVOD
- Localizing the site of the responsible lesion(s)
- Gauging disease severity and potential for healing

COMPONENTS OF THE NONINVASIVE EXAMINATION

- Segmental limb pressures
- Ankle-brachial index
- Treadmill stress testing
- Plethysmography
- Doppler waveform analysis and duplex Doppler imaging
- TCO_2 determination

COMPONENTS OF THE NONINVASIVE EXAMINATION

Aside from a thorough history and physical examination, perhaps the simplest test for evaluating a patient with PVOD symptoms is the ankle-brachial index (ABI). A complete nonivasive examination, however, goes well beyond solely obtaining ABI's. In fact, there are six components to the complete examination, each one imparting different information, and each one having advantages and limitations. The components of the complete noninvasive examination are segmental limb pressures, ABI's, treadmill stress testing, plethysmography, Doppler waveform analysis and duplex Doppler imaging, and transcutaneous oxygen tension (TCO_2) measurement. Some vascular laboratories perform all six parts routinely on all patients, whereas others will perform a select few, based on the clinical setting.

SEGMENTAL LIMB PRESSURES

Segmental limb pressure measurements aid in establishing the presence of vascular disease and in localizing disease to a given side or vascular segment within the leg. The segments interrogated are the upper thigh, lower thigh, calf, and ankle. Four pneumatic cuffs are placed on the leg at the four segments with the patient in the supine position. The cuffs should be inflated sequentially above systolic pressure (SBP) at the respective segment. The systolic pressure at each segment is then measured by deflating the cuffs and noting the pressures at which the return of flow was detected by the continuous wave (CW) Doppler probe distal to the cuff. A brachial systolic pressure measurement is also obtained. These pressure measurements are stored in a patient database and can be represented graphically on a patient data sheet **(Fig. 1)**.

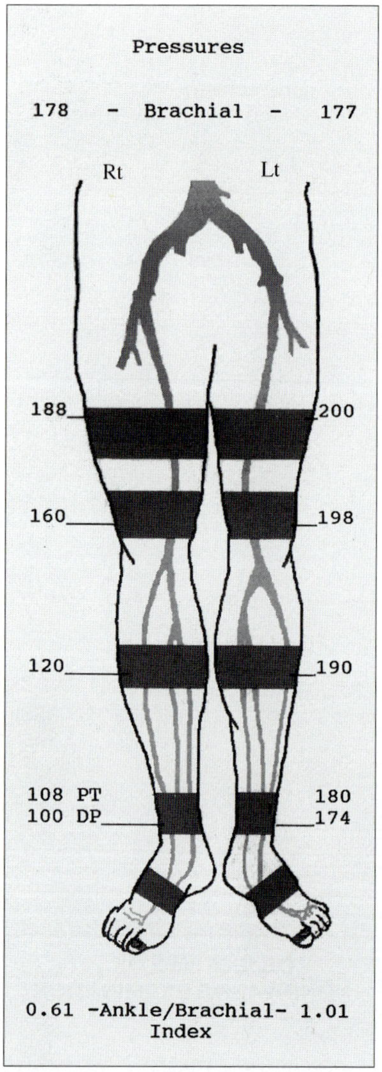

Figure 1. *Patient data sheet demonstrating the positions of the pneumatic cuffs and the resultant segmental pressure obtained at each cuff level.*

The segmental limb pressure exam has the advantage of being readily available and easy to perform, even by operators without extensive experience in vascular testing. The interpretation of the data is usually straightforward; however, certain pitfalls may complicate the interpretation of the results. For example, significant medial arterial calcification may cause vascular noncompliance, resulting in a falsely elevated segmental systolic pressure measurement. This is an important pitfall in patients with chronic renal failure and diabetes mellitus. In addition, the segmental limb pressure exam may be insensitive to mild stenoses. This exam may also inaccurately localize disease when there are multiple abnormalities in parallel or in series in a single extremity. Finally, the segmental pressure examination allows only intrapatient comparisons, with no way to compare the data to an established standard.

Despite these pitfalls, there are some useful rules of thumb for predicting the possibility of wound healing based on segmental pressures. Healing of foot or ankle ulcers requires an ankle pressure of > 50 mm Hg in nondiabetic patients and > 80 mm Hg in diabetic patients. Healing of a below-knee amputation requires a calf pressure of 50-70 mm Hg.

NORMAL LIMB PRESSURE GRADIENTS

- Ipsilateral (vertical) pressure gradients:
  - < 20 mm Hg between adjacent segments
  - < 30 mm Hg from thigh to ankle
- Right-left (horizontal) pressure gradients:
  - < 20 mm Hg

ANKLE-BRACHIAL INDEX

The ABI utilizes segmental and brachial pressure data in a way that normalizes the results within and between patients. To compute the ABI, CW Doppler is used to measure the ankle systolic pressure, after a pneumatic cuff is inflated and released, by detecting the return of flow in either the posterior tibial or dorsalis pedis arteries. A brachial pressure is obtained and a simple ratio of the two is calculated, with the brachial pressure in the denominator. This index can also be obtained for other sites as well, e.g., thigh-brachial index, toe-brachial index, finger-brachial index. A normal ankle brachial index is ≥ 1.0. Normal thigh pressures are higher than brachial pressures, yielding a thigh-brachial index > 1.0. Toe-brachial indices (TBI) normally range from 0.7 to 0.9, while finger-brachial indices normally range from 0.95 to 1.0 **(Table 1)**.

PITFALLS OF SEGMENTAL PRESSURES

- Only intrapatient comparison is possible
- Inaccuracies may occur with mild and multifocal PVOD
- Significant vascular calcification may falsely elevate pressures

TABLE 1: CORRELATION OF ANKLE-BRACHIAL AND TOE-BRACHIAL INDICES WITH CLINICAL SIGNS AND SYMPTOMS

Symptom/Sign	Ankle Pressure (Systolic - mm Hg)	ABI	TBI
Intermittent claudication	70-100	0.5-0.8	0.35 ± 0.15
Rest pain			
Nondiabetics	35-55	0.4-0.5	0.11 ± 0.10
Diabetics	55-80		
Ulcer/Gangrene	< 30	< 0.3	

NORMAL PRESSURE INDICES

- Ankle-brachial index ≥ 1.0
- Thigh-brachial index > 1.0
- Toe-brachial index: 0.7 – 0.9
- Finger-brachial index: 0.95 – 1.0

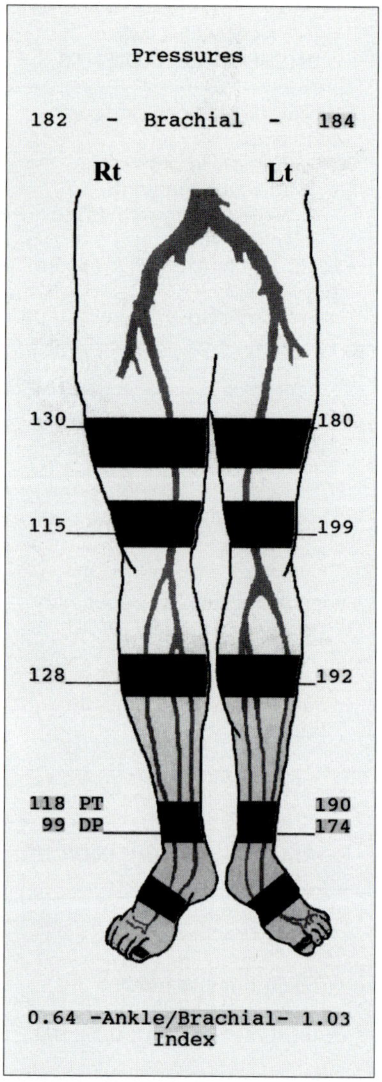

Figure 2. *Noninvasive data sheet of a patient with right leg claudication. Note the abnormal resting ABI of 0.64 on the right, compared to the normal ABI of 1.03 on the left.*

An ankle-brachial index of < 0.5 indicates vascular disease in more than one major artery. An example of an ABI measurement in a patient with claudication is shown in **Figure 2**.

ABI measurement is plagued by the same pitfalls as segmental pressure measurements, because the same physiologic data are used for both. ABI's may be falsely elevated in patients with extensive vascular calcification, such as in diabetics and patients with chronic renal failure. Furthermore, ABI's may be normal or near normal in patients with impending bypass graft occlusion.

Although simple pressure measurements are valuable, they are rarely the only component of the vascular exam. The other five elements of the exam either help to increase the sensitivity of pressure measurements and indices, or use completely different physiologic parameters to noninvasively evaluate the patient.

Treadmill Stress Test

Poiseuille's equation defines the relationship between blood flow, resistance, and pressure gradient through a cylindrical structure or blood vessel as follows:

$$Q = \frac{\pi P r^4}{8 \nu L},$$

where Q = blood flow rate, P = pressure gradient across the vessel, r = inner radius of vessel, ν = fluid viscosity, and L = length of vessel segment being evaluated. The resistance across a vessel is given by the equation:

$$R = \frac{L}{r^4}.$$

This relationship expresses how the resistance to blood flow would increase across a fixed stenosis. Poiseuille's equation can therefore be rewritten as:

$$Q = \frac{\pi P}{8 \nu R}.$$

Assuming a constant blood viscosity and a constant resistance to blood flow across a fixed stenosis, the following relationship can be derived.

Blood flow across a stenosis α Pressure gradient across a stenosis.

As blood flow to a muscle bed increases with exercise, the pressure drop across a stenosis proximal to the muscle bed increases. Therefore, exercise effectively increases the sensitivity of a segmental pressure measurement. This is the principle behind treadmill stress testing (TST).

FLOW-PRESSURE RELATIONSHIP

- The blood flow across a stenosis is proportional to the drop in pressure across the stenosis

The technique of TST begins with obtaining a baseline resting ABI followed by having the patient walk on a treadmill with a 12° incline at a speed of approximately 2 miles per hour. The patient is urged to walk as long as possible, up to 5 minutes, or until ischemic symptoms prohibit further exercise. The examination is terminated immediately for symptoms of angina, syncope, respiratory distress, or significant weakness. At the end of the exercise period, the ABI is obtained immediately and at 1-minute intervals thereafter, up to 5 minutes. From these measurements, a curve is plotted to demonstrate the dynamic response of the ABI to the stress test.

A normal stress test should result in an initial, postexercise ABI greater than 80% of the resting value. By 1 minute postexercise, the ABI should be at least 90% of the resting value. An abnormal stress test is defined as a drop in ankle pressure greater than 20 mm Hg along with a delayed return to baseline **(Fig. 3)**.

While treadmill stress testing increases the sensitivity for PVOD detection in the presence of mild disease or normal resting pressures and ABI's, it is limited by a patient's ability to ambulate, their exercise tolerance, and any premorbid conditions that would make the stress test prohibitively difficult or dangerous.

TREADMILL STRESS TESTING

Benefits
- Increases sensitivity of pressure measurements for detecting mild PVOD
- May suggest PVOD in the presence of normal resting pressures

Limitations
- Inability of patient to ambulate
- Poor patient exercise tolerance
- Premorbid conditions which prohibit exercise

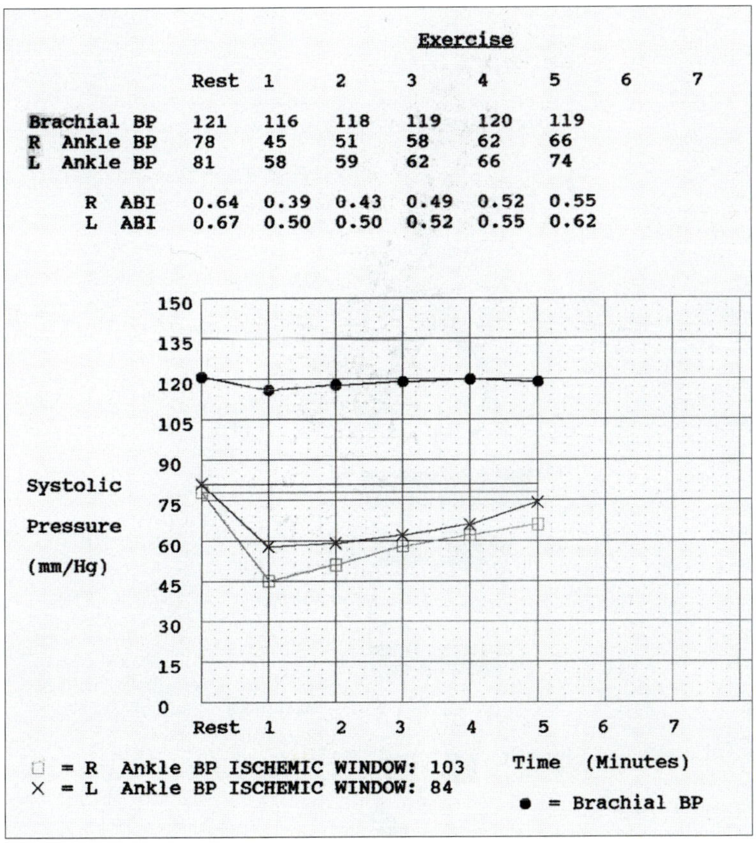

Figure 3. *Treadmill stress test demonstrating a further decline in ABI's bilaterally with exercise. There is also a delayed return to baseline.*

PLETHYSMOGRAPHY

Plethysmography, also known as pulse-volume recording (PVR), is one of two waveform analysis techniques incorporated into the noninvasive vascular exam. PVR measures the sum total of both arterial and venous blood volume in the leg at different levels. Partially inflated pneumatic cuffs are placed at the upper thigh, lower thigh, calf, ankle, and metatarsal levels, which detect the dynamic changes in blood volume. The data from the cuffs is then transduced into a waveform which reflects the arterial pressure pulse.

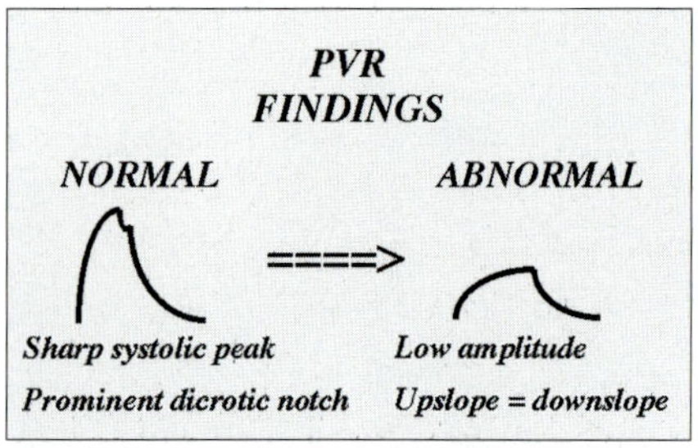

Figure 4. *Schematic representation of the effects of progressive peripheral vascular occlusive disease on the plethysmographic waveform.*

A normal PVR waveform will possess a sharp systolic peak and a prominent dicrotic notch. Progression of vascular disease is denoted by a progressive smoothing of the pulse contour and a decline in amplitude **(Fig. 4)**. Another key finding in an abnormal PVR study is the loss of the normal augmentation response that occurs at the calf. This indicates the presence of femoropopliteal disease **(Fig. 5)**. As a rule of thumb, a pulsatile PVR waveform in the calf and metatarsals is required for healing of a below-knee amputation and transmetatarsal amputation respectively.

Plethysmography in combination with the pressure-dependent components of the noninvasive exam produce a sensitivity for vascular disease of over 90%. Plethysmography is not limited by the presence of vascular calcification. However, despite these advantages, PVR has the pitfalls of decreased sensitivity for lesions distal to significant proximal disease as well as an inability to reliably distinguish short, high-grade stenoses from long occlusions. This is an important limitation, as it decreases the ability to discern if a lesion is amenable to percutaneous transluminal angioplasty versus bypass surgery. Because of such limitations, angiography is frequently needed to precisely define the disease morphology.

PLETHYSMOGRAPHY

- Progression of vascular disease is indicated by a progressive smoothing of the pulse contour and a decline in amplitude. Loss of the normal augmentation response that occurs at the calf indicates the presence of femoropopliteal disease.

PLETHYSMOGRAPHY

Benefits
- Increases sensitivity of the noninvasive detection of PVOD
- Not limited by the presence of vascular calcification

Limitations
- Decreased sensitivity for lesions distal to other sites of significant disease
- Inability to reliably distinguish short, high-grade stenoses from long occlusions

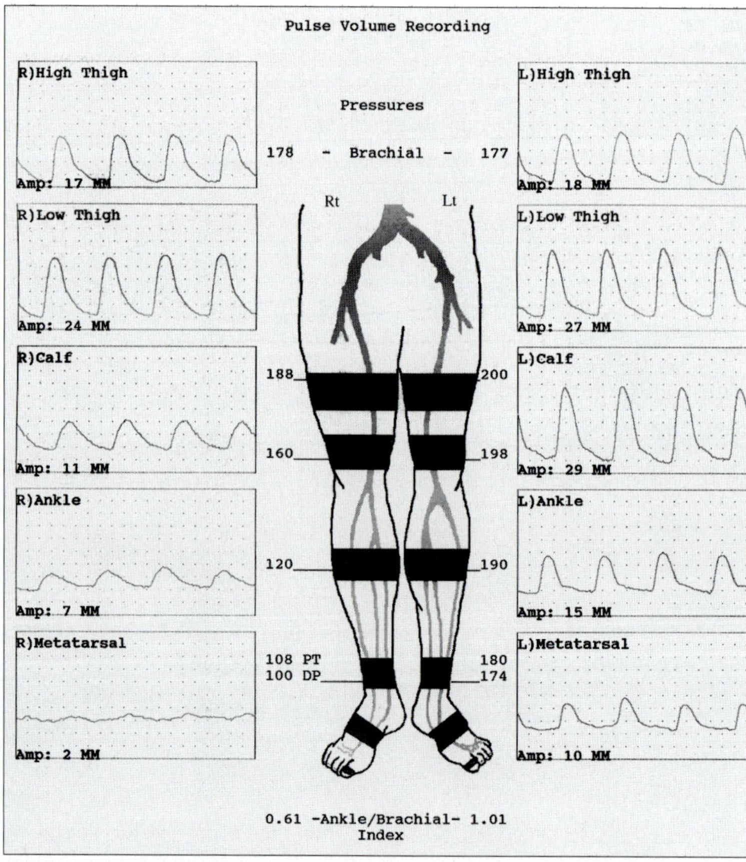

Figure 5. *The decreased amplitude in the PVR waveform in the right calf reflects the presence of right-sided femoropopliteal disease.*

DOPPLER WAVEFORMS AND DUPLEX DOPPLER IMAGING

Doppler waveform analysis and duplex Doppler imaging help to overcome some of the limitations of plethysmography. A Doppler waveform can be obtained at segments throughout the entire leg by the simple application of the CW Doppler probe, which is a standard feature of most hemodynamic measurement stations. The waveforms are then displayed on the patient data sheet **(Fig. 6)**.

The normal Doppler waveform in a native vessel is triphasic. It consists of an early forward systolic component (above the baseline), an end-systolic reversal of flow (below the baseline), and slower forward flow later in diastole (above the baseline). As the severity of PVOD increases the waveform distal to the lesion(s) dampens to biphasic and, ultimately, monophasic patterns. It should be noted, however, that the normal Doppler waveform in patent bypass grafts is biphasic **(Fig. 7)**. Two additional Doppler waveform patterns are spectral broadening, which is caused by turbulent flow due to a stenosis, and a nonpulsatile waveform, which reflects significant dimunition of blood flow distal to a high grade stenosis or occlusion **(Fig. 8)**.

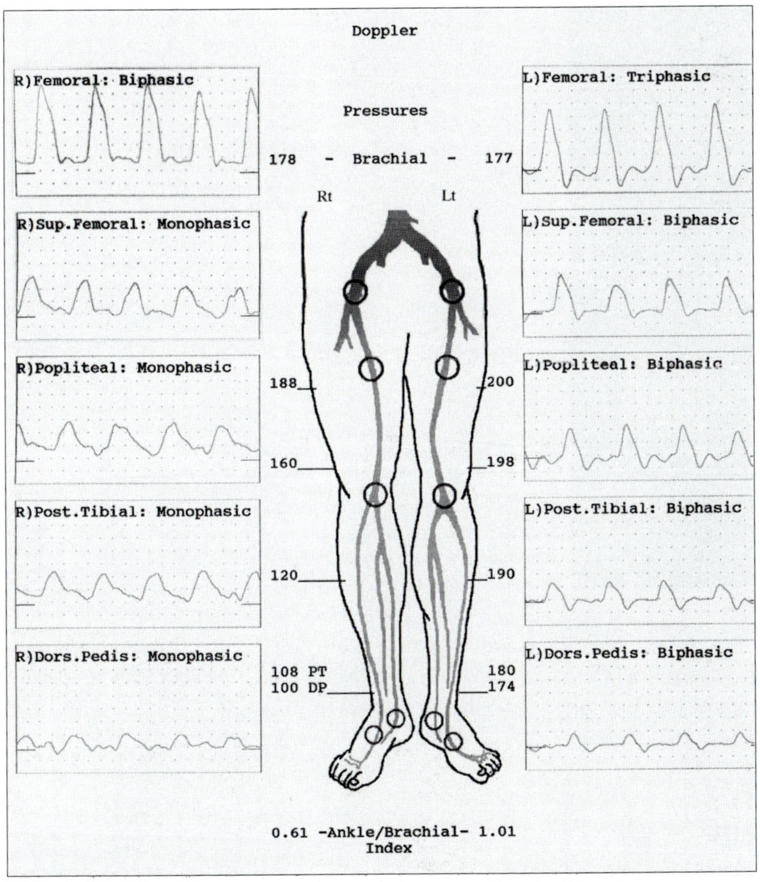

Figure 6. Noninvasive patient data sheet demonstrating the Doppler waveforms throughout both legs.

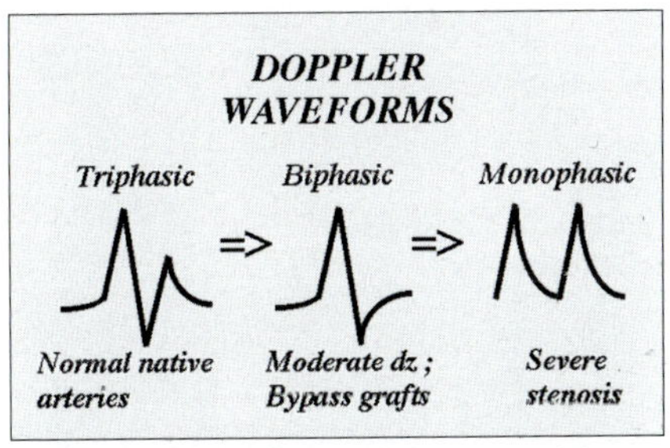

Figure 7. Schematic representation of the effects of progressive peripheral vascular occlusive disease on the Doppler waveform.

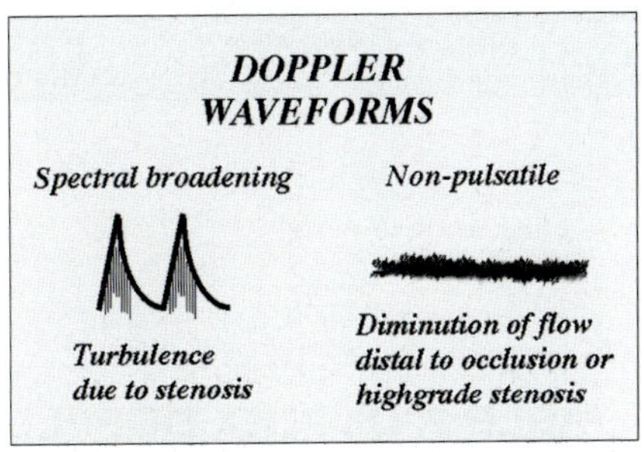

Figure 8. Schematic representation of the effects of arterial stenosis and complete occlusion on the Doppler waveform.

In addition to waveform analysis, a dedicated ultrasound unit should possess the capacity for duplex Doppler evaluation. Doppler evaluation offers the additional advantages of vascular imaging and velocity measurement **(Fig. 9)**.

The dependence of the velocity of blood flow through a vessel upon cross-sectional area is given by the following relationship:

$$Q = A * V \text{, which can be rewritten as } V = \frac{Q}{A} \text{, or } V \,\alpha\, 1/d^2 \text{,}$$

where Q = the flow rate of blood, A = cross-sectional area of the vessel, d = vessel diameter, and V = the peak systolic velocity of blood flow through the vessel. As cross-sectional area varies with the vessel diameter squared, the velocity of flow across a stenosis increases due to the decreased vessel diameter at that point. Conversely, measurement of the change in velocity allows deduction of the change in vessel diameter at the site of a stenosis.

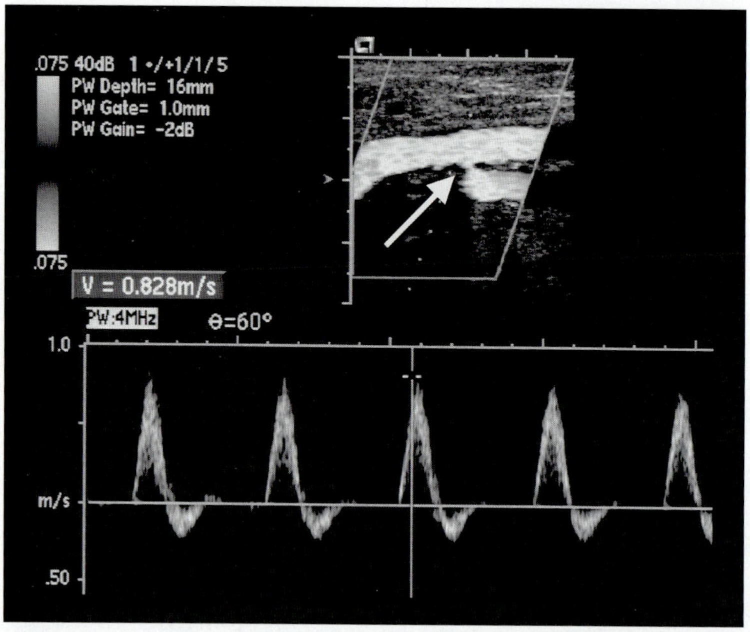

Figure 9. *Duplex Doppler image demonstrating normal multiphasic flow in the common femoral artery as well as a stenosis in the proximal profunda femoris artery (arrow).*

The ratio of the peak systolic flow velocity at a stenosis (Vs) to the velocity just proximal to the stenosis (Vp) is known as the velocity shift (Vs/Vp). The relationship of velocity shift to percent stenosis is charted in **Table 2**.

TABLE 2: DUPLEX DOPPLER VELOCITY SHIFT VERSUS DEGREE OF STENOSIS

Velocity shift	% Stenosis
1.5	< 20%
1.5 – 2.5	20 – 50%
2.5 – 3.5	50 – 75%
> 3.5	> 75%

The ability to ascertain flow velocities is a great asset in the surveillance of lower extremity vascular bypass grafts, as compromised graft flow may be asymptomatic and still result in a normal ABI until the point of actual graft failure by thrombosis. Therefore, duplex Doppler bypass graft surveillance plays a central role in the postoperative follow-up of the vascular patient. In the typical graft surveillance study, segmental limb pressures and plethysmography are omitted to avoid subjecting the graft to compressive forces. It has been shown that a peak systolic velocity of 0.45 meters per second or less in a bypass graft is a harbinger of graft failure. Often a focal stenosis is responsible for the decreased flow and, indeed, this may be manifested as a focal velocity shift on the duplex Doppler study. A positive graft surveillance study often initiates arteriography to diagnose and, potentially, treat the offending lesion(s) percutaneously or surgically.

DUPLEX DOPPLER GRAFT SURVEILLANCE

- A peak systolic velocity of 0.45 meters/second or less in a graft is a sign of impending graft failure, despite lack of symptoms and normal ankle-brachial index.

Pitfalls of duplex Doppler imaging include the inability to penetrate through heavily calcified vessels and extensive adipose tissue. This modality is also heavily operator-dependent.

PITFALLS OF DUPLEX DOPPLER IMAGING

- Inability to penetrate through heavily calcified vessels and extensive adipose tissue
- Heavily operator-dependent

TRANSCUTANEOUS OXYGEN TENSION

Transcutaneous oxygen tension (TCO_2) is a useful parameter for predicting the probability of wound healing in the PVOD patient. While this test may not be standard equipment in all noninvasive laboratories, it is complementary to the tests already covered, and it can help sway the decision to intervene on a patient or not.

The TCO_2 probe is used to measure the oxygen tension at the thigh, the calf, and the foot. A normal TCO_2 is 60 mm Hg. In order to heal a below-knee amputation or transmetatarsal amputation, the TCO_2 must measure at least 40 mm Hg in both cases. Healing of a foot ulcer requires a TCO_2 of at least 25 mm Hg.

TCO_2 CRITERIA FOR WOUND HEALING

- Below-knee amputation: $TCO_2 > 40$ mm Hg
- Transmetatarsal amputation: $TCO_2 > 40$ mm Hg
- Foot ulcer: TCO_2 at least 25 mm Hg

REFERENCES

Bird CE, Criqui MH, Fronek A, Denenberg JO, Klauber MR, Langer RD. Quantitative and qualitative progression of peripheral arterial disease by noninvasive testing. Vasc Med 1999; 4(1):15-21.

Criqui MH, Denenberg JO, Bird CE, Fronek A, Klauber MR, Langer RD. The correlation between symptoms and noninvasive test results in patients referred for peripheral arterial disease testing. Vasc Med 1996; 1(1):65-71.

Criqui MH, Denenberg JO, Langer RD, Fronek A. The epidemiology of peripheral arterial disease: importance of identifying the population at risk. Vasc Med 1997; 2(3):221-226.

Duprez D, Missault L, Van Wassenhove A, Clement DL. Comparison between ankle and toe index in patients with peripheral arterial disease. Int Angiol 1987; 6(3):295-297.

Foldes MS. Postoperative lower extremity bypass surveillance: beyond ankle arm blood pressures. J Vasc Nurs 1995; 13(3):75-78.

Hanna GP, Fujise K, Kjellgren O, et al. Infrapopliteal transcatheter interventions for limb salvage in diabetic patients: importance of aggressive interventional approach and role of transcutaneous oximetry. J Am Coll Cardiol 1997; 30(3):664-669.

Kalani M, Brismar K, Fagrell B, Ostergren J, Jorneskog G. Transcutaneous oxygen tension and toe blood pressure as predictors for outcome of diabetic foot ulcers. Diabetes Care 1999; 22(1):147-151.

Kazmers A, Koski ME, Groehn H, et al. Assessment of noninvasive lower extremity arterial testing versus pulse exam. Am Surg 1996; 62(4):315-319.

Piecuch T, Jaworski R. Resting ankle-arm pressure index in vascular diseases of the lower extremities. Angiology 1989; 40(3):181-185.

Pilcher DB, Ricci MA. Vascular ultrasound. Surg Clin North Am 1998; 78(2):273-293.

Ramsey DE, Manke DA, Sumner DS. Toe blood pressure. A valuable adjunct to ankle pressure measurement for assessing peripheral arterial disease. J Cardiovasc Surg (Torino) 1983; 24(1):43-48.

Sykes MT, Godsey JB. Vascular evaluation of the problem diabetic foot. Clin Podiatr Med Surg 1998; 15(1):49-83.

Ubbink DT, Spincemaille GH, Reneman RS, Jacobs MJ. Prediction of imminent amputation in patients with nonreconstructible leg ischemia by means of microcirculatory investigations. J Vasc Surg 1999; 30(1):114-121.

Vogt MT, McKenna M, Wolfson SK, Kuller LH. The relationship between ankle brachial index, other atherosclerotic disease, diabetes, smoking and mortality in older men and women. Atherosclerosis 1993; 101(2):191-202.

Zierler RE. Vascular surgery without arteriography: use of Duplex ultrasound. Cardiovasc Surg 1999; 7(1):74-82.

Zierler RE, Zierler BK. Duplex sonography of lower extremity arteries. Semin Ultrasound CT MR 1997; 18(1):39-56.

CHAPTER 8 PREVIEW ANSWERS

Case 1

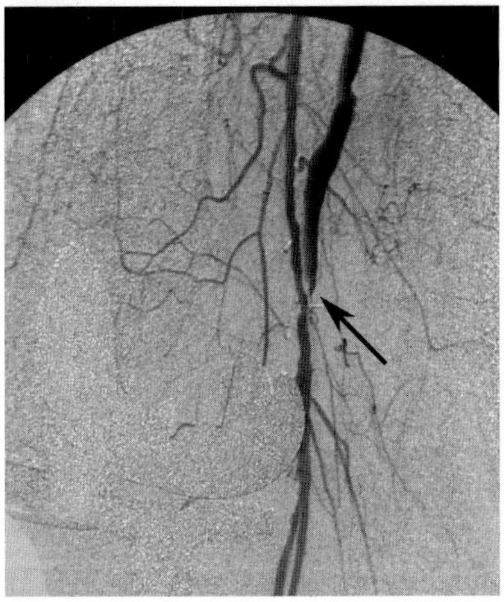

a.

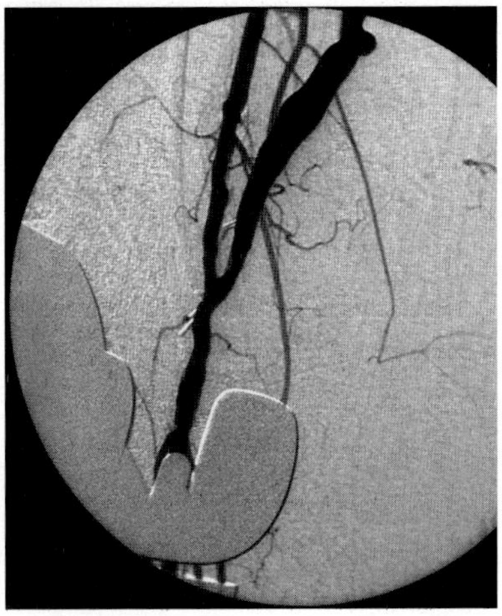

b.

Findings: The velocity map (*Preview Case 1a*) of this bypass graft survey demonstrates a focal velocity shift to a rate of 6.02 meters/sec, also shown on the duplex Doppler scan (*Preview Case 1b*). This velocity shift at the site of the distal anastomosis indicates a significant stenosis. a) Angiography confirms this stenosis (arrow). b) The stenosis responded to angioplasty.
Diagnosis: Distal anastomotic graft stenosis, amenable to percutaneous transluminal angioplasty.

Case 2

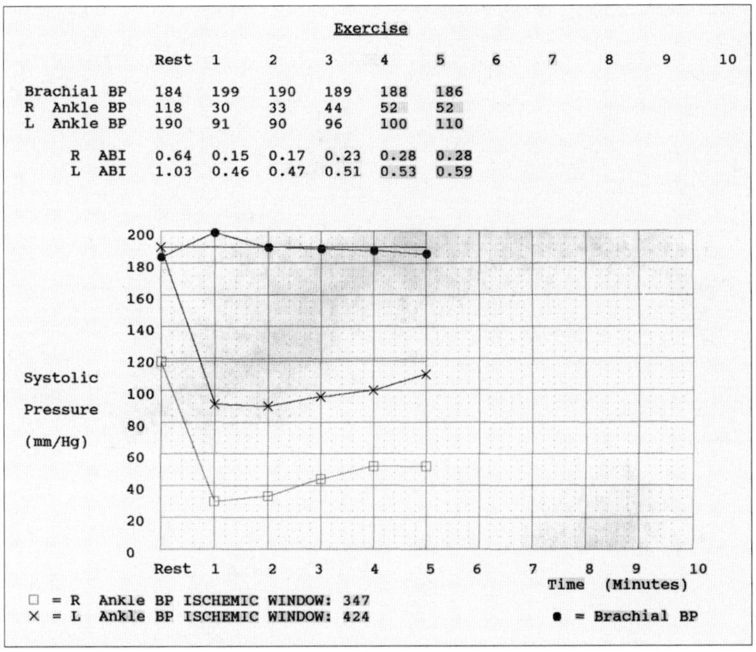

	Rest	1	2	3	4	5	6	7	8	9	10
Brachial BP	184	199	190	189	188	186					
R Ankle BP	118	30	33	44	52	52					
L Ankle BP	190	91	90	96	100	110					
R ABI	0.64	0.15	0.17	0.23	0.28	0.28					
L ABI	1.03	0.46	0.47	0.51	0.53	0.59					

a.

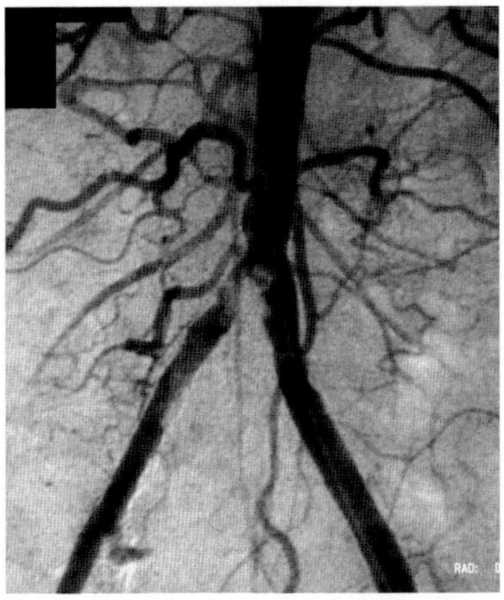

b.

Findings: Plethysmography demonstrates decreased amplitudes throughout the right leg (*Preview Case 2a*). The Doppler waveform data sheet demonstrates monophasic flow throughout the right leg and biphasic flow throughout the left leg (*Preview Case 2b*). The ABI on the left is normal at rest, while the ABI on the right is significantly impaired. No vertical segmental pressure gradients exist in either leg; however, a significant horizontal gradient is present with right-sided impairment. a) Given that the patient has bilateral symptoms, a treadmill stress test is performed to evoke hemodynamic changes in the left leg. The decrease in ABI's with exercise and impaired return to baseline indeed suggests the presence of bilateral disease, the right side more severe than the left. Given the lack of any significant vertical segmental pressure gradients in either leg, the lesions likely involve the iliac vessels. b) Angiography confirmed bilateral iliac artery stenoses.
Diagnosis: Bilateral common iliac artery stenoses, with the right side more severe than the left.

VASCULAR INTERVENTIONS

CHAPTER 9 PREVIEW

Case 1

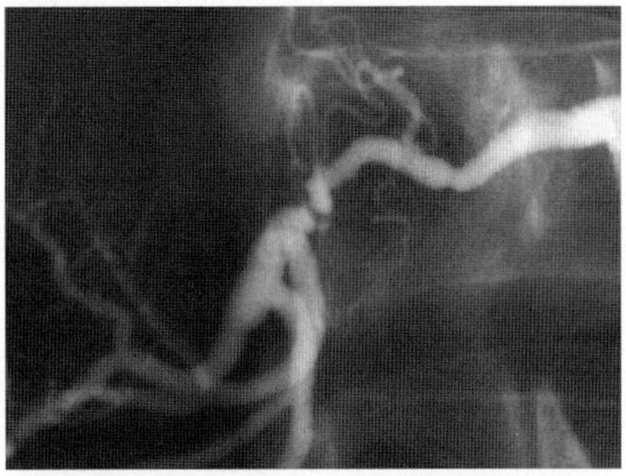

History: 42-year-old woman with hypertension requiring two medications for adequate control.
Question: Is this renal artery pathology suitable for balloon angioplasty?

Case 2

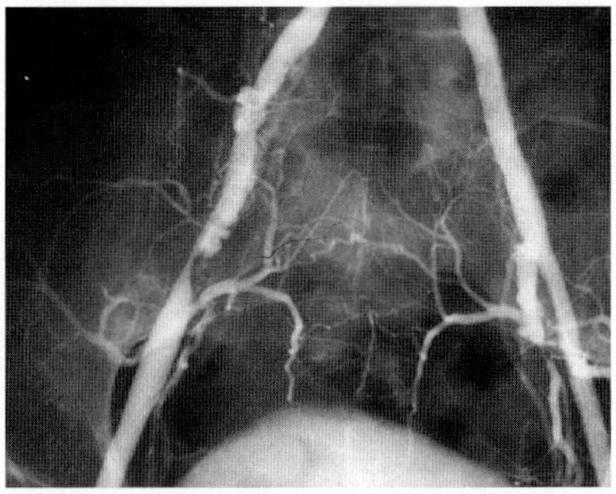

History: 56-year-old man with two-block claudication.
Question: Should this right external iliac artery stenosis be treated with angioplasty?

Case 3

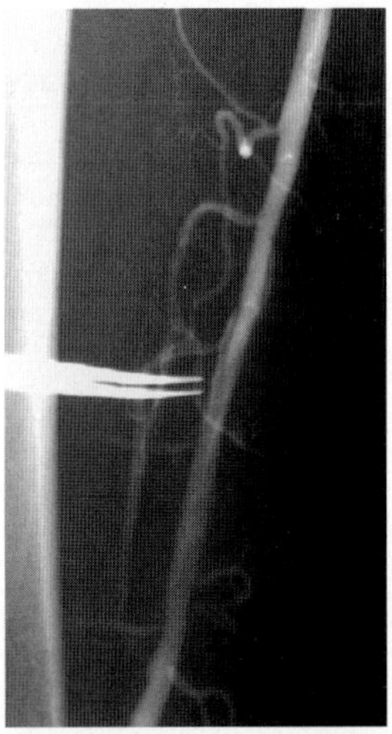

History: A 60-year-old man underwent balloon angioplasty of the distal right superficial femoral artery.
Question: Is this an acceptable postangioplasty arteriogram or should a stent be placed?

Case 4

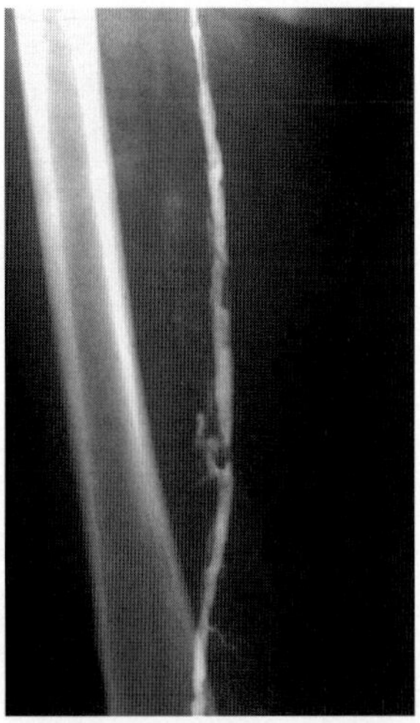

History: A 64-year-old man underwent balloon angioplasty of the distal right superficial femoral artery.
Question: Does this postangioplasty arteriogram show anything that might warrant further therapy?

Chapter 9
Angioplasty

Robert K. Kerlan, Jr., MD

CHAPTER OUTLINE

- Pathophysiology of balloon angioplasty
- Indications
  - Lesion classification
  - Brachiocephalic angioplasty
  - Renal arteries
  - Visceral arteries
  - Infrarenal abdominal aorta
  - Iliac arteries
  - Femoropopliteal arteries
  - Infrapopliteal arteries
  - Bypass grafts
- Technique
  - Preprocedural evaluation
  - Procedural care
  - Hemodynamic assessment
  - Pharmacologic agents
  - Balloon selection
  - Crossing the lesion
  - Balloon inflation and deflation
  - Postangioplasty assessment
  - Postprocedural care
- Results
- Complications

INTRODUCTION

Since its original description by Dotter and Judkins more than 30 years ago, percutaneous transluminal angioplasty **(Fig. 1)** has become a very common procedure and the cornerstone of vascular interventions. In the 1960's, large, tapered dilators were used to enlarge the vessel. However, the technique of angioplasty did not gain widespread acceptance until the balloon dilation catheter was developed by Andreus Gruntzig, allowing angioplasty to be performed through a smaller-sized arterial puncture. Angioplasty is commonly used to treat coronary artery disease, renal artery narrowings and peripheral vascular insufficiency. This chapter will review the pathophysiology, indications, technique and results of balloon angioplasty. A number of adjunctive devices have been developed to increase the immediate technical success as well as the durability of percutaneous vascular interventions including lasers, atherectomy catheters and stents. These devices will be discussed in the following chapter.

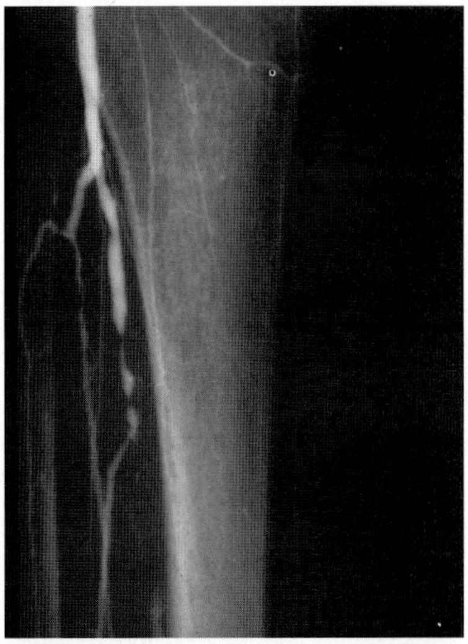

a

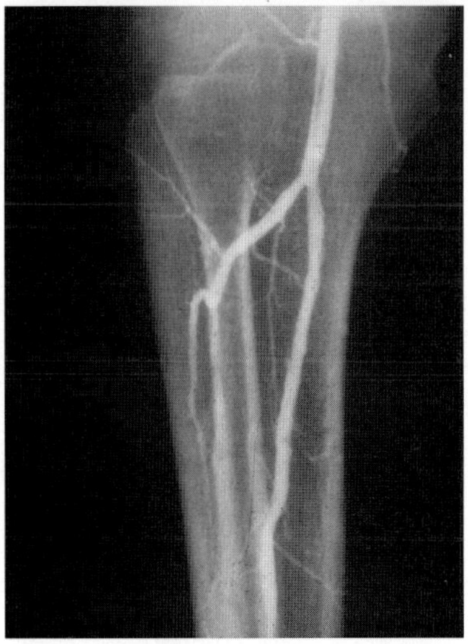

b

Figure 1. *76-year-old diabetic man with advanced ischemia of the right third and fourth toes. a) Arteriogram reveals extensive atherosclerotic narrowing of the tibioperoneal trunk. b) Following angioplasty with a 4-mm diameter balloon, a normal luminal caliber is restored.*

PATHOPHYSIOLOGY OF PERCUTANEOUS TRANSLUMINAL ANGIOPLASTY

Arteries and veins are comprised of three distinct layers: intima, media and adventitia. The intima extends from the vascular lumen lined with endothelial cells to the internal elastic lamina (IEL) and is composed of the endothelial cells, basement membrane and a subendothelial connective tissue layer. The media extends from the IEL to the external elastic lamina (EEL) and is composed principally of smooth muscle cells. The adventitia extends outward from the EEL and is composed of connective tissue, vasa vasorum, neural and lymphatic structures.

ARTERIES: HISTOLOGY

Intima
- From endothelium to internal elastic lamina (IEL)

Media
- From IEL to external elastic lamina (EEL)

Adventitia
- From EEL blending into perivascular connective tissue

The vast majority of arterial narrowings encountered are of atherosclerotic origin. The basic lesion develops from fibrofatty depositions within the intima. This is usually associated with thinning and fibrosis of the media. These basic or simple lesions may become complicated by the deposition of calcium, ulceration, intraplaque hemorrhage or thrombosis. Complicated plaques can lead to significant luminal narrowing and require treatment. The treatment required depends upon the extent of disease but the two most common options are surgical bypass or balloon angioplasty.

When the disease is diffuse or results in long-segment occlusions, surgical bypass with vein or prosthetic material may be performed. More focal disease without long-segment occlusions is often treated with balloon angioplasty.

Potential mechanisms by which balloon angioplasty enlarges the arterial lumen include:

1) Arterial wall stretching
2) Plaque fracture and local dissection
3) Plaque compression and remodeling
4) Plaque micropulverization/embolization

Arterial wall stretching results in enlargement of the lumen. This stretching is observed histologically as disruption of intercellular smooth muscle cell junction. Subsequently there is smooth muscle cell necrosis and fibrosis.

Plaque fracture and local dissection clearly play an important role. The plaque usually fractures at its thinnest point. These tears usually extend to the IEL but often extend into the media. Other segments of the plaque remain attached to the vessel wall preventing embolization of the dissected fragment.

Plaque compression and remodeling was originally thought to be the primary mechanism of balloon angioplasty. Histologic studies failed to support this mechanism. However, recent intravascular ultrasound studies (Losordo et al) suggest that plaque compression may contribute significantly, and may account for up to 30% of the improved luminal diameter.

Plaque micropulverization/embolization is an undesirable mechanism by which to enlarge the vessel. Fortunately, this event is unusual, being clinically detectable in less than 5% of cases. In most cases, this mechanism does not appear to contribute significantly to the angioplasty result.

INDICATIONS

The Standards of Practice Committee of the Society of Cardiovascular and Interventional Radiology published recommendations for angioplasty based on lesion morphology and location in 1990. This work will be briefly summarized.

Lesions can be classified as follows:

- *Category* 1: Balloon angioplasty alone is the procedure of choice. Treatment of lesions of this type have a very high likelihood of a durable clinical success.
- *Category* 2: Lesions which are well-suited (though not ideal) for angioplasty. This category includes lesions dilated prior to a surgical procedure to treat multilevel disease.

• *Category* 3: Lesions which are amenable to percutaneous dilatation, but have only a moderate chance of durable improvement (compared to surgical bypass) due to disease extent, location or severity. PTA may be performed if the patient is a poor surgical risk or lacks veins suitable for autogenous bypass.
• *Category* 4: Usually associated with extensive vascular disease, the lesions have a low technical success rate, a low long-term clinical benefit or are associated with a high procedural complication rate.

Specific categories with reference to location are as follows:

BRACHIOCEPHALIC ARTERIES

Category 1

- Focal stenoses (< 3 cm) of the subclavian or brachiocephalic, not involving the right carotid or either vertebral artery origin **(Fig. 2)**

Category 2

- Intermediate length stenoses (> 3 cm) of the subclavian or brachiocephalic, not involving the right carotid or either vertebral artery origin **(Fig. 3)**
- Stenoses of the brachiocephalic or subclavian dilated to provide inflow for an arterial bypass graft
- Brachiocephalic or subclavian bypass graft stenoses with a low risk of cerebral embolization
- Carotid artery lesions done for reasons of location with open surgical carotid access
- Focal, proximal stenoses of vertebral arteries requiring minimal manipulation

Category 3

- Short-segment occlusions (< 5 cm) of the subclavian or brachiocephalic which may involve the vessel origin **(Fig. 4)**

Category 4

- Stenoses that involve the origin of the right carotid or vertebral arteries
- Long segment (> 5 cm) occlusions

Recently, there has been considerable interest in carotid angioplasty, usually associated with stent placement. Currently, several trials are underway to evaluate the potential clinical usefulness of this technique.

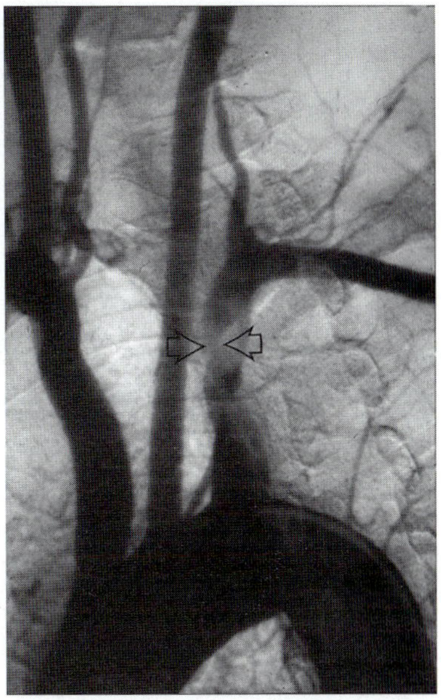

Figure 2. *Short, focal stenosis (arrows) of the proximal left subclavian artery not involving the origin of the left vertebral artery. This would be considered a category 1 lesion and is well suited for balloon angioplasty.*

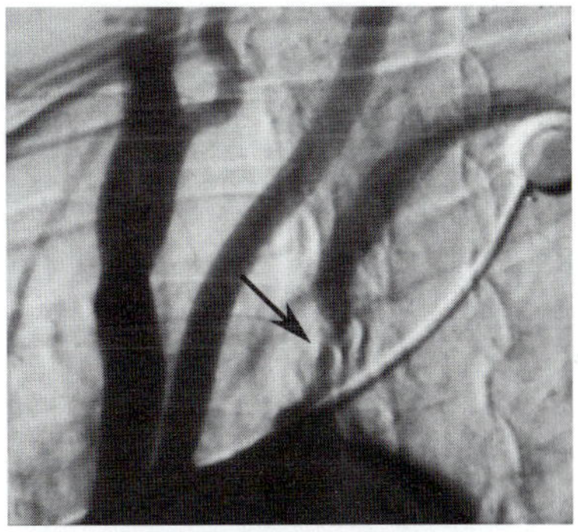

Figure 3. *Longer (approximately 3 cm) atherosclerotic narrowing of the proximal left subclavian (arrow), suitable but not ideal for balloon angioplasty. This would be considered a category 2 lesion.*

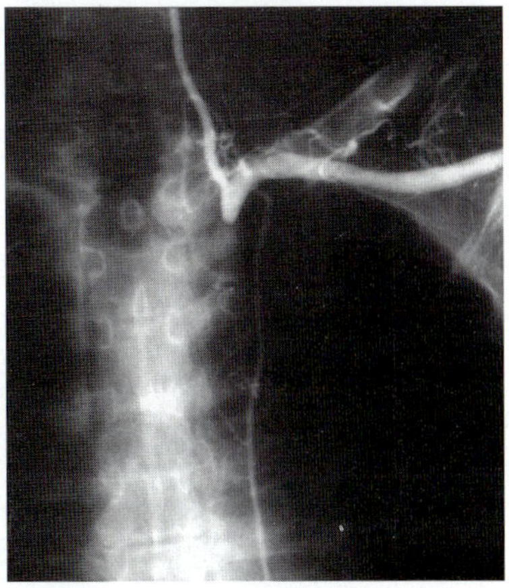

Figure 4. *Moderate length (4 to 5 cm) left subclavian occlusion would be considered either category 3 or 4 lesion. Though angioplasty may be attempted, surgical alternatives may be more appropriate.*

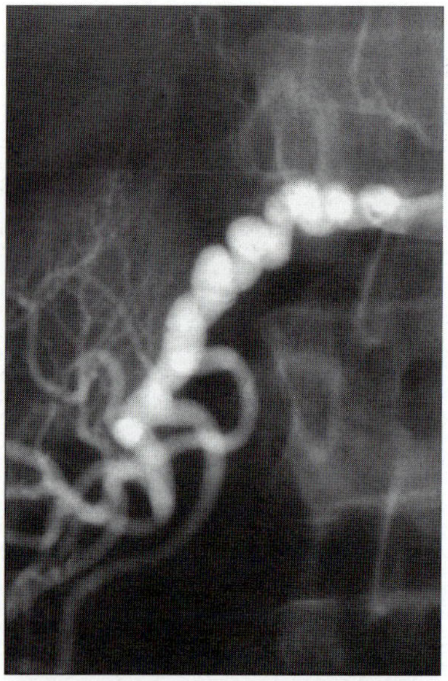

Figure 5. *Typical beaded appearance of medial fibroplasia subtype of fibromuscular dysplasia. Balloon angioplasty is the treatment of choice in the symptomatic patient (category 1).*

RENAL ARTERIES

Category 1

- Nonostial, short-segment (< 3 cm) atherosclerotic stenoses
- Stenoses secondary to fibromuscular dysplasia **(Fig. 5)**
- Stenoses in renal transplants

Category 2

- Bilateral, nonostial atherosclerotic stenoses
- Stenoses associated with surgical bypass grafts **(Fig. 6)**
- Stenoses associated with worsening renal function in patients with serum creatinine < 3.0 mg/dl

Category 3

- Ostial atherosclerotic stenoses **(Fig. 7)**
- Nonatheromatous lesions involving the proximal renal arteries
- Stenoses associated with worsening renal function (serum creatinine > 3.0 mg/dl)
- Renal artery occlusions

Category 4

- Stenoses which arise from aneurysmal or severely diseased aortas
- Stenoses associated with renal artery aneurysms **(Fig. 8)**

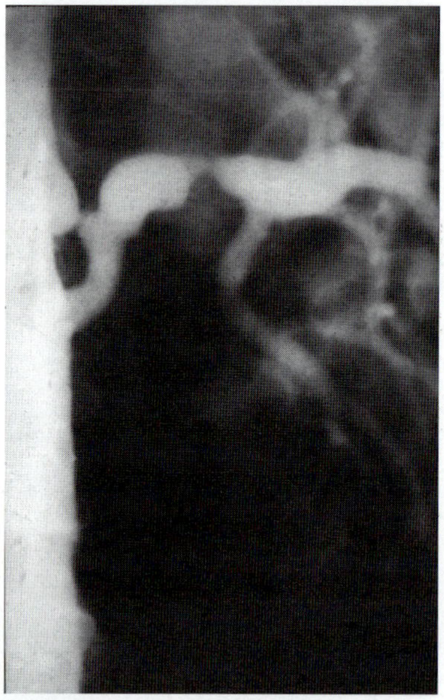

Figure 6. *A stenosis distal to an aortico-left renal artery vein graft would be considered a category 2 lesion.*

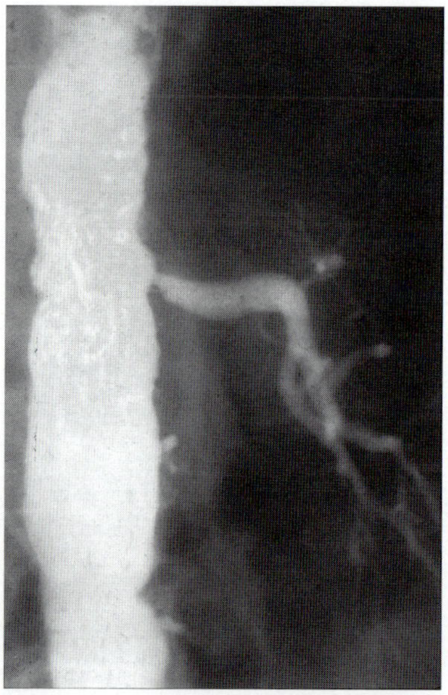

Figure 7. *High-grade ostial stenosis at the origin of the left renal artery in a patient with a single kidney (category 3).*

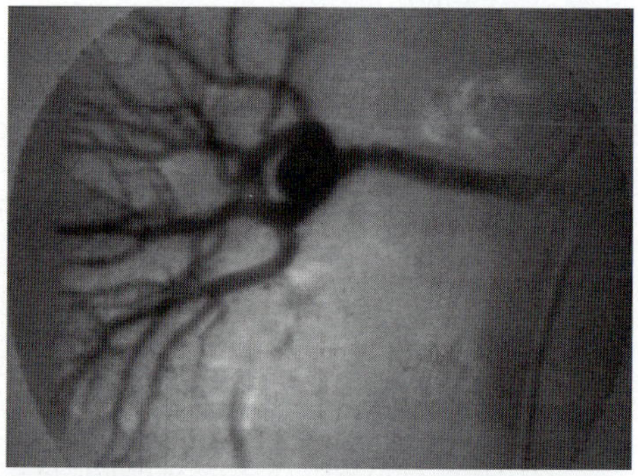

Figure 8. *Distal right main renal artery stenosis associated with an aneurysm at the point of bifurcation. Balloon dilatation of this lesion would carry a risk of rupturing the aneurysm (category 4).*

VISCERAL ARTERIES

Category 1

- There are no lesions for which angioplasty is clearly the procedure of choice

Category 2

- Short-segment (< 3 cm) stenoses of the celiac or SMA that do not involve the origin of the vessel **(Fig. 9)**

Category 3

- Long-segment (> 3 cm) stenoses of the celiac or SMA that do not involve the origin of the vessel
- Short stenoses involving the celiac or SMA orifice **(Fig. 10)**

Category 4

- Celiac or SMA occlusions **(Fig. 11)**
- IMA stenoses or occlusions
- Lesions in patients with clinical evidence of acute mesenteric ischemia

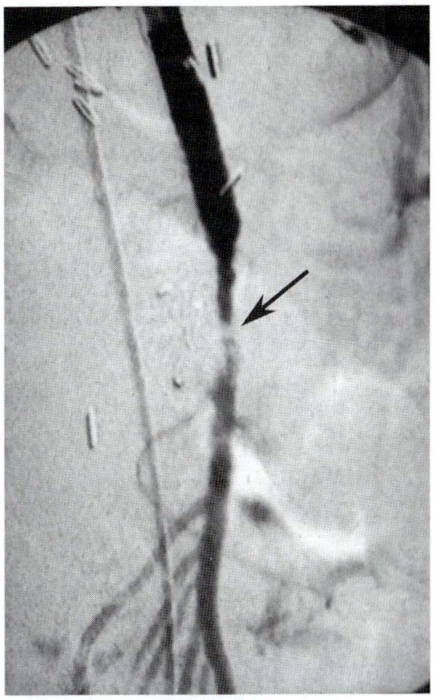

Figure 9. *Short-segment stenosis (arrow) immediately beyond the distal anastomosis of an aortosuperior mesenteric bypass graft (category 2).*

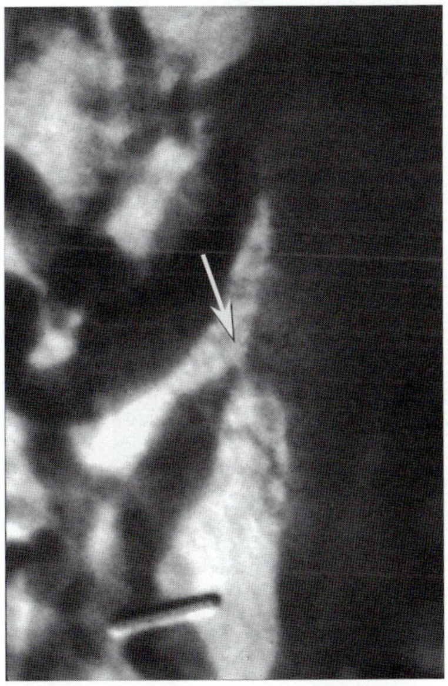

Figure 10. *Tight stenosis at origin of superior mesenteric artery (arrow). This would be considered a category 3 lesion.*

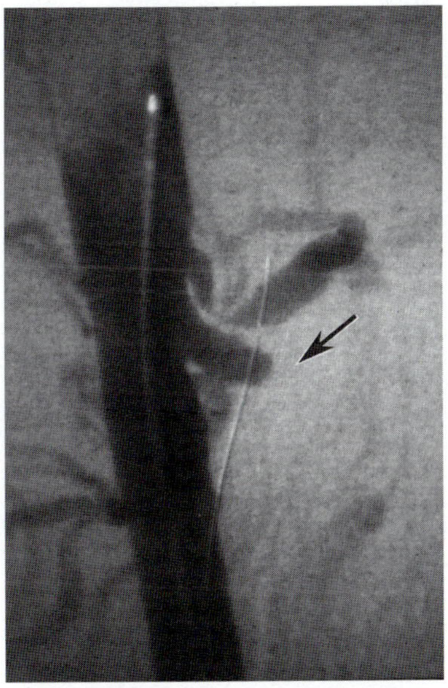

Figure 11. *Occlusion of the superior mesenteric artery (arrow) associated with dorsal compression of the celiac trunk. These lesions would not be considered suitable for angioplasty (category 4).*

INFRARENAL ABDOMINAL AORTA

Category 1

- Short-segment (<2 cm) stenoses with minimal disease elsewhere (in the infrarenal abdominal aorta) **(Fig. 12)**

Category 2

- Medium length stenoses (2-4 cm) with mild disease elsewhere **(Fig. 13)**

Category 3

- Long-segment stenoses (> 4 cm)
- Medium length stenoses (2-4 cm) with moderate to severe disease elsewhere

Category 4

- Aortic occlusions **(Fig. 14)**
- Aortic stenoses associated with aortic aneurysms

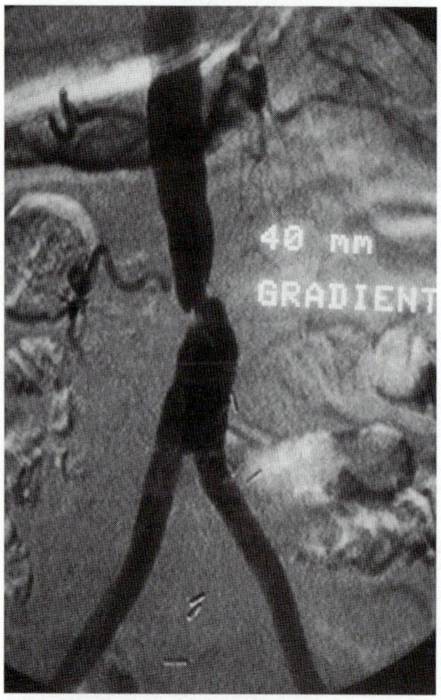

Figure 12. *Focal aortic stenosis immediately above the proximal anastomosis of an aortobifemoral bypass graft. Balloon angioplasty would be an excellent therapeutic alternative for this patient (category 1).*

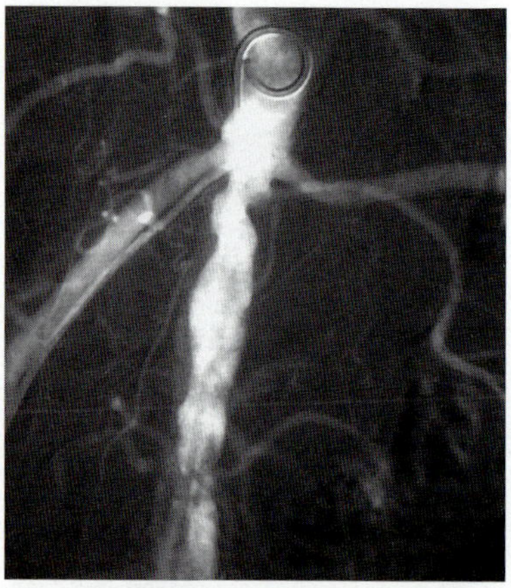

Figure 13. *Moderate stenosis of infrarenal abdominal aorta just below the renal arteries. Disease throughout the infrarenal aorta, however, qualifies this lesion as category 2.*

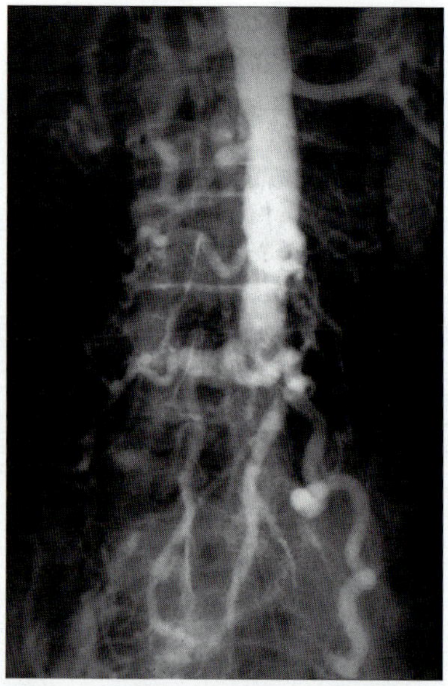

Figure 14. *Chronic aortic occlusion with well-developed collaterals (category 1).*

ILIAC ARTERIES

Category 1

- Short-segment (< 3 cm), noncalcified, concentric stenoses **(Fig. 15)**

Category 2

- Medium length (3-5 cm) stenoses
- Calcified or eccentric short-segment stenoses **(Fig. 16)**

Category 3

- Long-segment (5-10 cm) stenoses
- Chronic occlusions < 5 cm in length following fibrinolytic therapy **(Fig. 17)**

Category 4

- Very long stenoses (> 10 cm)
- Chronic occlusions > 5 cm in length following fibrinolytic therapy **(Fig. 18)**
- Extensive, bilateral, aortoiliac disease
- Iliac stenoses associated with abdominal aortic aneurysm or other lesion requiring aortic surgery

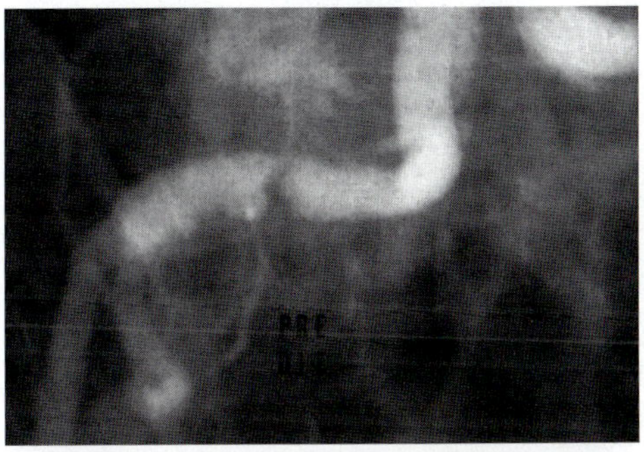

Figure 15. *Focal atherosclerotic web of the right common iliac artery ideally suited for percutaneous transluminal angioplasty (category 1).*

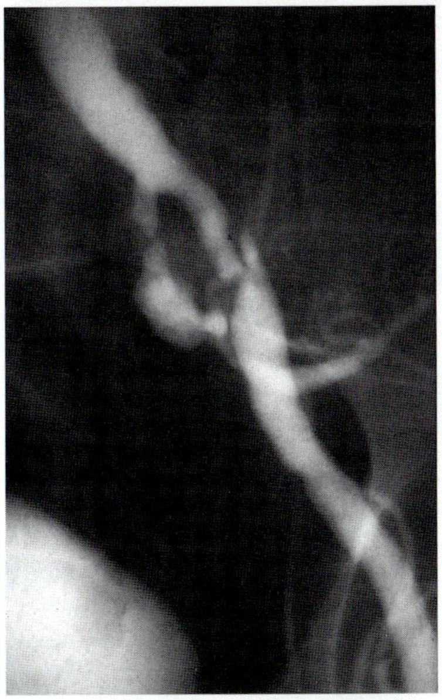

Figure 16. *Short-segment eccentric stenosis of the left external iliac artery. An associated stenosis of the internal iliac artery is also present (category 2).*

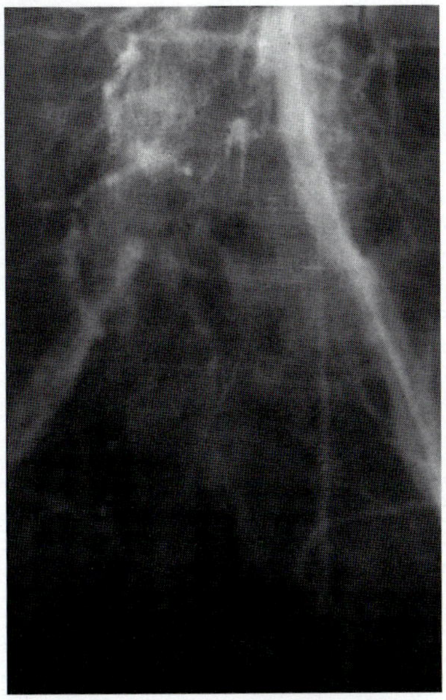

Figure 17. *Short-segment occlusion of the right common iliac artery with reconstitution of the external iliac artery. This would be considered a category 3 lesion.*

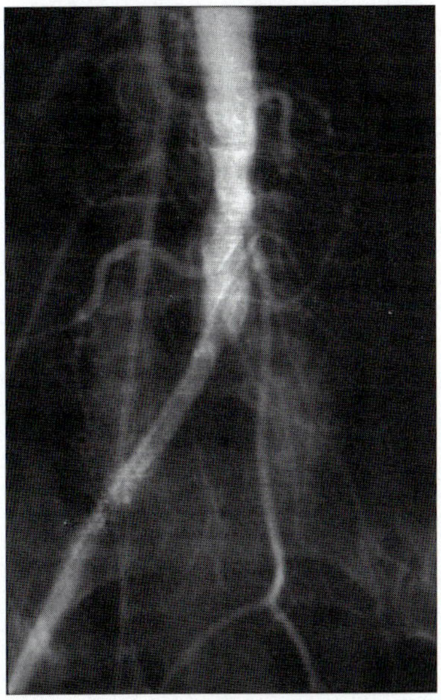

Figure 18. *Long-segment occlusion of the left common and external iliac artery following attempted fibrinolytic therapy. Though angioplasty is possible, the surgical alternatives may be more appropriate. A focal stenosis of the right common iliac/external iliac junction is also present (category 4).*

FEMOROPOPLITEAL ARTERIES

Category 1

- Short-segment (< 3 cm) stenoses or occlusions not involving the SFA origin or distal popliteal artery immediately proximal to the origin of the calf vessels **(Fig. 19)**

Category 2

- Medium length stenoses or occlusions (3-10 cm) not involving the distal portion of the popliteal artery **(Fig. 20)**
- Heavily calcified short-segment stenoses
- Multiple short-segment lesions **(Fig. 21)**
- Single or multiple lesions in which there is no continuous tibial runoff to improve inflow for a distal surgical bypass graft

Category 3

- Single, medium length lesions involving the distal popliteal artery **(Fig. 22)**
- Multiple, calcified short-segment lesions
- Long-segment (> 10 cm) stenoses or occlusions

Category 4

- Complete common and/or SFA occlusions **(Fig. 23)**
- Complete popliteal occlusions
- Proximal trifurcation occlusions

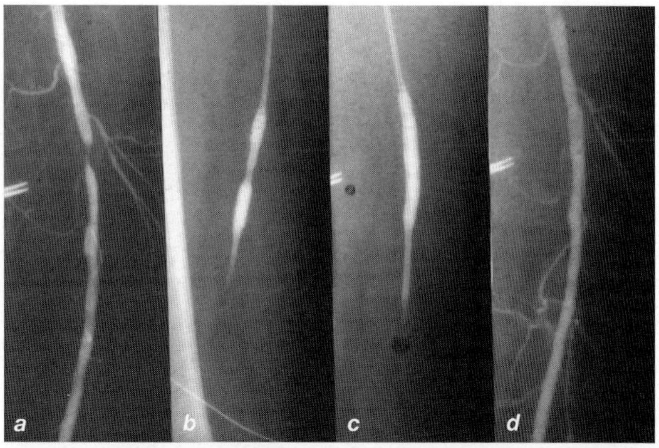

Figure 19. *Category 1 lesion of the right superficial femoral artery in a 58-year-old man. a. Arteriogram depicts a short-segment stenosis. b. Angioplasty balloon is inflated across the lesion. A deformity in the balloon at the site of stenosis is apparent during inflation. c. Angioplasty balloon completely inflates effacing the lesion. d. Arteriogram following the angioplasty confirms the successful result.*

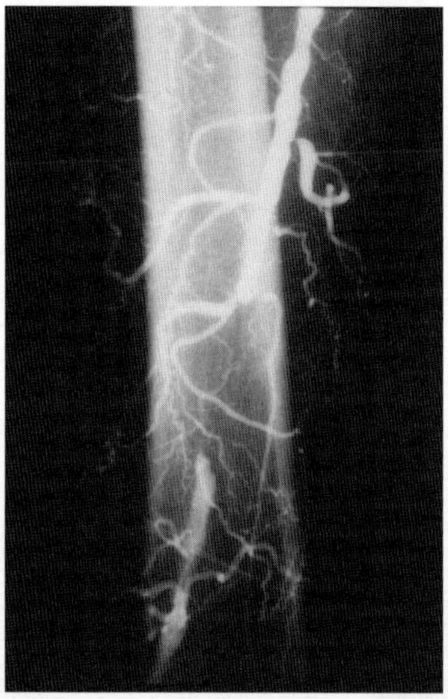

Figure 20. *Focal occlusion of the right proximal popliteal artery that is suitable for angioplasty in the symptomatic patient (category 2).*

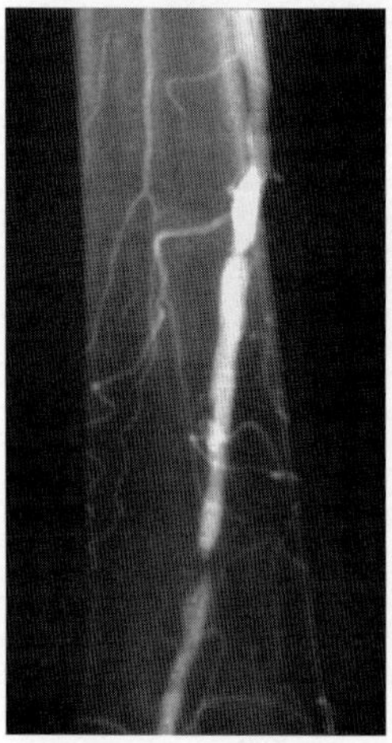

Figure 21. *Sequential focal stenoses in the right superficial femoral and popliteal arteries would be considered a category 2 lesion.*

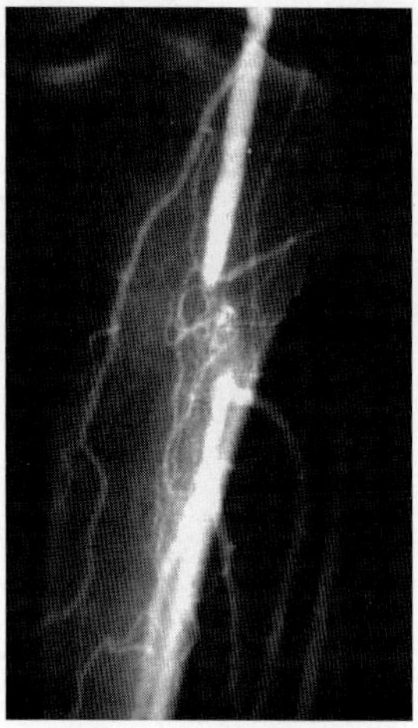

Figure 22. *Focal occlusion of distal popliteal artery. Though angioplasty is possible, if dissection is created extending into the trifurcation, marked extremity ischemia could be precipitated (category 3).*

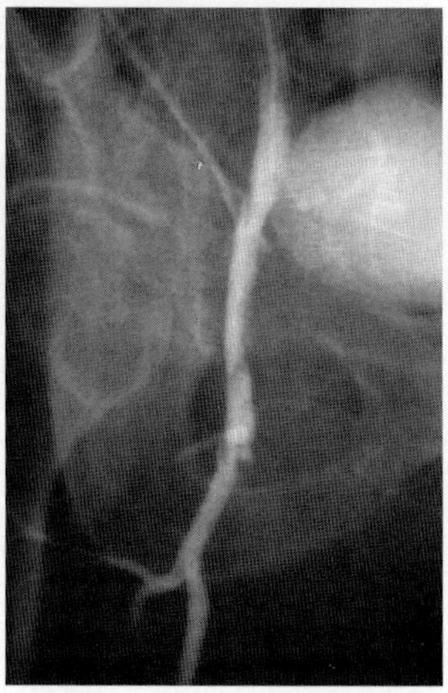

Figure 23. *Right superficial femoral artery occlusion extending to the origin of the vessel. Surgical bypass would be a better alternative for the symptomatic patient (category 4).*

INFRAPOPLITEAL ARTERIES

Category 1

- Short-segment (< 1 cm) stenoses **(Fig. 24)**

Category 2

- Multiple short-segment stenoses
- One or two short-segment stenoses at the vessel origin **(Fig. 25)**
- Stenoses treated during femoropopliteal bypass surgery

Category 3

- Medium-length (1-4 cm) stenoses
- Medium-length (1-2 cm) occlusions
- Extensive disease at the trifurcation

Category 4

- Occlusions longer than 2 cm
- Diffusely diseased vessels **(Fig. 26)**

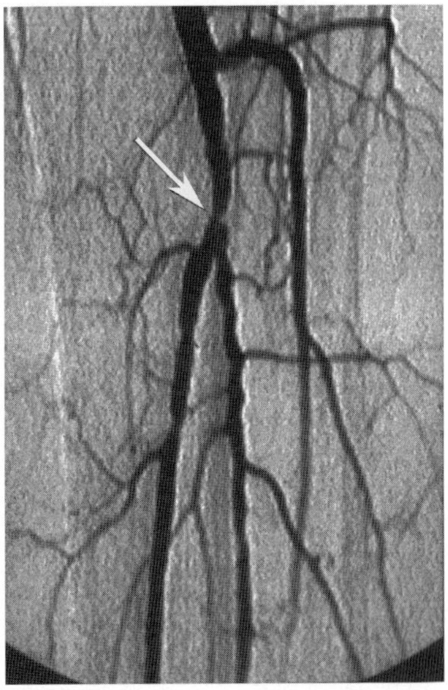

Figure 24. *Weblike stenosis of the distal tibioperoneal trunk (arrow) is well-suited to balloon angioplasty in the symptomatic patient (category 1).*

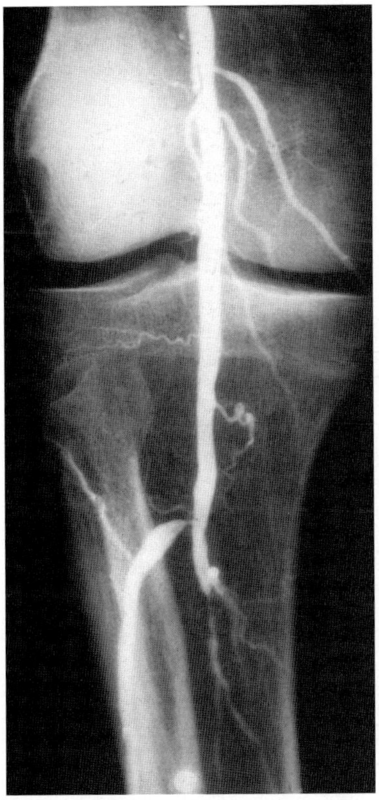

Figure 25. *Focal stenosis at the origin of the anterior tibial artery (category 2).*

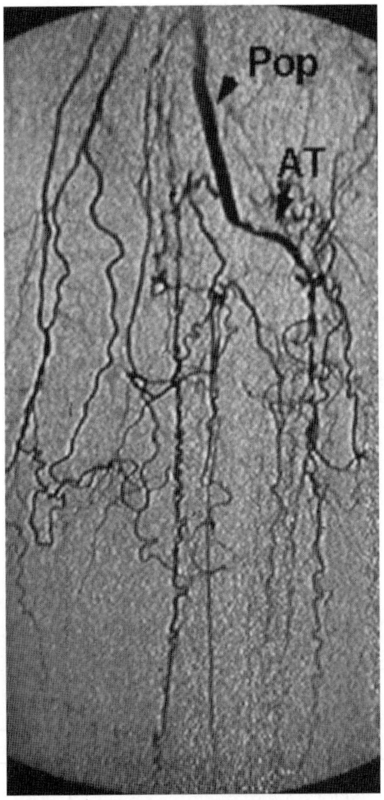

Figure 26. *Extensive occlusions of the trifurcation vessels which would not be considered suitable for angioplasty in most patients (category 4).*

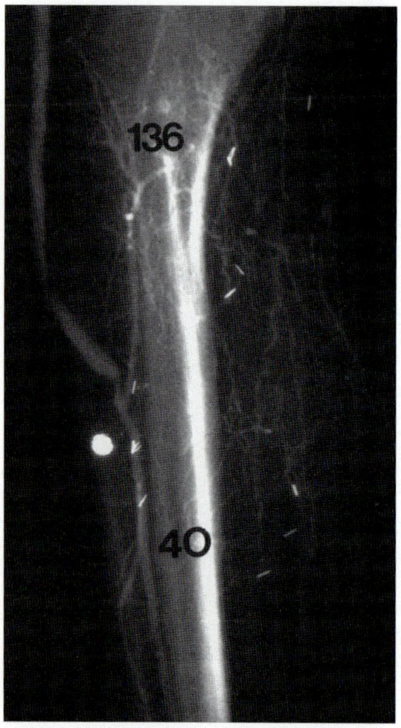

Figure 27. *High-grade stenosis at the distal anastomosis of a femoroanterior tibial saphenous vein bypass graft. A 96 mm Hg pressure gradient was measured across the lesion (category 1).*

BYPASS GRAFTS

Category 1

- Focal stenoses at distal anastomosis of femoropopliteal or femorotibial saphenous vein bypass graft **(Fig. 27)**
- Focal stenoses at proximal or distal aspect of saphenous vein aortorenal bypass graft

Category 2

- Focal stenoses of proximal anastomosis, femoropopliteal or femorotibial saphenous vein bypass graft **(Fig. 28)**
- Stenoses < 5 cm occurring within vein bypasses
- Stenoses associated with aortofemoral or aortoiliac bypass grafts **(Fig. 29)**
- Stenoses associated with prosthetic, extraanatomic bypass grafts

Category 3

- Medium-length (5-10 cm) stenoses of vein bypass grafts

Category 4

- Long-segment (> 10 cm) stenoses in vein bypass grafts
- Stenoses associated with anastomotic aneurysms

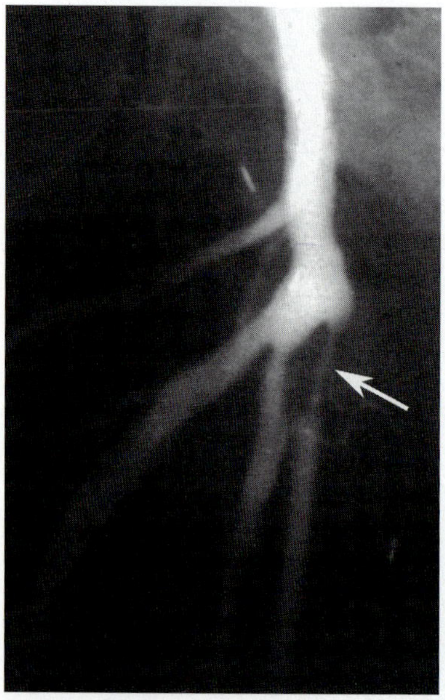

Figure 28. *Focal stenosis (arrow) at the proximal anastomosis of a right femoropopliteal saphenous vein bypass graft (category 2).*

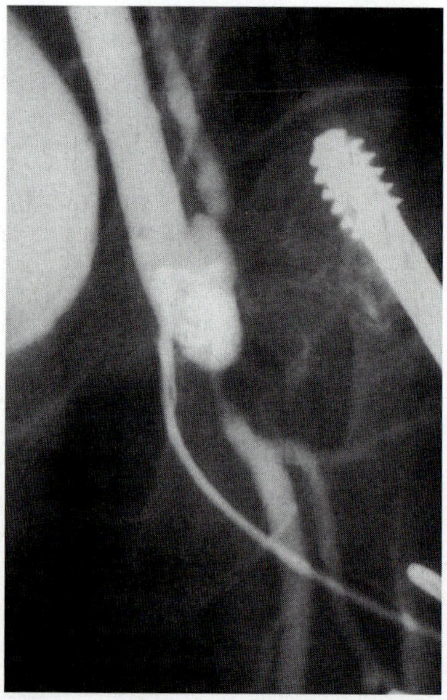

Figure 29. *Critical stenosis immediately distal to aorto-left femoral anastomosis. Balloon angioplasty may be performed, but the extensive scar tissue may limit the durability of the result (category 2).*

TECHNIQUE

Preprocedural Evaluation

The Standards of Practice Committee of the Society of Cardiovascular and Interventional Radiology recommends that patients undergoing angioplasty should have documentation of the following:

1) A written medical history including pertinent medical and surgical procedures, a list of current medications, and allergic history;
2) A physical examination focused upon the vascular system but adequate to exclude a significant, concurrent condition;
3) Complete diagnostic arteriography prior to intervention;
4) Informed consent;
5) Relevant laboratory information.

Procedural Care

During the procedure, all patients undergoing angioplasty should have:

1) Continuous cardiac monitoring and intermittent blood pressure measurement;
2) Intravenous access;
3) Pulse oximetry if sedation is used.

Hemodynamic Assessment

Angiography may underestimate the degree of arterial narrowing if images are obtained only in a single plane. Morphologically, a 50% decrease in luminal diameter is usually considered significant and is associated with more than a 75% reduction in cross-sectional area. If a question exists whether a lesion is hemodynamically significant or not, pressure measurements can be made on both sides of the lesion. A 10 mm Hg systolic pressure gradient is considered significant at rest. However, if a significant gradient is not present at rest, it may be revealed with vasodilatation of the distal vascular bed. Pharmacologic agents used to create vasodilation include nitroglycerine 50-100 µg IA and tolazoline 25 mg IA. A systolic gradient of 20 mm Hg or more is considered significant following vasodilatation.

Pharmacologic Agents

Anticoagulation with heparin, 3,000 to 10,000 units, is recommended for most angioplasties, although focal stenoses in high-flow regions, such as the iliac artery, are sometimes performed without heparin. Vasodilators are useful to prevent spasm and are strongly recommended for infrapopliteal and renal angioplasty. Vasodilating agents include nifedipine 10 mg SL 20 minutes prior to the procedure, and nitroglycerine 50 µg to 100 µg IA or tolazoline 25 mg IA during the procedure. Aspirin 325 mg po daily is recommended for patients with atherosclerotic lesions following the angioplasty.

MEDICATIONS COMMONLY USED WITH BALLOON ANGIOPLASTY

- Heparin 3,000-10,000 IV or IA
- Nitroglycerine 50 µg-100 µg IA
- Aspirin 65mg-325mg po

Selection of Balloon

Except when dilating the abdominal aorta, most investigators recommend selecting a balloon 10%-20% larger than the anticipated normal diameter of the vessel. Therefore, balloon diameter is usually selected on the basis of measuring the size of the adjacent or contralateral nonstenotic artery with a calibrated digital measuring system or calibrated catheter. Conventional film-screen angiograms may also be used for the measurement. Balloon length should be selected to span the lesion, but not to extend more than 1 cm on either side of the lesion.

BALLOON SIZE

- Balloon diameter 10%-20% larger than normal, unaffected vessel is considered optimal
- Balloon length which spans the lesion and does not extend more than 1 cm on either side

Selection of balloon material should be based on the type and anatomic location of the lesion. Atherosclerotic lesions will usually dilate with low-pressure balloons; intimal hyperplasia (postsurgical stenoses) require higher pressures to dilate. Densely calcified lesions require balloons which are puncture resistant to prevent rupture. Compliant balloons will enlarge with increasing pressure. Matsumoto recently reviewed these balloon parameters which are summarized in **Table 1**.

TABLE 1: CHARACTERISTICS OF DIFFERENT TYPES OF ANGIOPLASTY BALLOONS

Balloon Material	Compliance	Burst Pressure	Puncture/Scratch Resistance
Polyvinyl chloride	5+	2+	No literature
Polyethylene	3+	3+	2-3+
Polyolefin copolymers	4+	3+	No literature
Polyethylene teraphthalate	1+	5+	1+
Nylon derivatives	2+	4+	2-3+
Nylon-reinforced PU	1+	4-5+	4-5+

(From Matsumoto AH, 1994.)

Other considerations when selecting a balloon are related to wall tension, which is described by the Law of Laplace:

$$T\ (\textit{tension}) = P\ (\textit{pressure})\ x\ R\ (\textit{radius})$$

As is apparent from this relationship, more dilating force (and risk of vessel rupture) is present when larger balloon sizes are used. The practical significance of this relationship is that large vessels, such as the abdominal aorta, should be dilated with caution. This is the one circumstance where selecting a balloon size which will "underdilate" the lumen may be appropriate.

Crossing the Lesion

Safely crossing the lesion with a guidewire is the most critical step in performing a successful angioplasty. Usually, the lesion is crossed with a torquable, floppy tipped guidewire with fluoroscopic guidance augmented by injection of contrast media or digital roadmapping. After the floppy guidewire has been advanced safely through the lesion, it is followed with the diagnostic catheter. The floppy guidewire is then removed and replaced with a stiffer, exchange guidewire. The diagnostic catheter is then removed and replaced with the angioplasty balloon catheter. Coaxial balloon systems are also available which are extremely useful for renal and visceral angioplasty.

BALLOON INFLATION

- 10 ml or 12 ml syringe
- Mechanical inflators with pressure gauges

Balloon Inflation and Deflation

Balloons may be inflated either by hand or with mechanical devices with pressure gauges. Increasing amounts of pressure are applied until the area of narrowing is completely effaced. In most cases, the balloon is inflated with a 10 ml or 12 ml syringe filled with contrast material diluted 50% with saline flush. More rapid balloon deflation is achieved by attaching a 20 ml or 30 ml aspirating syringe. No scientific information is available regarding the appropriate length of time for balloon inflation.

Postangioplasty Assessment

An angioplasty is considered technically successful when the residual stenosis is 30% or less **(Fig. 30)**. Restenosis and clinical failure is much more common when there is a 30% or more residual stenosis. Therefore, if a significant residual stenosis is present, additional intervention may be warranted (larger balloon, stent or atherectomy).

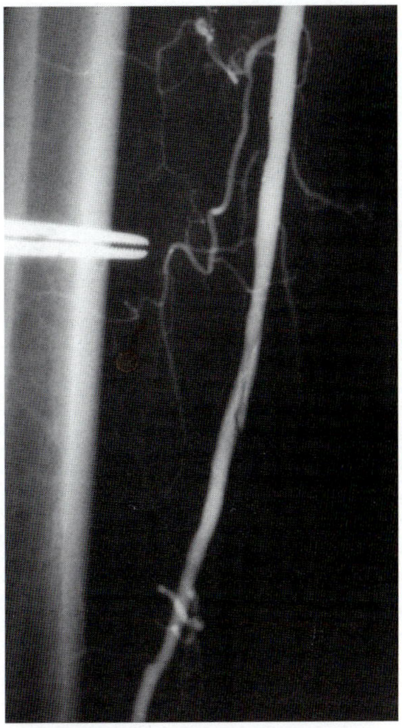

Figure 30. *SFA angiogram following balloon dilatation. A linear defect is present at the angioplasty site, which is a common appearance following dilatation. No further treatment is necessary in the presence of adequate flow.*

Postprocedural Care

The Standards of Practice Committee of the SCVIR recommends the following postprocedural care regimen:

1) All patients should be observed at bed rest for a variable period of time depending upon the size of the arteriotomy and the presence of risk factors for bleeding. These risk factors include hypertension, coagulopathy or the presence of a hematoma following the procedure. (A specific time period is not designated.)
2) During the initial postprocedural period, the patient should be monitored by skilled nurses or other trained professionals.
3) Hemodynamic monitoring is required to detect evidence of hemorrhage.
4) The urine output should be recorded.
5) Initial ambulation should be supervised.
6) In patients undergoing renal angioplasty, close blood pressure monitoring is required during the initial postprocedural period. IV access should be maintained in case fluid resuscitation is required to maintain the blood pressure.
7) The operating physician or her/his designee should evaluate the patient following the procedure and be available for continuing care during hospitalization and after discharge.

RESULTS

The SCVIR defines technical success as the substantial relief of stenosis or occlusion with residual narrowing of 20% or less, significant hemodynamic improvement and no major morbidity. Clinical success is defined as the complete relief of or substantial improvement in presenting symptoms. In the reported literature, patency is usually defined on clinical grounds based upon the lack of symptom recurrence and noninvasive measurements which are improved compared with baseline measurements. **Table 2** is a summary of the results of balloon angioplasty in the treatment of lower extremity ischemia as presented in the SCVIR *Peripheral Vascular Interventions* Syllabus:

TABLE 2: RESULTS OF BALLOON ANGIOPLASTY FOR PERIPHERAL VASCULAR DISEASE

Site	Number	% Technical Success	Short-Term Patency % (Yrs)	Long-Term Patency % (Yrs)
Infrarenal aorta	NA	92	95% (1-year)	80% (5-year)
Iliac	>6000	92	90% (1-year)	72% (5-year)
Femoropopliteal	~1500	89-98	59%-81% (1-year)	17%-58% (5-year)
Infrapopliteal	~300	75-100	56% (1-year)	44%-83% (2-year)

The results of renal angioplasty are more difficult to quantify due to the variability of reporting criteria. The SCVIR has defined a clinical *cure* of renovascular hypertension as diastolic blood pressure of < 90 mm Hg without antihypertensive agents. *Improvement* is defined as diastolic blood pressure < 90 mm Hg with the same or reduced dosages of antihypertensive medication or a diastolic blood pressure between 90 and 110 mm Hg, but at least a 15 mm Hg decrease in diastolic pressure from preprocedural measurements. *Failure* is defined as not meeting the above criteria for either cure or improvement. On the basis of these definitions, the results of renal angioplasty for the treatment of renovascular hypertension can be summarized in **Table 3**.

TABLE 3: RESULTS OF RENAL PTA IN PATIENTS WITH RENOVASCULAR HYPERTENSION

Etiology	Number	% Cured	% Improved	% Failed
Atherosclerosis (all)	1031	17%	47%	36%
Fibromuscular disease	314	43%	42%	15%
Renal artery graft stenosis	116	8%	47%	45%
Stenosis due to vasculitis	31	55%	29%	16%

(Adapted from Martin LG, 1994.)

Considerable debate regarding balloon dilatation of stenoses which involve the origin of the renal artery (ostial lesions) versus dilatation of lesions within the renal artery itself (nonostial lesions) exists. Most investigators have found dilatation of ostial lesions to have a lower technical success and a higher recurrence rate than nonostial stenoses. Some investigators recommend placing stents for all ostial stenoses in hopes of gaining a longer-term success.

Clinical success for the treatment of renal failure is defined as a decrease in serum creatinine to normal levels or 20% below levels obtained before angioplasty. In a summary of 177 patients from several series as reported in the SCVIR *Peripheral Vascular Interventions* Syllabus, 44% of patients were considered to have benefited from the procedure.

COMPLICATIONS

Complications and incidence associated with angiography/angioplasty are shown in **Table 4**.

TABLE 4: COMPLICATIONS OF ANGIOPLASTY

Complication	% Incidence
Puncture site hematoma	2% - 8%
Puncture site pseudoaneurysm	< 1%
Puncture site A-V fistula	Rare (clinically significant)
PTA site dissection (extensive)	Not uncommon
PTA site acute thrombosis	5% - 8%
PTA site rupture	0.5%
Distal embolization	1% - 5%
Cholesterol emboli	Very rare
Renal failure	NA

(Adapted from Picus D, 1994.)

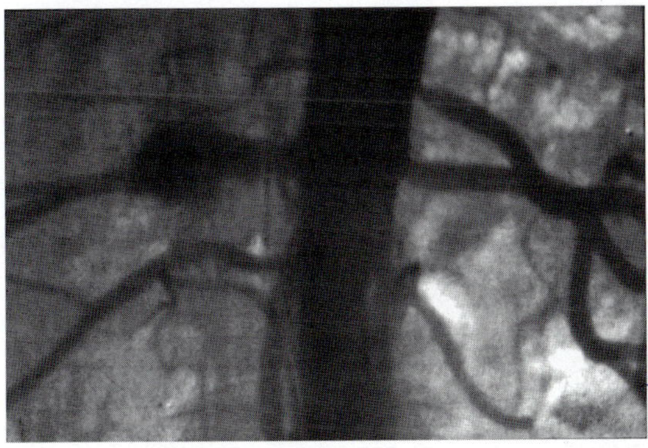

Figure 31. *Extravasation at site of right renal artery angioplasty. The artery should be occluded with a balloon and the patient taken directly to the operating room.*

Puncture site complications can usually be managed with supportive measures and ultrasound-guided compression. PTA site dissection does not require treatment when limited to the region of angioplasty in the presence of adequate flow. When the dissection is more extensive and flow is reduced, the flap is usually addressed with prolonged balloon inflation or stent placement. PTA site rupture in the abdomen or pelvis **(Fig. 31)** should be managed by balloon tamponade of the arterial defect and immediate surgical consultation (in addition to resuscitative measures). Rupture of a superficial femoral artery, popliteal artery or tibial vessel is often self-limited and may be observed for continued hemorrhage. Distal embolization **(Fig. 32)** should be managed with surgical embolectomy if limb-threatening ischemia is present. Fibrinolytic therapy or aspiration embolectomy may be attempted if ischemia is not limb-threatening. Cholesterol emboli has no effective therapy and may be fatal. Renal failure secondary to contrast media will usually resolve with time, though supportive dialysis may be necessary acutely.

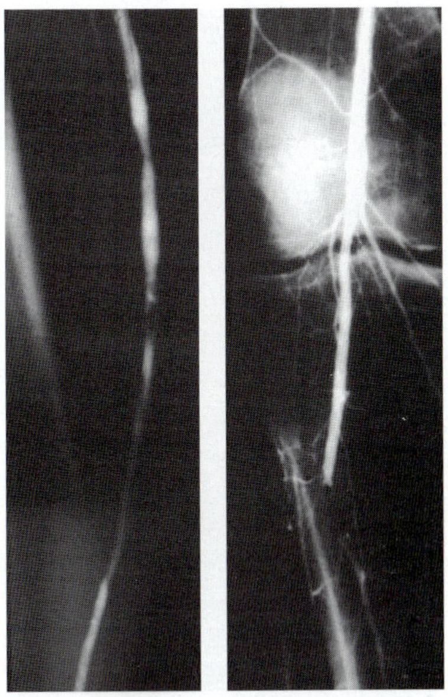

Figure 32. *An embolus occludes the distal right popliteal artery following superficial femoral artery angioplasty.*

REFERENCES

Bakal CW. Subtrifurcation percutaneous transluminal angioplasty: techniques and results. In: Darcy MD, LaBerge JM, eds. SCVIR Syllabus: Peripheral Vascular Interventions. San Francisco: Society of Cardiovascular & Interventional Radiology, 1994; 65-72.

Becker, GJ, Katzen BT, Dake MD. Noncoronary angioplasty. Radiology 1989; 170:921-940.

Cole PE. Pathophysiology of percutaneous transluminal angioplasty. In: Darcy MD, LaBerge JM, eds. SCVIR Syllabus: Peripheral Vascular Interventions. San Francisco: Society of Cardiovascular & Interventional Radiology, 1994; 3-12.

Losordo DW, Rosenfield K, Pieczek A, et al. How does angioplasty work? Serial analysis of human iliac arteries using intravascular ultrasound. Circulation 1992; 86:1845-1858.

Martin LG. Renal angioplasty: patient selection and results. In: Darcy MD, LaBerge JM eds. SCVIR Syllabus: Peripheral Vascular Interventions. San Francisco: Society of Cardiovascular & Interventional Radiology, 1994; 73-83.

Marx MV. Superficial femoral and popliteal percutaneous transluminal angioplasty: technique and results. In: Darcy MD, LaBerge JM, eds. SCVIR Syllabus: Peripheral Vascular Interventions. San Francisco: Society of Cardiovascular & Interventional Radiology, 1994; 50-64.

Matsumoto AH. Peripheral percutaneous transluminal angioplasty: balloon catheters and guidewires. In: Darcy MD, LaBerge JM, eds. SCVIR Syllabus: Peripheral Vascular Interventions. San Francisco: Society of Cardiovascular & Interventional Radiology, 1994; 24-35.

Picus D. Aortic and iliac angioplasty. In: Darcy MD, LaBerge JM, eds. SCVIR Syllabus: Peripheral Vascular Interventions. San Francisco: Society of Cardiovascular & Interventional Radiology, 1994; 36-49.

Picus D. Complications of percutaneous transluminal angioplasty. In: Darcy MD, LaBerge JM, eds. SCVIR Syllabus: Peripheral Vascular Interventions. San Francisco: Society of Cardiovascular & Interventional Radiology, 1994; 107-118.

Standards of Practice Committee of the Society of Cardiovascular and Interventional Radiology: Guidelines for percutaneous transluminal angioplasty. Radiology 1990; 177:619-626.

Standards of Practice Committee of the Society of Cardiovascular and Interventional Radiology: Angioplasty standard of practice. JVIR 1992; 3:269-271.

Vesely TM. General percutaneous transluminal angioplasty techniques: clinical aspects. In: Darcy MD, LaBerge JM, eds. SCVIR Syllabus: Peripheral Vascular Interventions. San Francisco: Society of Cardiovascular & Interventional Radiology, 1994; 13-23.

CHAPTER 9 PREVIEW ANSWERS

Case 1

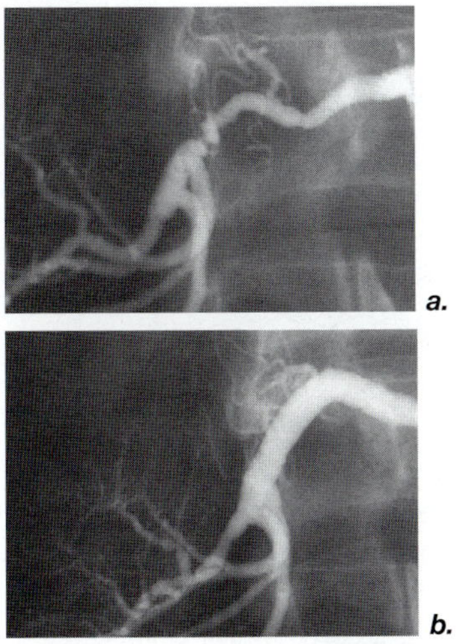

Findings: This 42-year-old woman has hypertension secondary to fibromuscular dysplasia. Symptomatic fibromuscular dysplasia of the main renal artery not associated with aneurysm formation is an ideal (category 1) lesion for balloon dilatation. a) Right renal arteriogram reveals typical changes of the medial fibroplasia subtype of fibromuscular hyperplasia. b) Following balloon dilatation, there is a marked cosmetic improvement without evidence of residual stenosis.

Case 2

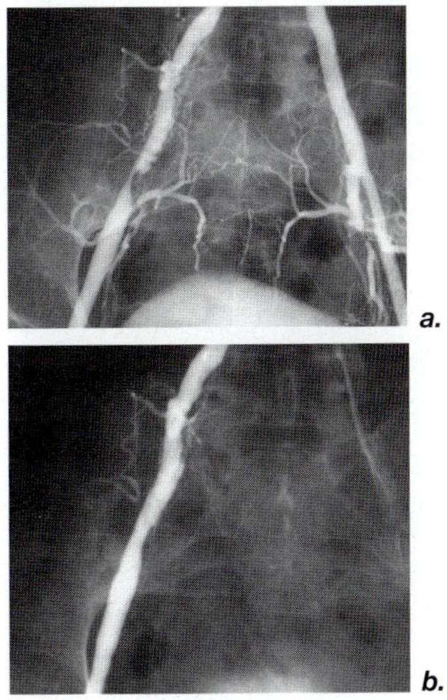

Findings: This 56-year-old man has a focal eccentric stenosis of the external iliac artery associated with atherosclerotic webs and would be considered a Category 2 lesion, well-suited for balloon angioplasty. a) Preangioplasty arteriogram. b) Postangioplasty arteriogram shows no significant residual stenosis with only minimal persistent narrowing at the angioplasty site.

Case 3

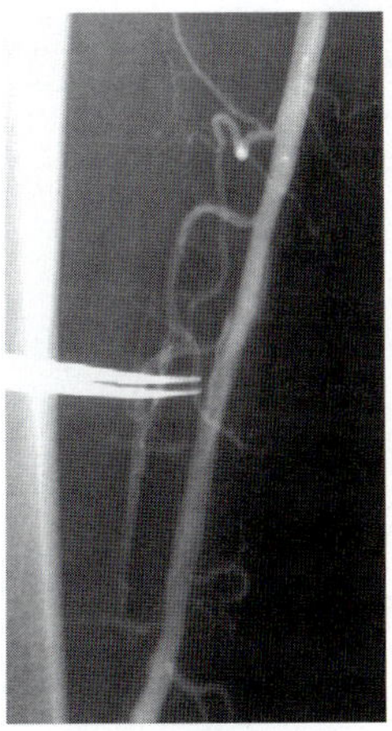

Findings: The postangioplasty arteriogram demonstrates a nonocclusive intimal crack. This angiographic appearance is frequently observed following a successful angioplasty. A stent should not be placed in this location in the presence of adequate flow.

Case 4

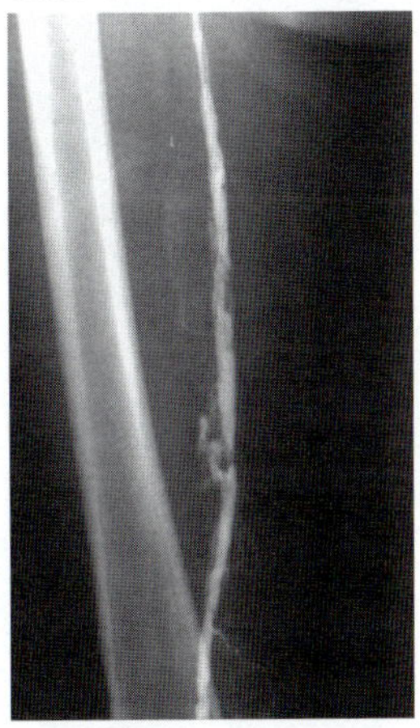

Findings: The postangioplasty arteriogram demonstrates extravasation from the superficial femoral artery. The patient needs to be followed closely for a developing hematoma or pseudoaneurysm formation. If extravasation occurs in this location, close clinical observation is warranted. If a similar finding were to be observed following an iliac or renal angioplasty, a balloon would be inflated at the angioplasty site followed by another intervention such as a covered stent or surgery.

CHAPTER 10 PREVIEW

Case 1

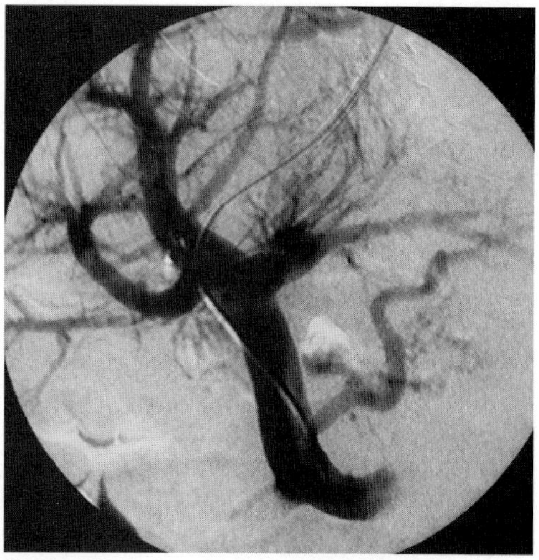

Figure 1.

History: A 48-year-old man presented with bleeding esophageal varices.
Question: What type of stent is most commonly used to create a TIPS?

Case 2

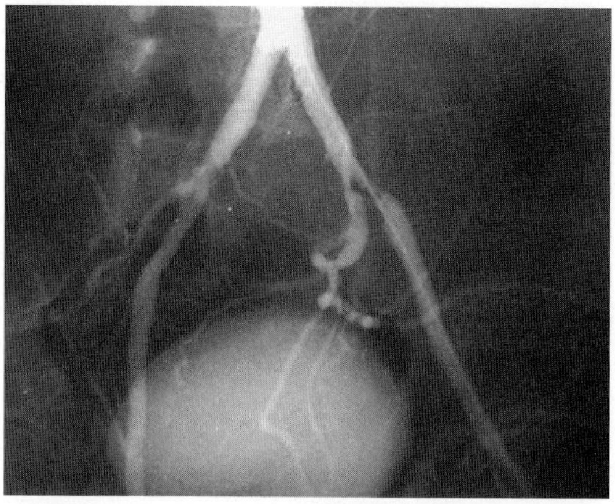

Figure 2.

History: A 58-year-old man with claudication was noted to have a stenosis of the left external iliac artery.
Question: What are the therapeutic options in this patient?

Case 3

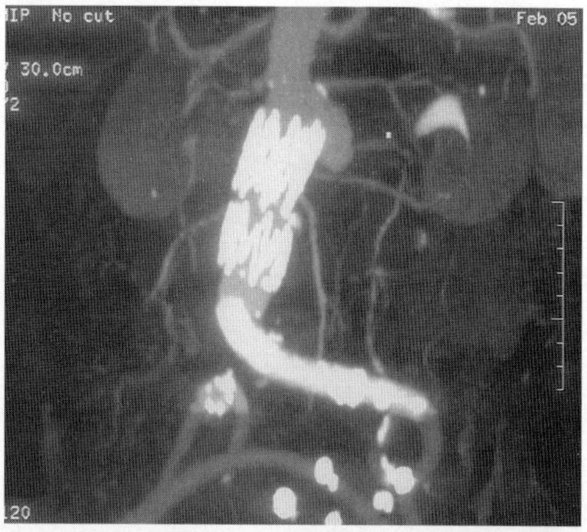

Figure 3. *Maximal-intensity projection image (MIP).*

History: A 75-year-old man underwent endovascular stent-graft repair of an infrarenal abdominal aortic aneurysm. One year later he developed progressive back pain. A contrast-enhanced abdominal CT scan was performed and 3-D reconstructions were obtained.
Questions: What kind of stent-graft is this? What is your diagnosis?

Case 4

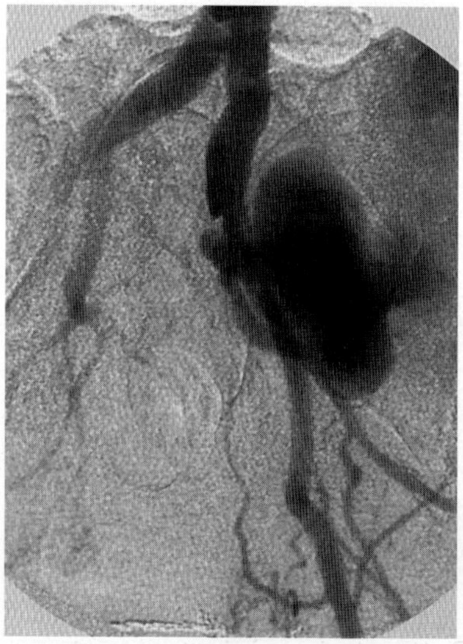

Figure 4. *Pelvic arteriogram, right posterior oblique projection.*

History: A 55-year-old woman complained of weakness following pelvic surgery. On exam she was found to be hypotensive and anemic. CT revealed a large left-sided retroperitoneal pelvic hematoma. Angiography was performed.
Question: What is your diagnosis?

Chapter 10
Stents and Stent-Grafts

Robert K. Kerlan, Jr., MD,
Jeanne M. LaBerge, MD, and
Timothy A. Chuter, MD

CHAPTER OUTLINE

Vascular stents
- History
- Stent design and deployment
  - balloon expandable
  - self-expandable
- Clinical applications
  - arterial
    - iliac
    - femoropopliteal
    - renal
    - visceral
    - abdominal aorta
    - brachiocephalic
    - cerebrovascular
  - venous
    - thoracic
    - abdominal and pelvic
    - upper extremity
    - transjugular portosystemic shunts
    - splanchnic venous system

Stent-grafts
- History
- Stent design
- Clinical results

VASCULAR STENTS

History

A vascular stent is a device that is inserted through a catheter to maintain patency of a blood vessel or a transparenchymal tract. At the University of Oregon in the mid-1960's, Charles Dotter designed the first device to keep dilated arteries open. The device was a cylindrical coil that was inserted through a catheter as a straight wire, resuming its cylindrical shape following deployment. Unfortunately, in animal experiments, the device caused thrombosis and this original stent was never used in humans.

Subsequently, during the initial investigation of portosystemic shunts at the same institution, Rösch used short segments of Teflon tubing to create a conduit from the portal venous system to the hepatic vein. This plastic stent also had an extremely low patency rate in animal experiments and was never used clinically.

In the late 1980's, Julio Palmaz developed a metal stent designed to be deployed on an angioplasty balloon. In a series of dog experiments, this stent did not cause vascular thrombosis and would remain patent for prolonged periods of time. At the same time, Gianturco and Rösch designed a stent that would self-expand following insertion through a catheter and not require an angioplasty balloon. These two basic designs, balloon-expandable and self-expandable, have been adapted to a number of specific devices that are currently in clinical use. This chapter will describe these devices and review their clinical applications.

Stent Design and Deployment

Physical characteristics vary considerably among the different stents available. As previously mentioned, there exists two basic stent designs: balloon expandable and self-expandable. Stents also differ with regard to potential diameter, length, flexibility and hoop strength. Though most stents are available in a variety of diameters and lengths (which are selected to fit with the individual patient's anatomy), flexibility and hoop strength are also extremely important parameters.

METALLIC STENT CHARACTERISTICS

Basic design
- Balloon expandable
- Self-expanding

Material
- Stainless steel
- Nitinol
- Other biomedical alloy

Sizing
- Maximal and minimal diameter
- Length (expanded and unexpanded)

Flexibility refers to the ease with which an undeployed stent can be advanced around bends and corners to the desired location. Moreover, a flexible stent is more likely to conform to the patient's anatomy if deployed in a vessel which is not straight. Hoop strength (which would be better termed tube strength) is the outward force provided by the stent to maintain the desired internal diameter. For example, an extremely tight arterial narrowing may not be kept open by a stent of low hoop strength.

As a general rule, vascular stents are always inserted through a vascular access sheath to prevent dislodgment or damage at the percutaneous entry site.

Balloon-Expandable Stents

Balloon-expandable stents are constructed of malleable metal, which deforms under the pressure of a dilating balloon. Though a diameter range is usually preselected, the precise diameter of the expanded stent depends on what diameter balloon is used to dilate it.

Palmaz Stent: The Palmaz stent (Johnson & Johnson) is the most common stent used for the treatment of peripheral vascular disease. It is composed of a cylinder of stainless steel wires intermeshed in a rectangular design connected at the midportion of each rectangular side **(Fig. 1)**. The device is securely "crimped" on an angioplasty balloon either by squeezing it with the thumb or index finger, or using a specialized "crimping tool." The device should not be placed on low-profile balloons or balloons with antifriction coating, due to the possibility of dislodging the stent during deployment. This stent may also be purchased premounted on a balloon catheter. At the time of this writing, the Palmaz stent is the only balloon-expandable stent approved by the United States Food and Drug Administration for the treatment of iliac occlusive disease.

The range of stent diameter and length of stent must be selected. Very small stents ranging in size from 2.0 mm to 5.0 mm in diameter are used for coronary arteries. For peripheral applications, either a stent designed to have an expanded diameter from 4 mm to 9 mm or one designed to range from 8 mm to 12 mm is selected. The 4-mm to 9-mm stent is mounted on a 5-F balloon and inserted through a 6-F or 7-F sheath. The 8-mm to 12-mm balloon is designed to be deployed on a 7-F angioplasty balloon and is usually placed through a 9-F or 10-F sheath.

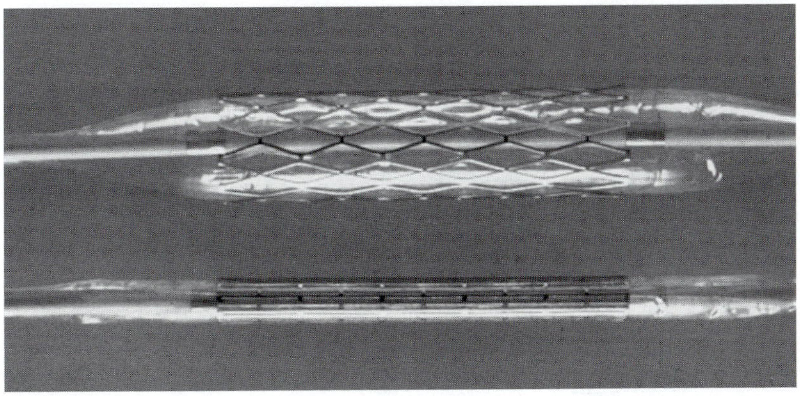

Figure 1. *Palmaz stent (Cordis). The stent is expanded by a dilating balloon and remains expanded after the balloon is deflated.*

Due to physical constraints, as a stent is dilated, it shortens. The amount of shortening depends upon the diameter the stent is dilated to. This is the reason that Palmaz stents must be selected on the basis of a desired diameter range. Specifically, if a stent which is designed to be from 4 mm to 9 mm in final diameter is dilated with a 12-mm-diameter angioplasty balloon, it would shorten excessively (to less than 1 cm in length) and not be a useful clinical device.

As a rule, Palmaz stents have very low flexibility and high hoop strength. The lack of flexibility precludes advancement over the aortic bifurcation, and therefore lesions in the lower extremity must be approached ipsilaterally. Moreover, these stents should not be placed across areas which bend, such as the hip joint or knee joint, as the stent would potentially be deformed by patient motion. However, Palmaz stents have splendid hoop strength and will more reliably maintain the desired luminal diameter when compared to other stents.

Corinthian Stent: The Corinthian stent (Johnson & Johnson) **(Fig. 2)** is a modified Palmaz stent with different lattice architecture and a different steel polymer. The stent was designed to be more flexible allowing advancement over the aortic bifurcation, as well as easier deployment into the renal arterial system.

Self-Expandable Stents

As a general rule, self-expandable stents are more flexible and have diminished hoop-strength compared to balloon expandable stents. Moreover, as the final release and detachment of self-expandable stents often occurs abruptly, precise positioning is usually more difficult compared to a balloon-expandable stent. In addition, and in contrast to balloon expandable stents, the stent diameter cannot be further increased by balloon dilatation. Therefore, if the selected diameter is too small, there is no way to correct the error.

Figure 2. *Corinthian stent (Johnson & Johnson). The design of this stent makes it more flexible than the original Palmaz stent.*

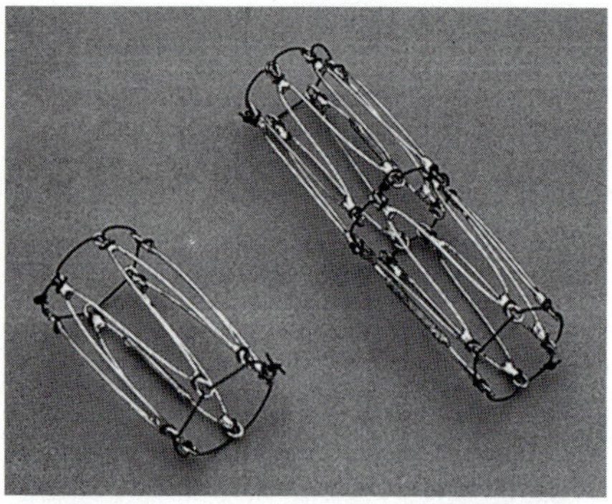

Figure 3. *Gianturco-Rösch Z-stent (Cook). This was the first self-expandable stent but it is seldom used intravascularly.*

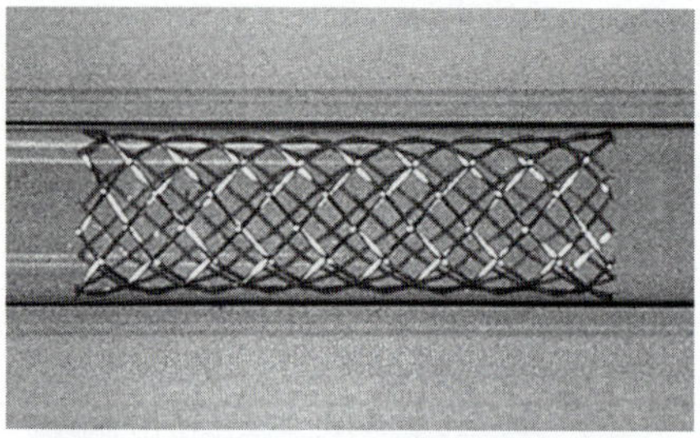

Figure 4. *Wallstent (Boston Scientific). This is currently the most common self-expanding stent used in the United States.*

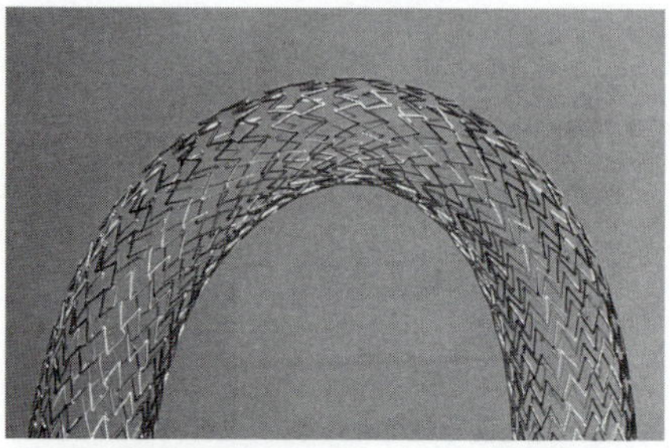

Figure 5. *Smartstent (Johnson & Johnson). This self-expanding nitinol stent shortens minimally during deployment and is currently being evaluated for vascular applications.*

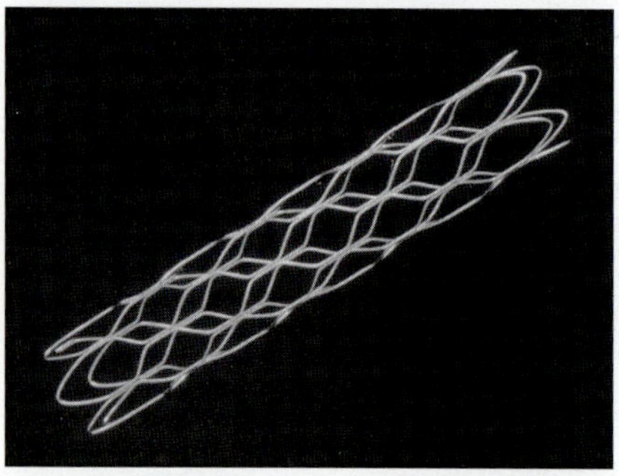

Figure 6. *Symphony stent (Boston Scientific). This self-expanding nitinol stent has the widest stent interstices. It is currently being evaluated for vascular applications.*

Gianturco-Rösch Z-Stent: The Gianturco-Rösch Z-stent (Cook Incorporated, Bloomington, Ind.) **(Fig. 3)** was the first available self-expanding stent. As the name implies, the stent is constructed by a series of straight metal struts connected in a serial "Z" configuration. Suture connects the apices of the metallic struts at both ends of the stent, determining its final diameter. The inherent spring of the metal affords the self-expanding quality. This stent was originally designed for biliary strictures. Though it is seldom used at the current time for vascular applications in its original form, the "Z-stent" design has become a popular component of many stent-grafts.

Wallstent: The Wallstent (Boston Scientific, Watertown, Mass.) **(Fig. 4)** is the most common self-expanding stent used for peripheral vascular applications. At the time of this writing, the Wallstent is the only self-expandable stent approved by the United States Food and Drug Administration for the treatment of iliac occlusive disease and TIPS.

The Wallstent comes in a variety of diameters ranging from 5 mm to 24 mm, and lengths ranging from 10 mm to 90 mm. These stents come premounted on catheter delivery systems. The delivery catheters range in size (depending on the size of the stent) from 7-F to 10-F. These deployment devices have outer plastic constraining membranes (or sheaths) which keep the stents bound to the catheters. When the delivery catheter has been appropriately positioned, this constraining sheath is gradually withdrawn, allowing the stent to expand. When the sheath is pulled beyond the proximal margin of the stent, the stent detaches from the catheter and cannot be retrieved.

The major advantage of the Wallstent is its flexibility, allowing insertion through tortuous vessels or over the aortic bifurcation into the contralateral lower extremity. The chief disadvantage is the considerable shortening which occurs during deploying making precise positioning (particularly the proximal end) difficult.

Other Self-Expanding Stents: A number of other self-expanding stents similar to the Wallstent have been released by the FDA for insertion into the biliary system. Some investigators are using these devices in the vascular system as an "off-label" application. These stents are made of nitinol, which is a polymer of nickel and titanium. The unique feature of this alloy is that it is flexible at room temperature but reforms rigidly into its predetermined shape at body temperature. As a result, there is less shortening when these stents are released from their constraining sheaths. Examples of self-expanding nitinol stents include the Smartstent (Johnson & Johnson) **(Fig. 5)**, the Memotherm stent (Bard), and the Symphony stent (Boston Scientific) **(Fig. 6)**.

Clinical Applications

Arterial

Stents were originally designed to be placed into arteries for the treatment of atherosclerotic narrowing. This remains true today as the majority of metallic stents are deployed in coronary arteries by cardiologists for the treatment of atherosclerotic coronary artery disease. However, stents are also used frequently in the management of peripheral vascular disease.

Iliac Artery: The iliac artery is the most common noncoronary site for stent deployment. This was the first location approved by the FDA for intraarterial stent deployment. Initially, stents were used following balloon angioplasty of the iliac artery for three indications: 1) significant flow-limiting post-angioplasty dissection, 2) significant residual pressure gradient, or 3) recurrent stenosis at a previously angioplastied site. Though some investigators continue to adhere to these guidelines, many interventional radiologists now place stents primarily (without first performing a balloon angioplasty) for the treatment of iliac artery occlusive disease.

The wisdom of this approach is supported by Bosch and Hunink's metaanalysis of over 2,000 patients who underwent either balloon angioplasty or iliac stent placement. These investigators concluded that balloon angioplasty and iliac stent placement had similar complication rates but the technical success rate was higher and long-term failure rate lower when stents were inserted. The same conclusion was drawn by Henry et al who reported primary patencies of iliac stents to be 94%, 91%, 86%, and 86% at 1-, 2-, 3-, and 4-year follow-up using life table analysis in a group of 310 patients..

Currently, Palmaz stents ranging in diameter from 8 mm to 12 mm are most commonly used to treat focal iliac stenoses less than 3 cm in length **(Fig. 7)**. Longer segments of disease (> 3 cm) are often treated with Wallstents **(Fig. 8)**. Self-expanding nitinol stents may also become commonplace in the treatment of longer segment iliac disease when FDA approval is granted.

Iliac stents are also routinely placed when completely occluded iliac arteries are recanalized.

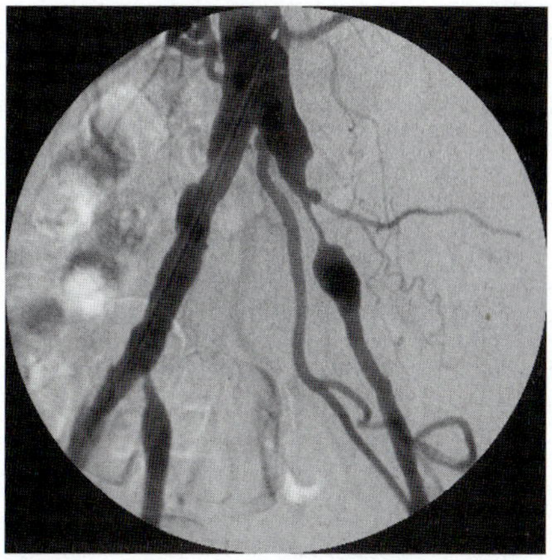

a.

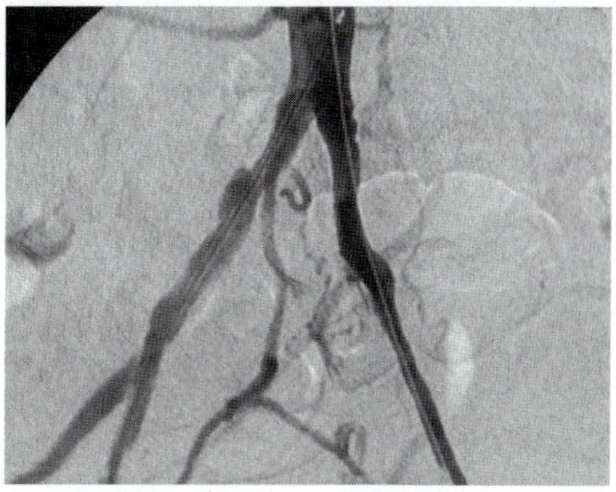

b.

Figure 7. *63-year-old woman with exertional pain in the left thigh and calf. a) Pelvic arteriogram shows a focal stenosis of the left common iliac artery. b) A Palmaz stent has been deployed, completely effacing the area of stenosis.*

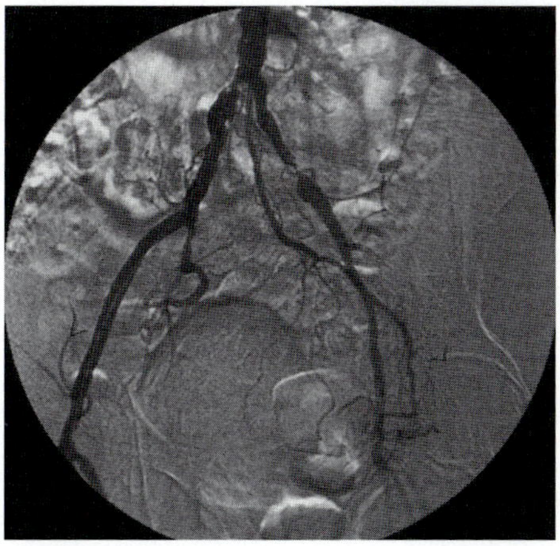

a.

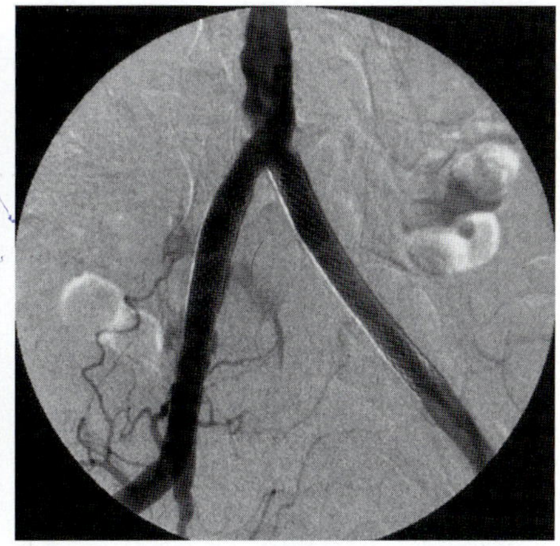

b.

Figure 8. *55-year-old woman with bilateral lower extremity exertional pain. a) Pelvic arteriogram shows a focal stenosis of the common iliac arteries bilaterally. b) Wallstents have been deployed at the aortic bifurcation. No residual stenosis is present.*

Femoropopliteal Arteries: Though stent deployment is exceedingly common for the treatment of iliac artery stenoses, similar enthusiasm has not been engendered for femoropopliteal disease. Specifically, though stents are not uncommonly placed for angioplasty failures **(Fig. 9)**, the long-term results do not parallel that of iliac stent insertion. In a series of 188 Palmaz stent insertions in 126 patients reported by Henry, a 4-year primary patency of 65% in the superficial femoral artery and 50% in the popliteal artery was noted.

The reasons for the diminished patency of femoropopliteal stents compared with iliac stents is uncertain but likely relates to lower flow, smaller vessel size, external compressibility, and tendency of femoro-popliteal atherosclerosis to be more diffuse and rapidly progressive.

Renal Artery: The role of stent placement into the renal arteries **(Fig. 10)** has been a little more difficult to define. Analysis of the data is encumbered by variable definitions of success and failure. Clearly, stents are being placed frequently into renal arteries following angioplasty.

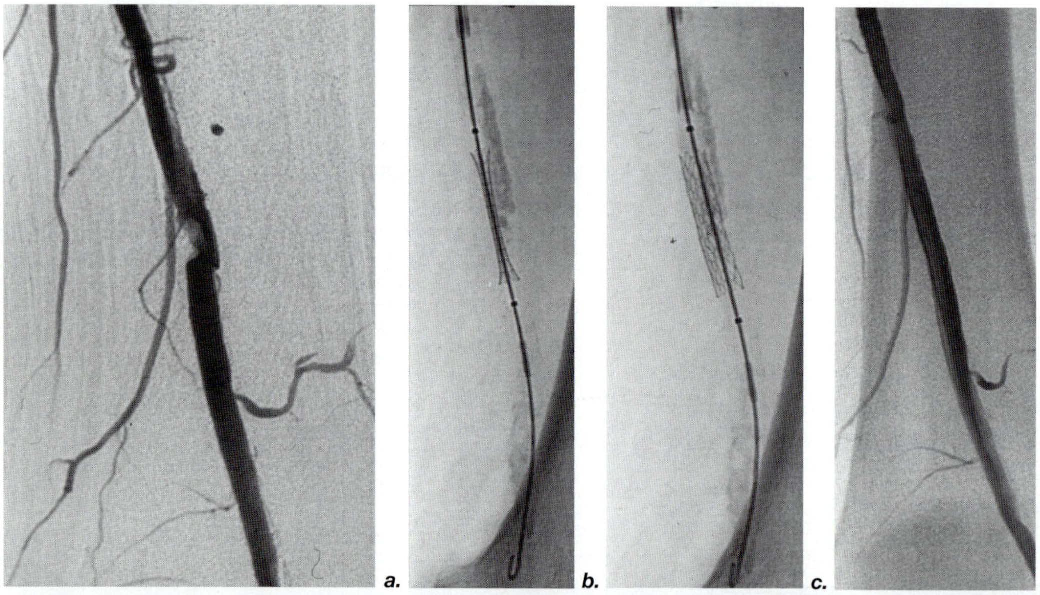

a. b. c. d.

Figure 9. *Palmaz stent placement at the femoropopliteal junction. a) Significant residual stenosis is present following balloon angioplasty of calcific stenosis. b) A Palmaz stent is deployed in the region of residual narrowing. c) Palmaz stent fully deployed. d) Arteriogram following stent deployment shows no residual stenosis. (Case courtesy of Allan Bloom, MD, Hadassah Hospital, Jerusalem.)*

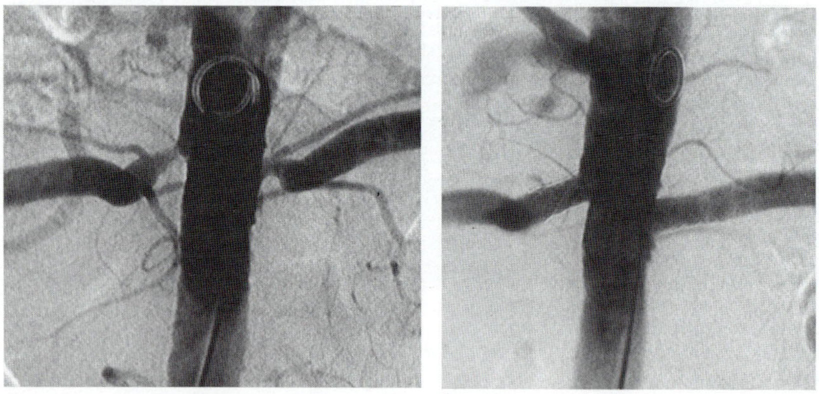

a. b.

Figure 10. *58-year old woman with uncontrolled hypertension. a) Abdominal aortogram confirms the presence of a tight, bilateral renal artery stenosis. b) Palmaz stents have been inserted bilaterally and no residual stenosis is present.*

Many practitioners reserve stent insertion for failed angioplasty (greater than 30% residual stenosis following balloon dilatation), significant angioplasty-induced dissection, or ostial lesions (which generally do not yield a durable result following PTA). However, other investigators are placing stents primarily in hopes of achieving a better long-term patency.

The indications for performing renal artery dilatation and stent placement have also been debated. Palmaz believes that intervention is clearly warranted in patients who have renal artery stenosis and significant renal dysfunction or flash pulmonary edema. Possible (but less clear) indications include bilateral renal artery stenosis with significant hypertension, renal artery stenosis in a patient with a single kidney, and unilateral renal artery stenosis with difficult to control blood pressure. Palmaz does not recommend renal artery intervention in patients with unilateral renal artery stenosis and easily controllable blood pressure or an asymptomatic renal artery stenosis.

Henry and Amor et al reported their experience of 259 balloon-expandable stents treating 244 renal artery stenoses in 210 patients. In this series, a five-year primary patency of 79% with a mean follow-up period of 25 months was noted. Clinically, of the 184 patients treated for hypertension, 19% were cured (no longer requiring treatment), 61% were improved (requiring diminished doses of medication), and 20% were unchanged. In 48 patients with renal insufficiency, function improved in 29%, remained stable in 67%, and deteriorated in 4%. Though these results are encouraging, the lack of comparison to a control group of matched patients diminishes the impact of the study.

In a prospective controlled trial of 85 patients, van de Ven et al compared primary stent placement versus balloon angioplasty (with stent placement only for inadequate results) in the treatment of ostial renal artery stenoses. At 6-month follow-up, restenosis occurred in 48% of patients who underwent balloon angioplasty compared to 14% of patients who received renal artery stents. Factors which appear to lead to recurrent stenosis in patients undergoing renal stenting include a residual stenosis of more than 30% at the time of the initial procedure and stents which are dilated to less than 6 mm.

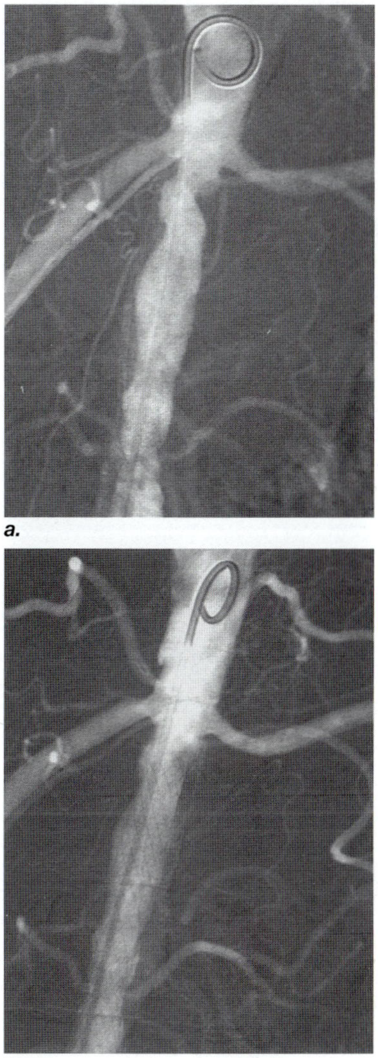

Figure 11. *62-year-old woman with recurrent bilateral lower extremity claudication following aortoiliac bypass. a) Aortogram reveals a focal stenosis in the infrarenal abdominal aorta. b) A Wallstent is inserted and no residual stenosis is evident.*

Visceral Arteries: Balloon expandable and self-expandable stents have also been used to successfully treat celiac and superior mesenteric arterial stenoses. Though the early results are encouraging, the number of cases is small and insufficient to make definitive comparisons to surgical revascularization. Sheeran et al reported their results with stenting 12 patients including three celiac and seven superior mesenteric artery stenoses and three short segment superior mesenteric artery occlusions. A technical success rate of 92% was encountered with a primary and primary-assisted patency of 74% at 18 months. There was one death due to bowel infarction in a successfully stented patient.

Abdominal Aorta: Stents have also been used to successfully treat stenoses of the abdominal aorta **(Fig. 11)**. Though the results of aortic stenting have been excellent for occlusive disease, balloon angioplasty in properly selected patients is also extremely effective. In a series of 25 patients by Westcott and Bonn, there was no significant difference in percent stenosis reduction, diminution of transstenotic gradient, or in immediate clinical outcome when balloon angioplasty was compared to stent placement. However, potential differences in long-term durability have yet to be evaluated.

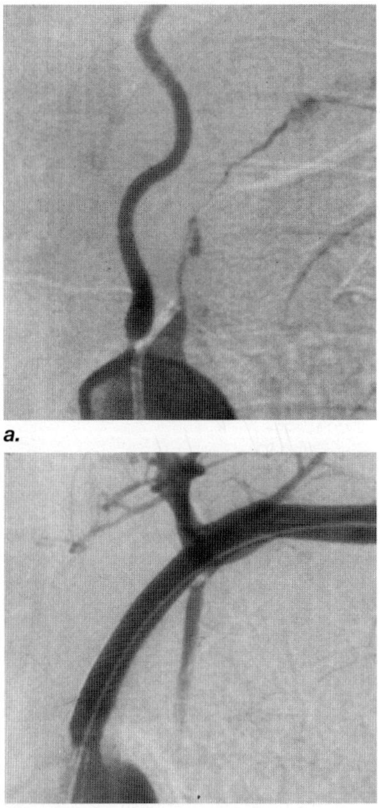

a.

b.

Figure 12. 68-year-old woman with exertional pain in her left arm. a) Thoracic aortogram shows a long-segment narrowing of the left subclavian artery. b) A Wallstent is deployed across the stenotic segment restoring pulsatile arterial flow to the left arm.

Brachiocephalic Vessels: Stent placement has also been used to treat narrowings of the brachiocephalic vessels **(Fig. 12)**. Indications include arm ischemia, vertebrobasilar insufficiency secondary to proximal obstruction or subclavian steal, as well as improving flow to left internal mammary coronary artery bypass grafts.

Rodriguez-Lopez et al reported their results in the treatment 53 subclavian stenoses and 17 subclavian occlusions with balloon dilatation followed by stent placement. These authors concluded that stent placement provided improved durability compared to simple balloon angioplasty; however, the trial was uncontrolled.

In a smaller series of seven patients, Sueoka reported 100% patency at a mean follow-up of 12 months when stents were used to treat unsuccessful balloon angioplasty (> 30% residual narrowing or intimal dissection) of the subclavian arteries in patients with subclavian steal.

Cerebrovascular Arteries: Over the past decade, there has been a tremendous amount of interest generated in treating stenoses of the carotid and vertebral arteries with balloon angioplasty and stent placement. Though stents have been successfully used for this purpose in a considerable number of patients, debate continues to rage over the appropriateness of this form of therapy. Proponents point to the minimally invasive nature of arterial stent placement coupled with its high success rate. However, it should be recognized that surgical carotid endarterectomy is well tolerated by even extremely ill individuals and has both established durability and a "track record" of safety with stroke occurring in 4% or less of patients.

In a literature review of 3,129 carotid stent placements by Wholey et al, a 2.49% minor stroke, 1.37% major stroke, and 0.96% procedure-related mortality rate was reported. Long-term durability has yet to be compared to traditional surgical alternatives. Currently, a number of prospective controlled, randomized trials are underway to analyze carotid stenting which should help clarify the precise role of this procedure in clinical practice.

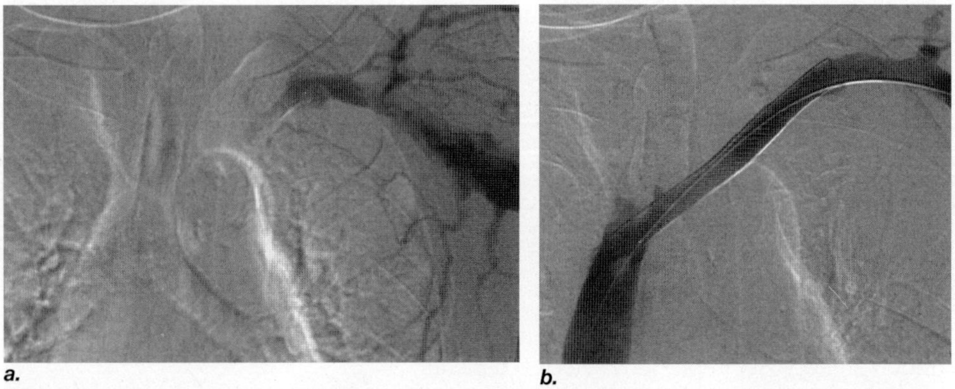

a. b.

Figure 13. 74-year-old woman with increased venous pressures during dialysis. a) Venogram discloses the presence of an innominate vein occlusion. b) Patency is reestablished following deployment of a Wallstent across the occluded segment.

Venous Stenting

Thorax: Stents have been used to treat narrowings and occlusions of the subclavian and innominate veins as well as the superior vena cava. Thoracic venous obstruction may be secondary to benign processes such as repeated venous access (particularly in dialysis patients) **(Fig. 13)**, mediastinal fibrosis due to histoplasmosis, radiation therapy, or other inflammatory processes. Malignant venous obstruction is usually secondary to lung or mediastinal tumors, but may also result from mediastinal metastases.

Both self-expandable and balloon expandable stents have been used. When complete occlusion is present, fibrinolytic therapy is often administered prior to stent deployment.

Qanadli et al reported successful recanalization of 12/12 patients with benign superior vena caval occlusion. Self-expanding Wallstents were used in all patients to reestablish patency. One patient developed a restenosis 2 months after treatment, but the remaining patients remained asymptomatic during the mean follow-up period of 11 months.

Nicholson et al evaluated stent placement in 26 patients with malignant superior vena caval obstruction and compared this group retrospectively to 25 patients who underwent radiation therapy for similar conditions. These investigators concluded that patients undergoing stent placement had significantly more durable palliation compared to radiation therapy and proposed stent placement as the treatment of choice for this condition.

In a mixed group of 51 patients with benign and malignant disease, Kee et al reported a 95% success rate in reestablishing superior vena caval patency.

Abdomen and Pelvis: Over the past 10 years, the treatment of iliofemoral and iliocaval thrombosis has substantially changed. Prior to the advent of fibrinolytic therapy and stents, patients with iliofemoral and caval thrombosis were treated with chronic anticoagulation. A substantial percentage of these patients developed chronic venous insufficiency with marked lower extremity edema, which interfered considerably with activities of daily living. Patients with extensive acute venous occlusion leading to the syndrome of phlegmasia cerulea dolens would even undergo venous thrombectomy or venous bypass in an attempt to alleviate their symptomatology.

With the advent of fibrinolytic therapy and stent placement, management of these patients has changed dramatically. In the majority of patients, patency can be reestablished and the precipitating stenosis treated.

Technically, fibrinolytic infusion catheters can be placed by the internal jugular, contralateral femoral or ipsilateral popliteal approach **(Fig. 14)**. When thrombus extends into the common or superficial femoral vein, the ipsilateral popliteal vein is the preferred site of access. This vein is punctured under ultrasound guidance and a multiside-holed infusion catheter is placed into the thrombus. Following fibrinolysis, a venogram is performed to ascertain the underlying cause of the thrombosis. In most patients, a stenosis is detected which can be durably corrected with stent placement.

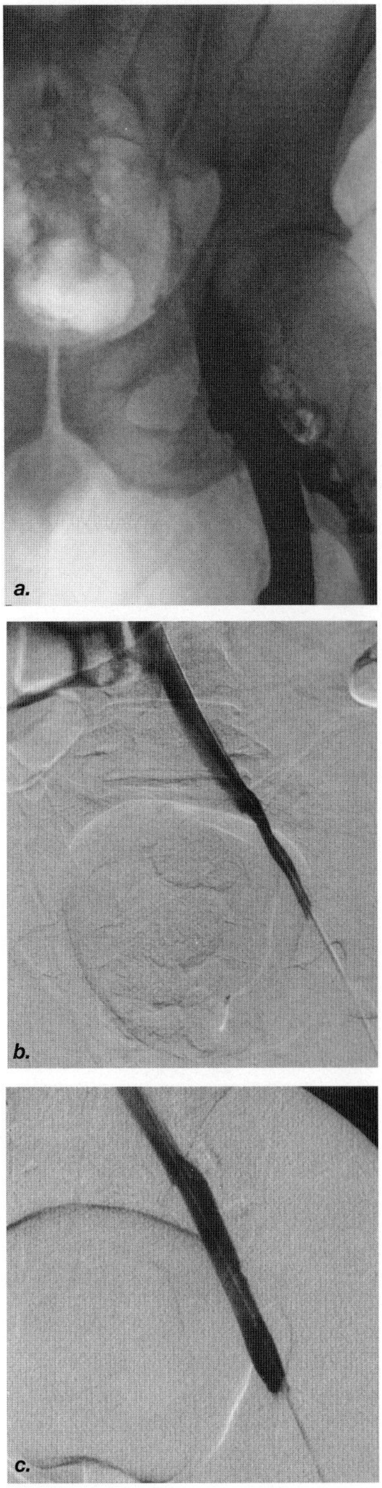

Figure 14. *50-year-old man with marked swelling of the left leg. a) Diagnostic venogram reveals occlusion of the left external and common iliac veins. b) Following fibrinolytic infusion from the ipsilateral popliteal approach, the external and common iliac veins are now patent, but the external iliac vein is diffusely narrowed. c) A Wallstent is inserted improving the diameter of the iliac vein.*

Binkert et al reported a 100% 3-year primary patency rate in eight women with benign left common iliac vein compression (May-Thurner syndrome) using Wallstents in all patients. Carlson et al also reported a 100% success rate at 6-month follow-up for patients with iliac vein thrombosis secondary to pelvic malignancies treated with stenting.

Upper Extremity: Venous stents are not infrequently used adjacent to dialysis access grafts to treat resistant venous outflow stenoses. Both balloon expandable and self-expandable stents have been used for this indication; however, less migration has been noted with the latter.

Quinn et al reported their results with stent placement in 40 dialysis patients with venous outflow stenoses. The primary patency rate of 27% at six months and 11% at one year was essentially the same as the 31% at six months and 10% at one year in 47 patients treated with balloon angioplasty. Though these results appear dismal, the secondary patency rates at the same time intervals were more encouraging with 64% and 64% for patients treated with stents compared to 80% and 71% for patients treated with balloon angioplasty. Moreover, the options for these patients are limited. However, as the data indicates, stents should be reserved for salvaging unsuccessful angioplasties, as these devices do not provide increased durability.

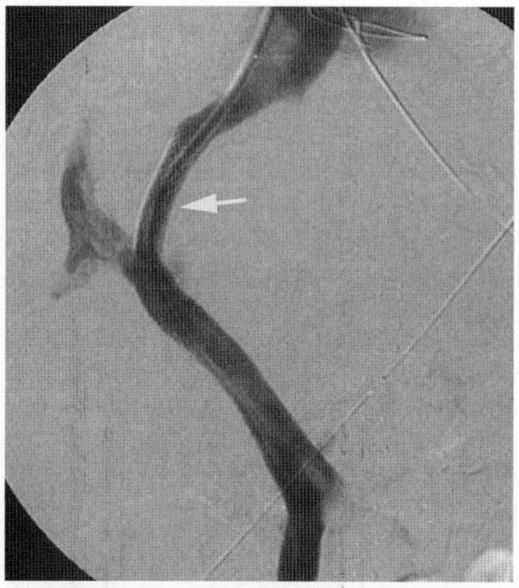

Figure 15. *50-year-old man with bleeding esophageal varices. A shunt (TIPS) has been created by the deployment of a Wallstent (arrow) bridging the hepatic parenchymal tract between the portal and hepatic veins.*

Transjugular Intrahepatic Portosystemic Shunts: The creation of a transjugular intrahepatic portosystemic shunt (TIPS) could not be accomplished without vascular stents. In this procedure, a tract is created between the hepatic vein and portal vein to provide decompression of the portal venous system. Colapinto et al attempted creation of a hepatic parenchymal tract by prolonged inflation of angioplasty balloons during the early 1980's. Unfortunately, a durable connection between the portal and hepatic venous system could not be established. However, when Richter performed a similar procedure, but used a stent to bridge the veins, prolonged patency was achieved. Currently, flexible self-expanding stents are routinely used to successfully establish these shunts **(Fig. 15)**.

Splanchnic Veins: Stents have also been used to treat portal, splenic, and superior mesenteric venous thrombosis **(Fig. 16)**. These veins can be recanalized using guidewire and catheter techniques from either a transhepatic or transjugular/transhepatic approach. Stein and Link were able to achieve an 86% technical success rate using flexible Wallstents to reopen eight portal veins, three splenic veins, and ten occlusions involving the portal, splenic, and superior mesenteric veins. A 63.5% primary patency rate was achieved at a mean follow-up of 43 months. Though procedures such as this should not be considered routine, significant palliation can be provided in the appropriately selected patient.

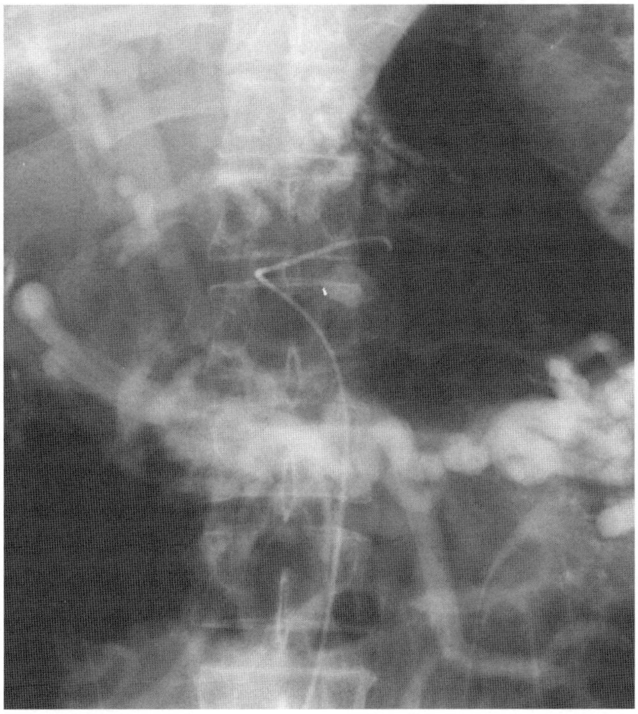

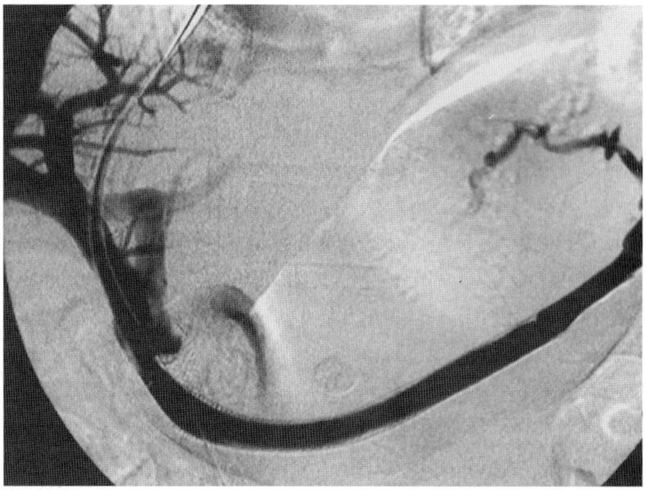

b.

a.

Figure 16. 58-year-old man with bleeding varices secondary to splenic and superior mesenteric venous occlusion. a) Preliminary arterial portogram shows collaterals around the splenic vein and main portal vein. b) Wallstents have been inserted from a transjugular approach into both the splenic and superior mesenteric veins, decompressing the portal venous system.

STENT-GRAFTS

History

With the success and rapid adoption of intravascular metallic stents, endovascular technology quickly evolved to embrace fabric coverings. Bare metal or "uncovered" stents worked well to treat occlusive disease of the aorta and lower extremities but were not appropriate for the management of aneurysmal disease. In treating aneurysms, the goal is to exclude the aneurysm from the circulation and thereby prevent subsequent rupture **(Fig. 17)**. To accomplish this task with stents requires the addition of an impermeable graft material to the metal stent framework. The resulting hybrid product is referred to as a "stent-graft."

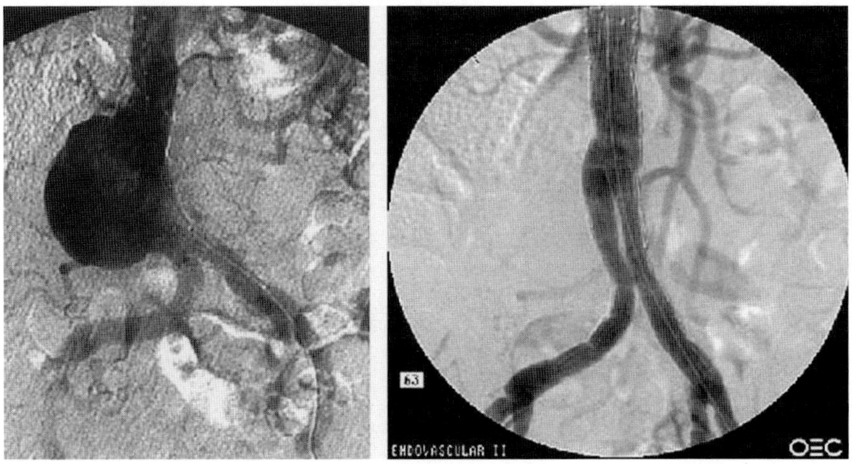

a. b.

Figure 17. Endovascular repair of abdominal aortic aneurysm. a) Preprocedure angiogram demonstrates a large eccentric infrarenal aortic aneurysm and a tortuous right iliac artery. b) Angiogram following deployment of a bifurcated stent-graft shows successful exclusion of the aneurysm.

The first clinical use of stent-grafts was in the treatment of abdominal aortic aneurysm (AAA). In 1991, Parodi reported the treatment of five patients with tubular stent-grafts and established that endovascular treatment of AAAs was possible. However, the widespread application of this procedure has been limited by device design problems. To date, no stent-grafts are approved for use by the U.S. Food and Drug Administration (FDA) although many are in clinical trials **(Table 1)**. In this section, we will concentrate on the application of stent-grafts for AAA since this is the application for which there is the greatest clinical experience.

TABLE 1. STENT-GRAFTS UNDER INVESTIGATION IN THE U.S.

- AneuRx (Medtronix)
- EVT (Guidant)
- Excluder (Gore)
- Talent (World Medical)
- Vanguard (Boston Scientific)
- Zenith (Cook)

Design

The first technical challenge of stent-graft design was building a device small enough to fit through the iliac arteries. Since the iliac arteries are generally 8 mm to 10 mm in diameter, the stent-graft introducer device must be less than 24-F to 30-F. In the past decade, a wide variety of stent graft devices have been devised that can be delivered through introducers of this size. These devices differ in several key features such as the type of metal stent, the type of synthetic material, and the overall deployment design **(Table 2)**.

The three most common design configurations for abdominal aortic stent-grafts are 1) simple tube graft, 2) unilateral aortoiliac graft with contralateral fem-fem graft or 3) a bifurcated design. Experience has shown that simple tube grafts are not appropriate for most patients because AAAs almost always extend all of the way to the aortic bifurcation. So in practice, most grafts are of the latter two designs **(Figs. 18, 19)**. Synthetic materials that have been used for these grafts include PTFE, dacron or polycarbonate urethane.

The supporting metal framework may be made of stainless steel or nitinol and may be attached to the synthetic material at the ends only (termed partially supported) or over the entire length of the graft (termed fully supported) **(Fig. 20)**. Unsupported stent-grafts have a tendency to kink and often need to be reinforced by the coaxial deployment of an inner metal stent. For this reason, most new stent-grafts have a fully supported design.

TABLE 2. STENT-GRAFT DESIGN CHARACTERISTICS

Materials
- Metal: nitinol, stainless steel
- Fabric: dacron, PTFE, polycarbonate urethane

Attachment mechanism

Delivery system

Anatomic configuration
- Tube graft
- Unilateral iliac
- Bifurcated

Fully supported vs not fully supported

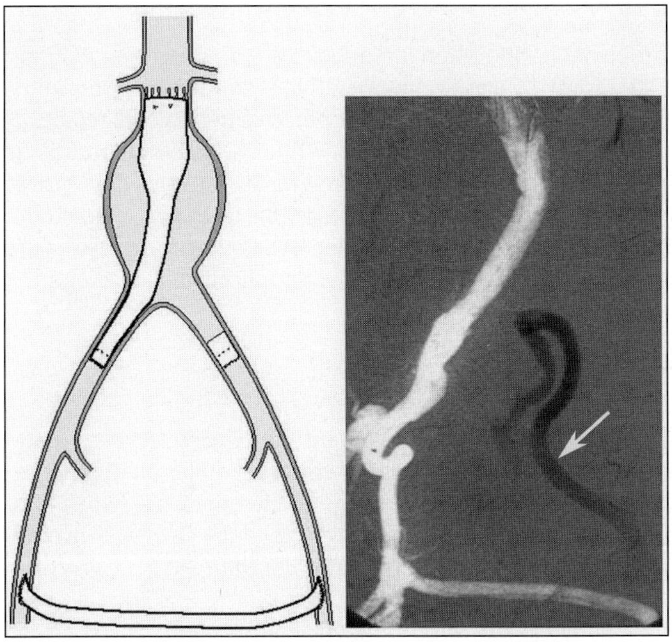

a. b.

Figure 18. *Aorto-unilateral iliac stent graft design. a) Schematic showing a graft extending from the infrarenal aorta to the right common iliac artery. The contralateral iliac artery is occluded by an "occluder stent" and the left leg is perfused via a fem-fem crossover graft. b) Completion angiogram on a patient with this type of stent. Note reflux into the left external (arrow) and internal iliac arteries.*

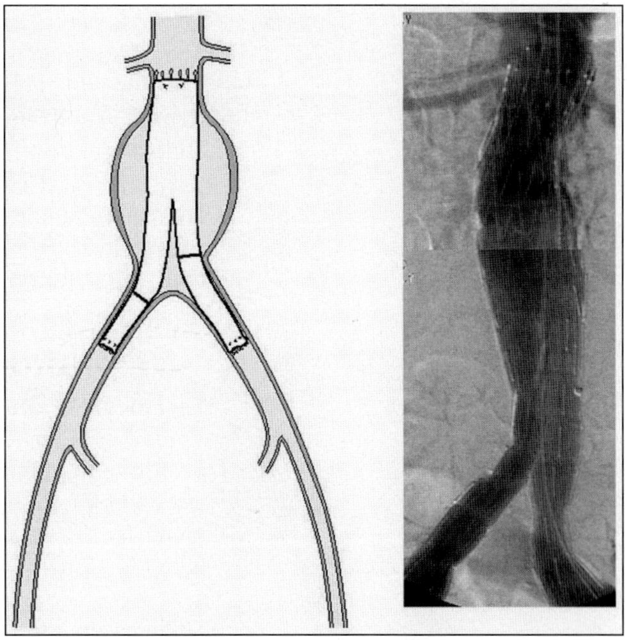

a. b.

Figure 19. *Bifurcated aortic stent-graft. a) Schematic showing the aortic stent-graft extending into both limbs. The main body of the device is placed through the right iliac artery and an extender graft is added to incorporate the left iliac artery. b) Completion angiogram on a patient with this type of stent.*

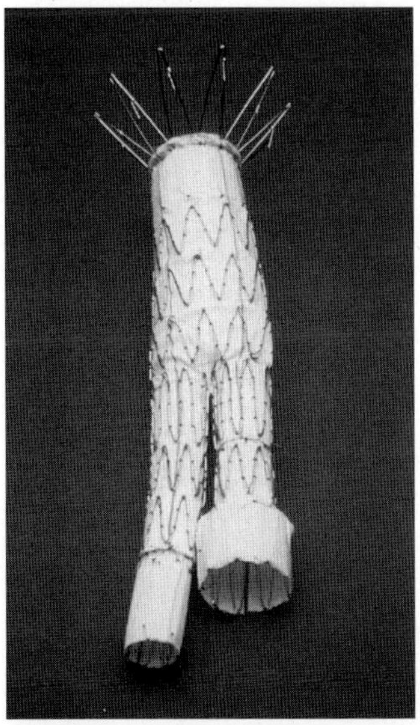

a.

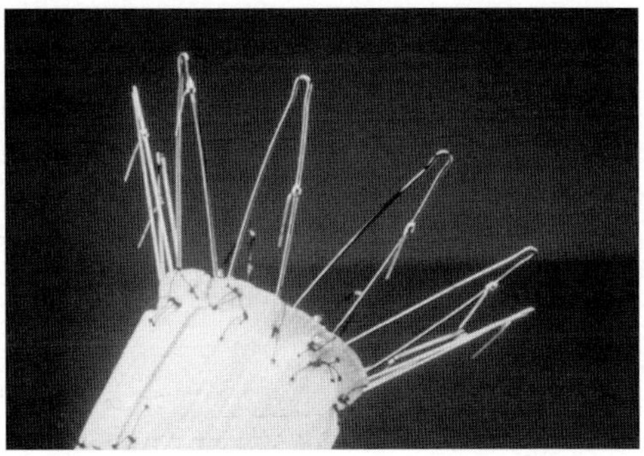

b.

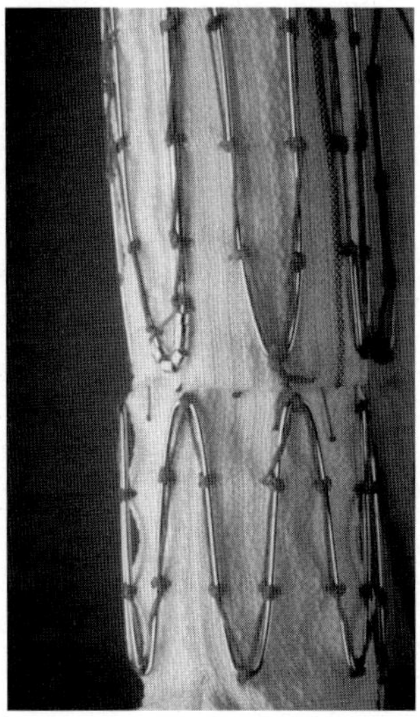

c.

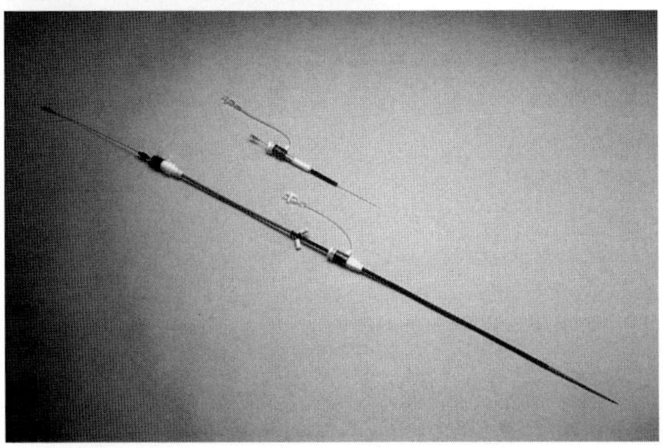

d.

Figure 20. *Photograph of a fully-supported, bifurcated stent-graft with the attachment stent wires outside of the graft fabric (Zenith Endovascular Graft, Cook Inc., Bloomington, Ind.) a) Bifurcated device. b) Proximal attachment barbs. c) Supporting stent on the outside of the fabric. d) The device is inserted through a 24-F introducer.*

Clinical Results

Aneurysm repair is a prophylactic operation intended to eliminate the risk of death from rupture. It is easy to study conventional operative repair because mortality, morbidity and failure occur in the immediate postoperative period. The results of endovascular aneurysm repair are harder to assess. While short-term morbidity is low, mid- and long-term morbidity are not known. Moreover, the field of endovascular aneurysm repair is in a state of constant flux as device characteristics are modified and improved. In this environment, even short-term results are hard to establish. Nevertheless, some of these preliminary clinical results are presented below.

Technical Success

A number of small series have now shown that stent-grafts can be inserted and positioned reliably in most patients. Extensive preoperative planning is necessary to determine the best approach and to determine the appropriate measurements for the stent-graft. But, success can be achieved in over 90% of well-imaged patients. Conversion to open surgery is required in fewer than 1%.

COMPLICATIONS OF ENDOVASCULAR AAA REPAIR

- Distal embolization
- Buttock claudication
  - With internal iliac artery occlusion
- Colonic ischemia (rare)
- Local groin complications
  - Lymphocele, hematoma, infection
- Postoperative fever

When compared to open surgery, endovascular stent-graft placement appears to offer lower morbidity. The benefit of the endovascular approach is the result of lower renal, cardiovascular and pulmonary toxicity. Complications of endovascular treatment include distal embolization of mural thrombus, ischemia from aortic side-branch or internal iliac artery occlusion, local groin complications (lymphocele, hematoma, necrosis, and infection), and self-limited postoperative fever. Distal embolization can be a devastating complication but has become less common with improved angiographic technique and improved device performance. Ischemia from occlusion of aortic side branches, leading to paralysis or colon ischemia, is quite rare. However, buttock claudication resulting from internal iliac artery occlusion is fairly common.

Anatomic Considerations

Proximal Attachment: The proximal stent performs two functions: fixation and sealing. Fixation is important because aortic flow generates large forces that may result in stent migration. Self-expanding stents do not exert enough pressure on the aortic wall to prevent migration and must be reinforced with barbed attachments. Balloon-expanded stent-grafts currently are not barbed and rely on tissue ingrowth to prevent migration. Overall migration rates in most series in follow-up are less than 5%.

The proximal neck of the aneurysm should be long enough to accommodate the proximal stent struts and to allow a secure seal. In order to accomplish this task, it may be necessary to partially cover the renal arteries with the stent struts. Although the long-term effects are not known, there does not appear to be any short-term ill effect from covering the renal artery orifice.

Distal Attachment: The initial aortic stent grafts were tube grafts attached to the aorta by just a proximal stent. These grafts invariably resulted in "leakage" or continued aneurysm perfusion around the distal graft. AAAs rarely have an uninvolved distal segment long enough or straight enough for distal stent implantation. For this reason, most stent grafts are implanted into the iliac arteries.

Implantation into the iliac arteries provides several different anatomic options. First the grafts may be unilateral or bilateral (bifurcated) grafts. Bifurcated grafts may have the iliac segments sewn into the fabric of the graft (these are the so-called unibody grafts) or the bifurcation is constructed in situ by connecting multiple stent graft components. Unilateral aortomonoiliac stent grafts have the simplest deployment system but require a concomitant femoral-femoral crossover graft and contralateral common iliac occluder.

Iliac Aneurysm: Approximately 30% of patients with aortic aneurysm have concomitant common iliac aneurysm, and most of these common iliac aneurysms extend all the way to the aortic bifurcation, leaving no suitable site for stent-graft implantation in the common iliac artery. The external iliac artery is rarely involved and can be used as the site for distal stent-graft implantation in almost all patients. However, this approach has several disadvantages, the most important of which is internal iliac artery occlusion with the potential consequences of impotence, buttock claudication, colon ischemia and paralysis. In our experience, buttock claudication is the most common of these problems and occurs in up to 50% of patients. Fortunately, collaterals develop over time and in most cases, symptoms resolve within six months to one year.

The presence of bilateral common or internal iliac artery aneurysms poses a special problem. Occluding both internal iliac arteries places the patient at increased risk for complications. Alternatively, we have used a staged approach that includes coil embolization of both internal iliac artery aneurysms followed by stent-graft placement **(Fig. 21)**. We first embolize one internal iliac artery then wait several months before embolizing the other, in the hope that collateral development will provide some protection from pelvic ischemia.

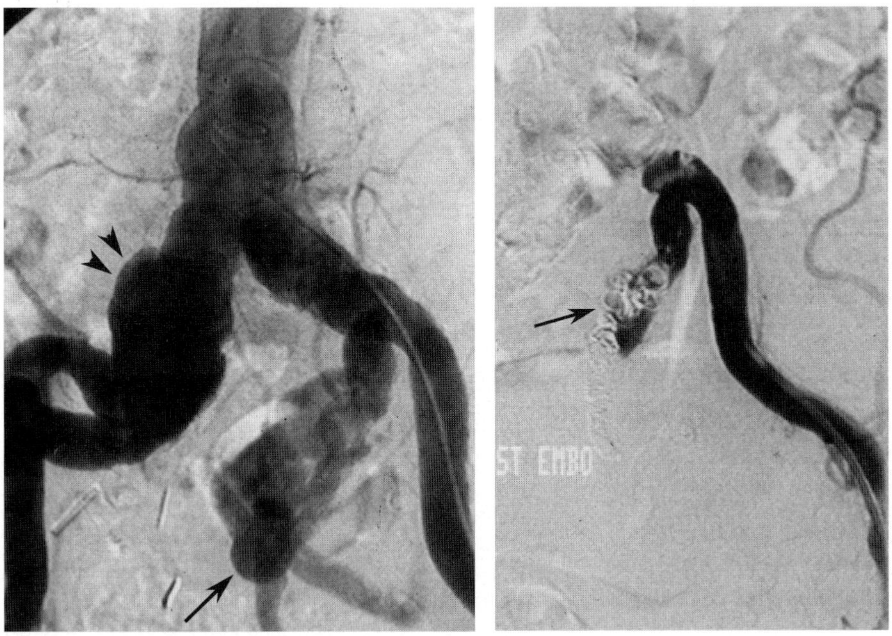

a. **b.**

Figure 21. *Endovascular management of AAA with concomitant iliac artery aneurysms. a) LPO pelvic arteriogram shows a large left internal iliac artery aneurysm (arrow) as well as a right common iliac artery aneurysm (arrowheads). b) The left internal iliac artery was embolized with coils (arrow). Subsequently the right internal iliac artery was embolized and a bifurcated stent-graft was placed that extended into the external iliac arteries.*

IDEAL ANATOMY FOR ENDOVASCULAR AAA REPAIR

- Neck greater than 15 mm
- Neck angulation less than 60 degrees
- Absence of iliac artery aneurysm
- No severe iliac artery tortuosity

Indications and Contraindications Based on Anatomy

Ideal anatomy includes an aneurysm neck longer than 15 mm with an angulation of less than 60 degrees. The iliac arteries should be free of aneurysms and free of severe tortuosity. However, patients without ideal anatomy may still be amenable to endovascular repair. In fact, about 25% of patients treated in our series at UCSF had anatomy that was not ideal as defined by the approved criteria.

Definite anatomic contraindications have not yet been defined but there are some patients who by virtue of their anatomy are poor candidates for this procedure. Some patients lack suitable sites for secure homeostatic stent-graft implantation. Others with small or very tortuous iliac arteries lack an acceptable route for transfemoral delivery system insertion. Despite these anatomic constraints, about 70% of patients with AAA can be treated with endovascular repair.

Follow-up

The most important endpoint in follow-up of endovascular AAA repair is aneurysm rupture. Short-term results suggest that this complication occurs uncommonly (< 1% of patients). Follow-up imaging evaluation is recommended to assess aneurysm size and perfusion. Serial CT evaluation at 3 and 6 months is standard. Aneurysm diameter usually stabilizes or declines following endovascular repair.

"ENDOLEAK" CATEGORIES

- I. Direct leak at the ends of the stent-graft
- II. Indirect, retrograde flow through side branches
- III. Indirect, through graft fabric
- IV. Indirect, at modular component joint

Although rupture is rare, evidence of persistent perfusion of the aneurysm is common. These so-called "endoleaks" are so common that they have been classified into four types:

I. around the distal end of the stent-graft
II. side-branch perfusion via the lumbars or IMA
III. and IV. through the stent-graft fabric either in the body or at the anastomosis of a modular component.

We believe that type I endoleaks should be treated by stent-graft revision and placement of additional stent-grafts. The natural history of type II leaks is not known and some investigators feel that they may resolve spontaneously and do not put the patient at great risk for rupture. However, if the leak appears large or if the aneurysm diameter increases, these endoleaks may be occluded by side-branch embolization **(Fig. 22)**.

Clinical Indications

Endovascular repair of AAA remains an investigational procedure for which clear indications and contraindications have not yet been established. Surgical risk may be shown to be an important determinant of therapy. It is thought that patients at high risk for open surgery may benefit most from endovascular repair. Low-risk patients with long life expectancy may be better treated by standard surgical repair. Unfortunately, the actual benefits and risks of the endovascular approach in terms of long-term morbidity and costs just is not known.

One of the chief objectives of endovascular AAA repair is low morbidity compared to open surgical repair. This is particularly true in the case of traumatic, inflammatory, anastomotic and ruptured AAAs. In these circumstances surgical exposure to affected arteries may be difficult or impossible and in these cases endovascular repair is particularly advantageous. Unfortunately, in the acute setting, preoperative imaging and stent planning is not possible. Under these conditions, off-the-shelf prefabricated grafts must be available in the appropriate size for the emergency patient.

Other Applications

Stent-graft technology has also been used to treat a variety of other arterial conditions including trauma, occlusion and dissection in a variety of other locations. Information on these applications is fairly sparse. Readers are encouraged to review a recent article by Dake et al on the thoracic applications of stent grafts. The promising application of stent-grafts is discussed in the chapter on TIPS.

Short small-bore covered stents have been used sparingly in the peripheral vessels. Vessels with small diameters are not particularly well suited to this application because of the bulky nature of current devices. However, in some situations, particularly for emergency use, PTFE-covered stents have proven to be quite valuable.

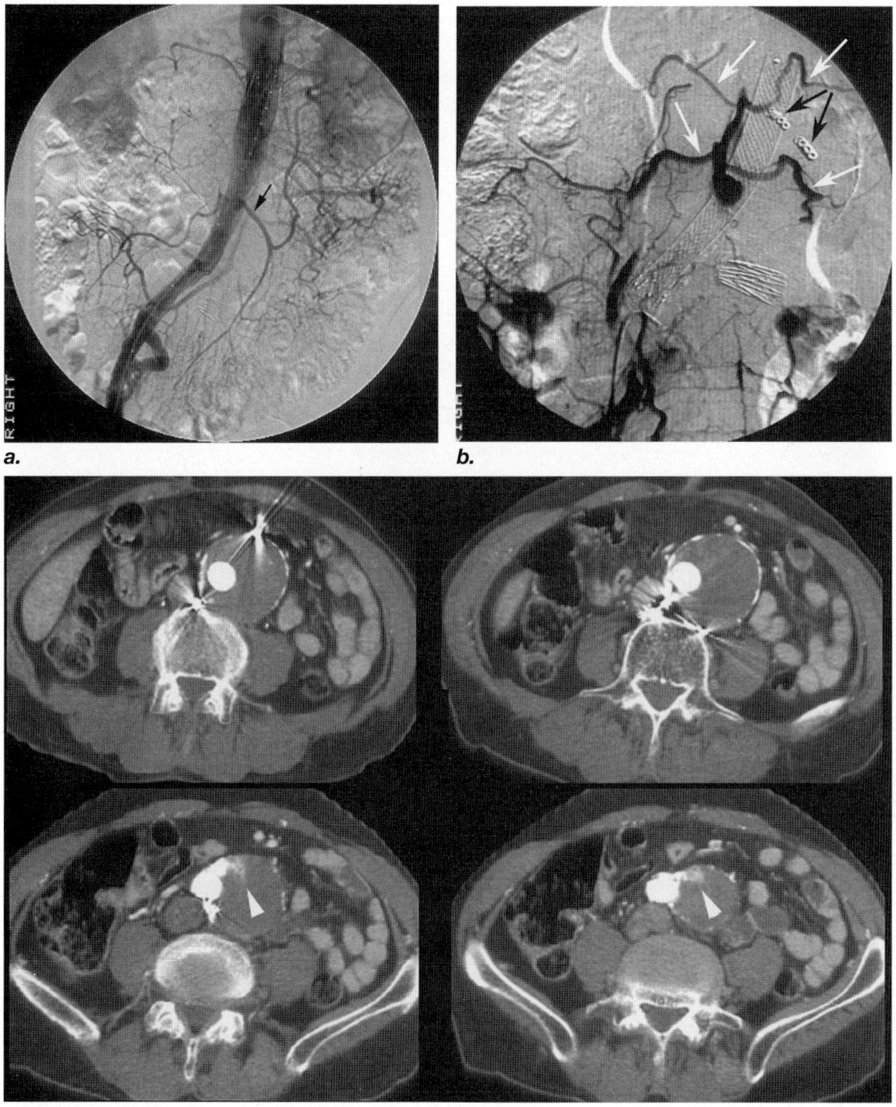

a. ***b.*** ***c.***

Figure 22. *Type II (side-branch) endoleak in a patient who underwent aortounilateral right iliac stent-graft. a) Late arterial phase image from midstream aortogram performed 2 months after endovascular repair reveals perfusion of the perigraft space from retrograde flow in the inferior mesenteric artery (arrow). The IMA origin was embolized with coils and then a selective right internal iliac arteriogram was obtained. b) Late arterial phase image from a right internal iliac artery injection shows perfusion of the perigraft space from multiple lumbar collaterals (white arrows). Note coils in the IMA (black arrows). c) CT following embolization of both the IMA and lumbar arteries reveals persistent perfusion of the aneurysm (arrowheads). The patient was treated conservatively and follow-up CT at 1 year showed no evidence of endoleak.*

REFERENCES

Vascular Stents

Binkert CA, Schoch E, Stuckmann G, et al. Treatment of pelvic venous spur (May-Thurner syndrome) with self-expanding metallic endoprostheses. Cardiovasc Intervent Radiol 1998; 2:22-26.

Bosch JL, Hunink MG. Meta-analysis of the results of percutaneous transluminal angioplasty and stent placement for aortoiliac occlusive disease. Radiology 1997; 204:87-96.

Carlson JW, Nazarian GK, Hartenbach E, et al. Management of pelvic venous stenosis with intravascular stainless steel stents. Gynecol Oncol 1995; 56:362-369.

Colapinto RF, Stronell RD, Gildiner M, et al. Formation of intrahepatic portosystemic shunts using a balloon catheter: preliminary clinical experience. AJR 1983; 140:709-714.

Henry M, Amor M, Ethevenot G, et al. Palmaz stent placement in iliac and femoropopliteal arteries: primary and secondary patency in 310 patients with 2-4 year followup. Radiology 1995; 197:167-174.

Henry M, Amor M, Henry I, et al. Stents in the treatment of renal artery stenosis: long-term followup. J Endovasc Surg 1999; 6:42-51.

Palmaz JC. The current status of vascular interventions in ischemic nephropathy. JVIR 1998; 9:539-543.

Richter GM, Noeldge G, Palmaz JC, et al. Transjugular intrahepatic portacaval stent shunt: preliminary clinical results. Radiology 1990; 174:1027-1030.

Rodriguez-Lopez JA, Werner A, Martinez R, Torruella LJ, Ray LI, Diethrich EB. Stenting for atherosclerotic occlusive disease of the subclavian artery. Ann Vasc Surg 1999; 13:254-260.

Sheeran SR, Murphy TP, Khwaja A, Sussman SK, Hallisey MJ. Stent placement for treatment of mesenteric artery stenoses or occlusions. JVIR 1999; 10:861-867.

Stein M, Link DP. Symptomatic spleno-mesenteric-portal venous thrombosis: recanalization and reconstruction with endovascular stents. JVIR 1999; 10:363-371.

Sueoka BL. Percutaneous transluminal stent placement to treat subclavian steal syndrome. JVIR 1996; 7:351-356.

van de Ven PJ, Kaatee R, Beutler JJ, et al. Arterial stenting and balloon angioplasty in ostial atherosclerotic renovascular disease: A randomised trial. Lancet 1999; 353:282-286.

Westcott MA, Bonn J. Comparison of conventional angioplasty with the Palmaz stent in the treatment of abdominal aortic stenoses from the STAR registry. SCVIR Transluminal Angioplasty and Revascularization. JVIR 1998; 9:225-231.

Stent-Grafts

Blum U, Voshage G, Lammer J, et al. Endoluminal stent-grafts for infrarenal abdominal aortic aneurysm. N Engl J Med 1997; 336:13-20.

Chuter TA, Faruqui RM, Sawhney R, et al. Endoleak following endovascular repair of abdominal aortic aneurysm. J Vasc Surg (in press).

Chuter TA, Gordon RL, Reilly LM, et al. Abdominal aortic aneurysm in high-risk patients: short- to intermediate-term results of endovascular repair. Radiology 1999; 210:361-365.

Dake MD, Kato N, Mitchell RS, et al. Endovascular stent-graft placement for the treatment of acute aortic dissection. New England J Med 1999; 340:1546-1552.

Kato N, Dake M, Semba C, et al. Treatment of aortoiliac aneurysms with use of single-piece tapered stent-grafts. JVIR 1988; 9:41-49.

Mialhe C, Amicabile C, Becquemin J. Endovascular treatment of infrarenal abdominal aneurysms by the stentor system: preliminary results of 79 cases. J Vasc Surg 1997; 26:199-209.

Parodi JC, Palmaz JC, Barone HD. Transfemoral intraluminal graft implantation for abdominal aortic aneurysms. Ann Vasc Surg 1991; 5:491-499.

CHAPTER 10 PREVIEW ANSWERS

Case 1

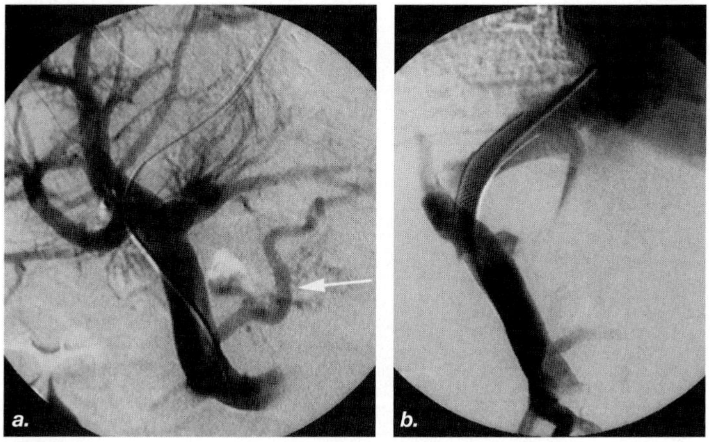

Figure 1.

Findings: a) Portal venography performed from a transjugular approach shows an enlarged coronary vein (arrow). b) A shunt from the hepatic vein to the portal vein is created by the deployment and dilatation of a self-expanding Wallstent within the parenchymal tract. The varices no longer fill.

Diagnosis: Bleeding esophageal varices successfully managed by creation of a TIPS, using a Wallstent.

Case 2

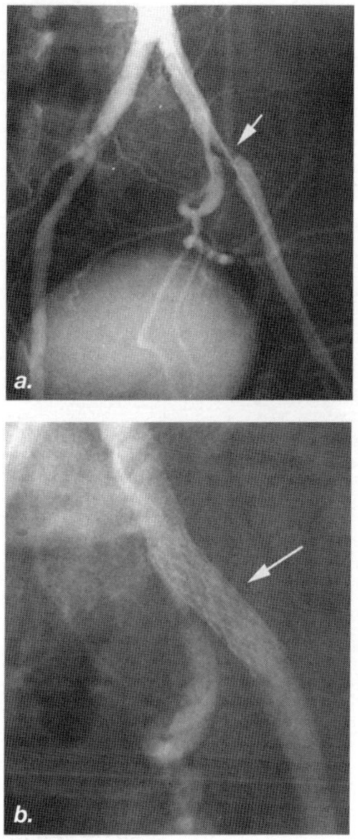

Figure 2.

Findings: a) Pelvic angiography shows a stenosis of the left external iliac artery (arrow). b) A Palmaz stent (arrow) has been deployed across the stenosis, improving the blood flow to the left lower extremity.

Diagnosis: Left external iliac stenosis secondary to atherosclerosis successfully treated by the placement of a Palmaz stent.

Case 3

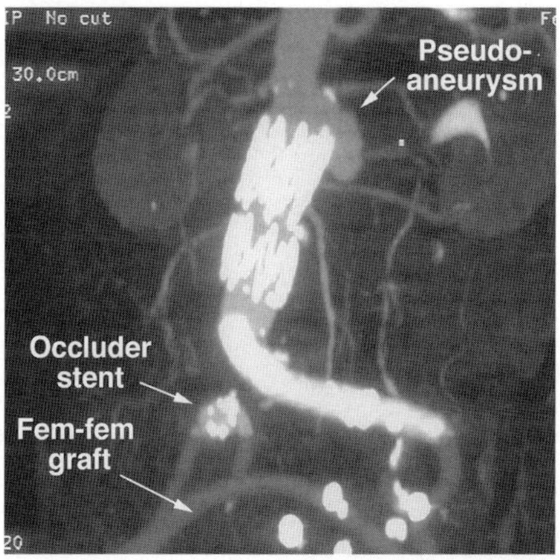

***Figure 3.** Maximal-intensity projection image (MIP).*

Findings: This is a partially-supported, unilateral aortoiliac stent-graft. The graft material is supported proximally and distally by Gianturco Z-stents. The contralateral iliac artery is occluded by placement of a "occluder stent" in the common iliac artery and the right leg is perfused via a fem-fem crossover graft. There is a pseudoaneurysm at the proximal anastomosis. In this case the pseudoaneurysm was due to infection, a very rare complication affecting stent-grafts.

Diagnosis: Mycotic pseudoaneurysm of aorta at the proximal end of a unilateral aortoiliac stent graft.

Case 4

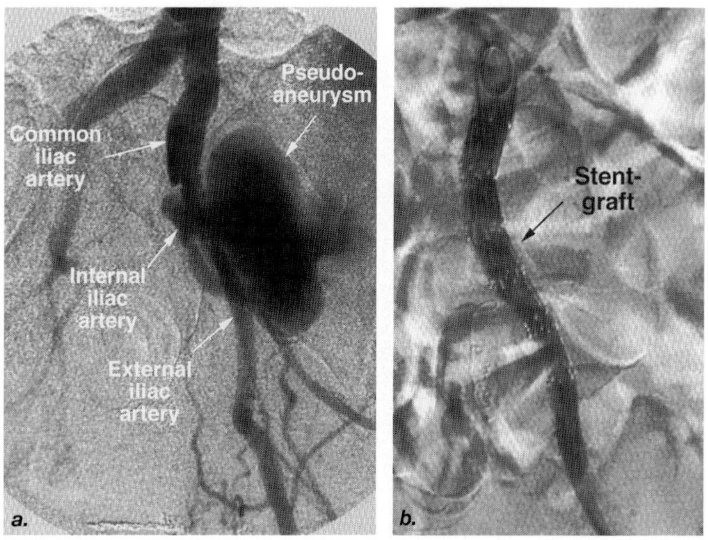

***Figure 4.** a) Pre-treatment arteriogram. b) Post-treatment arteriogram.*

Findings: a) There is a large pseudoaneurysm at the bifurcation of the left common iliac artery. The patient was considered at high risk for operative repair. b) A covered stent was placed across the pseudoaneurysm extending from the common iliac artery to the external iliac artery. The internal iliac artery was occluded with coils prior to stent-graft deployment.

Diagnosis: Postoperative pseudoaneurysm at the bifurcation of the left iliac artery successfully treated by stent-graft.

CHAPTER 11 PREVIEW

Case 1

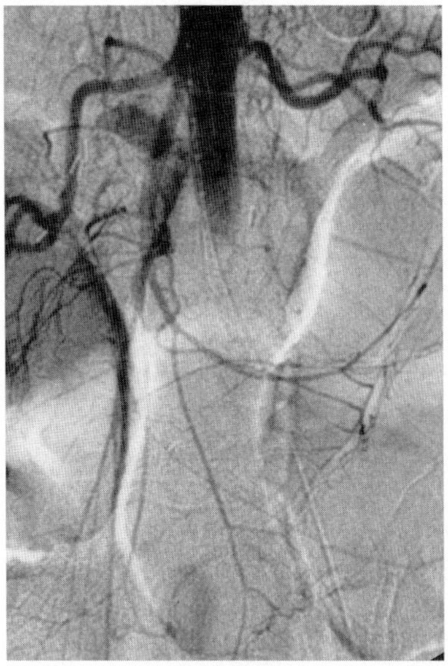

History: This 58-year-old woman previously had undergone an iliofemoral bypass and femoral-femoral bypass. She now presents with pain in both legs.
Question: Can this patient be treated with fibrinolytic infusion?

Case 2

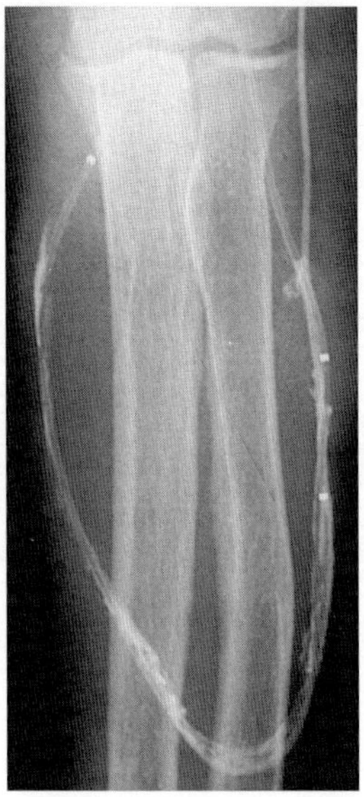

History: Occluded dialysis graft.
Question: What are the options for reopening the graft?

Case 3

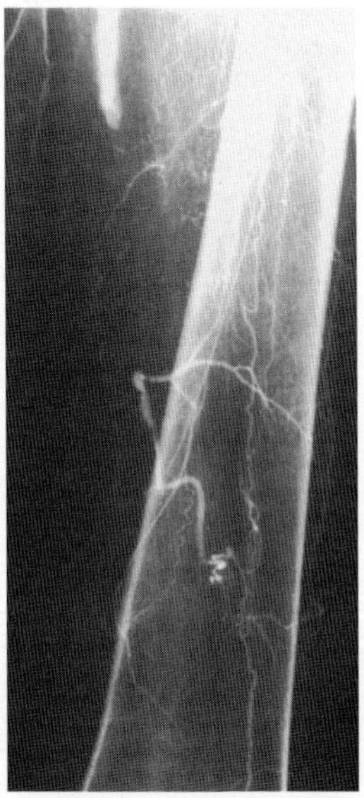

History: A 58-year-old man presents with numbness and an inability to move his lower leg.
Question: Should this patient be treated with a fibrinolytic infusion?

Case 4

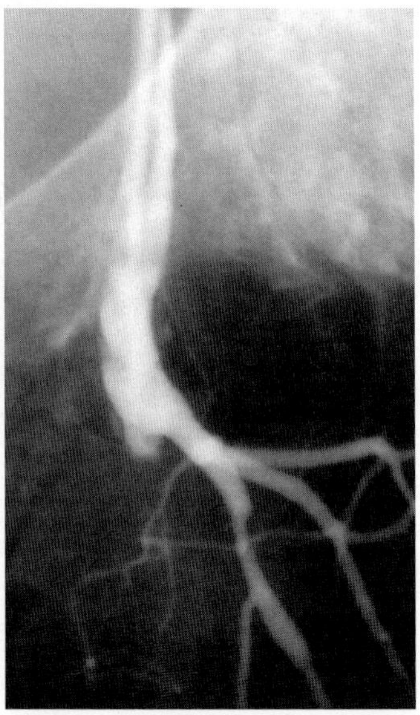

History: 55-year-old man undergoing fibrinolytic infusion to reopen a left femoral-popliteal bypass graft.
Question: Is the appearance of the infusion catheter in the left common femoral artery normal?

Chapter 11
Fibrinolytic Therapy

Robert K. Kerlan, Jr., MD

CHAPTER OUTLINE

- Fibrinolytic Agents
  - Streptokinase
  - Urokinase
  - Tissue Plasminogen Activator
- Technique
- Indications
- Contraindications
- Results
- Complications

INTRODUCTION

Since the concept of regional thrombolytic therapy was put forth by Charles Dotter in the early 1960's, catheter-directed fibrinolytic therapy has evolved into a common and accepted procedure for the management of acute thrombosis. The objective is to achieve a regional therapeutic thrombolytic effect without inducing a systemic fibrinolytic state. Despite more than thirty years of experience, no one agent or technique has become universally accepted. Different agents and dose recommendations make analysis of the existing data difficult. This chapter will focus upon currently available agents and examine their clinical applications from the viewpoint of the practicing radiologist.

THROMBOLYTIC THERAPY GOAL

- Rapid dissolution of thrombus
- Minimal systemic effect

FIBRINOLYTIC AGENTS

Three fibrinolytic agents are currently in clinical use: streptokinase, urokinase, and tissue plasminogen activator (tPA) (although at the time of this writing, urokinase is no longer distributed by the manufacturer). Though all of these agents use a different molecular mechanism, the ultimate result is the conversion of plasminogen to plasmin. The generated plasmin is then available to cleave fibrin molecules into fibrin split products, which is clinically recognized as dissolution of thrombus.

STREPTOKINASE

- Source: Lancefield group C beta-hemolytic streptococci
- Mechanism:
  - Combines with a molecule of circulating plasminogen to form an activated complex
  - This activated complex converts an additional plasminogen molecule to plasmin
- Half-life: 23 minutes

Streptokinase

Although it was the first agent used for selective fibrinolytic therapy, streptokinase is seldom used by interventional radiologists. As a bacterial byproduct from Group C beta-hemolytic streptococci, this agent is relatively inexpensive to manufacture. However, associated antigenicity creates problems with allergy and resistance in patients who have experienced previous streptococcal infections. Moreover, the indirect mechanism of action which requires the agent to combine with a free plasminogen molecule to create an activator complex leads to a non-linear dose-response curve. The half-life is 23 minutes and the usual dose is 5,000 units/hour.

STREPTOKINASE

- Advantages
  - Inexpensive and readily available
- Disadvantages
  - Circulating antibody neutralization
  - Antigenesis
  - Anaphylaxis/Serum sickness

UROKINASE
• Source: Tissue culture, human fetal renal cells, endothelium • Mechanism: Nonspecific serine protease which converts plasminogen to plasmin • Half-life: 16 minutes

Urokinase

Previously, the most common agent used due to a combination of safety, efficacy and marketing. In 1999, the United States Food and Drug Administration curtailed distribution of urokinase because of concerns regarding the methodology of production which potentially could lead to viral contamination. As urokinase is a human protein, problems with allergy and resistance are not encountered. The half-life is 16 minutes and the usual dose ranges from 60,000 units/hour to 250,000 units/hour.

UROKINASE
• Advantages — Safety — Nonantigenic • Disadvantages — Expensive — Not currently available

Tissue Plasminogen Activator

A genetically engineered product of the biotech industry, it was hoped that tPA would limit its activity to plasminogen within the targeted thrombus. Though this hope was not realized in clinical practice, in many centers tPA has become the agent of choice for regional thrombolytic therapy. This agent is extremely potent and despite readjustments of dosage regimens, systemic effects and hemorrhagic complications have been observed. Diminishing the dose to lower levels is limited by the stability of tPA when diluted. The half-life of tPA is 6 minutes and the dose ranges frcm 0.5 mg to 5 mg per hour.

tPA
• Source: Product of melanoma cells which allowed isolation and DNA coding. Subsequent insertion into Chinese ovary cells and *E. coli* carrier • Mechanism: Binds to plasminogen and fibrin to form a three-part complex. Plasminogen is then activated to plasmin • Half-life: 6 minutes (plasma)

TECHNIQUE

A wide variety of techniques for fibrinolytic therapy have been described. Though strict scientific comparison of these methods has not been performed, certain principles appear to have merit. The infusion catheter is usually inserted through a vascular sheath. This facilitates catheter exchanges and may allow angiographic imaging while the infusion catheter is in place.

It is clearly advantageous to imbed the catheter within the thrombus rather than leaving the tip proximal to the clot. A number of different catheters and catheter systems have been developed specifically for intrathrombotic fibrinolytic infusion. These systems include catheters with multiple side-holes **(Figs. 1, 2)** with and without tip-occluders, coaxial catheters with adjustable infusion lengths, as well as hollow infusible guidewires.

tPA
• Advantages — Selective fibrinolysis with diminished serum fibrinogenolysis — Availability • Disadvantages — Dilution — Expense

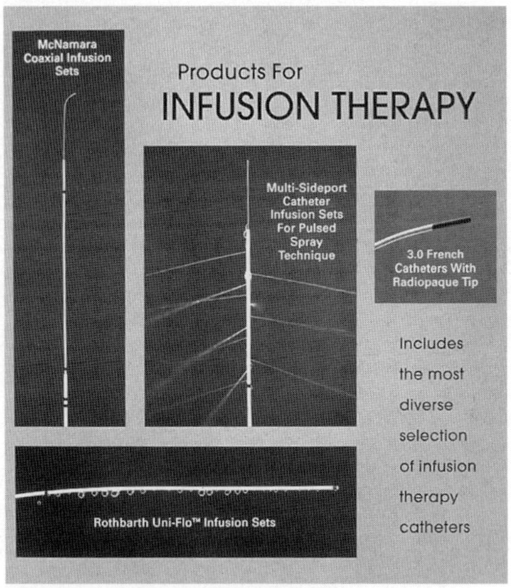

Figure 1. *A wide variety of catheters have been specifically designed for fibrinolytic infusion. (Figure courtesy of Cook Inc., Bloomington, Ind.)*

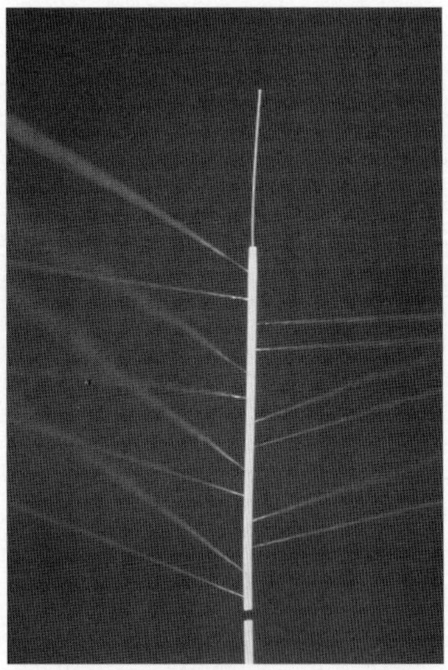

Figure 2. *Multiple side-hole catheter with a tip occluder is a common way to deliver fibrinolytic therapy. (Figure courtesy of Cook Inc., Bloomington, Ind.)*

Lacing the thrombus with a concentrated dose of fibrinolytic agent prior to the initiation of the infusion may shorten the lysis time. During this maneuver in the arterial system, care must be used not to dislodge clot distally. Dislodgment of clot in the venous system is usually of no clinical consequence.

Dosage regimens vary from institution to institution and are dependent upon which agent is being used. Typical ranges per hour of infusion are described in the following table.

TABLE 1: DOSAGES OF FIBRINOLYTIC AGENTS

Agent	Dosage Range
Streptokinase	2,500 units to 10,000 units/hour
Urokinase	50,000 units to 250,000 units/hour
tPA	0.5 mg to 5 mg/hour

Infusions are generally performed in an intensive care unit or transitional unit where continuous nursing supervision is feasible. Infusions generally last from a few hours to a maximum of 2 or 3 days depending upon the amount of thrombus and dosage regimen used.

During the infusion, patients must be watched carefully for evidence of bleeding. All stools should be checked for the presence of occult blood. No intramuscular injections are given.

A problem which can develop during the course of a fibrinolytic infusion is the formation of pericatheter thrombus surrounding the delivery catheter proximal to the infusion ports. To prevent this complication, many investigators advocate systemic heparinization while the infusion catheter is in place.

Baseline laboratory evaluations should include a blood count, platelet count, fibrinogen level, and prothrombin time. Partial thromboplastin times are warranted for patients undergoing concomitant heparin therapy. The laboratory values should be repeated at 6- to 12-hour intervals. A significant drop in the blood count and platelet count may indicate occult systemic bleeding. A drop in the fibrinogen level below 100 mg/dl usually indicates systemic fibrinolytic effect and places the patient at an elevated risk for a hemorrhagic complication. Most investigators recommend discontinuing the fibrinolytic infusion if the fibrinogen drops below 100 mg/dl.

REGIONAL THROMBOLYTIC THERAPY DELIVERY SYSTEMS

- Single end-hole catheter
- Coaxial catheter/injectable guidewire systems
- Multiple side-hole regional systems with "pulsed-spray"

MULTIPLE SIDE-HOLE INTRATHROMBUS INFUSION

- Intrathrombus injection of fibrinolytic agent which isolates the enzyme to the clot
- Diminished systemic exposure to agent
- Simultaneous treatment of the entire thrombus reducing treatment time

As an alternative to continuous infusion of fibrinolytic agents, a short course of high-dose fibrinolytic therapy may be administered in the angiographic suite using the "pulse-spray" technique. With this technique, a highly concentrated solution of fibrinolytic agent is forcibly injected through a multiple side-holed catheter with a tip-occluder, mechanically disrupting the clot and increasing the surface area exposed to the pharmacologic agent. This technique is almost always used for occluded dialysis access grafts on an outpatient basis. Typical parameters would be 0.2 ml to 0.4 ml of urokinase at a concentration of 25,000 units/ml injected every 30 seconds for 15 to 20 minutes.

INDICATIONS

Thrombolytic therapy may be used for acute arterial or venous thrombotic occlusions. Fibrinolytic therapy is also used in selected cases of embolic disease, though surgery may be more expedient when the underlying vessel is not extensively diseased. **Table 2** summarizes the current applications of selective fibrinolytic infusions.

TABLE 2: INDICATIONS FOR FIBRINOLYTIC INFUSION

Accepted Indications

- Acute occlusion of saphenous vein femoral-popliteal arterial bypass graft
- Acute occlusion of prosthetic femoral-popliteal arterial bypass graft
- Acute occlusion of dialysis access fistula
- Restoration of venous access catheter patency
- Acute occlusion of one-limb of aortofemoral bypass graft
- Acute segmental occlusion of native iliac or femoral artery
- Acute embolic occlusion of popliteal or brachial artery
- Acute occlusion of subclavian vein
- Acute thrombosis of the superior vena cava
- Acute iliac or femoral venous thrombosis

Investigational Applications

- Acute cerebrovascular thrombotic or embolic occlusion
- Acute visceral and renal artery occlusion

Acute Occlusion of Saphenous Vein Femoral-Popliteal Arterial Bypass Graft

As thrombosed vein grafts cannot be effectively declotted with Fogarty embolectomy balloons, and autogenous veins are clearly the best bypass conduit, thrombolytic therapy is almost always indicated when a vein graft occludes. After restoration of patency, a causative lesion is usually uncovered which, in most cases, is amenable to balloon angioplasty **(Fig. 3)**. Moreover, if the bypass extends into one of the tibial vessels, there may not be a reasonable surgical alternative.

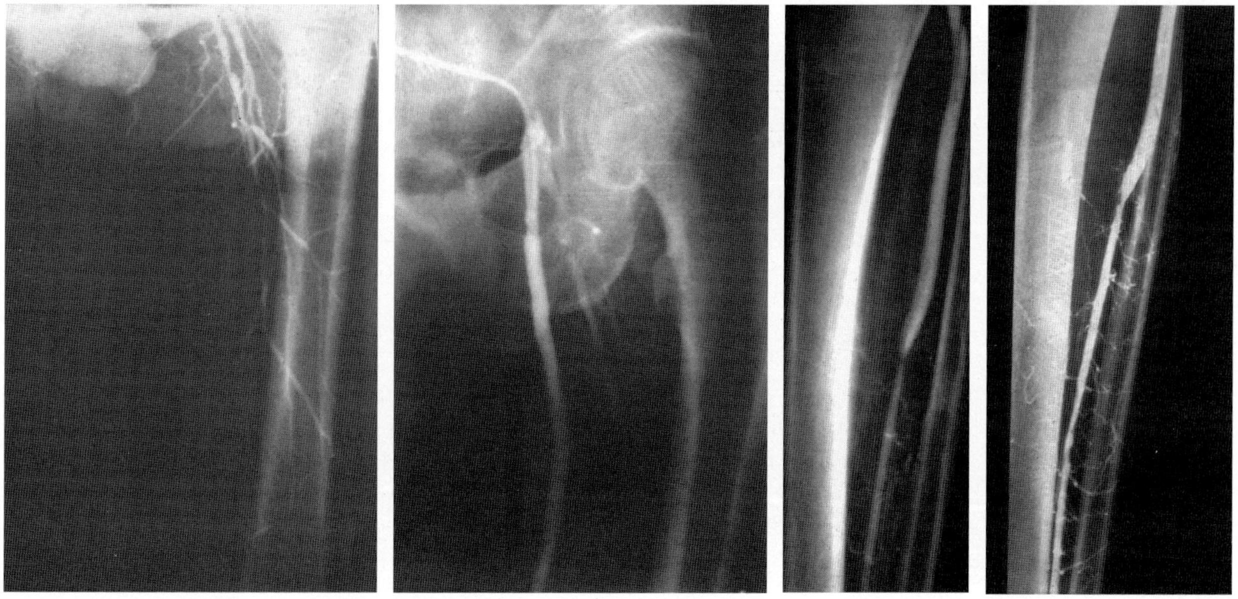

a. b. c. d.

Figure 3. *55-year-old man with acute ischemia of left lower extremity 6 months following left femoral to anterior tibial bypass graft. a) Left thigh arteriogram confirms occlusion of the reversed saphenous vein graft. b) 24 hours following the initiation of a urokinase infusion via an antegrade left femoral puncture, the graft is reopened. c) Lower extremity arteriogram through the infusion catheter reveals a distal anastomotic stenosis as the underlying cause of the thrombosis. d) Balloon dilatation was performed, with an acceptable result.*

Acute Occlusion of Prosthetic Femoral-Popliteal Arterial Bypass Graft

Prosthetic bypass grafts, usually made from PTFE or Dacron, are more amenable to Fogarty embolectomy compared to saphenous vein conduits. However, selective fibrinolytic therapy is often attempted to reopen prosthetic bypasses as endoluminal therapy can usually relieve the underlying cause of the thrombosis, avoiding the need for an open surgical procedure.

Acute Occlusion of Dialysis Access Grafts

The management of occluded dialysis grafts **(Fig. 4)** has evolved into the most common indication for fibrinolytic therapy in centers dealing with dialysis patients. Though the therapeutic efficacy of this form of management has never been prospectively compared to surgical thrombectomy in a randomized controlled trial, thrombolytic therapy is often used as the initial form of therapy to reestablish graft patency. Many of the techniques for fibrinolytic therapy have been developed in this group of patients.

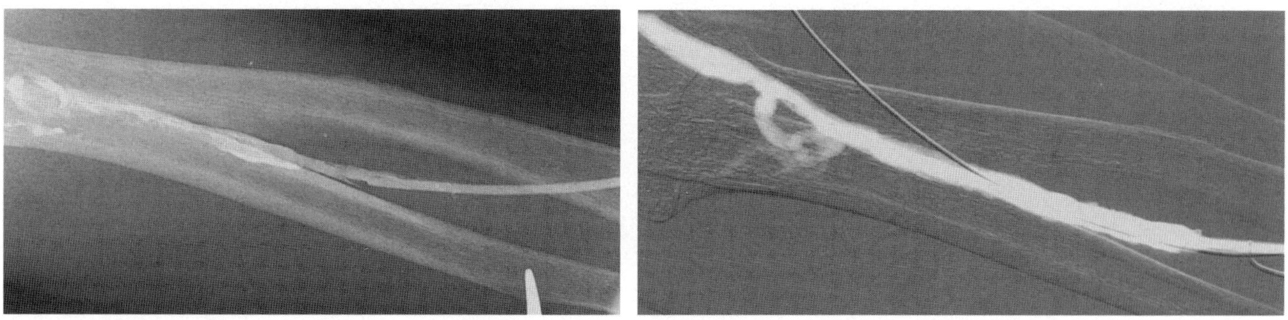

a. b.

Figure 4. *62-year-old woman with occluded left upper arm dialysis access graft. a) Minimal contrast is injected confirming thrombosis of left forearm dialysis fistula. b) Graft is reopened with pulse-spray fibrinolysis.*

Typically, two multiple side-holed catheters are placed in an overlapping "criss-cross" configuration. Pulse-spray fibrinolytic agents are administered simultaneously through each catheter. A dilatation balloon catheter is used to dilate the venous anastomosis, which is by far the most common cause of the graft thrombosis. A Fogarty balloon is then used to dislodge the platelet plug adjacent to the arterial anastomosis which is usually resistant to fibrinolysis.

Though fibrinolytic agents have frequently been used to reopen occluded grafts, mechanical thrombectomy devices offer a viable alternative and are used by many practitioners for this indication.

Restoration of Venous Access Catheter Patency

Venous access catheters which become occluded with thrombus can often be restored by the instillation of thrombolytic agents. Concentrated low-volume solutions containing between 5,000 and 100,000 units of urokinase are usually used for this purpose. Streptokinase (5,000 to 50,000 units) or tPA (1 mg to 5 mg) may also be used when urokinase is not available.

Acute Occlusion of Aortofemoral Bypass Limb

Occluded limbs of aortofemoral bypass grafts may be treated with fibrinolytic therapy **(Fig. 5)**. The infusion is usually done from the contralateral femoral approach through a Simmons or Sos-type catheter. A coaxial infusion catheter or wire can be used to lyse the occluding thrombus. A stenosis is usually revealed at the distal anastomosis as the cause for the graft occlusion, which may be amenable to balloon angioplasty. However, dense scarring may limit the durability of the angioplasty result and surgical revision may be required.

The large volume of thrombus makes distal embolization of fragments a frequent occurrence during lysis of an aortofemoral graft. Though these fragments are generally lysed with additional infusion, the transient ischemia encountered can be quite disturbing. Moreover, indwelling catheters through the contralateral side increase the possibility of infection or thrombosis of the patent limb.

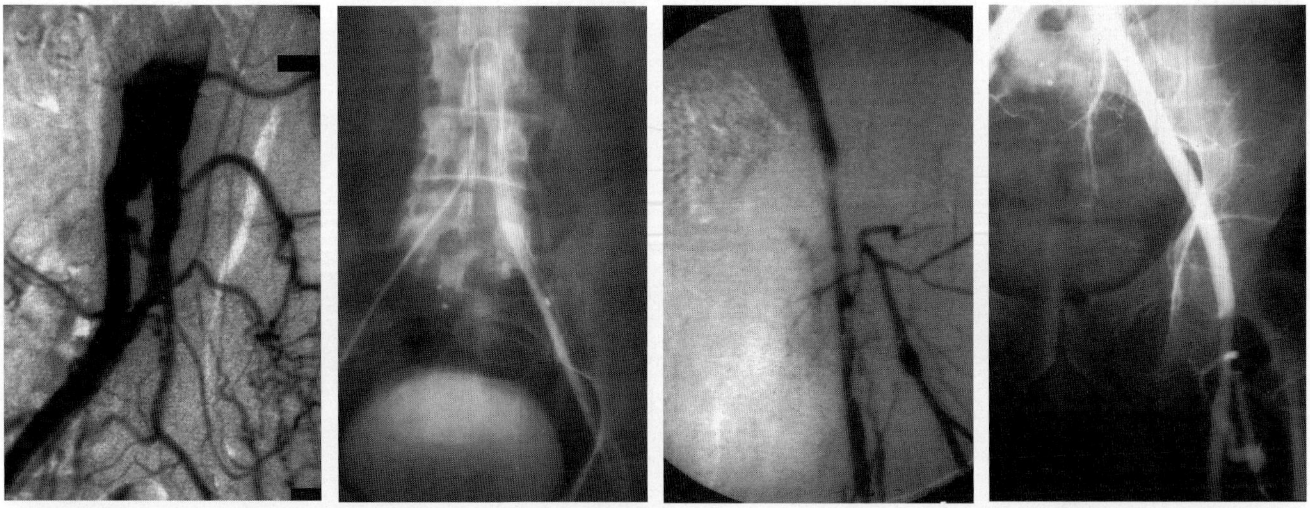

a. b. c. d.

***Figure 5.** 60-year-old man with acute left lower extremity ischemia 14-months following aorto-bifemoral bypass graft (AFBG) placement. a) Abdominal aortogram confirms occlusion of the left limb of the AFBG. b) An infusion catheter has been placed into the left limb from the right femoral approach, and partial lysis has been achieved. c) After additional fibrinolytic infusion, an arteriogram shows a stenosis at the distal anastomosis, occlusion of the left profunda femoral origin, and residual thrombus in the left common femoral and proximal left superficial femoral arteries. d) Following balloon dilatation of the distal anastomosis and additional fibrinolytic infusion, the distal anastomosis and left common femoral artery are widely patent.*

Acute Segmental Occlusion of Native Iliac and Femoral Arteries

Acute thrombosis of iliac or femoral arteries may not be recognized clinically. This nonrecognition can be attributed to the fact that thrombosis usually occurs at the site of a preexistent high-grade stenosis and that collaterals have developed circumventing the occluded segment. On occasion, these occlusions are encountered and recognized by a recent change in clinical symptomatology or the ease of guidewire passage through the occluded segment during attempted angioplasty. Under these circumstances, fibrinolytic therapy may be performed followed by angioplasty to correct the underlying stenosis **(Figs. 6, 7)**.

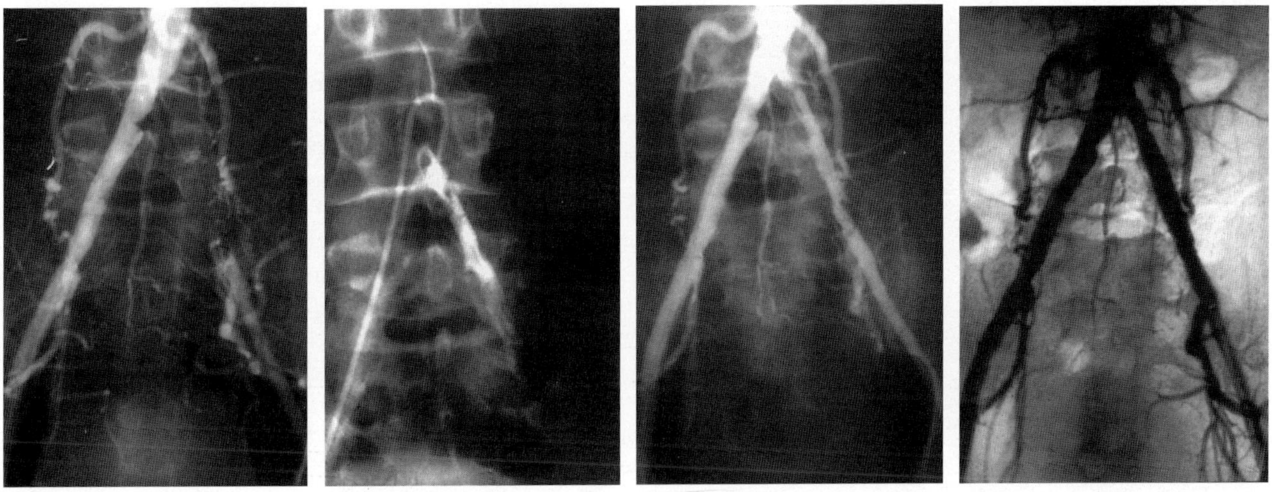

a. b. c. d.

Figure 6. *53-year-old man with left leg and buttock claudication which was recently more pronounced. a) Abdominal aortogram depicts left common iliac occlusion. The left internal iliac artery fills through lumbar collaterals and reconstitutes the left external iliac artery. b) A urokinase infusion is initiated through a catheter which crosses the aortic bifurcation into the occluding thrombus. c) Follow-up arteriogram discloses a critical stenosis in the proximal left common iliac artery. d) Following balloon angioplasty from an ipsilateral left-sided approach, the left common iliac artery is widely patent. (Case courtesy of Barry T. Katzen, MD; Miami Vascular Institute.)*

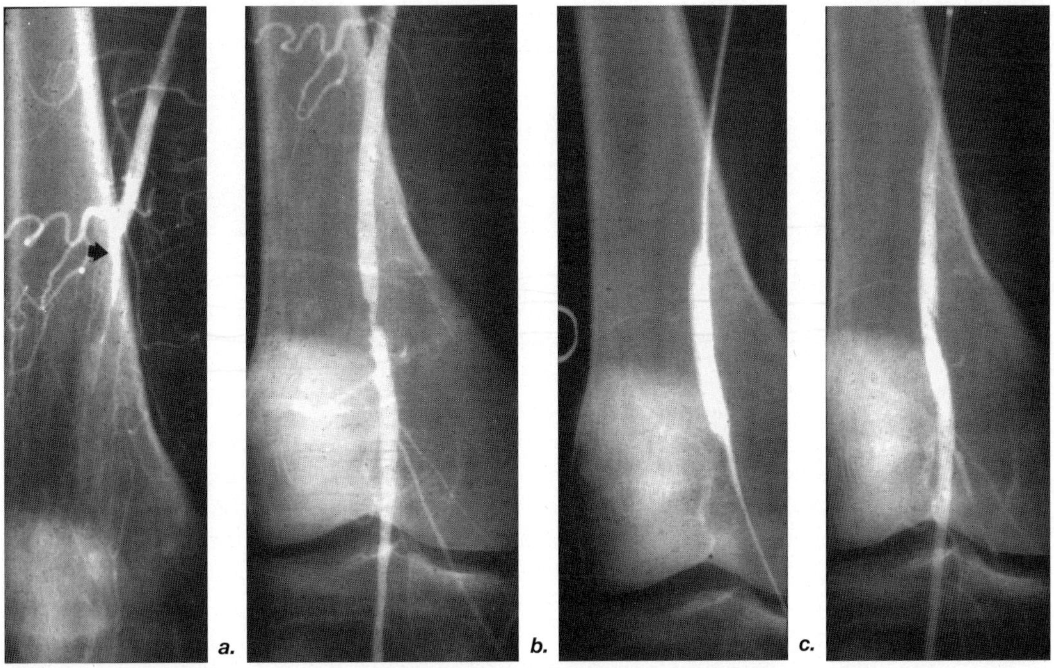

a. b. c. d.

Figure 7. *52-year-old man with acute ischemia of the right calf. a) Right leg angiogram reveals occlusion of the right popliteal artery. The tapered termination of the vessel (arrow) suggests the presence of acute thrombosis. b) An overnight urokinase infusion lyses the thrombus and an underlying atherosclerotic stenosis is revealed. c) Balloon angioplasty of the lesion is performed. d) Postangioplasty arteriogram demonstrates minimal residual narrowing.*

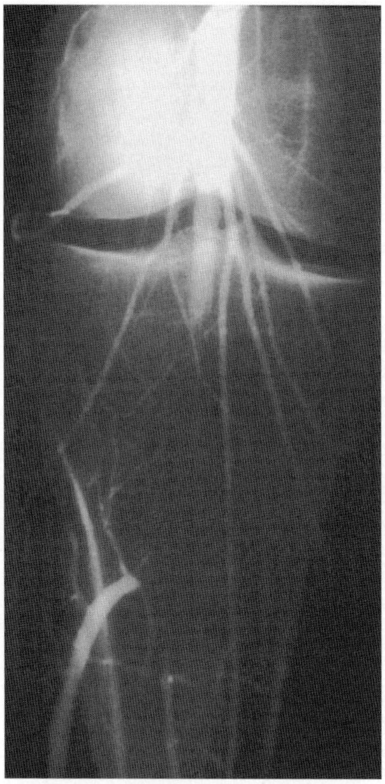

Figure 8. *Acute embolic occlusion of the right popliteal artery. Note typical location and lack of associated atherosclerotic disease.*

Embolic Occlusion of the Popliteal or Brachial Artery

Emboli **(Fig. 8)** can be treated with selective fibrinolytic infusion. As most large emboli originate from mural thrombus in the heart, if a significant systemic fibrinolytic effect is engendered, there exists a theoretical risk of additional embolization. It should be recognized that most patients do not develop measurable systemic fibrinolytic effects from regional infusion, making lysis of embolic clots safe in the majority of patients. However, emboli frequently create profound ischemia as no preexistent collaterals are present. Therefore, surgical removal may be the most expeditious course of management.

Acute Occlusion of the Subclavian Vein

Subclavian vein thrombosis is a difficult clinical problem, usually encountered in young and middle-aged patients. The thrombosis may develop secondary to a repetitive or strenous activity involving the upper extremity and is termed "effort thrombosis" or Paget-Schroetter syndrome. Subclavian vein thrombosis is one of the complications of the thoracic outlet syndrome, the others being arterial compression and brachial plexus nerve compression, with the latter being the most common manifestation. The underlying cause is usually an obstructing band of connective tissue crossing the vein in the costoclavicular space. Though a substantial percentage of these patients do reasonably well with anticoagulation alone, about half of these individuals experience considerable chronic impairment from venous stasis. Therefore, aggressive treatment with fibrinolytic agents is usually warranted **(Fig. 9)**. During the fibrinolytic infusion, therapeutic anticoagulation with heparin is recommended to diminish pericatheter thrombosis which can be problematic during venous infusions. Following lysis of the clot, surgery is usually performed to address the extrinsic obstructing lesion at the thoracic inlet which is invariably present. Angioplasty may be used to augment the result if a residual stenosis is present. Stent placement is usually not performed in this clinical situation due to the lack of long-term durability.

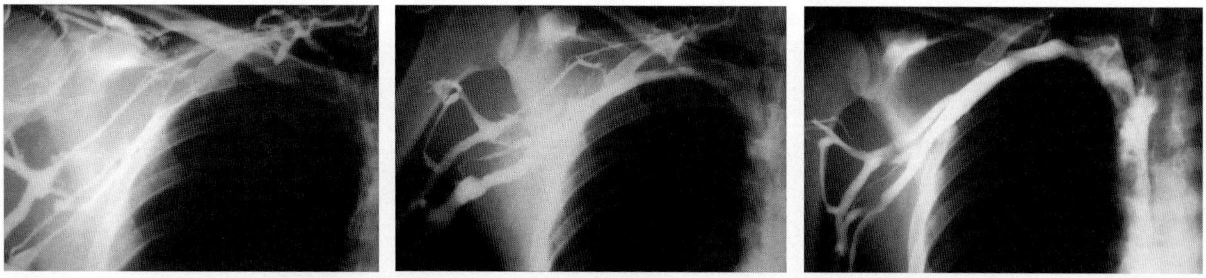

a. **b.** **c.**

Figure 9. *31-year-old man with edema of the right arm. a) Right upper extremity venogram demonstrates occlusion of the right axillary and subclavian vein. b) 24-hours following a urokinase infusion, the thrombus has lysed. Narrowing of the medial subclavian vein is present. c) A follow-up venogram following surgical removal of obstructing bands at the thoracic inlet shows a nearly normal appearance.*

Acute Thrombosis of the Superior Vena Cava

Superior vena caval thrombosis is a disabling clinical problem characterized by facial and bilateral upper extremity edema, headache, unclear mentation, difficulty swallowing and shortness of breath. Though often a complication of metastatic malignancy which has spread to the mediastinum, this syndrome may also be observed as a complication of indwelling central venous lines, or benign fibrosing mediastinitis related to granulomatous disease.

When a significant acute thrombotic component is present, fibrinolytic therapy **(Fig. 10)** can be used to reestablish patency. Depending on the etiology of the thrombosis, adjunctive angioplasty or stent placement may be necessary to maintain patency following successful clot lysis.

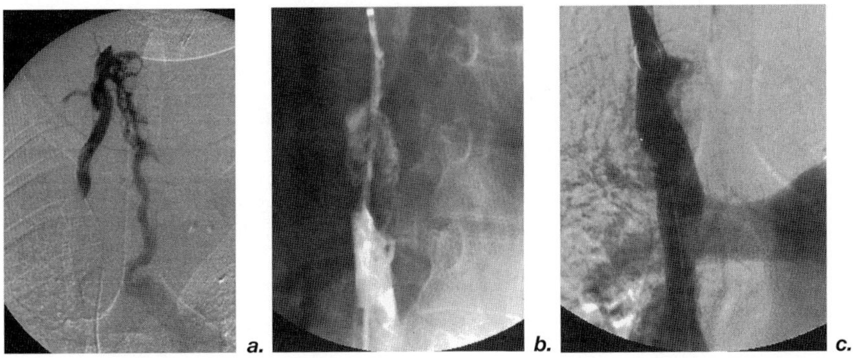

Figure 10. *Superior vena caval thrombosis. a) Superior vena cavogram reveals caval occlusion. The catheter has been advanced through thrombus. b) An infusion catheter is positioned across the thrombus. c) Following 20 hours of fibrinolytic infusion, the patency of the vena cava is restored.*

Iliofemoral Venous Occlusion

Over the past several years, fibrinolytic therapy has been used for the treatment of iliofemoral venous thrombosis. Patients with iliofemoral venous thrombosis usually present with unilateral leg and thigh edema. When there is extensive obstruction of the venous system, coupled with limited collateral flow, the edema is severe and the circulation may be compromised. This syndrome of a painful, edematous, and cyanotic extremity is termed phlegmasia cerulea dolens.

The etiology of iliofemoral venous thrombosis may be idiopathic, secondary to a hypercoagulable state, or due to compression of the iliac vein by tumor or scar tissue. When the left common iliac vein is narrowed due to compression by the overlying right common iliac artery, the condition is termed the May-Thurner syndrome.

Prior to the advent of regional thrombolytic therapy, the majority of patients with iliofemoral venous thrombosis were treated with bed-rest and anticoagulation with heparin, followed by long-term coumadin therapy. A significant percentage of these patients would suffer from chronic venous insufficiency, marked by persistent edema, trophic skin changes and ulcerations.

Because of the potential for these disabling chronic symptoms, patients with acute iliofemoral venous thrombosis are often treated with regional fibrinolytic therapy **(Fig. 11)**. The most common approach is to insert a multiple side-holed infusion catheter into the clot from an ipsilateral popliteal approach. The ipsilateral popliteal vein is punctured under ultrasound guidance with the patient in the prone position. As an alternative, iliac vein thrombosis can be treated by a catheter inserted retrograde from a contralateral femoral, or more frequently, a jugular venous approach. It should be noted, however, that treatment of femoral vein thrombosis from a retrograde approach can be difficult due to valves encountered in the retrograde direction.

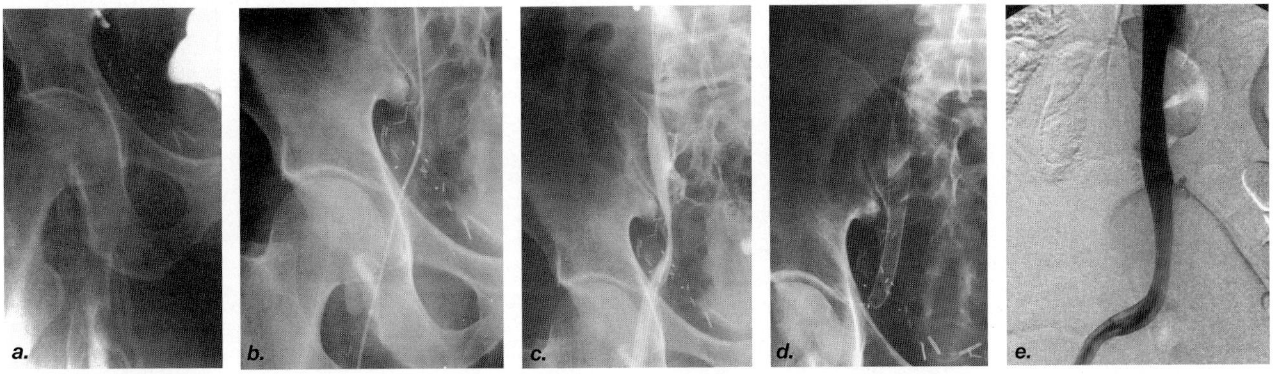

Figure 11. *Iliofemoral venous thrombosis. a) Right leg venogram shows iliofemoral venous thrombosis. b) An infusion catheter is placed through the thrombus from an internal jugular approach. c) The vein is reopened but stenoses are evident in the right external and right common iliac vein. d) Wallstents are used to bridge the stenotic areas. e) Following stent placement, no residual stenosis is evident.*

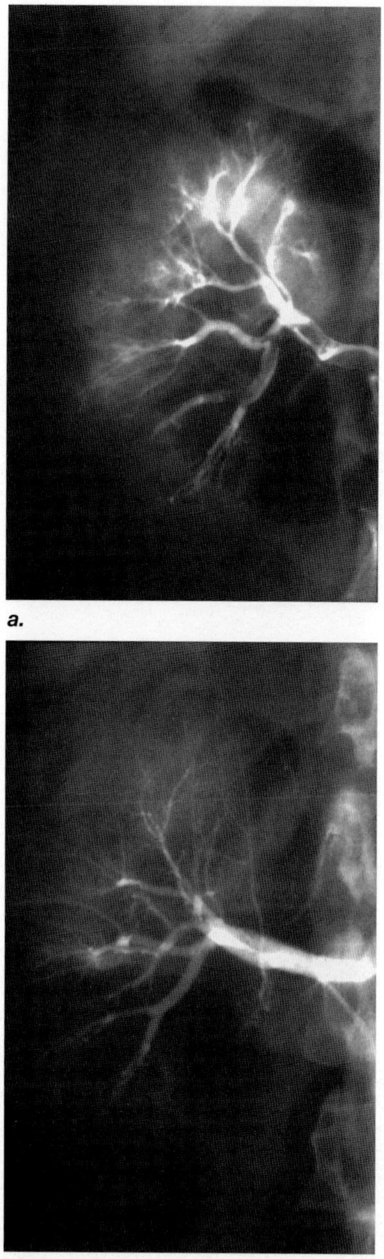

a.

b.

Figure 12. *62-year-old woman with atrial fibrillation and right flank pain. a) Selective right renal arteriogram shows an embolus within the right main renal artery extending into an upper pole branch. b) Following 3 hours of intraarterial urokinase infusion at 250,000 units/hour, the embolus has lysed and flow has been restored.*

Fibrinolytic therapy is continued for 1 to 3 days, until the majority of the thrombus is lysed. The majority of patients require iliac venous angioplasty often coupled with stent placement to treat the causative iliac venous narrowing.

Surprisingly, the occurrence of symptomatic pulmonary emboli during the course of iliofemoral venous thrombolysis is infrequent. Though occasional cases of pulmonary emboli have been reported, the placement of prophylactic inferior vena caval filters is usually not performed.

Acute Stroke

The use of systemic and regional fibrinolytic therapy in the management of acute cerebrovascular ischemia is currently being evaluated in multicentered controlled trials. All patients must undergo computed tomographic imaging to exclude the presence of intracerebral hemorrhage before fibrinolytic agents are administered. The results of these investigational trials are not currently available.

Acute Mesenteric Ischemia

Fibrinolytic therapy is not routinely used for patients with acute mesenteric ischemia due to the possibility of associated bowel necrosis. Most patients with emboli to the mesenteric circulation undergo surgical embolectomy and resection of compromised bowel. Patients with acute thrombosis secondary to an obstructing atherosclerotic lesion generally undergo mesenteric revascularization by either endarterectomy or bypass.

Recently, limited series have been reported using fibrinolytic therapy for selected patients with acute mesenteric ischemia. Patients were considered candidates if there was no physical examination evidence of peritonitis and no imaging evidence of irreversible bowel ischemia. In this subgroup of patients, selective fibrinolytic infusion, often followed by balloon angioplasty, was used. Approximately 60% of patients had resolution of their abdominal pain and did not require surgery. The reported experience, however, is quite small and conclusions regarding the wisdom of this therapeutic strategy await the report of a larger series of patients.

Acute Renal Ischemia

The majority of patients who occlude their renal artery do not come to immediate clinical attention, as the thrombosis occurs in the background of chronic atherosclerotic obstruction. However, patients who have acute embolic occlusion of the main renal artery (usually from a cardiac source) may present with flank pain. Unfortunately, by the time the diagnosis is made, most patients will have undergone irreversible ischemic damage to the kidney.

In patients who are diagnosed quickly, arteriography followed by selective fibrinolytic infusion **(Fig. 12)** may be useful to salvage some function of the involved kidney. Due to the sensitivity of the renal parenchyma to ischemic damage, short-term high-dose infusions (250,000 units/hour or greater) are preferable to lower-dose, longer-duration infusions.

TABLE 3: CONTRAINDICATIONS TO FIBRINOLYTIC THERAPY
Absolute
• Acute or subacute stroke
• Recent neurologic surgery
• Known CNS neoplasm
• Active gastrointestinal bleeding
Relative
• Recent abdominal or thoracic surgery
• Uncontrolled hypertension
• Remote history of stroke
• Remote history of GI bleeding

CONTRAINDICATIONS

Contraindications to regional fibrinolytic infusion are listed in **Table 3**.

When a patient is predisposed to a significant hemorrhagic event, the potential risks and benefits of fibrinolytic therapy must be carefully weighed. Though hemorrhagic complications occur infrequently, the results can be tragic when systemic bleeding is encountered. However, when fibrinolytic therapy offers the only viable therapeutic alternative, the risk may be warranted, even in patients who have had recent surgery.

RESULTS

Multiple studies have documented the efficacy of thrombolytic therapy and compared the efficacy of the different agents. A summary from these series comparing the different types of fibrinolytic agents is presented in **Table 4**.

TABLE 4: RESULTS OF REGIONAL FIBRINOLYTIC INFUSION COMPARING DIFFERENT FIBRINOLYTIC AGENTS

	SK	*UK*	*tPA*
Initial success	52%	81%	90%
Total complications	40%	26%	--
Bleeding complications:			
Total	34%	8%	27%
Major	14%	4%	8%
Infusion duration	44 h	19 h	7 h

(Sullivan KL, 1994.)

The clinical success of regional thrombolysis depends upon the type of occlusion being treated. A summary of series comparing the results of catheter directed fibrinolytic therapy in the treatment of different types of thrombotic or embolic occlusion is presented in **Table 5**.

COMPLICATIONS

As with any invasive therapeutic option, complications are occasionally encountered. Using fibrinolytic agents, both local and systemic complications may be observed. In addition to complications attributable directly to the fibrinolytic agent, complications related to catheterization as well as to radiographic contrast media may also be observed. Complications of fibrinolytic therapy vary in each individual series. Approximate rates of complications are listed in **Table 6**.

TABLE 5: RESULTS OF FIBRINOLYTIC THERAPY

Area Treated	(n)	Clinical Success (%)
Native lower extremity artery	222	77%
Lower extremity bypass graft	217	79%

(Cynamon J, 1994.)

TABLE 6: COMPLICATIONS OF REGIONAL FIBRINOLYTIC THERAPY

Complication	Frequency
Bleeding (req. transfusion or surgery)	5-15 %
Pericatheter thrombosis (req. surgery)	3 %
Stroke (intracranial hemorrhage)	2 %
Renal failure	2 %
Myocardial infarction	< 1 %
Death	< 1 %
Graft infection	< 1 %

REFERENCES

Cynamon J. Results of thrombolysis in peripheral arterial occlusive disease. In: Darcy MD, LaBerge JM, eds. SCVIR Syllabus: Peripheral Vascular Interventions. San Francisco: Society of Cardiovascular & Interventional Radiology, 1994; 140-152.

Holden RW. Plasminogen activators: pharmacology and therapy. Radiology 1990; 174:993-1001.

LeBlang SD, Becker GJ, Benenati JF, Zemel G, Katzen BT, Sallee SS. Low-dose urokinase regimen for the treatment of lower extremity arterial and graft occlusions: experience in 132 cases. J Vasc Interv Radiol 1992; 3:475-483.

McNamara TO, Fischer JR. Thrombolysis of peripheral arterial graft occlusions: improved results using high-dose urokinase. AJR 1985; 144:769-775.

Mewissen MW, Seabrook GR, Meissner MH, Cynamon J, Labropoulos N, Haughton SH. Catheter-directed thrombolysis for lower extremity deep venous thrombosis: report of a national multicenter registry. Radiology 1999; 211:39-49.

Meyerovitz MF, Goldhaber SZ, Reagan K, et al. Recombinant tissue-type plasminogen activator versus urokinase in peripheral arterial and graft occlusions: a randomized trial. Radiology 1990; 175:75-78.

Ouriel K, Vieth FJ, Sasahara AA. A comparison of recombinant urokinase with vascular surgery as initial treatment for acute arterial occlusion of the legs. N Engl J Med 1998; 338:1105-1111.

Ouriel K, Vieth FJ, Sasahara AA. Thrombolysis or peripheral arterial surgery, phase 1 results. J Vasc Surg 1996; 23:64-73.
Semba CP, Dake MD. Catheter-directed thrombolysis for iliofemoral venous thrombosis. Seminars in Vascular Surgery 1996; 9:26-33.
Shortnell CK, Francis CW. Thrombolytic therapy for arterial thrombosis. Current Opinion in Hematology 1999; 6:309-313.
Simó G, Echenagusia AJ, Camunez F, Turégano F, Cabrera A, Urbano J. Superior mesenteric arterial embolism: local fibrinolytic treatment with urokinase. Radiology 1997; 204:775-779.
Sullivan KL. Thrombolytic agents: pharmacology, dose regimens, and comparison of agents. In: Darcy MD, LaBerge JM, eds. SCVIR Syllabus: Peripheral Vascular Interventions. San Francisco: Society of Cardiovascular & Interventional Radiology, 1994; 119-125.
Sullivan KL, Gardiner GA Jr, Shapiro MJ, Bonn J, Levin DC. Acceleration of thrombolysis with a high-dose transthrombus bolus technique. Radiology 1989; 173:805-808.
The STILE Investigators: results of a prospective randomized trial evaluating surgery versus thrombolysis for ischemia of the lower extremity. Ann Surg 1994; 200:251-268.

CHAPTER 11 PREVIEW ANSWERS

Case 1

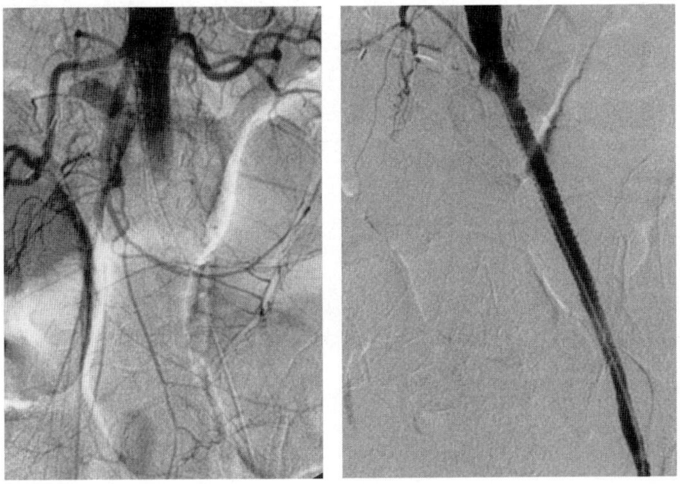

a. b.

Findings: a) The aortogram performed through the occluded graft confirms thrombosis of the iliofemoral bypass. b) The graft is reopened following an 18-hour infusion of tissue plasminogen activator (tPA).

Case 2

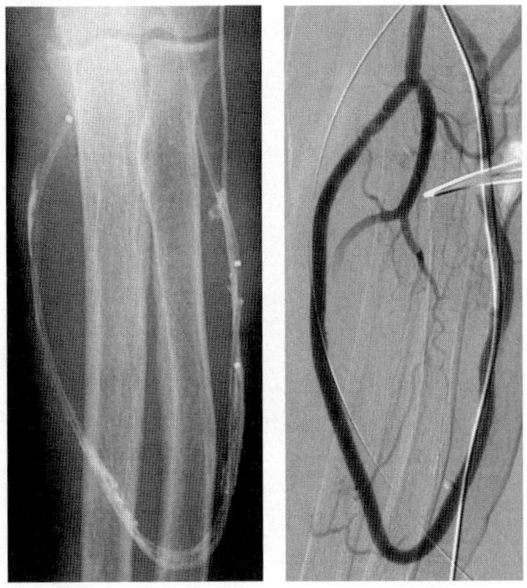

a. b.

Findings: a) The options for treating the occluded graft include fibrinolytic therapy, mechanical thrombectomy, and surgical thrombectomy. b) Urokinase infusion through criss-crossed catheters was used to reopen the dialysis graft.

Case 3

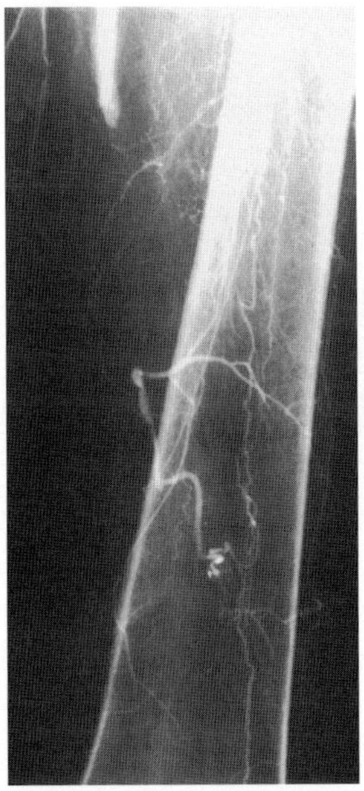

Findings: Patients with neurologic symptoms need to undergo emergent surgical revascularization rather than fibrinolytic therapy.

Case 4

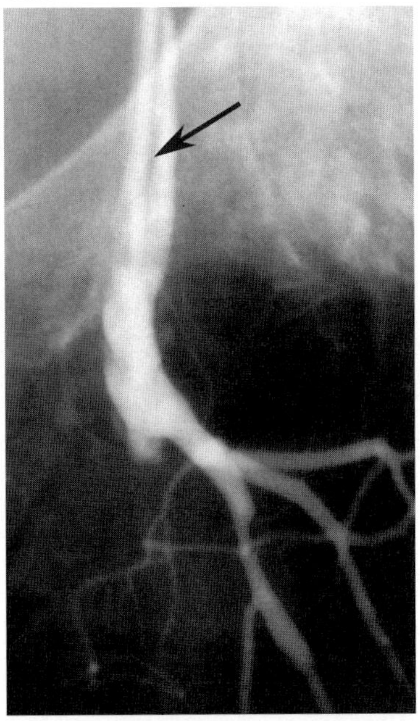

Findings: The infusion catheter has a surrounding sheath of pericatheter thrombus (arrow) and has pulled proximal to the femoral popliteal graft. The fibrinolytic material is not reaching the graft; rather it is flowing away through the patent profunda femoral artery.

CHAPTER 12 PREVIEW

Case 1

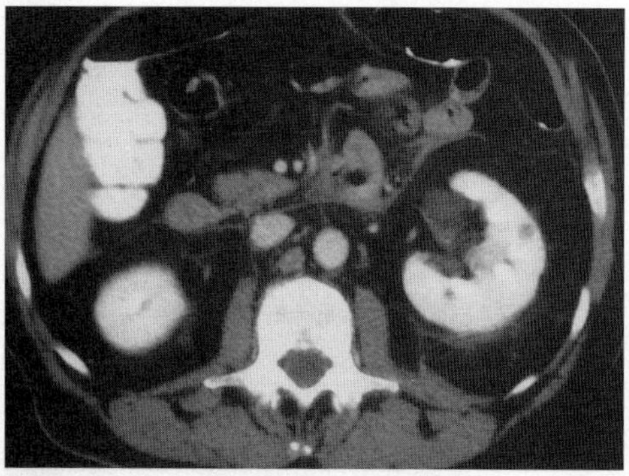

a.

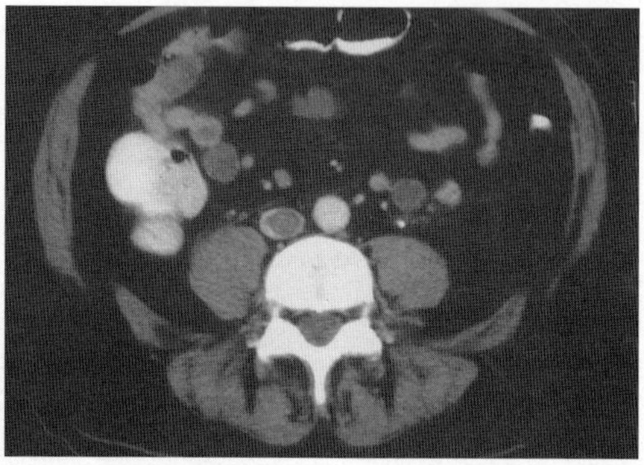

b.

History: Two contrast-enhanced CT scans of the abdomen in a patient with a pulmonary embolus.

Question: Is the inferior vena cava normal?

Case 2

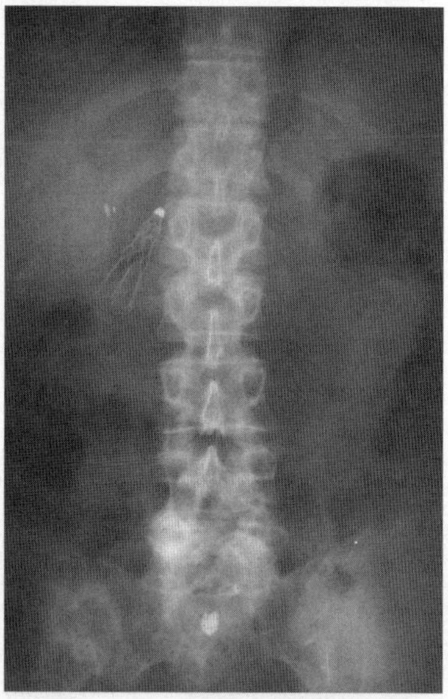

History: This patient had a filter placed from the jugular approach.

Question: Do you diagnose a problem?

Case 3

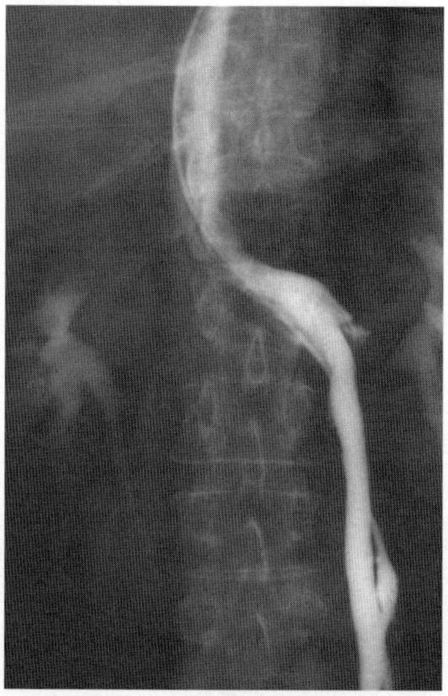

History: A venogram was performed from a jugular approach.

Question: What do you see?

Chapter 12
Inferior Vena Cava Filters

Roy L. Gordon, MD

CHAPTER OUTLINE

- Indications for filter placement
- Contraindications to filter placement
- Available filters
  - Vena Tech or LGM
  - Bird's Nest
  - Titanium Greenfield
  - Over-the-wire stainless steel Greenfield
  - Simon Nitinol
- Filter performance
- Complications of filter placement
- Technique of filter placement
  - Right femoral approach
  - Level of renal veins
  - Caval diameter
  - Venous anomalies
  - Position of filter
  - Suprarenal placement
  - Filter configuration
  - Left femoral approach
  - Right internal jugular approach
  - Left internal jugular approach
  - Right external jugular approach
  - Other approaches
- Management of complications
- Future developments

INTRODUCTION

Some 75% to 90% of pulmonary emboli (PE) originate in the veins of the leg or pelvis. The treatment of choice is adequate anticoagulation. However, there is a subset of patients at risk for PE who cannot be adequately anticoagulated in whom partial interruption of the inferior vena cava (IVC) is indicated to prevent emboli from reaching the lungs. This can be achieved by the use of percutaneously inserted filters placed in the IVC. The goal of the filter is to trap any thrombi passing up the IVC to the lungs without causing complete occlusion of the IVC. It is estimated that some 30,000 to 40,000 filters are placed each year in the United States.

This chapter addresses the use of these filters and focuses on five filters in widest use in the USA today, including Titanium Greenfield (TG), over-the-wire-stainless steel Greenfield (SG-OTW), Vena Tech (VT) (also called the LGM filter), Bird's Nest (BN) and Simon Nitinol (SN). There is some debate whether clinically significant pulmonary emboli originate in the venous bed that drains into the superior vena cava (SVC). Thrombosis in these veins is increasingly common as a result of the widespread use of these vessels for central venous access and dialysis. Although a few recent papers have reported safe placement of filters in the SVC, this is not a commonly performed procedure because of a lack of consensus that the procedure is either needed or safe. In the remainder of this chapter all discussions are therefore about filters in the IVC.

COMMONLY USED IVC FILTERS

- Titanium Greenfield
- Over-the-wire-stainless steel Greenfield
- Vena Tech
- Bird's Nest
- Simon Nitinol

INDICATIONS FOR FILTER PLACEMENT (Table 1)

The three most common indications for IVC filter placement are:

1) Contraindication to anticoagulation in patients with PE or deep venous thrombosis (DVT) of major veins such as the IVC, pelvic or leg veins;
2) Failure of anticoagulation therapy due to progression of DVT, recurrent PE, or noncompliance; and
3) Complications from anticoagulation.

Less common indications include:

- Massive PE;
- Free-floating thrombus in iliac veins or IVC;
- Chronic, recurrent PE with pulmonary hypertension and cor pulmonale; and
- Conditions predisposing patients to high risk of PE (e.g. post brain surgery, post hip surgery, post pelvic trauma, prior to major pelvic surgery).

Some authors have suggested that the ease of percutaneous filter placement has led to overutilization, particularly in patients who are terminally ill with malignant disease.

TABLE 1: FILTER INDICATIONS

- PE/DVT anticoagulation contraindicated
- Failed anticoagulation
  - Progressive DVT
  - Recurrent PE
  - Complication with anticoagulation
- Thrombus in IVC or iliac veins
- After massive PE
- Recurrent PE with cor pulmonale or pulmonary hypertension
- High risk following brain or hip surgery
- High risk following pelvic trauma or surgery

CONTRAINDICATIONS TO FILTER PLACEMENT

There are no absolute contraindications to filter placement. However, the potential benefits and risks must be carefully evaluated and balanced in each patient. The following situations demand special consideration. If a patient has untreated and uncontrolled bacteremia, immediate and appropriate antibiotic treatment should be started before filter placement. Filter placement in pediatric patients is especially controversial because of the uncertain long-term effects of filters and the long-term durability of these filters. Equally controversial is the use of filters in pregnant women.

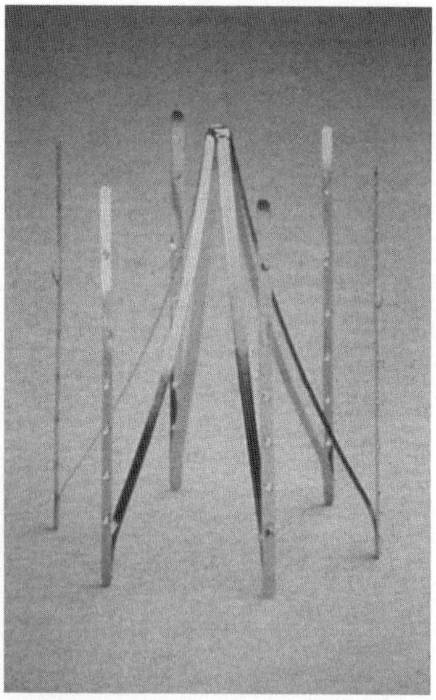

Figure 1. *Vena Tech Filter.*

AVAILABLE FILTERS

Vena Tech (VT) *or* LGM

This is a cone-shaped filter **(Fig. 1)** that uses side legs to achieve axial orientation and central location. It is easy to insert from either a right femoral or a right internal jugular approach using an introducer with an outer diameter of 14.6-F. An advantage of the VT is that a single system can be used from either the jugular or femoral approach reducing the cost of holding supplies of equipment. Although not advised, nor always possible, placement from a left femoral vein puncture can often be successfully performed. Made of an alloy named Phynox, the VT lacks ferromagnetic properties, has no detectable motion in an MRI magnetic field, and causes only a minimal imaging artifact.

The legs of the VT do not appear to penetrate the IVC wall. Problems with the VT are that it does not always spring open completely on deployment **(Fig. 2)**. It is not clear if this has any clinical importance in terms of filtering efficiency. To minimize this problem, the delivery system should be well-flushed with heparin saline to minimize thrombus formation and delivery of the filter should be performed rapidly. In cases where the filter does not open completely, we make gentle attempts with a guidewire to encourage complete opening which are generally successful **(Fig. 3)**. Other interventionalists avoid this for fear of causing migration and because they doubt whether it is necessary. No hard data are available on these issues.

VENA TECH FILTER	
Sheath	14.6 F
Maximum diameter	28 mm
Length	38 mm
Material	Phynox
MR movement	No

Follow-up studies on the VT are more complete than for many of the early filters. Recurrent pulmonary emboli rates are in the 3% range which is in the accepted range for most filters, but IVC thrombosis seems to be more common than for other filters. A progressive decrease in patency has been reported with a patency rate of only 66.8% at 9 years follow-up. This high thrombosis rate may be a result of longer and more diligent follow-up of VT filters than for other filters. Minor caudal migration has been reported, but we have observed no clinical problems with migration. The VT is currently the filter we use for most patients with cavae not wider than 28 mm in diameter.

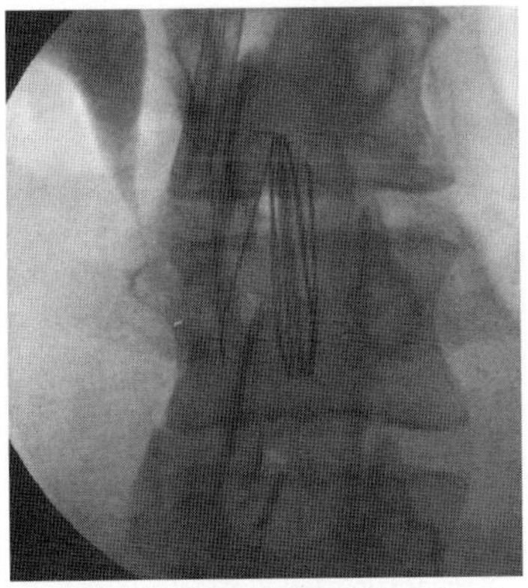

Figure 2. *The legs of this Vena Tech filter are incompletely open.*

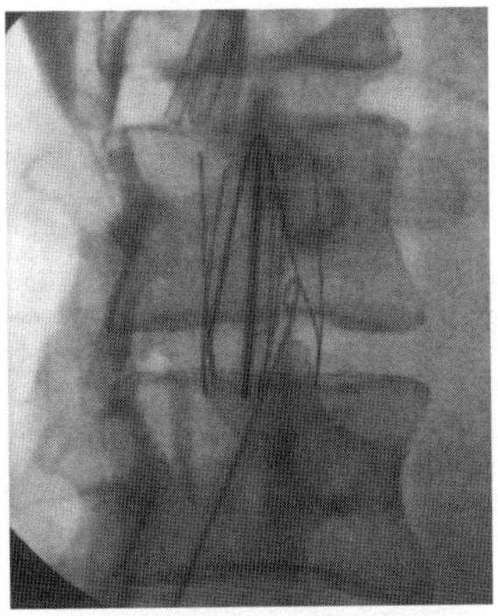

Figure 3. *After gentle manipulation with a Rosen guidewire, the legs are now open and they are in a good position.*

Bird's Nest (BN)

The BN **(Fig. 4)** consists of an irregular tangle or "nest" of wires within the lumen of the IVC. It is a little more difficult to insert than the VT or TG, but with experience, deployment is straightforward. The introducer outer diameter is 14-F, but it is more flexible than the VT or TG introducers. Separate sets are needed for jugular or femoral placements. The BN is made of stainless steel and is subject to torque and movement in the MRI field. A period of 6 weeks should elapse to prevent movement after placement before scanning a patient in an MRI machine. Considerable imaging artifacts are usual. The major advantage of the BN is that it can be used in cavae with diameters larger than 28 mm. Such large cavae are uncommon, being reported in 2% to 20% of patients studied. Right heart outflow impedance or obstruction, such as tricuspid regurgitation, constrictive pericarditis or cor pulmonale may be associated with large cavae.

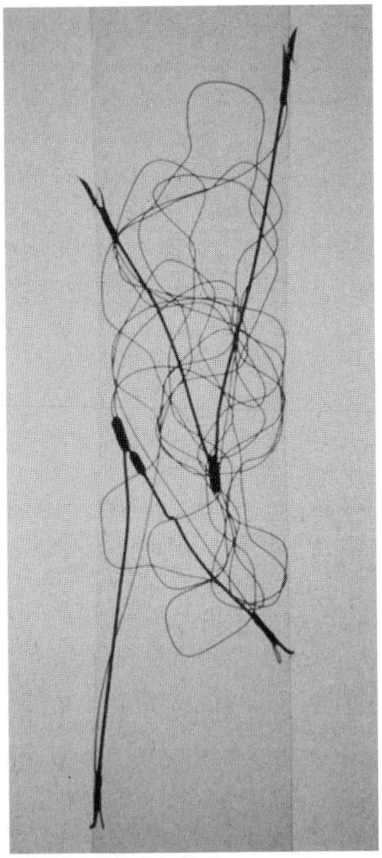

Figure 4. *Bird's nest filter.*

BIRD'S NEST FILTER	
Sheath	14 F
Maximum diameter	40 mm
Length	80-110 mm
Material	Stainless
MR movement	Yes

It is not always possible to deploy the BN so that its wires will form a tight nest. Prolapse of the wires in a cephalad direction is common. The clinical significance of this is not documented, although in vitro studies show no loss in filtering efficiency because of prolapse or from a less-tight "nest" in a wide cava. The BN is the longest of the filters once deployed (80mm-110mm) and the distance between the iliac vein confluence and the entry of the lowest renal vein may be too short to accommodate a BN filter. Recurrent pulmonary emboli rates are in the standard range of about 3%; migration is not a problem, but there can be some penetration of the caval wall by the four fixation prongs. IVC thrombosis rates are 3% to 19%.

Titanium Greenfield (TG)

This is a cone-shaped filter **(Fig. 5)** derived from the original stainless steel Kimray-Greenfield filter. It is easy to insert using a somewhat inflexible introducer (outer diameter 14.3-F) from the right side only **(Figs. 6, 7)**. Separate jugular or femoral systems are required. The TG has no torque and no image artifact in the MRI magnet. The deployment of the TG from the femoral approach is an uncontrolled procedure in that the filter springs open from its constrained position within its carrier (less than 14-F) to the full diameter of the IVC in an abrupt nongradual fashion. This may cause the filter to be tilted in position, to have crossed legs and to cause partial penetration of the caval wall. Recurrent pulmonary emboli and caval thrombosis rates are standard, being 4% and 9% respectively.

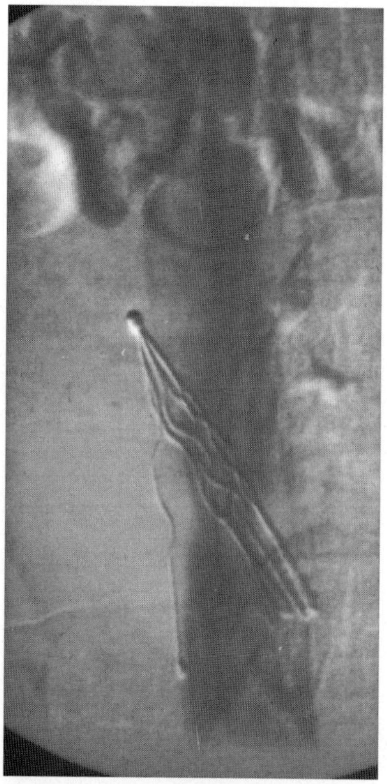

Figure 5. *A Titanium Greenfield filter badly tilted in the IVC with crossed filter legs.*

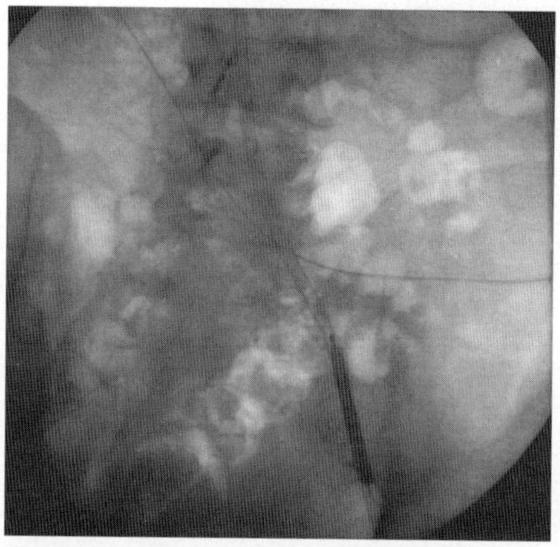

Figure 6. *An unsuccessful attempt was made at inserting a Titanium Greenfield filter from the left femoral approach. The rigid delivery capsule cannot be advanced up the introducer sheath.*

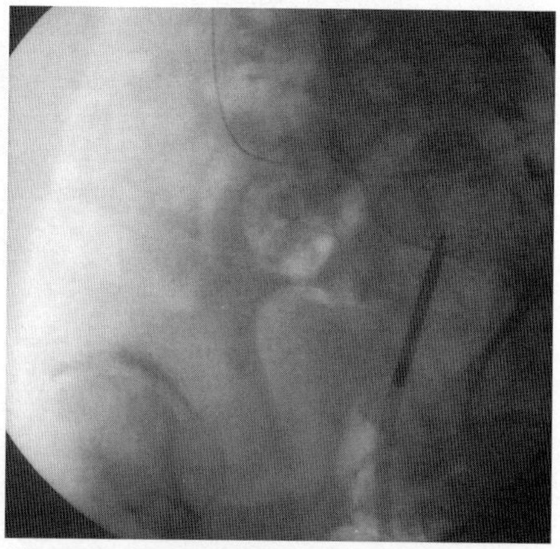

Figure 7. *An oblique view shows that the sheath is kinked because of the natural pelvic curve of the left iliac vein.*

TITANIUM GREENFIELD	
Sheath	14.3 F
Maximum diameter	30 mm
Length	47 mm
Material	Titanium
MR movement	No

Over-the-Wire Stainless Steel Greenfield
This is an alternative to the TG. The guidewire is intended to overcome the frequency of tilted positioning and asymmetric leg position seen with the TG. There is debate as to the effectiveness of this modification. The introducer has an outer diameter of 15.6-F. The filter is minimally susceptible to torque in the MR, but does cause a considerable imaging artifact. Performance is otherwise similar to the TG.

Simon Nitinol (SN)
The SN **(Fig. 8)** is manufactured from a nickel-titanium alloy that has a temperature-dependent memory. At body temperature, the filter assumes its memorized configuration, but at room temperature it can be straightened out and delivered through a 9-F introducer. When deployed, it forms an umbrella-shape with conical self-centering legs. The unique advantage of the SN is that the delivery system is only 9-F and flexible enough to allow insertion from right and left approaches and even from the brachial veins. The SN is not subject to movement and only causes a minimal artifact in the MRI magnet. Recurrent pulmonary emboli rates and IVC thrombosis rates are 2% and 10% respectively. There is, however, concern regarding higher migration rates, caval penetration rates and the durability of the memory of the shape of the filter.

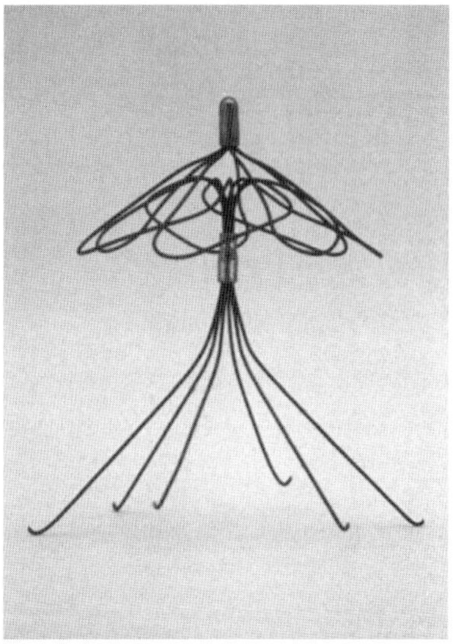

Figure 8. *Simon Nitinol filter.*

SIMON NITINOL	
Sheath	9 F
Maximum diameter	28 mm
Length	38 mm
Material	Nitinol
MR movement	No

MR MOVEMENT	
Vena Tech	No
Simon Nitinol	No
Titanium Greenfield	No
Greenfield stainless steel	Yes
Bird's Nest	Yes

FILTER PERFORMANCE

Do filters prevent PE? No prospective randomized studies have been performed to compare the efficacy of filter placement versus no treatment for the prevention of PE. Since such studies cannot be ethically performed, the best estimate of efficacy is gleaned from follow-up studies of patients who have undergone filter placement. Follow-up in many of these studies is incomplete and based largely on clinical assessment alone. Recurrent PE is reported in the range of 0% to 6%, with 5% being a reasonable estimate. It should be remembered that this estimate is only for symptomatic PE suspected on clinical grounds and probably will represent an underestimate because asymptomatic PE will be missed, as will unsuspected PE in patients who die without autopsy. There does not appear to be a significant difference between the various types of filters in their ability to prevent PE.

If new clot is found in the cava on follow-up study, does this mean that the filter has caused the complication of caval thrombosis, or is this just an indication that the filter is performing its intended role, i.e. of capturing distal thrombi and preventing their passage to the lungs? The best that can be reported is the incidence of caval thrombosis. Objective studies of the cava such as duplex sonography, contrast-enhanced CT, magnetic resonance angiography, and contrast vena cavography provide a true estimate of the presence of clot. Clinical evaluation for leg swelling is not a reliable indicator of caval obstruction. Studies by CT have shown filters causing complete caval obstruction in patients who remain asymptomatic. Twenty-eight patients underwent filter placement and later CT scan at the Mayo Clinic. Caval thrombosis was found in nine patients, six of whom were totally asymptomatic even on retrospective review. The longer the follow-up, the higher the rate of thrombosis; although in some cases clot captured by the filter will undergo lysis, restoring the cava to normal over time. Considering these problems, it is not surprising that reported thrombosis rates vary widely from 5% to 30%, making it difficult to compare different filter types.

COMPLICATIONS OF FILTER PLACEMENT

- New or worse DVT
- IVC thrombosis
- Caval penetration
- Filter migration
- Filter fracture

COMPLICATIONS OF FILTER PLACEMENT

The reported incidence of complications varies widely and depends not only on filter type, but more importantly on the methods used to assess complications and on the duration of follow-up. Representative figures are given in parenthesis, but as previously explained there are great variations in reported numbers. Reported complications include 1) New or worsened DVT may be the result of progression of the underlying condition for which the filter was used, but may also be a result of filter placement caused by decreased venous flow or access vein trauma (22%). 2) IVC thrombosis, partial or complete. This can be evidence of excellent clot capturing performance or of interference with caval flow leading to thrombosis (19%). 3) IVC wall penetration which is difficult to assess and even when present is very infrequently of clinical significance (9%). 4) Filter migration of minor degrees is not of great importance, but migration of the filter to the heart or misplacement of the filter in the right renal vein are serious complications (6%). 5) Fracture of filter components have been seen over time, but adverse clinical sequelae are rare (2%). Indirect complications include entrapment of guidewires or intravenous catheters that are passed through the filter.

IVC FILTER PLACEMENT PREPARATION

- Review CT, US, MR
  - Evaluate presence of IVC or iliac vein clot
- Choose access
- Perform venogram

TECHNIQUE OF FILTER PLACEMENT

We do not request interruption or reversal of anticoagulant therapy prior to filter placement since filters are placed via a venous puncture and postplacement hemostasis at the puncture site can usually be easily achieved by standard manual compression even in anticoagulated patients.

Available current imaging studies such as CT, US and MR are reviewed in conjunction with the patient's history and physical examination to help in choosing the best access route. Does the patient have jugular or femoral lines in place? Have these veins been used or obliterated by previous use? Is there evidence of DVT in one or the other leg? If an ultrasound examination of the legs was performed, was there clot in the femoral vein at the groin and was it possible to image any of the external iliac vein? Clot in both femoral veins or IVC seen on CT, US or MR would dictate a jugular approach.

PURPOSE OF PRELIMINARY CAVOGRAM

- Exclude clots in IVC
- Identify renal vein level
- Measure IVC diameter
- Detect anomalies

Right Femoral Approach

In the absence of prior evidence of clot, we puncture the right femoral vein and insert a 5-F dilator a short distance up the vein. A small contrast injection viewed fluoroscopically or digitally recorded is performed to exclude clots in the iliac vein before advancing a catheter to the confluence of the iliac veins **(Fig 9)**. An inferior vena cavogram is then performed in order to address four primary issues: 1) to exclude the presence of clots in the IVC, 2) to locate the level of the renal veins, 3) to measure the diameter of the IVC just below the entry of the renal veins, and 4) to exclude the presence of anomalous venous anatomy.

Level of Renal Veins

The defect caused by inflow of nonopacified renal venous blood into the IVC usually makes it easy to identify the level of the renal veins. In case of doubt, repeating the cavogram during a Valsalva maneuver may cause reflux of contrast into the renal veins. Selective catheterization of the renal veins may occasionally be necessary to establish their position. There may be more than one vein draining each kidney.

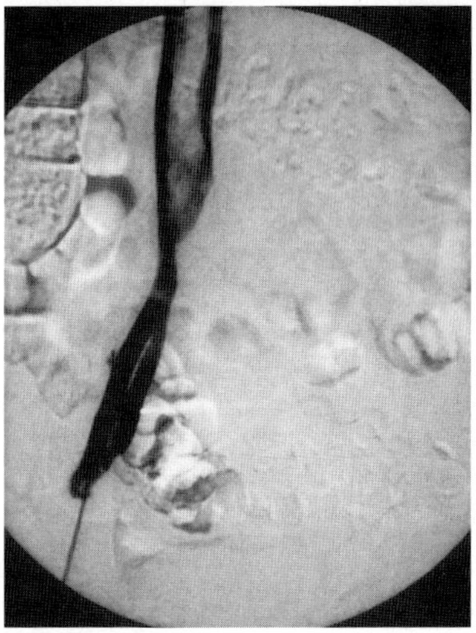

Figure 9. *A preliminary venogram performed through a dilator in the right external iliac vein demonstrates thrombus in the IVC. The filter placement therefore had to be performed from the jugular approach with the filter positioned above the thrombus.*

Caval Diameter

The diameter of the cava should be no larger than 28 mm if a TG, a SN or a VT filter is to be used. For wider cavae, the filter of choice is the BN which is suitable for use in cavae up to 40 mm in diameter. The IVC diameter is most accurately measured by using a marker pigtail catheter. A study of caval diameters in 65 patients at the time of filter placement found that the mean infrarenal caval diameter was 20 mm, range 13 mm to 30 mm, standard deviation 3 mm. Only two cavae (3%) were more than 28 mm in diameter.

Venous Anomalies

Venous anomalies must be recognized prior to filter placement. Three anomalies are of particular interest:

1) Double inferior vena cava (0.2%-3.8% incidence) is the result of persistence of both the right and left cardinal veins. The cavae may be of equal size but often the right is larger. The left cava joins the left renal vein and they cross anterior to the aorta to join the right-sided cava. This anomaly can be overlooked when the right cava is large. A Valsalva maneuver may help in the diagnosis which is best seen on bilateral femoral vein puncture. Filters should be placed in both cavae **(Figs. 10, 11)**.

VENOUS ANOMALIES OF INTEREST FOR FILTER PLACEMENT

- Double IVC
  - Right larger than left
  - Treat with two filters
- Circumaortic renal vein
  - Place the filter low
- Left-sided IVC

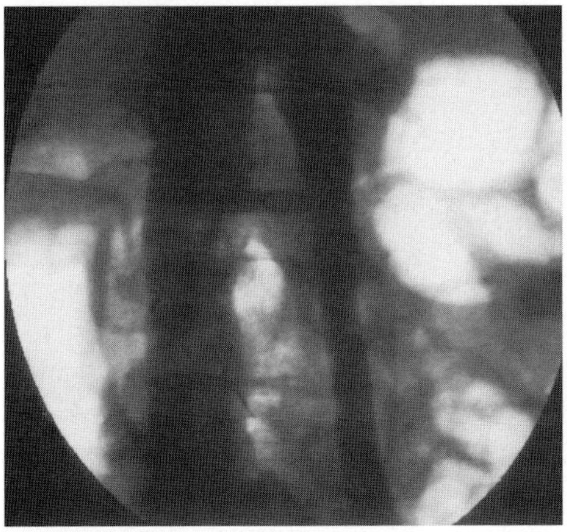

Figure 10. *An example of a double IVC, which occurs with an incidence of about 0.25% – 3.8%. A Bird's Nest filter has been placed in the right-sided cava.*

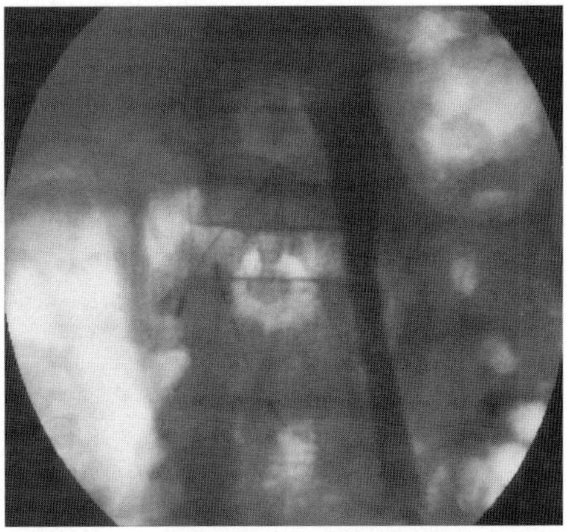

Figure 11. *After placement of the Bird's Nest filter in the right IVC, a Vena Tech filter was placed in the smaller left IVC of this patient with a double IVC.*

2) Circumaortic renal vein occurs in 8.7% of the population. In this anomaly, the anterior half-ring of the left renal vein joins the IVC at the usual level, but the posterior half crosses behind the aorta and joins the IVC at a lower level. In patients with this anomaly, the filter should be placed below the entry level of all renal vein branches.

3) Left-sided IVC has a prevalence of 0.2-0.5%. The cava lies along the left side of the spine, receives the left renal vein, and crosses in front of the aorta to receive the right renal vein, and then continues in the normal location to reach the heart.

Position of Filter

Ideally, the top of the filter should be positioned just below the entry of the lowest renal vein. When older, highly thrombogenic filters such as the Mobin-Uddin were used, great care was taken to accurately place the filter so that there was no dead space between the top of the filter and the renal veins. The fear was that clot could form above the filter in this space and be a source of emboli. With modern, less thrombogenic filters, this consideration is less important. Similarly, modern filters can be placed in the suprarenal cava, if necessary, because of clots in the distal cava, with less fear of caval thrombosis. If the suprarenal cava has a diameter larger than 28 mm, a BN will be required.

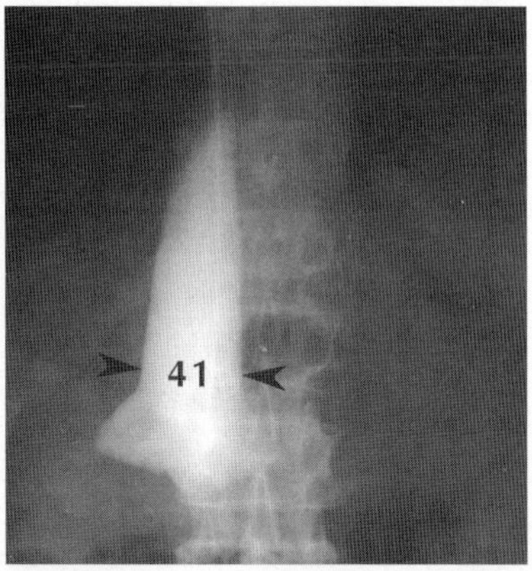

Figure 12. *This patient has thrombus in the infrarenal IVC. A cavogram performed from the jugular approach shows that the suprarenal cava is large with a diameter of 41 cm. A Bird's Nest filter is therefore indicated.*

Suprarenal Placement

May be indicated **(Fig. 12)** occasionally in the following circumstances:

a) Renal vein thrombosis;
b) IVC thrombosis extending above the level of the renal veins (including thrombus secondary to tumor such as renal cell carcinoma);
c) During pregnancy and in women anticipating pregnancy;
d) Pulmonary embolism subsequent to left ovarian vein thrombosis;
e) Pulmonary embolism after placement of an infrarenal filter when an extremity source has been ruled out;
f) Unsatisfactory configuration of an infrarenal filter.

There is concern that the suprarenal placement may result in migration of the filter cephalad as the cava is wider at this level. If a BN is used, will the wires prolapse into the atrium with deleterious effects? Will the filter cause caval thrombosis leading to renal vein thrombosis? Hard data are not available, but different interventionalists often hold strong preferences. If needed, I would use a VT in a supracaval location with a cava less than 28 mm in diameter or, in a larger cava, a BN.

Filter Configuration

Filters do not always assume the intended configuration upon deployment, and each of the commonly used filters has specific problems. The BN filter may have marked cephalad prolapse of its nest of wires. This yields a less dense concentration of wires, but recent in vitro studies suggest that the effectiveness of clot trapping is not impaired by this prolapse. No clinical correlation with these laboratory studies is available as is the case with other in vitro studies discussed below. The wires may prolapse into a renal vein with as-yet undocumented consequences. The TG filter may be deployed with a tilt in its axis relative to the IVC. Laboratory studies show impaired filtration with increasing tilt off-axis. The over-the-wire Greenfield is better in this respect than earlier models. Asymmetric opening of the Greenfield may also be a problem.

The VT may deploy with incompletely opened struts, yielding a decreased trapping efficiency in vitro. Much debate surrounds the safety versus the necessity of trying to manipulate the suboptimally placed filter following deployment. VT recommends a rapid deployment technique, and all manufacturers recommend thorough heparinized-saline irrigation of all delivery systems to limit problems of clot formation that may bind the filter legs and impair proper deployment.

Left Femoral Approach

Manufacturers of the Greenfield and VT filters warn against using this approach because of tortuosity of the veins, leading to possible kinking and perforation of the insertion sheath. If other routes were unavailable, we have deployed the VT successfully, passing up the left iliac under close fluoroscopic control. The BN and SN have more flexible delivery systems that pass up the left side with less problem.

Right Internal Jugular Approach

This is the route favored by some workers, and is indicated when there is suspected clot in the cava or iliac veins. As the delivery system is passed retrograde down the cava, the right renal vein may be inadvertently entered and the filter deployed in the renal vein **(Fig. 13)**. Careful technique with cavography is essential to avoid this preventable complication. Rarely, other veins joining the cava, such as a hepatic vein, may be entered.

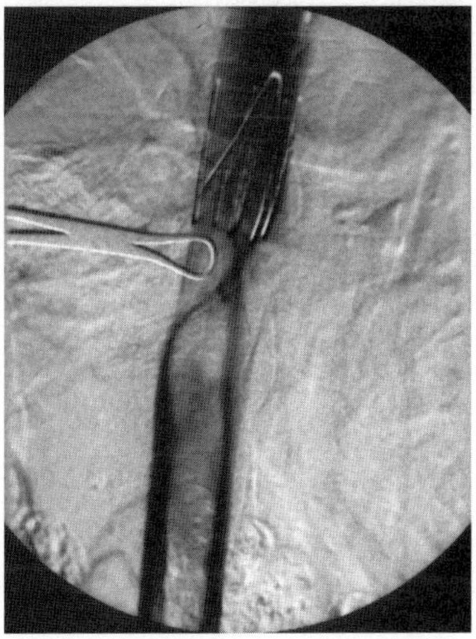

Figure 13. *A VT filter has been placed from the right internal jugular approach because of previously identified thrombus in the lower IVC. The filter is positioned above the thrombus and is somewhat tilted.*

Left Internal Jugular Approach

Should be avoided for the deployment of the Greenfield or VT for the same reasons as the left femoral: tortuosity of the vein and inflexibility of the delivery systems. This route can be used for the SN and BN.

Right External Jugular Approach

Has been successfully used, but the internal jugular is preferable. The SN can be used via this approach.

Other Approaches

Because of its flexibility and small-diameter delivery system, the SN can be inserted via either external jugular or antecubital vein, in addition to the conventional routes.

MANAGEMENT OF COMPLICATIONS

Rarely, a filter will end up in the heart because of migration or technical error. Case reports have described percutaneous retrieval, otherwise surgical removal is indicated. In cases where filters are suboptimally, albeit securely deployed, additional filters can be inserted caudally either in the cava or in both iliac veins. If the deployment appears unstable, a cephalad placement of an additional filter may be necessary to prevent migration and provide effective filtration.

FUTURE

Work continues to develop the perfect filter and new filter designs are being evaluated. In addition, there is interest in the development of a temporary or removable filter. Potential applications include: short-term prophylaxis for trauma patients; for patients with a short-term contraindication to anticoagulation; for patients undergoing thrombolytic therapy for lower extremity DVT; and for prophylactic treatment of patients after PE who have poor cardiopulmonary reserve.

A temporary filter remains attached to a catheter or wire that projects from the insertion site, while a retrievable filter has no external tethering, but has a mechanism such as a hook that allows for transvenous removal. These filters are in clinical use in Europe, although not yet available in the USA. The use of carbon dioxide or gadolinium as an alternate contrast agent has been described for patients who are either allergic to conventional contrast agents or who suffer from impaired renal function. Initial reports have appeared on the bedside insertion of filters in patients who are too sick to transport to the radiology department.

REFERENCES

Alexander JJ, Yuhas JP, Piotrowski JJ. Is the increasing use of prophylactic percutaneous IVC filters justified? Am J Surg 1994; 168:102-106.

Brown DB, Labuski MR, Cardella JF, Singh H, Waybill PN. Determination of inferior vena cava diameter in the angiography suite: comparison of three common methods. J Vasc Interv Radiol 1999; 10:143-147.

Carson JL, Kelley MA, Duff A, et al. The clinical course of pulmonary embolism. N Engl J Med 1992; 326:1240-1245.

Crochet DP, Brunel P, Trogrlic S, Grossetête R, Auget J-L, Dary C. Long-term follow-up of vena tech-LGM filter: Predictors and frequency of caval occlusion. J Vasc Interv Radiol 1999; 10:137-142.

Dorfman GS. Evaluating the roles and function of vena caval filters: will data be available before or after these devices are removed from the market? Radiology 1992; 185:15-17.

Engmann E, Asch MR. Clinical experience with the antecubital Simon Nitinol IVC filter. J Vasc Interv Radiol 1998; 9:774-778.

Ferris EJ, McCowan TC, Carver DK, McFarland DR. Percutaneous inferior vena caval filters: follow-up of seven designs in 320 patients. Radiology 1993; 188:851-856.

Grassi CJ. Inferior vena caval filters: analysis of five currently available devices. AJR 1991; 156:813-821.

Greenfield LJ, Proctor MC, Cho KJ, Cutler BS, et al. Extended evaluation of the titanium Greenfield vena caval filter. J Vasc Surg 1994; 20:458-465.

Hicks ME, Malden ES, Vesely TM, Picus D, Darcy MD. Prospective anatomic study of the inferior vena cava and renal veins: comparison of selective renal venography with cavography and relevance in filter placement. J Vasc Interv Radiol 1995; 6:721-729.

Johnson SP, Raiken DP, Grebe PJ, Diffin DC, Leyendeckere JR. Single institution prospective evaluation of the over-the-wire Greenfield vena caval filter. J Vasc Interv Radiol 1998; 9:766-773.

Katsamouris AA, Waltman AC, Delichatsios MA, Athanasoulis CA. Inferior vena cava filters: in vitro comparison of clot trapping and flow dynamics. Radiology 1988; 166:361-366.

Kaufman JA, Geller SC. When to use an inferior vena cava filter. AJR 1995; 164:256-257.

Matchett WJ, Jones MP, McFarland DR, Ferris EJ. Suprarenal vena caval filter placement: follow-up of four filter types in 22 patients. J Vasc Interv Radiol 1998; 9:588-593.
McCowan TC, Ferris EJ, Carver DK, Molpus WM. Complications of the nitinol vena caval filters. J Vasc Intern Radiol 1992; 3:401-408.
Millward SF. Temporary and retrievable inferior vena cava filters: current status. J Vasc Intern Radiol 1998; 9:381-387.
Prince MR, Novelline RA, Athanasoulis CA, Simon M. The diameter of the inferior vena cava and its implications for the use of vena caval filters. Radiology 1983; 149:687-689.
Reed RA, Teitelbaum GP, Taylor FC, Pentecost MJ, Roehm JOF. Use of the Bird's Nest filter in oversized inferior venae cavae. J Vasc Interv Radiol 1991; 2:447-450.
Rosen MP, Porter DH, Kim D. Reassessment of vena caval filter use in patients with cancer. J Vasc Interv Radiol 1994; 5:501-506.
Savader SJ. Inferior vena cava filters. In: Trerotola SO, Savader SJ, Durham JD, eds. SCVIR Syllabus: Venous Interventions. San Francisco: Society of Cardiovascular & Interventional Radiology, 1995; 276-292.
Shellock FG, Morisoli S, Kanal E. MR Procedures and biomedical implants, materials, and devices: 1993 update. Radiology 1993; 189:587-599.
Shlansky-Goldberg R, Wing CM, LeVeen RF, Cope C. Effectiveness of a prolapsed Bird's Nest filter. J Vasc Interv Radiol 1993; 4:505-511.
Spence LD, Gironta MG, Malde HM, Mickolick CT, Geisinger MA, Dolmatch BL. Acute upper extremity deep venous thrombosis: safety and effectiveness of superior vena caval filters. Radiology 1999; 210:53-58.
Stoneham GW, Burbridge BE, Millward SF. Temporary inferior vena cava filters: in vitro comparison with permanent IVC filters. J Vasc Interv Radiol 1995; 6:731-736.
Venbrux AC. Temporary inferior vena cava filters. In: Trerotola SO, Savader SJ, Durham JD, eds. SCVIR Syllabus: Venous Interventions. San Francisco: Society of Cardiovascular & Interventional Radiology, 1995; 293-299.

CHAPTER 12 PREVIEW ANSWERS

Case 1

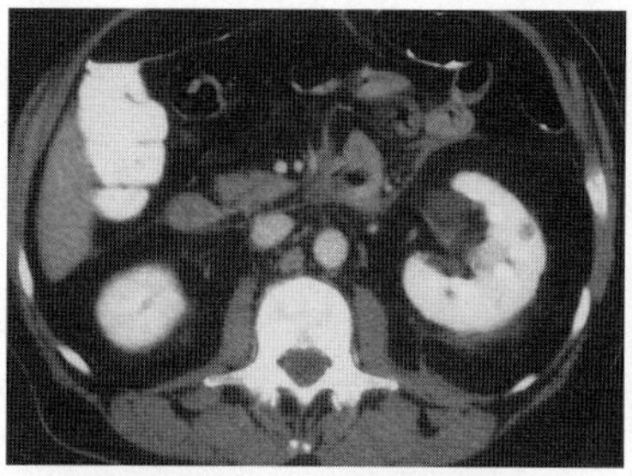

a.

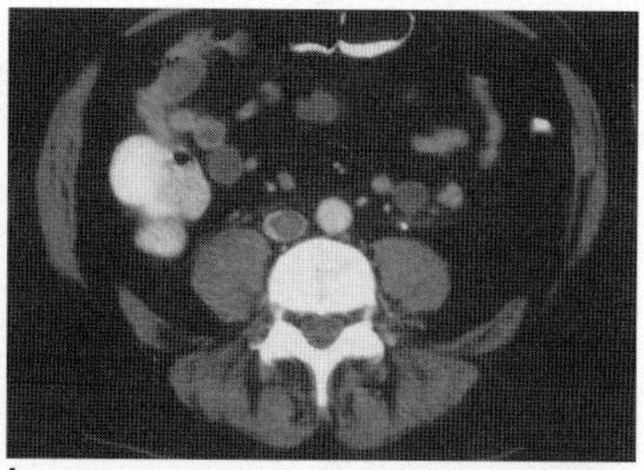

b.

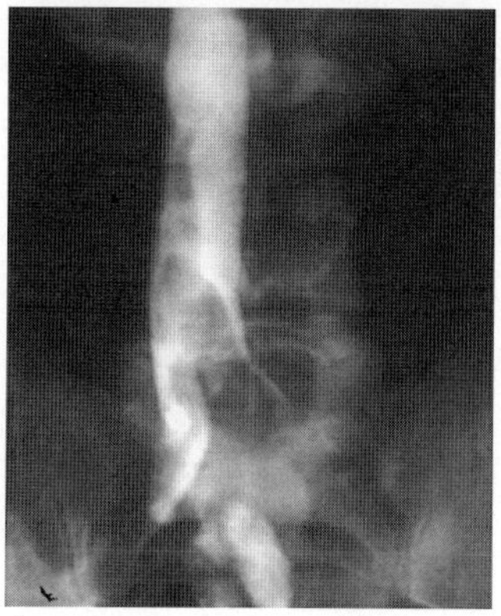

c.

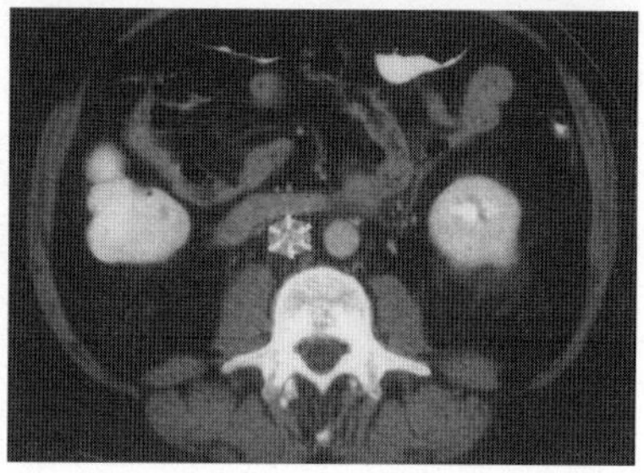

d.

Findings: a) At the level of the kidneys both the aorta and the IVC are normal and filled with contrast agent. b) At a lower level, the IVC is filled with thrombus with only a ring of contrast seen peripherally. c) A venacavogram from the jugular approach confirms the CT finding of thrombus in the infrarenal IVC. d) A follow-up CT scan shows that a VT filter has been placed in the IVC above the clot.
Diagnosis: Thrombus in the IVC treated by transjugular filter placement.

Case 2

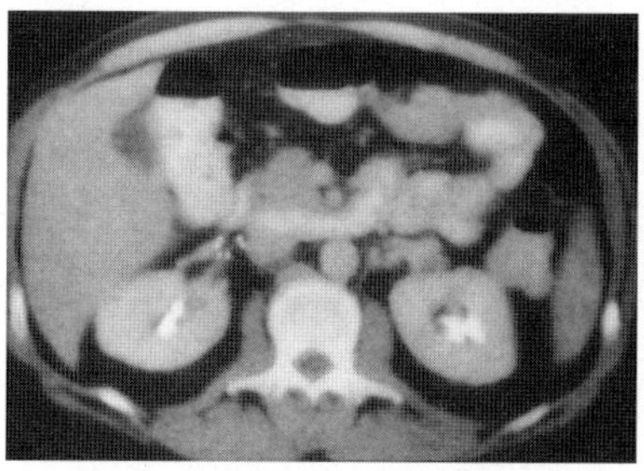

Findings: The CT scan shows that a Greenfield filter has been inadvertently placed in the right renal vein. This is a recognized pitfall when using the jugular route.
Diagnosis: Filter in right renal vein.

Case 3

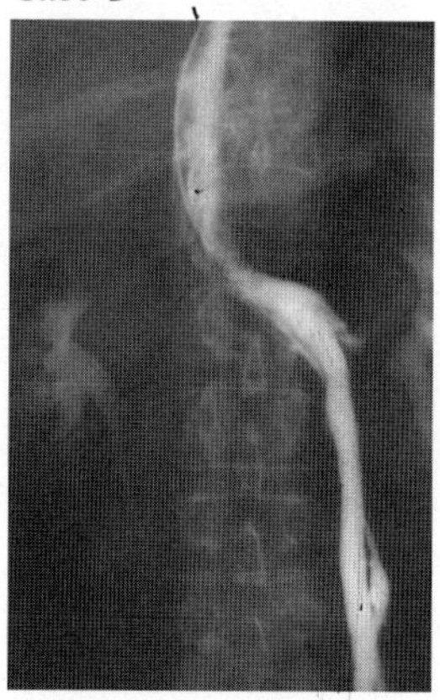

Findings: This is an example of a left-sided IVC which is found in about 0.2%-0.5% of patients.
Diagnosis: Left-sided IVC.

CHAPTER 13 PREVIEW

Case 1

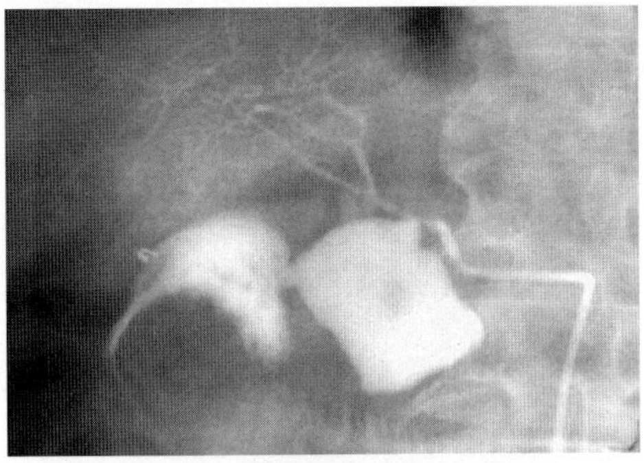

History: Gastrointestinal bleeding following a Whipple procedure.
Question: Could this patient be treated with transcatheter embolization?

Case 2

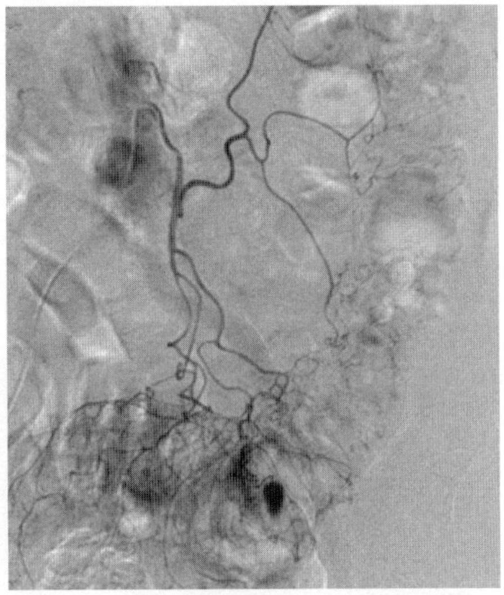

History: Lower gastrointestinal bleeding in an 81-year-old man.
Question: Could this patient be treated with transcatheter embolization?

Case 3

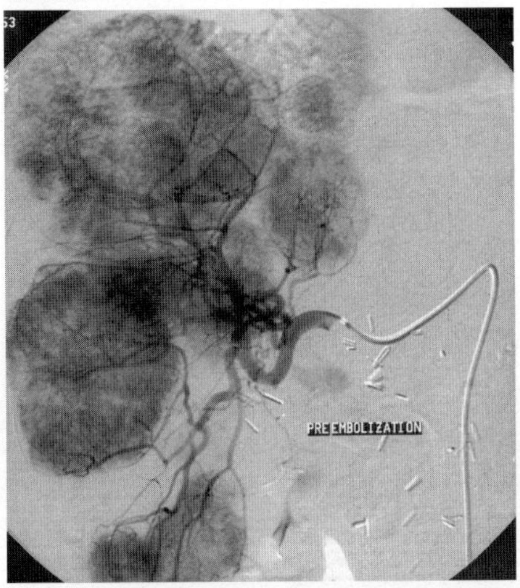

History: Right upper quadrant pain and diarrhea.
Question: Does embolization play any role in this patient with carcinoid metastases to the liver?

Chapter 13
Transcatheter Embolization

Robert K. Kerlan, Jr., MD

OUTLINE OF EMBOLIZATION

- Embolization technique
- Indications and results
  - Hemorrhage
    - Musculoskeletal trauma
    - Gastrointestinal hemorrhage
    - Massive hemoptysis
    - Iatrogenic trauma
    - Hemorrhage secondary to pancreatitis
    - Postpartum hemorrhage
  - Palliation of malignancy
    - Neoplasms
    - Hepatic
    - Renal
    - Pelvic malignancy
    - Skeletal metastases
  - Miscellaneous
    - Splenic embolization for thrombocytopenia
    - Varicoceles
    - Uterine fibroids

INTRODUCTION

Since its original clinical inception in the early 1970s, transcatheter embolization has evolved into an extremely useful clinical tool for the control of hemorrhage and palliation of malignancy. Recent technical advances, including the development of coaxial microcatheter systems, has expanded the potential applications of transcatheter therapy. Moreover, the number of embolic agents available has increased substantially, allowing selective blockade at different levels of the arterial bed. This discussion will focus upon transcatheter embolotherapy including technique, indications, and results.

EMBOLIC AGENTS

- Coils and balloons
- Particulates
- Liquids

TECHNIQUE

Embolic material can be divided into three groups: 1) coils and balloons, 2) particulate material, and 3) liquids.

Coils and balloons are designed for focal occlusion of large blood vessels. If the vessel is an end artery, occlusion of the distal vascular bed usually occurs. If the artery is not a terminal branch, collateral vessels will usually lead to reconstitution of the distal arterial bed.

COILS

- Focal
- Permanent
- Block vessels 1 mm to 20 mm in diameter

COILS

- Stainless steel coils
- Platinum microcoils

PARTICULATES

- Multifocal to diffuse
- Transient or permanent
- Blocks 50-micron arterioles to 5-mm arteries

Coils **(Fig. 1)** are made of either platinum or stainless steel. These coils are constructed from pieces of preshaped guidewire with attached dacron fibrils to enhance thrombogenicity. Preselected sizes ranging from 1 mm to 20 mm in diameter are available. Platinum microcoils are designed to be inserted through 3-F microcatheters. Standard stainless steel coils have a .035-inch or .038-inch wire thickness and require 5-F or larger catheters for placement. Stainless steel coils should not be deployed through catheters made of polyurethane, as the coil may become entrapped in the distal tip of the catheter. Polyethylene and braided nylon catheters are suitable for coil deposition.

Specialized coils have also been developed which can be inserted through the catheter and retrieved if the coil is improperly positioned (Guglielmi coils). When these coils are placed in the appropriate location, an electric current is applied and the coil is detached from the insertion device.

Detachable balloons are available from 8 mm to 20 mm in diameter and require specialized coaxial catheter systems for deployment. These devices are not commonly used outside of the cerebral circulation. The specific method of deployment is dependent upon the type of balloon being used.

Particulate materials include gelatin sponge, polyvinyl alcohol particles and collagen microfibers. Particulates are used to occlude vessels 50 microns to 5 mm in diameter and can be delivered through 3-F microcatheters as well as standard 5 to 7-F angiographic catheters.

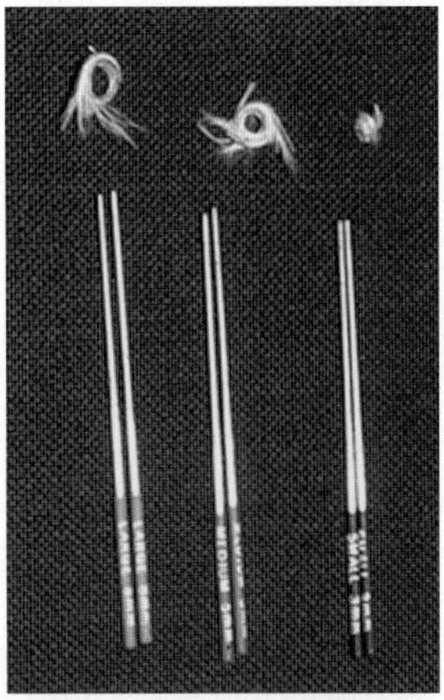

Figure 1. Gianturco coil spring emboli and loading cartridges.

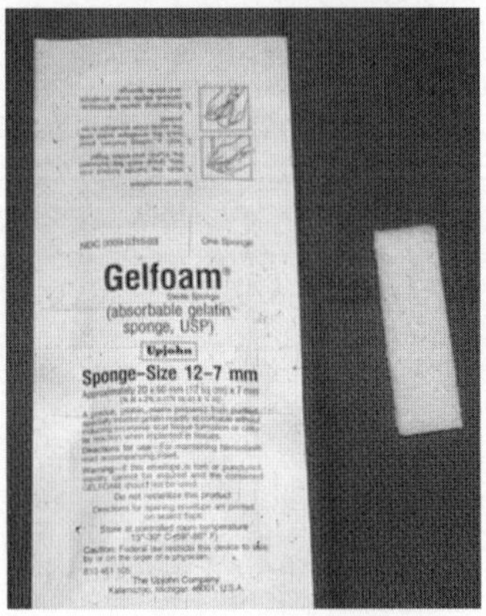

Figure 2. Gelfoam sponge.

Gelatin sponge (Gelfoam, Upjohn, Kalamazoo Mich.) **(Fig. 2)** is the most common particulate agent used for non-CNS embolization. The material is gradually absorbed after injection leading to temporary occlusion of the embolized vessel. Recanalization of the vessel occurs within weeks of the embolization procedure. Pledgets of any imaginable size (which will fit through the catheter being used) can be cut from pieces of gelfoam sponge. These pledgets can either be placed into a syringe and injected with dilute contrast media, or placed directly into the hub of the catheter. Gelfoam pledgets will obstruct arteries and arterioles ranging in size from 500 microns to almost 5 mm depending on the size of the particle.

Powdered Gelfoam is also available for embolization. This material contains particles ranging from 50 to 300 microns and will obstruct arterioles, potentially leading to tissue necrosis. The use of powdered gelfoam is usually restricted to the palliation of hepatic and renal malignancies.

Polyvinyl alcohol (Ivalon, Contour) is commercially available in preselected sizes ranging from 150 to 1,000 microns. Supplied as sterile particles, polyvinyl alcohol must be mixed with diluted contrast media prior to injection. Suspension is improved by agitated mixing between two syringes connected with a stopcock. In contrast to gelfoam, polyvinyl alcohol particles are not degraded over time and provide permanent embolic occlusion.

Microfibrillar collagen has recently been used for particulate embolization. As with polyvinyl alcohol, the size of the occlusive particles can be preselected and range from 100 to 1,000 microns. Collagen is the most temporary particulate used for embolization. Degradation begins soon after injection and recanalization of the targeted vessel occurs within hours to days.

Excessive force should not be applied to the syringe during injection of particulate material, as this may create reflux and nontarget embolization.

Liquid embolic materials include dehydrated ethanol and isobutyl-cyanoacrylate derivatives. Dehydrated ethanol is cytotoxic and leads to coagulative necrosis of the injected vascular bed. Both arterial and venular occlusion develop rapidly. The level of blockade extends from the capillary bed to the larger arteries. Dehydrated alcohol is usually injected through 7-F occlusion balloon catheters to prevent reflux into nontargeted vascular beds. When used in the venous circulation, dilution by blood rapidly dissipates the alcohol, diminishing the cytotoxic and sclerosing effect.

Isobutyl-cyanoacrylate derivatives require extensive experience in microcatheter techniques for appropriate application. These agents may be mutagenic and are only available on an experimental basis. The level of vascular occlusion induced by these substances is dependent upon the rate of polymerization, which in turn is related to the amount of tantalum or pantopaque added to the compounds. Isobutyl-cyanoacrylate derivatives polymerize when exposed to ionic solutions such as blood. This polymerization leads to a risk of gluing the catheter into the patient.

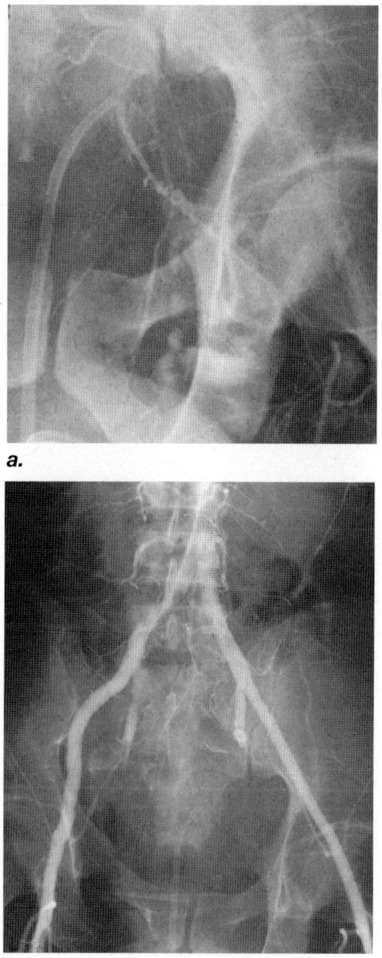

a.

b.

Figure 3. *38-year-old man status post high-speed motor vehicle accident presents with continued hemodynamic instability and a negative peritoneal lavage. a) Late arterial image from pelvic arteriogram reveals extravasation from branches of the left obturator artery. b) Successful occlusion of the internal iliac arteries with Gelfoam and coils controls the bleeding.*

INDICATIONS AND RESULTS

Hemorrhage

Musculoskeletal Trauma

The first clinical use of transcatheter embolization was to control massive hemorrhage from a pelvic fracture. Autologous clot was injected through the catheter, successfully occluding the internal iliac artery. Since its initial application in the early 1970s, this technique has been used extensively for patients with pelvic fractures who are unstable or who have continued bleeding **(Fig. 3)**.

The selection of patients for arteriography and embolization should be based upon the mechanism of injury as well as the clinical status of the patient. In general, 7%-11% of patients with pelvic fractures require embolization. When subdivided in accordance to mechanism of injury, only 1.7% of patients with lateral compression injuries require embolization, compared to 20% of patients with anteroposterior compression, vertical shear and combined mechanical injury patterns.

The decision to perform pelvic embolization depends upon the patient's stability, rate of blood loss, and results of peritoneal lavage. In some circumstances, imaging information provided by computed tomography is valuable in excluding associated intraabdominal injuries, depicting the position of pelvic fracture fragments, assessing the size of the retroperitoneal hematoma, and potentially revealing a focus of active extravasation. Patients requiring continued blood transfusion, with or without hemodynamic instability, who have no evidence of significant intraperitoneal bleeding are prime candidates for angiography and embolization. Most patients should have pelvic stabilization with an external fixator prior to angiography. The application of the fixator will promote tamponade and hemostasis, making angiography unnecessary in many patients. However, precious time should not be wasted placing an external fixator in a grossly unstable patient, when a swift angiographic procedure followed by embolization could potentially arrest the hemorrhage.

A pelvic arteriogram is obtained through a pigtail catheter positioned in the lower abdominal aorta to demonstrate the location of the bleeding vessel. However, even if no extravasation is identified, internal iliac embolization should be performed if the patient is unstable or the source of continued blood loss is suspected to be the bony pelvis. Gelfoam pledgets ranging in size from 1 mm to 4 mm followed by stainless steel coils ranging in size from 3 mm to 8 mm are appropriate for most patients. Gelfoam powder and polyvinyl alcohol fragments less than 500 microns in size are contraindicated due to the possibility of creating an ischemic injury to the sciatic nerve and pelvic viscera. Bilateral embolization of both the anterior and posterior divisions of the internal iliac artery is well tolerated and appropriate for most patients. Unilateral embolization will often be unsuccessful as cross-pelvic anastomoses will usually lead to continued hemorrhage.

Technical success of internal iliac embolization is greater than 90%. Atherosclerotic disease of the iliac vessels will occasionally preclude successful catheterization. Complications are rare, but non-target embolization due to reflux of particulate material has been observed. The clinical results of pelvic embolization may be dramatic with almost immediate stabilization of the patient. Average blood requirements should drop precipitously after occlusion of the internal iliac arteries.

Gastrointestinal Hemorrhage

Transcatheter embolization for gastrointestinal bleeding has diminished in frequency as endoscopic techniques for direct control have evolved and preventative pharmacotherapy has improved. Unfortunately, the patients now referred to angiography for gastrointestinal bleeding are sicker, with nongastric bleeding sites, and usually have coagulopathies associated with multiple transfusions.

To combat the added degree of difficulty, coaxial microcatheter systems have emerged as a critical tool for superselective catheterization and embolization. Platinum microcoils can be delivered into arteries less than a millimeter in diameter. The high fluoroscopic visibility of these coils allows precise positioning of these occluding devices.

Gastric bleeding occurs from branches of the left gastric artery in over 85% of patients. Therefore, if a gastric source of hemorrhage has been adequately documented, it is reasonable to occlude the left gastric artery even if the specific bleeding site is not identified. Gastric bleeding from Mallory-Weiss tears, gastritis, or gastric ulcer can be controlled in over 80% of patients with embolotherapy. However, in the presence of a significant coagulopathy, a considerably lower success rate may be anticipated.

Bleeding from the duodenum is more difficult to control **(Fig. 4)**. This difficulty is engendered by the extensive inflammation associated with many duodenal ulcers as well as the rich arterial anastomoses in the duodenal region. These arterial arcades which extend between branches of the gastroduodenal artery and the superior mesenteric artery often require embolization both proximal and distal to the bleeding site to achieve hemostasis. For these reasons, angiographic interventions lead to successful control in only 50% to 70% of patients with bleeding duodenal ulcers.

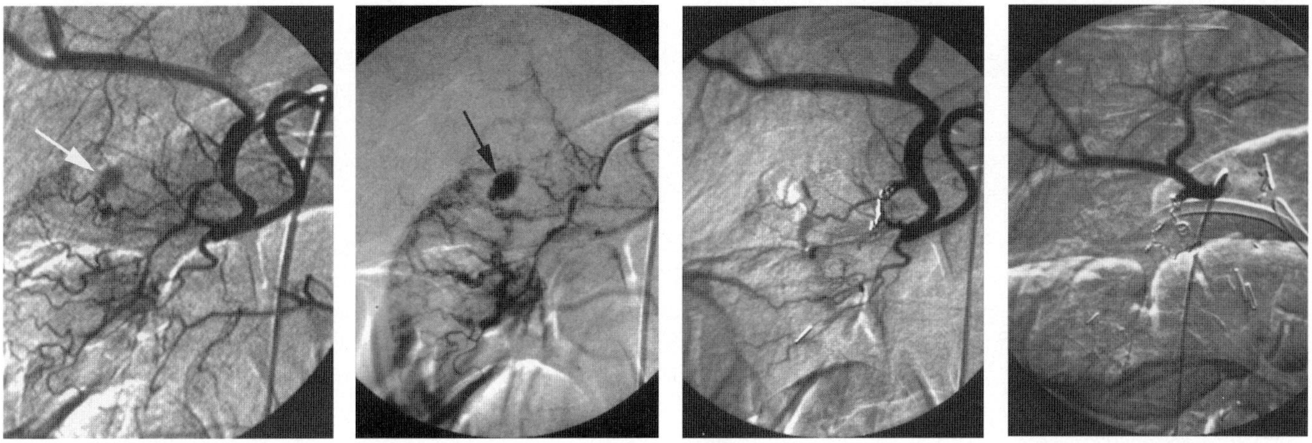

a. b. c. d.

Figure 4. *54-year-old man with bleeding duodenal ulcer. The bleeding could not be controlled with endoscopic techniques. a) Gastroduodenal arteriogam reveals a focus of contrast extravasation in the descending duodenum (arrow). b) Extravasation is more apparent when a superselective arteriogram of a branch of the gastrodudodenal artery is performed. c) Superselective embolization of the duodenal branch is unsuccessful in controlling the patient's bleeding. d) Following embolization of the entire gastroduodenal artery with microcoils, the extravasation is no longer seen.*

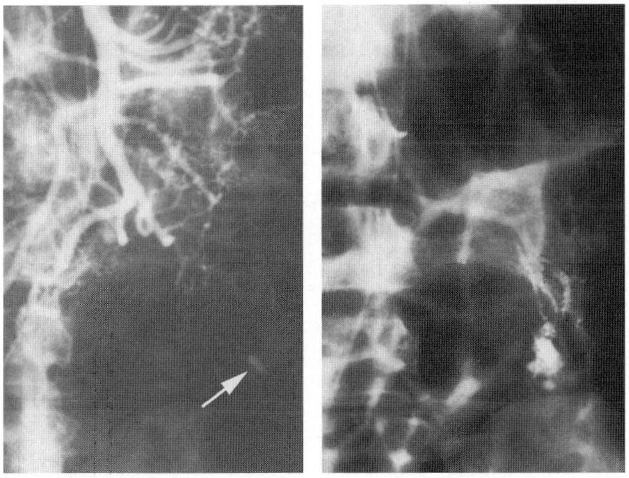

a. b.

Figure 5. *48-year-old woman with melena and a positive radionuclide scan for small bowel bleeding. a) Selective superior mesenteric arteriogram reveals extravasation in a distal jejunal or ileal branch (arrow). b) A 3-F coaxial microcatheter is manipulated into the bleeding artery. Contrast is injected confirming the appropriate position. The patient was taken to the operating room where methylene blue was injected through the catheter, identifying the bleeding site. A 15-mm carcinoid tumor was found in the surgical specimen.*

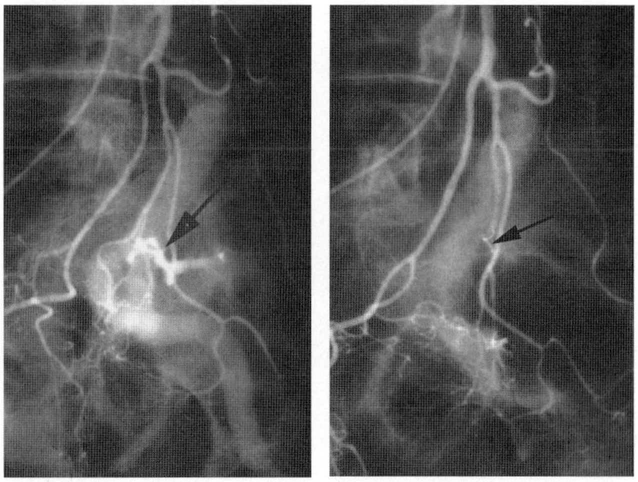

a. b.

Figure 6. *49-year-old man with massive rectal hemorrhage following endoscopic polypectomy. a) Inferior mesenteric arteriogram shows extravasation from a sigmoid branch (arrow). b) Following embolization with microcoils (arrow), no further extravasation is visible.*

Embolization to control bleeding in the small bowel **(Fig. 5)** and colon **(Fig. 6)** remains controversial. There clearly exists a higher risk of inducing bowel infarction when embolization of superior or inferior mesenteric arterial branches is performed. Infarction is precipitated in 5% to 20% of cases and is usually evident on clinical grounds. Though this necessitates surgical intervention, if a bleeding site has been identified, the surgical procedure is often identical to that which would have been performed to control the hemorrhage. A more important issue is delaying the diagnosis of a potentially curable neoplasm. Therefore, small bowel and large bowel embolization should only be performed when there is reasonable confidence regarding the etiology of the hemorrhage. It is possible, however, to control small and large bowel bleeding in up to 80% of patients when the bleeding site is identified.

Some practitioners use superselective embolization exclusively to control gastrointestinal hemorrhage, and no longer use the traditional infusion of pitressin. A controlled clinical trial comparing these two treatment strategies has not been performed.

Massive Hemoptysis

Selective catheterization of bronchial arteries **(Fig. 7)** followed by embolic occlusion has been established as an effective technique to control massive hemoptysis. Polyvinyl alcohol particles ranging in size from 250 microns to 750 microns are the preferred embolic material. Platinum microcoils may be used when arteriovenous shunting is present. Bronchoscopy is useful prior to embolization to ascertain whether the bleeding is originating from the right or left side. Moreover, selective intubation of the uninvolved bronchus can provide an important temporizing measure.

MASSIVE HEMOPTYSIS: ETIOLOGY

- Bronchiectasis
- Tuberculosis
- Cystic fibrosis
- Carcinoma
- Aspergillosis

Two types of patients typically present with massive hemoptysis: 1) patients with chronic inflammatory disease and 2) patients with bronchogenic neoplasms.

Chronic inflammatory diseases include bronchiectasis, cystic fibrosis and tuberculosis. The bronchial arteries in these patients are typically enlarged and are not difficult to catheterize. An acute success rate of approximately 85% can be anticipated. In comparison, patients with bronchogenic neoplasms usually have normal-sized bronchial arteries **(Fig. 8)** which can be technically difficult to catheterize, resulting in a lower success rate.

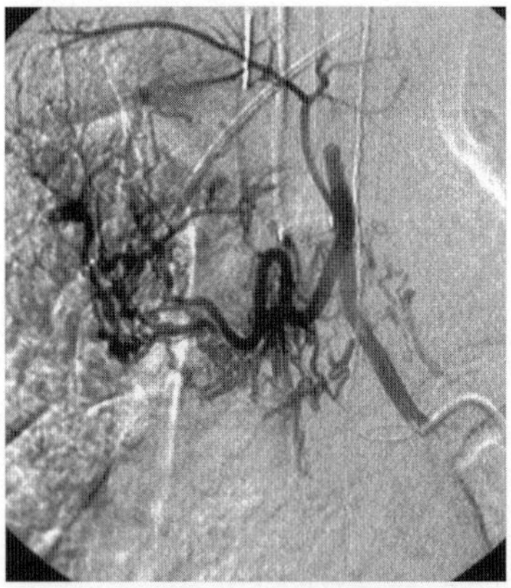

a.

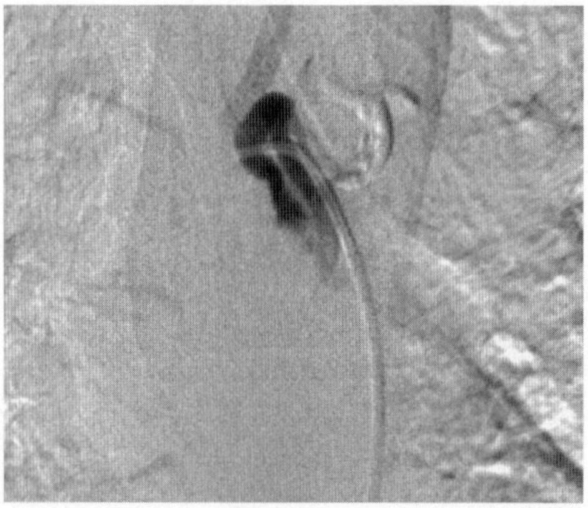

b.

Figure 7. *19-year-old man with cystic fibrosis and recurrent hemoptysis. a) Bronchial artery injection shows abnormal proliferation of vessels consistent with chronic inflammation. b) Successful occlusion of the bronchial artery is achieved with polyvinyl alcohol particles.*

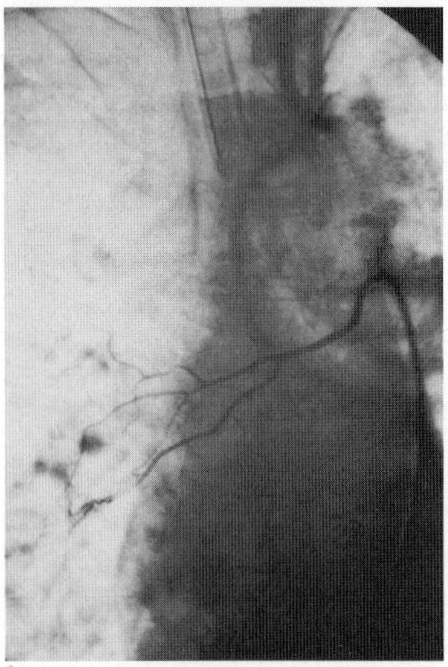

Figure 8. *60-year-old man with unresectable bronchogenic carcinoma and recurrent hemoptysis. Selective right bronchial arteriogram shows no significant abnormality; however, the vessel was embolized with polyvinyl alcohol particles.*

Complications related to bronchial artery embolization are unusual, but care must be employed to avoid embolization of branches supplying the spinal cord. Though this complication has been reported, fewer than 10 cases have been documented in the literature.

Bleeding Secondary to Iatrogenic Trauma

Bleeding from the liver is not an infrequent complication of liver biopsy or percutaneous biliary drainage procedures. In most cases, the bleeding is self-limited and does not require intervention. However, if bleeding is massive or persistent, arteriographic evaluation is warranted. A focal arterial injury is often identified, which can be treated with superselective embolization. Because of the rich anastomotic network of intrahepatic arterial collaterals, embolic material should be deposited as close as possible to the bleeding site **(Fig. 9)**. When possible, microcoils should be placed both distal and proximal to the arterial injury. If this is not possible, injection of Gelfoam or polyvinyl alcohol particles followed by an occluding coil is recommended. As the liver receives perfusion from both the hepatic artery and portal vein, infarction is unusual following embolization.

Bleeding from the kidney **(Fig. 10)** may also be observed following percutaneous biopsy or nephrostomy tube placement. Angiography may reveal an arterial defect or an arteriovenous fistula. In contrast to the liver, renal arteries seldom have significant intraparenchymal anastomoses. As infarction can be expected following embolization, occlusion of the smallest arterial branch possible is recommended. Coils are the agent of choice.

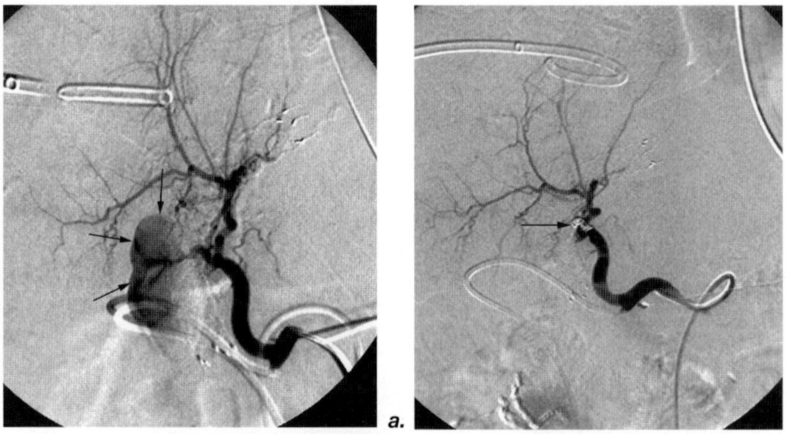

a. b.

Figure 9. *64-year-old man with recurrent bleeding following an attempted transhepatic biliary drainage. a) Hepatic arteriogram shows extravasation from a right hepatic arterial branch (arrows). b) The bleeding branch is successfully occluded with platinum microcoils (arrow).*

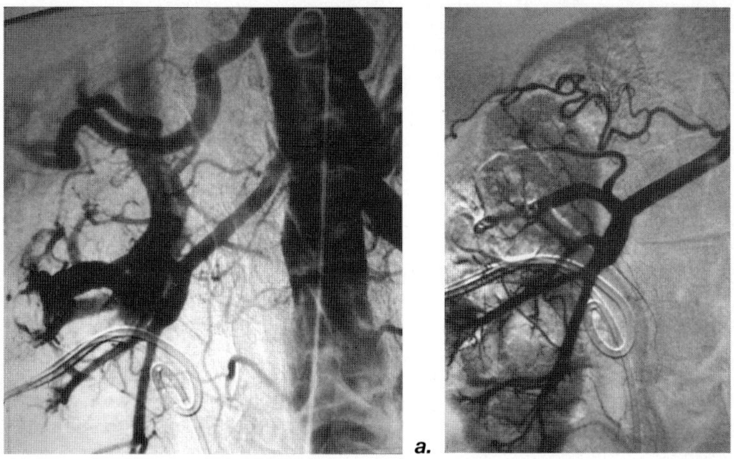

a. b.

Figure 10. *78-year-old man with massive hematuria following percutaneous nephrostomy a) Abdominal aortogram shows a large arteriovenous fistula involving the mid-right kidney. b) Successful embolization with standard-sized coils controls the hemorrhage.*

Hemorrhage Secondary to Pancreatitis

Arterial complications are being recognized with increasing frequency secondary to pancreatitis. A retrospective review of 104 cases reported by Boudghene et al revealed involvement of the splenic (42%), gastroduodenal (22%), and the small pancreatic arteries (25%) by either formation of a false aneurysm or erosion into a pseudocyst. Embolization was successful in controlling hemorrhage acutely in all 32 patients where this form of therapy was used; however, a 37% rebleeding rate was encountered. Embolization was considered useful in this series to stabilize patients prior to pancreatic débridement. Arteriography was essential for diagnosis as many of these lesions were small and not visible with ultrasound or computed tomography.

Postpartum Hemorrhage

Uncontrolled postpartum hemorrhage may be life threatening. Bilateral internal iliac artery embolization can play a critical role in the management of these patients. Yamashita et al reported successful management of 15 women with massive postpartum hemorrhage. No significant complications were reported. If the embolization is performed with gelfoam, the uterine arteries will recanalize, leaving open the possibility of future pregnancies.

Palliation of Malignancy

Neoplasms in which embolotherapy has been beneficial include primary and secondary hepatic tumors, hypervascular renal adenocarcinomas prior to resection, bleeding uterine malignancies and palliation of hypervascular skeletal metastases.

PALLIATE MALIGNANCY

- Liver
  - Hepatocellular carcinoma
  - Metastatic carcinoid
  - Metastatic islet cell tumors
  - Metastatic sarcomas
- Kidney
- Bleeding uterine malignancies
- Osseous metastases

Hepatic Neoplasms

Hepatic neoplasms responsive to embolotherapy include hepatocellular carcinomas, metastatic carcinoid tumors, metastatic islet cell adenocarcinomas and hypervascular metastatic sarcomas. The embolic material is usually gelfoam powder or polyvinyl alcohol particles combined with chemotherapeutic agent. This procedure is known as chemoembolization **(Fig. 11)**. This technique delivers highly concentrated chemotherapeutic agents to the tumor, and then arrests arterial blood flow. Intratumoral chemotherapeutic levels can be achieved which are 10 to 25 times higher than those observed following intravenous or even intraarterial infusion.

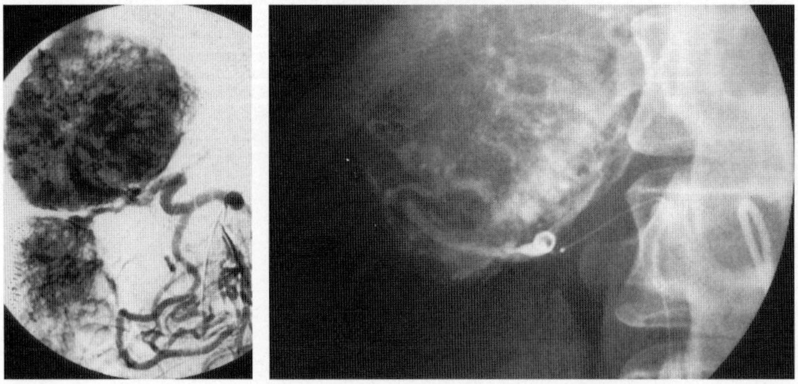

a. **b.**

Figure 11. *40-year-old woman with metastatic hemangiopericytoma to the liver. a) Superior mesenteric arteriogram fills the hepatic artery through pancreaticoduodenal collaterals, due to compression of the celiac trunk by a median arcuate ligament. Two large hypervascular liver lesions are present. b) Dense staining of the tumors persists following chemoembolization.*

Following chemoembolization, patients generally experience abdominal pain and fever which is self-limited. Prophylactic antibiotics are recommended to diminish the incidence of postembolization abscess formation.

Using chemoembolization for unresectable hepatocellular carcinoma, an average survival of 2.7 years has been reported in uncontrolled trials. This compares favorably to the 6- to 8-month average survival observed in untreated patients. Patients with coexistent cirrhosis or multifocal tumors do not respond as well.

Response rates following hepatic embolization for metastatic islet cell carcinomas and carcinoid tumors are also encouraging. Significant necrosis of tumor in 60% to 80% of patients has been observed with the majority of patients being relieved of symptoms related to hormone production.

Unfortunately, metastases from the breast, lung and gastrointestinal tract do not consistently respond to this form of therapy.

Renal Neoplasms

Blood loss at the time of resection of hypervascular renal adenocarcinomas can be reduced by complete preoperative embolization with alcohol **(Fig. 12)**. It should be noted, however, that incomplete embolization may actually increase perioperative transfusion requirements. As alcohol is extremely cytotoxic, it must be used with extreme caution. Balloon catheters are recommended to prevent reflux of alcohol into the aorta, adrenal and gonadal arteries.

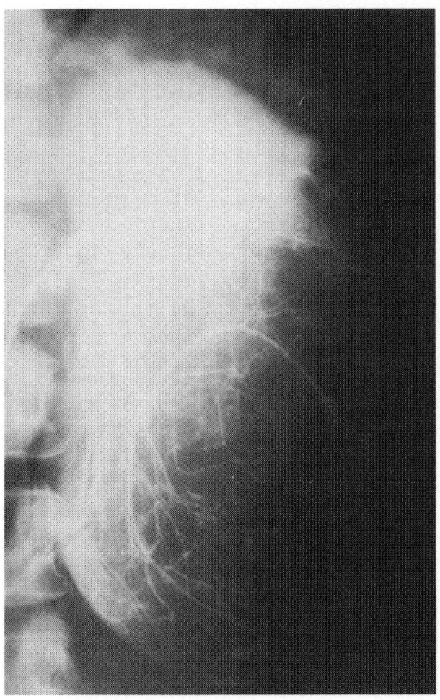

a.

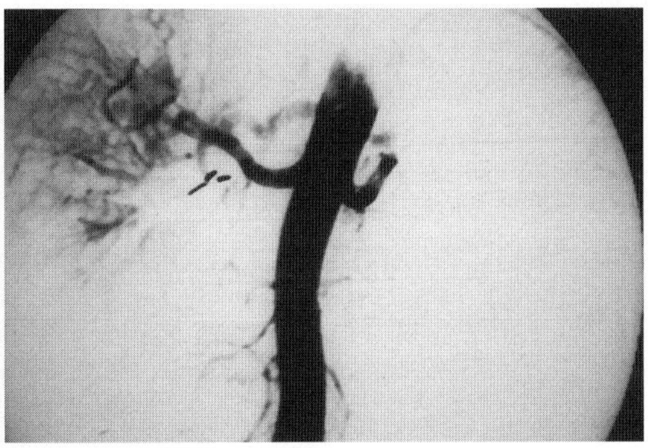

b.

Figure 12. *38-year-old man with left-sided renal cell adenocarcinoma. a) Late arterial phase from selective left renal arteriogram reveals a large tumor in the lower pole of the kidney. b) Following embolization with ethanol, an abdominal aortogram confirms that the arterial flow to the left kidney has been successfully occluded.*

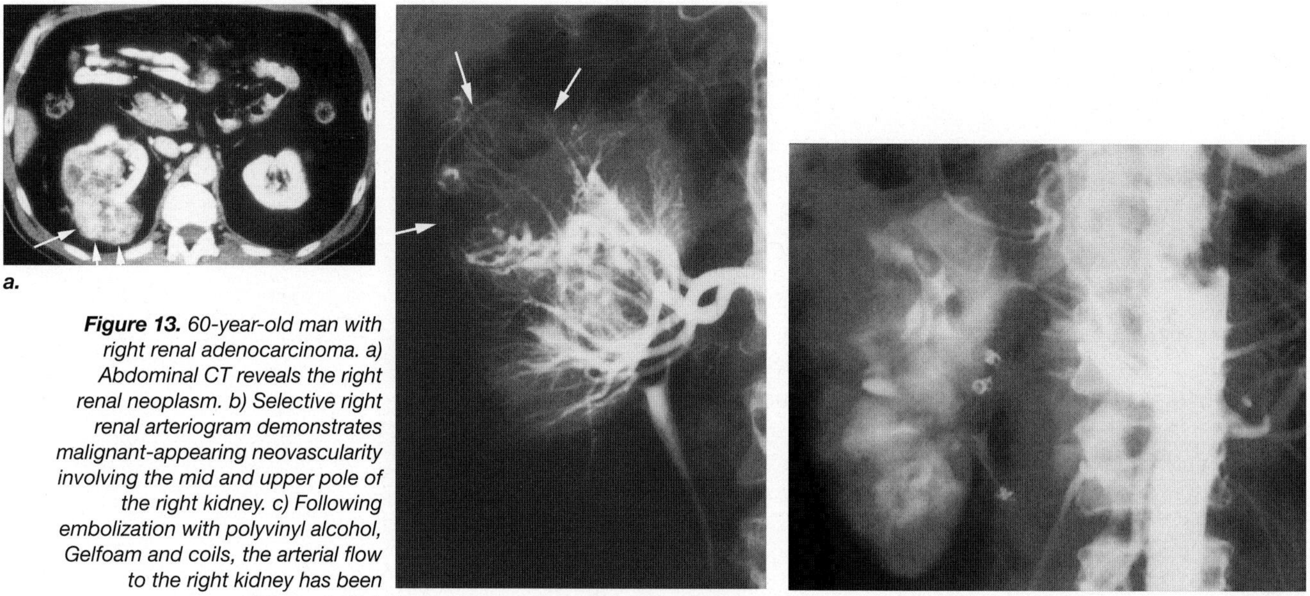

a.

Figure 13. 60-year-old man with right renal adenocarcinoma. a) Abdominal CT reveals the right renal neoplasm. b) Selective right renal arteriogram demonstrates malignant-appearing neovascularity involving the mid and upper pole of the right kidney. c) Following embolization with polyvinyl alcohol, Gelfoam and coils, the arterial flow to the right kidney has been successfully occluded. b. c.

Though dehydrated alcohol has been demonstrated to be effective for preoperative renal embolization, superiority compared to particulate agents has not been unequivocally established. Particulate materials used include both Gelfoam pledgets and powder, as well as polyvinyl alcohol **(Fig. 13)**. If coils are used for preoperative renal embolization, they must be deposited peripherally to avoid the possibility of inadvertent dislodgment during surgery.

Pelvic Malignancy

Unresectable or recurrent carcinoma of the uterine cervix often leads to recurrent vaginal hemorrhage which may be difficult to control. Bilateral embolization of the anterior division of the internal iliac artery with polyvinyl alcohol particles can be useful in reducing the rate of bleeding. A report of 17 patients by Yamashita et al confirmed that this technique is highly effective in acutely controlling the hemorrhage. Unfortunately, a recurrence rate of 40% was observed. Complications are surprisingly infrequent despite the fact that many of the patients have previously undergone radiation therapy. However, focal skin ulceration and transient leg numbness have been observed. Care should be employed to avoid embolization of the inferior gluteal artery which supplies the sciatic nerve.

Skeletal Metastases

Embolization of hypervascular metastases can be palliative in patients with pain, or can be performed preoperatively to reduce blood loss during surgery. Metastatic renal cell adenocarcinoma and metastatic thyroid carcinoma are the most common types of tumors treated. Polyvinyl alcohol is the agent of choice due to the permanent occlusive nature of the material. In a study by Gellad et al, operative blood loss for palliative spinal surgery was reduced from 3,500 ml to 1,850 ml when preoperative embolization was performed.

Miscellaneous Indications for Transcatheter Embolization

Splenic Artery Embolization

Splenic artery embolization has been advocated as a potential therapy for marked thrombocytopenia related to splenomegaly. Either the proximal splenic artery is occluded with coils leaving the distal circulation intact, or a partial embolization of intrasplenic segmental vessels with particulates or coils is performed. Embolization is only considered when platelet counts fall well below 50,000/ml. Early enthusiasm for this technique was tempered by the frequent appearance of post-embolization splenic abscess. Broad spectrum parenteral antibiotic prophylaxis, as well as soaking the embolic material in antibiotic solution, was recommended to diminish the incidence of this complication.

More recent work indicates that postembolization splenic abscess can be avoided in most patients by only partially embolizing the spleen or placing coils in the main splenic artery. With increasing frequency, splenic artery embolization is performed prior to surgical splenectomy. If the surgery is performed soon after embolization, concern over causing a splenic abscess is removed and a more aggressive embolization performed. Though a dramatic increase in the number of circulating platelets is seldom observed following the procedure, an elevation above 50,000/ml is generally achieved. At this level, the safety of a surgical splenectomy is improved.

Complete splenic artery embolization may also be useful to facilitate surgical splenectomy in patients with thrombocytopenia or splenic vein occlusion. In these patients, the entire artery is blocked immediately prior to the operation **(Fig. 14)**.

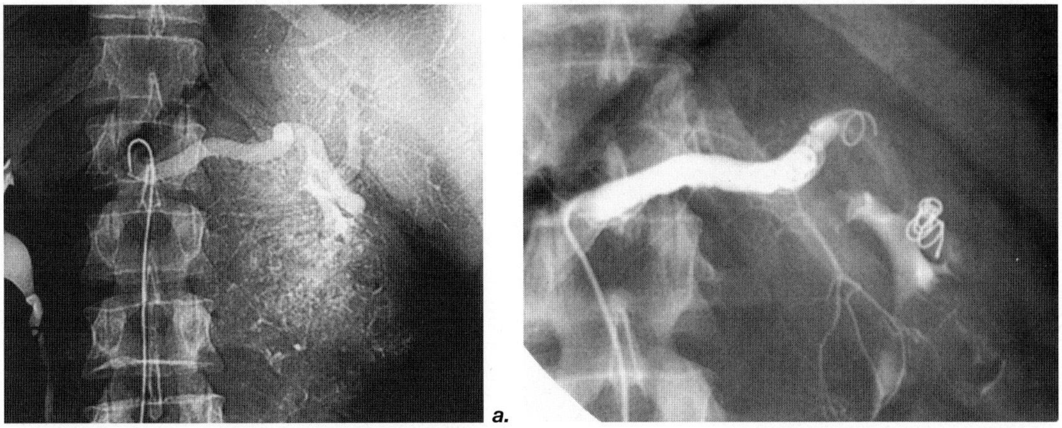

a. b.

Figure 14. *38-year-old man with pancreatic mass and splenic vein occlusion. a) Splenic arteriogram depicting the vascular mass in the pancreatic tail. b) The splenic artery is occluded by coil spring emboli.*

Proximal splenic artery embolization may also be useful in patients with advanced pancreatic carcinoma. These tumors commonly occlude the splenic vein and erode into the stomach and cause bleeding. Only limited surgical and endoscopic options are available to these patients. Transcatheter occlusion of the splenic artery will diminish both arterial and venous pressure and may reduce the rate of bleeding to a manageable level. Controlled delivery of coil spring emboli may be facilitated by placement of an occlusion balloon catheter.

Embolization of Testicular Varicoceles

Transcatheter occlusion of the testicular vein is not uncommonly performed in patients with varicoceles. Most often, these patients have had a prior surgical ligation but a recurrence has developed. When transcatheter occlusion of testicular veins was first performed, detachable balloons were used. More recently, coils have become the most common occluding agent.

Uterine Fibroids

Recently, considerable interest has developed regarding transcatheter embolization for the management of symptomatic uterine fibroids. Though only short-term results are available, it appears that this procedure is effective in reducing menorrhagia as well as pressure symptoms related to the bulky tumors in more than 90% of patients. The procedure is performed by selectively injecting the uterine arteries with polyvinyl alcohol particles bilaterally **(Fig. 15)**. Though significant complications are unusual, occasionally infection following embolization necessitates hysterectomy.

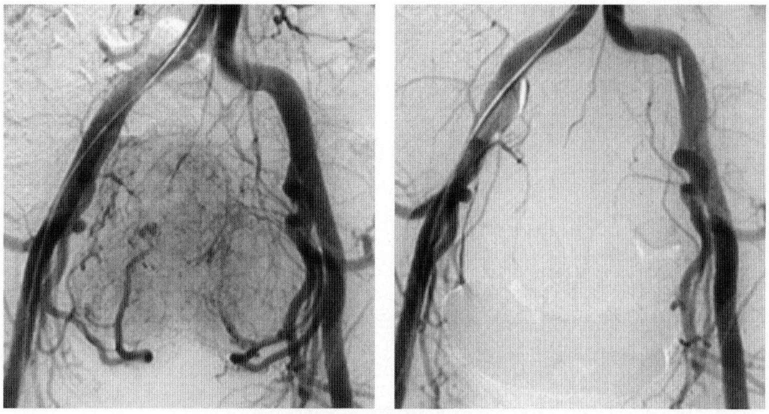

a. ***b.***

Figure 15. *44-year-old woman with recurrent menorrhagia. a) Pelvic arteriogram reveals a hypervascular myomatous uterus. b) Following bilateral uterine artery embolization, arterial flow to the fibroid uterus is no longer visualized.*

CONCLUSION

The development of coaxial catheters and microcoils has facilitated the growth of arterial embolization as a therapeutic modality. In the management of uncontrolled hemorrhage, these techniques are safer for the patient and more cost-effective than analogous surgical procedures. It is anticipated that the clinical applications of these techniques for the management of bleeding will continue to increase as the benefits are attractive to both the patient and referring physician.

The utility of arterial embolization for the palliation of malignancy is less well established. Currently, this technique is only appropriate for specific types of tumors. More clinical trials are needed to determine the benefit and cost-effectiveness of these techniques in comparison to more conventional forms of therapy in the treatment of patients with advanced malignancies.

REFERENCES

Bakal CW, Cynamon J, Lakritz PS, Sprayregen S. Value of preoperative renal artery embolization in reducing blood transfusion requirements during nephrectomy for renal cell carcinoma. J Vasc Interv Radiol 1993; 4:727-731.

Ben-Menachem Y, Coldwell DM, Young JWR, Burgess AR. Hemorrhage associated with pelvic fractures: causes, diagnosis, and emergent management. AJR 1991; 157:1005-1014.

Boudghene F, L'Hermine C, Bigot JM. Arterial complications of pancreatitis: diagnostic and therapeutic aspects in 104 cases. J Vasc Interv Radiol 1993; 4:551-558.

Burgess AR, Eastridge BJ, Young JWR, et al. Pelvic ring disruptions: effective classification system and treatment protocols. J Trauma 1990; 30:848-856.

Burn P, McCall J, Chinn R, Healy J. Embolization of uterine fibroids. Brit J Radiol 1999; 72:159-161.

Cornud F, Belin X, Amar E, Delafontaine D, Helenon O, Moreau JF. Varicocele: strategies in diagnosis and treatment. Eur Radiol 1999; 9:536-545.

Encarnacion CE, Kadir S, Beam CA, Payne CS. Gastrointestinal bleeding: treatment with gastrointestinal arterial embolization. Radiology 1992; 183:505-509.

Gellad FE, Sadato N, Numaguchi Y, Levine AM. Vascular metastatic lesions of the spine: preoperative embolization. Radiology 1990; 176:683-686.

Goodwin SC, Walker WJ. Uterine artery embolization for the treatment of uterine fibroids. Curr Opin Obstetrics and Gyn 1998; 10:315-320.

Guy GE, Shetty PC, Sharma RP, et al. Acute lower gastrointestinal hemorrhage: treatment by superselective embolization with polyvinyl alcohol particles. AJR 1992; 159:521-525.

Hayakawa K, Tanaka F, Torizuka T, et al. Bronchial artery embolization for hemoptysis: immediate and long-term results. Cardiovasc Interv Radiol 1992; 15:154-158.

Kam J, Jackson H, Ben-Menachem Y. Vascular injuries in blunt pelvic trauma. Radiol Clin North Am 1981; 19:171-176.

Kerlan RK Jr., Pogany AC, Burke DB, Ring EJ. Angiographic management of upper gastrointestinal hemorrhage. AJR 1986; 147:1185-1188.

Lang EK. Transcatheter embolization in management of hemorrhage from duodenal ulcer: long-term results and complications. Radiology 1992; 182:703-770.

Malisch TW, Guglielmi G, Vinuela F, et al. Intracranial aneurysms treated with Guglielmi detachable coil: midterm clinical results in a consecutive series of 100 patients. J Neurosurg 1997; 87:176-183.

Moertel CG, Johnson CM, McKusick MA, et al. The management of patients with advanced carcinoid tumors and islet cell carcinomas. Ann Int Med 1994; 120:302-309.

Peene P, Wilms G, Baert AL. Embolization of iatrogenic renal hemorrhage following percutaneous nephrostomy. Urol Radiol 1990; 12:84-87.

Soulen MC. Chemoembolization of hepatic malignancies. Oncology 1994; 8:77-84.

Taniguchi K, Nakata K, Kato Y, et al. Treatment of hepatocellular carcinoma with transcatheter arterial embolization. Analysis of prognostic factors. Cancer 1994; 73:1341-1345.

Yamashita Y, Harada M, Yamamoto H, et al. Transcatheter arterial embolization of obstetric and gynecological bleeding: efficacy and clinical outcome. Brit J Radiol 1994; 67:530-534.

CHAPTER 13 PREVIEW ANSWERS

Case 1

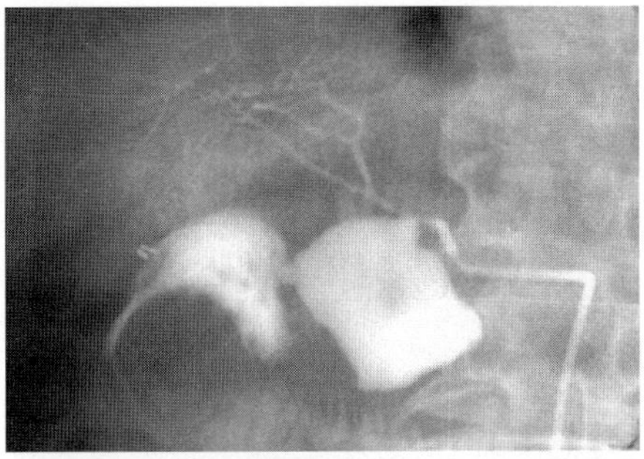

a.

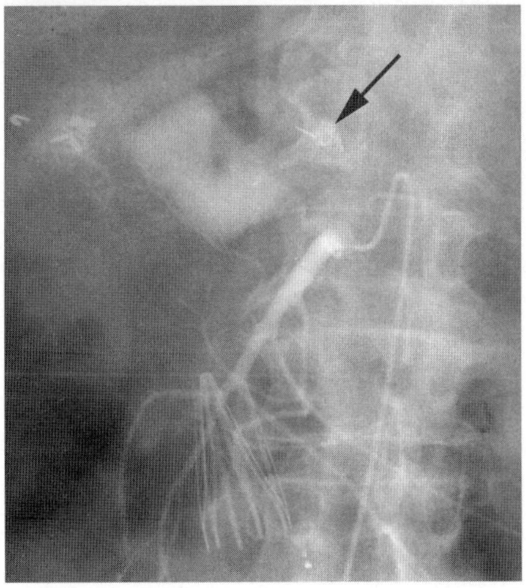

b.

Findings: a) Extravasation from common hepatic artery can be treated with percutaneous embolization. A patent portal vein will usually prevent hepatic infarction. b) Following deployment of coils (arrow) to occlude the common hepatic artery, a superior mesenteric arteriogram is performed showing no collateral flow to the area of extravasation. Residual extravasated contrast is present from prior injections.
Diagnosis: Postoperative bleeding from gastroduodenal arterial stump.

Case 2

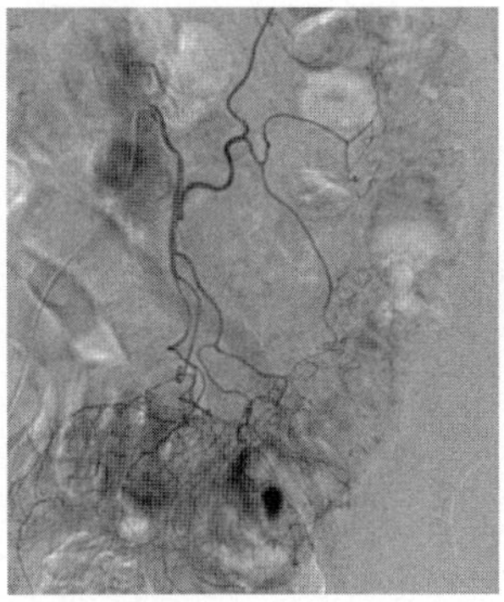

a.

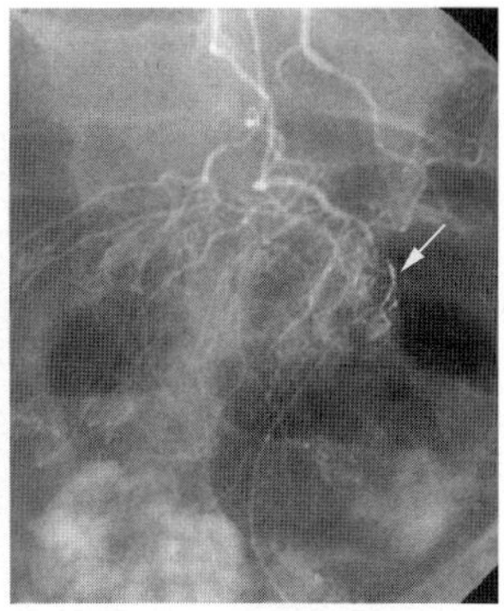

b.

Findings: a) Inferior mesenteric arteriogram reveals extravasation of contrast into the sigmoid colon. b) Superselective coil embolization (arrow) is used to control the bleeding.
Diagnosis: Lower gastrointestinal bleeding from sigmoid diverticulosis.

Case 3

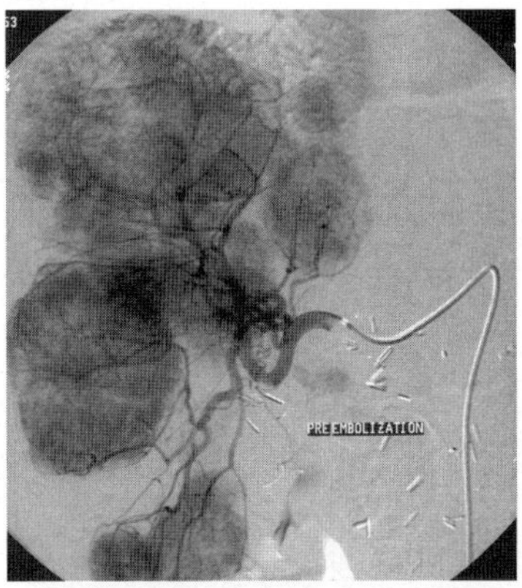

a.

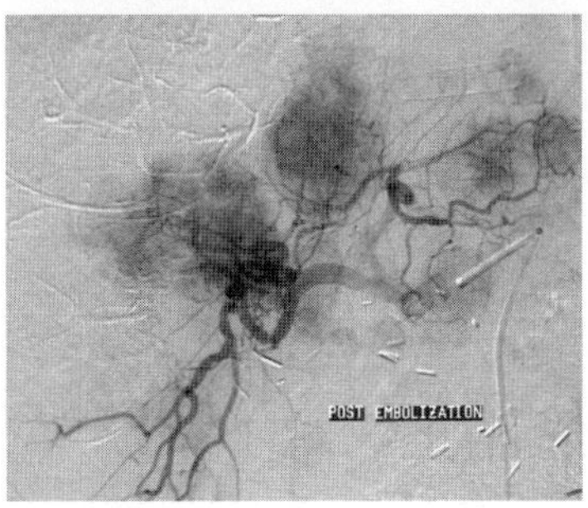

b.

Findings: a) Hepatic arteriogram shows multiple hypervascular hepatic metastases. b) A considerable reduction in the hypervascularity is present following embolization with Gelfoam powder.
Diagnosis: Metastatic carcinoid tumor.

CHAPTER 14 PREVIEW

Case 1

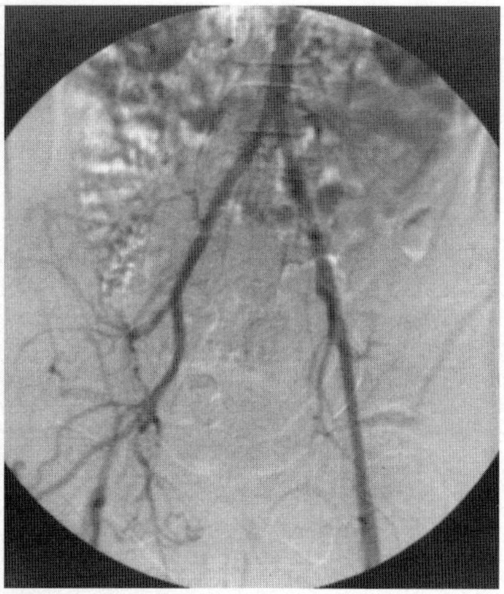

Figure 1. *Intravenous digital subtraction angiogram (IV DSA) with image intensifier centered over the pelvis.*

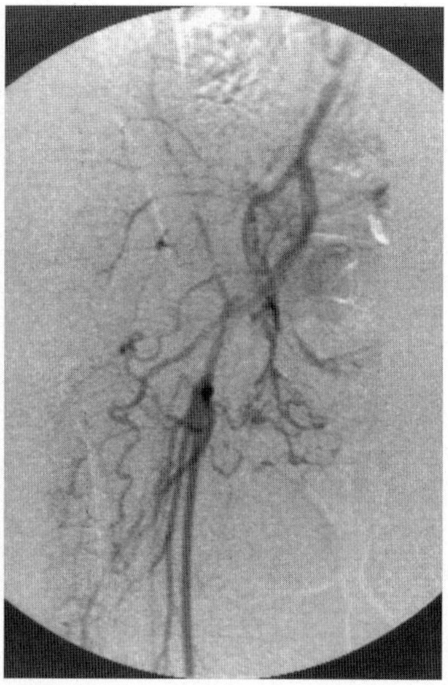

Figure 2. *IV DSA with image intensifier centered over the right hip.*

History: A 10-year-old girl was evaluated for leg length discrepancy (left leg longer than right). On physical exam, the right femoral pulse was palpable but diminished. Her past medical history was significant for congenital heart disease. She had undergone cardiac catheterization at age 1 year. An intravenous digital subtraction arteriogram was performed with imaging over the pelvis and right hip **(Figs. 1, 2)**.
Questions: What is your diagnosis? Can you describe the collateral circulation pattern shown here?

Case 2

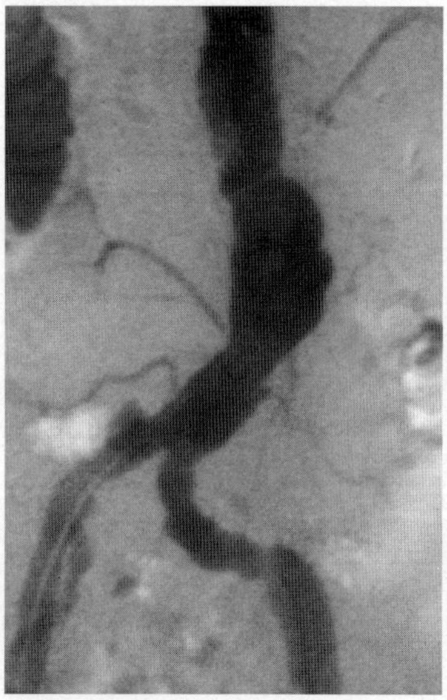

Figure 3. *Lower abdominal aortogram, subtraction film AP projection.*

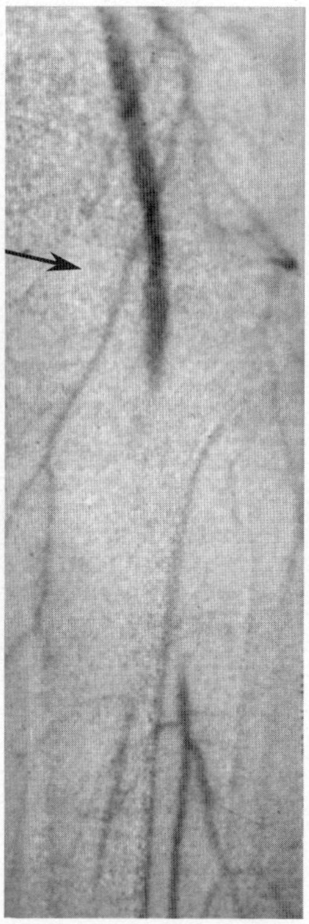

Figure 4. *Selected image from left lower extremity arteriogram, DSA image centered over the left knee (arrow marks the level of the knee joint).*

History: A 55-year-old man with hypertension underwent aortography **(Fig. 3)**. A left renal artery stenosis was detected (not shown) and the following day the patient underwent angioplasty of the left renal artery. Initial catheterization of the left renal artery was difficult and during the procedure the patient complained of pain in his left leg and foot. Arteriography of the left leg was performed. A selected image from that study is shown **(Fig. 4)** (arrow marks the level of the knee joint).
Questions: What is your diagnosis? How should this patient be treated?

Chapter 14
Angiographic Complications

Susan D. Wall, MD, and
Jeanne M. LaBerge, MD

CHAPTER OUTLINE

Puncture site complications
- Hematoma
- Pseudoaneurysm
- Arteriovenous fistula
- Dissection
- Spasm
- Acute thrombosis

Vascular complications remote to the puncture site
- Arterial dissection
- Acute thrombosis
- Spasm
- Arterial rupture
- Distal embolization
- Cholesterol emboli

Systemic reactions
- Syncope, vasovagal response
- Contrast reactions, immediate and delayed
- Hemodynamic instability, hypo- and hypertension

INTRODUCTION

Public interest in medical complications has been heightened by a recent report from the Institute of Medicine indicating that medical errors kill close to 100,000 patients each year in the United States. Moreover, 3%-4% of all hospitalized patients suffer an iatrogenic injury during their hospital stay. These startling statistics are of particular interest to interventional radiologists because 35% of all medical errors involve the performance of a procedure or operation, and procedures are the essence of interventional radiology.

In this chapter, we review procedural complications associated with diagnostic angiography and vascular interventions. The chief aim of this chapter is to provide an organizational framework for assessing and treating vascular complications.

ADVERSE EVENTS RELATED TO ANGIOGRAPHY

- Systemic
  - allergic
  - syncopal
- Catheter-induced
  - at the puncture site
  - at the intervention site
  - in distal vessels (embolic)

INCIDENCE

Adverse events associated with angiographic procedures may be due to systemic reactions or catheter-induced vascular injury. Complications due to catheterization may be observed at the puncture site, at the site of vascular intervention, or in distal vascular beds due to embolization.

Fortunately, complications associated with diagnostic and therapeutic angiography are uncommon. Overall, adverse events occur in 2.0% of procedures. Serious complications occur in approximately 0.5% of cases with fatal events in only 0.03%-0.06% of procedures.

INCIDENCE OF PROCEDURAL COMPLICATIONS

- Overall 2.0%
- Fatal 0.03%-0.06%
- Puncture site morbidity by location:
  - Femoral 1.7%
  - Axillary 3.3%
  - Brachial 7.0%
  - Translumbar 3.0%

DIMINISHED PULSE AT PUNCTURE SITE: DIFFERENTIAL DIAGNOSIS

- Hypotension (vasovagal reaction)
- Spasm
- Local injury to access vessel (dissection, thrombosis)
- Compromised flow due to the size of the catheter or sheath
- Embolic occlusion
- Hematoma

HEMATOMA RISK FACTORS

- Obesity
- Hypertension
- Anticoagulation or coagulopathy
- Large catheter or sheath size
- "Low" puncture site

Loss or Decrease in Pulse at the Puncture Site

It is important to pay close attention to the quality of the pulse at the puncture site. A decrease in pulse should be recognized promptly and the cause should be treated appropriately. Differential diagnosis includes: hypotension (vasovagal reaction), spasm, local injury to the punctured vessel (dissection or thrombosis), obstruction by the catheter **(Figs. 3, 4)**, embolus, and hematoma.

Hematoma

This is the most common arterial access site complication. Obesity, hypertension, and coagulopathy are important risk factors. Large catheter size and "low" puncture site may also contribute to the problem. To minimize the risk of hematoma, the patient's blood pressure should be controlled so that the diastolic pressure remains below 90 mm Hg. Adequate control of blood pressure is especially important at the time of catheter insertion and catheter removal.

Normal coagulation parameters are an essential ingredient for maintaining and sustaining hemostasis at the puncture site. Coagulopathic patients should be treated so that laboratory parameters (INR, PT, PTT, platelets) are normal prior to the procedure. When intravenous heparin is used during the procedure, the PTT and activated coagulation time (ACT) should be normalized prior to catheter removal.

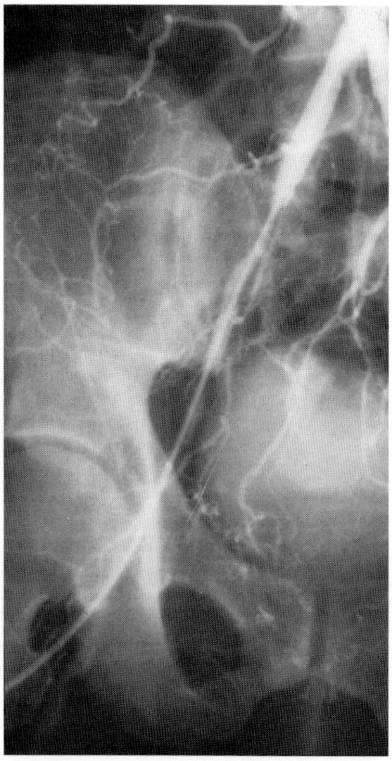

Figure 3. *Right iliac arteriogram following ipsilateral catheterization. The right iliac vessels are diffusely stenotic and the catheter further compromises flow to the right lower extremity.*

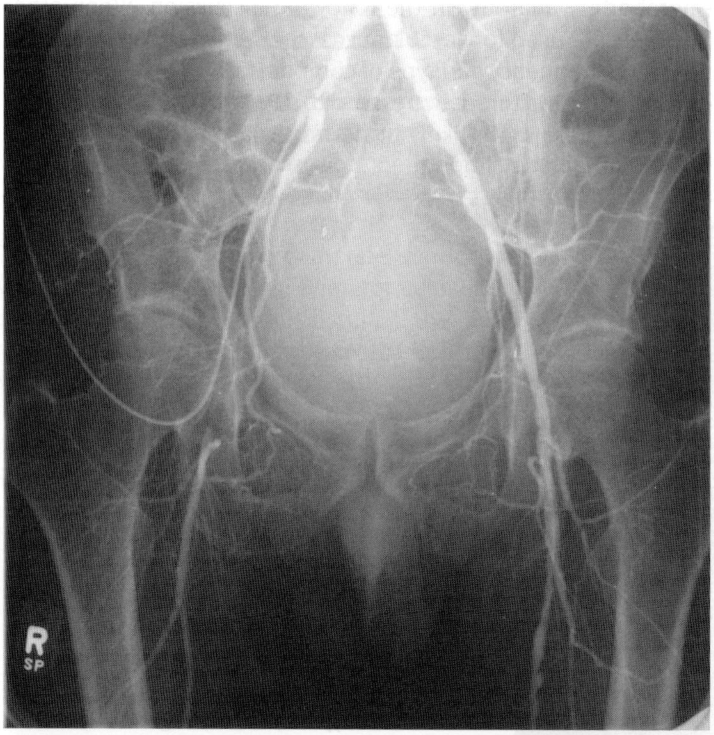

Figure 4. *A patient with right leg claudication underwent arteriographic evaluation. The pelvic arteriogram demonstrates absent filling of the external iliac artery. The right profunda femoris artery is opacified through collaterals. In this case, the diagnostic catheter had completely obstructed flow in the severly stenotic external iliac artery. The patient was anticoagulated with 4,000 units of intravenous heparin immediately after catheterization.*

We believe that durable hemostasis at the puncture site is best achieved by manual compression. (A variety of percutaneous closure devices are now available as an alternative to manual compression but we restrict their use to selected patients.) Because the midpoint of the femoral artery usually overlies the femoral head, the favored site for vascular entry is the midaspect of the common femoral artery. Vascular entry at this point allows for effective compression of the puncture site using the femoral head. When a puncture site is selected inferior to the femoral head, effective compression may be difficult to achieve and hematoma or pseudoaneurysm may be the consequence.

Pseudoaneurysm

Pseudoaneurysm of the common femoral artery results from inadequate manual compression of the puncture site following catheter removal. This complication occurs more frequently when access is "low" because of an increased likelihood of puncturing the superficial femoral artery rather than the common femoral artery **(Fig. 5)**. When the puncture site is caudal to the head of the femur, manual compression is relatively less effective and a pseudoaneurysm may develop. Periprocedural anticoagulation with intravenous heparin also increases the risk of pseudoaneurysm formation.

PSEUDOANEURYSM

- Results from inadequate compression
- Associated with "low" puncture site into SFA
- May expand and rupture
- Treat with ultrasound-guided compression
- May require thrombin injection or surgical repair

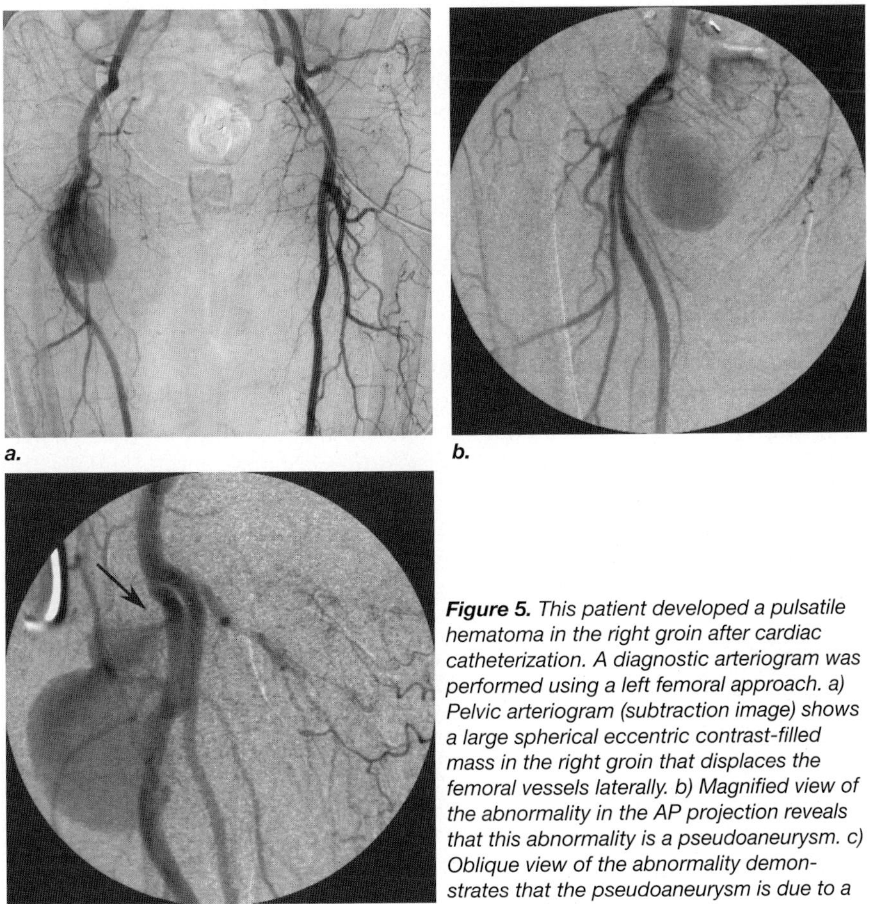

a. b. c.

Figure 5. *This patient developed a pulsatile hematoma in the right groin after cardiac catheterization. A diagnostic arteriogram was performed using a left femoral approach. a) Pelvic arteriogram (subtraction image) shows a large spherical eccentric contrast-filled mass in the right groin that displaces the femoral vessels laterally. b) Magnified view of the abnormality in the AP projection reveals that this abnormality is a pseudoaneurysm. c) Oblique view of the abnormality demonstrates that the pseudoaneurysm is due to a defect in the wall of the SFA (arrow).*

Pseudoaneurysm should be suspected when postprocedure examination of the access site reveals a pulsatile hematoma. Ultrasound is the preferred method for diagnosing this problem and evaluating the effectiveness of treatment **(Fig. 6)**. Initial treatment with ultrasound-guided compression is successful in 55%-75% of cases. Ultrasound-guided introduction of thrombin provides a second-line method of treatment when compression is unsuccessful. Surgical repair may be necessary in a minority of patients in order to avoid expansion and rupture.

Retroperitoneal Hemorrhage

This is an uncommon but potentially life-threatening complication of femoral access. It occurs when the puncture site is "too high," that is, when the puncture site is above the inguinal ligament. In this circumstance, manual compression may not be effective in promoting hemostasis and as a consequence retroperitoneal hemorrhage may ensue. This problem may not be immediately apparent since the hematoma does not involve the soft tissues of the groin and is not visible. Rather, bleeding is occult and may not be noticed until volume loss precipitates cardiovascular changes such as hypotension and tachycardia.

RETROPERITONEAL HEMORRHAGE

- Secondary to "high" puncture site, above inguinal ligament
- Occult until vital signs reflect significant hemorrhage
- Perform immediate noncontrast CT if suspected
- Surgical repair necessary

Attention to anatomic landmarks may reveal an inadvertently high puncture. If the catheter entry site is near or above the origins of the deep iliac circumflex or inferior epigastric arteries, the angiographer should assume that he has entered the external iliac artery. If retroperitoneal hematoma is suspected, a noncontrast pelvic CT scan should be performed. Scan coverage should extend into the groin and proximal thigh. If a large amount of blood is detected around the distal external iliac or common femoral arteries, a vascular surgeon should be notified. The patient should be taken immediately to the operating room or observed closely in the intensive care unit for signs of cardiovascular change indicating further bleeding.

Arteriovenous Fistula

Arteriovenous fistula most commonly results from "low" arterial entry, an entry below the common femoral artery. This complication is explained by the spatial relationship of the femoral artery and the vein. At the level of the common femoral artery, the femoral vein lies medial to the artery. Whereas, more caudally, the distal common femoral artery or superior femoral artery lies anterior to the vein. Therefore, simultaneous puncture of the artery and vein is more common with lower punctures than with higher punctures. Also, adequate compression is difficult to obtain when catheter entry occurs caudal to the femoral head.

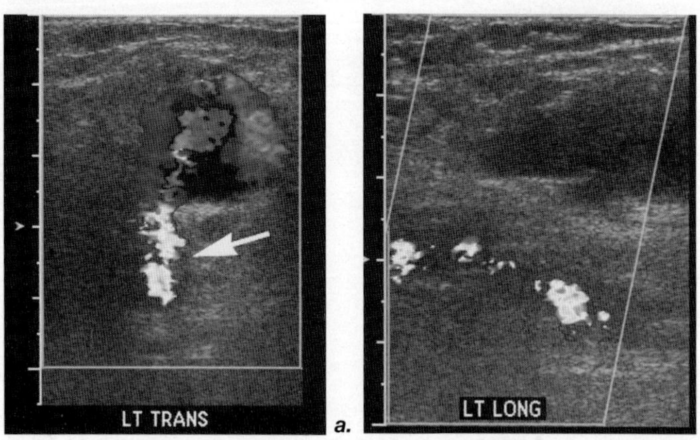

a. b.

Figure 6. *An 85-year-old woman underwent iliac stent placement from a left transfemoral approach. Three hours later, examination of the left groin revealed a large area of ecchymosis and a pulsatile hematoma. Sonography was performed. a) Transverse color Doppler image reveals a large pseudoaneurysm with a well-defined neck (arrow). b) Longitudinal image, following 20 minutes of directed manual compression, shows resolution of flow in the pseudoaneurysm while flow in the femoral artery is maintained.*

Small arteriovenous fistulas may not cause symptoms; however, large fistulas may cause symptoms by diverting or "stealing" arterial blood away from the lower extremity. As a result, patients may experience symptoms of arterial insufficiency that mimic those caused by arterial stenosis.

VASCULAR COMPLICATIONS REMOTE TO THE PUNCTURE SITE

Injury of vessels remote to the puncture site is uncommon but a variety of problems have been reported. Vascular injury may be induced by a catheter or guidewire during diagnostic angiography or may result from trauma inflicted on the wall of a blood vessel during an intervention. The spectrum of these nonpuncture site injuries is similar to those observed at the puncture site and includes: arterial dissection, acute thrombosis, spasm, arterial rupture and distal embolization.

Arterial Dissection

This can occur when a guidewire is forcibly directed along a subintimal path **(Fig. 7)**. Rough handling of the guidewire or inattention to the tactile sensation of resistance can initiate subintimal passage. Use of guidewires with very floppy tips may limit this complication. In addition, road-mapping techniques provide a valuable aid to atraumatic guidewire passage across a difficult arterial stenosis.

Limited arterial dissection is an expected outcome of percutaneous transluminal angioplasty (PTA). The benefit of PTA comes from the restructuring of the arterial lumen that is derived from a controlled intimal dissection and rupture of the media. Dissection induced by PTA can be counterproductive when it is too extensive and causes narrowing of the vessel lumen rather than enlargement. Complications of excessive dissection include occlusion, thrombosis and/or distal embolization **(Fig. 8)**. Hence, it is helpful to retain guidewire access across a lesion undergoing angioplasty until a satisfactory outcome has been achieved.

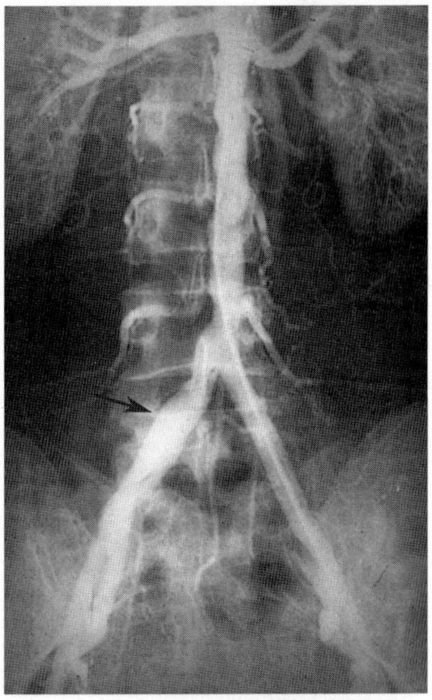

a.

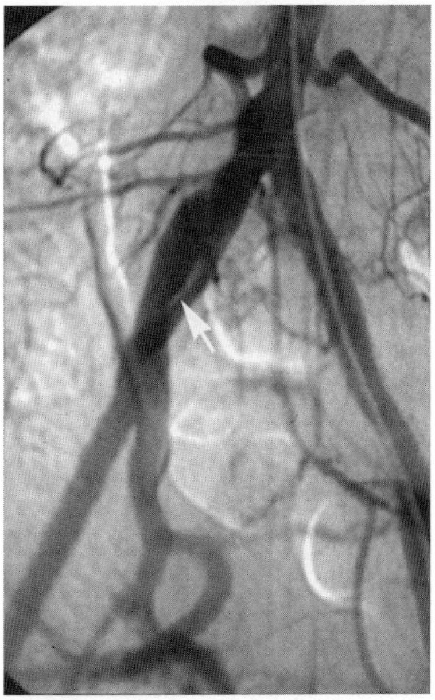

b.

Figure 7. *This patient with hyperlipidemia developed right leg claudication following cardiac catheterization. a) Aortogram performed via left transfemoral catheterization reveals an abnormality of the right common femoral artery (arrow). The diameter of the vessel is too large. b) RPO digital subtraction arteriogram reveals an intimal flap in the vessel (arrow). These findings are consistent with post-traumatic dissection, likely secondary to catheterization.*

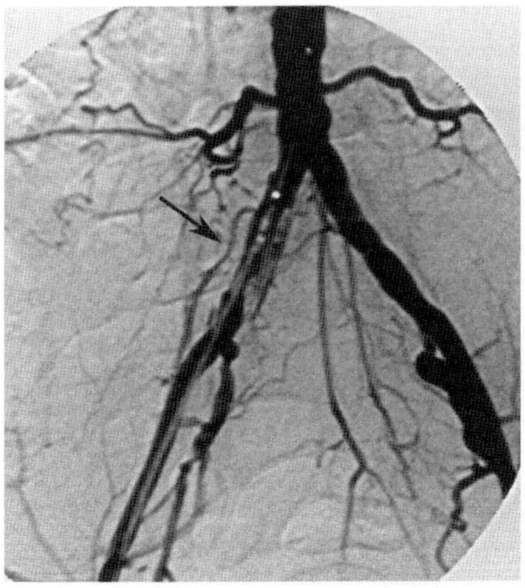

a.

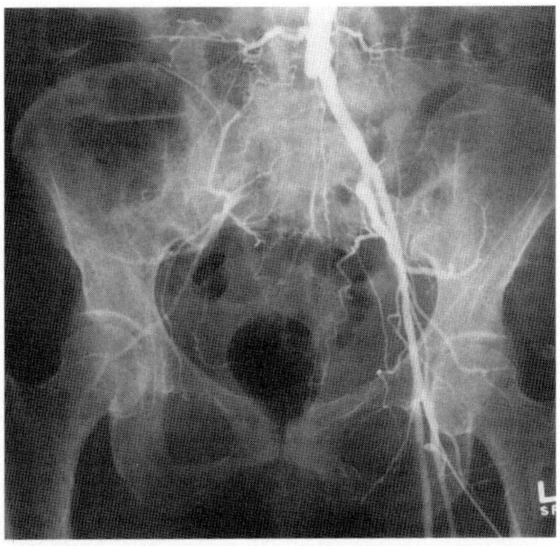

b.

Figure 8. *Right iliac angioplasty was performed in this patient with 2-block right leg claudication. a) Postangioplasty iliac arteriogram shows an extensive dissection (arrow). There was poor flow through the vessel which persisted despite repeat dilatation and heparinization. (Procedure performed before the advent of endovascular stents.) b) Two weeks later, the patient continued to complain of claudication slightly more severe than before PTA. Repeat pelvic angiogram reveals right iliac occlusion.*

A flow-limiting dissection observed during PTA must be treated immediately. Sometimes the intimal flap can be "tacked down" by repeat balloon inflation. Deployment of a metallic stent provides a more reliable treatment and is usually a very effective means of restoring flow at the angioplasty site. If these efforts are not effective, surgical repair may be necessary.

RISK FACTORS FOR ARTERIAL RUPTURE WITH PTA

- Steroids
- Calcified vessels
- Large balloons
- High-pressure balloons

Arterial Rupture

Arterial rupture occurs in less than 0.5% of angioplasty procedures, but it is a serious complication and can be catastrophic. Factors that predispose a patient to rupture associated with PTA include steroid use, calcified vessels, and the use of large and/or high-pressure balloons. This complication should be suspected if the patient complains of severe pain that persists after balloon deflation. When the angioplasty site is in an accessible vessel (such as the superior femoral artery), tamponade can be achieved by manual compression. When the rupture site is inaccessible to external compression (as with the iliac or renal arteries), hemostasis should be accomplished by balloon inflation proximal to or at the site of rupture. Surgical intervention must be performed immediately. Alternatively, placement of a covered stent may provide an immediate solution to the problem.

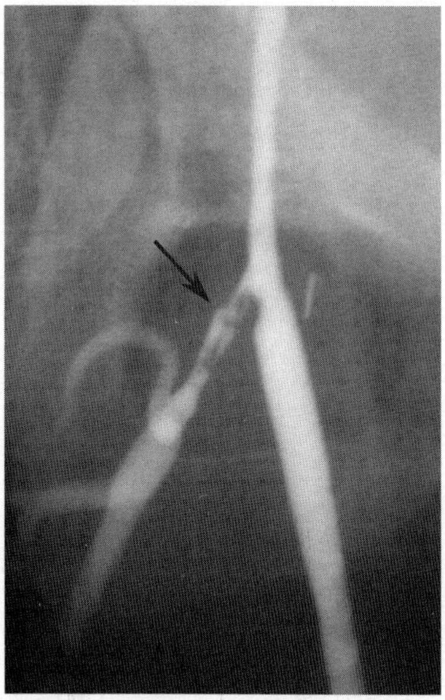

Figure 9. *A patient with a history of multiple prior vascular procedures underwent angioplasty to treat a stenosis of the right profunda femoris artery origin. This right femoral arteriogram obtained following PTA reveals thrombus at the PTA site (arrow). Thrombus resolved following administration of urokinase.*

Acute Thrombosis

This complication occurs when blood flow is slowed due to spasm, dissection or catheter occlusion of a narrowed artery. Treatment involves heparin for anticoagulation and direct introduction of a thrombolytic agent such as urokinase or tPA **(Figs. 9, 10)**.

Spasm

Spasm results from irritation of the vessel wall by a guidewire or catheter. It is easier to prevent than it is to treat. Antispasmodic medications such as nitroglycerine can be introduced intraarterially prior to balloon angioplasty and into a venous bypass graft prior to guidewire/catheter introduction. If spasm occurs and intraarterial antispasmodic medication does not provide successful treatment, balloon dilatation may be beneficial. Sublingual administration of a calcium channel blocker such as nifedipine may also be useful.

COMPLICATIONS DISTAL TO THE SITE OF INTERVENTION

Distal Embolization

This is an uncommon complication of angioplasty. It is more common in the setting of an acute occlusion **(Fig. 11)**, especially when the occlusion involves a large vessel such as the common or external iliac artery. There does not seem to be an increased incidence of distal embolization after recanalization of chronic occlusions when PTA follows thrombolysis.

Embolization may be symptomatic and result in diminished peripheral pulses. However, most distal emboli are clinically silent **(Fig. 12)**. A Doppler-derived ankle systolic pressure that has decreased more than 20 mm Hg compared to the preangiographic value is indicative of iatrogenic embolic occlusion. Small emboli may lyse spontaneously, but large emboli will require treatment with a thrombolytic agent, aspiration embolectomy or surgical thrombectomy.

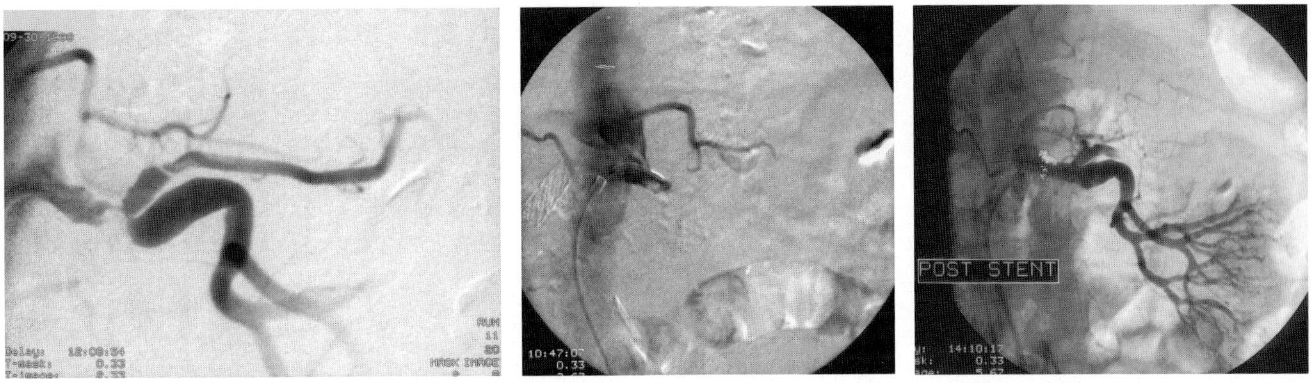

a. b. c.

Figure 10. *This hypertensive man underwent renal angioplasty to treat a left renal artery stenosis. a) Initial renal arteriogram reveals a complex severe stenosis of the left renal artery. b) Initial angioplasty of the lesion resulted in occlusion of the vessel. c) Post-PTA thrombosis was successfully treated by urokinase infusion and placement of a Palmaz stent.*

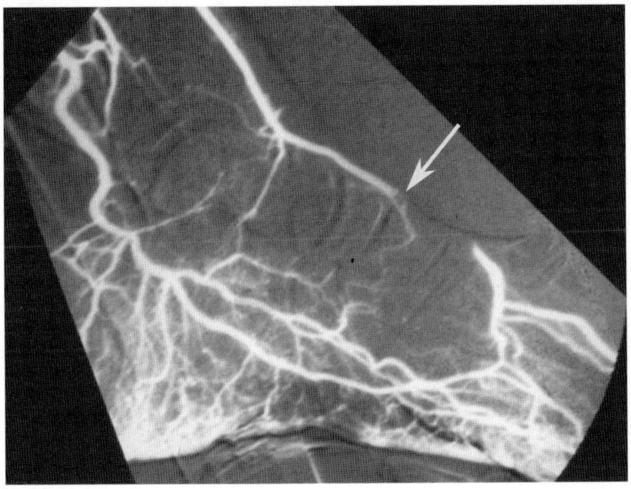

Figure 11. *Lateral foot arteriogram obtained following popliteal artery angioplasty. A small filling defect is identified in the middorsalis pedis artery (arrow). This likely represents embolized debris from the PTA. It resolved following the administration of urokinase.*

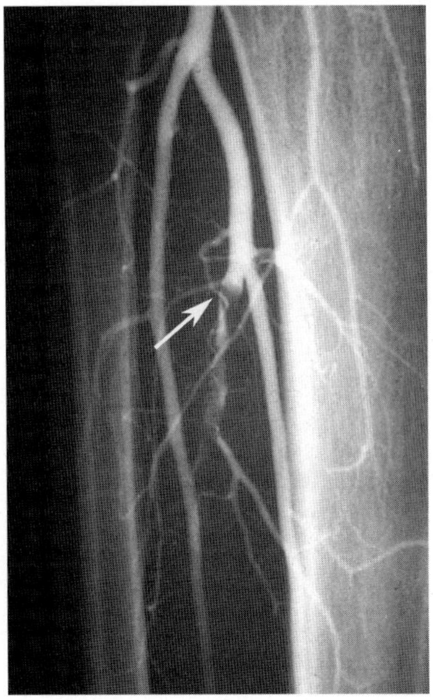

Figure 12. *Arteriogram of the right leg in a patient undergoing fibrinolytic infusion to lyse a proximal surgical graft. The anterior tibial and posterior tibial arteries are widely patent and the foot pulses were normal. The filling defect in the peroneal artery (arrow) represents clot embolized from the graft. An embolus in this location may be clinically silent.*

Cholesterol Emboli

Cholesterol emboli occur when small cholesterol crystals spontaneously break off the arterial wall, flow distally, and occlude small downstream vessels. If the source vessel is the ascending aorta or aortic arch, stroke can result. When the lower extremity is involved, physical examination will reveal palpable distal pulses, but painful small cyanotic patches on the skin will be noted. Cholesterol emboli from the abdominal aorta can result in renal failure and/or bowel infarction. This is a very serious complication with a mortality of approximately 50%-80%. Treatment of cholesterol emboli is difficult as anticoagulation is often ineffective. Surgical resection may be necessary in cases of severe end-organ ischemia.

SYSTEMIC REACTIONS

Syncope

Syncope caused by a vasovagal reaction may result in bradycardia and hypotension. Treatment includes elevation of the lower extremities to enhance blood return to the central venous system and a bolus of intravenous fluids. If symptoms persist, atropine (0.6 mg-1.0 mg) can be given intravenously.

Contrast Reactions

These are experienced by about 3% of patients undergoing arteriography or venography. Minor reactions occur three times more often in patients with a history of prior reaction than in those with none. Similarly, a history of asthma increases the risk of allergic reaction threefold. Major nonfatal contrast reactions occur less often during venography (0.09% of procedures) than during arteriography (0.14% of procedures). But, fatal reactions occur more often during venography (0.02%) than during arteriography (0.004%). Laryngeal edema with intraarterial use of iodinated contrast occurs in less than 0.06% of procedures.

Patients with a history of contrast allergy should be premedicated to reduce the risk of allergic reaction. Our treatment algorithm is as follows. Minor reactions call for the use of Benadryl (50 mg) by mouth the evening before and a 25 mg-50 mg intravenous dose at the beginning of the procedure. Premedication for a prior moderate reaction (e.g. bronchospasm) includes Benadryl plus hydrocortisone 100 mg every 6 hours for three doses before the procedure. If severe reaction has occurred in the past (laryngospasm, anaphylaxis, shock, cardiopulmonary arrest), contrast exposure is avoided, if possible. If angiography is necessary, patients are premedicated with Benadryl, hydrocortisone and cimetidine (Tagamet). Importantly, anesthesia stand-by is available to assist in the treatment of a life-threatening reaction. Should a serious reaction occur, the airway should be maintained, cardiovascular resuscitation with fluid and epinephrine (0.2-0.4 mg IV) should be implemented and if ventricular tachycardia/fibrillation occurs, electric cardioversion may be necessary.

Despite proper screening and premedication, patients may experience contrast reactions. Treatment of these reactions should be tailored to the individual patient and is determined by the type of reaction. A suggested treatment algorithm is shown in **Table 2**.

TABLE 2: TREATMENT OF CONTRAST REACTIONS

Reaction	Treatment
Urticaria	Benadryl 50 mg IV
Bronchospasm	Epinephrine 0.3-0.4 mg IV slowly Benadryl 25-50 mg IV Hydrocortisone 100 mg IV if severe
Laryngospasm/edema	Epinephrine 0.3-0.4 mg IV slowly Benadryl 25-50 mg IV Cimetidine 300 mg over 20 min Benadryl 50 mg IV Intubation
Cardiopulmonary arrest	Cardiopulmonary resuscitation

Contrast Nephropathy

This is defined as an increase in serum creatinine by at least 20% or 0.3 mg/dl above the preangiographic level. Predisposing factors include: preexisting renal disease, marginal or abnormal renal function, diabetes mellitus, multiple myeloma, hyperuricemia, proteinuria, dehydration, advanced age, nephrotoxic antibiotics, and prior recent intravenous contrast study. Creatinine elevation usually is transient with return to baseline in days to weeks. Occasionally temporary or permanent dialysis is required. Normal hydration should precede and follow angiographic and venous administration of iodinated contrast.

RISK FACTORS FOR CONTRAST NEPHROPATHY

- Preexisting renal disease
- Diabetes mellitus
- Multiple myeloma
- Hyperuricemia
- Proteinuria
- Dehydration
- Advanced age
- Nephrotoxic antibiotics
- Recent exposure to iodinated contrast

SUMMARY

The Standards of Practice Committee of the Society of Cardiovascular and Interventional Radiology has published guidelines for diagnostic angiography. Suggested threshold levels for complications are shown in **Table 3**. Physicians who perform angiographic procedures should be mindful of these potential complications and should document the rates of adverse events in their practice.

TABLE 3: SCVIR COMPLICATION THRESHOLDS

Indicators	Threshold
• Puncture site	
— Hematoma	3.0 %
Delays discharge	
Requires transfusion	
Requires surgery	
— Occlusion	0.5%
— Pseudoaneurysm	0.5%
— Arteriovenous fistula	0.1%
• Nonpuncture site	
— Distal emboli	0.5%
— Subintimal dissection	2.0%
— Occlusion of selected vessel	2.0%
— Subintimal injection	1.0%
• Cerebral angiography	
— All neurological deficits	4.0%
— Permanent neurological deficits	1.0%

REFERENCES

Almgren B, Karacagil S, Nybacka O. Arteriovenous fistula following transfemoral arteriography. Aust NZ J Surg 1990; 60:549-550.

Altin RS, Flicker S, Naidech HJ. Pseudoaneurysm and arteriovenous fistula after femoral catheterization: associations with low femoral punctures. AJR 1989; 152:629-631.

Committee on Quality Health Care in America. To err is human: building a safer health system. Institute of Medicine, National Academy of Sciences. National Academy Press, 1999.

Cragg AH, Nakagawa N, Smith TP, Berbaum KS. Hematoma formation after diagnostic angiography: effect of catheter size. J Vasc Interv Radiol 1991; 2:231-233.

Dawson P, Strickland N. Thromboembolic phenomena in clinical angiography: role of materials and technique. J Vasc Interv Radiol 1991; 2:125-132.

Fellmeth BD, Roberts AC, Bookstein JJ, et al. Postangiographic femoral artery injuries: nonsurgical repair with US-guided compression. Radiology 1991; 178:671-675.

Kent KC, McArdle CR, Kennedy B, Baim DS, Anninos E, Skillman JJ. A prospective study of the clinical outcome of femoral pseudoaneurysms and arteriovenous fistulas induced by arterial puncture. J Vasc Surg 1993; 17:125-133.

Leape LL, Brennan TA, Laird N, et al. The nature of adverse events in hospitalized patients. N Engl J Med 1999; 324:377-384.

Lilly MP, Reichman W, Sarazen AA, Carney WI. Anatomic and clinical factors associated with complications of transfemoral arteriography. Am Vasc Surg 1990; 4:264-269.

Quint LE, Holland D, Dorobkin M, Cascade PN. Role of femoral vessel catheterization and altered hemostasis in the development of extraperitoneal hematomas: CT study in 44 patients. AJR 1993; 160:855-863.

Rupp SB, Bogelzang RL, Nemcek AA, Yungbluth MM. Relationship of the inguinal ligament to pelvic radiographic landmarks: anatomic correlation and its role in femoral arteriography. J Vasc Interv Radiol 1993; 4:409-413.

Spies JB. Complications of diagnostic arteriography. Seminars in Interv Radiol 1994; 11:93-101.

Spies J, Bakal C, Burke D, et al. Standard for disgnostic arteriography in adults. J Vasc Intervent Radiol 1993; 4:385-395.

Waugh JR, Sacharias N. Arteriographic complications in the DSA era. Radiology 1992; 182:243-246.

CHAPTER 14 PREVIEW ANSWERS

Case 1

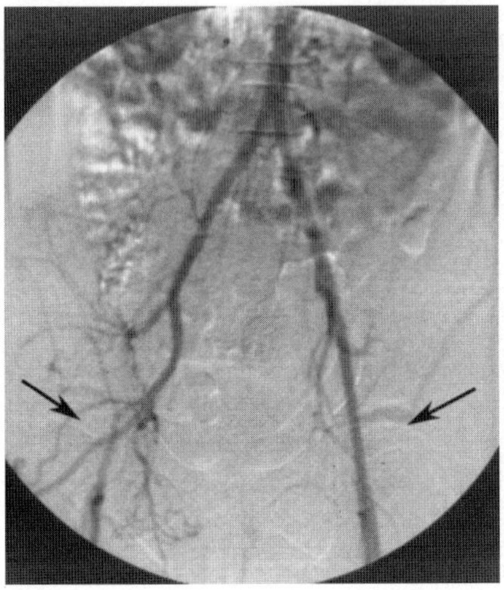

Figure 1. *IV DSA with image intensifier centered over the pelvis (arrows mark the hip joints for reference).*

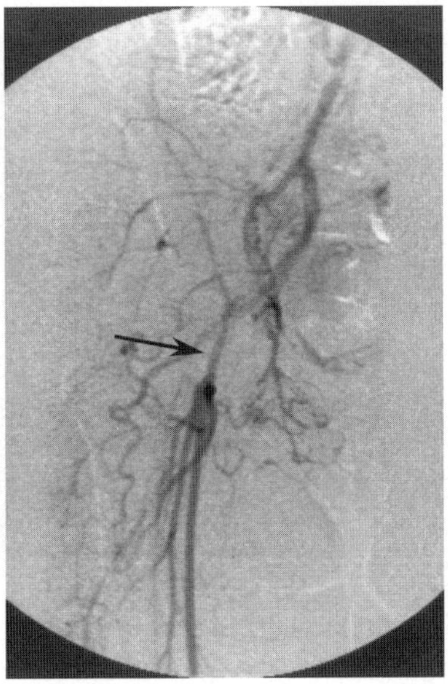

Figure 2. *IV DSA with image intensifier centered over the right hip (arrow marks the right hip joint).*

Findings: Intravenous digital subtraction angiogram **(Fig. 1)** displays marked asymmetry of the pelvic arteries. The left side appears normal. The right-sided vessels are abnormal. Close-up view over the right hip **(Fig. 2)** reveals that the external iliac artery is missing. Large internal iliac artery collaterals anastomose with branches of the profunda femoris artery to restore flow to the right leg. Note that the common femoral artery (arrow) remains patent.

Diagnosis: External iliac artery occlusion, likely a complication of prior transfemoral cardiac catheterization and spasm of the external iliac artery resulting in vascular occlusion.

Anatomy of collateral flow: The anterior division of the internal iliac artery—the internal pudendal, obturator and inferior gluteal arteries—anastomose with the lateral and medial femoral circumflex branches of the profunda.

Case 2

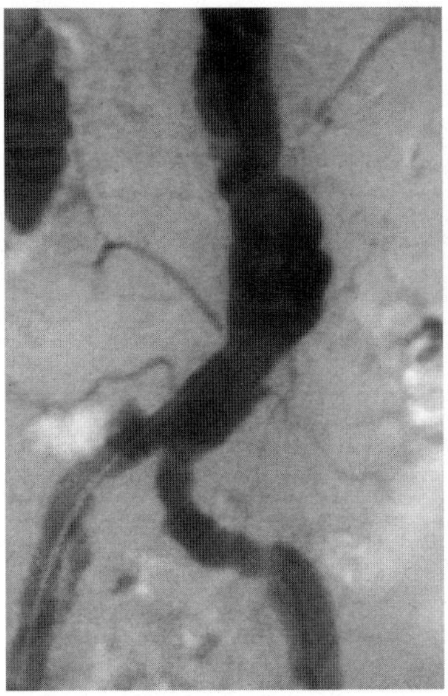

Figure 3. *Lower abdominal aortogram subtraction film, AP projection.*

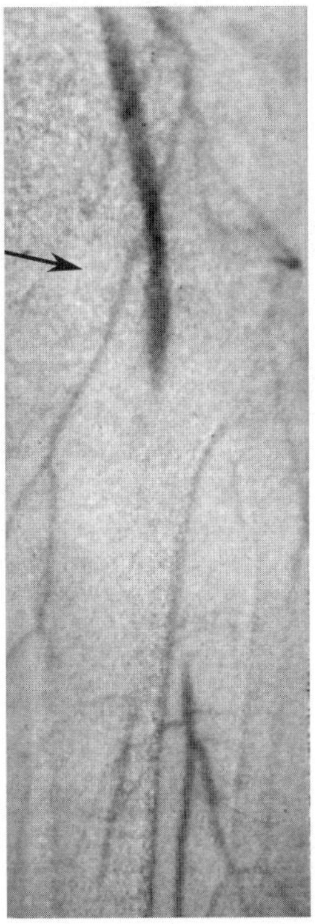

Findings: Abdominal aortography **(Fig. 3)** reveals a small aneurysm of the infrarenal aorta. A selected image from a left leg runoff centered over the left knee **(Fig. 4)** shows abrupt occlusion of the popliteal artery without evident collaterals. This appearance is typical of acute embolic occlusion. In this case, the embolus was a complication of arterial catheterization. A fragment of clot or plaque from the aortic aneurysm likely broke off and lodged in the left popliteal artery trifurcation.

Diagnosis: Acute embolic occlusion of the left popliteal artery secondary to arterial catheterization.

Treatment: The patient underwent surgical embolectomy with restoration of flow to the left lower extremity.

Figure 4. *Selected image from left lower extremity arteriogram; digital subtraction image centered over the left knee (arrow marks the level of the knee joint).*

CHAPTER 15 PREVIEW

Case 1

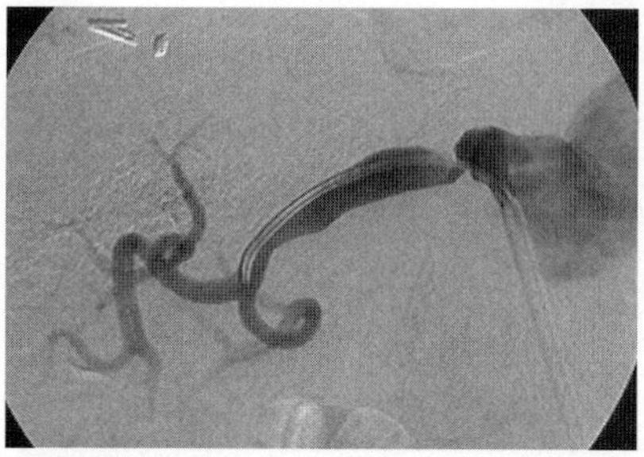

a.

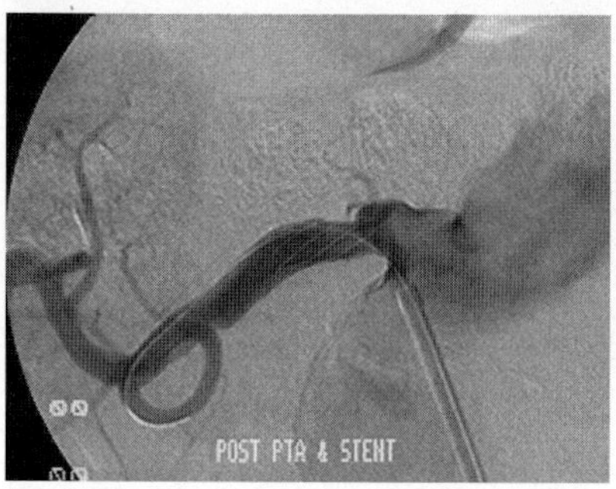

b.

History: A 62-year-old man with hypertension underwent a) selective right renal arteriogram, and b) stent placement.
Question: Which pharmacologic agents might be used in the performance of this procedure?

Case 2

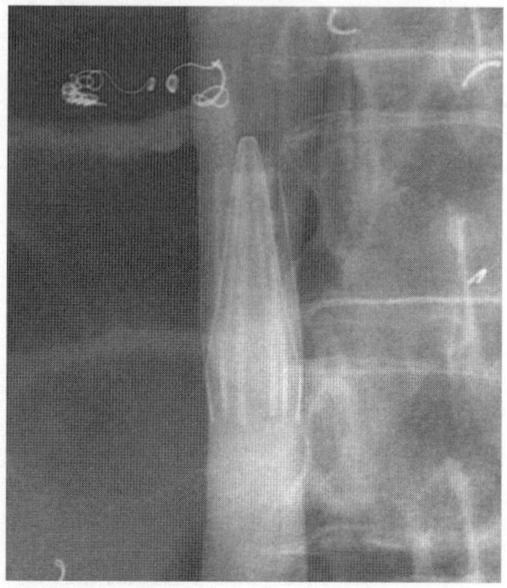

History: A 60-year-old woman with recurrent pulmonary emboli underwent IVC filter placement.
Question: Should prophylactic antibiotics be administered prior to IVC filter insertion?

Case 3

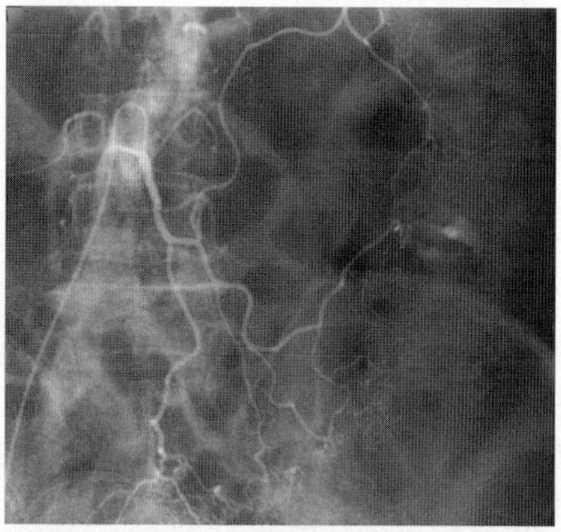

History: A 78-year-old man with lower GI bleeding underwent IMA arteriography.
Question: Which pharmacologic agent might be useful to treat his bleeding?

Case 4

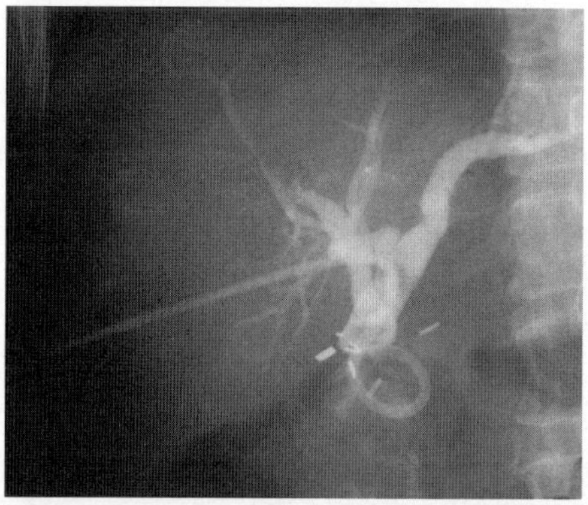

History: A 44-year-old woman presented with abnormal liver function tests following hepaticojejunostomy.
Question: Should prophylactic antibiotics be administered prior to percutaneous transhepatic biliary drainage?

Chapter 15
Pharmacologic Therapy

Robert K. Kerlan, Jr., MD, and Susan D. Wall, MD

CHAPTER OUTLINE

- Antibiotics
  - Antibiotic prophylaxis
  - Bacterial endocarditis prophylaxis
  - Antibiotic therapy
- Conscious sedation
  - Basic principles
  - Agents
    - narcotics
    - anxiolytics
    - neuroleptics
  - Reversal agents
- Preprocedural treatment of hypertension
- Pharmacoangiography
  - Vasodilators
  - Vasoconstrictors
- Anticoagulants

INTRODUCTION

The medications commonly used in angiography and interventional radiology are reviewed here. These medications include antibiotics, sedatives, antihypertensives, vasodilators, vasoconstrictors and anticoagulants. Fibrinolytic agents are discussed in the chapter devoted to thrombolytic therapy.

ANTIBIOTICS

Complications related to bacterial infections are not uncommonly encountered by the practicing interventional radiologist. Strict adherence to sterile technique can minimize infectious complications, but the nature of our subspecialty requires catheter manipulation within contaminated spaces on a daily basis. As such, antibiotics are among the most common drugs used by interventional radiologists.

Antibiotic Prophylaxis

Antibiotics are given in two clinical situations: 1) to treat an existing infection (antibiotic therapy), and 2) to prevent a clinical infection from developing (antibiotic prophylaxis). Interventional radiologists commonly encounter patients undergoing antibiotic therapy, but the selection of agents has usually been made by the referring clinician. Conversely, the decision to administer, as well as the selection of prophylactic antibiotics, is usually made by the interventional radiologist.

ANTIBIOTIC PROPHYLAXIS: SURGERY

- Goal
  To prevent infections of the surgical wound following the procedure

ANTIBIOTIC PROPHYLAXIS: PERCUTANEOUS DRAINAGE

- Goal
  To prevent life-threatening sepsis during and immediately following the procedure

PROCEDURES FOR WHICH USE OF PROPHYLACTIC ANTIBIOTICS IS INDICATED

- Percutaneous biliary drainage
- Percutaneous nephrostomy
- Percutaneous abscess drainage
- Embolization procedures resulting in tissue necrosis

PROCEDURES FOR WHICH USE OF PROPHYLACTIC ANTIBIOTICS IS NOT INDICATED

- Diagnostic arteriography
- Balloon angioplasty
- Routine nephrostomy tube change

Though an extremely common clinical practice, the usefulness of administering prophylactic antibiotics prior to the performance of interventional radiologic procedures has seldom been evaluated with controlled clinical trials.

Of the available studies, Cronan et al conducted a prospective evaluation of 104 nephrostomy tube changes in 74 patients and detected an 11% incidence of bacteremia. The incidence of bacteremia was nearly identical in the group receiving preprocedural antibiotics compared to the group who did not.

In a retrospective review of 388 interventional biliary tract procedures, Clark et al noted seven bacteremias. Five of these seven (71%) had received antibiotics; in two cases, the organisms were sensitive to the antibiotics given.

In a retrospective analysis of 494 hepatic chemoembolization procedures, Reed et al concluded that prophylactic antibiotics decreased the incidence of post-procedural hepatic abscess formation. It should be noted that the statistical significance of this conclusion was disputed as only one of nine patients who did not receive prophylactic antibiotics developed an infectious complication. Similar evaluations for other interventional radiologic procedures have not been performed.

The initial recommendations for antibiotic prophylaxis for interventional radiologic procedures was put forth by Spies et al in 1988. These recommendations were derived by applying standards for open surgical procedures to procedures performed percutaneously. The open surgical standards in the prevention of wound infection were put forth by the National Academy of Sciences/National Research Council (NAS/NRC). The NAS/NRC classified procedures into four categories: clean, clean-contaminated, contaminated and dirty. Clean procedures were those in which spaces potentially containing bacteria were not transgressed. If a noninflamed space containing bacteria was entered (gastrointestinal, biliary, genitourinary, or respiratory tracts) the procedure was considered "clean-contaminated." Contaminated indicated that a space containing inflammation had been transgressed and the classification "dirty" was applied when pus or free spillage of contaminated material was present. Antibiotics are recommended for open surgical procedures based upon which category the procedure is classified.

Spies et al recommended using this classification in deciding whether prophylactic antibiotics are necessary prior to a percutaneous intervention. However, it must be recognized that marked differences exist between open surgical and percutaneous procedures. Surgical prophylaxis is directed at preventing infection of the wound by infected fluid or skin organisms. In contrast, the small incision made during percutaneous drainage procedures seldom serves as a site for a clinically important infection. The infectious risk of percutaneous drainage procedures is that contaminated fluid under pressure may be entered with a needle or catheter creating a potential communication with the infected contents and the bloodstream. In these procedures, prophylactic antibiotics are intended to diminish the impact of the bacteremia. Therefore, data from studies of surgical wound prophylaxis may not be directly applicable to interventional radiologic procedures performed on the same organ.

TABLE 1: PROPHYLACTIC ANTIBIOTICS FOR INTERVENTIONAL RADIOLOGIC PROCEDURES

Procedure	Organism	Agent	Dose
Diagnostic angiography	None	None	
PTA	None	None	
Vascular stent	None	None	
IVC filter	None	None	
Chemoembolization: liver	E. coli Klebsiella Enterobacter Enterococcus Clostridium	Ampicillin Gentamicin	2 g IV and 1.5 mg/kg
PTC/PTBD	Klebsiella Enterobacter E. coli Enterococcus	Ceftriaxone	1 g IV
Gastrostomy Antegrade pyelography Nephrostomy	None	Cefazolin	1 g IV
Nephrostomy tube change	E. coli P. mirabilis Enterococcus Pseudomonas	None	
Abdominal fluid aspiration (uninfected ascites, lymphocele, simple hepatic or renal cyst)	None	None	

The current recommended practice is to use antibiotic prophylaxis for interventional radiologic procedures which are not classified as clean. The selected antibiotic **(Table 1)** is based upon activity against likely organisms, toxicity, and cost. The administration of prophylactic antibiotics with extremely broad spectrums of coverage should be discouraged as this practice promotes cultivation of resistant organisms. Also, the emergence of drug resistant enterococci has focused attention on the indiscriminate use of vancomycin as a prophylactic agent.

ANTIBIOTIC PROPHYLAXIS

- Designed to cover most likely, but not all, potential pathogens

Bacterial Endocarditis Prophylaxis

Indicated for all invasive procedures which fall into the "clean-contaminated," "contaminated," or "dirty" classifications in patients who have the following conditions:

- Prosthetic cardiac valves
- Previous bacterial endocarditis
- Most congenital cardiac malformations
- Rheumatic heart disease
- Hypertrophic cardiomyopathy
- Mitral valve prolapse with valvular regurgitation

Bacterial endocarditis prophylaxis is not indicated for invasive procedures in patients with the following conditions:

- Previous coronary artery bypass graft surgery
- Mitral valve prolapse without valvular regurgitation
- Physiologic, functional, or innocent heart murmurs
- Previous Kawasaki disease without valvular dysfunction
- Previous rheumatic fever without valvular dysfunction
- Cardiac pacemakers and implanted defibrillators

The recommended regimen for endocarditis prophylaxis for patients undergoing procedures in the genitourinary system, gastrointestinal tract, biliary tract, peritoneal cavity or potentially contaminated spaces in the retroperitoneum is listed in **Table 2**.

TABLE 2: RECOMMENDED REGIMEN FOR BACTERIAL ENDOCARDITIS PROPHYLAXIS

Ampicillin	2.0 g IV or IM	30 min before
Gentamicin	1.5 mg/kg (not > 80 mg)	30 min before
Amoxicillin	1.5 g orally	6 hrs after above

(Adapted from Dajani AS, et al 1990.)

For patients with significant allergy to penicillin, vancomycin 1 g IV (to be infused over 1 hour) is substituted for ampicillin. In addition, rather than giving the oral dose of amoxicillin 6 hours later, the vancomycin and gentamicin are repeated 8 hours following the initial dose.

Antibiotic Therapy

Antibiotic therapy when clinical evidence of infection exists should be initiated prior to the procedure then modified according to culture and sensitivity results. If culture data is unavailable, therapy may be initiated in accordance to the guidelines outlined in **Table 3**.

TABLE 3: RECOMMENDED THERAPEUTIC ANTIBIOTICS FOR PATIENTS WITH CLINICAL EVIDENCE OF INFECTION

Procedure	Organisms	Agent	Dose
PTC / PTBD	Klebsiella Enterobacter E. coli Enterococcus	Ampicillin and gentamicin	2 g IV q 6 h 1.5 mg/kg IV q 8h
Antegrade pyelography Percutaneous nephrostomy Nephrostomy tube change	E. coli P. mirabilis Enterococcus Pseudomonas	Ampicillin and gentamicin	2 g IV q 6 h 1.5 mg/kg IV q 8h
Abdominal fluid aspiration or drainage (possible infection)	Gram negative aerobes, Anaerobes	Cefoxitin or cefotetan or gentamicin and metronidazole or gentamicin and clindamycin	2.0 g IV q 6h 1.0 g IV q 12h 1.5 mg/kg IV q 8h 500 mg IV q 8h 1.5 mg/kg IV q 8h 900 mg IV q 8h

SEDATION

Sedative medications are critical to allow invasive procedures to be performed on patients who are not anesthetized. However, these medications also are a potential source of devastating complications. In most interventional radiology laboratories, a combination of agents is used to achieve an adequate comfort level for the patient. The commonly used agents include narcotics, benzodiazepenes, and butyrophenones. In most circumstances, these drugs are given intravenously in small dosage increments, adjusted to the individual patient's size and tolerance.

CONSCIOUS SEDATION

- Primary goals
  - Alleviation of anxiety
  - Pain control
- Secondary goals
  - Postprocedure sedation (vascular interventions)
  - Amnesia

Basic Principles

Five levels of sedation are generally recognized:

Level 1: Awake and alert
Level 2: Drowsy but easily arousable
Level 3: Asleep but arousable
Level 4: Deep sedation; asleep and difficult to arouse
Level 5: Anesthetized; asleep and unarousable.

CONSCIOUS SEDATION

- Undesirable Effects:
  - Respiratory depression
  - Aspiration
  - Hypotension

CONSCIOUS SEDATION: AGENTS

- Anxiolytic: benzodiazepine
- Analgesia: narcotic
- Neuroleptic: butyrophenone

Different procedures require different levels of sedation. Procedures which involve minimal discomfort such as arteriography are usally performed with level 1 or 2 sedation. More painful procedures such as biliary drainage require level 3 or 4 sedation.

Monitoring

To perform conscious and unconcious sedation safely, the following monitoring is required:

1) A nurse to continually monitor the patient on a one-to-one basis
2) Intermittent (preferrably automated) blood pressure monitoring
3) Pulse oximetry.

Patient Support

The following supportive measures are required for safe conscious sedation:

1) A reliable intravenous access for administration of medications
2) Oxygen
3) Sedative medications (as described below) and reversal agents
4) A crash cart
5) Personnel competent to provide basic life support measures in the event of a cardiac arrest.

Agents

<u>Narcotics</u>

All narcotics are opiate derivatives. These powerful analgesics activate central nervous system receptors to diminish the sensation of pain. These agents have many properties in common including respiratory depression and diminishing the level of consciousness. At the usual therapeutic dosages, narcotics have little or no effect on myocardial contractility or blood pressure. Blood pressure reductions observed with these agents are generally due to alleviation of anxiety.

NARCOTIC ANALGESICS

Agent	Dose
• Fentanyl (Sublimaze)	12.5 µg-150 µg
• Morphine	1 mg-15 mg
• Meperidine (Demerol)	12.5 mg-150 mg

Morphine: Morphine is the longest acting of the narcotic analgesics, lasting 4 to 6 hours. The usual dose ranges from 1 mg to 15 mg IV given in increments of 1 mg to 4 mg. Common side effects include itchiness and mild nausea from histamine release. Respiratory depression may be profound in frail and elderly individuals.

Demerol: A synthetic narcotic with an intermediate duration of action (3 to 4 hours). The usual dose ranges from 12.5 mg to 150 mg IV generally given in increments of 12.5 mg to 25 mg. Nausea is a common side effect and is more common with this agent than morphine. Severe reactions may be noted when given to patients taking monoamine oxidase (MAO) inhibitors.

Fentanyl: The shortest acting of the commonly used agents, lasting from 15 minutes to 30 minutes. However, if repeated large doses are given, the redistribution (which limits its duration of action) is overcome, leading to a prolongation of the duration of action. The usual dose ranges from 12.5 µg to 150 µg, usually given in 12.5 µg increments. This is the agent of choice in patients with hepatic dysfunction and has less respiratory depression in comparison to the other agents. Moreover, fentanyl does not cause histamine release, diminishing hypotensive side effects.

Anxiolytics

Anxiolytics are agents which diminish the anxiety associated with having an invasive procedure. The most common agents used are benzodiazepines. The mechanism of action of these agents is potentiation of gamma aminobutyric acid (GABA) receptors and there is less respiratory depression than with narcotic analgesics. Mild hypotension due to peripheral vasodilatation is noted in some patients. Benzodiazepines affect cognition less than motor function so the patient may perceive unaltered mentation but have difficulty ambulating.

BENZODIAZEPINES

Agent	Dose
• Lorazepam (Ativan)	0.5 mg-2.0 mg
• Diazepam (Valium)	5 mg-10 mg
• Midazolam (Versed)	0.5 mg-5 mg

Midazolam: The most commonly used agent in conscious sedation. The usual dose is 1 mg to 5 mg IV given in 0.5 mg to 1 mg increments. Mild hypotension may be encountered due to peripheral vasodilation. Though repiratory depression is not a major feature in most patients, profound depression may be noted in patients with COPD as the hypoxic drive is severely depressed. Midazolam is an excellent amnestic agent.

Diazepam: Diazepam (Valium) was frequently used before midazolam became available. However, problems with intravenous solubility and venous irritation have essentially removed it from common practice. The dose is 2 mg to 10 mg IV in 2 mg increments.

Lorazepam: Lorazepam (Ativan) is a longer acting hypnotic with excellent amnestic qualities. Due to the 6 or more hour duration of action, it is usually not used for periprocedural sedation.

Neuroleptics

Droperidol: This agent is used for anxiety and leads to a dissociative response. The usual dose is 1.25 mg to 5 mg given in 1.25-mg increments. Droperidol causes an alpha-adrenergic blockade which may lead to vasodilatation and hypotension. A reflex tachycardia may result. Sedation and unconsciousness may be engendered by higher doses.

Reversal Agents

Opiate Reversal

The respiratory depression caused by overadministration of narcotic analgesics can be reversed with naloxone (Narcan). The dosage is .05 mg to 0.2 mg given IV every 2 to 3 minutes. Rebound hypertension, agitation, and tachycardia may be observed. Congestive heart failure may also be precipitated by diminished venous compliance. In patients who have received both opiates and benzodiazepines, the opiates should be reversed first.

REVERSAL AGENTS

Agent	Dose
• Naloxone (Narcan)	0.4 mg-0.8 mg
• Flumazenil (Mazicon)	1 mg-5 mg/10 min

Benzodiazepine Reversal

Flumazenil (Mazicon) is an agent developed to reverse the sedative and respiratory depressive effects of benzodiazepines. The usual dose is 1 mg over 1 to 3 minutes up to 5 mg in a 10-minute period. Seizures can be precipitated in patients who are habituated to benzodiazepines.

PREPROCEDURAL MANAGEMENT OF HYPERTENSION

Hypertension is common among patients undergoing angiography and interventional radiologic procedures. As the presence of significant hypertension increases the chance of a hemorrhagic complication, it is optimal to diminish the systolic pressure below 180 mm Hg and diastolic to below 90 mm Hg prior to angiography, biopsy or a percutaneous drainage procedure.

Patients who have a history of hypertension are usually on antihypertensive medication. Though patients are usually NPO prior to an invasive procedure, they should be allowed to take their usual antihypertensive medications with small sips of water on the day of the examination.

If hypertension is present despite the usual medication regimen, often the intravenous sedation with anxiolytics and narcotic analgesics will be sufficient to lower the blood pressure into the acceptable range. If hypertension persists despite the sedative agents, additional measures are warranted.

TREATMENT OF HYPERTENSION

• Nitroglycerine paste	1- 3 inch
• Nifedipine	10 mg sublingual
• Labetalol	5-10 mg slow IV
• Metoprolol	5 mg over 2 min

The initial agent used in our lab is nitroglycerine paste. The paste is absorbed transdermally. The initial dose is 0.5 inch to 1.0 inch which can be increased up to 3.0 inches. If the blood pressure falls too far, the nitropaste is simply wiped off and the effects should dissipate within 5 minutes.

If nitropaste is insufficient, it may be necessary to use other agents. Though the next agent administered varies in accordance to the individual practitioner, we prefer using sublingual nifedipine. Nifedipine (Procardia) prevents smooth muscle contraction by selectively inhibiting calcium ion influx across the cellular membrane.

The usual dose is 10 mg sublingually **(Fig. 1)**. With sublingual administration, sensitive patients may experience a precipitous drop in blood pressure, therefore, close monitoring is warranted. The blood pressure lowering is also potentiated when nitroglycerine has been simultaneously administered.

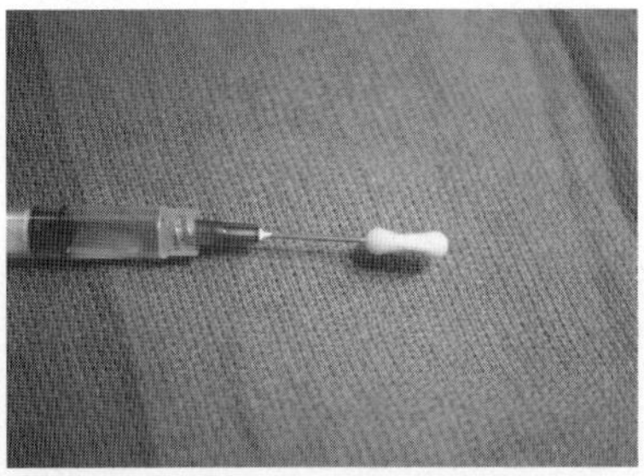

Figure 1. *Nifedipine solution is drawn from a 10-mg pavule into a syringe.*

Nifedipine may also be used as an antispasmodic prior to renal or tibial angioplasties. When used for this indication, 10 mg is given orally at least 1 hour prior to the procedure.

Other agents used for pre- and intraprocedural blood pressure lowering include labetalol and metoprolol. Labetalol blocks both alpha and beta-1 adrenergic receptors and is usually given at a dose of 5 mg to 10 mg intravenously. Labetalol should not be given with calcium channel blockers as this may cause unacceptable depression of myocardial contractility.

Metoprolol is another beta-blocker which can be used in the management of hypertension. The mechanism of action is prinicipally beta-1 blockade, resulting in depression of myocardial contractility. The dose is 5 mg intravenously administered over a period of 2 minutes.

PHARMACOANGIOGRAPHY

Vasodilators

Vasodilators are used commonly during diagnostic and therapeutic arteriography. In diagnostic arteriography, these agents are used to augment potential pressure gradients in the analysis of lesion significance in patients with peripheral vascular disease. Vasodilators may also be used to improve visualization of the portal vein during transarterial portography, and enhance visualization of runoff vessels during peripheral arteriography, particularly in the upper extremity. These agents are also crucial in inducing an erection during pharmacoangiography of the penis in the evaluation of impotence.

Vasodilators are given prophylactically to prevent spasm during renal and lower extremity angioplasty. Therapeutically, vasodilators have been administered to treat nonocclusive mesenteric ischemia and occasionally to treat vasospasm associated with Raynaud's disease.

VASODILATORS

- Nitroglycerine
- Tolazoline
- Papaverine
- Phentolamine
- Prostaglandin E-1
- Reserpine

Nitroglycerine: Nitroglycerine is a short-acting vasodilator whose mechanism of action is direct smooth muscle relaxation. Most interventional radiologists consider nitroglycerine to be the agent of choice for vasodilation due to its short half-life, safety, and established effectiveness. Clinical indications include prevention of spasm prior to angioplasty, treatment of catheter (or angioplasty) induced spasm, as well as dilatation of the distal vascular bed when measuring pressures across a lesion of questionable hemodynamic significance.

The usual dose is 50 to 100 μg administered as a bolus diluted to a concentration of 10 μg/ml. Special plastic (polycarbonate) or glass syringes should be used to administer this drug as the agent adheres to the walls of conventional polyvinylchloride plastic syringes. It should not be used in patients who are hypotensive or patients with constrictive pericarditis. The patient's blood pressure should be monitored closely after administration of this agent.

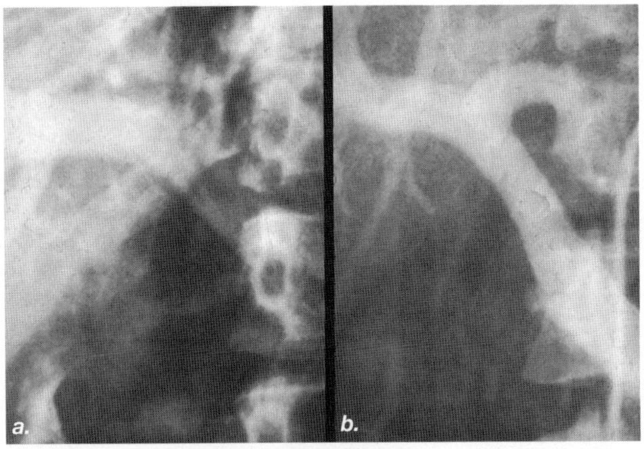

Figure 2. *Arterial portography in a 48-year-old man with hepatitis C. a) Arterial portogram obtained during the venous phase of a superior mesenteric arterial injection shows modest opacification of the portal vein but no morphologic detail. b) The superior mesenteric injection was repeated with the same rate and volume following the injection of 25 mg tolazoline into the SMA. The venous phase reveals the portal venous morphology in considerably greater detail.*

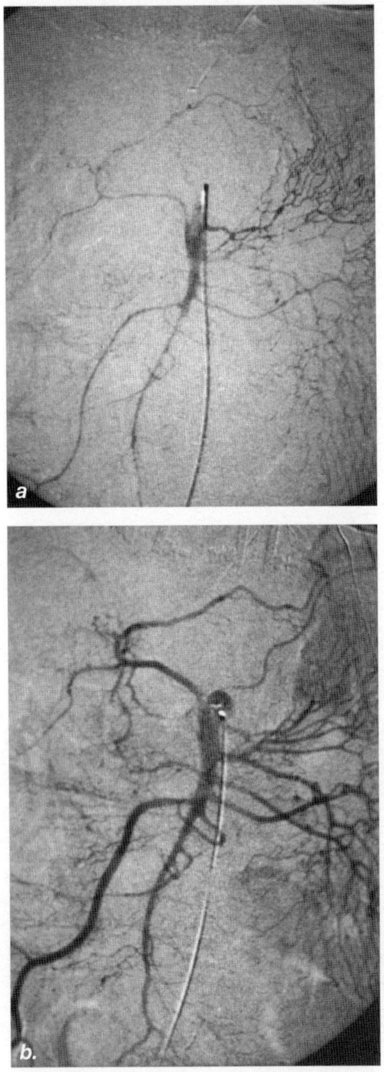

Figure 3. *74-year-old woman in chronic congestive heart failure presents with moderate to severe abdominal pain. a) Superior mesenteric arteriogram shows extensive vasoconstriction throughout the mesenteric arterial bed. b) Following 24 hours of a papaverine infusion, a repeat superior mesenteric arteriogram shows considerably less vasoconstriction.*

Tolazoline: Tolazoline (Priscoline) is a potent vasodilator which acts by direct smooth muscle relaxation in doses used in interventional radiology. Tolazoline is most often used to enhance visualization of the portal vein during arterial portography by dilating the mesenteric arterial bed **(Fig. 2)**. Other indications include enhancement of distal peripheral vascular beds for improved opacification during upper and lower extremity arteriography as well as dilatation of the distal vascular bed during pressure measurements across a lesion of questionable hemodynamic significance. The usual dose is 12.5 mg to 25 mg injected as a bolus, generally diluted in 5 to 10 ml of normal saline. Commonly, a blotchy rash develops when tolazoline is injected into an artery supplying the cutaneous vascular bed. This rash is generally limited to the region where the drug has been injected. Hypotension and arrhythmias are also occasionally observed.

Papaverine: Papaverine is a vasodilator which, like the previously discussed agents, acts by direct smooth muscle relaxation. The clinical applications of this agent include intracorporeal injection in males to induce an erection and infusion into the superior mesenteric artery to treat nonocclusive mesenteric ischemia.

The dose to induce an erection is 30 mg to 60 mg. The dose for the treatment of nonocclusive mesenteric ischemia **(Fig. 3)** is a 30 mg to 60 mg bolus followed by an infusion at a rate of 1 mg/min. The agent forms a precipitate when combined with Ringer's lactate, urokinase and heparin. Therefore, the agent is generally administered in a solution of nonheparinized saline. Papaverine depresses arteriovenous node conduction and is contraindicated in patients with heart block. Like other vasodilators, hypotension may also be induced and close blood pressure monitoring is mandatory.

Phentolamine: Phentolamine (Regitine) is a short-acting alpha-adrenergic blocker that is rarely used to assess the neurogenic component of vasospastic disorders. In the past, this drug was used prophylactically in patients with suspected pheochromocytoma to prevent precipitation of a hypertensive crisis during arteriography. However, modern imaging techniques have made this application obsolete. Intraarterially it is given as a 50-μg to 100-μg bolus.

Prostaglandin E-1: Prostaglandin E-1 is a potent direct smooth muscle relaxant. Continuous infusion is used in patients with pulmonary hypertension. However, for diagnostic and therapeutic angiography, no clear advantages over other agents have emerged. Intraarterially it is given as a 5.0-μg to 7.5-μg bolus.

Reserpine: Reserpine depletes catecholamine stores and therefore acts by diminishing alpha-adrenergic tone. In the past, this agent was used in 0.5 mg to 1.0 mg intra-arterial boluses to treat vasoconstrictive disorders such as Raynaud's disease. Currently this agent is seldom used.

Vasoconstrictors

Vasopressin: Vasopressin (Pitressin, antidiuretic hormone) is a direct smooth muscle constrictor. The principle application of this agent is in the treatment of gastrointestinal bleeding by continuous infusion **(Fig. 4)**. Now that embolization has become an acceptable alternative in the treatment of bleeding in branches of the celiac axis, the main application of vasopressin is in the treatment of bleeding originating from branches of the superior mesenteric or inferior mesenteric arteries. It is not used, nor is it effective, for treatment of arterial bleeding from other sources.

VASOCONSTRICTORS

- Vasopressin
- Epinephrine (seldom used)

Vasopressin should be infused into the main trunk of the superior or inferior mesenteric artery. If it is given into small branches, intense vasospasm resulting in infarction may be precipitated. This drug is given by continuous infusion with a starting dose of 0.2 u/min. After a 20-minute infusion, the arteriogram is repeated. If there is evidence of continued extravasation, the rate of infusion is increased to 0.4 u/min. If the bleeding continues after an additional 20 minutes at the higher dose level, the patient is treated with either embolization or surgery.

If the bleeding is controlled, the infusion is continued for 24 to 48 hours with a tapering dose regimen reducing the rate by one-half every 12 hours. After 1 u/min has been infused for 12 hours, the infusion is changed to normal saline. If the bleeding does not recur after 12 hours of normal saline infusion, the catheter is removed.

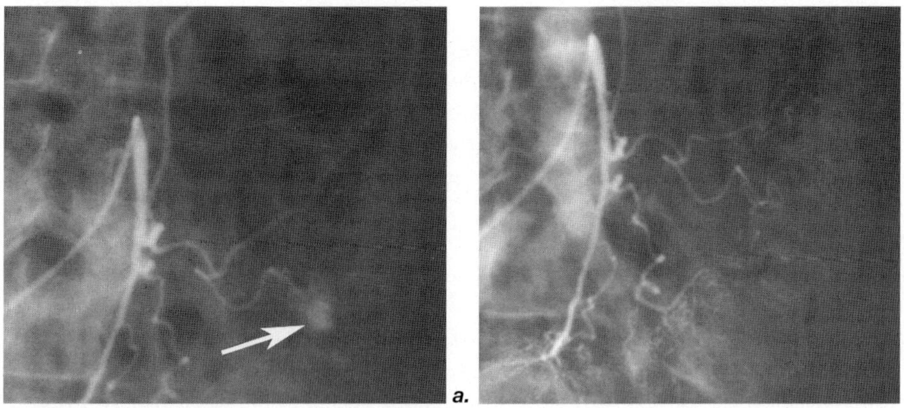

Figure 4. *68-year-old man with recurrent hematochezia requiring repeated transfusions. a) Selective inferior mesenteric arteriogram shows extravasation into the distal descending colon (arrow). b) Following 20 minutes of an intraarterial vasopressin infusion at a dose of 0.2 u/min, the extravasation is no longer present.*

Vasopressin infusion is often associated with transient abdominal cramping during the first few hours of infusion. If abdominal pain persists, the possibility of bowel ischemia or infarction must be considered. Other complications associated with vasopressin infusion include myocardial ischemia due to coronary artery constriction and peripheral vascular arterial insufficiency. Peripheral vascular constriction has led to digital and extremity necrosis. Also, as vasopressin is an antidiuretic hormone, water retention and hyponatremia may be observed.

Epinephrine: Prior to the advent of cross-sectional imaging, vasoconstrictors were used with some frequency in the performance of diagnostic pharmacoangiography, differentiating normal blood vessels from tumor vessels. This was touted to be of particular value in the detection of subtle renal cell carcinomas where 5 μg to 10 μg of epinephrine would be injected directly into the renal artery to reveal subtle neovascularity. Moreover, elegant renal venography can be performed when the arterial blood flow is markedly reduced by renal arterial epinephrine injection. However, modern imaging techniques have made these techniques obsolete.

ANTICOAGULANTS

Heparin

Heparin is a naturally occurring mucopolysaccharide which acts at multiple sites in preventing the conversion of prothrombin to thrombin. Heparin is the most commonly used anticoagulant during the course of diagnostic and therapeutic vascular procedures, being administered whenever there is a significant concern regarding the formation of blood clot. Heparin is routinely given as a bolus prior to the performance of angioplasty, stent placement, atherectomy or in any situation in which a catheter is partially occluding the arterial flow. A continuous infusion is generally used during prolonged infusions of fibrinolytic agents to prevent the formation of pericatheter thrombosis. It is routinely added to catheter flush solutions and, in some institutions, to contrast material.

ANTICOAGULANTS

- Heparin
- Enoxaparin
- Coumadin

The usual dose is a 3,000- to 5,000-unit bolus. When a continuous infusion is performed, a typical starting dose is 800 to 1,000 units per hour with subsequent adjustments made on the basis of the partial thromboplastin time. Heparin may also be administered subcutaneously for prophylaxis against deep venous thrombosis. Heparin must always be given parenterally. The plasma half-life is dose-dependent but is approximately 1 hour in the therapeutic range.

Heparin is contraindicated in patients who are actively bleeding. Moreover, heparin may cause thrombocytopenia in sensitive patients. A mild, transient thrombocytopenia is not uncommonly observed. However, a specific syndrome, heparin-induced thrombocytopenia (HITs), is observed in a small percentage of patients. This syndrome is characterized by massive platelet aggregation due to the presence of an antiplatelet antibody which causes platelet clumping in the presence of heparin. This clumping can lead to massive intravascular thrombosis in both arteries and veins leading to death. The treatment of HITs is to discontinue heparin and administer antiplatelet agents.

The anticoagulant effect of heparin can be reversed by the administration of protamine sulfate. One mg of protamine sulfate neutralizes 100 units of heparin but must be adjusted in accordance to the time of heparin administration, allowing for reduction in dose due to physiologic clearance. Giving an excessive dose of protamine should be avoided as this may also lead to anticoagulation. Protamine sulfate is given slowly (1 to 3 minutes) by the intravenous route. Transient hypotension is not unusual due to peripheral vasodilatation induced by the drug.

Low-molecular Weight Heparin
Low-molecular weight heparin (enoxaparin, danaparoid, dalteparin, ardeparin) is an effective agent for antithrombotic prophylaxis. It must be given subcutaneously but does not require monitoring of the partial thromboplastin time. This agent is associated with less hemorrhagic complications compared with conventional heparin therapy for the same degree of antithrombotic prophylaxis. Currently, this agent is not used in conjunction with the performance of vascular interventional procedures.

Warfarin
Warfarin (Coumadin) is a vitamin K antagonist used for long-term anticoagulation in patients at high risk for thrombotic or embolic complications. The drug has a relatively low therapeutic index and must be monitored closely by periodic assessment of the patient's prothrombin time. Because the mechanism of action is depletion of coagulation factors II, VII, IX, and X, to achieve therapeutic anticoagulation takes 2 or 3 days. The therapeutic dose ranges from 2.5 mg to 10 mg p.o. daily. The therapeutic range of the international normalized ratio (INR) is 1.5 to 2.5. Higher degrees of anticoagulation should be avoided due to the risk of hemorrhagic complications.

If hemorrhagic complications develop, the anticoagulant effect of warfarin can by acutely reversed by repletion of the clotting factors with fresh frozen plasma. If such rapid reversal is unnecessary, vitamin K (phytonadione, AquaMephyton) can be given and the coagulation status will revert to normal within 12 to 24 hours.

Warfarin is commonly encountered in patients undergoing vascular interventional procedures. The decision to perform an intervention on a patient undergoing warfarin therapy is dependent on the degree of anticoagulation as measured by the INR and the type of intervention to be performed.

ANTIPLATELET AGENTS

Aspirin
Aspirin is the most common antiplatelet agent used in the United States. Aspirin works by irreversibly inhibiting platelet cyclooxygenase, creating a platelet defect which lasts for the life of the platelet (about 10 days). Some interventionalists begin aspirin prior to angioplasty and stent placement. Most interventionalists recommend long-term platelet therapy following angioplasty or stent placement.

ANTIPLATELET AGENTS

- Aspirin
- Dipyridamole (Persantine)
- Ticlopidine (Ticlid)
- Clopidogrel (Plavix)
- Abciximab

Apart from interventional procedures, aspirin is commonly used as prophylaxis in patients with coronary artery and cerebrovascular disease. The decision to discontinue aspirin therapy prior to an interventional procedure is dependent on the type of intervention performed. Most interventionalists prefer discontinuance of aspirin therapy for a week preceding a large-bore core biopsy or any type of lung biopsy. The dose is between 65 mg and 325 mg p.o. daily. Reversal is seldom necessary but would require platelet transfusion.

Dipyridamole
Dipyridamole (Persantine-DuPonte Pharmaceuticals) has a similar antiplatelet activity to aspirin, but has only a limited anticoagulant effect in the usual clinical dose range. In addition to antiplatelet activity, dipyridamole has a vasodilatory effect. The usual dose is 75 mg to 100 mg every 6 to 8 hours. Bleeding complications are infrequently observed. Dipyridamole may increase the risk of contrast-induced acute renal failure and the amount of contrast administered to these patients should be kept to an absolute minimum.

Ticlopidine

Ticlopidine (Ticlid-Roche Pharmaceuticals) diminishes platelet aggregation through a different mechanism than aspirin. The drug irreversibly binds to ADP preventing ADP dependent platelet aggregation. The dose is 250 mg p.o. b.i.d. and this drug is most often used for patients who are intolerant of aspirin. Though nausea, dyspepsia and diarrhea are observed in up to 20% of patients, the most feared side effect is bone marrow depression resulting in leukopenia and thrombocytopenia, which is seen in 1% of patients. Regular monitoring of the white blood count during the first 3 months of therapy is recommended to detect this potentially serious complication.

Clopidogrel

Clopidogrel (Plavix, Bristol-Myers Squibb Co) diminshes platelet aggregation using the same mechanism as ticlopidine, binding to ADP. However, clopidogrel is not associated with bone marrow depression and is used for platelet inhibition in patients who cannot take aspirin. The usual dose is 75 mg p.o.daily.

Abciximab

Abciximab works by a different mechanism than aspirin, ticlopidine or clopidogrel. Abciximab is a monoclonal antibody that blocks glycoproteins on the platelet membrane (GPIIb-IIIa) preventing platelet aggregation. This agent has been approved by the FDA for patients undergoing high-risk coronary angioplasty and atherectomy. The dose is 0.25 mg/kg IV bolus given 10 minutes prior to the procedure followed by an infusion of 10 μg/min for 12 hours. This agent is not commonly used in peripheral angioplasty.

REFERENCES

Clark CD, Picus D, Dunagan WC. Bloodstream infections after interventional radiologic procedures in the biliary tract. Radiology 1994; 191:495-499.

Cronan JJ, Horn DL, Marcello A, et al. Antibiotics and nephrostomy tube care: preliminary observations. Radiology 1989; 172: 1043-1045.

Dajani AS, Bisno AL, Chung KJ, et al. Prevention of bacterial endocarditis: recommendations by the American Heart Association. JAMA 1990; 264:2911-2919.

Gorelick PB, Born GVR, D'Agostino RB, Hanley DF, Moye L, Pepine CJ. Aspirin revisited in light of the introduction of Clopidogrel. Stroke 1999; 30:1716-1721.

Kadir S. Diagnostic Angiography. Philadelphia: B.C. Decker Inc., 1992.

Kim D, Orron DE. Peripheral vascular imaging and intervention. St. Louis: Mosby Year Book, 1992.

Page CP, Bohnen JMA, Fletcher JR, McManus AT, Solomkin JS, Wittmann DH. Antimicrobial prophylaxis for surgical wounds. Arch Surg 1993; 128:79-88.

Spies JB, Rosen RJ, Lebowitz AS. Antibiotic prophylaxis in vascular and interventional radiology: a rational approach. Radiology 1988; 166:381-387.

Topol EJ, Byzova TV, Plow EF. Platelet GPIIb-IIIa blockers. Lancet 1999; 353:227-231.

Wojtowycz M. Handbook of interventional radiology and angiography. 2nd edition. St. Louis: Mosby, 1995.

CHAPTER 15 PREVIEW ANSWERS

Case 1

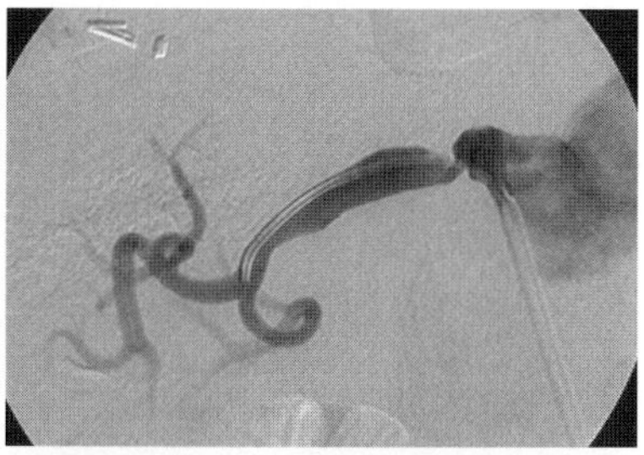

a.

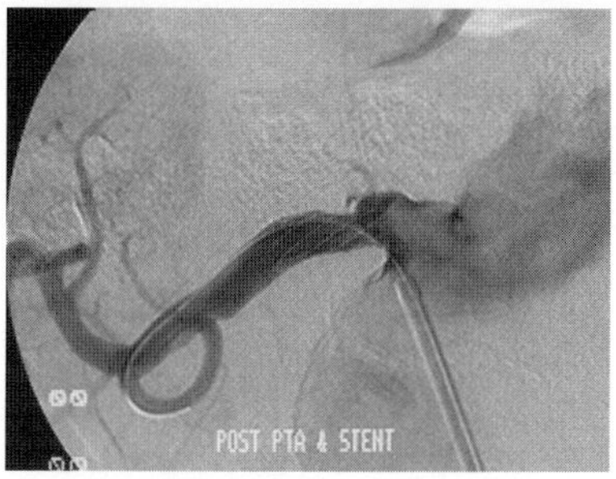

b.

Findings: Renal angioplasty and stent placement. a) Arteriogram performed through guiding catheter confirms location of high-grade right renal artery stenosis. b) Following angioplasty and stent placement the narrowing is improved. Medications used were nitroglycerine 100 mg IA, nifedipine 10 mg sl, and heparin 5,000 units IA.

Case 2

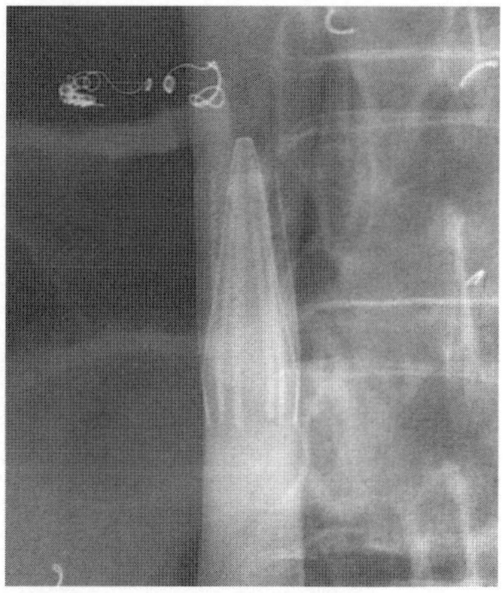

Findings: Antibiotics are usually not given prior to insertion of IVC filters.

Case 3

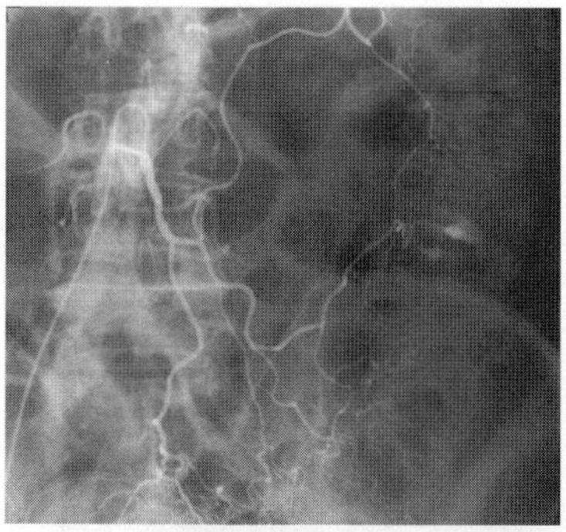

a.

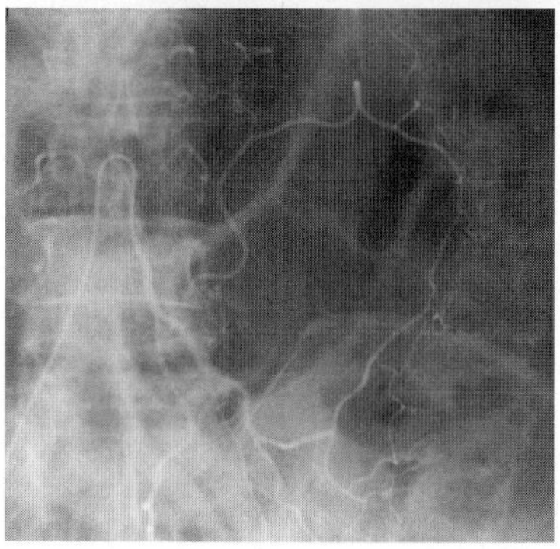

b.

Findings: a) Inferior mesenteric arteriogram shows extravasation from the mid-descending colon. b) A vasopressin infusion was started through the catheter at a dose of 0.2 u/min. The repeat arteriogram after 20 minutes of infusion shows no further extravasation.

Diagnosis: Descending colonic hemorrhage secondary to diverticulosis.

Case 4

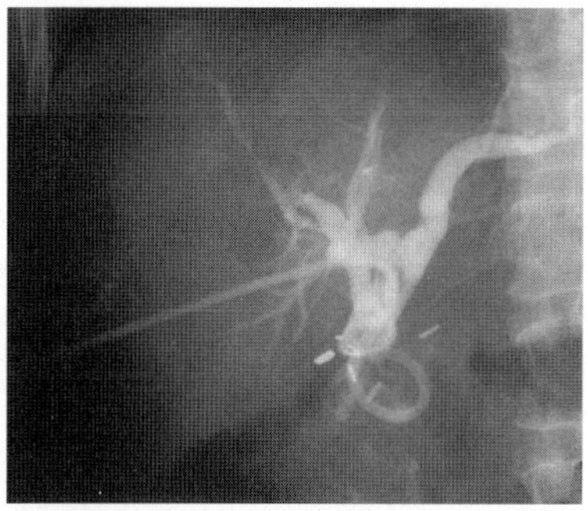

Findings: Prophylactic antibiotics should be used prior to transhepatic biliary drainage. We routinely administer a 3rd generation agent such as ceftriaxone 1 gm IV.

CHAPTER 16 PREVIEW

Case 1

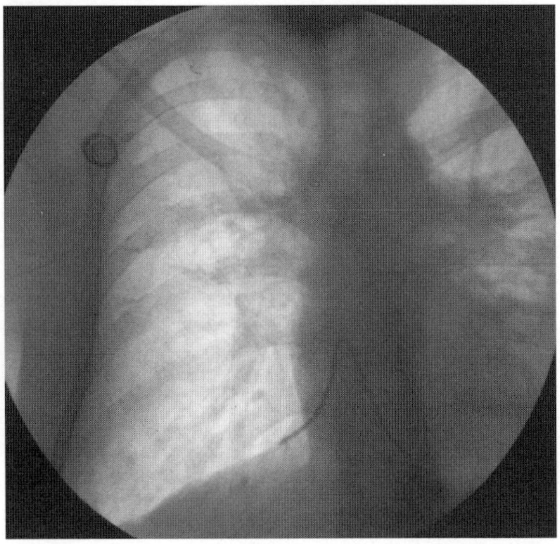

Question: What do you observe on the chest radiograph of this cancer patient?

Case 2

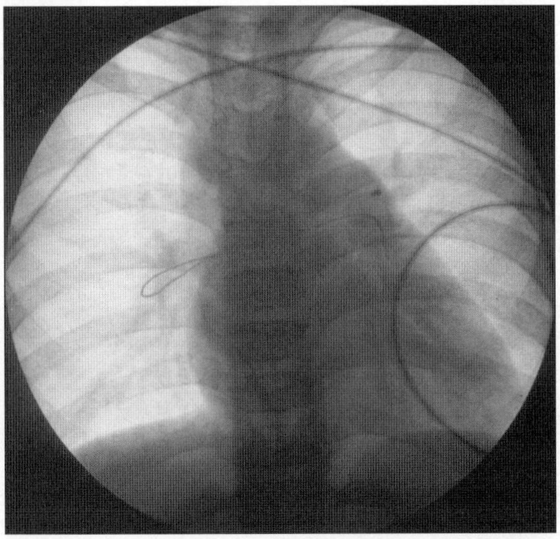

Question: What do you observe on this chest radiograph?

Chapter 16
Percutaneous Removal of Intravascular Foreign Bodies

Roy L. Gordon, MD

CHAPTER OUTLINE

- Indications
- Technique
- Equipment
- Catheter and wire retrieval
- Misplaced caval filters and stents
- Misplaced occlusion coils
- Bullets and the like
- Additional intravascular maneuvers

INTRODUCTION

Removing foreign bodies from the vascular system involves many of the elements of interventional radiology that make interventionalists most happy. Each patient represents an individual challenge often requiring the use of various devices adapted to the problem at hand. There is much scope for improvisation. Foreign body extraction, using the techniques of interventional radiology, usually involves no more than a percutaneous vascular puncture under local anesthesia and sedation. This is in marked contrast to the conventional open surgical techniques that may involve major surgery under general anesthesia in patients who are often in very poor condition. Percutaneous foreign body removal is successful in the great majority of patients treated and there is a low complication rate. Not only does a successful extraction make the interventionalist happy, but even more so is it a cause for relief for the referring physician who may have caused the foreign body problem during a venous catheter placement. The most common clinical problems involve catheter fragments or guidewires lodged in the right heart or pulmonary artery. Myocardial perforation, pericardial tamponade, dysrhythmias, endocarditis and pulmonary emboli are possible consequences of failure to remove foreign bodies.

INDICATIONS

Intravascular foreign bodies may be lodged in the arterial or the venous lumen. In some cases the foreign body may be partially within the vessel lumen and also extend partly into the adjacent tissues or into a pseudoaneurysm. Although the most common types of foreign bodies are venous catheters and wires, other foreign bodies include misplaced intravascular occlusion coils, vena cava filters, vascular stents and bullet fragments. In general all intravascular foreign bodies should be removed because they pose a significant risk to the patient. In a 1978 review by Fisher and Ferreyro, 71% of 220 patients suffered serious morbidity or death as a result of an intravascular foreign body. The extraction should be performed as soon as safely possible after diagnosis of the problem. However, as in all interventions, the risk of the procedure must be weighed against the risk of the foreign body. Foreign bodies are sometimes found incidentally, without clear evidence of how long they have been in the vascular system. It is not possible to predict whether an individual foreign body will, over time, have become endothelialized or incorporated, making removal difficult or impossible. We will usually make an attempt at extraction even some years after the foreign body embolization is believed to have occurred.

POTENTIAL COMPLICATIONS OF FOREIGN BODIES

- Intravascular septic nidus
- Endocarditis
- Cardiac arrhythmia
- Pulmonary emboli
- Vascular occlusion
- Perforation
- Cardiac tamponade

TECHNIQUE

Prior to starting the extraction, previous imaging studies should be reviewed to localize the foreign body as accurately as possible, to identify any potentially complicating anatomical features, and to allow selection of the best extraction plan. It may be necessary to perform additional imaging studies with oblique views **(Fig. 1)**. If needed, angiograms or venograms can be performed during the course of extraction. The patient's cardiac status should be carefully reviewed before starting the procedure. If extraction involves working with transcardiac catheters, arrhythmias may be induced and one should be well prepared to deal with them.

Important steps common to all retrievals are:

a) Placement of a vascular sheath large enough to allow passage of the foreign body. This must be done ahead of time, because once the foreign body has been captured it is too late to upsize the extraction sheath. With some large awkwardly shaped foreign bodies, surgical cut-down may be a necessary part of the treatment plan following intravascular retrieval. The sheath is usually placed in a femoral vessel, as these are large, easy to work with and permit safe passage of large devices.

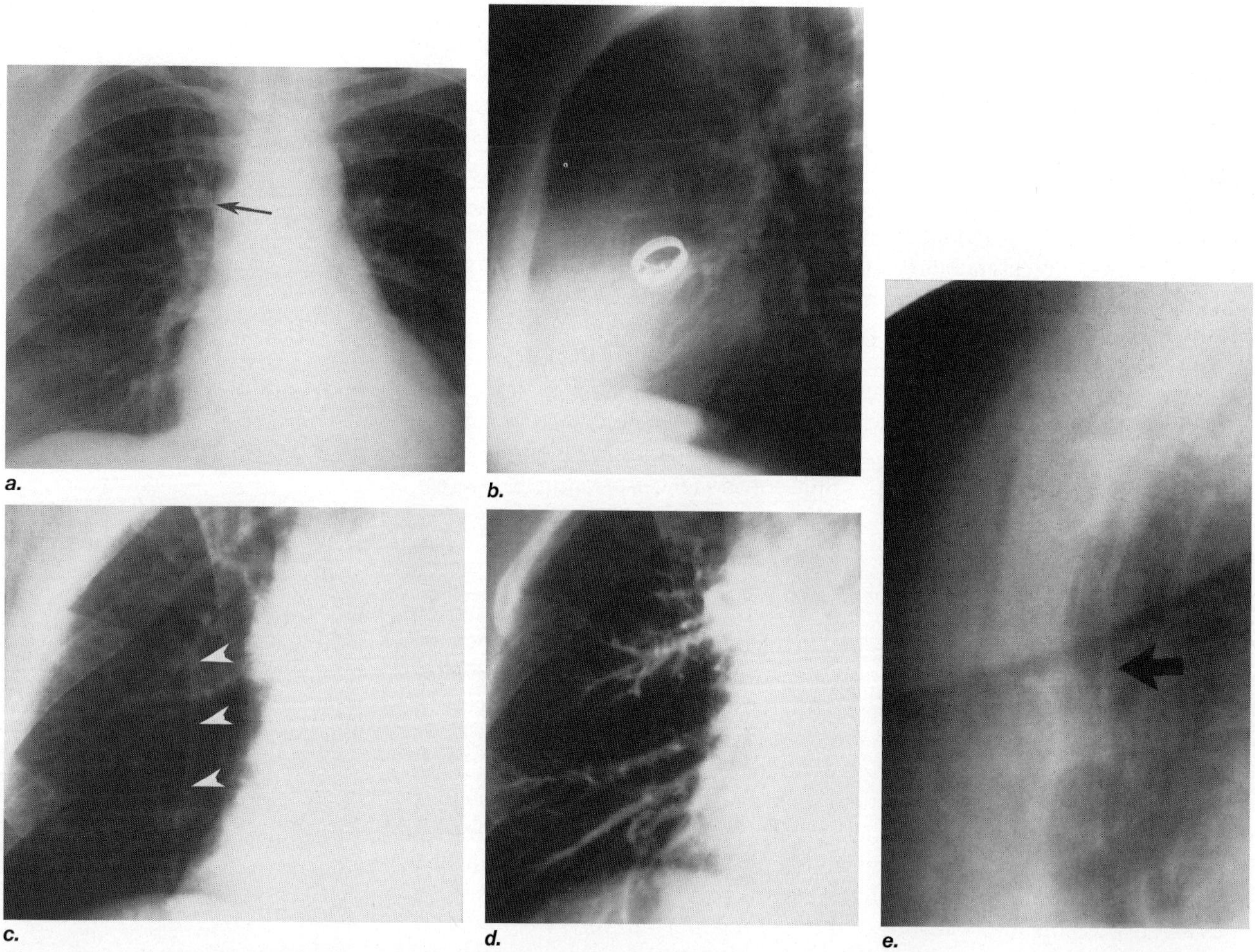

Figure 1. *a) A routine chest radiograph on a patient one year after cardiac surgery for valve replacement shows a catheter fragment near the mediastinum on a frontal film (arrow). b) The catheter is not seen on the lateral chest radiograph. c) An oblique view shows the fragment oriented in the long axis of the body (arrowheads). d) A pulmonary angiogram shows clearly that the vessels and the fragment are aligned at almost 90 degrees to each other and so the fragment is obviously not within a pulmonary artery. e) A just off lateral chest radiograph shows the catheter fragment to be lying just below the sternum and not accessible for intravascular retrieval.*

b) Good oblique or lateral views obtained by rotating the patient or using a C-arm to define the location of the foreign body **(Fig. 2)**.

c) If there is a free end, a snare or basket is suitable; otherwise some hooking device must be used.

d) When working through the heart, continuous cardiac monitoring is used to detect arrhythmias provoked by catheter manipulations. The possibility of vessel perforation, damage to papillary muscles or valves must be remembered and retrieval attempts should be tempered with suitable caution.

POTENTIAL COMPLICATIONS OF FOREIGN BODY RETRIEVAL

- Cardiac arrhythmia
- Cardiac perforation
- Valve or chordae damage
- Endothelial injury with thrombosis
- Vessel perforation and bleeding
- Distal embolization of foreign body
- Fragmentation of foreign body

EQUIPMENT

A vascular sheath of large enough caliber to accept the foreign body is required. The common femoral vein in adult patients is large enough to accept large sheaths so that this is usually not a problem. However, in children or when performing an arterial retrieval, sheath diameter may have to be limited. Sheaths should not be of the peel-away type, because they may split when the foreign body is dragged down into the lumen of the sheath. Snares are most useful for foreign body retrieval.

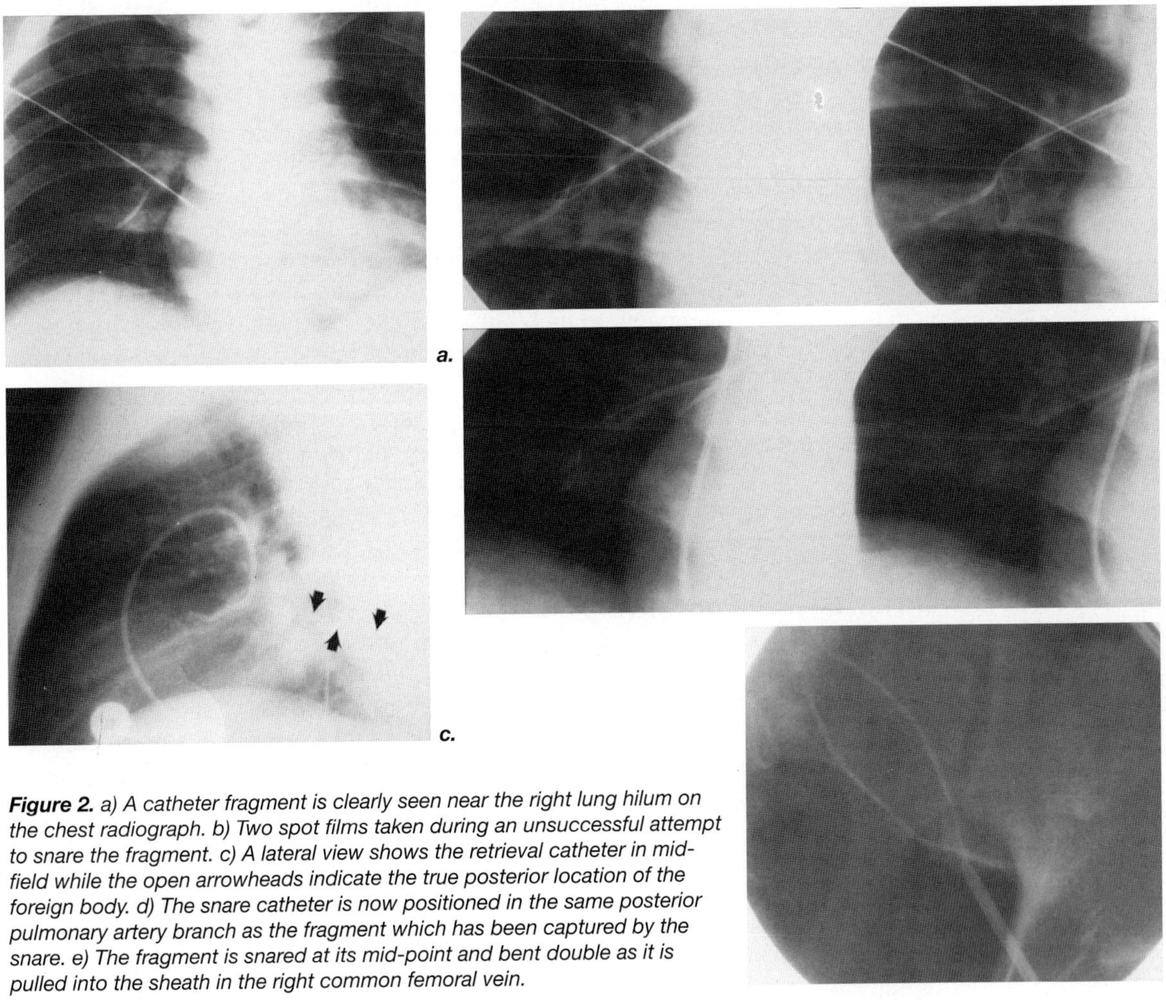

a. b. c. d. e.

Figure 2. *a) A catheter fragment is clearly seen near the right lung hilum on the chest radiograph. b) Two spot films taken during an unsuccessful attempt to snare the fragment. c) A lateral view shows the retrieval catheter in mid-field while the open arrowheads indicate the true posterior location of the foreign body. d) The snare catheter is now positioned in the same posterior pulmonary artery branch as the fragment which has been captured by the snare. e) The fragment is snared at its mid-point and bent double as it is pulled into the sheath in the right common femoral vein.*

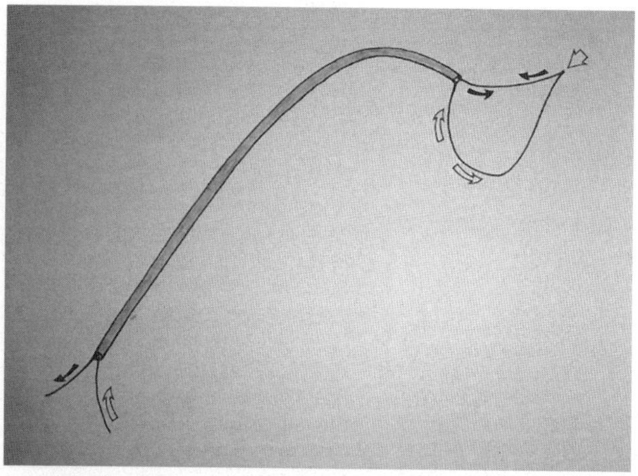

Figure 3. Diagram of a homemade intravascular retrieval snare. A guidewire (.021- or .025-inch diameter) is bent in half with an acute bend at its apex (open arrowhead). The two free ends are fed through a curved 6-F angiographic catheter. The shape and size of the snare loop are controlled by in or out adjustment of the free ends of guidewire (arrows). Further control of the loop is achieved by rotation of the curved angiographic catheter.

Home-made snares can be put together in the angiography room using a long .021 or .025-inch guidewire doubled back on itself through a curved angiographic catheter **(Fig. 3)**. Commercially available snares such as the nitinol Amplatz gooseneck snare (Microvena, White Bear Lake, MN) are very convenient and effective **(Fig. 4)**. Snare loop diameters of 15 mm, 25 mm, and 35 mm are packaged with 6-F catheters, loops of 5 mm and 10 mm go through 4-F catheters and microsnares with loops of 2 mm, 4 mm, and 7 mm go through 3-F microcatheters. Stone baskets of the Dormia type are sometimes useful for extracting catheters or wires **(Fig. 5)**, where there is no free end to snare and for extracting other foreign bodies such as bullet fragments. They can be used through 5-F or 6-F catheters. Standard angiographic catheters such as pigtails or small hooks such as the Sos or shepherd's crook are useful for hooking and pulling catheters from the heart to the IVC, particularly when no free ends are available **(Fig. 6)**. They can be used in conjunction with coaxial tip-deflecting wires. Their main role is in repositioning a fragment to make available a free catheter end for snaring. They usually are not suitable for performing a complete extraction procedure. Small grasping forceps and biopsy forceps are available for transcatheter applications, but should be used as a last resort only. They can cause endothelial damage and do not usually afford a secure grip on the foreign body. Intravascular balloons can be placed prior to retrieval attempts to block central embolization of the foreign body. In some cases, a balloon can be used to reposition a foreign body that cannot be entrapped.

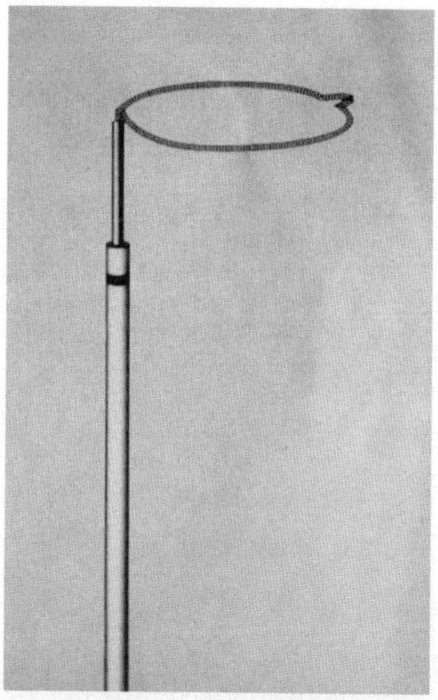

Figure 4. Diagram of a nitinol Amplatz goose-neck snare. The snare is shown fully opened protruding in its normal position at right angles through the 6-F delivery catheter. This catheter has a radiopaque marker band near its tip. The catheter has a gentle curve at its end similar to the catheter in Figure 3. However, in this diagram, it is shown as a straight catheter.

CATHETER AND WIRE RETRIEVAL

The most common foreign bodies that require retrieval are catheters or wires that have embolized to the central veins, the right atrium or the pulmonary artery. If there is a free end of the foreign body in the inferior vena cava (IVC) or superior vena cava (SVC), it can be snared directly and removed via a femoral vein sheath. A catheter bouncing around in the heart or main pulmonary artery should first be hooked using a pigtail or Sos-type catheter. The hooking catheter can be rotated on itself a few times, after engaging the foreign body which twists upon itself, thus making the link with the hooking catheter more secure. A tip-deflecting wire can be used coaxially through the hooking catheter to reinforce its hook. The hooking catheter is then drawn down bringing the foreign body into the IVC. A snare is then used to grab the foreign body in the IVC. Capture of the foreign body is easier in the less capacious and nonpulsatile IVC than in the heart or pulmonary artery. There are fewer arrhythmias and less chance of hooking any cardiac chordae. The hook technique may also be of value for catheters lying more distally in the pulmonary arterial tree. Oblique views are essential to ensure that the retrieval device and the foreign body are in the same plane. A basket can also be used, but the less traumatic snare should be used whenever possible.

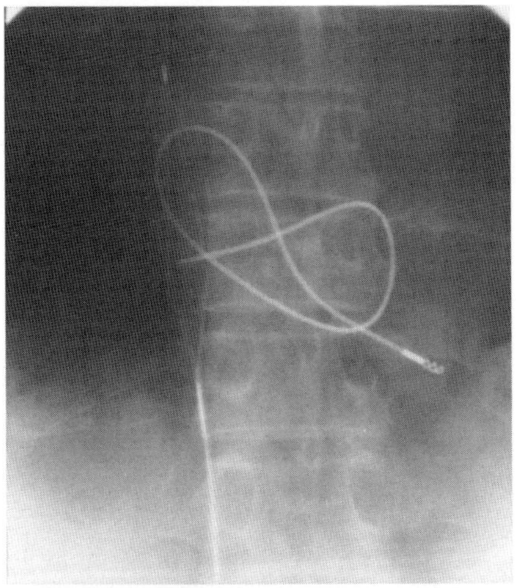

a.

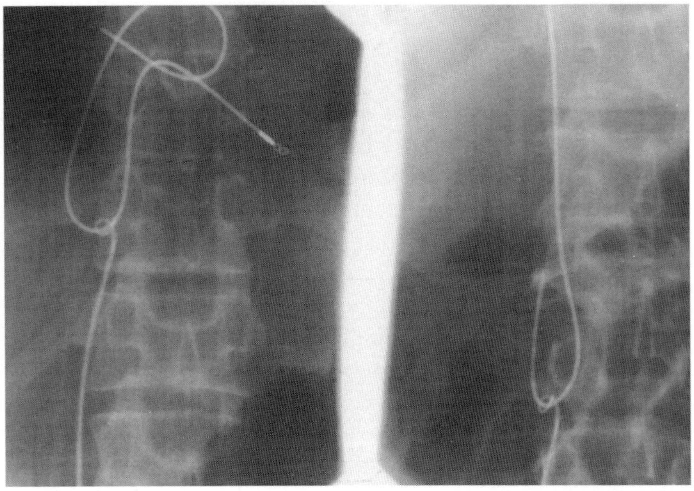

b.

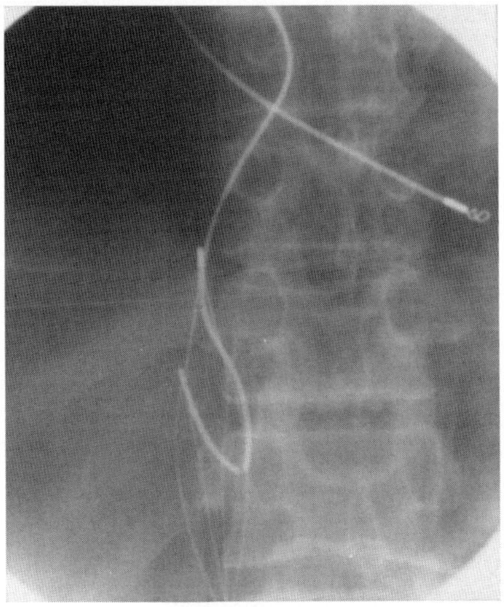

c.

Figure 5. *a) A patient with broken-off transvenous pacer leads knotted in the heart. b) A tip-deflector wire was hooked around the leads, undoing the knot and drawing a free end into the IVC. c) The leads were firmly grasped by a basket in the IVC. The leads could not be extracted, because of severe arrhythmias induced during attempts to pull the leads free from their cardiac implantation site.*

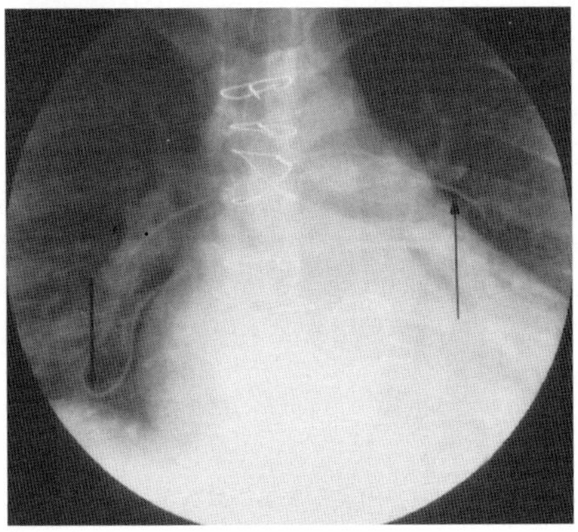

a.

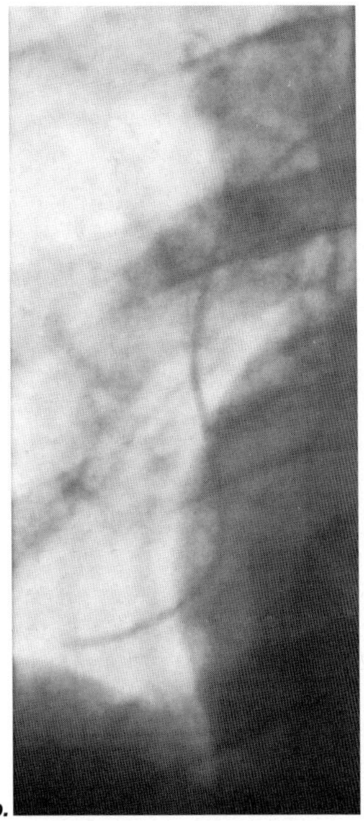

b.

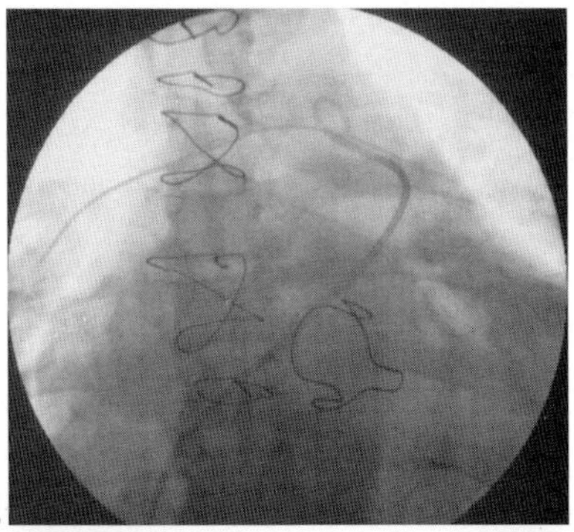

c.

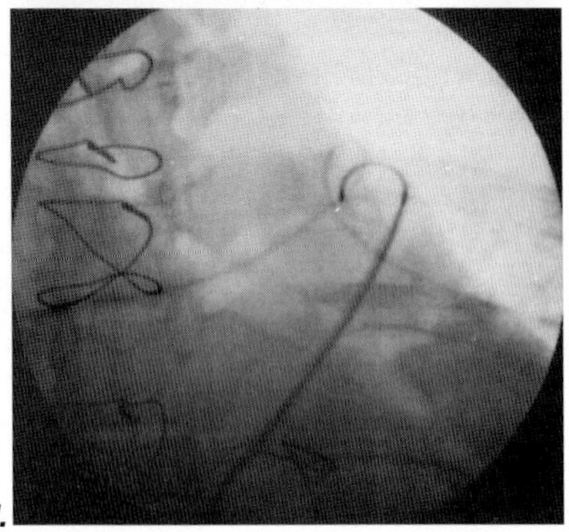

d.

Figure 6. *a) A long piece of a broken PICC line has embolized centrally and lies across the pulmonary artery with the free ends located in a not easily accessible location (arrows). b) Closeup of the catheter end on the left side. c) A pigtail catheter has been passed through the right side of the heart to the main pulmonary artery where it hooks the mid-part of the PICC line. d) The pigtail is reinforced with a tip-deflector wire. The PICC line has been engaged by the pigtail which is rotated on itself prior to drawing the PICC into the IVC where it was snared and removed via a sheath in the femoral vein.*

MISPLACED CAVAL FILTERS AND STENTS

Filters and stents are designed to expand and stay securely in position after intravascular delivery. Once expanded they are not only many times larger than the system through which they were delivered, but may also be equipped with sharp hooks to prevent movement. Retrieval of these devices can be technically very difficult and traumatic so that careful consideration must be given to the need for retrieval. For example, can a patient with a less than optimal filter placement be best treated by leaving the original filter in place and positioning another filter above it which will both secure the original filter in its suboptimal location and also provide better filtration? Can a misplaced stent be expanded in place and another stent placed in the desired location? Clearly a free filter or stent in the heart or pulmonary circulation represents a very great risk and urgent removal is required. The key maneuver is to position one or more snares around the filter or stent so that it can be compressed into its pre-delivery size. It may be helpful to have access from more than one site, for example from both a femoral and a jugular vein or from right and left femoral veins. Once the device has been collapsed, it is drawn into a suitably large sheath and then extracted. Long sheaths should be used so that the compressed leading edge of the device can be safely lodged within the lumen of the sheath close to the point of capture. It is then moved as a unit with the sheath through the venous system. A surgical cut-down on the femoral vein may be required for final extraction, but this is a much smaller procedure than the alternative of cardiothoracic surgery.

MISPLACED OCCLUSION COILS

Coils may lodge in unintended locations during a therapeutic embolization procedure. If the problem is recognized immediately, it is often possible to extract the coil using a small snare. Of particular interest are the microsnares which can be deployed through a 3-F microcatheter. This should be done immediately before the coil induces thrombosis in its undesired location. This can be a quite straightforward procedure or it can turn out to be very difficult and traumatic. Fortunately, a single coil will oftentimes not induce significant occlusion on its own and due consideration should be given to leaving the coil alone. The procedure should not be worse for the patient than leaving the aberrant coil in place. It is sometimes hard for a determined interventionalist to put his ego aside and realize that the patient's interests are best served by leaving the coil alone.

BULLETS AND THE LIKE

Foreign bodies such as bullets or fragments can sometimes be extracted using baskets. Even if the foreign body is too large to extract entirely by the percutaneous route, intravascular capture may allow the foreign body to be positioned in a location where a simple surgical cut-down can easily and safely be performed. Balloons may be of great value in maneuvering and relocating a foreign body that cannot be captured.

ADDITIONAL INTRAVASCULAR MANEUVERS

The intravascular techniques described above have proved of value in other settings as listed in brief below:

a) Repositioning of central venous catheters or LeVeen shunts, using wires, hooks and snares **(Fig 7)**.

b) Stripping of fibrin sheaths from central venous catheters, using snares **(Fig. 8)**.

c) Unknotting of vascular catheters. Important principles include: avoiding pulling on the catheter as this tightens the knot, always using guidewires with care to open the knot but avoiding puncturing the catheter, use of an additional vascular entry site and tip-deflector hooks to ease open the knot.

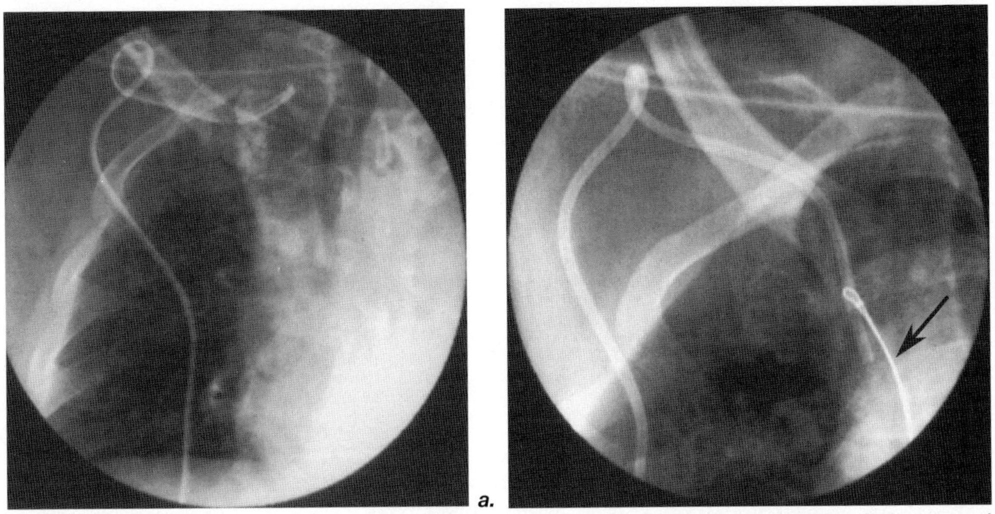

a. b.

Figure 7. a) This tunneled catheter has pulled back and is in a suboptimal position with its tip facing upwards. Fluid could be infused, but no blood could be aspirated. b) Using a snare (arrow) inserted via the femoral vein, the tip of the catheter was pulled down into a satisfactory position in the SVC.

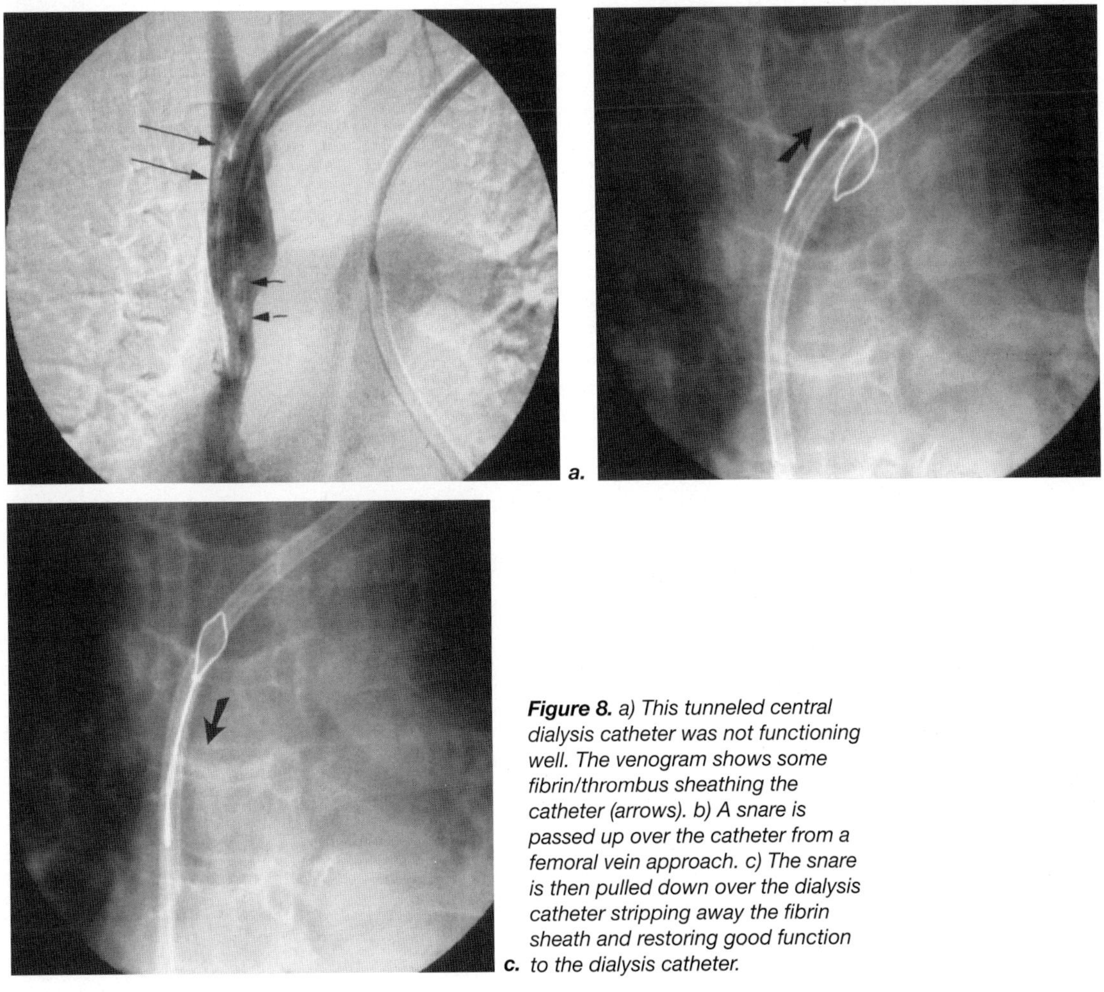

a. b. c.

Figure 8. a) This tunneled central dialysis catheter was not functioning well. The venogram shows some fibrin/thrombus sheathing the catheter (arrows). b) A snare is passed up over the catheter from a femoral vein approach. c) The snare is then pulled down over the dialysis catheter stripping away the fibrin sheath and restoring good function to the dialysis catheter.

REFERENCES

Carrasco CH, Richli WR, Charnsangavej C, Wallace S. Technical note: repositioning misplaced central venous catheters. Cardiovasc Intervent Radiol 1987; 10:234-236.

Cassidy FP, Zajko AB, Bron KM, Reilly JJ, Peitzman AB, Steed DL. Noninfectious complications of long-term central venous catheters: radiological evaluation and management. AJR 1987; 149:671-675.

Cekirge S, Weiss JP, Foster RG, Neiman HL, McLean GK. Percutaneous retrieval of foreign bodies: experience with the Nitinol goose-neck snare. J Vasc Interv Radiol 1993; 4:805-810.

Chomyn JJ, Craven WM, Groves BM, Durham JD. Percutaneous removal of a Gianturco coil from the pulmonary artery with use of flexible intravascular forceps. J Vasc Interv Radiol 1991; 2:105-106.

Cohen HR, Deutsch AM, Ryvicker MJ, Schatz SL. Reduction of catheter knots. Radiology 1980; 134:243-245.

Deutsch LS. Percutaneous removal of intracardiac Greenfield vena cava filter. AJR 1988; 151:677-679.

Egglin TKP, Dickey KW, Rosenblatt M, Pollak JS. Retrieval of intravascular foreign bodies: experience in 32 cases. AJR 1995; 164:1259-1264.

Furui S, Yamauchi T, Makita K, et al. Intravascular foreign bodies: loop-snare retrieval system with a three-lumen catheter. Radiology 1992; 182:283-284.

Gaylord GM, Johnsrode IS. Split 24-F Amplatz dilator for percutaneous extraction of an intravascular bullet: case report and technical note. Radiology 1989; 170:888-889.

Grabenwoeger F, Dock W, Pinterits F, Appel W. Fixed intravascular foreign bodies: a new method for removal. Radiology 1988; 167:555-556.

Gordon DH, Glanz S, Sclafani SJ. Prevention of embolization of a catheter fragment to the lower extremity by compression of the femoral artery. Cardiovasc Intervent Radiol 1982; 5:117-118.

Hartnell GC, Wilde PH. Nonsurgical repositioning of central venous catheters. Radiology 1988; 168:581.

Knelson MH, Hudson ER, Suhocki PV, Payne CS, Sallee DS, Newman GE. Functional restoration of occluded central venous catheters: new interventional techniques. J Vasc Interv Radiol 1995; 6:623-627.

McCowan TC, Ferris EJ, Reifsteck JE, Liu GC, Baker ML. Retrieval of dislodged birdsnest inferior vena caval filters. J Vasc Interv Radiol 1988; 3:179-183.

McSweeney WJ, Schwartz DC. Retrieval of a catheter foreign body from the right heart using a guidewire deflector system. Radiology 1971; 100:61-62.

Olcott EW, Gordon RL, Ring EJ. The injection technique for repositioning central venous catheters: technical note. Cardiovasc Intervent Radiol 1989; 12:292-293.

Roff R, Alspaugh JP, Peterman S, et al. Percutaneous basket removal of an intravascular bullet. J Vasc Interv Radiol 1986; 129-131.

Roizental M, Hartnell GG. The misplaced central venous catheter: a long loop technique for repositioning. J Vasc Interv Radiol 1995; 6:263-265.

Rossi P. "Hook catheter" technique from transfemoral removal of foreign body from the right side of the heart. AJR 1970; 109:101-106.

Saeed Mohsin, Knowles Jr HJ, Brems JJ, Takiff H. Percutaneous retrieval of a large Palmaz stent from the pulmonary artery. J Vasc Interv Radiol 1993; 4:811-814.

Schneider PA, Bednakiewicz M. Percutaneous retrieval of Kimray-Greenfield caval filter. Radiology 1985; 156:547.

Smith PL. An improved method for intra-arterial foreign body retrieval. Radiology 1982; 145:539.

Thomas HA, Sievers RE. Nonsurgical reduction of arterial catheter knots. AJR 1979; 132:1018-1019.

Uflacker R, Lima S, Melichar AC. Intravascular foreign bodies: percutaneous retrieval. Radiology 1986; 160:731-735.

Vaeusorn N, Burodom N, Harnchonboth A, Noikland P, Viranuvatti J. Modified loopsnare for percutaneous removal of intravascular catheter fragments. Radiology 1982; 145:839-840.

Wholey MH, Zikria EA, Boyle B. Inadvertent atrial fixation of a Swan-Ganz catheter by suture and a method for its percutaneous removal. Cardiovasc Intervent Radiol 1987; 10:171-174.

Yakes WF. Percutaneous retrieval of Kimray-Greenfield filter from right atrium and placement in inferior vena cava. Radiology 1988; 169:849-851.

CHAPTER 16 PREVIEW ANSWERS

Case 1

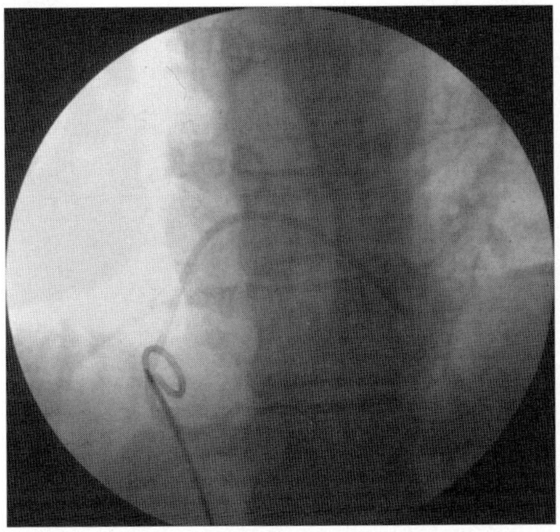

a.

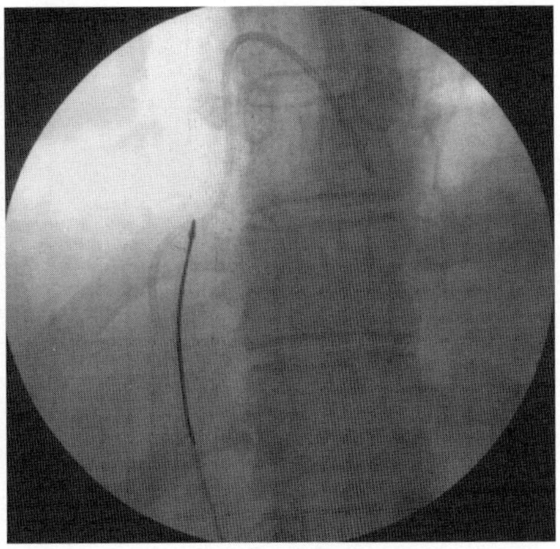

b.

Answer: a) There is a right-sided implanted chest port. The catheter has broken off and embolized to the right atrium. A pigtail with coaxial tip-deflecting wire was used to pull the catheter fragment into the IVC. b) The catheter was snared in the IVC prior to percutaneous transvenous retrieval.

Case 2

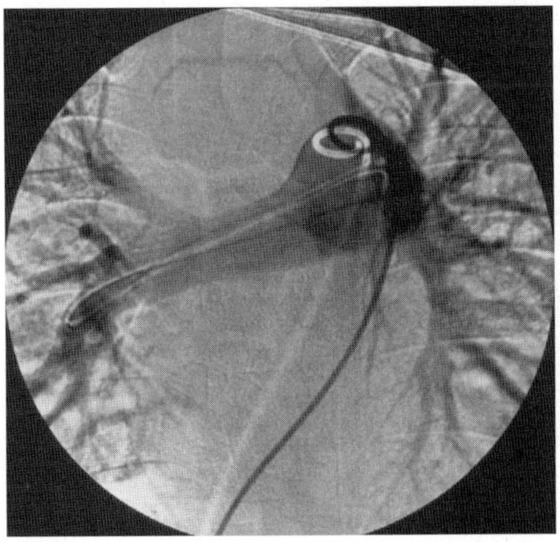

a.

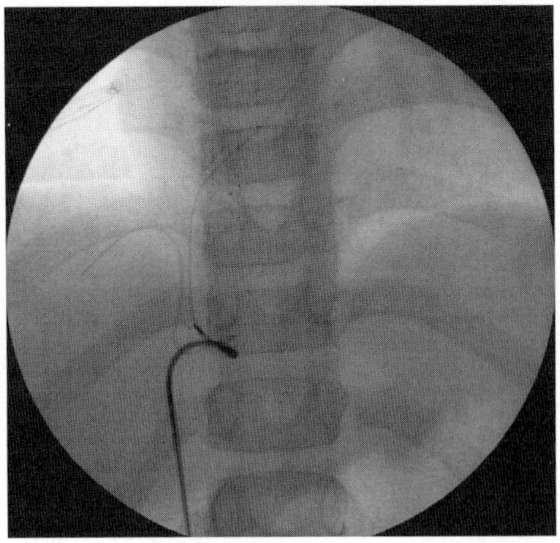

b.

Answer: a) There is a long catheter fragment lying across the pulmonary artery. A pigtail in the pulmonary artery has been used to perform a pulmonary angiogram, nicely demonstrating the position of the fragment. b) The pigtail catheter was used to pull the fragment down into the IVC where it was snared and extracted.

CHAPTER 17 PREVIEW

Case 1

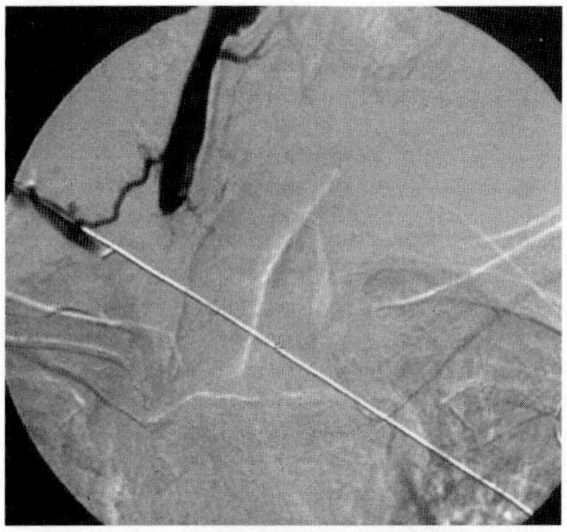

Figure 1. *Venogram of the right internal jugular vein.*

History: A 53-year-old woman developed acute renal failure and required temporary dialysis. The right internal jugular vein (IJ) was punctured in order to place a 13-F dual lumen dialysis catheter into the superior vena cava. However, obstruction of the central IJ was noted and venography was performed **(Fig. 1)**.
Questions: What is your diagnosis? What are options for insertion of a dialysis catheter in this patient?

Case 2

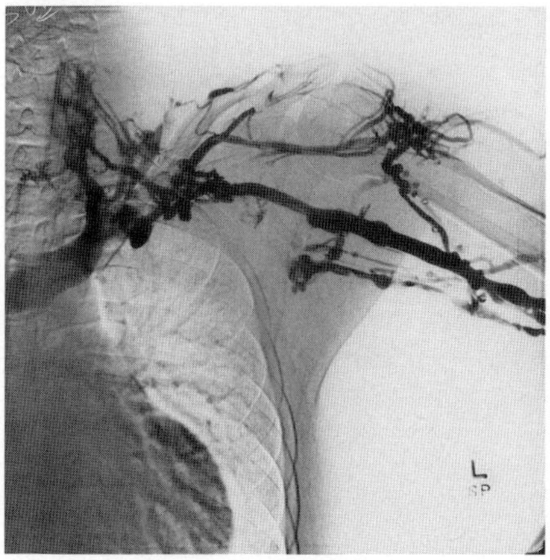

Figure 2. *Left arm venogram, subtraction film.*

History: A 60-year-old man with osteomyelitis required intravenous antibiotics for 6 weeks. He was referred to Interventional Radiology for PICC line placement. The basilic vein of the left arm was catheterized without difficulty but a guidewire could not be advanced beyond the axillary vein. A left arm venogram was performed **(Fig. 2)**.
Questions: What is your diagnosis? How could this patient be treated?

Case 3

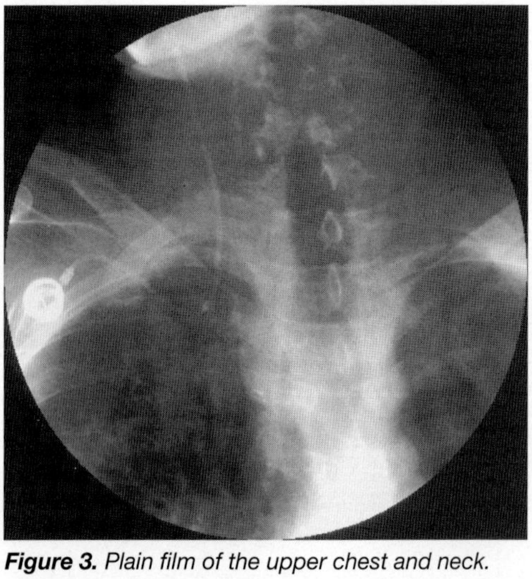

Figure 3. *Plain film of the upper chest and neck.*

History: A 32-year-old-man with cystic fibrosis underwent placement of a chest port to facilitate the periodic administration of intravenous antibiotics. The port worked well for 6 months but then the referring physician noticed difficulty aspirating blood through the port. A chest film was obtained **(Fig. 3)**.
Questions: What is your diagnosis? How could this patient be treated?

Case 4

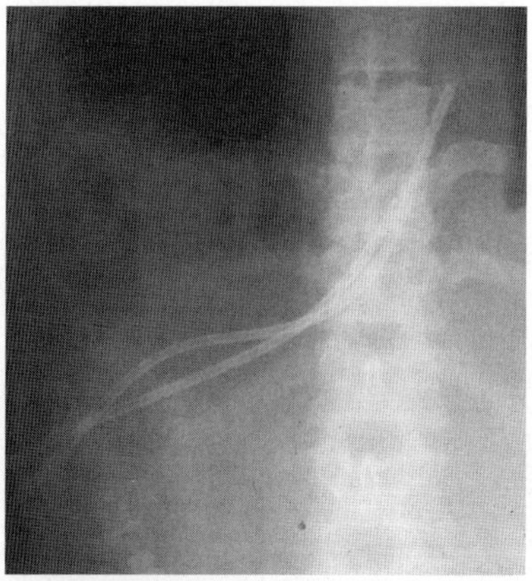

Figure 4. *Plain film of the right upper quadrant following an IR procedure.*

History: A middle-aged man with a long history of end-stage renal disease required dialysis. Prior renal transplantation had failed due to rejection. Multiple arm arteriovenous access grafts had failed. Recently, upper extremity venography revealed bilateral subclavian vein occlusion. In addition, the inferior vena cava was occluded to the level of the renal veins. This patient was referred to Interventional Radiology for placement of a long-term dialysis catheter. A spot-film from that procedure is shown **(Fig. 4)**.
Question: What is the route of venous access used in this case?

Chapter 17
Central Venous Access

Jeanne M. LaBerge, MD,
and Rajiv Sawhney, MD

CHAPTER OUTLINE

- Anatomic consideration for venous access
- Indications
- Devices: nontunneled and tunneled catheters; ports
- Procedural technique
- Complications and management
- Results

INTRODUCTION

At the beginning of the 21st century, for better or worse, the field of vascular access rests principally within the purview of the interventional radiologist. This is for the betterment of patient care because interventional radiologists are optimally suited to wrestle with the technical and anatomic challenges of complex venous access problems. But some might argue that it is to the possible detriment of radiologists because the demands of vascular access have placed a huge burden on inverventional services. Certainly, this activity is occupying an increasingly large percentage of the interventionalist's practice and has become an important part of our job.

In this chapter, we present an organized review of central venous access (CVA) as it is practiced in our section. We begin the discussion with a review of anatomy relevant to these procedures.

ANATOMIC CONSIDERATIONS FOR VENOUS ACCESS

Upper Extremity Veins **(Fig. 1)**

The veins of the upper arm consist of a superficial and deep system. The principal superficial veins are the cephalic, basilic, and median cubital. The paired brachial and axillary veins form the deep system.

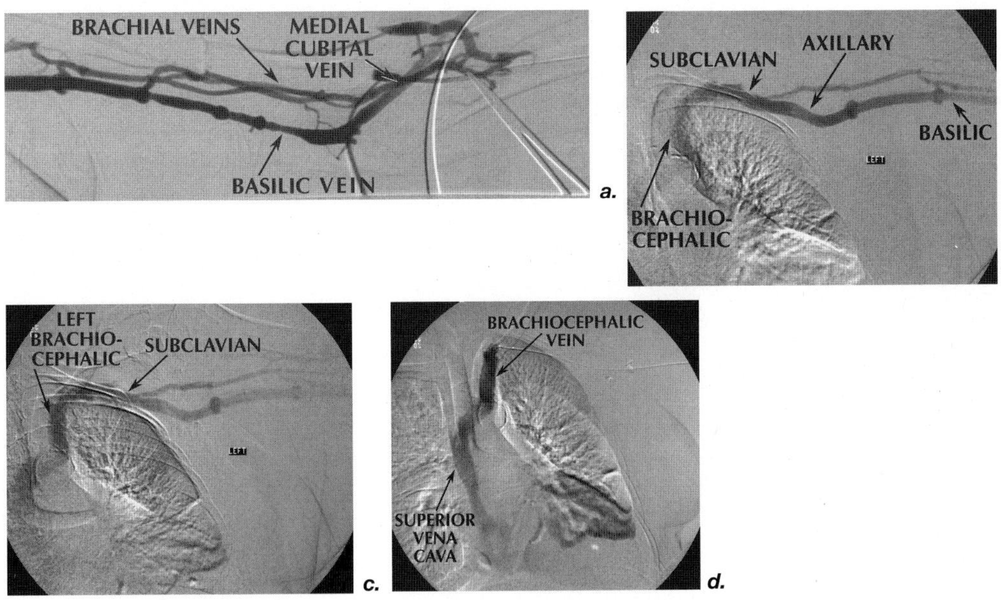

Figure 1. *Dialysis fistulogram showing the veins of the left upper extremity. a) Anatomy at the elbow: median cubital vein, basilic vein and deep brachial veins. b) The basilic vein drains into the axillary and then the subclavian vein. c) The central veins of the chest are depicted here: the subclavian, left brachiocephalic veins. d) A catheter is placed centrally and venography reveals the proximal brachiocephalic vein and superior vena cava.*

UPPER EXTREMITY VEINS

- Superficial:
  Cephalic, basilic, median cubital
- Deep:
  Paired brachial, axillary

The cephalic vein originates along the radial aspect of the forearm, courses across the elbow and along the ventralateral aspect of the biceps, and then enters the chest through the infraclavicular fossa passing behind the pectoralis major muscle. The cephalic vein is usually smaller than the basilic but is noteworthy since it is the intended outflow vein for the Cimino fistula and should be preserved in patients who may require dialysis access.

The basilic vein is the largest vein draining the upper arm. It originates along the ulnar aspect of the forearm, passes over the ventral, medial aspect of the elbow, and then pierces through the deep fascia in the mid-upper arm. It joins with the paired brachial veins at the border of the teres major muscle to become the axillary vein. In most cases, the basilic and cephalic veins are connected at the elbow by the median cubital vein.

The axillary vein begins at the inferior border of the teres major muscle at the junction of the brachial and basilic veins and extends to the lateral aspect of the first rib.

Subclavian Vein **(Fig. 2)**

The subclavian vein extends from the lateral border of the first rib to the medial aspect of the anterior scalene muscle. It is approximately 3 cm to 4 cm long and 1 cm to 2 cm in diameter. It begins as the continuation of the axillary vein at the lateral border of the first rib, crosses over the first rib, and passes in front of the anterior scalene muscle, behind the subclavius muscle to enter the chest. At the medial border of the anterior scalene muscle, the subclavian vein joins the IJ to form the brachiocephalic vein. Of note, the thoracic duct left enters the superior margin of the left subclavian vein near the IJ junction.

AXILLO-SUBCLAVIAN VEINS

- Axillary vein:
  Begins at the junction of the basilic and brachial veins at the teres major muscle and extends to the lateral aspect of the first rib
- Subclavian vein:
  Begins at the lateral aspect of the first rib and extends to the junction with the IJ, at the medial aspect of the anterior scalene muscle

Puncture site considerations: The subclavian vein has traditionally been punctured using visible landmarks observed on physical exam. With this technique, the vein is punctured at or just medial to the costoclavicular junction. A skin puncture site is chosen approximately 1 cm below the clavicle at the junction of the medial and middle thirds of the clavicle. The needle is directed underneath the clavicle toward the sternal notch or superior aspect of the sternal end of the clavicle.

We now routinely use ultrasound guidance for subclavian vein catheterization, and our puncture site has been modified slightly. With ultrasound, our skin entry point is slightly lateral to the standard location and the point of venous entry is the axillary vein just lateral to the costoclavicular space.

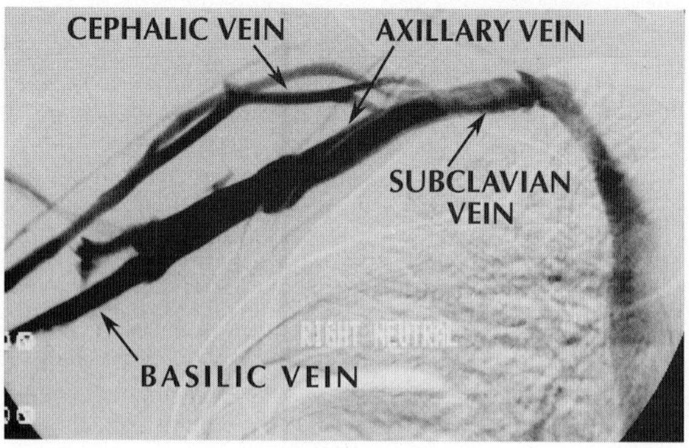

Figure 2. *Right upper extremity venogram. The right basilic, axillary, subclavian and brachiocephalic veins are opacified. The opacified cephalic vein drains into the subclavian vein.*

Internal Jugular Vein **(Fig. 3)**

The IJ vein is the continuation of the sigmoid sinus as it emerges from the base of the skull. It courses inferiorly from the jugular foramen to the clavicle. The IJ maintains consistent relationships to both the sternocleidomastoid muscle (SCM) and the carotid artery. The upper third of the IJ lies medial to the SCM. The lower two-thirds of the IJ lies beneath the SCM and exits the neck between the two heads of the SCM. The IJ drains into the brachiocephalic vein just below the clavicle.

INTERNAL JUGULAR VEIN

- Preferred route of access for dialysis catheters
- Extends from the base of the skull to the clavicle
- Maintains a consistent relationship to the SCM muscle and carotid artery

Throughout its length, the IJ is located lateral to the carotid artery. Superiorly, it is slightly posterolateral to the carotid, and inferiorly, it is slightly anterolateral to the carotid. The IJ becomes larger as it descends in the neck. As a result of this gradual widening, the IJ is a considerably larger target low in the neck than it is high in the neck.

Puncture site considerations: The right IJ is favored over the left IJ for most central venous access procedures because the right IJ forms a straight line with the brachiocephalic vein, and because the thoracic duct courses through the left lower neck and upper chest. However, when the right IJ is not available, the left is quite an acceptable alternative for central venous access.

The three common traditional approaches to IJ puncture are the central, posterior and anterior approaches as defined by the location of puncture relative to visible SCM landmarks. The puncture site for the anterior approach is located 5 cm below the angle of the mandible and 5 cm above the clavicle. At this point, the IJ is just lateral to the common carotid artery and just medial to the border of the SCM. The central puncture site is located at the apex of the triangle formed by the two heads of the SCM (Sedlow's triangle). The puncture site for the posterior approach is located lateral to the SCM about 5 cm above the clavicle, where the external jugular crosses the lateral margin of the SCM.

Interestingly, our standard approach to IJ puncture has been revised with the introduction and common usage of ultrasound guidance for this technique. We now routinely use a low anterior (central) approach and choose the puncture site based on ultrasound visualization rather than physical exam and visible landmarks.

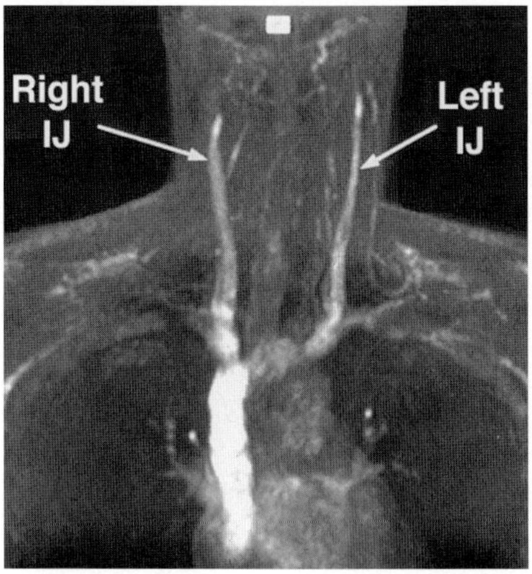

Figure 3. *Coronal MRA of the neck and chest. This image demonstrates the course of the right and left internal jugular veins as they drain into the chest. Note the straight course of the right IJ as it drains into the brachiocephalic and SVC.*

External Jugular Vein **(Fig. 4)**
The external jugular (EJ) vein is formed below the ear and behind the angle of the mandible where a branch of the posterior facial vein joins the posterior auricular vein. The EJ then passes downward and obliquely backward across the surface of the SCM, pierces the deep fascia of the neck just above the middle of the clavicle, to end in the subclavian vein lateral to the anterior scalene muscle.

Puncture site considerations: The EJ is usually visible in the neck (particularly with the patients in Trendelenburg position) and standard IV techniques can be used for cannulation. It should be noted that the EJ is usually a small vessel and may not accommodate the large central venous access catheters. However, large EJ veins can usually accommodate 10-F or even 12-F catheters. Care must also be taken when advancing large catheters from the EJ into the subclavian vein. The bend at this junction can be acute and excessive force may tear the vein in this location.

The Large Veins of the Chest: Brachiocephalic Veins and Superior Vena Cava **(Fig. 5)**
The brachiocephalic vein is formed by the union of the IJ vein and the subclavian vein. The right is usually about 2.5 cm long and the left is about 6 cm long. The brachiocephalic veins join to form the superior vena cava. The superior vena cava (SVC) is about 7 cm in length. It enters the superior posterior aspect of the right atrium. The lower half of the SVC lies within the pericardium. The azygos vein drains into the SVC below the brachiocephalic vein confluence.

A double SVC may be present in less than 1% of the population. A single left SVC has been reported in 0.4%; in this situation the left SVC drains into the coronary sinus or right atrium.

CENTRAL VEINS

- Brachiocephalic veins:
  Right- 2.5 cm; left- 6 cm
- Superior vena cava:
  Length 7 cm
- Variants:
  Duplicated SVC, left-sided SVC

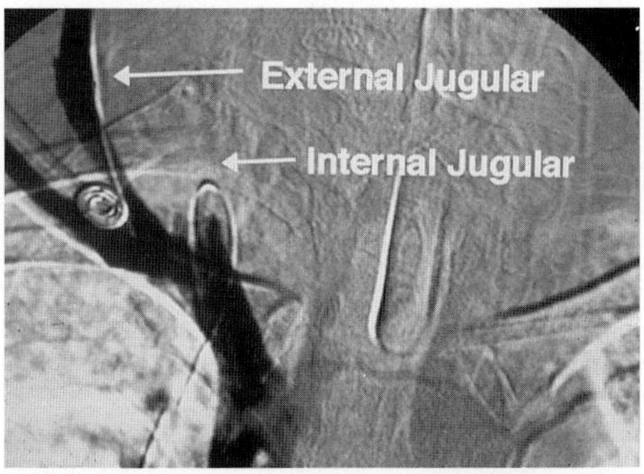

Figure 4. *External jugular venogram obtained in a patient with a Leveen shunt. The EJ vein drains into the subclavian vein and was used as an alternate means of vascular access in this patient whose right IJ vein was not available for access.*

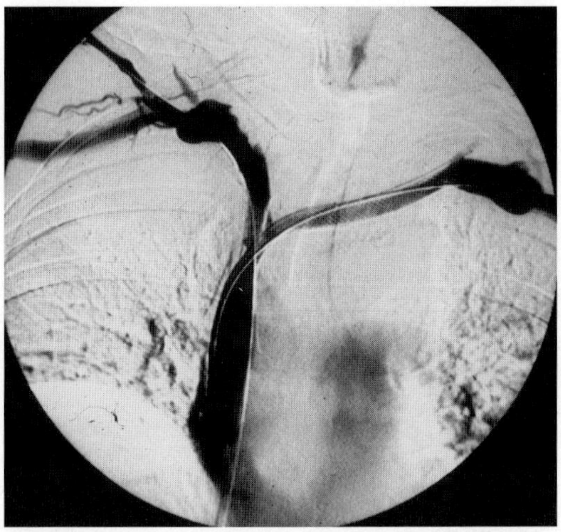

Figure 5. *Bilateral arm venograms demonstrating the course of the brachiocephalic veins which drain into the SVC. Note that Wallstents have been placed in the right left brachiocephalic veins in this patient with lung cancer who had developed SVC syndrome.*

INDICATIONS FOR CENTRAL VENOUS ACCESS

Indications for central venous access (CVA) include difficult intravenous (IV) access, need for prolonged IV access, the use of irritating medications, outpatient IV therapy, and requirements for dialysis and plasmapheresis. At UCSF, the most common reasons for performing CVA procedures not related to dialysis are for the administration of antibiotics, chemotherapy, blood products, and hyperalimentation.

CVA can be broadly divided into the following three groups based on the duration of the access requirement: 1) temporary (days to weeks), 2) prolonged (weeks to months) and 3) permanent (months to years). In addition, the requirements of venous access related to dialysis are so different from other applications, one could categorize CVA as 1) related to dialysis or 2) not related to dialysis. Such categorization is helpful in determining the devices appropriate for an individual patient.

Contraindications to placement of venous access devices are few but include ongoing sepsis and uncorrectable coagulopathy. Patients with ongoing sepsis should be treated with an appropriate course of antibiotics and have negative blood cultures before a device is placed. Coagulopathy should be corrected with blood products (fresh frozen plasma or platelets) prior to the procedure.

INDICATIONS FOR CENTRAL VENOUS ACCESS

- Dialysis
- Nondialysis
  - Antibiotics
  - Chemotherapy
  - Blood product administration
  - Hyperalimentation
  - Poor veins

TEMPORAL REQUIREMENT FOR CENTRAL VENOUS ACCESS

- Temporary (days to weeks)
- Prolonged (weeks to months)
- Permanent (months to years)

CONTRAINDICATIONS FOR DEVICE PLACEMENT

- Ongoing sepsis
- Uncorrectable coagulopathy

DEVICES (Figs. 6-9)

Numerous CVA devices are currently available from a variety of commercial vendors. These catheters can be inserted peripherally in the extremities or centrally through the tissues of the chest wall. Catheters may be inserted using a direct percutaneous entry or they may be buried under the skin to reduce the incidence of infection (e.g., tunneled catheters and ports).

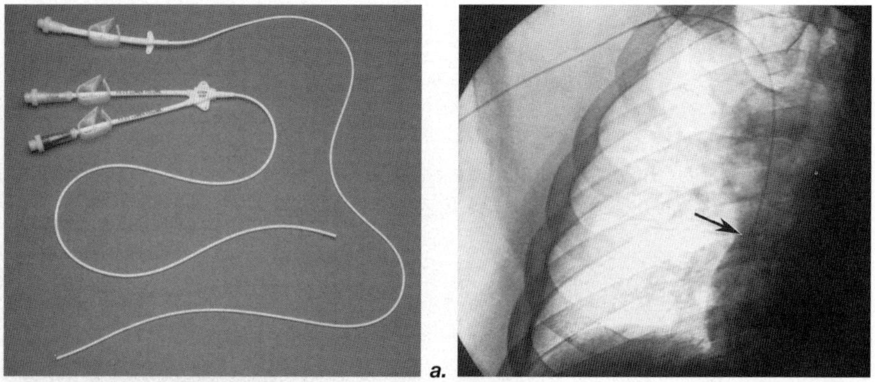

a. b.

***Figure 6.** Peripherally inserted central venous catheter. a) Catheters may be single or double lumen. We prefer the 5-F single lumen PICC line. b) The tip of the catheter is positioned in the SVC at the junction with the right atrium (arrow).*

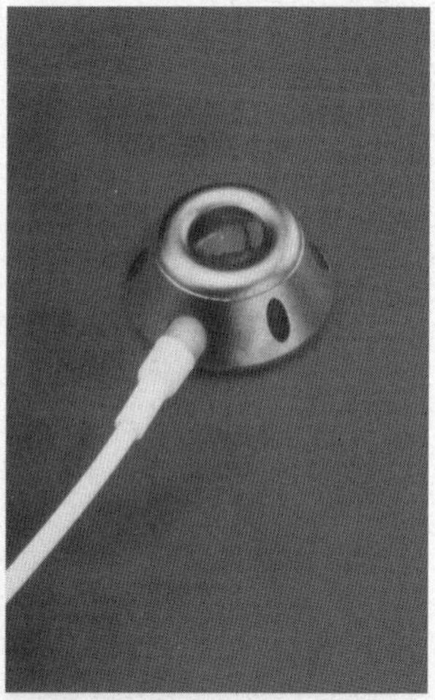

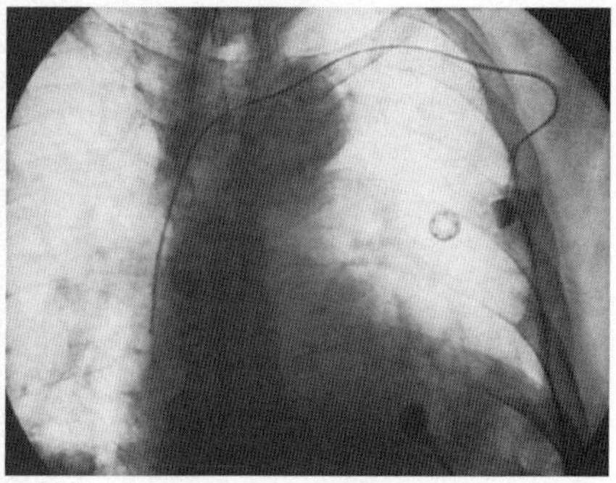

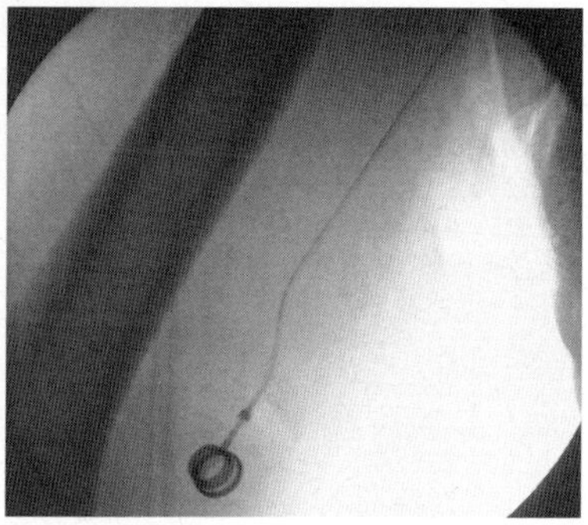

a. b. c.

***Figure 7.** Ports. a) This is the appearance of a typical device. It is inserted into a subcutaneous pocket in the chest or arm. b) Chest radiograph demonstrating the typical appearance of a left chest port. The venous entry of the catheter is the left subclavian vein. c) Radiograph of a right arm port (Pas-port) implanted in the upper arm.*

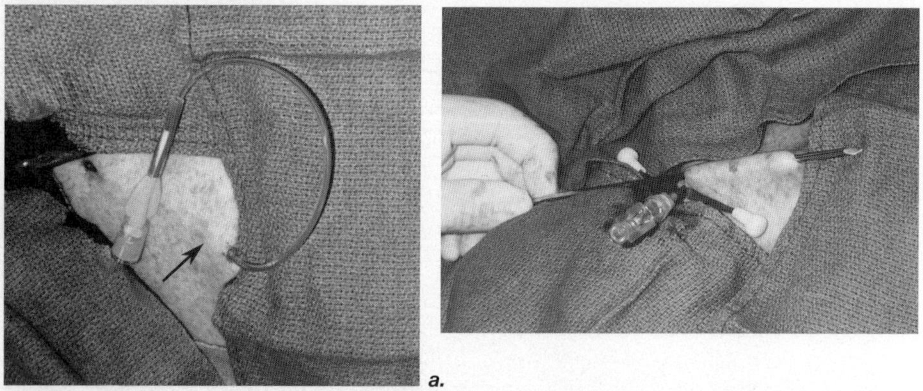

a. b.

***Figure 8.** Tunneled catheter. a) Photograph of the exit site of a single-lumen tunneled Groshong catheter. The catheter's bacteriostatic cuff is positioned just underneath the skin at the entry site (arrow). This cuff acts as an anchor for the catheter to prevent inadvertent movement and also impedes the spread of infection from the entry site. b) Photograph showing the tunneling process. The sheath shown here enters the subclavian vein and tunneling is directed inferiorly along the anterior chest wall.*

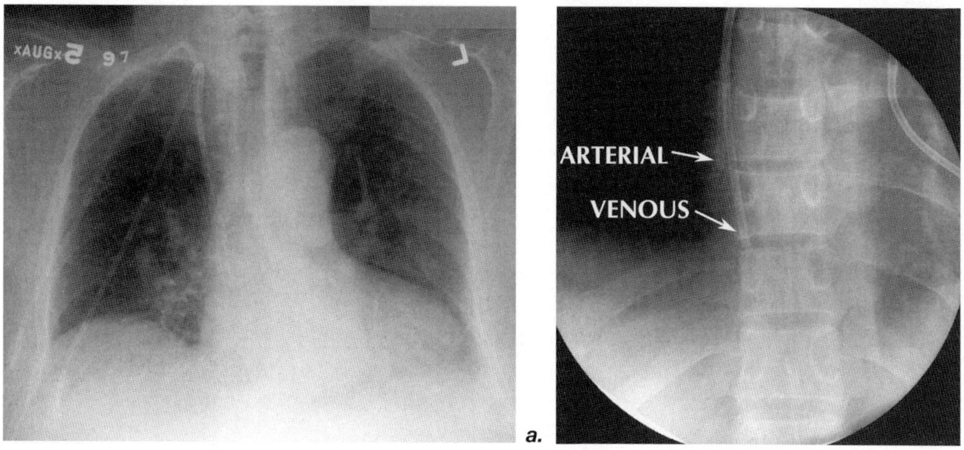

a. b.

Figure 9. *Dialysis catheter. a) Chest radiograph showing tunneled dual-lumen Tesio catheter. b) The catheter enters via the left internal jugular vein. c) The tips of the catheters are positioned in the SVC or right atrium. In this case, the arterial catheter is in the SVC and the venous catheter is in the right atrium.*

Device selection is determined by a number of factors including the type of therapy, length of therapy, frequency of therapy, patient preference, physician preference, and nursing preference. A list of generic CVA device options and their expected duration of use is provided below **(Table 1)**.

Catheters are made of medical grade silicone or polyurethane. They are available in various sizes and may be single lumen or multi-lumen in design. Catheter tips may have an end-hole, valved-tip (Groshong design), or dual lumen with a staggered-tip design.

TABLE 1: CENTRAL VENOUS ACCESS DEVICES

Category	Device Options—Anticipated Duration of Use
Peripheral catheters **(Fig. 6)**	PICC lines—weeks to months
Subcutaneous ports **(Fig. 7)**	Peripheral ports (arm)—months to years
	Central ports (chest wall)—months to years
Chest wall catheters **(Fig. 8)**	Nontunneled—weeks to months
	Tunneled—months to years
Dialysis catheters **(Fig. 9)**	Temporary, nontunneled—days
	Permanent, tunneled—months to years

PROCEDURAL TECHNIQUE

Device selection is made with input from the patient and the referring physician. Informed consent is obtained and coagulation parameters and platelet levels are checked. Strict adherence to sterile technique is crucial. This includes appropriate room cleansing, sterile patient preparation and draping. IV sedation and appropriate cardiopulmonary monitoring are recommended. While the use of prophylactic antibiotics has been debated, we administer a single dose of cefazolin (1 g, IV) to patients receiving tunneled catheters or subcutaneous ports.

The initial venous puncture is routinely performed under ultrasound guidance. For peripheral access, the basilic vein is the vein of first choice. The cephalic vein is usually small and less able to accommodate venous catheters. The brachial veins lie deep and in close proximity to the brachial artery. For central venous access, the axillary/subclavian vein or IJ vein is punctured. Subclavian vein puncture can also be performed using fluoroscopy and appropriate landmarks or using fluoroscopy and contrast infusion via a peripheral vein. When central access via the subclavian or jugular veins is not possible, alternative venous access can be achieved using a transhepatic approach into the hepatic vein or using a translumbar approach into the infrahepatic inferior vena cava.

CENTRAL VENOUS ACCESS: IMMEDIATE COMPLICATIONS

- Bleeding
- Contrast reaction
- Pneumothorax
- Air embolism

CENTRAL VENOUS ACCESS: DELAYED COMPLICATIONS

- Infection (skin, tunnel/pocket, systemic)
- Wound dehiscence
- Catheter dysfunction (fibrin sheath, catheter tip malposition, catheter thrombosis, catheter fracture)
- Venous abnormality (stenosis, thrombosis)

CAUSES OF CATHETER DYSFUNCTION

- Catheter tip malposition
- Fibrin sheath occluding tip
- Catheter thrombosis
- Venous thrombosis
- Catheter fracture: "pinch-off" syndrome

We routinely use a 21-gauge micropuncture access set for initial venous access (Cook, Inc., Bloomington, Ind.) and feel that it adds to the safety of the puncture. Once venous access has been obtained, a wire is advanced into the venous system and the vascular entry site is dilated. The remaining technical steps for device placement depend on the type of device being inserted. In general, we prefer to position all CVA catheters such that the tip lies at the junction between the SVC and right atrium.

For nontunneled applications, the catheter is advanced over a wire or through a peel-away sheath and the tip of the catheter is positioned using fluoroscopic guidance. For tunneled catheters, a subcutaneous tunnel is first created between the venotomy site and the catheter exit site using a tunneling device. The catheter is then pulled through the subcutaneous tunnel, advanced through a peel-away sheath, and positioned within the venous system.

For ports, a subcutaneous surgical pocket is created, the catheter is tunneled from the venotomy site to the pocket, the catheter is positioned and attached to the port. The port is sutured to the tissue within the pocket and the incision is closed.

COMPLICATIONS AND MANAGEMENT

Immediate complications include bleeding and hematoma, contrast reaction, pneumothorax and air embolism (for central venous punctures). Delayed complications include infection, wound dehiscence, catheter dysfunction (fibrin sheath formation, catheter thrombosis, catheter malposition), and venous injury (venous stenosis and venous thrombosis).

A fibrin sheath develops along the surface of all venous catheters. When excessive, it may occlude the catheter end-hole. Wire manipulation and catheter stripping using a snare device from the femoral approach can be used to salvage these catheters **(Fig. 10)**. Alternatively, infusion of a thrombolytic agent may be helpful **(Fig. 11)**. This problem can also be treated by changing the catheter over a wire for a new catheter. This exchange helps disrupt the fibrin sheath. In some cases, balloon disruption of the fibrin sheath and placement of the catheter tip deeper into the right atrium may be required.

Malpositioned catheters can be exchanged over a wire and the new catheter positioned appropriately. Alternatively, the tip of the malpositioned catheter can be repositioned using wires placed through the catheter or snares manipulated from the femoral approach. "Pinch-off" syndrome may occur when the subclavian vein is punctured too centrally and as a result the catheter is compressed by the subclavius tendon or costoclavicular ligaments. Catheter malfunction and even catheter fracture, requiring removal and foreign body removal using a snare device, may ensue **(Fig. 12)**. (*These techniques are described in greater detail in the chapter on foreign body retrieval.*)

Catheter-related venous thrombosis is reported in up to 10% of patients. Management depends on the site of thrombosis, symptoms, and need to maintain the access site. In general, if continued venous access is required and the patient is symptomatic, anticoagulation with heparin is appropriate. In symptomatic patients without a compelling requirement for continued access, the device is removed and heparin is administered. In patients with symptoms of SVC syndrome, anticoagulation and fibrinolytic therapy may be indicated.

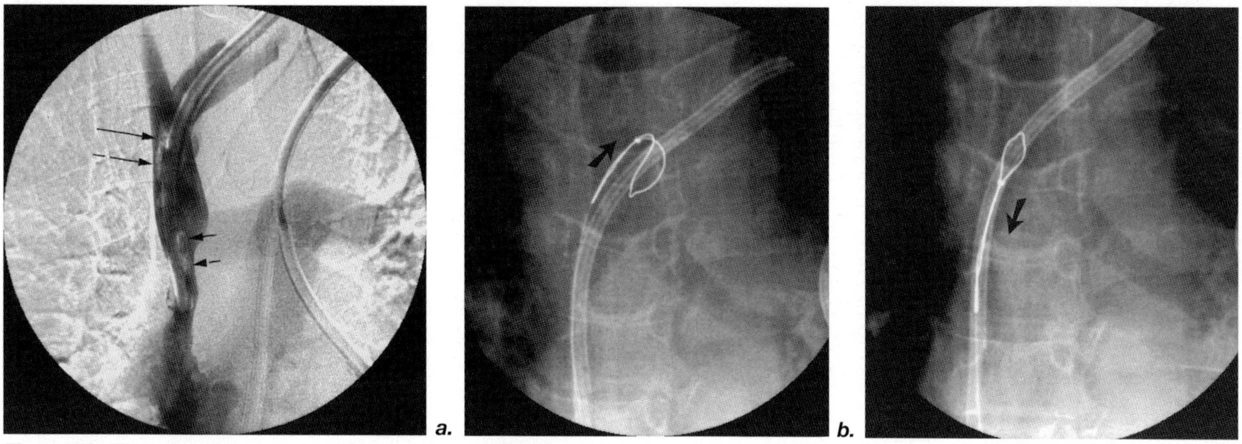

a. b. c.

Figure 10. Fibrin sheath removal. a) Venography revealed that a fibrin sheath was the cause of catheter malfunction in this patient. b) A loop snare was advanced into the SVC from a femoral venous approach and the fibrin strip was stripped off the catheter. The catheter worked well for 2 months following this procedure but it was ultimately replaced over a wire.

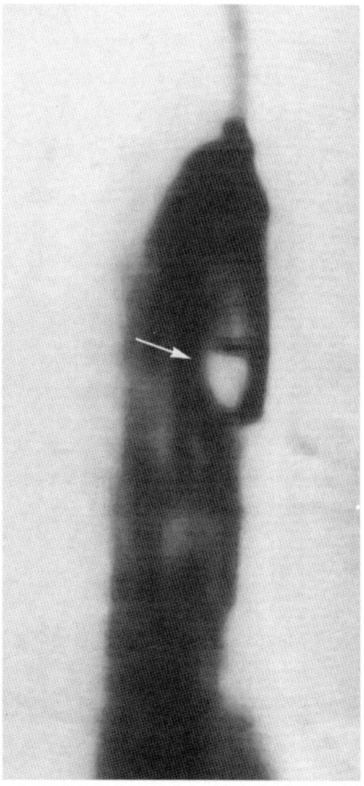

Figure 11. *Angiogram through existing chest port catheter. The referring physician could not aspirate blood through this catheter. The angiogram demonstrates a filling defect (clot) at the tip of the catheter (arrow). Urokinase was infused through the catheter, the clot resolved and catheter function was restored.*

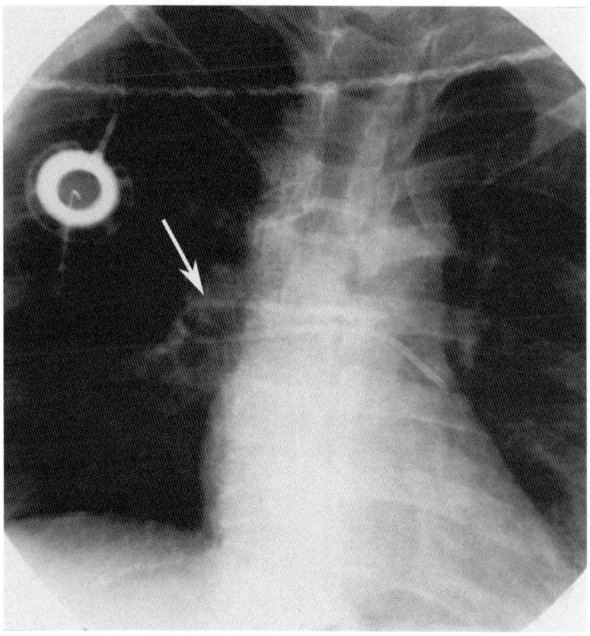

Figure 12. *Chest port catheter fragment. This chest port catheter has separated from the port and has embolized to the main pulmonary artery (arrow). It was retrieved with a loop snare using a femoral venous approach and then the port was replaced.*

TREATMENT OF DEVICE-RELATED INFECTION

- Superficial exit site infections
  - Local care, oral antibiotics
  - Mild tunnel, pocket infections
  - Local care, oral antibiotics
- More severe exit site, tunnel, pocket infections
  - Device removal
  - Incision and drainage
  - Antibiotics

CATHETER-RELATED VENOUS STENOSIS/THROMBOSIS

- Venous stenosis: balloon dilatation
- Venous thrombosis: symptomatic
  - Need to maintain access: Heparin, leave device intact
  - Low need to maintain access: Heparin, remove device
  - SVC syndrome: Heparin or lytic therapy
  - Septic thrombosis: Remove device, local therapy, heparin, antibiotics

Wound or exit site infections can be successfully treated with local care and oral antibiotics. Mild tunnel or pocket infections can also be treated with local wound care and antibiotics. More severe infections require device removal, and incision and drainage of infected material along with antibiotic therapy. Patients with catheter-related sepsis can be treated aggressively with IV antibiotics without immediate removal of the device. If the patient does not respond or if the symptoms recur after the IV antibiotics have been stopped, the device must be removed. A new device should not be placed until after an appropriate antibiotic course has been completed and negative blood cultures are documented. Removal of access devices should be performed by the interventional team. Tunneled catheters can be removed after minimal blunt dissection. Ports are removed with incision, blunt dissection, and wound closure. If there is a pocket infection, the wound should be left open **(Fig. 13)**.

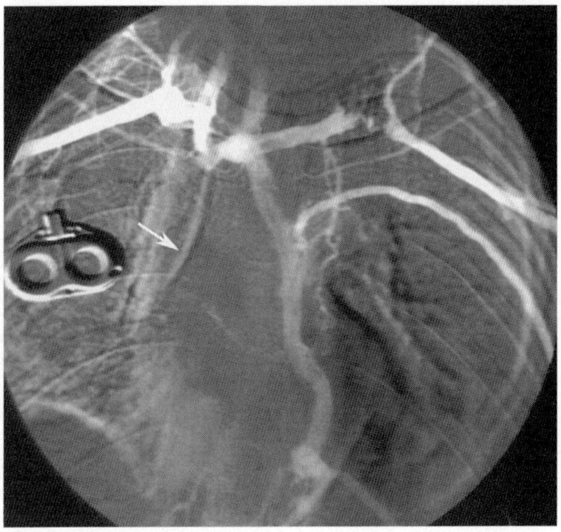

Figure 13. *SVC syndrome was induced in this hypercoaguable patient by catheter-induced thrombosis of the brachiocephalic veins. Arm venography demonstrates occlusion of the brachiocephalic vein and a large hemiazygos chest collateral. The right double-lumen port is visible on this subtraction film. An arrow marks the tip of the chest port catheter in the thrombosed superior vena cava.*

RESULTS

Temporary Venous Catheters

The technical success of catheter insertion approaches 100% with the use of ultrasound and fluoroscopy. Complications vary depending on the site of access. A brief summary of expected immediate results for various access routes is provided below.

PICC lines: Several large series (> 1,000 patients) have been published. Techniques have varied from bedside insertion by nurses to radiologic insertion by interventionalists using ultrasound and fluoroscopy. Results of two series are shown below **(Table 2)**. Ng et al reported results with nonimage-guided insertion by physicians in which cut-down was used if percutaneous insertion failed. Cardella et al compared nursing bedside insertion to radiologic insertion using fluoroscopy.

Internal Jugular Vein Catheters: Goldfarb and Lebrec reported successful jugular catheterization in 99% of 1,000 patients using visible landmarks without ultrasound guidance. Immediate complications were uncommon **(Table 3)**.

Additional immediate complications that have been reported by others include air embolism, bacteremia, sepsis, and hemopericardium. Neurologic and hemorrhagic complications of IJ puncture can be readily explained by the position of the jugular within the carotid sheath. Besides the carotid artery and jugular vein, the other structures in the carotid sheath are the vagus nerve and the cervical sympathetic nerves.

TABLE 2: PICC LINE RESULTS IN TWO LARGE SERIES

	Ng et al*	Cardella et al**
• Number of catheters	1,000	1,273 (552 by radiologists)
• Technical success	96% 14% cutdown	98% by radiologists in fluoro 82% by nurses at bedside
• Completed therapy with single device	69%	75%
• Duration of PICC	40 days 19% > 90 days	28 days
• Requirement for change or new catheter		
— Dislodgement	8%	9%
— Mechanical failure	5%	10%
— Occlusion	4%	5%
— Infection	4%	2%
— Thrombophlebitis	-	3%

(**Mayo Clinic Proceedings* 1997; 72:225-233. ***JVIR* 1996; 7:5-13.)

TABLE 3: COMPLICATIONS OF INTERNAL JUGULAR VEIN CATHETERIZATION*

Complication	Incidence
• Neck hematoma	1%
— With airway compromise	0.1%
• Arterial puncture	7%
— With sequelae of same	rare
• Transient Horner's syndrome	0.2%
• Pneumothorax	0.2%

(**Anesthesiology* 1982; 56:321-323.)

Venous thrombosis is a delayed complication of IJ catheterization. However, because IJ thrombosis is not usually clinically apparent, the exact incidence of thrombosis is unknown. Technical success appears to be higher and IJ catheterization is associated with fewer complications when ultrasound guidance is used for IJ puncture. Gordon et al reported 99.9% success and 2.3% complication rate with real-time ultrasound guidance for IJ puncture.

TABLE 4: RESULTS OF SUBCLAVIAN LINE PLACEMENT

	Mansfield*	Openshaw**
• Number of patients	821	100
• Technical success	88%	100%
• Complications		
— Pneumothorax	1.5%	0%
— Mediastinal hematoma	0.6%	0%
— Arterial puncture	3.4%	2%
— Air embolus without sequelae	-	1%

(*N Eng J Med 1994; 331:1735-1738. **JVIR 1994; 5:111-115.)

Subclavian Vein Catheters: Two large series have been recently published documenting the immediate results of subclavian venous catheter placement. The results are shown in **Table 4**. In the Mansfield study, the procedure was performed without fluoroscopy or real-time ultrasound guidance. In the Openshaw study, fluoroscopy was used, either alone or with real-time ultrasound guidance.

Long-term Catheters: Tunneled Catheters and Ports

Immediate technical results for placement of long-term catheters are dependent on the site of venous access and are similar to those reported above. The most important results concerning long-term catheters are durability and long-term complications. Specifically, device patency, device infection and venous thrombosis are the parameters of greatest interest.

In general, infection rates are lower for subcutaneous ports than for tunneled catheters. Groeger and colleagues reviewed the rate of infection in two large groups of cancer patients and found a significantly lower infection rate with ports (.02 infections per 100 catheter days) than with tunneled lines (.27 infections per 100 catheter days).

Recent literature has also shown that infection rates for permanent catheters placed in the interventional radiology suite are similar to the infection rates for catheters placed in the operating suite.

Dialysis Catheters

The requirements for tunneled dialysis catheters are similar to other tunneled devices in that a low infection rate and high patency rate is desired. In addition, a special requirement of dialysis catheters is the ability to sustain high blood flow rates. Flow rates of 300 ml/min or greater are desired. Consequently, fairly large bore catheters (12-F to 14-F) are required to provide adequate flow rates.

In the past, large bore dialysis catheters have been placed through either the subclavian or the IJ vein. However, now IJ approach is used almost exclusively. This is because of the high incidence of subclavian vein stenosis following large-bore catheterization of the subclavian vein. It is crucial to preserve the venous outflow for future arm fistulas or grafts.

Immediate technical results of catheter insertion have been similar to that reported for smaller bore catheters through the same venous approach. Although the potential for air embolism is greater due to the larger size of dialysis catheters, studies to date have not shown a markedly increased risk for this complication. Similarly, the risk of bleeding and pneumothorax are similar to those reported for smaller catheters.

Infection and catheter occlusion continue to be a major problem limiting the extended use of tunneled dialysis catheters. Catheter design may play a significant role in determining follow-up performance. The two largest studies to date have been reported by Lund and Trerotola. The results of these two studies are shown in **Table 5**. Note that predicted patency at 1 year ranges from 25% to 50% and the percentage of catheters removed for infection varied from 6.8% to 11.2%.

At UCSF, we favor the Tesio catheter system and use it as our primary device for long-term dialysis access. This twin catheter system is composed of two 10-F catheters, each with a detachable hub. The detachable hub facilitates catheter insertion and the independent double-catheter design affords a large lumen that potentially maximizes flow rates. We have recently reported our UCSF results with the Tesio system and these are summarized in **Table 6**.

TABLE 5: TUNNELED DIALYSIS CATHETER RESULTS

	Lund, 1996*	Trerotola, 1997**
• Number of catheters	237 (inserted)	250
• Numbers of patients	140	175
• Location	Subclavian	Right IJ
• Length of follow-up	16,240 catheter days 2-497 (73d)	21,572 catheter days 2-643(87d)
• Major complications		
— pneumothorax	6/237(2 chest tubes)	0
— venous thrombosis	2/222	0
• Infection		
— No. of catheters	32/222 catheter (14.4%)	34/250 (13.6%)
— No. of episodes/100 catheter days at risk	.2	.16
— No./% removed	25/222 (11.2%)	17/250 (6.8%)
• Patency		
— 6 months	44%	62%
— 1 year	25%	48%

(*Radiology 1996; 198:467-472. **Radiology 1997; 203:489-495.)

TABLE 6: TESIO RESULTS—UCSF

• Number of catheters	78
• Location	IJ
• Follow-up	65 days (mean) 4,367 (total)
• Complications	2.5%
• Infection	.09 (per 100 catheter days)
• Malfunction	4.5%
• Flow rate	300 (mean)

REFERENCES

PICC *Lines*:

Cardella JF, Cardella K, Bacci N, Fox PS, Post JH. Cumulative experience with 1,273 peripherally inserted central catheters at a single institution. J Vasc Interv Radiol 1996; 7:5-13.

Ng PK, Ault MJ, Ellrodt AG, Maldanado L. Peripherally inserted central catheters in general medicine. Mayo Clin Proced 1997; 72:225-233.

IJ *Lines*:

Goldfarb G, Lebrec D. Percutaneous cannulation of the internal jugular vein: an experience based on 1,000 attempts. Anesthesiology 1982; 56:321-323.

Gordon AC, Saliken JC, Johns D, Owen R, Gray RR. US-guided puncture of the internal jugular vein: complications and anatomic considerations. J Vasc Interv Radiol 1998; 9:333-338.

Subclavian lines:

Mansfield PF, Hohn DC, Fornage BD, Gregurich MA, Ota DM. Complications and failures of subclavian-vein catheterization. N Engl J Med 1994; 331:1735-1738.

Openshaw KL, Picus D, Hicks ME, Darcy MD, Vesley TM, Picus J. Interventional radiologic placement of Hohn central venous catheters: results and complications of 100 consecutive patients. J Vasc Interv Radiol 1994; 5:111-115.

Chest ports and tunneled catheters:

Damascelli B, Patelli G, Frigerio LF, et al. Placement of long-term central venous catheters in outpatients: study of 134 patients over 24,596 catheters days. AJR 1997; 168:1235-1239.

Foley MJ. Radiologic placement of long-term central venous peripheral access system ports (PAS Port): results in 150 patients. J Vasc Interv Radiol 1995; 6:255-262.

Groeger JS, Lucas AB, Thaler HT, et al. Infectious morbidity associated with long-term use of venous access devices in patients with cancer. Ann Intern Med 1993; 119:1168-1174.

Simpson KR. Interventional radiologic placement of chest wall ports: results and complications in 161 consecutive placements. J Vasc Interv Radiol 1997; 8:189-195.

Dialysis catheters:

Lund GB, Trerotola SO, Scheel P Jr, et al. Outcome of tunneled hemodialysis catheters placed by radiologists. Radiology 1996; 198:467–472.

Perini S, LaBerge JM, Pearl JM, et al. Tesio catheter: Radiologically guided placement, mechanical performance, and adequacy of delivered dialysis. Radiology 2000; 215:129-137.

Trerotola SO. Outcome of tunneled hemodialysis catheters placed via the right internal jugular vein by interventional radiologists. Radiology 1997; 203:489-495.

CHAPTER 17 PREVIEW ANSWERS

Case 1

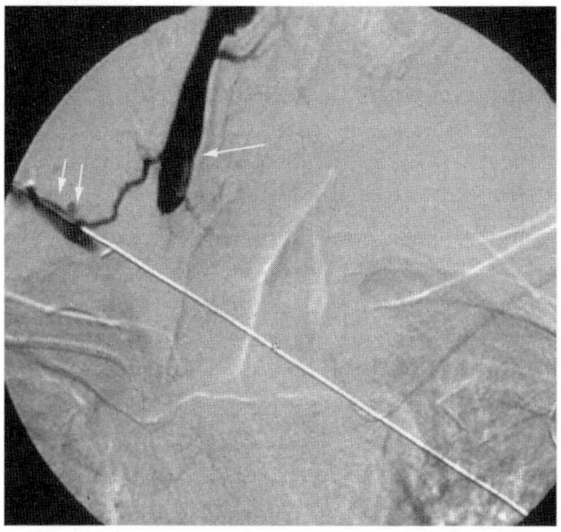

Figure 1. *Hand-injected venogram of the right internal jugular vein.*

Findings: A catheter has been placed in the peripheral aspect of the IJ. Injection of contrast **(Fig. 1)** reveals occlusion of the central portion of the vein (arrow). A small collateral is seen that is filling the external jugular vein (double arrow).

Diagnosis: Central IJ occlusion.

Treatment: If the left IJ is patent, the dialysis catheter can be inserted into the SVC from the left neck. Insertion of dialysis catheters through the subclavian veins is to be avoided because of the risk of catheter-induced stenosis.

Case 2

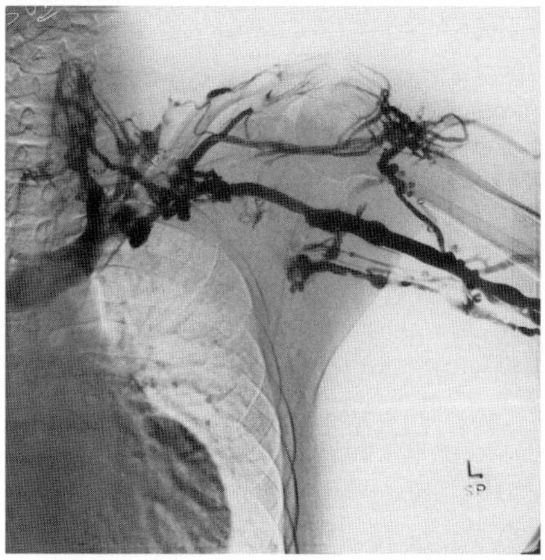

Figure 2. *Left arm venogram, subtraction film.*

Findings: Left arm venography **(Fig. 2)** reveals occlusion of the left subclavian vein. The axillary vein is patent and prominent collaterals are seen about the shoulder to reconstitute flow in the left brachiocephalic vein.

Diagnosis: Left subclavian vein occlusion with collateral reconstitution of the left brachiocephalic vein.

Treatment: The easiest option is to place a PICC line in the right arm. Leaving a PICC line in the left axillary vein puts the patient at risk for developing further thrombosis in the axillary vein.

Case 3

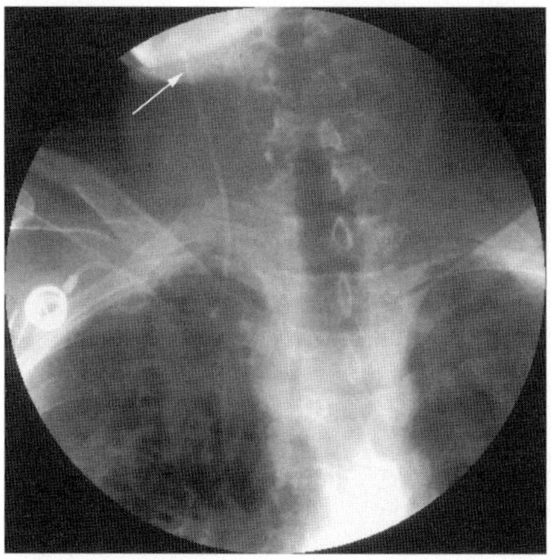

Figure 3. *Plain film of the upper chest and neck.*

Findings: Plain film **(Fig. 3)** shows that the right chest port catheter has migrated so that the tip is in the right internal jugular vein (arrow).

Diagnosis: Port catheter migration from the SVC to the right internal jugular vein.

Treatment: Several treatment options may be considered. The catheter could be repositioned using a transfemoral approach to snare the catheter and pull it down into the SVC. Alternatively, the catheter could be replaced over a wire. (This would require cut-down over the port.)

Case 4

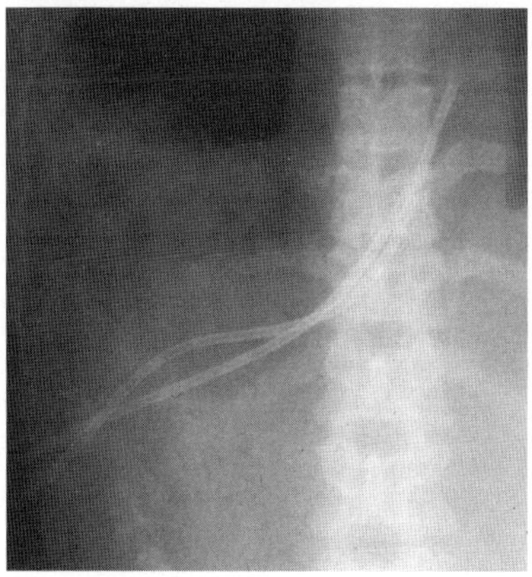

Figure 4. *Plain film of the right upper quadrant following an IR procedure.*

Findings: Plain film of the right upper quadrant reveals two catheters entering from a right lateral abdomen ending in the right atrium. In this case, a Tesio twin catheter system has been inserted using a transhepatic appoach into the right hepatic vein. The catheters were then tunneled along the anterior abdominal wall for use during dialysis.

Diagnosis: Transhepatic insertion of Tesio dialysis catheters in a patient with extensive venous occlusions.

CHAPTER 18 PREVIEW

Case 1

Bilirubin: 14.2 mg/dl (normal 0.1-1.2)
Albumin: 2.6 g/dl (normal 3.4-5.0)
Prothrombin time (PT): 18.7 sec. (normal 10.4-12.8)
Hematocrit: 28% (normal 36-46)
Platelet count: 60K. (normal 140K-450K)

Figure 1. *Laboratory data on admission.*

History: A 39-year-old man with alcoholic liver disease and hepatitis presented with massive variceal bleeding. He had a history of prior variceal bleeding treated by sclerotherapy. His laboratory data on admission are shown above. On physical exam, he was not encephalopathic but he had a moderate amount of abdominal ascites.

Questions: What is the Child's-Pugh score for this patient and what are the treatment options to control his bleeding?

Case 2

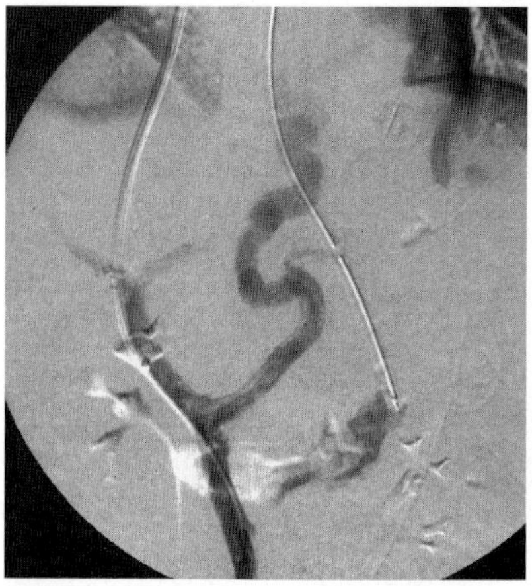

Figure 2. *Transjugular portal venogram.*

History: A 32-year-old woman with hepatitis C presented with variceal bleeding refractory to sclerotherapy. TIPS was requested. The initial pre-TIPS portosystemic pressure was 18 mm Hg.

Questions: Does this patient have portal hypertension? What is the desired portosystemic gradient following TIPS?

Case 3

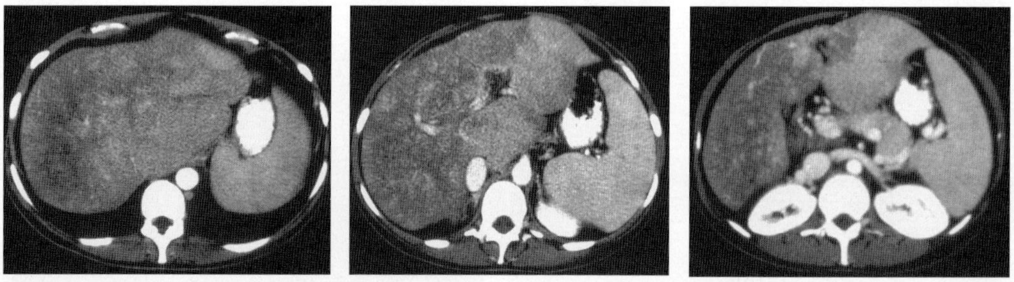

Figure 3. *Contrast-enhanced abdominal CT.*

History: A 34-year-old woman complained of abdominal fullness and right upper quadrant pain. Physical exam revealed hepatosplenomegaly. Laboratory evaluation was notable for a slight elevation in the hematocrit (50%), platelet count (377K), PT (14.5 seconds) and bilirubin (1.7 mg/dl). Sonography revealed ascites in the pelvis. An abdominal CT was obtained and then liver biopsy was performed.

Questions: What is your diagnosis? How could this patient be treated?

Chapter 18
Portal Hypertension and Transjugular Intrahepatic Portosystemic Shunts

Jeanne M. LaBerge, MD

CHAPTER OUTLINE

- Background information on the diagnosis and treatment of portal hypertension
- Review of basic TIPS technique and expected complications
- Summary of current TIPS results
- Accepted indications and contraindications

INTRODUCTION

The transjugular intrahepatic portosystemic shunt (TIPS) procedure is a novel interventional procedure that is now widely accepted as a treatment for patients with refractory variceal bleeding or ascites **(Fig. 1)**. A thorough understanding of this procedure and its role in the management of patients with portal hypertension requires an understanding of the pathophysiology of portal hypertension and current alternative treatment options. In this chapter we provide background information on portal hypertension and present the TIPS procedure in the context of other available treatments.

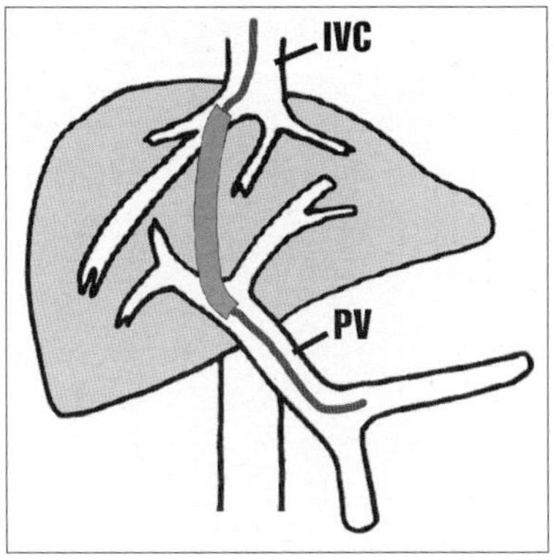

Figure 1. *Schematic representation of TIPS. A stented channel is created within the liver from the portal vein to the hepatic vein in order to decompress the portal system in patients with portal hypertension.*

BACKGROUND INFORMATION

Physiology of Portal Hypertension

In the fasting state, the portal vein conducts approximately one liter of blood to the liver per minute. Since hepatic artery flow is only 350 ml/min in the fasting state, portal vein flow accounts for 75%-80% of total hepatic blood flow. The superior mesenteric and inferior mesenteric veins contribute approximately 80% of portal vein flow, and the splenic vein, 20%. Portal vein flow may increase by a factor of eight after eating. Almost all of the postprandial increase in portal vein flow is mesenteric in origin.

The normal pressure in the portal vein varies between 5 and 10 mm Hg. The pressure difference between the portal vein and the inferior vena cava, known as the portosystemic gradient (PSG), is normally 3-6 mm Hg. Portal hypertension is defined as an elevation of the absolute portal pressure or the portosystemic gradient. Quantitatively, portal hypertension exits when the absolute portal pressure is greater than 11 mm Hg or when the portosystemic gradient is greater than 6 mm Hg. **(Fig. 2)**.

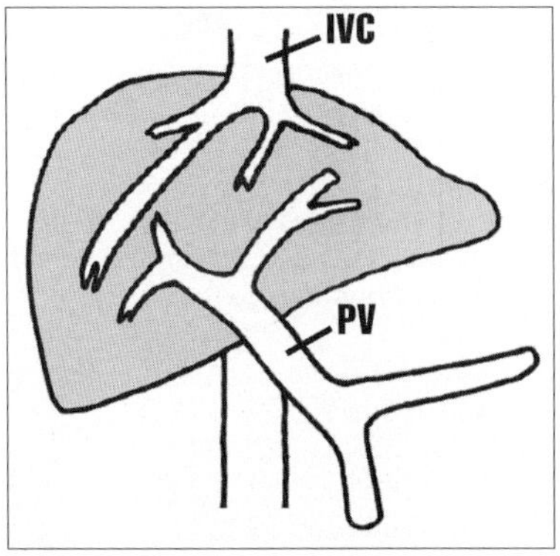

Figure 2. *The gradient between the portal vein (PV) and the inferior vena cava (IVC) is normally < 6 mm Hg.*

A brief review of the hepatic microcirculation will help explain the common etiologies of portal hypertension. Blood flows into the hepatic sinusoid from the hepatic artery and the portal vein, and exits the sinusoid from the central hepatic vein. Obstruction to blood flow through the sinusoid will result in elevation of pressure in the portal vein. In addition, any obstruction to inflow from the portal vein or outflow from the hepatic vein may also result in portal hypertension.

CAUSES OF PORTAL HYPERTENSION

- Liver cirrhosis
  - Alcoholic
  - Biliary
  - Infectious, viral
- Noncirrhotic portal fibrosis
- Budd-Chiari syndrome
- Schistosomiasis

COMPLICATIONS OF PORTAL HYPERTENSION

- Variceal bleeding
- Ascites
- Encephalopathy
- Hypersplenism

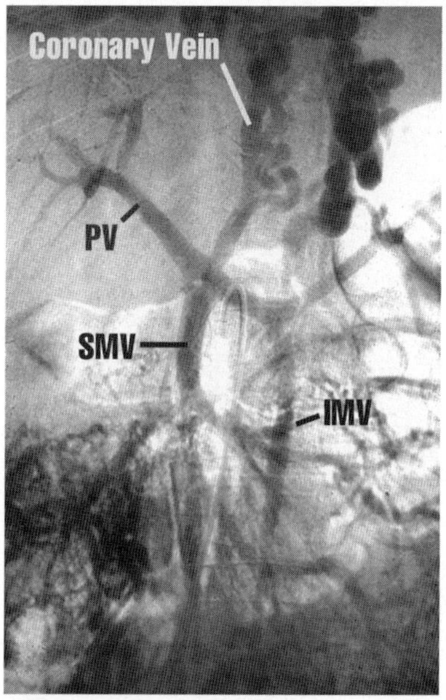

Figure 3. Arterial portography in a patient with variceal bleeding. An SMA injection opacifies the SMV and the main and intrahepatic portal veins. However, there is also opacification of the coronary vein and IMV indicating the presence of portal hypertension.

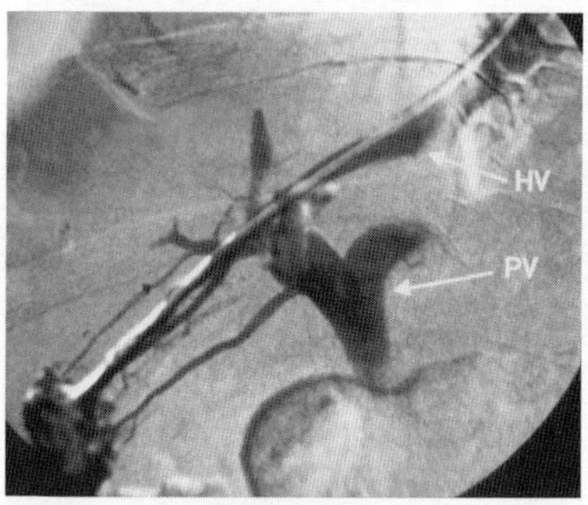

Figure 4. Wedged hepatic venogram. In this case a catheter was advanced into a peripheral right hepatic vein and contrast injection opacifies the main portal vein. Pressure measurement through this catheter revealed an absolute pressure of 20 mm Hg.

The most common causes of portal hypertension are listed below:

1) liver cirrhosis (alcoholic, biliary, viral);
2) noncirrhotic portal fibrosis;
3) Budd-Chiari syndrome (hepatic venous or inferior vena caval obstruction);
4) schistosomiasis (intrahepatic sinusoidal portal venous obstruction).

Patients with portal hypertension may be asymptomatic or they may manifest symptoms of any of the following problems:

1) variceal bleeding;
2) ascites;
3) encephalopathy;
4) hypersplenism.

Variceal hemorrhage occurs most frequently (in up to 85% of cases) from varices at the gastroesophageal (GE) junction. Varices at the GE junction are superficial and are more likely to protrude into the esophageal lumen than varices elsewhere in the esophagus. The risk of variceal hemorrhage increases with the size of the varix. That is, large varices are much more likely to rupture than small varices. The likelihood of rupture is also dependent on the amount of surrounding supporting tissue. Varices with large diameters and thin walls are at greatest risk for rupture.

Diagnosis of Portal Hypertension

A definite diagnosis of portal hypertension is obtained by direct measurement of the portal venous pressure and portosystemic gradient. However, the diagnosis is more commonly made by indirect evaluation. Most commonly, the diagnosis of portal hypertension is made endoscopically by visual identification of gastroesophageal varices. The risk of hemorrhage may be assessed by the size and appearance of the varix.The larger the varix, the greater the risk of bleeding. In addition, the "red wale" sign suggests an increased risk of hemorrhage. Gastric varices may be present in conjunction with esophageal varices in patients with portal hypertension. However, if gastric varices are present in the absence of esophageal varices, splenic vein occlusion should be suspected.

Patients with portal hypertension may also be evaluated by sonography or arterial portography. Sonographic signs of portal hypertension include splenomegaly, ascites and the presence of numerous small tortuous vessels reflective of varices. Sonography is frequently used to assess the size and patency of the portal vein prior to TIPS or liver transplantation.

Arterial portography is used to clarify the anatomy of the portal vein when Doppler sonography is inconclusive. Arterial portography is accomplished by contrast injection into the splenic or superior mesenteric arteries **(Fig. 3)**. When portal hypertension is severe, arterial portography may not adequately display the portal venous anatomy because of the magnitude of portal flow diversion through varices or because of total reversal of flow in the main portal vein. When arterial portography is inconclusive, imaging of the portal venous system can be accomplished by direct portal catheterization or by wedged hepatic venography **(Fig. 4)**. An indirect measure of portal pressure can be obtained by wedged hepatic pressure measurement. In this technique, the free hepatic vein pressure is subtracted from the wedged pressure to yield the "corrected sinusoidal pressure (CSP)." A normal CSP is less than or equal to 5 mm Hg.

Treatment of Portal Hypertension and its Complications

The initial approach to patients with portal hypertension and its attendant complications of variceal bleeding, ascites, encephalopathy and hypersplenism is to conduct a thorough medical evaluation to establish a diagnosis and to assess the patient's underlying medical condition. As the initial medical evaluation is being conducted, patients who are acutely ill will undergo emergency medical treatment. Resuscitation with fluid and blood products and admission to the intensive care unit (ICU) are standard treatments for patients with acute variceal bleeding. Thorough evaluation of the patient's medical condition is important because the prognosis and treatment options are very much dependent on the patient's underlying medical condition **(Fig. 5)**.

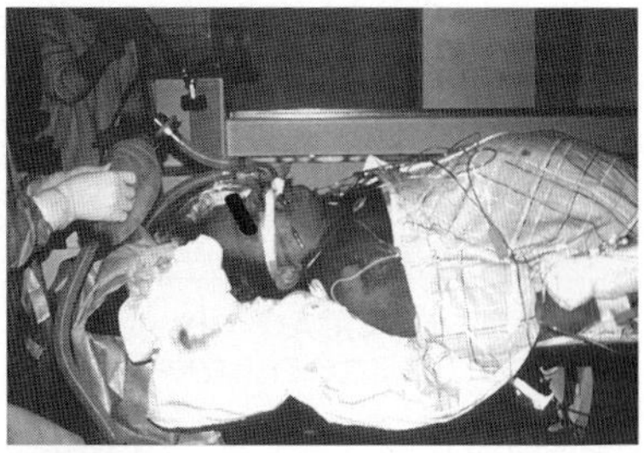

Figure 5. *This patient with Child's C cirrhosis suffered from two complications of portal hypertension: acute variceal bleeding and ascites. Treatment of acute variceal hemorrhage required admission to the ICU, fluid resuscitation and endotracheal intubation.*

The Child's-Pugh scoring system is a means of categorizing the severity of liver disease. The modified scoring system shown on **Table 1** was described by DiMagno et al and is the scoring system used at UCSF.

INITIAL THERAPY FOR VARICEAL BLEEDING

- Resuscitation
- Intensive care
- Early endoscopy

TABLE 1: MODIFIED CHILD'S-PUGH SCORING SYSTEM*

Variable	*Points for Severity of Each Variable* ***(1)***	***(2)***	***(3)***
Encephalopathy	none	moderate	severe
Ascites	none	moderate	severe
Bilirubin (mg/dl)	<2	2-3	>3
Albumin (g/dl)	≥3.5	2.8-3.4	<2.8
Prothrombin time (secs)	≤14	15-17	≥18

*(*Points are summed for all five categories to determine Child's-Pugh score and Child's class as follows: A=5, 6; B=7, 8, or 9; C=10-15.)*

(DiMagno EP et al, Mayo Clinic Proceedings, 1985.)

Once the patient's medical condition has been stabilized and a diagnosis of portal hypertension has been made, definitive treatment options may be considered. For the purposes of this chapter, the discussion of therapeutic options will be limited to those used in the treatment of variceal bleeding. These include:

1) medical therapy: sclerotherapy, variceal band ligation, administration of pharmacologic medications that reduce portal pressure, and balloon tamponade;
2) surgery including devascularization procedures and portosystemic shunts;
3) liver transplantation;
4) TIPS.

THERAPEUTIC OPTIONS FOR VARICEAL BLEEDING

- Medical management
- Endoscopic sclerotherapy
  - band ligation
  - sclerotherapy
- Surgical shunting
  - portacaval
  - selective
- Liver transplantation
- TIPS

A brief discussion of each of these therapies including their advantages and disadvantages follows.

<u>Sclerotherapy</u>

Over the past decade, endoscopic sclerotherapy has become the primary treatment modality for variceal bleeding. It is highly effective; nearly 90% of acute variceal hemorrhage can be stopped by this approach. In more than 75% of patients, the variceal channels can be obliterated with repeated injections.

SCLEROTHERAPY

- Primary treatment modality
- 90% success in acute setting
- 20% complication rate

A variety of sclerosants are available. While sodium morrhuate has been used extensively in the past, ethanolamine oleate is now more commonly employed because it appears to result in a lower incidence of ulceration and scarring. The sclerosant is injected into the varix or into the adjacent tissue. The injection is performed under direct endoscopic vision. The intended histologic effect of sclerotherapy is variceal thrombosis.

COMPLICATIONS OF SCLEROTHERAPY

- Pain
- Fever
- Pleural effusions
- Chemical mediastinitis
- Bleeding
- Sepsis
- Ulcerations
- Stricture
- Perforation

Complications occur in 20% of patients undergoing sclerotherapy. The majority of complications are minor and involve pain, fever, and reactive pleural effusion. Postinjection back-bleeding is usually minor but can be quite severe in patients with severe portal hypertension. Esophageal ulceration is a major complication that can result in persistent bleeding or frank ulceration. A sequela of ulceration is stricture formation, which may require subsequent dilation.

Variceal Band Ligation

Variceal band ligation is a treatment alternative that avoids the systemic complications of sclerotherapy. In this procedure, a rubber band is placed around the varix in an attempt to strangulate the varix and cause congestion, stasis, and thrombosis. The most suitable candidate for variceal band ligation is the patient with very large varices.

Pharmacologic Control

If sclerotherapy or banding fails to control bleeding, then pharmacologic or mechanical means are employed. The pharmacologic agents used to control variceal hemorrhage include vasopressin, nitroglycerine and somatostatin. These agents reduce portal pressure and thus reduce variceal wall tension. Bleeding can be reliably controlled by reducing the portal pressure gradient to below 12 mm Hg.

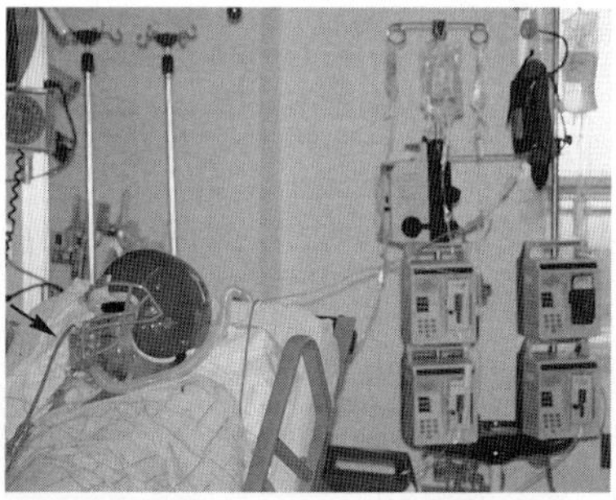

Figure 6. This patient with acute variceal hemorrhage was admitted to the ICU and treated with balloon tamponade (arrow). An octreotide infusion was administered in order to lower portal pressure.

Vasopressin, which lowers portal pressure by producing splanchnic vasoconstriction, has been the principal pharmacologic agent used to control bleeding. However, side effects such as systemic hypertension and cardiac ischemia limit its application. Concurrent use of nitroglycerine may decrease the incidence of these side effects and potentiate portal pressure reduction. Somatostatin may be effective in controlling acute bleeding by reducing splanchnic blood flow and reducing portal pressure. Octreotide, a synthetic analog of somatostatin, is now preferred over somatostatin because of its longer half-life.

The prophylactic treatment of varices and portal hypertension remains controversial. beta-blocker therapy appears to be beneficial in preventing the initial variceal hemorrhage. The nonselective beta-blockers (propranolol and nadolol) appear to be more effective than the selective beta-blockers in decreasing splanchnic blood flow. Alcoholic patients with large varices respond best to this therapy.

BALLOON TAMPONADE

- Complications in 20%
  - Aspiration
  - Ulceration
  - Perforation

Balloon Tamponade

Balloon tamponade is a very effective means of stopping acute bleeding **(Fig. 6)**. The balloons may be inflated to independently occlude the esophageal or gastric varices. The most common balloon catheter used at our institution is the Minnesota tube, which has side-holes for aspiration in both the esophageal and gastric segments. The catheter must be secured under tension to ensure adequate variceal compression.

Although balloon tamponade is a very effective means of controlling acute bleeding, it is associated with a 10%-20% morbidity, including aspiration, esophageal perforation, and ulceration.

Surgery

Most patients with variceal hemorrhage are successfully treated with either sclerotherapy, pharmacotherapy, tamponade or a combination of these modalities. However, when bleeding continues despite these means, a variety of surgical therapeutic alternatives are available. These procedures can be broadly classified into two categories: nonshunt operations and portosystemic shunts. Nonshunt operations include expedited transplantation, devascularization procedures and splenectomy. Portosystemic shunt options include shunts which substantially divert portal blood flow, such as the portacaval or mesocaval shunt, and shunts which preserve portal blood flow, such as the distal splenorenal shunt.

The plethora of surgical alternatives for the management of variceal bleeding attest to the limitations of each these therapies. Each is associated with a greater procedural morbidity and mortality than sclerotherapy and many are associated with delayed morbidity from the adverse consequences of portosystemic shunting, namely, encephalopathy and liver failure. Consequently, none of these alternatives is appropriate for bleeding prophylaxis and most are employed only when endoscopic or pharmacologic therapy are unsuccessful.

Devascularization Procedures

Devascularization procedures can be performed in the emergent and nonemergent setting. Esophageal transection with a staple gun is the simplest emergency devascularization procedure. Simple variceal ligation with esophageal transection is an effective means of controlling acute bleeding, but bleeding recurs in 20%-50% of patients due to the prompt development of additional varices. Improved long-term control of bleeding has been reported with the Sugiura operation, a more extensive procedure consisting of transthoracic paraesophageal devascularization, esophageal transection, splenectomy, esophagogastric devascularization, pyloroplasty and vagotomy.

Surgical Shunts

A wide variety of surgical shunts has been devised for control of variceal bleeding. The three most common shunt procedures **(Fig. 7)**, are the portacaval, mesocaval and distal splenorenal shunts. All three of these procedures divert blood flow from esophagogastric varices. The portacaval and mesocaval shunts are nonselective shunts in that they divert portal blood flow from the liver as well as the varices. The distal splenorenal shunt is a selective shunt that only diverts flow from esophagogastric varices and does not interfere with mesenteric portal venous perfusion of the liver. Surgical decompression of esophagogastric varices results in more effective control of bleeding than can be achieved with sclerotherapy. But this improvement in control of bleeding is achieved at the cost of increased procedural morbidity and mortality, as well as shunt-related complications such as encephalopathy.

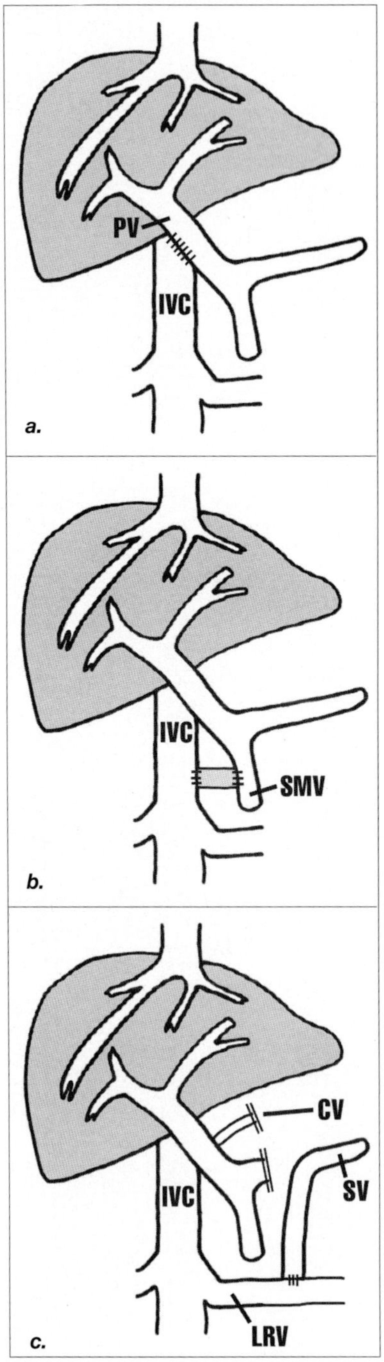

Figure 7. *Most common surgical shunts. a) Side-to-side portacaval shunt. A 20-mm anastomosis is made directly between the PV and the infrahepatic IVC. b) Mesocaval shunt. A synthetic graft is inserted to connect the IVC and the SMV. c) Distal splenorenal shunt. The splenic vein (SV) is connected to the left renal vein (LRV) and the coronary vein (CV) is ligated.*

Selection of a shunt based on expected clinical outcome is fraught with controversy because randomized trials comparing outcome with different shunts have produced conflicting results. Stiegmann has recently averaged the results of six prospective randomized trials comparing nonselective shunts to selective shunts with respect to operative mortality, rebleeding and encephalopathy (SCVIR *Syllabus: Portal Hypertension*, 1995). These results are summarized in **Table 2** below.

TABLE 2: RESULTS OF SURGICAL DECOMPRESSIVE SHUNTS

	Nonselective Shunts	Selective Shunts
Operative mortality	8%	10%
Rebleeding	6%	13%
Encephalopathy	40%	26%

(Modified with permission of Van Stiegman G, 1995.)

Operative mortality was similar for both types of shunts. Variceal rebleeding rates were slightly higher following selective shunt than following nonselective shunt. Encephalopathy rates were lower following selective shunt than following nonselective shunt. Although the data differs somewhat among the studies, many experts feel strongly that the distal splenorenal shunt is associated with a lower incidence of shunt-related encephalopathy than nonselective shunts; and, it is for this reason that they suggest that the distal splenorenal shunt is preferable to the portacaval or mesocaval shunt.

Expedited Transplantation

Transplantation is the only therapy that cures the underlying cause of variceal bleeding—portal hypertension. By restoring normal liver function, transplantation eliminates portal hypertension and thereby definitively controls acute bleeding and prevents future bleeding. However, the indiscriminate use of transplantation as a primary treatment for variceal hemorrhage is inadvisable because of the limited availability of donor organs, the morbidity and cost of transplantation, and the morbidity of long-term immunosuppression. Transplantation is a therapy usually reserved for patients with symptomatic end-stage liver disease and its use in the management of patients with good hepatic reserve is limited. Nevertheless, expedited transplantation is an attractive means of controlling chronic recurrent bleeding in patients with poor liver disease who are otherwise suitable candidates for transplantation.

All of the therapeutic options for the management of variceal hemorrhage listed above have shortcomings. Sclerotherapy may be initially unsuccessful at controlling bleeding in 10% of patients; and bleeding recurs within the first year in up to 50% of patients who were initially controlled by sclerotherapy. Variceal bleeding is well-controlled by surgical shunts but this benefit is achieved at the price of a high operative morbidity and significant shunt-related complications such as encephalopathy. Transplantation is an effective means of curing portal hypertension but is impractical in the emergency setting and is inappropriate in patients whose liver disease is not severe enough to warrant transplantation.

Transjugular Intrahepatic Portosystemic Shunts (TIPS)
The search for a better therapy for portal hypertension and variceal bleeding led to the development of a nonoperative portosystemic shunt. The hope was that, because TIPS could be performed percutaneously without the need for open surgery, morbidity would be lower for TIPS than for surgical shunts. Moreover, because the TIPS shunt is narrower than the standard surgical shunt, it was hoped that the rate of shunt-related encephalopathy would be lower for TIPS than for surgical shunts.

THERAPEUTIC SHORTCOMINGS

- Sclerotherapy
  - 10-30% fail initially
  - 50% rebleed
- Surgical shunts
  - high 30-day mortality
  - Delayed encephalopathy
- Transplantation
  - Impractical acutely
  - May not be necessary

Considering the currently available data, TIPS appears to be competitive with both sclerotherapy and surgery in the management of recurrent variceal bleeding. The reported rate of rebleeding varies from 7% to 31% following TIPS. This rebleeding rate is lower than the rate achieved with sclerotherapy but higher than obtained following surgical shunting. Procedure-related morbidity and mortality appear to be lower for TIPS than for conventional surgical shunts. Thirty-day mortality rates after TIPS range from 3% to 15%. Encephalopathy occurs in about 25%-30% of patients after TIPS creation. In most cases, the symptoms are mild and can be readily managed with dietary protein restriction and lactulose therapy. In less than 5% of patients, uncontrolled encephalopathy poses a major clinical problem.

Although procedure-related morbidity and mortality appear to be lower for TIPS than for conventional surgical shunts, poor shunt patency has emerged as a major problem. One-year primary patency rates ranging from 25% to 66%, emphasize the importance of a close surveillance program and prompt percutaneous reintervention when necessary.

TIPS TECHNIQUE AND COMPLICATIONS

Technique
The technique of TIPS is well established. The basic steps of this procedure are illustrated in **Figure 8** and summarized below.

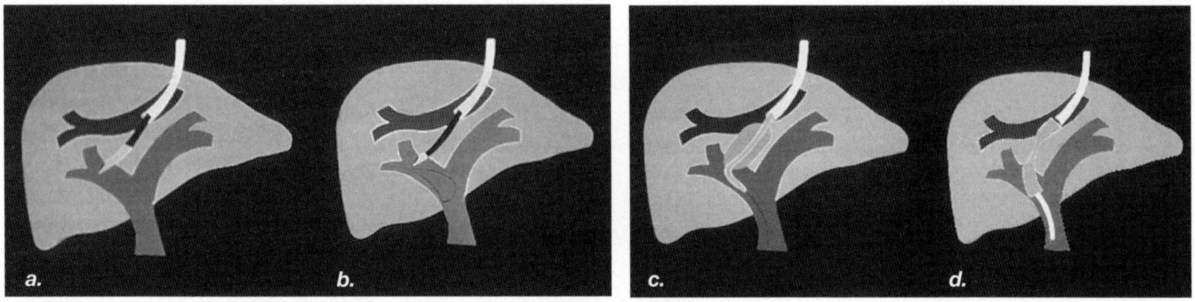

Figure 8. *Schematic representation of the TIPS procedure. a) A needle is advanced from the right hepatic vein into a branch of the right portal vein. b) A wire is threaded into the portal system; c) A balloon is advanced over the wire, positioned within the parenchymal tract and dilated; d) A metal stent is advanced over the wire and deployed to bridge the tract from the portal vein to the hepatic vein.*

TIPS Technique: Step-by-Step

1) Patients receive IV broad spectrum antibiotics before the procedure. (A first generation cephalosporin is recommended.) IV sedative and analgesic medication is administered. General anesthesia is not required in most cases.

2) The preferred approach for performing TIPS is the right internal jugular vein. If the right internal jugular vein is occluded, the procedure can be performed from the left internal jugular vein or the external jugular veins.

3) Using Seldinger technique, a 9-F 40-cm angiographic sheath with a 30-degree distal angle (Cook Inc., Bloomington, Ind.) is placed into the infra-hepatic inferior vena cava. Initial pressure in the inferior vena cava is measured.

4) The 9-F sheath is retracted to the atrial-caval junction, and a curved angiographic catheter is manipulated under fluoroscopic guidance into a hepatic vein. Selection of an appropriate hepatic vein is of critical importance. A vein of suitable diameter and appropriate angle greatly facilitates the procedure. The right hepatic vein is generally preferred, but TIPS can also be created from the middle and left hepatic veins.

5) Once the appropriate hepatic vein has been selected, the catheter is removed and replaced with a transjugular hepatic puncture needle. Two types of needles are commonly used: Colapinto and Rösch-Uchida. The Colapinto needle (Cook Inc., Bloomington, Ind.) has two parts: (1) a curved 16-gauge metal needle and (2) an outer 9-F catheter. The stiffness of the Colapinto needle enables a controlled puncture even through hard, cirrhotic liver parenchyma. The internal lumen of the needle allows passage of a 0.035-inch guidewire. The Rösch-Uchida transjugular needle (Cook Inc., Bloomington, Ind.) was designed to be more flexible and have a leading tip of smaller caliber compared with the Colapinto system. With the Rösch-Uchida system, a .038-inch flexible needle is used to puncture the portal vein.

6) Targeting of the portal vein for puncture remains a controversial topic. A high degree of success can be achieved using fluoroscopy alone to guide the needle into the anticipated region of the hepatic hilum without additional imaging. Prepuncture opacification of the portal vein can be easily achieved by wedged hepatic venography in most cases. Either contrast or carbon dioxide can be used for this purpose **(Fig. 9)**.

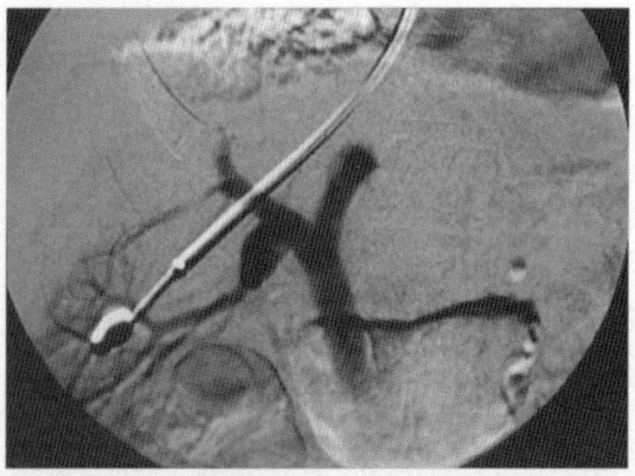

Figure 9. *Wedged hepatic venography may be used to opacify the portal vein and facilitate puncture of the portal vein. In this case, carbon dioxide was used as the contrast agent and full opacification of the intrahepatic portal system was achieved.*

7) The needle is rotated anteromedially and advanced caudally out of the right hepatic vein 4 -5 cm into the hepatic parenchyma. When a portal vein branch is entered, a guidewire is passed through the needle and manipulated down the main portal vein into the splenic or mesenteric vein.

8) The track is then dilated with a 5-F 8-mm diameter balloon. After track dilatation, a bridging stent is deployed. In the United States, most TIPS are performed with the Wallstent (Schneider USA, Minneapolis, Minn.) to bridge the hepatic parenchymal track. This wire mesh endoprosthesis has a predetermined unconstrained maximal diameter. Sizes typically used for TIPS range from 8 mm to 12 mm. The principal advantage of the Wallstent is its flexibility, which allows easy passage across the angles of the hepatic vein and portal vein entries. The Palmaz stent (Johnson & Johnson, Warren, N.J.) was the first stent used in TIPS and extensive experience is available with this device **(Fig. 10)**. The Gianturco-Rösch Z stent (Cook Inc, Bloomington, Ind.) and the Strecker stent (Medi-Tech, Boston Scientific Co., Watertown, Mass.) have also been used successfully to create TIPS.

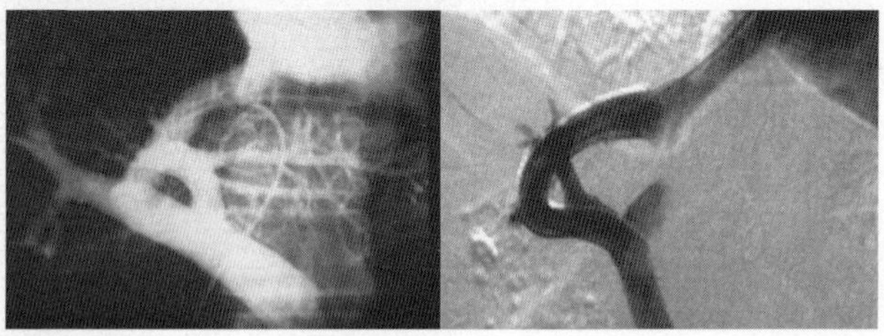

a. **b.**

Figure 10. *Stenting of the parenchymal tract for TIPS can be achieved with a variety of metal stents. a) The first stent used for this purpose was the Palmaz stent. b) Wallstents are more flexible and are the primary metal stent used for this application today.*

9) The Wallstent is initially dilated to 8 mm using an angioplasty balloon, as residual narrowing of the stent is invariably present despite predilatation of the track to this size. If the portal pressure gradient is 12 mm Hg or greater, or prominent filling of varices continues to be seen on the portal venogram, the stent is further dilated with a 10-mm-diameter balloon catheter. If varices are still filling briskly after dilating to 10 mm, the varices are selectively catheterized and embolized. In about 10% of patients, a 10-mm-diameter shunt does not produce sufficient reduction in the portosystemic gradient to prevent recurrent bleeding (< 12 mm Hg residual gradient). In these cases, further reduction in portal pressure could be obtained by performing a second parallel TIPS from a different hepatic vein to a portal vein branch in the opposite lobe of the liver. Alternatively, if a 12-mm Wallstent was selected, it can be dilated with a 12-mm balloon. A typical TIPS case is shown below **(Fig. 11)**.

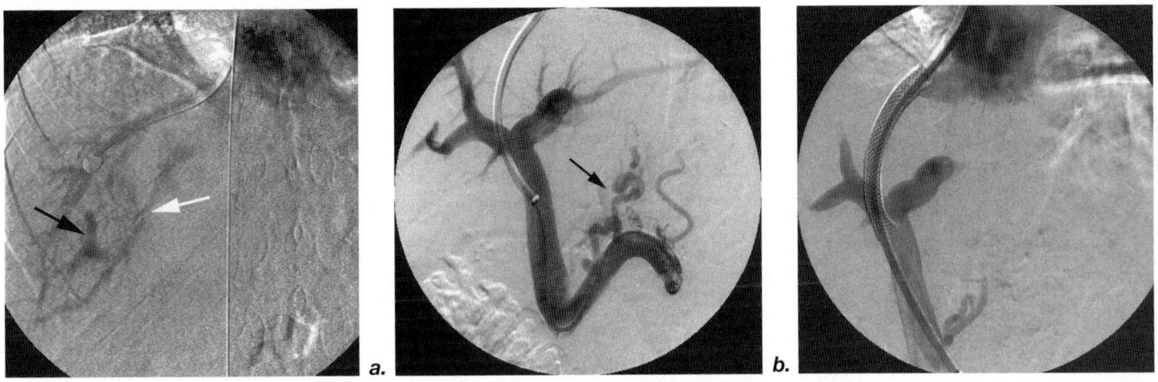
a. b. c.

Figure 11. *TIPS was performed in a 50-year-old man with intractable ascites. a) Wedged hepatic venogram was performed through an occlusion balloon catheter. Note opacification of the middle hepatic vein (white arrow) as well as a branch of the right portal vein (black arrow). b) After successful puncture of the right portal vein, a catheter was advanced into the splenic vein. Venography reveals varices (arrow). c) Completion venogram shows flow through the TIPS as well as continued flow into the intrahepatic portal vein branches (hepatopedal flow).*

Short-Term Complications

Procedural morbidity from TIPS occurs in up to 10% of cases as detailed in a thorough review of TIPS complications by Freedman et al (Radiographics, 1992.) Nonfatal procedure-related complications include self-limited intraperitoneal hemorrhage (1%-6%), hemobilia (1%-4%), sepsis, transient oliguric renal failure, and atrial arrhythmias. Increases in serum bilirubin, transaminases, prothrombin time and ammonia, that peak within the first week after TIPS and subsequently resolve, occur in 10%-20% of patients. Stent migration or marked stent misplacement have been encountered occasionally during the course of TIPS procedures.

PROCEDURE-RELATED COMPLICATIONS

	UCSF	Freiburg
• Bleeding	2	13
– Intraabdominal	1	6
– Hemobilia	1	4
– Subcapsular	-	3
• Fever	10	-
• Renal failure	3	-
• Stent migration	0	2

Very few direct procedure-related deaths have been reported. Procedural deaths from intraperitoneal hemorrhage occur only rarely, even when the liver capsule is traversed during the procedure. Deaths attributable to hepatic arterial injuries have been observed. Central venous pressure rises after TIPS, and heart failure has also been reported as a complication of the procedure. One patient at UCSF with severe underlying pulmonary hypertension died from acute cor pulmonale after TIPS placement. Death from hepatic failure has been observed in 3% to 7% of patients in the acute postprocedural period (first 30 days). Diminished hepatic perfusion from diversion of portal venous blood has been a suspected contributory factor.

Thirty-day mortality is variable and appears to be related to the underlying condition of the patient rather than the morbidity of the procedure itself. Thirty-day mortality rates after TIPS are from 3% to 15%. At UCSF, we have found the strongest predictor of early death is Child's-Pugh score > 12. Others have correlated early death with APACHE II score > 20.

Long-Term Complications

The chief long-term complications of TIPS are portosystemic encephalopathy (PSE), shunt stenosis and recurrent bleeding due to shunt dysfunction. The reported rates of these complications in nonrandomized and randomized trials are summarized in **Tables 3-5**. In general, shunt stenosis is common, occurring in as many as 50% of patients within the first 6 months to one year **(Fig. 12)**. Recurrent stenosis may lead to shunt dysfunction and recurrent bleeding. Recurrent variceal bleeding may occur in 10%-25% of patients at 1 year.

New or worsened encephalopathy occurs in 25%-30% of patients following TIPS. Patients at highest risk for this complication are elderly (age greater than 65 years). In most cases, symptoms are mild and readily managed with dietary protein restriction and lactulose therapy. In fewer than 5%, uncontrolled encephalopathy poses a major clinical problem. Intractable shunt-related encephalopathy can be treated by shunt revision. The TIPS can be occluded by temporarily placing a balloon occlusion catheter aross the shunt for 8-12 hours **(Fig. 13)**. We have successfully used this technique to relieve encephalopathy in five patients. The creation of an intentional intrashunt stenosis has also been reported.

Shunt Revision

Percutaneous shunt revision is an effective means of treating TIPS dysfunction and of maintaining shunt patency. Stenotic shunts may be treated by balloon dilatation alone **(Fig. 14)** or by restenting and dilatation **(Fig. 15)**. Technical success approaches 100%. Revision of TIPS occlusion is successful in over 80% of patients. Recanalized shunts are usually restented and then dilated. Acute thrombus may be redirected into varices using a balloon occlusion catheter. Extension of thrombus into the main portal vein is a serious but uncommon complication of shunt occlusion that can be effectively treated during shunt revision.

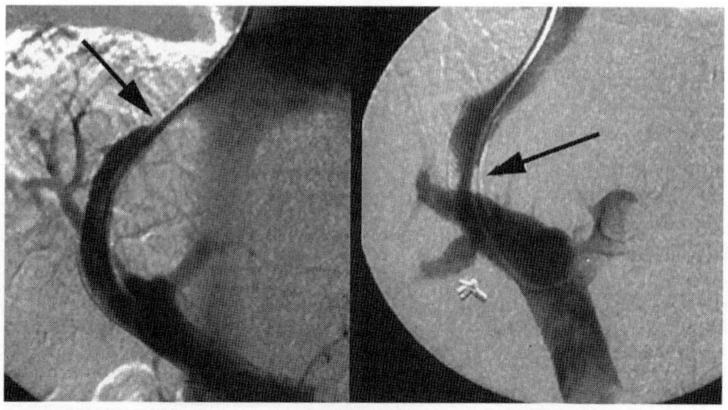

a. b.

Figure 12. Shunt stenosis following TIPS. a) The most common location for stenosis is in the hepatic vein (arrow). b) Less commonly stenosis occurs within the parenchymal tract (arrow).

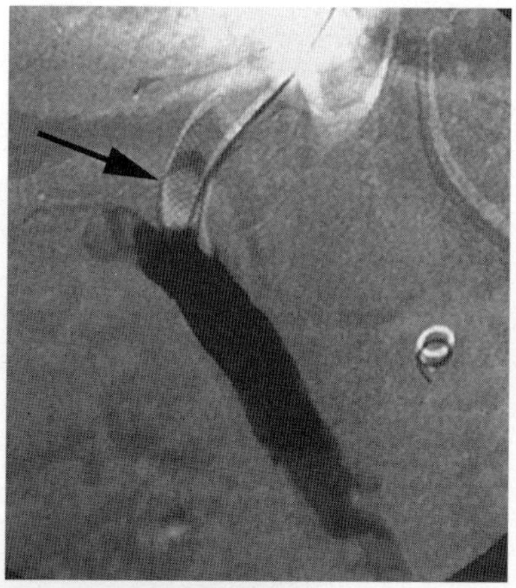

Figure 13. This patient developed uncontrollable encephalopathy following TIPS. The shunt was occluded by placing an occlusion ballloon (arrow) within the middle of the shunt and keeping it inflated overnight.

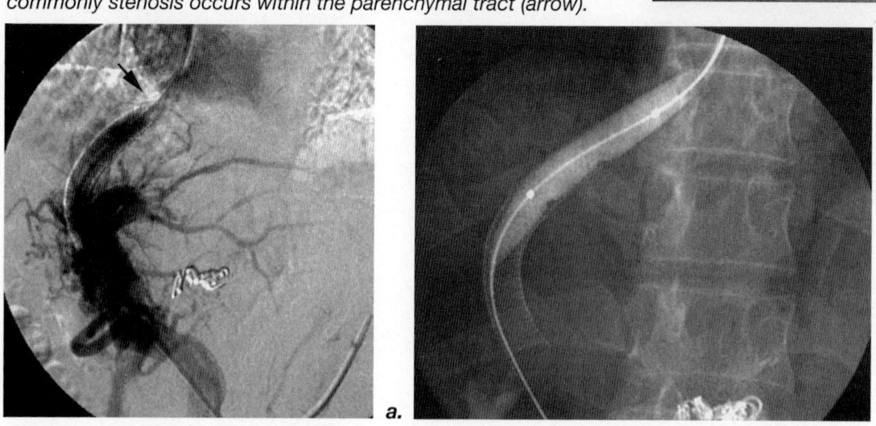

a. b.

Figure 14. This patient developed recurrent variceal bleeding 9 months after TIPS. a) Shunt venogram reveals a central hepatic vein stenosis (arrow). b) The stenosis was treated by balloon angioplasty performed from a transjugular approach.

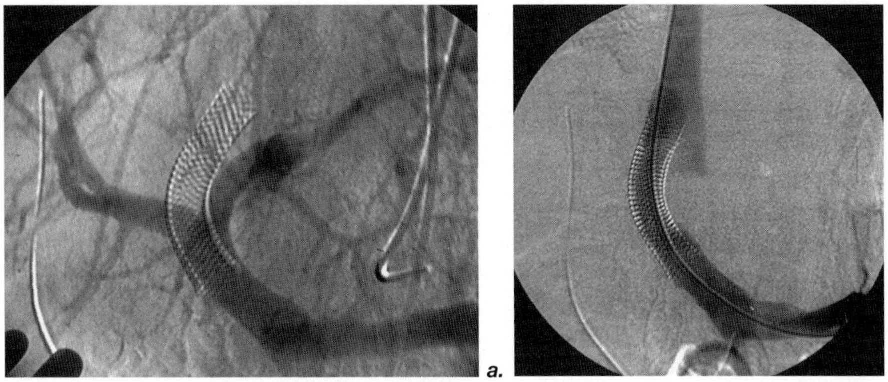

a. b. ***Figure 15.*** *This patient developed shunt occlusion and recurrent bleeding 1 year after TIPS. a) Arterial portogram reveals TIPS occlusion. b) The shunt was revised from a transjugular approach by deployment of a new stent within the occluded shunt.*

New and promising techniques for the prevention of TIPS stenosis include the use of PTFE-covered stents and the creation of direct portacaval shunts using covered stent technology.

Summary of Published TIPS Results

The most important clinical results have been mentioned above. **Tables 3-5** provide a summary of the results from the largest published series to date.

TABLE 3: EARLY RESULTS IN LARGE SERIES

Series	*LaBerge '93 (N=100)*	*Helton '93 (N=59)*	*Martin '93 (N=45)*	*Rossle '94 (N=100)*
Successful TIPS	96 (96%)	55 (93%)	45	93
Immediate clinical success	94 (94%)	48 (81%)	42 (93%)	93 (93%)
Procedural complications	9 (9%)	8 (14%)	0 (0%)	15 (15%)
Fatal procedural complications	0 (0%)	2 (3%)	0 (0%)	1 (1%)
Recurrent bleeding	17/88 (19%)	17/55 (31%)	3/44 (7%)	10 (18%)
New or worse encephalopathy	17/96 (18%)	10/59 (17%)	16/45 (36%)	23/93 (25%)

TABLE 4: TIPS PATENCY AT 1 YEAR

Series	*LaBerge '95*	*Haskal '94*	*Lind '94*
Primary	66%	50%	25%
Primary-assisted	83%	85%	83%

TABLE 5: ENDOSCOPIC SCLEROTHERAPY (ES) VERSUS TIPS FOR VARICEAL BLEEDING

Study	*PTS*		*Re-bled*		*PSE*		*Deaths*	
	ES	*TIPS*	*ES*	*TIPS*	*ES*	*TIPS*	*ES*	*TIPS*
Sanyal '97	39	41	27%	24%	13%	29%*	18%	29%*
Merli '96	38	37	50%	16%*	18%	54%*	18%	22%
Sauer '96	41	42	44%	9%*	10%	29%*	-	-
Cabrera '96	32	31	51%	23%*	13%	33%*	16%	16%
Rossle '97	65	61	41%	15%*	18%	36%*	12%	13%
Geaih '95	33	32	61%	41%	-	-	42%	50%
Cello '97	25	24	48%	13%*	44%	50%	32%	33%

*(Figures marked * are significantly different)*

TIPS *Indications and Contraindications*

Indications and contraindications for TIPS were evaluated at a conference sponsored by the National Digestive Diseases Advisory Board in February 1994. In their report, the board drew the following conclusions:

a) The two accepted indications for TIPS are: (1) acute variceal bleeding that cannot be successfully controlled with medical treatment including sclerotherapy and (2) recurrent variceal bleeding in patients who are refractory or intolerant to conventional medical management including sclerotherapy and pharmocologic therapy.

b) Additional indications which are unproven but promising uses for TIPS include: (1) therapy for refractory ascites and (2) portal decompression for patients suffering from hepatic venous outflow obstruction (Budd-Chiari syndrome).

c) TIPS is not recommended for: (1) initial therapy of acute variceal hemorrhage, (2) initial therapy to prevent initial or recurrent variceal hemorrhage, and (3) reduction of intraoperative morbidity during liver transplant surgery.

d) Absolute contraindications for TIPS placement are (1) right-sided heart failure with marked elevation of the central venous pressure, (2) polycystic liver disease and (3) severe hepatic failure. Relative contraindications include: (1) active intrahepatic or systemic infection, (2) severe hepatic encephalopathy poorly controlled by medical therapy, (3) hypervascular hepatic tumors, and (4) portal vein thrombosis. However, indications continue to evolve as more data emerge.

These 1994 recommendations remain generally accepted today.

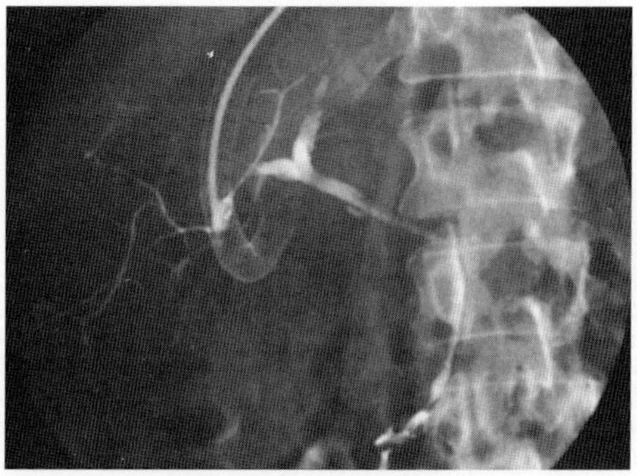

Figure 16. *Some investigators believe that shunt stenosis may be due, in part, to bile duct injury. In this case, injection of contrast within the occluded shunt resulted in opacification of the biliary tree indicating that the shunt had transected a bile duct.*

FUTURE DIRECTIONS

All clinical studies have shown a very high rate of TIPS failure as a result of restenosis or occlusion. The unresolved research issues are a) how to follow-up and detect shunt failure and b) how to prevent it from occurring or successfully treat it. Although many centers are working on these problems, no definitive solutions are yet available.

The possibility that bile duct injury may lead to TIPS stenosis **(Fig. 16)** has prompted the investigation of "covered" stents for TIPS. Metal stents can be covered with a variety of synthetic materials and inserted through transjugular sheaths only slightly larger than those required for delivery of "bare metal" stents. Theoretically, the synthetic covering prevents bile from entering the shunt lumen and causing stenosis. To date, the most promising results have been achieved with polytetrafluoroethylene (PTFE) covered stents **(Fig. 17)**.

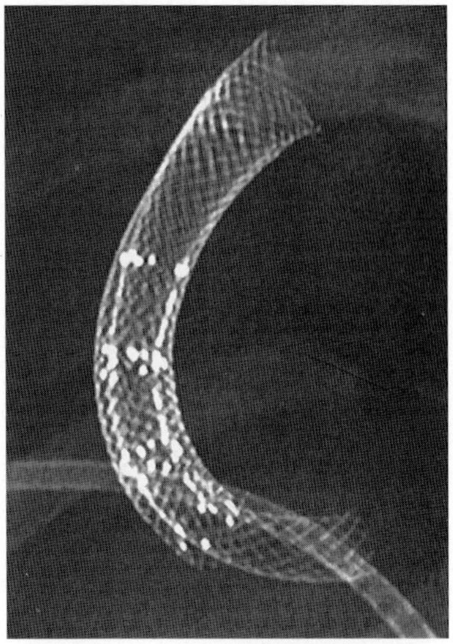

a.

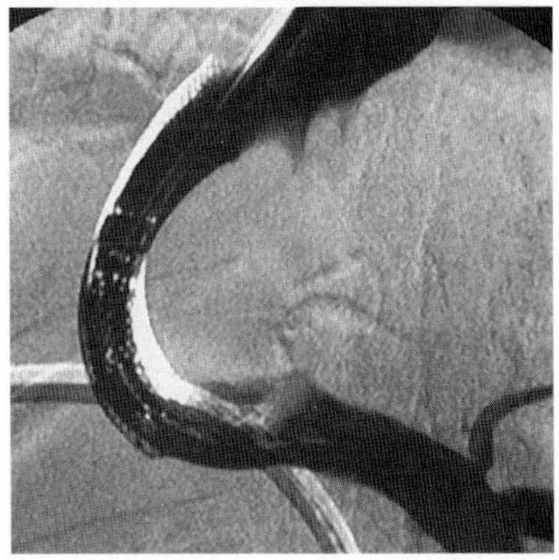

b.

Figure 17. *Covered stents may prevent shunt stenosis due to bile duct injry. In this case shunt occlusion was treated by placement of a PTFE-covered Gianturco stent within the parenchymal tract. a) The Gianturco stent is visible within the preexisting Wallstent. b) Venogram performed following revision shows good flow through the stent.*

REFERENCES

Cabrera J, Maynar M, Granados R, et al. Transjugular intrahepatic portosystemic shunt versus sclerotherapy in the elective treatment of variceal hemorrhage. Gastroenterology 1996; 110:832-839.

Cello JP, Ring EJ, Olcott EW, et al. Endoscopic sclerotherapy compared with percutaneous transjugular intrahepatic portosystemic shunt after initial sclerotherapy in patients with acute variceal hemorrhage: a randomized, controlled trial. Ann Intern Med 1997; 126:858-865.

DiMagno EP, Zinsmeister AR, Larson DE, et al. Influence of hepatic reserve and cause of esophageal varices on survival and rebleeding before and after the introduction of sclerotherapy: a retrospective analysis. Mayo Clinic Proceedings 1985; 60(3):149-157.

Freedman AM, Sanyal AJ, Tisnado J, et al. Complications of transjugular intrahepatic portosystemic shunt; a comprehensive review. Radiographics 1993; 13:1185-210.

Groupe D'Etude des Anastomoses Intra-Hepatiques. TIPS vs. sclerotherapy and propanolol in the prevention of variceal rebleeding: preliminary results of a multicenter randomized trial. Hepatology 1995; 22:297A.

Haskal ZJ, Pentecost MJ, Soulen MC, Shlansky-Goldberg RD, Baum RA, Cope C. Transjugular intrahepatic portosystemic shunt stenosis and revision. AJR 1994; 163:439-444.

Helton WS, Belshaw A, Althaus S, Park S, Coldwell D, Johansen K. Critical appraisal of the angiographic portacaval shunt (TIPS). Am J Surg 1993; 165:566-571.

LaBerge JM, Ring EJ, Gordon RL, et al. Creation of transjugular intrahepatic portosystemic shunts with the Wallstent endoprosthesis: results in 100 patients. Radiology 1993; 187:413-420.

LaBerge JM, Somberg KA, Lake JR, et al. Two-year outcome following transjugular intrahepatic portosystemic shunt for variceal bleeding: results in 90 patients. Gastroenterology 1995; 108:1143-1151.

Lind CD, Malisch TW, Chong WK, et al. Incidence of shunt occlusion or stenosis following transjugular intrahepatic portosystemic shunt placement. Gastroenterology 1994; 106:1277-1283.

Martin M, Zajko AB, Orons PD, et al. Transjugular intrahepatic portosystemic shunt in the management of variceal bleeding: indications and clinical results. Surgery 1993; 114:719-727.

Merli M, Riggio O, Capocaccia L, et al. TIPS vs. sclerotherapy for prevention of variceal rebleeding: results of a randomized controlled trial. Gastroenterology 1996; 110:A1265.

Rössle M, Deibert P, Haag K, et al. Randomised trial of transjugular-intrahepatic-portosystemic shunt versus endoscopy plus propranolol for prevention of variceal rebleeding. Lancet 1997; 349:1043-1049.

Rössle M, Haag K, Ochs A, et al. The transjugular intrahepatic portosystemic stent-shunt procedure for variceal bleeding. N Engl J Med 1994; 330:165-171.

Sanyal AJ, Freedman AM, Luketic VA, et al. Transjugular intrahepatic portosystemic shunts compared with endoscopic sclerotherapy for the prevention of recurrent variceal hemorrhage: a randomized, controlled trial. Ann Intern Med 1997; 126:849-857.

Sauer P, Theilmann L, Bent C, Roeren T, Richter G, Stremmel W, Stiehl A. Transjugular intrahepatic portosystemic stent shunt (TIPS) vs. sclerotherapy in the prevention of variceal rebleeding: a randomized study. Gastroenterology 1996; 110:A1313.

Shiffman ML, Jeffers L, Hoofnagle JH, Tralka TS. The role of transjugular intrahepatic portosystemic shunt (TIPS) for treatment of portal hypertension and its complications. Proceedings from National Digestive Disease Advisory Board Conference, February 1994.

Van Stiegman G. The role of surgery in the management of patients with variceal hemorrhage. In: LaBerge JM, Durham JD, eds. SCVIR Syllabus: Portal Hypertension: Options for Diagnosis and Treatment. 1st edition. San Francisco: Society of Cardiovascular & Interventional Radiology, 1995; 25.

CHAPTER 18 PREVIEW ANSWERS

Case 1

MODIFIED CHILD'S-PUGH SCORING SYSTEM*

Variable	***Points for Severity of Each Variable*** (1)	(2)	(3)
Encephalopathy	none	moderate	severe
Ascites	none	moderate	severe
Bilirubin (mg/dl)	<2	2-3	>3
Albumin (g/dl)	≥3.5	2.8-3.4	<2.8
Prothrombin time (secs)	≤14	15-17	≥18

(Points are summed for all five categories to determine Child's-Pugh score and Child's class as follows: A=5, 6; B=7, 8, or 9; C=10-15.)*

Figure 1. *Child's-Pugh scoring system.*

Findings: Based on the Child's-Pugh scoring system criteria listed above **(Fig. 1)** the patient would be awarded 1 point for no encephalopathy, 2 points for moderate ascites, 3 points for bilirubin > 3 mg/dl, 3 points for albumin < 2.8 g, and 3 points for PT > 18 seconds. These data yield total score of 12. Options for treatment include endoscopic therapy (vesical band ligation of sclerosis), surgical shunt (portocaval or mesocaval), or TIPS.

Diagnosis: Child's Class C with Child's-Pugh score of 12 successfully treated by TIPS.

Case 2

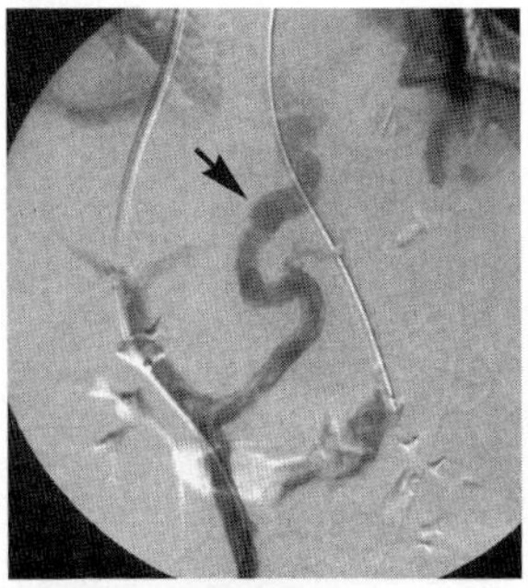

Figure 2. *Transjugular portal venogram, before TIPS.*

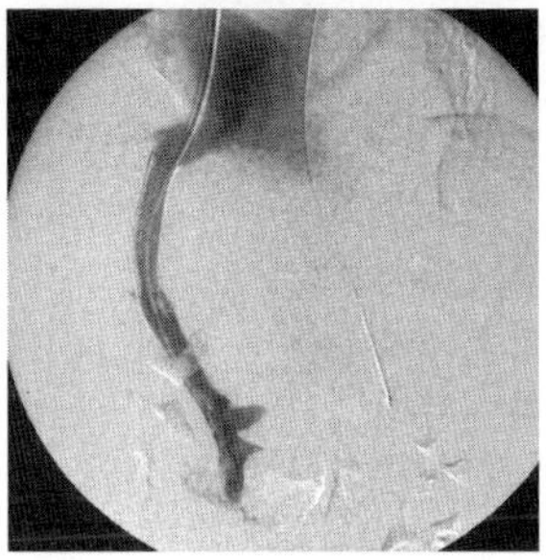

Figure 3. *Transjugular portal venogram, after TIPS.*

Findings: This patient does have portal hypertension. The PSG is greater than 6 mm Hg. and there is filling of portosystemic collaterals on the initial transjugular portal venogram **(Fig. 2)**. In this case, there is a large coronary vein (arrow). In order to effectively control variceal bleeding, TIPS should reduce the PSG to < 12 mm Hg. In this case the final post-TIPS PSG was 10 mm Hg and the patient stopped bleeding **(Fig. 3)**.

Diagnosis: Bleeding secondary to portal hypertension treated by TIPS.

Case 3

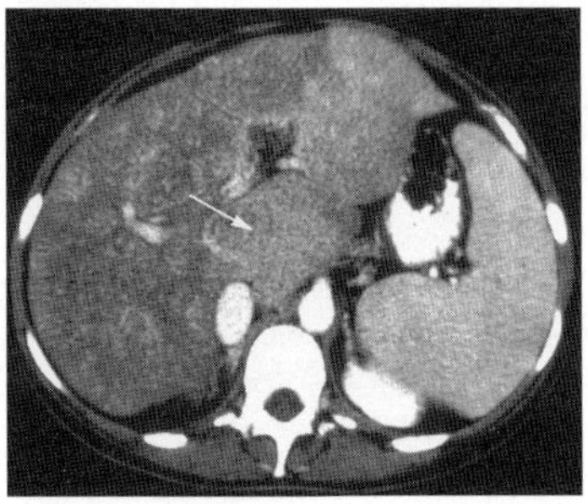

Figure 4. *Abdominal CT.*

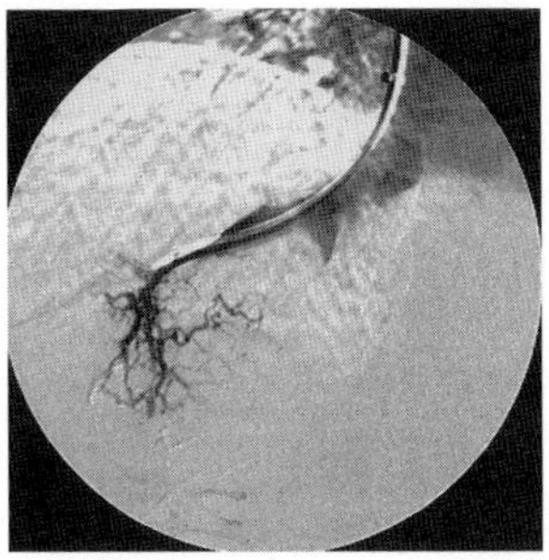

Figure 5. *Transjugular hepatic venogram.*

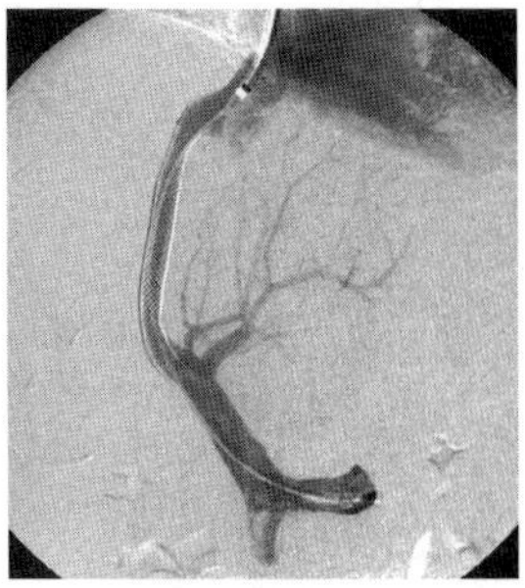

Figure 6. *Completion TIPS venogram.*

Findings: CT **(Fig. 4)** reveals hepatosplenomegaly with an abnormal inhomogeneous enhancement pattern in the right lobe and enlargement of the caudate lobe (arrow). These radiographic findings and the clinical presentation are compatible with Budd-Chiari syndrome (BCS) which is a disorder due to hepatic venous obstruction manifest by the clinical triad of hepatosplenomegaly, abdominal pain and ascites. Subsequently, hepatic venography **(Fig. 5)** revealed a narrow irregular peripheral hepatic vein. The goal of treatment in BCS is to palliate ascites and prevent liver failure. Portosystemic shunts and TIPS have been shown to be effective in relieving ascites. This patient was treated by TIPS **(Fig. 6)** and her ascites resolved.

Diagnosis: Budd-Chiari syndrome treated by TIPS.

CHAPTER 19 PREVIEW

Case 1

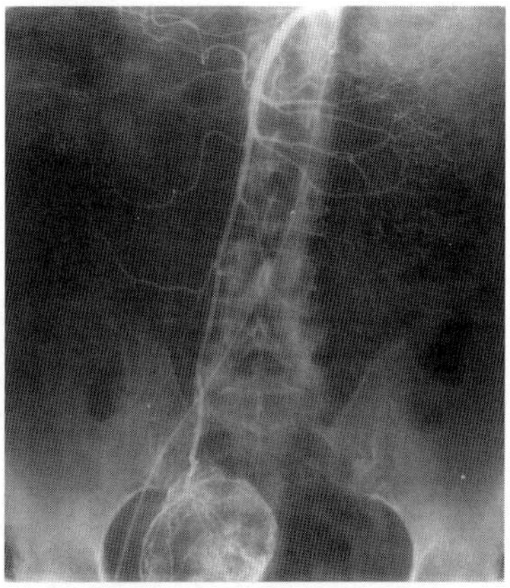

Figure 1a. *SMA arteriogram, early arterial phase.*

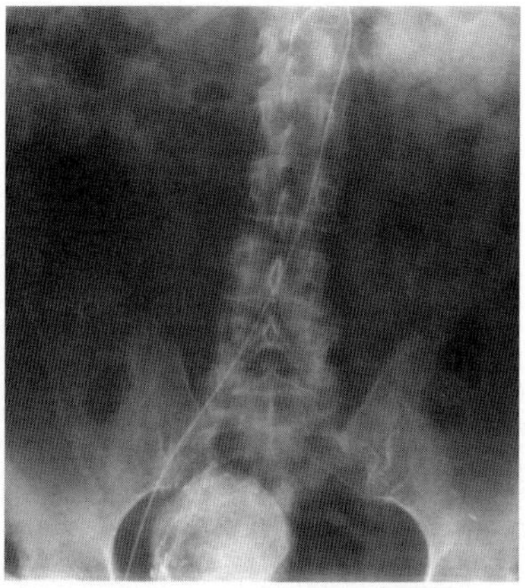

Figure 1b. *SMA arteriogram, late arterial phase.*

History: A 35-year-old woman presented with lower gastrointestinal tract bleeding. A visceral angiogram was performed. Early and late arterial phase images from a superior mesenteric artery (SMA) arteriogram are shown.

Question: What is your diagnosis?

Case 2.

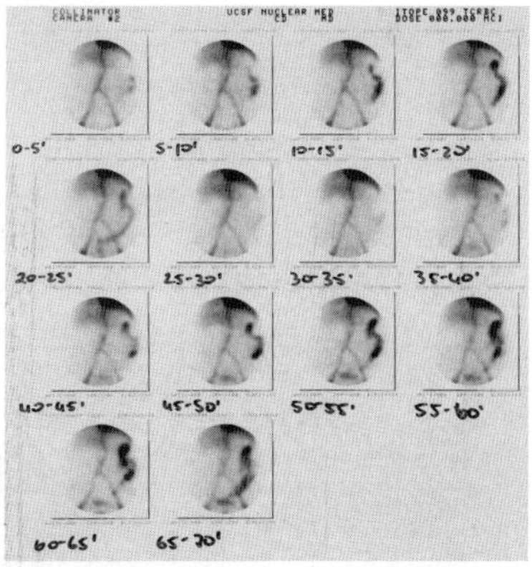

Figure 2a. *Nuclear medicine tagged-RBC study.*

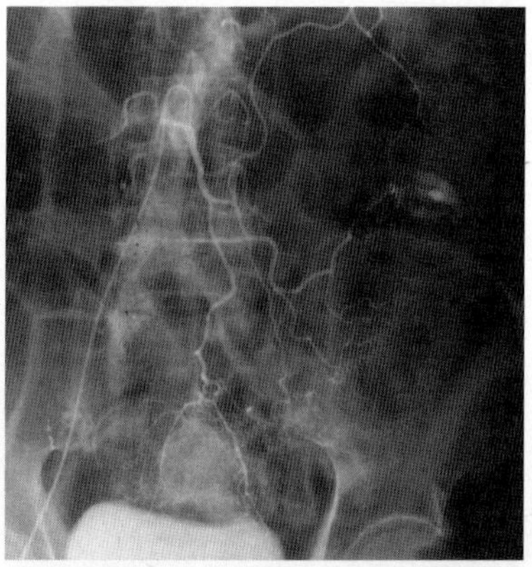

Figure 2b. *Selective IMA arteriogram, late arterial phase.*

History: A 70-year-old man with a history of diverticulosis presented with lower gastrointestinal tract bleeding. A nuclear medicine tagged-red cell scan was obtained and then a visceral angiogram was performed.

Questions: What is your diagnosis? How would you treat this patient?

Chapter 19
Management of Gastrointestinal Bleeding

Jeanne M. LaBerge, MD

CHAPTER OUTLINE

- Etiology of gastrointestinal bleeding
- Angiographic diagnosis
- Transcatheter therapy
  - Vasopressin infusion
  - Embolization

ETIOLOGY OF GASTROINTESTINAL BLEEDING

Gastrointestinal (GI) tract bleeding can be subdivided into the following clinically relevant categories:

1) Arterial
   - Upper GI (esophagus, stomach, duodenum)
   - Lower GI (small bowel, colon)
2) Venous
   - Gastroesophageal varices
   - Mesenteric varices

CATEGORIES OF GASTROINTESTINAL BLEEDING

- Arterial
  - Upper GI (esophagus, stomach, duodenum)
  - Lower GI (small bowel, colon)
- Venous
  - Gastroesophageal
  - Mesenteric

The focus of this chapter is on arterial causes of GI bleeding. The diagnosis and management of portal hypertensive bleeding has been covered in the previous chapter. It is important to note, however, that GI bleeding in patients with portal hypertension may be arterial in origin. In fact, an arterial source of bleeding may be seen in as many as 30% of bleeding patients with varices **(Fig. 1)**.

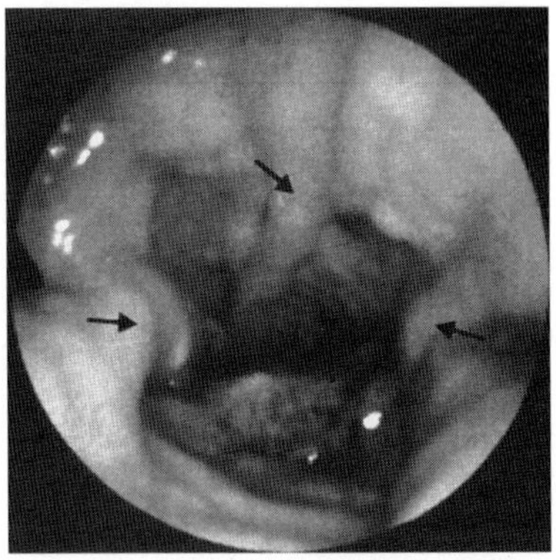

Figure 1. *Upper GI endoscopy in a patient with cirrhosis and recent upper GI bleeding revealed columnar varices (arrows). Complete upper GI endoscopy revealed a duodenal ulcer that was the source of the recent bleeding.*

The distinction between upper and lower GI bleeding is important. For the purposes of this chapter, upper gastrointestinal (UGI) tract bleeding is defined as bleeding from the esophagus, stomach or duodenum; while lower gastrointestinal tract bleeding is defined as bleeding originating from the small bowel (jejunum or ileum), or colon. The arterial supply to the UGI tract is derived chiefly from the celiac (left gastric, gastroduodenal [GDA], and splenic arteries) and the superior mesenteric artery (SMA) [via the inferior pancreaticoduodenal (IPD) artery]. The arterial supply to the lower GI tract is chiefly from the SMA (jejunal and ileal branches, right and middle colic arteries) and inferior mesenteric artery (IMA) (via the left colic and superior hemorrhoidal arteries) and occasionally the internal iliac artery (via the middle and inferior hemorrhoidal arteries).

UPPER GI TRACT: ARTERIAL SUPPLY

- Celiac
  - Left gastric, GDA, hepatic, splenic
- SMA
  - Inferior pancreatico-duodenal (IPD)

LOWER GI TRACT: ARTERIAL SUPPLY

- SMA
  - Jejunal, ileal, ileocolic, right colic, middle colic
- IMA
  - Left colic, superior hemorrhoidal
- Internal iliac
  - Middle hemorrhoidal

An important distinction between the blood supplies to these two regions is that the blood supply to the upper GI tract is redundant and there is considerable potential for collateral flow. The blood supply to the small bowel and colon, on the other hand, is not redundant and there is more limited potential for collateral flow. This anatomic distinction has important therapeutic implications. When embolizing upper GI tract vessels, one must be sure to block collateral inflow and thus it is important to embolize proximal and distal to the bleeding site. When treating lower GI tract bleeding it is important to embolize the bleeding vessel at a distal level (i.e. near the bleeding site) to avoid extensive bowel infarction.

Some common causes of upper and lower GI bleeding are listed in **Table 1**.

TABLE 1: CAUSES OF GI BLEEDING

UPPER GI BLEEDING

- Peptic ulcers: gastric and duodenal
- Mallory-Weiss tear
- Acute hemorrhagic gastritis
- Hemobilia: following transhepatic biliary drainage or sphincterotomy
- AIDS: Kaposi's sarcoma, lymphoma
- Aortoduodenal fistula
- Vascular neoplasm
- Gastroesophageal varices

LOWER GI BLEEDING

- Small Bowel
  - Tumor
  - Arteriovenous malformation
  - Diverticulum
  - Ulcer
  - Trauma
  - Aortoenteric fistula
  - Inflammatory bowel disease
  - Aneurysm
- Colorectal
  - Diverticulum
  - Angiodysplasia
  - Ischemic colitis
  - Tumor
  - Postpolypectomy

ANGIOGRAPHIC DIAGNOSIS OF GI BLEEDING

The goals of angiography in the evaluation of GI bleeding are to 1) confirm the diagnosis of arterial GI bleeding, 2) further localize the bleeding site, and 3) provide transcatheter therapy, if indicated.

GOALS OF ANGIOGRAPHY: EVALUATION OF GI BLEEDING

- Confirm diagnosis of bleeding
- Localize the bleeding site
- Provide transcatheter therapy

The vessel that is thought to be the source of bleeding is selectively catheterized and an angiogram is obtained **(Fig. 2)**. When UGI bleeding is suspected, standard angiographic evaluation includes celiac, gastroduodenal, left gastric and superior mesenteric artery injections. Lower gastrointestinal bleeding is usually evaluated by SMA and IMA injections. Arterial variants should be considered if a bleeding site is not identified on the standard injections (e.g., a middle colic artery originating from the celiac or dorsal pancreatic artery).

Conventional film screen technique is used, when available, for the initial angiographic runs. The detection of bleeding using digital imaging may be confounded by digital artifacts caused by bowel motion. Filming should be continued into the venous phase (for at least 30 seconds) so that extravasation of contrast can be detected. Bleeding at a rate of 0.5 to 1.0 ml/min can be detected with this technique.

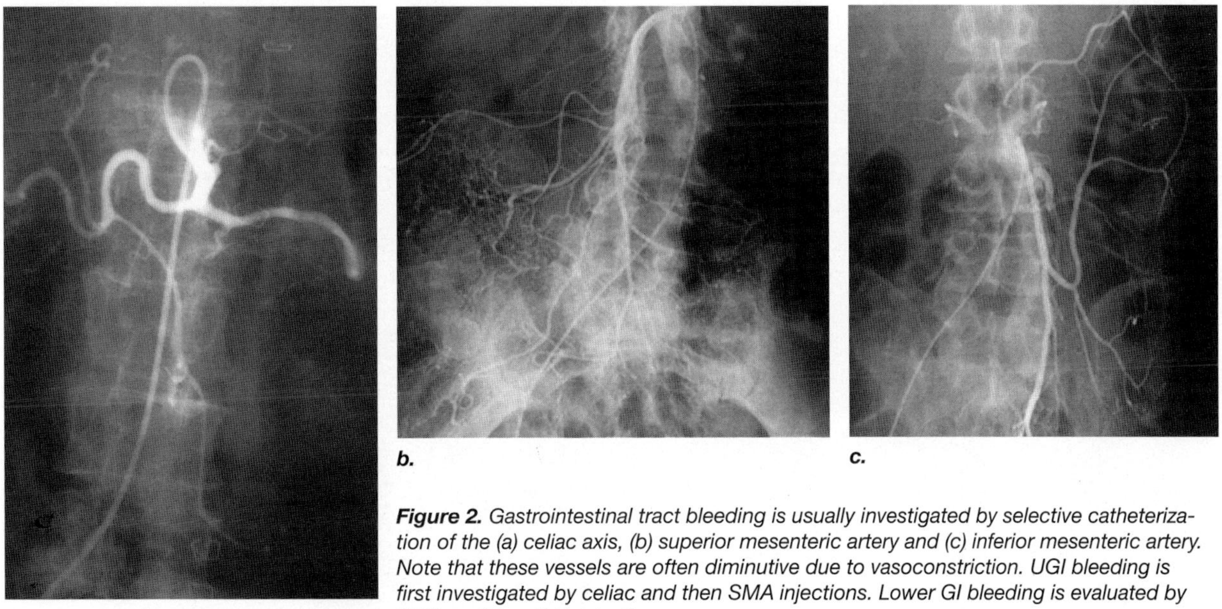

a. b. c.

Figure 2. Gastrointestinal tract bleeding is usually investigated by selective catheterization of the (a) celiac axis, (b) superior mesenteric artery and (c) inferior mesenteric artery. Note that these vessels are often diminutive due to vasoconstriction. UGI bleeding is first investigated by celiac and then SMA injections. Lower GI bleeding is evaluated by SMA and then IMA injections.

> **ANGIOGRAPHIC FINDINGS OF GI BLEEDING**
>
> - Extravasation into bowel
> - Extravasation into a confined space
> - Pseudoaneurysm, arterio-venous fistula
> - Other vascular abnormalities

Angiographic findings of GI bleeding include:

1) extravasation of contrast into the lumen of the bowel **(Fig. 3)**;
2) extravasation of contrast into a confined space outside of the vessel **(Fig. 4)**;
3) demonstration of a pseudoaneurysm (indirect sign);
4) demonstration of an AV fistula (indirect sign);
5) hyperemia, neovascularity or other vascular abnormality (indirect sign).

The angiographic findings in several specific entities are listed below:

• *Mallory-Weiss tear*: Linear collection of contrast at the gastroesophageal junction seen on left gastric arteriography.

• *Hemorrhagic gastritis*: Diffuse hyperemia of the stomach.

• *Hemobilia*: Extravasation of contrast filling the biliary tree. Look for abnormalities of the hepatic arteries to include pseudoaneurysm or abrupt cutoff of a vessel, particularly in the region of suspected trauma (e.g., biliary drainage catheter).

• *Diverticular bleeding*: This is the most common cause of lower GI bleeding in adults. Although left-sided diverticula are more common than right-sided diverticula, bleeding is more commonly observed in right-sided diverticula. Contrast may fill the bleeding diverticulum.

• *Angiodysplasia*: Angiographic findings include early and dense opacification of the ileocolic vein, as well as vascular tufts along the antimesenteric border of the cecum. Angiodysplasia is fairly common in older people and it may be an incidental finding rather than the source of active bleeding.

• *Vascular-enteric fistula*: The duodenum is the most likely site of bleeding. Findings on aortography include a false aneurysm at the site of the proximal anastomosis of an aortic bypass graft or a nipple-like outpouching in this region. Frank extravasation is rarely seen.

• *Small bowel tumors*: This is the most common cause of small bowel bleeding. Extravasation indicates active bleeding but tumor vascularity may also be present.

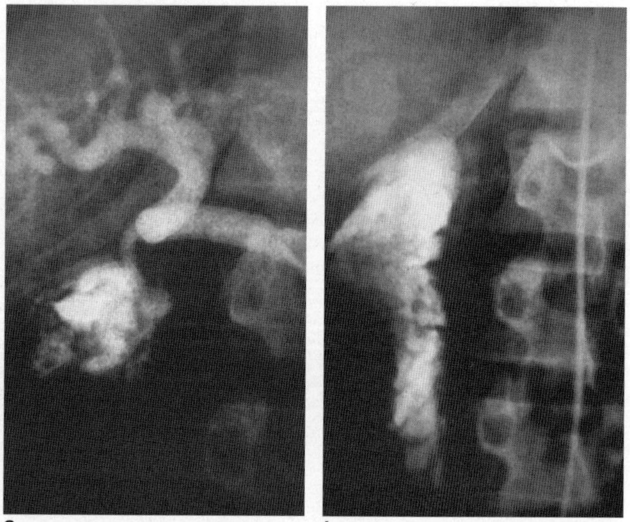

a. b.

Figure 3. Selective hepatic arteriography reveals the location of bleeding in a patient with duodenal ulcer disease. On the early film (a), there is a confined collection of contrast outside of the GDA. On the later film (b), there is frank extravasation of contrast into the lumen of the duodenum.

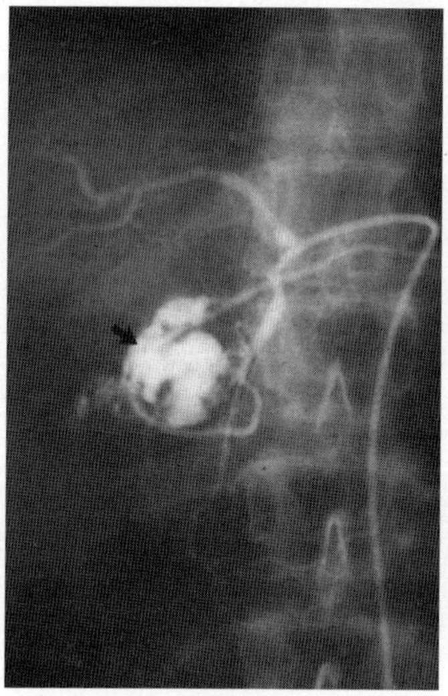

Figure 4. Selective hepatic arteriography shows a confined area of contrast extravasation from the branches of the GDA (arrow) in another patient with a bleeding duodenal ulcer.

Test your understanding of these findings by determining the most likely etiology of bleeding in **Figures 5-7**. (*The diagnoses are given at the bottom of the page.*)

If a bleeding source is not detected during the initial angiographic evaluation, the vessel most likely to be bleeding may be injected again. Some investigators have advocated the use of provocative agents (such as urokinase, tPA or heparin) to induce bleeding when the bleeding source is not apparent. The use of this approach is controversial.

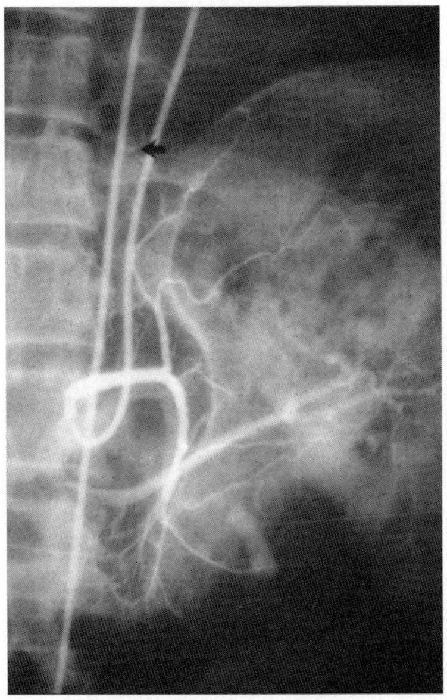

Figure 5. *Selective left gastric arteriogram demonstrates a punctate area of extravasation at the gastro-esophageal junction (arrow).*

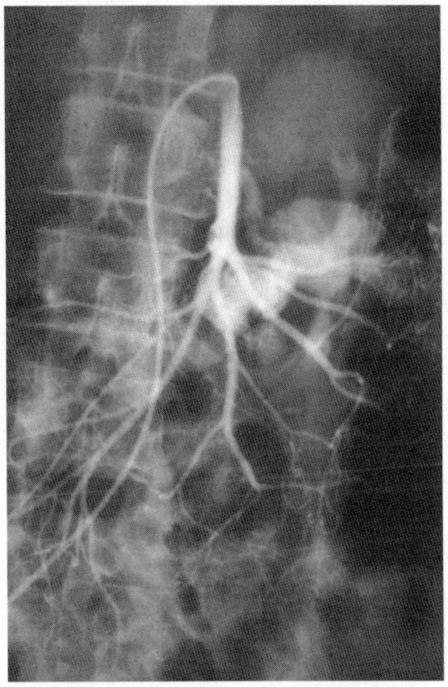

Figure 6. *Selective SMA arteriogram in a young woman with lower GI bleeding shows vascular enhancement of a dumbell-shaped mass and late extravasation of contrast into the proximal small bowel.*

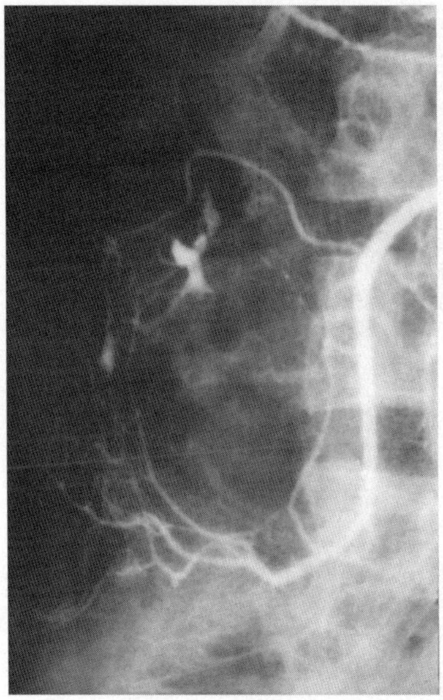

Figure 7. *Magnification view from a selective SMA arteriogram performed in an elderly man with lower GI bleeding shows extravasation from peripheral iliocolic branches.*

Diagnoses

Figure 5: *Mallory-Weiss tear*

Figure 6: *Neurofibroma of the small bowel*

Figure 7: *Actual diagnosis unknown; most likely angiodysplasia. (Could be diverticular disease. Whatever it was, it resolved with vasopressin infusion.)*

The second goal of angiography is to localize the bleeding site once it has been identified. Precise localization can be achieved by superselective injections or by the use of oblique imaging **(Fig. 8)**. If a patient is to undergo surgical resection, placement of a 3-F superselective catheter close to the bleeding site may facilitate identification of the bleeding bowel segment **(Fig. 9)**. At surgery, methylene blue dye can be injected through the superselective catheter, precisely localizing the bleeding site.

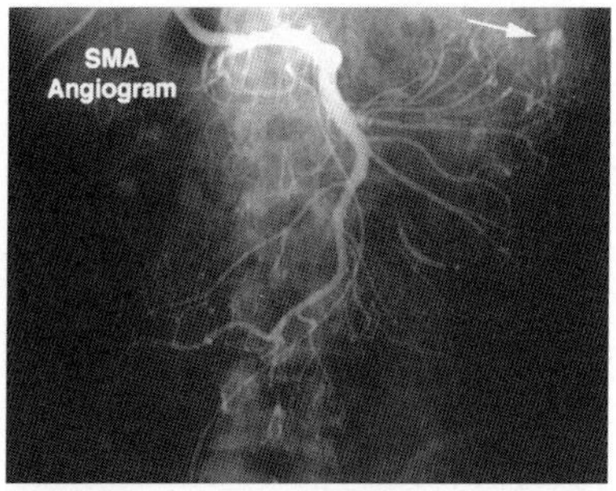

***a.** Early.*

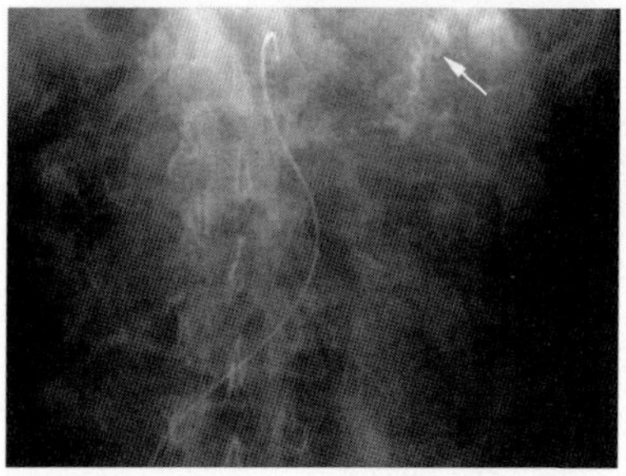

***b.** Late.*

***Figure 8.** SMA arteriogram (AP projection) performed in an elderly man with lower GI bleeding reveals a bleeding site in the left upper quadrant (arrow). An early film (a) shows extravasation and a late film (b) shows filling of the bowel lumen. Is this the small bowel or colon?*

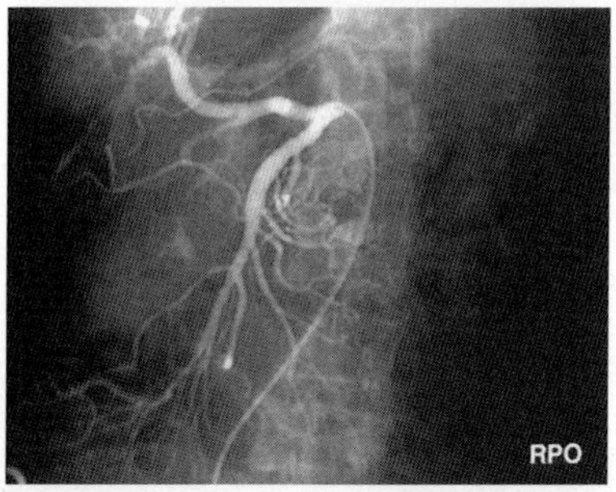

***c.** Early.*

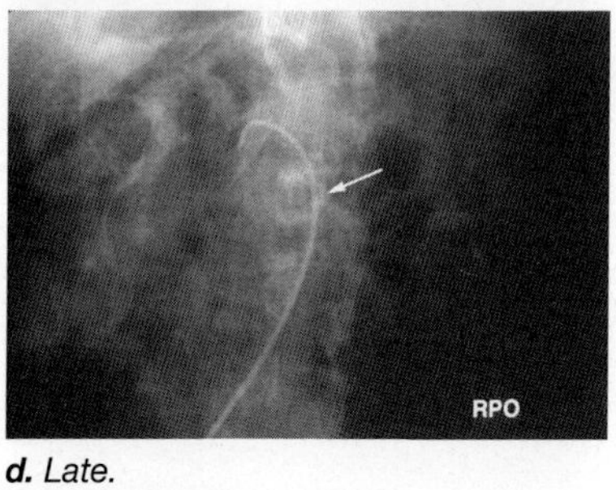

***d.** Late.*

***Figure 8, cont'd.** SMA injection with filming in an oblique projection (c,d) reveals extravasation in a bowel loop ventral to the splenic flexture of the colon. This patient was bleeding from a jejunal ulcer.*

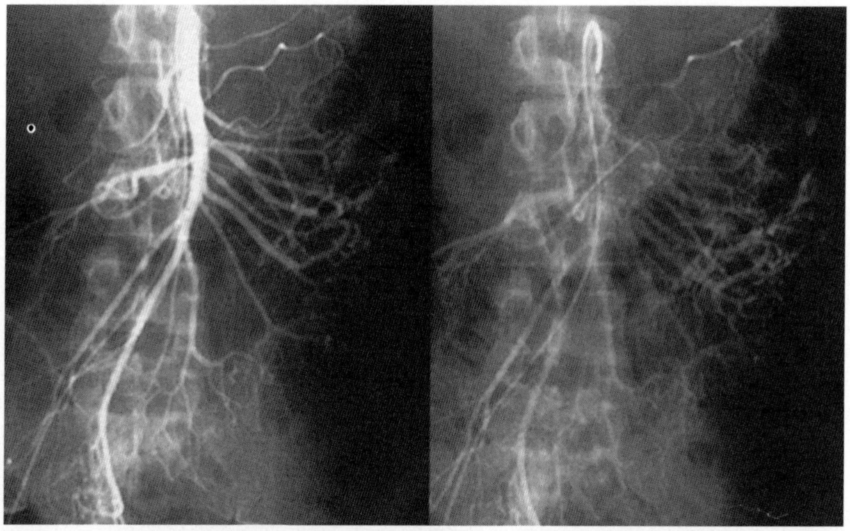

a. **b.**

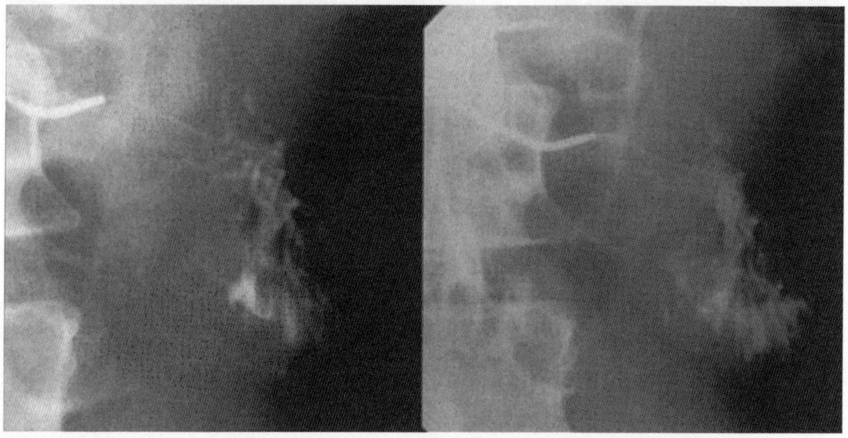

c. **d.**

***Figure 9.** Lower GI bleeding in a young man with AIDS. Initial SMA injection (a,b) suggested bleeding in the proximal small bowel. A 3-F catheter was selectively advanced into the bleeding jejunal artery. Contrast injection revealed extravasation into the small bowel (c,d). The patient was transferred to the operating room where the surgeons injected methylene blue dye through the catheter to reveal the bleeding segment of jejunum.*

TRANSCATHETER THERAPY

Vasopressin

Pharmacologic induction of vasoconstriction is a useful treatment option for select patients with GI bleeding. Embolization is the preferred therapy for UGI bleeding where plentiful collaterals permit this approach. The safety of embolization for lower GI bleeding is less well documented and some experts would argue that vasopressin infusion is preferable to embolization for lower GI bleeding. Selective infusion of vasopressin into the main trunk of the SMA or IMA can arrest bleeding in the majority of cases. A protocol for vasopressin infusion is shown in **Table 2**.

TABLE 2: VASOPRESSIN INFUSION PROTOCOL

1. Infuse 0.2 U/min for 20 minutes into main SMA or IMA. (Mix 100 U vasopressin in 500 ml normal saline and infuse with an arterial infusion pump at a rate of 60 ml/hr)
2. Repeat angiography and assess for extravasation or excessive vasoconstriction. If extravasation is identified, increase the dose to 0.4 U/min. If excessive constriction is identified, reduce the dose to 0.1 U/min. Repeat the angiography in 20 minutes. Do not increase the dose above 0.4 U/min.
3. Infuse at the starting dose for 24 hrs and then taper the dose over 24-48 hrs. Then infuse with saline for 12-24 hrs.
4. Patients must be monitored in the ICU for bleeding, cardiac toxicity (arrhythmias, infarction), systemic hypertension, electrolyte disturbances, bowel ischemia, and peripheral vascular ischemia.

Complications of vasopressin include: hypertension, bradycardia, coronary ischemia, arrhythmias, pulmonary edema, oliguria, hyponatremia, peripheral ischemia and bowel cramps. Rebleeding may occur in 20% of patients upon withdrawal of the infusion.

The advantages of vasopressin are: 1) it does not require subselective catheterization of distal branches and, therefore, it is technically less difficult than embolization, 2) drug-induced side effects are almost always reversible, and 3) bowel infarction occurs infrequently.

EMBOLIZATION OF GI BLEEDING: RESULTS

Etiology	Clinical Success
Mallory-Weiss	80%
Duodenal ulcer	50%-70%
Lower GI	? 75%

Embolization

Embolization is an acceptable nonoperative treatment for most UGI bleeds and some experts now argue that it is a viable nonoperative treatment for lower GI bleeding as well. The technique of transcatheter embolization is covered in detail in Chapter 13, and only a few additional comments will be included here. In general, gelfoam pledgets (Upjohn, Kalamazoo, Mich.) and coils are the agents that are used most frequently for this purpose at UCSF.

Gastric bleeding occurs from branches of the left gastric artery in over 85% of patients. Therefore, if a gastric source of hemorrhage has been adequately documented, it is reasonable to occlude the left gastric artery even if the specific bleeding site is not identified **(Fig. 10)**. Gastric bleeding from Mallory-Weiss tears, gastritis, or gastric ulcer can be controlled in over 80% of patients with embolotherapy. However, in the presence of a significant coagulopathy, a lower success rate may be anticipated.

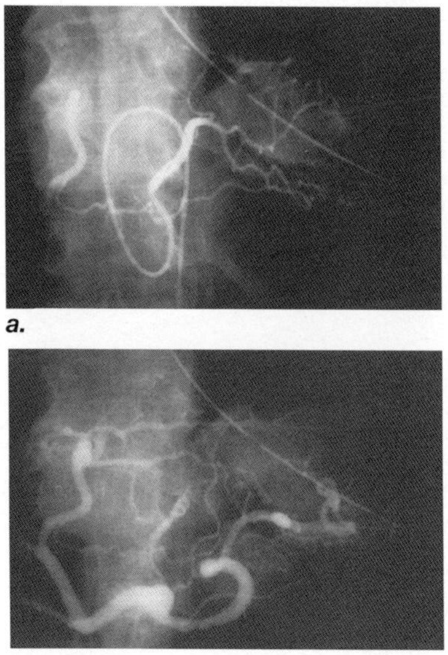

a.

b.

Figure 10. *A patient with a bleeding gastric ulcer by endoscopy underwent arteriography. No definite bleeding site was identified but the left gastric artery was preemptively embolized to prevent further bleeding. a) An enlarged left gastric artery was selectively catheterized and embolized with coils. b) The post-embolization celiac angiogram demonstrates occlusion of the left gastric artery.*

Bleeding from the duodenum is more difficult to control by embolization. This difficulty is engendered by the extensive necrosis associated with many duodenal ulcers as well as the rich arterial anastomoses in the duodenal region. The arterial arcades that extend between branches of the gastroduodenal artery and the superior mesenteric artery often require embolization both proximal and distal to the bleeding site to achieve hemostasis. For these reasons, angiographic interventions leads to successful control of bleeding in only 50% to 70% of patients bleeding from duodenal ulcers.

Embolization to control bleeding in the small bowel and colon remains controversial. At UCSF, superselective embolizaton has, for the most part, replaced vasopressin infusion to control lower gastrointestinal hemorrhage **(Fig. 11)**. Infusion of vasopressin has been reserved for patients in whom superselective catheterization cannot be performed. Coaxial microcatheter systems are used for superselective catheterization and embolization. Platinum microcoils are delivered into the bleeding artery (which is usually less than a millimeter in diameter). The high fluoroscopic visibility of these coils allows precise positioning of the occluding device. It is possible to control small and large bowel bleeding using this embolization technique in up to 80% of patients when the bleeding site is identified.

There is clearly a risk of inducing bowel infarction when embolization is performed in the branches of the superior or inferior mesenteric artery. Infarction may be precipitated in 5% to 20% of cases and is usually evident on clinical grounds. Though postembolization infarction necessitates surgical intervention, the surgical procedure is often identical to that which would have been performed to control the original bleeding. Another, perhaps more important issue is delaying the diagnosis of a potentially curable neoplasm. If tumor is suspected, operative resection of the affected bowel segment is advisable. Whenever embolization is performed, the cause of the bleeding should be investigated further once the bleeding has been controlled. Cross-sectional imaging, endoscopy or operative exploration may be necessary to uncover the underlying cause of bleeding.

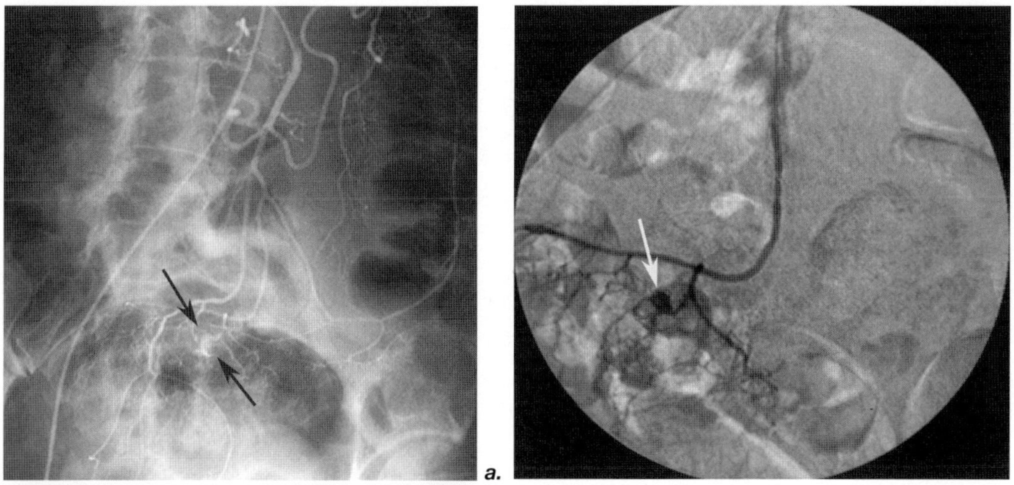

a. b.

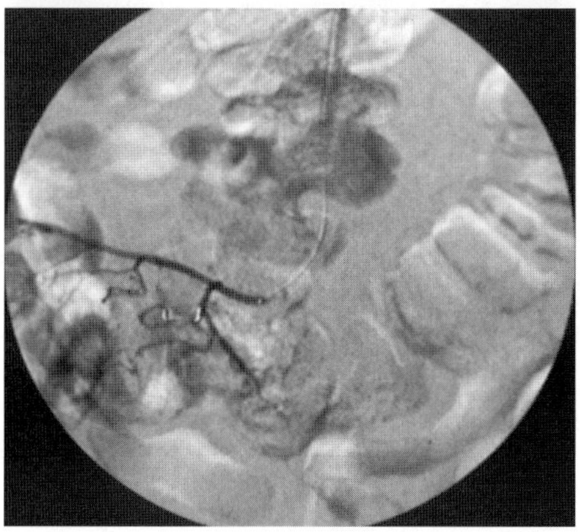

c.

Figure 11. *An elderly man presented with lower GI bleeding. A nuclear medicine tagged RBC study suggested rectosigmoid bleeding. a) Selective IMA injection revealed a bleeding site from a sigmoid branch of the IMA (arrows). b) Sub-selective injection into the bleeding vessel was obtained through a 3-F microcatheter. A small pseudoaneurysm is identified (arrow). c) Microcoils were delivered through the 3-F catheter to selectively embolize the bleeding vessel.*

REFERENCES

Gordon RL, Ahl K, Kerlan RK, et al. Selective arterial embolization for the control of lower gastrointestinal bleeding. Am J Surg 1997; 174:24-28.

Kerlan RK Jr., Pogany AC, Burke DB, Ring EJ. Angiographic management of upper gastrointestinal hemorrhage. AJR 1986; 147:1185-1188.

Rosen RJ, Sanchez G. Angiographic diagnosis and management of gastrointestinal hemorrhage: Current concepts. Radiologic Clinic of North America 1994; 32:951-967.

Wilcox CM, Clark WS. Causes and outcome of upper and lower gastrointestinal bleeding: the Grady experience. Southern Medical Journal 1999; 92:44-50.

CHAPTER 19 PREVIEW ANSWERS

Case 1

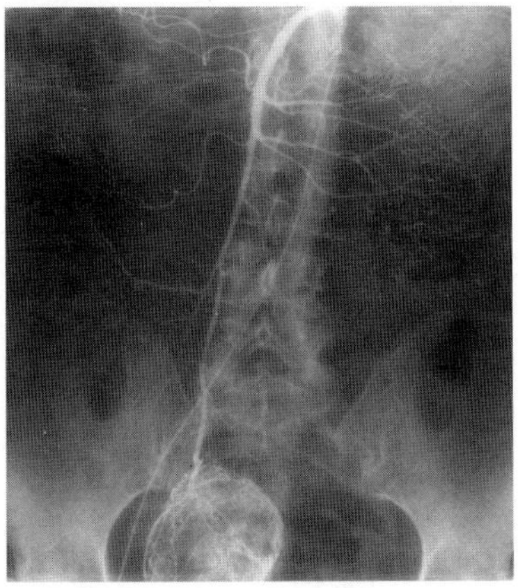

Figure 1a. *SMA arteriogram, early arterial phase.*

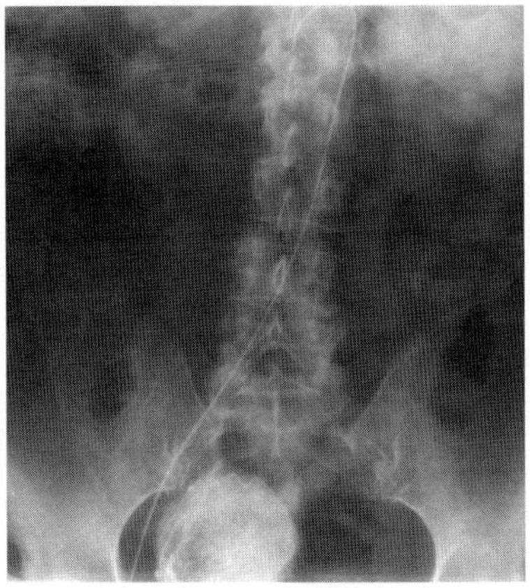

Figure 1b. *SMA arteriogram, late arterial phase.*

Findings: There is a vascular mass in the lower abdomen supplied by an ileal branch of the SMA. The mass is hypervascular and demonstrates dense staining on the late arterial phase image. Differential diagnosis includes small bowel neoplasm, particularly benign smooth muscle or neurogenic tumors.

Diagnosis: Leiomyoma of the small bowel.

Case 2

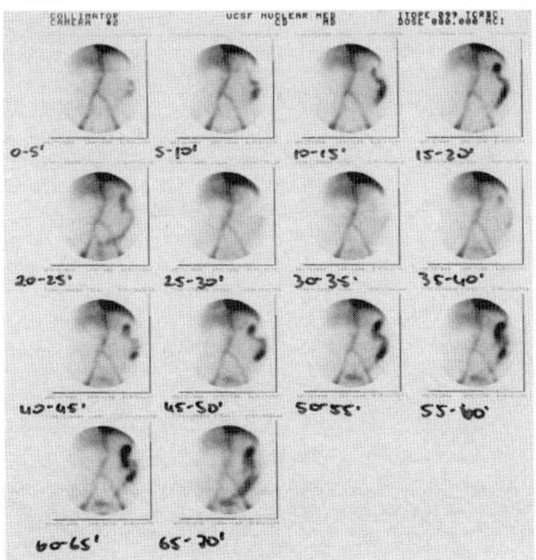

Figure 2a. *Nuclear medicine tagged-RBC study.*

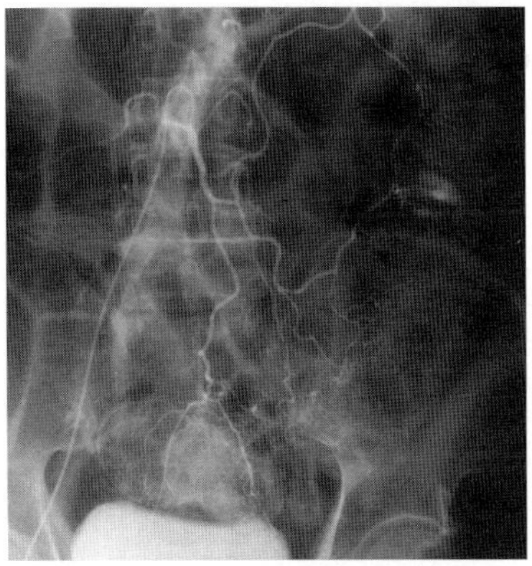

Figure 2b. *Selective IMA arteriogram, late arterial phase.*

Findings: The tagged RBC study shows increased activity in the left mid-abdomen. Over time, the activity is seen to conform to the anatomic configuration of the left colon. Selective IMA angiogram demonstrates extravasation of contrast from the left mid-descending colon. The bleeding vessel was successfully catheterized with a microcatheter and embolized. The patient stopped bleeding. Alternative therapies for this problem include vasopressin infusion or colonic resection.

Diagnosis: Lower gastrointestinal tract bleeding into left colon from left-sided diverticulum. Successful treatment by transcatheter microcoil embolization.

NONVASCULAR INTERVENTIONS

CHAPTER 20 PREVIEW

Case 1

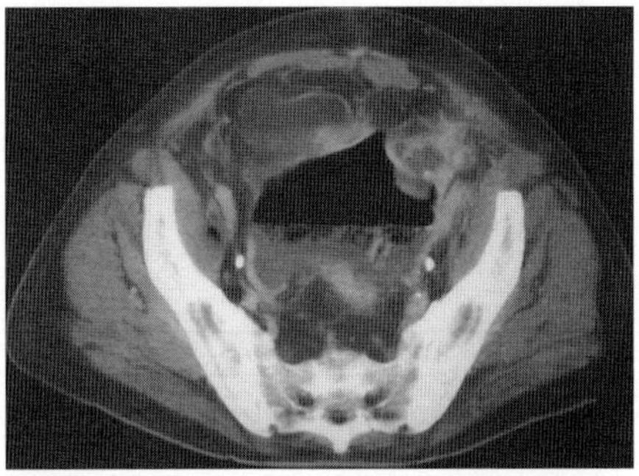

History: Fever and abdominal pain.
Question: Is there a safe approach to drainage of this pelvic abscess?

Case 2

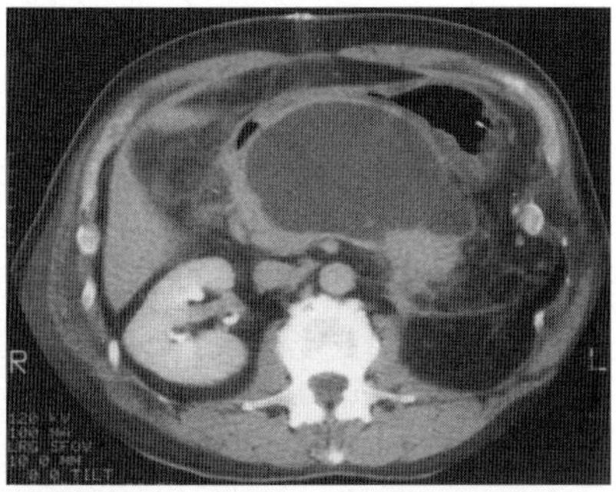

History: Pancreatic pseudocyst.
Question: How could this fluid collection be drained percutaneously?

Case 3

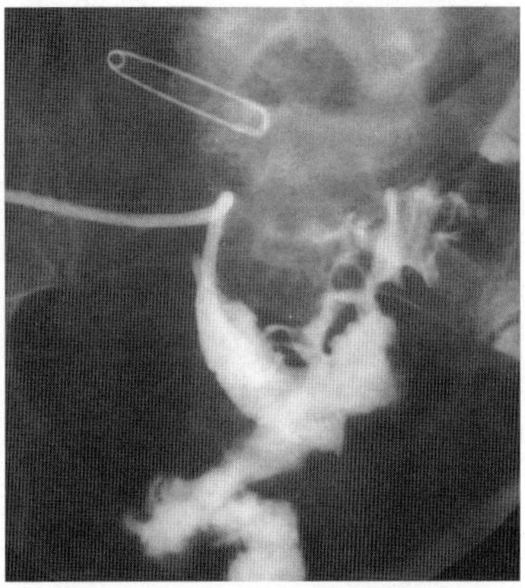

History: Contrast material was injected through this abscess drainage tube 3 days following the percutaneous drainage.
Question: What is the source of the infection?

Case 4

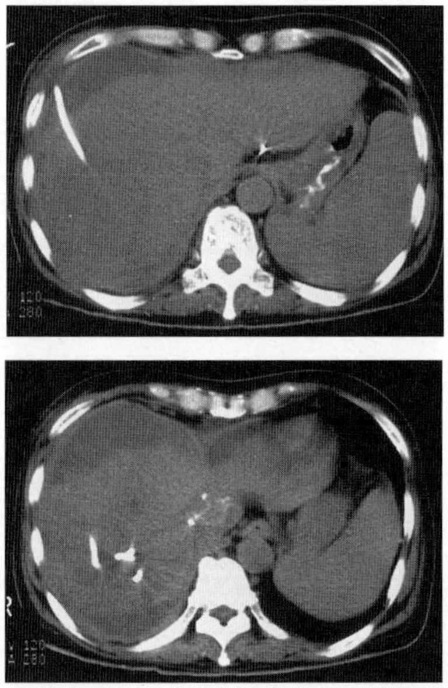

a.

b.

History: This liver transplant patient had percutaneous drainage of a liver abscess.
Question: What problems might be encountered in this patient following the drainage?

Chapter 20
Abscess Drainage

Robert K. Kerlan, Jr., MD

OUTLINE OF ABSCESS DRAINAGE

- Indications
- Contraindications
- Technique
  - Sedation and antibiotics
  - Localization of safe access route
  - Insertion of drainage catheter
  - Postprocedural care and follow-up
  - Assessment of results
  - Removal of tube
- Special considerations
  - Fluid collections complicating pancreatitis
  - Splenic abscesses
  - Hepatic abscesses
  - Pelvic abscesses
- Results
- Complications

INTRODUCTION

Percutaneous abscess drainage is one of the most valuable and frequently performed nonvascular interventions. Fluid collections in almost any region of the body can be approached including the peritoneal cavity, liver, spleen, retroperitoneum (including pancreas), thoracic cavity, mediastinum and neck. Personal experience and preference usually dictates the specific technique used to drain an abscess. A high degree of technical and clinical success can be anticipated.

Intraabdominal abscesses have been treated with both aspiration and drainage. Percutaneous aspiration is defined as evacuation of a fluid collection with either a needle or catheter, with removal of the needle or catheter immediately after withdrawing the fluid. Percutaneous drainage is defined as placement of a catheter to provide continuous drainage of a fluid collection over a period of time.

INDICATIONS

The indications for diagnostic aspiration and percutaneous fluid drainage have been stated by the Society of Cardiovascular & Interventional Radiology Standards of Practice Committee to be:

> "Presence of an abnormal fluid collection with suspicion that the fluid is infected, need for fluid characterization, or suspicion that the collection is producing symptoms sufficient to warrant drainage."

CONTRAINDICATIONS: RELATIVE

- Significant coagulopathy
- Presence of material which cannot be effectively evacuated through a drainage tube such as necrotic debris, hematoma or infected tumor

CONTRAINDICATIONS

The only absolute contraindication to percutaneous drainage is that the drainage catheter would, by necessity, traverse a vital structure (major vessel, heart, or neural structure) and likely cause harm to the patient. All other contraindications are relative and depend upon the suitability of surgical alternatives. These include:

1) Significant coagulopathy
2) Presence of material which cannot be effectively evacuated through a drainage tube such as necrotic debris, hematoma or infected tumor.

TECHNIQUE

The standard technique of percutaneous abscess evacuation includes:

1) Administering conscious sedation and appropriate antibiotic coverage
2) Localizing the fluid collection and safe access route
3) Insertion of a drainage catheter by either the trocar or Seldinger method
4) Postprocedural catheter care and follow-up
5) Assessment of results
6) Removal of tube.

Sedation and Antibiotics

Though superficial abscesses may be drained solely with local anesthesia, deeper and parenchymal abscesses are better addressed with the aid of conscious sedation. An intravenous narcotic in combination with a hypnotic sedative is appropriate. Standard physiologic monitoring must be applied.

Early reports suggested fluid aspiration be performed without antibiotic coverage for fear of altering the culture results. This practice is dangerous and has been abandoned. If suspicion of an infected collection exists, broad spectrum antibiotic coverage is recommended prior to the procedure in order to protect against life-threatening sepsis which can be caused by the procedure.

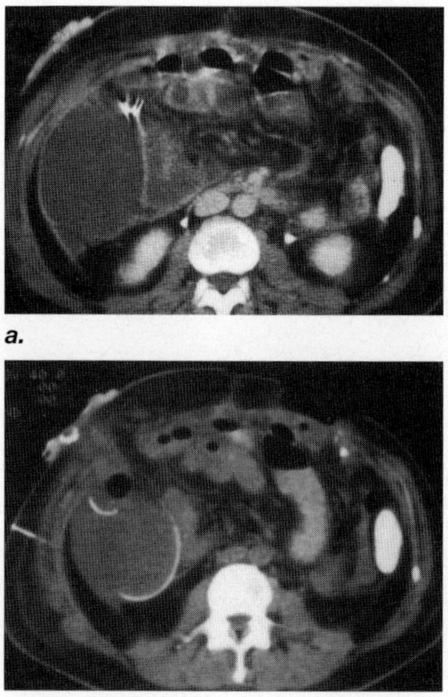

a.

b.

Figure 1. *58-year-old man with fever following colonic resection and construction of a colostomy. a) CT scan shows a fluid collection in right mid-abdomen where the surgery has been performed. b) Using CT guidance, the collection is punctured and a guidewire coiled within the collection. A drainage tube was subsequently placed.*

Localization

Imaging guidance should be used to localize the fluid and demonstrate a safe percutaneous access route **(Fig.1)**. The selection of ultrasound versus CT is generally reflection of operator preference and expertise. In the future, MR imaging may also play a significant role. Fluoroscopy may be helpful if the collection contains gas. Very large or palpable collections may not require imaging guidance.

Insertion of the Drainage Catheter

The skin over the selected entry site is sterilely cleansed and draped. A local anesthetic agent is used to numb the skin and a small skin incision is made with a #11 blade (Bard-Parker). Two basic techniques are used: 1) the trocar method and 2) the Seldinger method.

The trocar method uses the sharp stylet of the drainage tube to enter the collection and does not utilize a guidewire. A localizing needle is often placed prior to drainage catheter insertion for delineation of the appropriate angle and depth.

The Seldinger method involves placement of a guidewire through the needle. The needle is removed, the tract is dilated, and the drainage catheter is placed over the guidewire.

There is no scientific evidence that a specific catheter design or size is superior for the evacuation of abscesses. Tubes are available **(Fig. 2)** with and without parallel sump channels in both locking tip (Cope) and nonlocking tip designs. Catheter sizes range from 8-F to 30-F. The only rule that applies is that the catheter size should be large enough to drain the fluid into which it has been inserted. The 12-F and 16-F are the sizes used most frequently by our group **(Fig. 3)**.

The fluid is completely evacuated at the time of initial drainage. The fluid often will become blood tinged or bloody as the cavity collapses. Some investigators advocate irrigation of the cavity at the time of drainage with a sterile saline solution. There is no evidence to support or refute the advisability of this practice. Irrigation under pressure should never be performed as this may precipitate a serious septic episode.

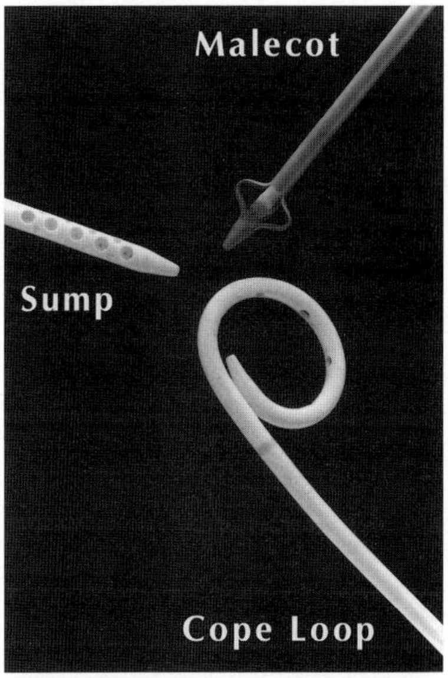

Figure 2. *A wide variety of tubes are available for abscess drainage.*

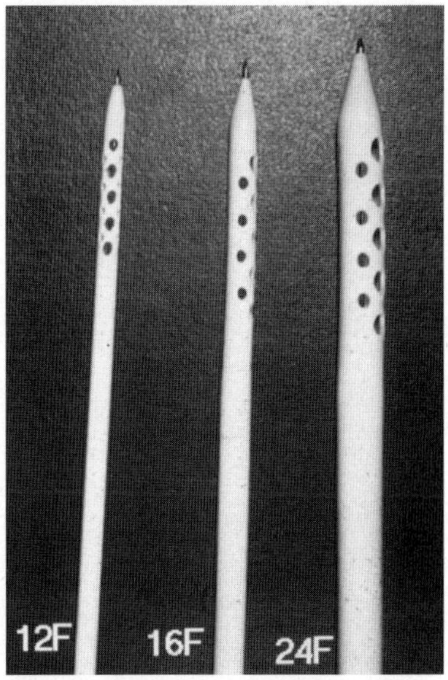

Figure 3. *The size of the drainage tube should be adequate to drain the fluid. A straight sump tube design is shown here.*

Postprocedure Care and Follow-up

If a sump tube has been inserted, the patient is returned to the ward and the catheter is attached to low-continuous wall suction. If a nonsump tube has been inserted, it is connected to a Jackson-Pratt (Bard) continuous suction bulb. The patient is seen a minimum of twice a day (once in the morning and once in the evening) and the catheter is flushed with 3 ml to 5 ml of sterile saline as needed to maintain patency.

Assessment of Results

Three clinical scenarios exist 24 hours to 48 hours following abscess drainage:
1) Minimal drainage (less than 20 ml/24 hours) with clinical improvement in patient's condition including decreased fever and WBC count.
2) Minimal drainage (less than 20 ml/24 hours) or continued drainage (more than 50 ml/24 hours) without clinical improvement or deterioration in the patient's condition.
3) Continued drainage of significant volumes of fluid (more than 50 ml/24 hours) with clinical improvement in the patient's condition.

Minimal drainage with clinical improvement is the anticipated result following drainage of uncomplicated abscesses. Almost any follow-up strategy will be successful in this circumstance. When the drainage becomes less than 10 ml in a 24-hour period, the patient is returned to the interventional suite and a sinogram is performed. If there is no significant residual cavity, the catheter is removed.

Minimal drainage or continued drainage without clinical improvement is the most distressing scenario. This indicates that the underlying problem has not been addressed. A repeat CT to assess the catheter position and to search for additional fluid collections should be performed. The catheter track must be inspected for potential iatrogenic injuries. If no additional fluid collections are identified, and the original collection has been evacuated, an alternative explanation for the patient's condition must be sought.

Continued drainage with clinical improvement usually indicates fistulous communication to the alimentary canal **(Fig. 4)**, bile duct, pancreatic duct, urinary tract or lymphatic system. A surprisingly high percentage (up to 60%) of abdominal abscesses are associated with these fistulous communications. The communications may be obvious by the nature of the effluent, or be occult requiring careful sinography for demonstration. Enterocutaneous fistulae may be divided into two groups: 1) low output and 2) high output.

Low-output fistulas drain less than 100 ml per day. In many cases, the fistula is not suspected until sinography is performed several days following the initial drainage procedure. When such a communication is identified, repositioning the drainage tube to a site adjacent to the leakage is recommended.

High-output fistulas may drain more than 1,000 ml per day. Management includes total parenteral nutrition as well as agents such as cimetidine and somatostatin to reduce fluid production from the bowel. Nasoenteric tubes as well as tubes traversing the bowel defect are warranted to diminish the quantity of effluent and promote healing.

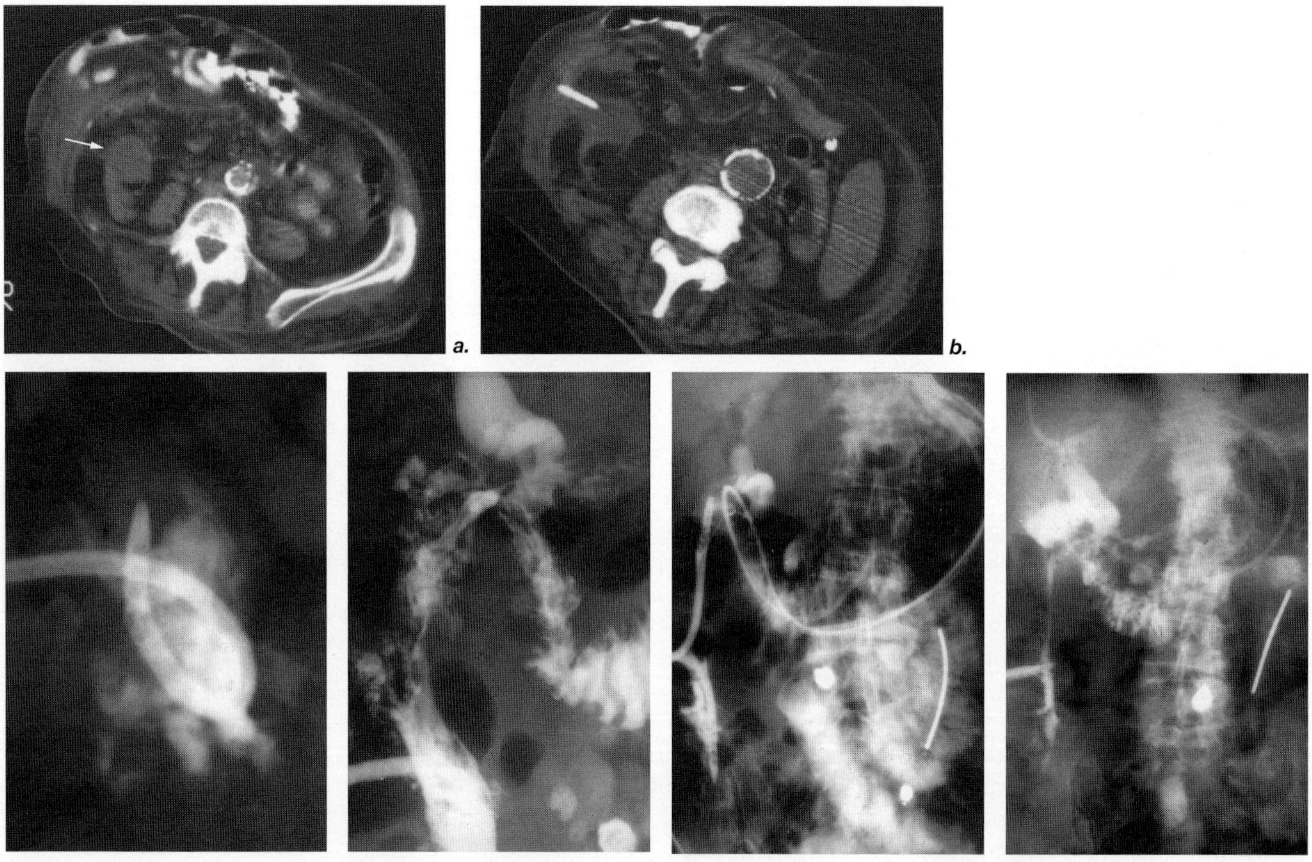

a. b. c. d. e. f.

Figure 4. *74-year-old woman with fever following an endoscopic sphincterotomy. a) CT scan discloses a right retroperitoneal fluid collection (arrow). b) A drainage catheter is inserted under CT guidance. c) On the following day, contrast material is injected through the tube showing an amorphous cavity without evidence of fistulous communication. d) The patient continued to drain approximately 200 ml of fluid per day and the tube injection was repeated 3 days later. A fistula extending to the descending duodenum and bile duct is now apparent. e) Drainage tubes were positioned adjacent to the site of fistulous communication as well as dependently in the abscess cavity. f) Two weeks later, the cavity has diminished in size. The drainage tubes were removed when the output decreased to less than 10 ml/day.*

Without surgical intervention, both low- and high-output fistulas can be cured in 60% to 80% of patients. Factors which preclude successful healing include distal obstruction, steroid therapy, significant colonic perforations or underlying bowel pathology such as ischemia or neoplasm. When these conditions are present, management with tube therapy alone is usually unsuccessful.

Removal of Tube

Drainage tubes are removed when the output is less than 10 ml in 24 hours and sinography demonstrates no significant residual cavity. Tube removal is often performed as an outpatient. Sedation is usually not required as the amount of discomfort (in most patients) is minimal.

SPECIAL CONSIDERATIONS

There are four areas which deserve special consideration as potential problems in the percutaneous management of abscesses. These include fluid collections complicating pancreatitis, splenic abscesses, hepatic abscesses and pelvic abscesses.

Fluid Collections Complicating Pancreatitis

Management of patients with severe, complicated pancreatitis is an extremely complex issue requiring a close cooperative effort between the internist, surgeon and interventional radiologist. There are conflicting reports in the literature regarding the success of percutaneous drainage in pancreatitis, which probably reflects the wide spectrum of severity of this disorder coupled with different patient populations encountered in these series. A somewhat controversial, but commonly accepted grading of the severity of pancreatitis on the basis of CT findings, was proposed by Balthazar et al **(Table 1)**.

TABLE 1: SEVERITY OF PANCREATITIS BY CT APPEARANCE

Grade A	Normal pancreas
Grade B	Focal or diffuse enlargement of the pancreas including contour irregularities, pancreatic ductal dilatation and foci of intrapancreatic fluid collections
Grade C	Intrinsic pancreatic abnormalities and indistinctness, haziness or streaking in the peri-pancreatic fat
Grade D	Single, ill-defined fluid collection or phlegmon
Grade E	Two or more poorly defined collections or presence of gas in or adjacent to the pancreas

The vast majority of patients encountered by the interventional radiologist have grade D and E disease. Fluid collections related to pancreatitis can be classified as unorganized peripancreatic fluid collections, pseudocysts that have a visible wall, and pancreatic abscesses that are any type of infected, loculated fluid within or adjacent to the pancreas. The dividing line between unorganized peripancreatic fluid collections and pancreatic pseudocyst is not well defined when the wall is developing.

The other important CT finding in acute pancreatitis is the presence of pancreatic necrosis. Kivisaari et al first reported that the lack of contrast enhancement of part or all of the pancreas on CT correlates closely with necrotic pancreatic tissue discovered at the time of surgical exploration. Patients with extensive necrosis will not be successfully treated with percutaneous therapy and require surgical débridement.

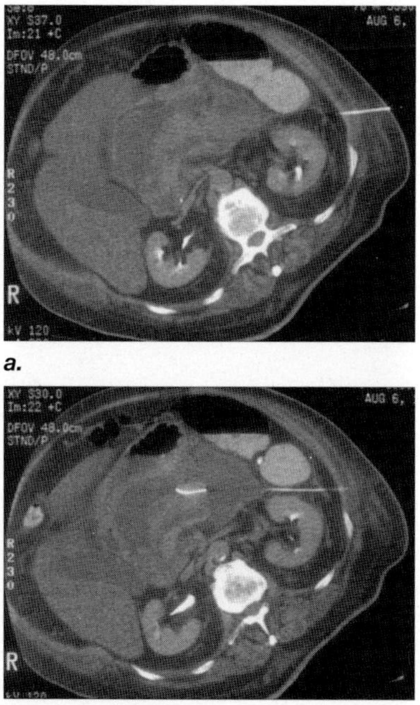

a.

b.

Figure 5. 44-year-old man with acute pancreatitis and fever. a) Peripancreatic fluid is drained under CT guidance by advancing a needle anterior to the left kidney and posterior to the descending colon. b) A slightly more caudal image confirms appropriate needle and wire position. A drainage catheter was subsequently placed; however, the patient subsequently required a surgical débridement.

Lee et al reviewed their experience with 59 fluid collections in 30 patients with severe, complicated pancreatitis who were treated in part or totally with percutaneous catheter drainage. Percutaneous drainage was successful in 14 patients, partially successful in eight patients and unsuccessful in eight patients. Sixteen of the 30 patients required surgery. Percutaneous therapy was successful in draining 28 of 49 peripheral (parahepatic, pararenal, and paracolic) collections. Percutaneous therapy was unsuccessful in nine of 10 central fluid collections **(Fig. 5)** (pancreatic bed and lesser sac). The authors recommend primary surgical drainage for central fluid collections and necrotic tissues, followed by percutaneous drainage if surgical evacuation is incomplete. Peripheral collections **(Fig. 6)** without associated pancreatic necrosis should be approached with percutaneous therapy.

Simple pseudocysts also represent a therapeutic challenge. Despite several reports touting the benefits of percutaneous drainage **(Fig. 7)**, the optimal management for this problem has not been established. Pseudocysts complicate approximately 10% of patients with acute pancreatitis. The majority of these collections will spontaneously resolve. Unless symptoms suggesting infection, obstruction of the bowel or bile duct, or unrelenting pain are present, most investigators recommend observation. It has been suggested that pseudocysts 6 cm or greater in diameter which have been present for 6 or more weeks require therapy, but more recent data discussed by Deziel et al indicate that expectant management is usually safe and results in a high rate of resolution.

When pseudocysts do not resolve and require drainage, the appropriate technique (percutaneous versus endoscopic versus surgical drainage) has not been established in a random, prospective trial. Advocates of percutaneous drainage claim that catheter therapy is well tolerated, and if unsuccessful, the patient can then undergo the more invasive surgical therapy. However, Rao et al have documented that following percutaneous interventions, operative morbidity is significantly increased. In their series, the operative morbidity following failed percutaneous or endoscopic therapy was twice that of patients who had not undergone previous intervention. Specifically, the incidence of postoperative abdominal abscess was four times greater in this group of patients.

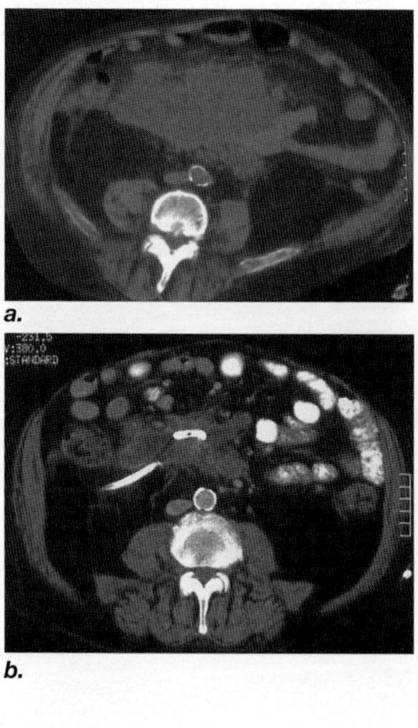

a.

b.

Figure 6. 66-year-old man with pancreatitis and fever. a) CT scan reveals a fluid collection within the mesentery caudal to the pancreas. b) The collection is successfully evacuated from a right postero-lateral retroperitoneal approach. The patient improved and was discharged from the hospital but required the drainage tube for more than 2 months.

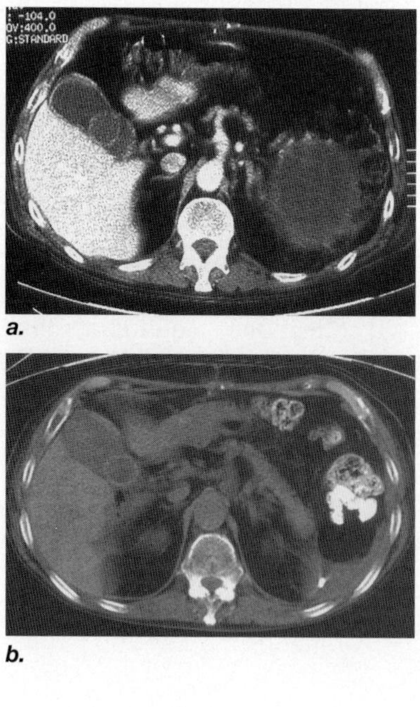

a.

b.

Figure 7. 70-year-old man with persistent pain attributable to a pancreatic pseudocyst. a) CT scan shows a large pseudocyst adjacent to the left hemidiaphragm. b) The pseudocyst is successfully drained from a posterior-subcostal approach.

If percutaneous drainage is elected as the appropriate therapeutic alternative, the procedure can be performed from a posterolateral retroperitoneal approach or an anterior intraperitoneal approach. The transgastric route may be used **(Fig. 8)** to create an internal cyst-gastrostomy. This transgastric approach has not been demonstrated to be superior to routes that do not transgress the stomach. Patients with pseudocysts complicating chronic pancreatitis often have impaired drainage from the pancreatic duct. Percutaneous drainage in these patients will likely result in a chronic pancreatic fistula.

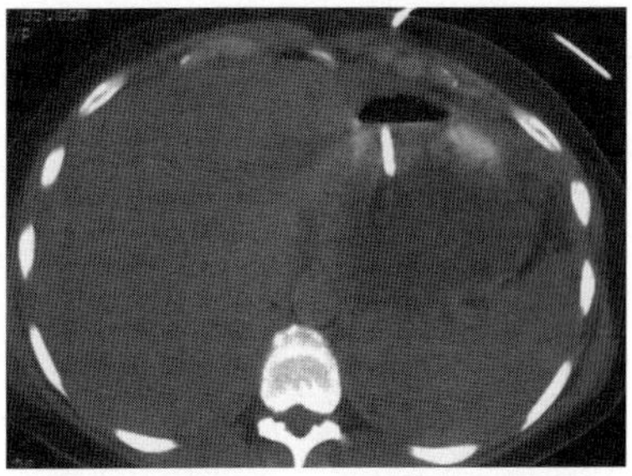

Figure 8. *A pancreatic pseudocyst is drained from an anterior transgastric approach in a 43-year-old woman.*

SPLENIC ABSCESSES

Numerous reports have appeared documenting the success of percutaneous aspiration and drainage of splenic abscesses. Schwerk et al reported curative needle aspiration in six of seven patients with pyogenic splenic abscesses and success in an additional patient with catheter drainage. Six patients had a solitary abscess and two patients had multiple abscesses. Despite these reports, a review by Levison suggests that splenectomy has not yet been replaced by percutaneous drainage.

PERCUTANEOUS DRAINAGE: SPLENIC ABSCESSES

- Percutaneous access must avoid pleural transgression
- Bleeding is a risk
- Ineffective evacuation is a concern
- Aspiration versus catheter drainage could be considered

Technically, splenic abscess drainage can be challenging. Transgression of the pleura may lead to a complicating empyema and hemorrhagic complications can be encountered. A minimal amount of intact splenic tissue should be violated and close observation for intraperitoneal hemorrhage following the procedure is warranted.

HEPATIC ABSCESSES

Controversy also exists regarding the clinical indications for imaging-guided percutaneous drainage of hepatic abscesses **(Fig. 9)**. Hansen and Vargish retrospectively reviewed 24 patients and concluded that open drainage with an associated definitive surgical procedure, such as cholecystectomy, was superior to percutaneous drainage alone. This conclusion was based upon the observation that a 25% mortality rate was evident among the percutaneous group, and no deaths were encountered with the open drainage group.

PERCUTANEOUS DRAINAGE: HEPATIC ABSCESSES

- Percutaneous vs surgical vs medical approaches
- Need for catheter drainage is disputed
- Need for drainage in amebic abscesses is controversial

A different conclusion was reached by Lambiase et al in a substantially larger group of 335 abscesses in 323 patients. A cure rate of 62% was documented with a failure rate of 9%. The remaining patients were felt to have benefited, but required additional therapy. Though a 14% 1-year mortality was found, only 5% were attributable to septic complications. Similar results were noted in a review by Dondelinger et al where a literature search comprising 252 patients revealed an overall success rate of 77% with a mortality rate of 6%.

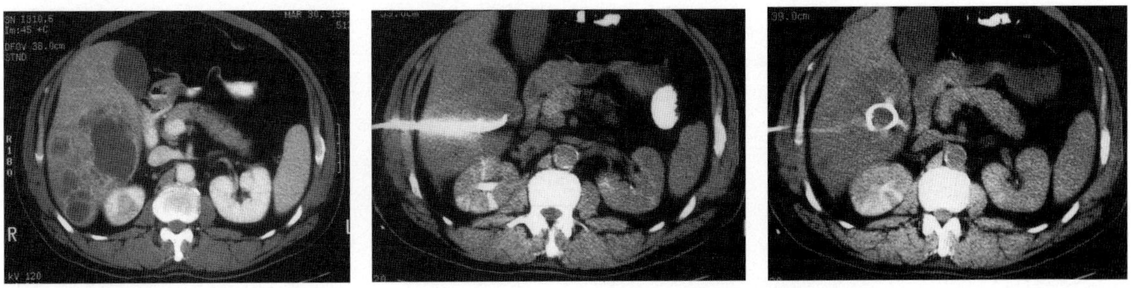

a. **b.** **c.**

Figure 9. *81-year-old man with history of diverticulitis presents with recurrent fever. a) Abdominal CT shows multiple low-density lesions in the liver with a large lesion medially within the posterior segment of the right lobe. b) The abscess is drained with CT guidance. c) A slightly more cephalic image shows the catheter tip coiled within the hepatic abscess.*

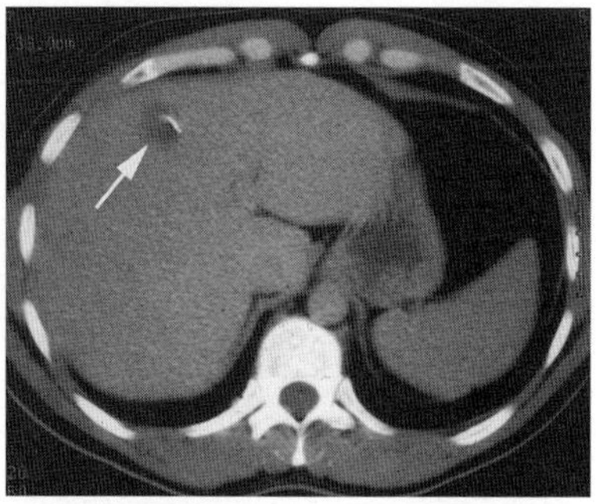

Figure 10. 24-year-old man with a history of amebic dysentery presents with right upper quadrant pain. An amebic abscess in the liver (arrow) is aspirated under CT guidance.

It is unclear whether prolonged catheter drainage is necessary to treat hepatic abscesses. Several series have documented equivalent results with repeated ultrasound-guided needle aspirations.

Controversy also persists regarding the management of amebic abscesses. Though a substantial number of these abscesses will resolve solely with amebocidal therapy, Filice et al reported a more rapid cure rate when percutaneous aspiration **(Fig. 10)** and/or catheter therapy was combined with medical treatment. We currently drain hepatic amebic abscesses when they are large, adjacent to the liver capsule or pericardium, if there is an associated pleural effusion, or if clinical symptoms persist despite amebocidal therapy.

PERCUTANEOUS DRAINAGE: PELVIC ABSCESSES

Access routes:

- Transabdominal
- Transgluteal
- Transvaginal
- Transrectal

PELVIC ABSCESSES

Pelvic abscesses represent a technical challenge due to the bony pelvis, bowel, genitourinary tract, and vascular structures. When a safe window exists anteriorly, it is the preferred route for percutaneous evacuation. However, due to gravity dependence, the majority of abscesses occur posteriorly in the pelvis. This may necessitate alternative entry routes including transgluteal, transrectal and transvaginal.

Transgluteal drainage **(Fig. 11)** traverses the sciatic notch. The safest route is generally adjacent to the sacrum as caudal as feasible to avoid both the sciatic nerve and internal iliac arterial branches. CT is the guidance method of choice.

Transrectal **(Fig. 12)** and transvaginal drainage **(Fig. 13)** should be performed under real-time ultrasound guidance with appropriately designed transducers. A protective sheath is required when introducing needles from this approach to avoid injury to both the patient and operator.

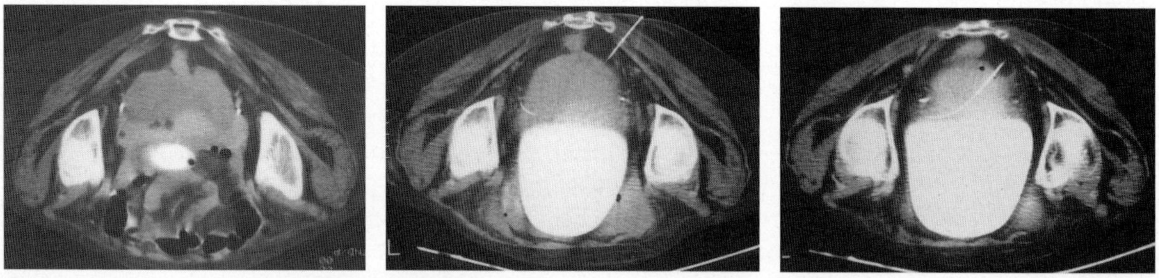

a. b. c.

Figure 11. 62-year-old man with postoperative fever. a) CT scan confirms the presence of a pelvic abscess. b) A needle and wire are placed into the collection from a right transgluteal approach with the patient in a prone position. Note that the needle has been placed close to the sacrum to avoid the sciatic nerve and arterial structures. c) A more caudal image confirms the wire to be securely coiled within the collection.

In a recent article by Kuligowska et al, an 88% success rate was reported in treating 24 patients with 33 pelvic abscesses by ultrasound-guided transrectal aspiration with an 18-gauge needle. Following aspiration of the purulent material, the cavity was lavaged with saline. Though no catheters were placed, their results were clearly comparable to those reported with catheter drainage.

RESULTS

Curative drainage of infected fluid collections is defined as complete resolution of infection requiring no further operative intervention. Partial success is defined as either adequate drainage of an abscess with surgery subsequently performed to repair an underlying problem or as a temporizing measure which stabilizes the patient prior to surgery. The rate of successful drainage is defined as the sum of both curative drainages and partially successful drainages.

The results are similar for both abdominal and chest collections. The following ranges of outcomes for drainage of infected fluid collections have been reported as appropriate by the Standards of Practice Committee of the Society of Cardiovascular & Interventional Radiology **(Table 2)**.

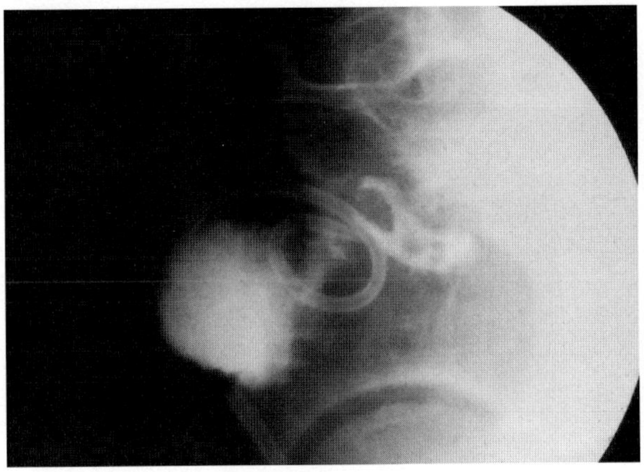

Figure 12. *Injection of an abscess drainage catheter that has been inserted from a transrectal approach to evacuate a pelvic abscess.*

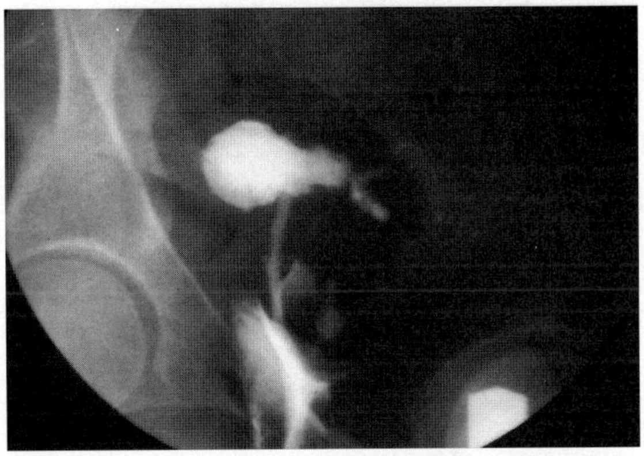

Figure 13. *Injection of an abscess drainage catheter inserted into the pouch of Douglas from a transvaginal approach.*

TABLE 2: PERCUTANEOUS ABSCESS DRAINAGE: RESULTS

• Curative drainage	80% - 85%
• Partial success	5% - 10%
• Successful drainage	85% - 95%
• Failure	5% - 10%
• Recurrence	5% - 10%

(SCVIR Standards of Practice Committee, JVIR 1995.)

Despite these impressive results, it is unclear that percutaneous aspiration or drainage of abscesses is appropriate for all patients. In a report of 91 patients by Levison and Zeigler, APACHE II scores (a quantitative assessment of the physiologic status of critically ill patients) were correlated with drainage technique (percutaneous versus open surgical) of abdominal abscesses and eventual outcome. Though not statistically significant, only 8% of patients with APACHE II scores equal to or more than 15 undergoing percutaneous drainage survived compared with 30% of patients who underwent open surgical drainage procedures.

COMPLICATIONS

Complications are encountered in approximately 10% of abscess drainages. After a comprehensive literature review, the Standards of Practice Committee of the SCVIR published these rates of specific complications following percutaneous abscess drainage **(Table 3)**.

TABLE 3: COMPLICATIONS OF PERCUTANEOUS ABSCESS DRAINAGE

Complication	Rate (%)	Threshold (%)
• Septic shock	1-2	4
• Bacteremia requiring significant new intervention	2-5	10
• Hemorrhage requiring transfusion	1	2
• Superinfection (includes infection of sterile fluid collection)	1	2
• Bowel transgression requiring intervention	1	2
• Pleural complication from abdominal procedure	1	2
• Pleural complication from thoracic procedure	2-10	20

(SCVIR Standards of Practice Committee, JVIR 1995.)

REFERENCES

Aassar OS, LaBerge JM, Gordon RL, et al. Percutaneous management of abscess and fistula following pancreaticoduodenectomy. Cardiovasc Intervent Radiol 1999; 22:25-28.

Balthazar EJ, Ranson JH, Naidich DP, Megibow AJ, Caccavale R, Cooper MM. Acute pancreatitis: prognostic value of CT. Radiology 1985; 156:767-772.

Deziel DJ, Prinz RA, Uzer MF, Matalon TAS. Pseudocysts in hepatobiliary and pancreatic disease: the team approach to management. Pitt HA, Carr-Locke DL, Ferrucci JT, eds. Boston: Little, Brown and Company, 1995; 342.

Dondelinger RF, Kurdziel JC, Gathy C. Percutaneous treatment of pyogenic liver abscesses: a critical analysis of results. Cardiovasc Intervent Radiol 1990; 13:174-182.

Filice C, DiPerri G, Strosselli M, et al. Outcome of hepatic amebic abscesses managed with three different therapeutic strategies. Digestive Diseases and Sciences 1992; 37:240-247.

Hansen N, Vargish T. Pyogenic hepatic abscess: a case for open drainage. American Surgeon 1993; 59:219-222.

Kivisaari L, Kalevi S, Standertskjold-Nordenstam CG, et al. A new method for the diagnosis of acute hemorrhagic-necrotizing pancreatitis using contrast enhanced CT. Gastrointest Radiol 1984; 9:27-30.

Kuligowska E, Keller E, Ferrucci JT. Treatment of pelvic abscesses: value of one-step sonographically guided transrectal needle aspiration and lavage. AJR 1995; 164:201-206.

Lambiase RE, Deyoe L, Cronan JJ, Dorfman GS. Percutaneous drainage of 335 consecutive abscesses: results of primary drainage with 1-year follow-up. Radiology 1992; 184:167-179.

Lee MJ, Rattner DW, Legemate DA, et al. Acute complicated pancreatitis: redefining the role of interventional radiology. Radiology 1992; 183:171-174.

Levison MA. Percutaneous versus open operative drainage of intra-abdominal abscesses. Infec Dis Clin North Am 1992; 6:525-544.

Levison MA, Zeigler D. Correlation of APACHE II score, drainage technique and outcome in postoperative intra-abdominal abscess. Surg Gynecol Obstet 1991; 172:89-94.

Rao R, Fedorak I, Prinz RA. Effect of failed computed tomography guided and endoscopic drainage on pancreatic pseudocyst management. Surgery 1993; 114:843-849.

Schwerk WB, Gorg C, Gorg K, Restrepo I. Ultrasound-guided percutaneous drainage of pyogenic splenic abscesses. J Clin Ultrasound 1994; 22:161-166.

Society of Cardiovascular and Interventional Radiology Standards of Practice Committee: Quality improvement guidelines for adult percutaneous abscess and fluid drainage. J Vasc Interv Radiol 1995; 6:68-70.

Wroblicka JT, Kuligowska E. One-step needle aspiration and lavage for the treatment of abdominal and pelvic abscesses. AJR 1998; 170:1197-1203.

CHAPTER 20 PREVIEW ANSWERS

Case 1

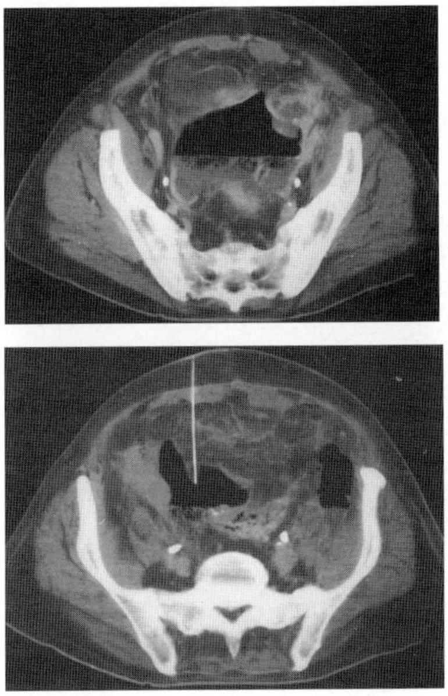

a. b.

Findings: a) CT shows a pelvic abscess without bowel interposed anteriorly. b) A needle is placed from an anterior approach to drain the abscess.
Diagnosis: Pelvic abscess successfully drained with anterior percutaneous catheter.

Case 2

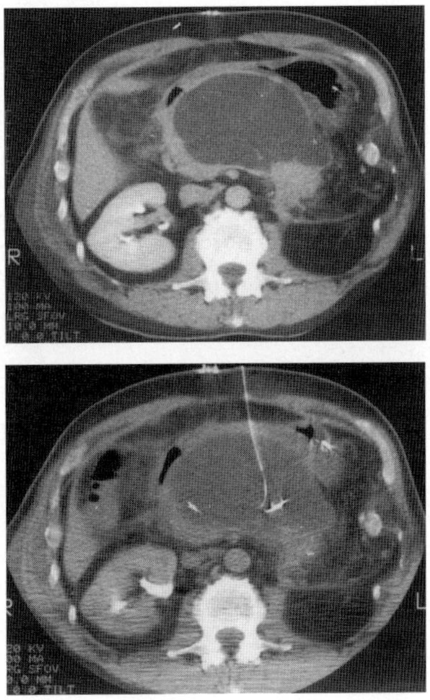

a. b.

Findings: a) A pancreatic pseudocyst is present in the lesser peritoneal sac, posterior to the stomach. b) A pseudocyst in this location can be drained from a transgastric approach.
Diagnosis: Pancreatic pseudocyst, drained from anterior transgastric approach.

Case 3

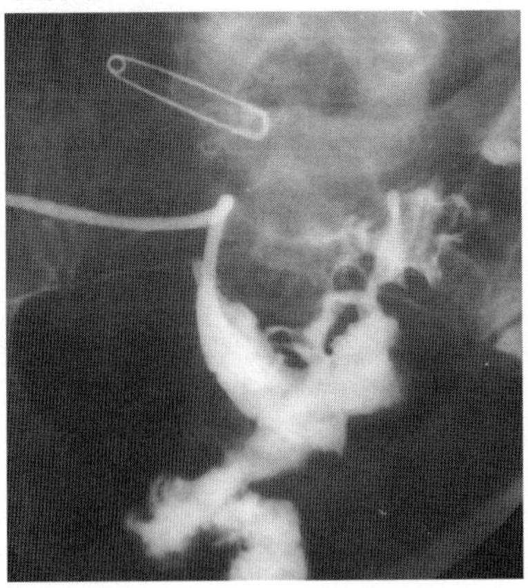

Findings: Contrast material injected through the drainage tube opacifies the sigmoid colon.
Diagnosis: This abscess is likely due to a ruptured sigmoid diverticulum.

Case 4

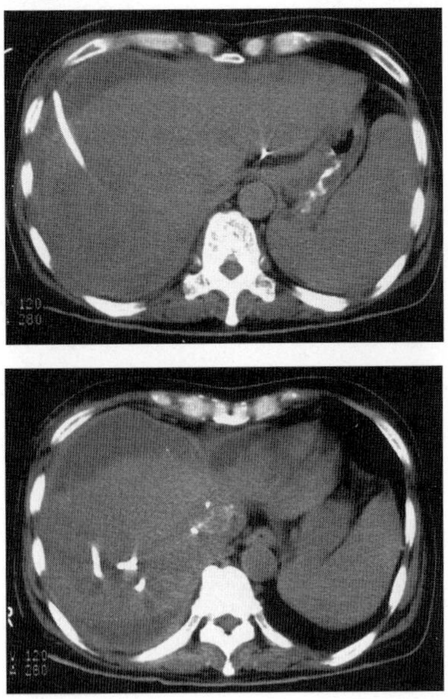

a. b.

Findings: a,b) Drainage of the liver abscess through the ascitic fluid. Drainage through uninfected ascitic fluid should be avoided as the ascites may become contaminated and infected.
Diagnosis: Hepatic abscess drainage tube placed through ascitic fluid.

CHAPTER 21 PREVIEW

Case 1

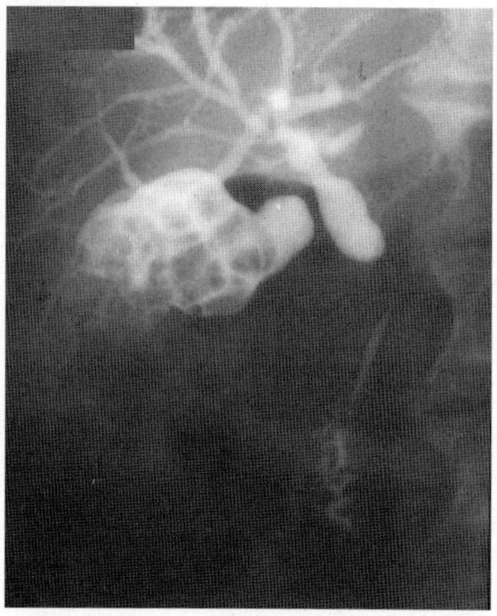

History: PTC of a 57-year-old woman with pain and jaundice.
Questions: What are the findings? What is a possible diagnosis?

Case 2

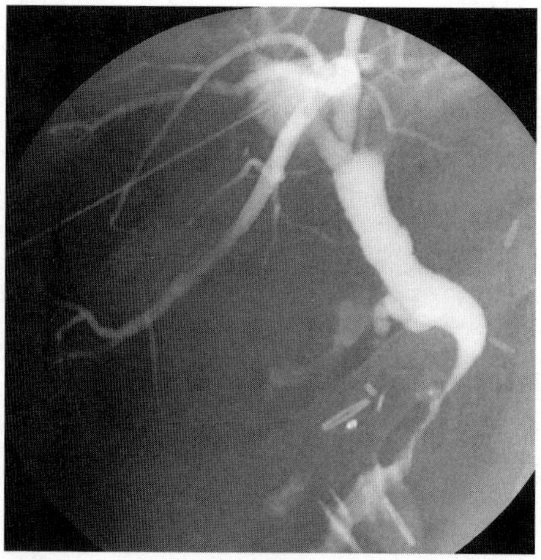

History: PTC of a patient following liver transplantation.
Question: Do you see the clinically suspected extrahepatic bile leak?

Case 3

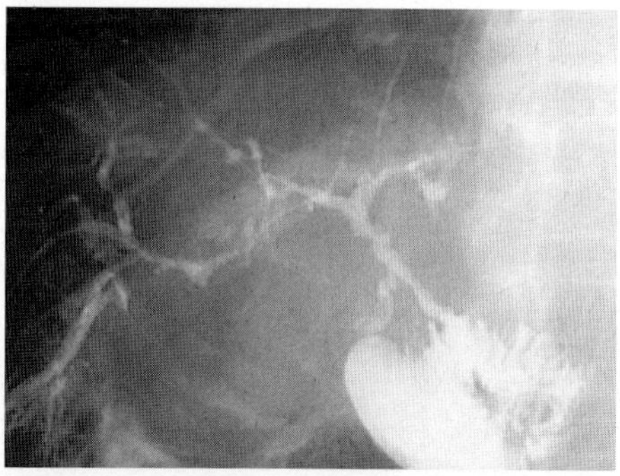

History: A patient underwent liver transplantation 4 years ago because of liver failure secondary to primary sclerosing cholangitis. She now has recurrent jaundice and fever resulting in a request for this PTC.
Question: What is your diagnosis?

Case 4

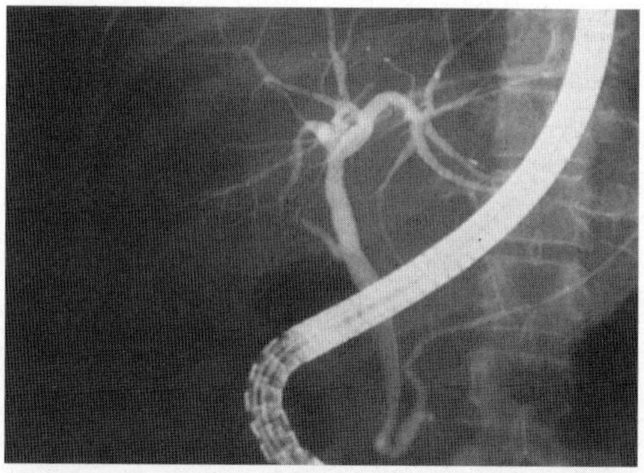

History: A 34-year-old woman had pain and fever following laparoscopic cholecystectomy and this is an ERCP performed 10 days after the operation.
Question: What do you see?

Chapter 21
Percutaneous Biliary Interventions

Roy L. Gordon, MD

TECHNIQUES

- Preprocedure evaluation
- Percutaneous transhepatic cholangiogram (PTC)
- Percutaneous transhepatic biliary drainage (PTBD)
  - benign biliary obstruction
- Percutaneous Roux-loop puncture
- Percutaneous cholecystostomy (PC)
- Procedures performed through surgical tracts

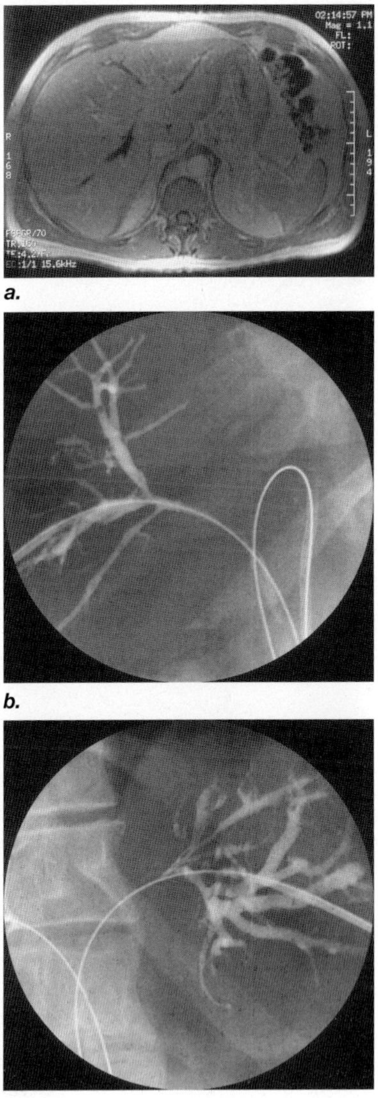

Figure 1. a) MR shows a Klatzkin tumor encasing the left and two main right biliary ducts, indicative of a non-operable condition. b) and c) are cholangiograms obtained as part of the procedure for placement of palliative metallic biliary stents. The cholangiograms do not add any clinically important diagnostic information that was not available on the prior MR.

INTRODUCTION

Access to the intrahepatic biliary tree can be reliably established by passing a thin needle percutaneously into the liver parenchyma and injecting contrast material while slowly withdrawing the needle under fluoroscopic observation. In many cases, real-time sonography can be used to guide a percutaneous needle into a dilated intrahepatic bile duct or into the gallbladder. These techniques are in routine and widespread clinical use. They have proved to be both efficacious and safe.

In this chapter we will consider percutaneous transhepatic cholangiography (PTC), percutaneous transhepatic biliary drainage (PTBD), percutaneous Roux-loop puncture, percutaneous cholecystostomy (PC) and access using existing routes such as T-tube tracks. The hallmarks of these techniques are that no surgical incision is required and that for the most part they can be performed under local anesthesia using conscious sedation without the need for general anesthesia. The retrograde endoscopic (ERCP) approach provides the alternative nonsurgical approach to the biliary tree and provides similar advantages.

INDICATIONS

Access to the biliary tree provides options for diagnosis and/or treatment.

Usually preliminary laboratory tests and noninvasive imaging techniques such as ultrasound (US), computed tomography (CT), magnetic resonance imaging (MRI), magnetic resonance cholangiography (MRC) or nuclear medicine studies will distinguish between medical nonobstructive jaundice and obstructive jaundice. They may also provide information about the anatomical level of the obstruction, the degree of obstruction and in many cases the cause of the obstruction **(Fig. 1)**.

Contrast cholangiography by means of PTC or ERCP is thus only indicated when: a) the above-mentioned tests do not yield adequate diagnostic information, or b) when the cholangiogram is the first step in a therapeutic intervention such as a drainage procedure or biopsy.

TECHNIQUES

Preprocedure Evaluation

Preliminary considerations are often very important and may include the following:

a) Review of relevant prior studies **(Fig. 2)**,

b) History of contrast allergies,

c) Coagulation status. Many patients have compromised liver function and may have problems with bleeding. In general, the platelet count should be above 50,000. Infusion of 1 unit of platelets should raise the count by 10,000 within 1 hour and usually the infusion of 6 units is advocated. PT and PTT should be within the institutional limits and levels above 16/40 may indicate the need for fresh frozen plasma infusion when viewed in conjunction with the platelet count. Some authors think the bleeding time is more useful and should be below 9 to 10 minutes. The International Normalized Ratio (INR) is in use in some centers to assess coagulation status. Vitamin K replacement therapy may be effective in the short term in some patients with cholestasis who are vitamin K deficient.

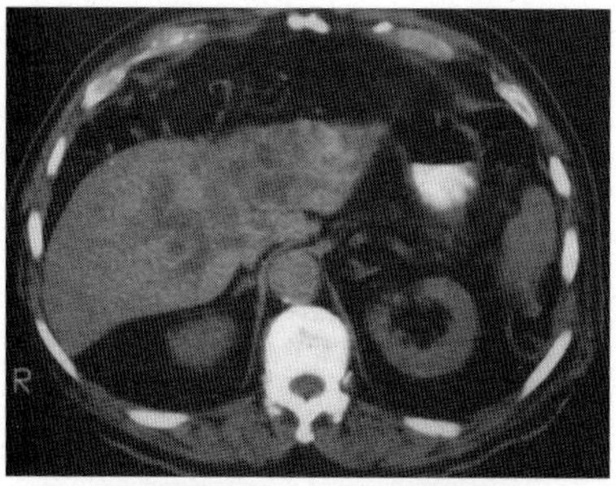

Figure 2. *Review of prior CT scans shows the high colon blocking access to most percutaneous approaches to the liver. A CT scan is not a prerequisite before PTC/PTBD, but thorough review of all prior imaging studies should always be done.*

d) Laboratory studies, including hemoglobin, hematocrit, WBC, liver panel, and serum creatinine, establish a preprocedure baseline to assess later progress or postprocedure complications such as sepsis or bleeding.

e) Prophylactic antibiotics are generally used for PTC, PTBD and most other biliary manipulations, although practices vary. Usually a broad spectrum cephalosporin is administered intravenously at the time of sedation. This drug will not cover enterococci so some authors routinely use even broader cover. In patients with heart valve problems, broad spectrum antibiotic cover is necessary with combined agents including ampicillin and gentamicin, or vancomycin, unasyn, or similar agents. This field changes rapidly and regular consultations with the institute's infectious disease group is necessary to ensure that the best drugs are used.

f) Sedation is a most important element in successful biliary work. Most centers do not have ready access to anesthesiologists to provide routine sedation and frequently interventional radiology nurses provide IV sedation and monitoring under the radiologists' supervision using agents with which they are familiar. ECG, vital signs and oximetry are routinely monitored. Some interventionalists use regional nerve blocks or occasionally have access to general anesthesia. It is important to establish adequate sedation and pain control before starting a procedure and to maintain adequate levels.

g) Inpatient versus outpatient treatment is a controversial topic in the era of health reform. No accepted guidelines are yet available, and individual practice may vary from hospital to hospital and patient to patient. For PTC, we plan a 23-hour admission for all patients. In a very few selected cases that proceed smoothly, who have been stable over six hours observation and who have an excellent home situation, we may consider early discharge home. All new PTBD's are inpatient procedures.

Percutaneous Transhepatic Cholangiography (PTC)

In most institutions, diagnostic PTC is used only when ERCP:

a) is unsuccessful or cannot be performed, e.g. in advanced pancreatic or duodenal lesions with complete obstruction of the distal ampulla of Vater, in some cases with duodenal or distal duct diverticulae, and in patients following upper gut surgery such as a Whipple procedure or a Bilroth II partial gastrectomy

b) does not provide sufficient delineation of the biliary tree, e.g. when the extrahepatic duct is completely obstructed and there is no filling of the intrahepatic ducts. **Table 1** lists the indications for PTC.

TABLE 1: INDICATIONS FOR PTC

- Delineate bile duct anatomy completely
- Define level and nature of obstruction in patients with dilated bile ducts
- Evaluate for presence of suspected bile duct stones or other filling defects
- Determine etiology of cholangitis
- Evaluate suspected bile duct inflammatory disorders
- Demonstrate site of bile duct leak

PTC is performed using a thin, flexible 22-gauge needle of the Chiba type. The puncture is made either from a right lateral intercostal approach or from an anterior approach, which is subcostal in the epigastric region. Review of prior imaging studies may be of value in selecting the best approach and in identifying any unsuspected problems such as ascites, loops of bowel interposed in front of the liver, hepatic cysts or other intrahepatic masses. The anterior approach is of value when the left ducts only are to be opacified, e.g., in hilar obstructions, or in children with split liver autografts. This approach has the disadvantage of making it more difficult to reduce x-ray exposure to the interventionalist's hands. Real-time sonographic guidance is of great value when puncturing dilated left-sided ducts and in avoiding puncture of adjacent left portal vein branches.

For most general diagnostic cholangiography, a right lateral intercostal entry is selected in the midaxillary line. A caudal site is employed to avoid puncturing the diaphragm or pleura. The chosen point of entry should be checked under fluoroscopy during deep inspiration to ensure that it is below the costophrenic angle. Despite these precautions, it is likely that the needle traverses the pleural space frequently. Fortunately, resultant complications such as pneumothorax are not common **(Fig. 3)**.

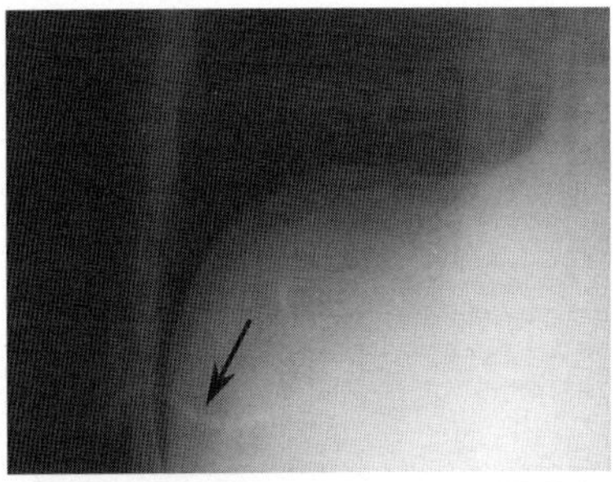

***Figure 3.** The entry site for this PTBD is too high and obviously crosses the pleural space (arrow). The chest radiograph shows no pleural fluid or other reaction and the patient was asymptomatic.*

The needle is passed through the lower half of the intercostal space to avoid damage to the neurovascular structures running on the inferior aspect of the rib. The first pass is made parallel to the table top and slightly angled towards the head along the course of the 11th rib, aiming for T-10 or T-11 and stopping at the right lateral margin of the vertebral bodies.

Contrast material is injected gently while the needle is slowly withdrawn. Forceful injection is avoided to minimize excessive parenchymal extravasation of contrast that interferes with subsequent imaging. Several structures may be identified as the contrast material is injected and the needle is withdrawn through the liver:

i) Portal veins: large, branching structures similar in appearance to the bile ducts, but easily differentiated from them because the intravascular contrast rapidly washes out. In patients with advanced portal hypertension, there may be the confusing picture of hepatofugal flow with central washout.

ii) Hepatic arteries: small, branching structures from which contrast washes out very quickly, causing contrast staining of the perfused hepatic parenchyma.

iii) Hepatic veins: large tubular structures without visible branches. The injected contrast material flows cephalad towards the right atrium.

iv) Lymphatics: small, irregular, beaded linear structures from which contrast does not wash out. There is no organized branching pattern, but rather a parallel grouping often headed towards the liver hilum.

v) Periportal extravasation: occasionally contrast material dissects around the portal structures, producing an appearance reminiscent of the bile ducts but without their sharp, well-defined margins. The portal structures are seen as a negative defect within the puddle of extravasated contrast material.

vi) Bile ducts: when the needle is in the lumen of a bile duct, injected contrast material moves slowly away from the needle tip, gradually filling up the ducts. If the injection is stopped for a moment, intraluminal contrast persists, unlike the washout seen when the needle is in a vascular structure.

Once in the bile duct, enough contrast material is injected to partially opacify the biliary tree and to demonstrate either an obstruction or free drainage. Bile cannot easily be drained through the 22-gauge needle so care should be taken to avoid injecting excessive volumes of contrast. When the biliary tree is obstructed, bacteria are frequently present in the bile, even when there is no clinical evidence of sepsis. Adding a large volume of contrast material to an already obstructed and distended ductal system can raise intraductal pressure above venous pressure and can result in bacteremia. The volume of contrast material needed to adequately demonstrate the anatomy is minimized by tilting the patient and redistributing the opaque material to the area of obstruction.

It is not uncommon that a bile duct is not punctured on the first needle pass and repeated passes are then performed. The ducts are most likely to be punctured in the hilum where they converge and where they attain their largest diameter. The major vascular structures also cluster in the hilum so that although central punctures may be helpful for diagnosis in nondilated systems, such central punctures then should not be converted into catheterization entry routes.

In some patients, particularly after bile duct injury following laparoscopic cholecystectomy or in others with high biliary obstruction, there will be a situation in which not all ducts will be in communication with all others. For complete delineation of the biliary tree, separate punctures of any isolated systems will be needed **(Fig. 4)**. Reference to available preliminary imaging studies, particularly MRC, may be helpful. The LPO projection best lays out the ductal system reducing overlapping and facilitating recognition of duct-deficient parts of the liver where there may be isolated nonopacified bile ducts.

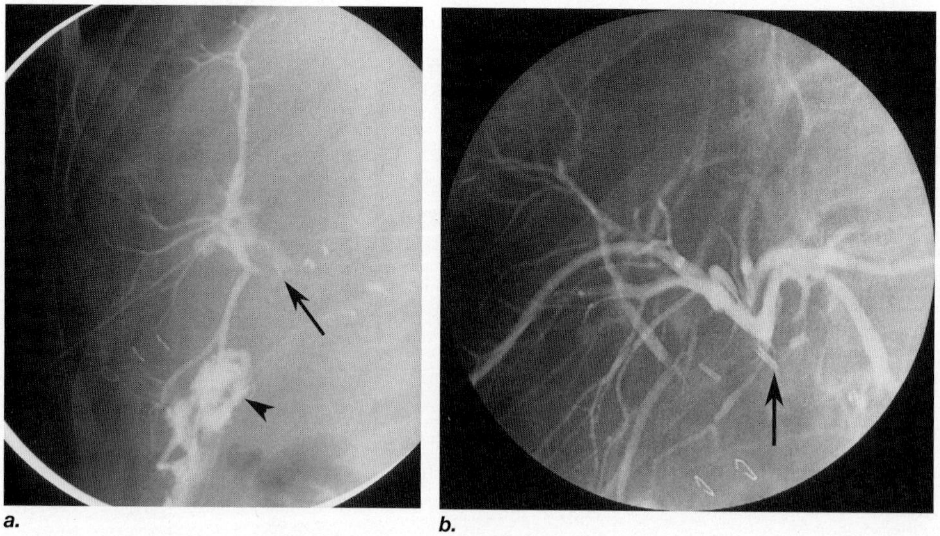

a. **b.**

Figure 4. *This patient sustained bile duct injuries during a laparoscopic cholecystectomy. a) The first pass of the PTC shows an isolated segment of the right biliary ductal system. There is a surgical clip occluding the duct (arrow) and there is a leak into the peritoneal space from a peripheral duct (arrowhead). This isolated right segment had a PTBD placed prior to surgery as well as a percutaneous abscess drain into the fluid collection resulting from the bile leak. b) A PTC and then drainage of the remainder of the biliary tree was also performed. There is also a surgical clip on this duct (arrow).*

The cholangiogram should be terminated when all the relevant anatomic information has been demonstrated, using as little contrast injection and as little distension of the system as possible. Current surgical thinking is that most patients will not require biliary drainage as a temporary measure after PTC. However patients with biliary obstruction and sepsis should undergo immediate percutaneous drainage, as should patients whose cholangiograms demonstrate small or large intrahepatic abscesses.

Results and Complications of PTC

The SCVIR threshold for success in opacifying the bile ducts by PTC is 95% and 65% for dilated and nondilated ducts respectively, with major complications reported at 2% and a suggested procedure complication threshold of 4%. A report on 450 PTC's by Mueller in 1980 from the Massachusetts General Hospital details a success rate of 99% for patients with dilated bile ducts and 74% for nondilated ducts. The major complication rate was 4.8% including one death for a 0.22% mortality rate. In the same year, Harbin reviewed the published literature and used a multiinstitutional questionnaire to evaluate a total of 3,596 PTC procedures. Success in patients with dilated ducts was 98% and overall major complications were reported in 3.28% of procedures. The details of these complications were: sepsis 1.84%, bile leakage 1.03%, intraperitoneal hemorrhage 0.28% and death 0.14%.

Anatomic Review

Hepatic lobar anatomy: The liver is composed of two lobes according to the distribution of its vascular structures. The right lobe is supplied by the right hepatic artery and the right portal vein; the left lobe is supplied by the left hepatic artery and the left portal vein. The boundary between the right and left lobes, the major lobar fissure, can be approximated by drawing a line between the gallbladder fossa and the inferior vena cava. Surgical dissection of the liver along this plane intersects the middle hepatic vein, which runs in the plane of the major fissure. Each lobe is further divided into two segments. The right lobe is composed of anterior and posterior segments while the left lobe is composed of medial and lateral segments.

Subsegmental anatomy: In 1957, Couinaud suggested that the liver could be further subdivided into eight subsegments based on the anatomic course of the portal and hepatic veins **(Fig. 5)**. The eight subsegments of Couinaud are:

1) Caudate
2) Posterior aspect, lateral segment of the left lobe
3) Anterior aspect, lateral segment of the left lobe
4) Medial segment, left lobe
5) Inferior aspect, anterior segment right lobe
6) Inferior aspect, posterior segment right lobe
7) Superior aspect, posterior segment right lobe
8) Superior aspect, anterior segment right lobe.

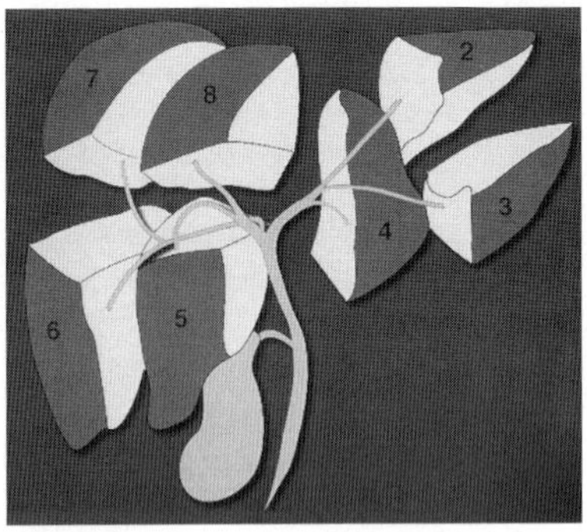

Figure 5. *Diagram showing the Couinaud subsegmental anatomy of the liver.*

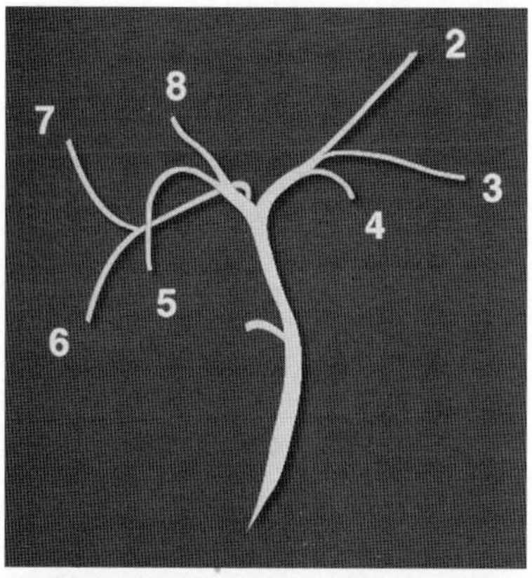

a.

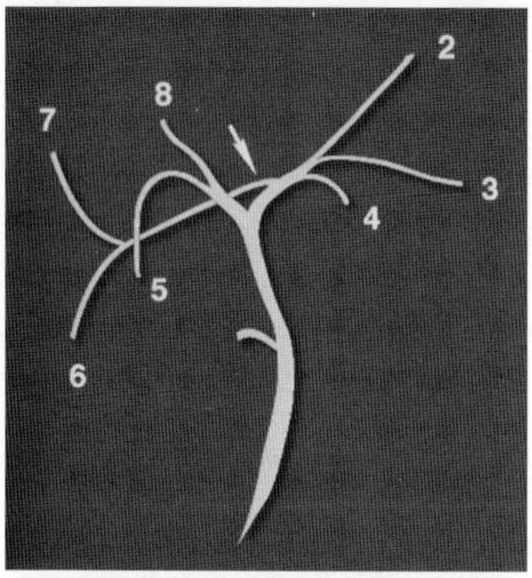

b.

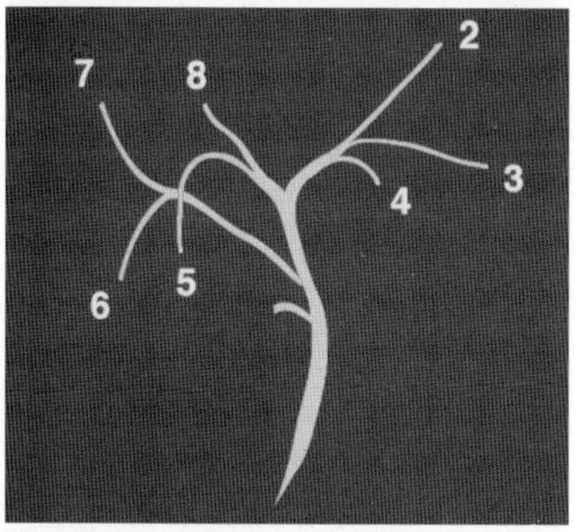

c.

Figure 6. *a) Normal bile duct anatomy. b) The posterior right duct from segments 6 and 7 drains into the left duct in about 20% of livers (straight arrow). c) The posterior right duct from segments 6 and 7 drains into the extra-hepatic common hepatic duct in about 2% to 6% of patients. It is at risk of being confused with the cystic duct and cut in error during cholecystectomy.*

Biliary anatomy: The left hepatic duct (LHD) drains the left lobe of the liver and is formed by the segmental ducts from Couinaud subsegments 2, 3 and 4. The main LHD generally lies anterior to the portal vein and is formed by inferior and superior horizontally oriented tributaries. The right hepatic duct (RHD) drains segments 5, 6, 7 and 8 and is formed by two main tributaries from the posterior and anterior segments of the right lobe. The RHD joins the LHD to form the common hepatic duct at the porta hepatis **(Fig. 6)**. The RHD is normally short and is subject to considerable normal variation. The most common intrahepatic biliary ductal variations are listed below:

1) Right posterior duct draining into the left duct: 20%
2) Trifurcation of the left and right anterior and posterior ducts: 12%
3) Aberrant segmental right posterior duct into the common hepatic or cystic ducts: 2-6%.

The common hepatic duct (CHD) is joined by the cystic duct to form the common bile duct (CBD). The cystic duct is 2-4 cm long and inserts along the right lateral border of the CHD. The most common anomalies are a low insertion into the CBD or a parallel course with the CHD.

Cholangiography: General Findings

The chief objectives of cholangiography are:

1) to clearly portray the intra- and extrahepatic biliary anatomy;
2) to evaluate the functional status of the biliary tree; specifically to evaluate for obstruction or bile leak **(Fig. 7)**;
3) to determine the level of biliary obstruction and, when possible, suggest the etiology **(Fig. 8)**.

Portraying biliary anatomy: In evaluating the intrahepatic biliary tree, it is important to optimize patient position to promote filling of all of the ducts. Imaging projection is the other critical factor. The LPO (RAO) projection is best for displaying the biliary confluence.

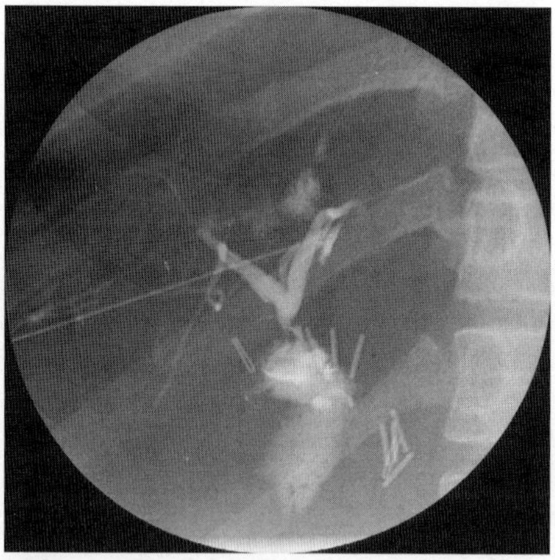

Figure 7. *A patient with a suspected biliary leak following a repair of an injury sustained during laparoscopic cholecystectomy. The fine needle PTC shows a leak at the bilo-enteric anastomosis. PTBD was requested. The diagnostic needle enters the left duct from a right intercostal approach. There are steep angles involved and this entry site should not be converted for the PTBD. A fresh puncture should be made peripherally in the right duct.*

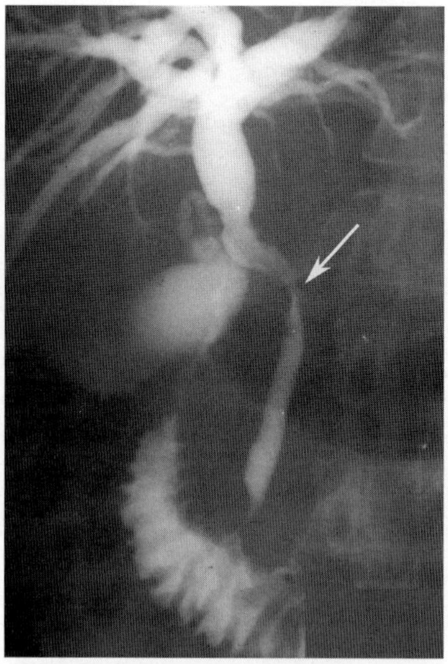

Figure 8. *A PTC of a patient with Hodgkin's lymphoma showing narrowing of the mid portion of the extrahepatic bile duct (arrow) caused by external pressure from enlarged lymph nodes.*

A common pitfall in cholangiographic interpretation is failure to recognize missing ducts. To this end, a thorough understanding of normal and variant biliary anatomy is required. Even so, it is important to realize that a dilated obstructed segmental duct may appear to occupy the territory of an entire lobe.

Evaluating functional status: Biliary obstruction is usually accompanied by bile duct dilatation, but this is not a universal finding. Patients with sclerosing cholangitis, patients status post liver transplantation or patients with early obstruction may have nondilated ducts. Some investigators have advocated measuring biliary pressures to determine degree of obstruction, but we have not found this useful.

Cholangiography: Specific Findings

Filling defects: Filling defects in the biliary system may be due to air, stones, blood clot or tumor. Air bubbles are round and occupy a nongravity-dependent position. Stones, on the other hand, usually move to a gravity-dependent position. They may be faceted and seen in association with gallbladder stones. Primary bile duct stones can have a variable appearance. They may appear as round discrete stones or they may fill the entire duct giving a cast-like appearance. Intraductal tumor is an uncommon finding but one that is characteristic of papillary ductal cholangiocarcinoma.

Obstruction: When obstruction is partial, the result is termed a "biliary stricture." Cholangiographic interpretation should include length and location of the stricture as well as description of the morphology. A distinction between benign and malignant causes of biliary obstruction can be inferred from the morphology of the obstruction. In general, benign obstruction results in partial obstruction with a smooth, tapered narrowing. Malignant obstruction is more likely to be complete with an abrupt, irregular border. However, this distinction between malignant and benign processes is not exact and pathologic evaluation is the only method for insuring a correct diagnosis.

In describing the location of a stricture (particularly benign strictures), it is common to use the Bismuth classification **(Fig. 9)**.

BISMUTH CLASSIFICATION

- Type 1: CHD > 2 cm
- Type 2: CHD < 2 cm
- Type 3: Hilar, confluence intact
- Type 4: Hilar, right and left ducts separate
- Type 5: Injury to aberrant right duct with or without CHD injury

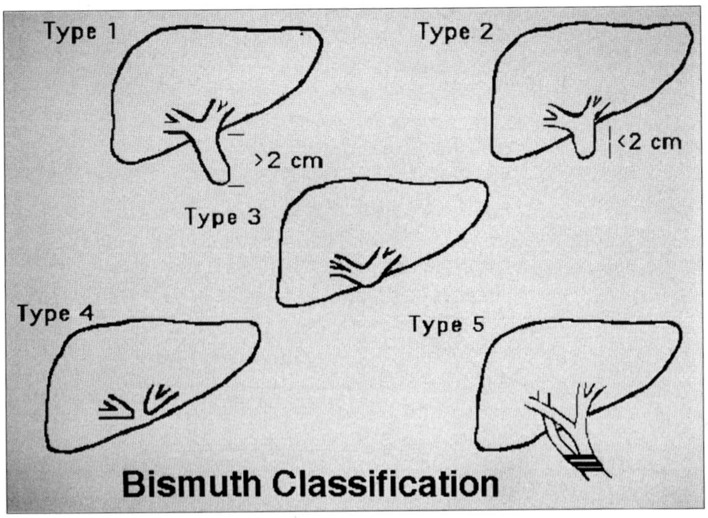

Figure 9. *Diagram of the Bismuth classification for level of bile duct invovlement.*

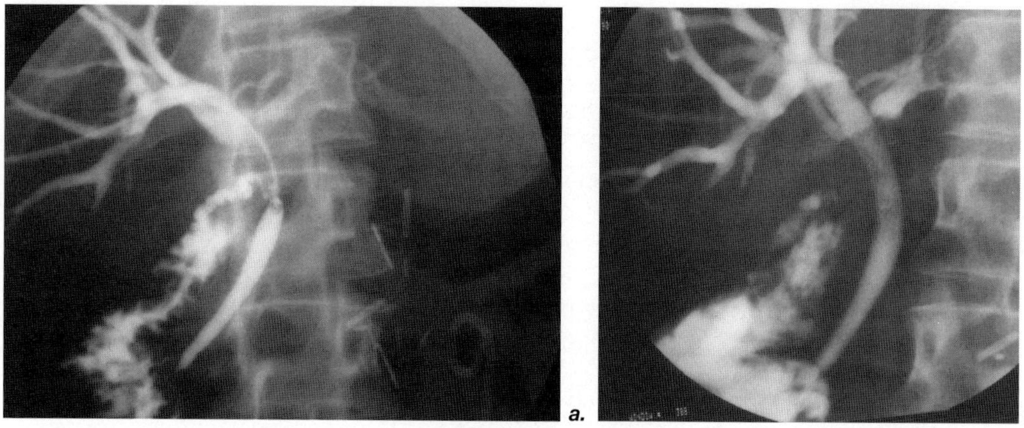

a. b. ***Figure 10.*** *a) The cholangiogram clearly demonstrates the extent and margins of this biopsy-proven malignant biliary obstruction. b) A Wallstent has been placed through the obstructed segment and is long enough to extend at least 3 cm on either side of the tumor margins. Excellent drainage is established.*

Brief Review of Biliary Pathology

Malignant biliary disease **(Table 2)**: Malignant biliary obstruction is most commonly due to tumors of the pancreas, bile duct, gallbladder, ampulla or metastatic disease with nodal compression of the duct **(Fig. 10)**.

Ampullary and pancreatic tumors usually result in mid to distal common bile duct obstruction. Metastatic disease and gallbladder carcinoma more typically affects the ducts at or below the hilum. Cholangiocarcinoma may produce obstruction at any level. In our population, cholangiocarcinoma most typically presents with hilar obstruction with strictures extending back into the intrahepatic ducts.

Benign biliary disease: In general, benign biliary disease is less common than malignant disease. **Table 3** lists a variety of benign diseases of the biliary tree. Of these, the benign processes that we see most commonly at UCSF are postoperative strictures, strictures following liver transplantation and sclerosing cholangitis.

TABLE 2: MALIGNANT BILIARY DISEASE

- Distal duct
  - Pancreatic carcinoma
  - Ampullary carcinoma
- Proximal duct (CHD, hilum)
  - Gallbladder carcinoma
  - Metastasis
  - Klatskin tumor
- Cholangiocarcinoma
  - Can occur anywhere in the duct

TABLE 3: BENIGN BILIARY DISEASE

- Postoperative strictures
  - Postcholecystectomy
  - Postliver transplantation
- Sclerosing cholangitis
- Congenital biliary disease
  - Biliary atresia
  - Choledochal cysts
  - Duodenal diverticula
  - Polycystic liver disease
- Stone-related strictures
  - Recurrent pyogenic cholangitis
  - Mirizzi syndrome
- Other acquired strictures
  - Pancreatitis
  - Papillary stenosis
  - AIDS cholangiopathy
  - Tuberculosis

Percutaneous Transhepatic Biliary Drainage (PTBD)

Indications

Percutaneous transhepatic biliary drainage (PTBD) may be indicated in a variety of settings.

These patients fall into two groups: those with biliary leaks and those with partial or complete biliary obstruction. In most institutions, PTBD is reserved for drainage of patients where retrograde endoscopic drainage cannot be performed. Endoscopic drainage is the first choice because of its lower complication rate and because it can often be performed on an outpatient basis and completed in a single intervention.

PTBD has the advantage of being successful in 96% to 100% of patients with dilated ducts and so is used when endoscopy fails. However, PTBD is not an outpatient procedure. PTBD may be favored in particular centers because of local differences in expertise or availability of ERCP. In patients with obstructive biliary sepsis, the percutaneous approach can be performed simply and rapidly on the supine patient and may be preferable to the endoscopic route. Relative contraindications to PTBD are coagulopathies, which should be addressed by suitable infusions of blood products; and significant ascites, which may require preliminary tapping.

INDICATIONS FOR PTBD

- Gain access to the biliary tree
- Decompress obstructed biliary tree
- Dilate biliary strictures
- Remove bile duct stones
- Divert bile away from and stent bile duct leaks

Biliary drainage can be established in three separate configurations:

a) External biliary drainage where the percutaneous catheter is positioned within the biliary tree and is attached to a collection bag outside the body to allow external drainage of bile. This is often a preliminary step to decompress an obstructed infected biliary tree. This allows the patient to recover from an acute septic state. It is also useful because it allows time for any clots or debris to clear and for the percutaneous tract to mature. External drainage is not suitable for long-term drainage because of the large fluid and electrolyte losses that occur over time. The diversion of all bile salts from the gut impairs the digestive processes and adversely affects the patient's general state of well-being.

b) Internal-external biliary drainage involves positioning the percutaneous drainage catheter through the bile ducts and down into the bowel. Side-holes in the catheter in the ducts above the obstruction drain bile through the catheter lumen to the bowel when the catheter is capped off outside the body **(Fig. 11)**.

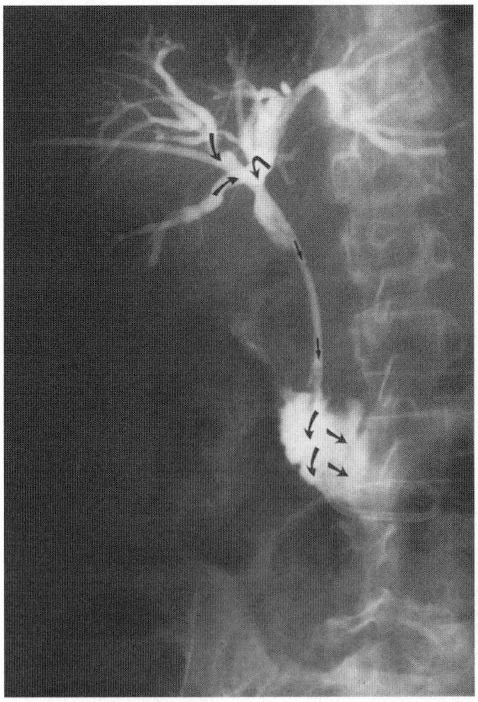

Figure 11. *A 65-year-old woman with gallbladder carcinoma. The cholangiogram shows an internal-external biliary drainage catheter which has been capped and allows drainage from the obstructed ducts through side-holes into the catheter and then down the lumen of the catheter to the duodenum.*

Advantages of the internal-external system are that fluid loss is eliminated and bile drains into the bowel. Such catheters are easily exchanged over a guidewire on an outpatient basis so that problems of catheter obstruction can be simply managed. Pain, infection and bile leakage at the skin site may be problems with this arrangement.

c) Implanted endoprostheses provide all the advantages of internal drainage without the problems associated with the percutaneous catheter exiting from the body wall. All endoprostheses are prone to obstruction over time and unblocking them involves a repeat percutaneous intervention in most cases. Endoprostheses can be plastic tubes or expandable metallic stents. Metallic stents are currently most often used.

Technique

After opacification of the ducts by PTC or ERCP, the cholangiogram is reviewed to select the best bile duct to puncture. The duct should be punctured in a peripheral location both to gain a sufficiently long length of duct above any obstructing lesion and also to avoid injury to the large portal venous and arterial structures, which are concentrated in the hilum of the liver. Left ducts are catheterized from an anterior subzyphoid or subcostal approach. Ultrasound guidance can often be helpful for dilated left ducts. For right lateral approaches, the skin entry site is intercostal and should be lower than the costophrenic angle. The entry site should be chosen to allow the shortest path to the target duct. A well- chosen path will facilitate subsequent maneuvers by minimizing any sharp angles along its course. For this reason the left ducts should not be catheterized from the right lateral approach **(Fig. 7)**.

In some cases the duct punctured at the time of the original diagnostic PTC provides a suitable site for cannulation, in which case a conversion system can be used. However it is important to resist the temptation to convert a suboptimal entry site at the time of PTC. Oblique views are of great help when puncturing the selected duct. A variety of needle systems are successfully used to gain access to the bile ducts. These systems are of two main types: those using a fine 21- or 22-gauge needle for the initial puncture followed by insertion of an .018" wire and subsequent conversion using systems such as the Cope, Neff or similar sets, and those where the puncture is performed using a 19-gauge needle covered by a flexible 4.8-F sheath. Once the needle is removed, the sheath will accept an .038" guidewire and the sheath can then be advanced over the guidewire into the bile duct. If the gallbladder is inadvertently punctured, as much bile as possible should be aspirated from the gallbladder before removing the needle. This is done to try to avoid bile leakage and peritonitis.

Once access to the bile duct has been achieved, a combination of torquable wires and catheters will in most cases allow fluoroscopically directed passage through the bile ducts, across any obstructions into the draining bowel. Successful negotiation through the obstructed duct is essential for antegrade drainage and subsequent endoprosthesis placement. Fortunately, catheterization through an obstruction is almost always feasible.

The next step is to place a stiff guidewire through the catheter into the bowel. The guidewire's rigidity smoothes out the angles to be negotiated and reduces buckling in the space between the liver and the lateral abdominal wall. The initial transhepatic tract will usually require predilatation before a drainage catheter can be inserted. This is particularly important if: the liver is hard; if there is ascites; if there are difficult angles; or if a large-diameter catheter is to be inserted. Semirigid plastic fascial dilators are deployed over the guidewire.

Dilators of increasing size are used consecutively until the tract is of sufficient diameter to admit the required drainage tube. Fluoroscopy is used to ensure that the dilator is being advanced along the course of the guidewire in order to avoid kinking the guidewire. Kinking is not uncommon and if it occurs, the dilator should be removed, and the wire exchanged for a new wire using a sheath as an intermediate step. A 5-F 4-mm balloon catheter will occasionally be needed when tract dilatation cannot be achieved with fascial dilators.

During introduction of the drainage catheter, the space between the liver and the abdominal wall is observed fluoroscopically. The catheter may buckle in this space, a phenomenon which is painful and which prevents successful passage of the catheter. Buckling can be avoided by advancing the catheter over the guidewire during the phase of respiration in which the liver tract aligns with the skin entry site. The metal stiffener supplied with the catheter is of great value, but must be used with care to avoid kinking the guidewire or damaging the liver parenchyma.

For external drainage, an 8-F or 10-F self-retaining loop catheter is usually selected. The larger caliber tube is used when the bile is thick or pus-like. The loop can be of the Cope type with a diameter of about 25 mm for use in a distended biliary tree. For less capacious systems, a Kumpe-type loop, which has a smaller diameter, is very helpful. The loop is positioned at the confluence of the right and left ducts. Side-holes are positioned along the inner curve of the loop to facilitate subsequent guidewire insertion. The catheter is fixed to the skin and attached to bag gravity drainage.

For internal-external drainage an 8-F,10-F, or 12-F self-retaining biliary loop catheter is used **(Fig. 11)**. The loop is positioned in the duodenum just below the ampulla and prevents the catheter from being pulled out. Skin fixation prevents forward displacement of the tube. Side-holes should be carefully positioned above the obstruction to drain the ducts. There should not be any side-holes in the liver parenchyma or along the tract towards the skin. This avoids venous bleeding into the catheter or bile leakage around the catheter. Additional side-holes may have to be made to tailor the commercially available catheters to fit the anatomy of the individual patient. Under fluoroscopic control, contrast material is injected and then withdrawn through the catheter to assure that the side-holes are correctly positioned for easy filling and emptying of the ductal system. If the drainage is heavily blood-tinged or if the patient has signs of infection, the tube is attached to temporary bag drainage; otherwise it may be capped externally.

If an endoprosthesis is to be used, it is ideally inserted at the same session as the initial drainage. The patient's hospital admission is thus kept short and the number of visits to the interventional department is reduced. This can be achieved in most patients particularly if an expandable metallic stent such as a Wallstent is used. The Wallstent is inserted using a delivery system of 7 F. It is deployed through a 7-F sheath of outer diameter of 8.5-F. The indwelling stent placement should be delayed in patients who a) have biliary sepsis, or b) are bleeding significantly following the initial drainage. They should be managed by temporary external biliary drainage until the sepsis or bleeding has resolved.

Both tubular plastic and expandable metallic stents are prone to obstruction and have a limited duration of patency. However, many patients with malignant obstruction die within 4 to 6 months and 12-F plastic stents often remain patent for the remainder of the patient's life providing good palliation. Metallic stents were anticipated to prolong stent patency significantly as they enabled a stent with a lumen of 10 mm to be introduced using an introducer system of only 8.5 F in outer diameter. In practice, it was disappointing to find that even metallic stents were very prone to occlusion as occurred in 24% of our series of 50 Wallstents.

Although the expected benefits of metallic stents have not been completely realized, prospective randomized trials have shown advantages of metallic over plastic stents which seem to offset the 8- to 10- fold greater cost of the metallic stents. In Lammer's comparative trial, the data for plastic versus metal stents are respectively 30-day mortality— 24%/10%; median survival— 98days/123days; median time to obstruction or death— 81 days/122days.

The cause of occlusion of metallic stents is not well established. Sludge formation is common and suggested approaches such as the use of choleretics and antibiotics have been tried with conflicting results. With low pancreatic lesions, stenting across the sphincter of Oddi is almost always needed, but theoretically, by leaving the stent above the sphincter and not breaching the integrity of the sphincter, reflux up the bile ducts can be minimized, thus reducing contamination of the biliary tree and reducing resultant sludge formation. Most agree that tumor overgrowth is an important factor and so using stents long enough to extend well beyond the tumor margins is used with some benefit **(Fig. 10)**. Tumor ingrowth through the mesh interstices of the metallic stent does occur, but its relative clinical importance is debatable as shown by two postmortem studies and by direct endoscopic inspection. In the first study, there was ingrowth in only one of 15 patients. The tumor was a hepatoma causing 50% narrowing of the stent, but not occlusion. In the other autopsy study of 24 patients, tumor ingrowth was found in only one patient and it was nonobstructive. Work continues in attempts to discover pharmacological stent coatings designed to inhibit sludge formation and tumor growth. Initial work with Wallstents covered by polyurethane has shown no improvement in patency compared to bare metal stents. There was considerable sludge formation and tumor overgrowth, and tumor was even found within the stent as a result of breakdown of the coating. Migration of coated stents has been a reported problem. In the USA, a limited pilot trial of silicone-coated Wallstents showed no advantage in using covered stents.

Disposable equipment costs for placement of expandable metallic stents are currently some 8 to 10 times higher than costs for a plastic stent. In most patients, the patency rates of either stent type are of adequate duration to cover the patients' short average survival time which is limited by the underlying obstructing malignancy. All types of stents are prone to obstruction over time, but over the long term, expandable metallic stents may have a better patency rate than plastic stents and, unlike plastic stents, have no tendency to migrate. They are certainly easier to insert than plastic stents, which are 12 F in caliber. If hospitalization times can be reduced by early placement of a metallic stents then this offsets their higher cost. Wallstents are currently the most common metallic stents used in the USA for treatment of malignant biliary obstruction. They have a flexible 7-F delivery system and are available in a variety of lengths and diameters. They are self-expanding and although they conform well to the contours of the biliary ducts, they have good hoop strength. Accurate placement of Wallstents requires experience in their use, because there is about a 40% shortening in length when the Wallstent goes from its fully constrained state to its fully expanded state. Wallstents have a reconstrainable property effective during the initial stages of deployment, which aids in accurate positioning.

Newer self-expanding stents, such as the Smart Stent, Memotherm and many others currently under development, have eliminated the shortening problem to a great extent. Experience is growing, particularly with their use outside of the United States. Balloon-expandable stents such as the Gianturco and Palmaz stents have the advantage that they can be more accurately positioned than Wallstents. For short lengths, they may prove less costly. Gianturco Z-stents are our stent of choice for benign biliary obstructions. Palmaz stents have not, as yet, seen wide use in the biliary tree despite their extensive use in the blood vessels. Many other balloon or self-expandable metallic stents are now available in the United States and experience is being gained in their use and performance.

Results

In several large series with malignant obstructions of different etiologies, average patencies for metallic stents were between 5.8 and 8.3 months and occlusion rates were from 7% to 24%. In Rossi's multicenter European study of 240 patients using life table analysis, the overall survival rates were 42% at 25 weeks and 16% at 50 weeks. The 30-day mortality rate was 14.6%. The 25-week patency rate for nitinol Strecker stents was 78% and for Wallstents was 67%.

Complications and Success Rates

Major complications associated with PTBD or percutaneous stent placement as reported in the Quality Improvement Guidelines of the Society of Cardiovascular & Interventional Radiology include: bleeding (2.5%), sepsis (2.5%), localized inflammatory problems (1.2%), pleural complications (0.5%) and death (1.7%). These rates are an average for all comers and vary in the reports from different centers. Bleeding from portal or hepatic veins is almost always self-limited. Bleeding from hepatic arteries is successfully treated by selective arterial embolization. The success rate for cannulation of the bile ducts after opacification by PTC is 95% for dilated ducts and 70% for nondilated ducts.

Benign Biliary Obstruction

The majority of benign biliary strictures occur after surgery and are short in length. Treatment options include surgical repair, balloon dilatation, stenting with plastic endoprostheses, and metallic stenting. Initial surgical repair has a success rate of only 78%-88% with recurrent stricture occurring in 15%-25% of patients. The success rate of repeat surgical repair decreases with the number of operations with good results in less than 61% of patients after the third operation. Interventional management with percutaneous transhepatic balloon dilatation can be successful, but there is a recurrent stricture rate of 27%-33% for anastomotic strictures and 12%-24% for nonanastomotic strictures at three years. Plastic endoprostheses can be placed retrograde via endoscopy or antegrade via a transhepatic approach. They are problematic because of a tendency to migrate, especially when placed across an anastomotic stricture. And, over time, all plastic endoprostheses require repeat endoscopic exchange because of occlusion with sludge and debris. Metallic stents may be indicated when all other treatments have failed. Long-term results are not available in a sufficient number of patients. Because patients with benign biliary stricture are likely to live for decades, treatment success must be measured accordingly.

The Bismuth classification of biliary obstruction is used to categorize the level of bile duct injury **(Figs. 9, 12)**. The surgical repair is increasingly difficult as the injury approaches the porta hepatis.In addition to the level of injury, surgical success is influenced by the experience of the surgeon, the time interval from injury to repair (worse with prolonged interval), and the presence of coexisting complications such as cholangitis, periductal infection, cirrhosis or portal hypertension. Restenosis occurs in 10%-30% of patients after surgical repair. Thereafter, subsequent surgical repair is even more complex. Interventional techniques have evolved in response to the need to address those patients who fail surgical repair, have nonoperable lesions (high in the porta hepatis or intrahepatic), or are limited by an unsatisfactory surgical risk.

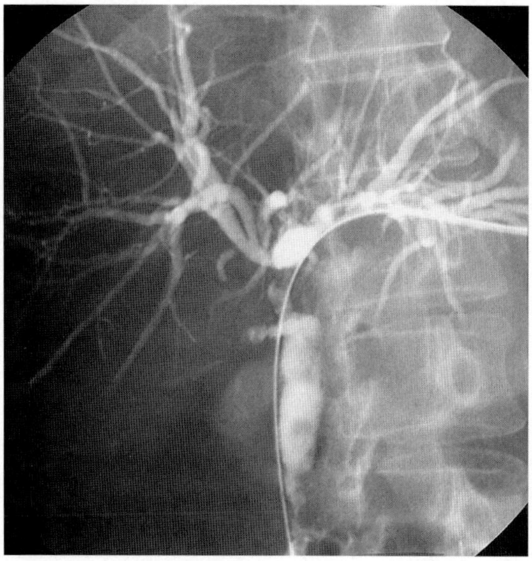

Figure 12. *The cholangiogram shows a high benign stricture that is of the Bismuth Type 3.*

Balloon Dilatation

Literature review of the success of interventional treatment of benign biliary stricture is difficult because reports include many variables regarding technique, and there is a mixing of etiology as well as the level of obstruction in most reports. Also, malignant and benign strictures frequently are reported together. Several reports indicate that overall an approximate 65% success rate of percutaneous balloon dilatation of benign biliary strictures can be expected, but repeated dilatations and prolonged stenting may be necessary to achieve this. Moore et al reported their experience in treating 18 patients with benign biliary strictures in 1987. In their series, 6 of 9 patients with choledochoenterostomy stricture were successfully dilated with only one dilatation, but one, three, and eight dilatations were required in the remaining three patients. They were unsuccessful in removing the external drainage catheter in two patients who died from other causes. The seven patients who became catheter-free were patent on follow-up at 22-55 months (mean, 35.4 months). Technical success was not achieved in two additional patients with choledochoenterostomy stricture. Five of six patients with postoperative biliary stricture (not bilioenteric) were successfully dilated with long-term follow-up available only in two patients who were patent at 42 and 55 months. One patient with an inflammatory common bile duct stricture was patent at 18 months.

Williams, et al, in 1987, reported their experience with balloon dilatation of benign biliary strictures in 74 patients. 67.5% were successfully relieved of obstruction on follow-up that ranged 1-58 months (mean = 28 months). Overall, 60% were asymptomatic and free of an internal-external transhepatic stent at 24 months. This was reduced to 38% at 36 months. Forty patients were strictured at a bilioenteric anastomosis and 24 had ductal stricture. Significantly, 72.5% (29/40) of patients with bilioenteric anastomotic strictures were successfully treated but this required repeated dilatations in all and an average of 8.8 months of percutaneous transhepatic stenting. 87% of patients with ductal strictures were successfully treated after repeated dilatations and an average of 4.7 months of stenting. Overall, ten patients never became stent-free, and 22% (16/64) of patients sustained recurrent stricture following stent removal (mean = 13.8 months, range =1-36 months, median = 8 months). Thirteen of 14 were successfully treated with repeated percutaneous dilatation and an average of 9.6 months of restenting.

Mueller et al reported a 67% patency rate in 73 patients with benign stricture treated with percutaneous balloon dilatation at a minimum follow-up of 36 months. It is unclear whether this is a primary or assisted patency rate, but they did indicate a 34% (25/73) recurrent stricture rate. Of these, 16 were successfully treated by repeated dilatation and nine by surgery. This was a retrospective multicenter study of 44 patients with bilioenteric anastomotic strictures, 28 with postoperative nonanastomotic strictures, and 17 with sclerosing cholangitis. The 36-month patency rates were 67%, 76%, and 42% respectively for each patient group. Data regarding the duration of stenting was not included. In summary, balloon dilatation of benign biliary stricture can be successful, but recurrent stricture is common and repeated dilatations with prolonged percutaneous transhepatic stenting often is necessary.

Surgery Versus Balloon Dilatation

Pitt et al reported their comparison of surgical repair of benign biliary stricture versus percutaneous balloon dilatation. Although this was not a randomized, prospective trial, the results listed in **Table 4** below are of some interest.

Their data reflects a greater success rate with surgical repair and a higher complication rate of hemobilia with balloon dilatation. The authors concluded that surgical treatment is preferable to percutaneous balloon dilatation unless the patient has a high risk for surgery or is unwilling to undergo surgery. Note that both patient groups endured an average of more than 13 months of percutaneous transhepatic biliary stenting.

TABLE 4: BENIGN STRICTURES – SURGERY VS DILATATION

	Surgical	**Balloon**
No. of patients	25	20
Clinical profile	same	same
Hosp days/cost	same	same
No. of procedures/pt (mean)	1	3.9
Procedural deaths	0	0
Hemobilia	4%	20%
Stenting	13.8	13.3
Follow-up/months (mean)	57 (7)	59 (6)
Success	88%	55% ($p < 0.02$)

(Pitt HA, et al 1989.)

Endoprostheses

Plastic indwelling endoprostheses were developed to address the problem of frequent stricture recurrence as well as the morbidity and cost of long-term transhepatic biliary stenting associated with both percutaneous balloon dilatation and surgical treatment. However, there are several problems associated with these 12-F plastic endoprostheses. For one, they have a tendency to migrate, especially when placed across a bilioenteric anastomosis. For another, and even more importantly, all eventually occlude. Endoscopic exchange can be performed in patients with normal gastroduodenal-biliary anatomy, but this must be done routinely about every 3 to 6 months. And, stents placed across a bilioenteric anastomosis cannot be reached endoscopically. Percutaneous transhepatic exchange can be accomplished, but this is not acceptable as a routine treatment for patients with benign disease who may have a life expectancy of decades.

Metallic Stents

Because of the above problems, expandable metallic stents have been used to treat biliary obstruction with the hope that their larger diameter (up to 12 mm) will provide longer patency and that migration will occur less frequently. There are two types of metallic stents. Those that have an intrinsic expanding force are "self-expandable." These include the Wallstent (Boston Scientifc), the modified Gianturco-Rosch "Z" Zigzag stent (Cook), Smart Stent (Cordis) and many others. The other types are mounted on an angiographic balloon catheter and deployed by balloon expansion, such as the Palmaz stent (Johnson & Johnson). Problems have related largely to stent obstruction due to mucosal hyperplasia and stone formation **(Table 5, Fig. 13)**.

Technique

When using a metallic stent, we choose the modified Gianturco Z stent for the following reasons. In benign disease, the stent is needed only as a scaffold. The Gianturco stent is in contact with less of the ductal mucosal surface than is the Wallstent, possibly stimulating less endothelial hyperplasia. It is more radiopaque and better seen fluoroscopically. The Gianturco stent is easier to place precisely, and finally, it does not shorten. Individual 15-mm long units can be linked in flexible chains to provide stenting over a variable distance. Diameters of 8, 10, and 12 mm are available. The upper and lower extent of the stricture must be defined precisely in order to select the best stent length and diameter and then to position the stent optimally.

Summary

We treat benign biliary strictures with the Gianturco Z stent only when other options have failed or are not possible. The first-line treatment of benign biliary strictures continues to be surgical repair. When surgery is not feasible or unsuccessful, our interventional approach begins with percutaneous transhepatic balloon dilatation. We leave a drainage catheter across the stricture for six weeks. This catheter is then exchanged for a catheter that is positioned above the stricture for a 1-2 week test period. If the stricture remains widely patent after this test period, the drainage catheter is removed. If the stricture recurs during this test period, it is redilated and a similar test period begins. If restenosis occurs after 2-3 dilatations, it is considered unsuccessful and we then consider placement of a metallic Gianturco Z stent. If restenosis occurs following placement of a Gianturco stent, treatment consists of débridement and balloon angioplasty or placement of another Gianturco stent. Metallic stents are reserved for intractable cases because they cannot be exchanged or removed either endoscopically or transhepatically and because their presence makes surgical repair difficult or impossible. Long-term results that match the duration of life expectancy in these patients with benign disease are not yet available.

TABLE 5: STENTING OF BENIGN BILIARY STRICTURE LITERATURE REVIEW

Author	Year	Pts	Months	Patency	Problems
Irving	1989	11	6-21	72.7%	Migration - 2 Technical failure - 1
Coons	1992	40	12	87%	(11 with PSC)
Maccioni	1992	18	30-41	55.5%	3 died, 5 recurrences Migration - 1 Stent obstruction - 3 Papillary stenosis - 1

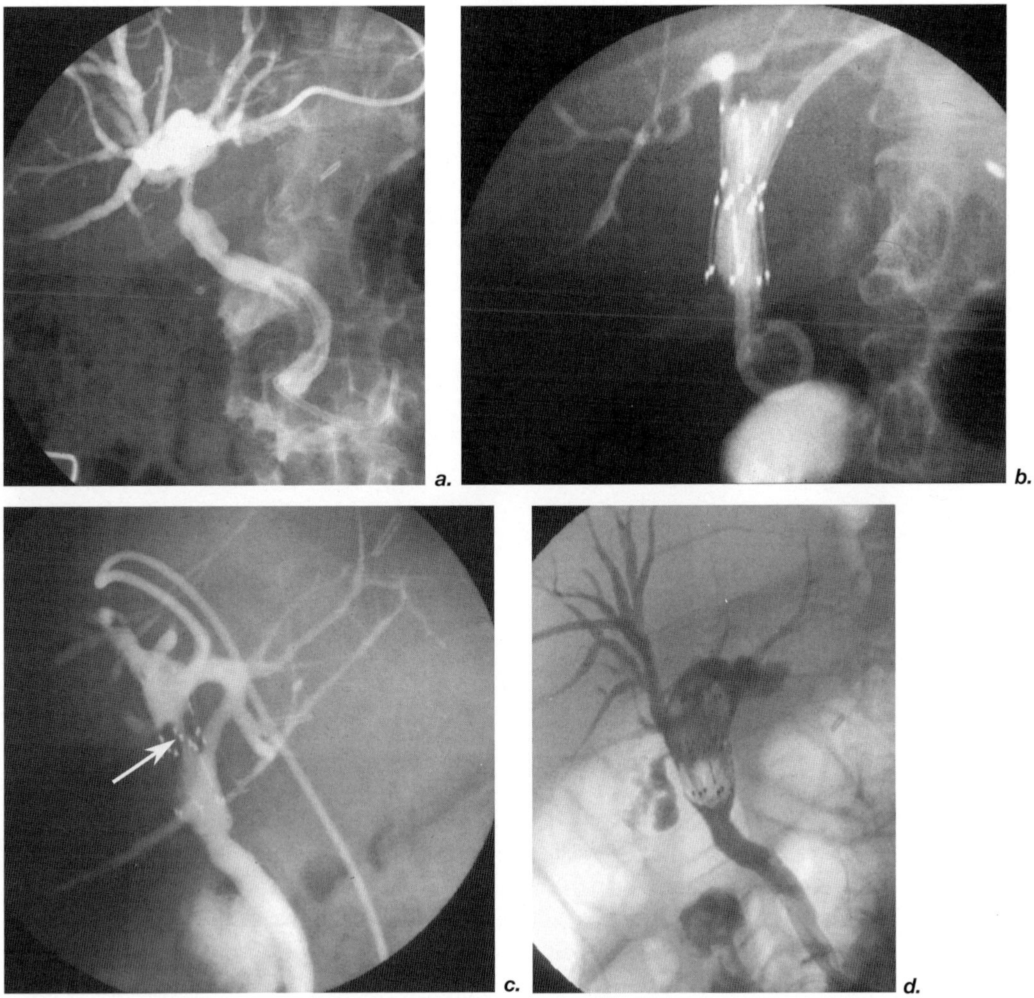

Figure 13. *a) PTBD in a liver transplant recipient. There is an ischemic stricture of the common hepatic duct high in the porta hepatis as a result of occlusion of his hepatic artery. b) This was successfully treated using a Gianturco stent. c) Six weeks later the stricture has recurred at the same site within the Gianturco stent (arrow). It was treated by placing a coaxial Palmaz stent. d) The patient did reasonably well, but 7 years later a PTC showed the stents encased in stone material causing almost complete obstruction. He was treated surgically.*

Percutaneous Roux Loop Puncture

A small group of patients with biliary problems will have had a surgically created Roux loop to which their bile ducts have been anastomosed. In cases where the loop lies adjacent to the abdominal wall it can be punctured percutaneously, provided there are no intervening loops of bowel or other organs. This route can then be used to gain retrograde access to the biliary tree without having to use the transhepatic route. It has the advantage of permitting access to both right- and left-sided ducts. This approach is of particular value in patients who require repeated dilations or stone extractions such as in recurrent strictures, sclerosing cholangitis and recurrent pyogenic cholangitis. In some patients, the Roux is constructed with the express intention of allowing repeated percutaneous access. The end of the Roux is tacked up to the underside of the abdominal wall and is marked with surgical clips. This is the so-called Hutson loop.

Percutaneous Cholecystostomy

Indications

Percutaneous cholecystostomy (PC) is indicated in the treatment of patients with suspected acute calculous or acalculous cholecystitis. Its role is to provide temporary drainage in patients who cannot undergo immediate surgical removal of the gallbladder because of serious comorbidities and/or contraindications to surgery. The drainage procedure may promote improvement in the patient's condition so that gallbladder excision at a later stage can be safely performed. Some patients are in such poor condition that they are managed permanently with the PC. In a few patients with acalculous cholecystitis, the PC can be removed once the acute problem has resolved. In patients with multi-organ problems, placement of a PC is of value to exclude the obstructed gallbladder as one possible source of sepsis. Some authors suggest that placement of a PC is indicated as a means of diagnostic cholangiography or biliary drainage; however, many disagree with this use. In rare cases, PC is used to gain access to the gallbladder for percutaneous stone extraction.

Technique

The gallbladder is usually punctured under direct real-time ultrasound guidance. This is most conveniently performed on the fluoroscopy table so that tube placement can be monitored by fluoroscopy. If necessary, in very sick patients, the complete procedure of percutaneous cholecystostomy can be achieved with ultrasound guidance performed at the patient's bedside. CT can also be used to guide gallbladder puncture. The route to the gallbladder can be transhepatic, in which case the chosen route should not traverse just a sliver of parenchyma that may be lacerated by the drainage tube, but should go through a reasonable thickness of liver tissue. The transhepatic route offers a secure placement with easy tube exchanges and potentially less bile leakage and tube displacement. Advocates of the alternate transperitoneal approach claim there is less danger of bleeding in patients with deranged coagulation when the transhepatic route is avoided. Hard data is not available to substantiate the superiority of one approach compared to the other. We favor the transhepatic approach where possible, but will puncture the gallbladder directly when this is the anatomically most favorable route. We generally use a Seldinger technique in the fluoroscopy suite starting with a rigid 19-gauge needle that will accept an 0.038-inch guidewire and placing a 10-F self-retaining loop catheter after suitable tract dilation. When the bile is very thick, a 12-F catheter should be used.

If the cholecystostomy is performed at the beside under ultrasound guidance, a single stick trocar, such as the McGahan device, is convenient. This eliminates the necessary exchanges of the Seldinger technique. Placement of a PC can be tricky because the thick-walled gallbladder may float away from the needle and wires have a tendency to coil in the pericholecystic space if care is not exercised. Removal of the PC should not be attempted sooner than six weeks after placement in order to allow enough time for a mature catheter tract to develop. A tube cholangiogram is essential prior to removal to demonstrate free passage of contrast agent down the cystic duct, to the common bile duct and down into the bowel. Before PC removal, a tractogram can be performed by exchanging the cholecystostomy tube for a side-arm sheath over the retained guidewire. Contrast injection will then demonstrate a formed tract without intraperitoneal bile leakage.

Results and Complications

PC should be successfully performed in about 98% of patients with an associated major complication rate of 9%. The major complications are hemorrhage and bile peritonitis.

Procedures Performed Through Surgical Tracts

Following a surgical operation, a surgeon may place a tube in the biliary tree to allow for external drainage of bile. An example is the T tube placed after a bile duct exploration. Since the advent of laparoscopic cholecystectomy, T tubes are much less frequently used. Access via the T-tube track may be used to percutaneously extract retained stones in the bile ducts. However, retained stones are usually extracted by the retrograde endoscopic route. This technique can be performed without having a T tube in place and within days after the cholecystectomy, as compared to a T tube extraction which necessitates a 5-week postsurgical delay to allow the track to mature. T tube extractions are indicated when the endoscopic approach has either failed or is impossible because of unfavorable anatomy. Some surgeons will use a temporary tube to stent complex bile duct repairs. These tubes provide a convenient portal of access to the biliary tree for the interventional radiologist to perform diagnostic studies or interventions. Perihepatic drains may in some instances also allow access to the biliary tree.

REFERENCES

Bismuth H. Postoperative strictures of the bile duct. In: Blumgart LH, ed. The biliary tract. Clinical Surgery International, Vol 5. Edinburgh: Churchill Livingstone, 1982; 207-218.

Blumgart LH. Benign biliary strictures. In: Blumgart LH, ed. Surgery of the liver and biliary tract, Vol 1. Edinburgh: Churchill Livingstone, 1988; 721-752.

Boguth L, Tatalovic S, Antonucci F, Heer M, Sulser H, Zollikofer CL. Malignant biliary obstruction: clinical and histopathologic correlation after treatment with self-expanding metal prostheses. Radiology 1994; 192:669-674.

Burke DR, Lewis CA, Cardella JF, et al. Quality improvement guidelines for percutaneous transhepatic cholangiography and biliary drainage. J Vasc Interv Radiol 1997; 8:677-681.

Coons, H. Metallic stents for the treatment of biliary obstruction: a report of 100 cases. Cardiovasc Intervent Radiol 1992; 15:367-374.

Culp WC, MCCowan TC, Lieberman RP, Goertzen TC, LeVeen RF, Heffron TG. Biliary strictures in liver transplant recipients: treatment with metal stents. Radiology 1996; 199:339-346.

Dick BW, Gordon RL, LaBerge JM, Doherty MM, Ring EJ. Percutaneous transhepatic placement of biliary endoprostheses: results in 100 consecutive patients. J Vasc Interv Radiol 1990; 1:97-100.

Gordon RL, Ring EJ, LaBerge JM, Doherty MM. Malignant biliary obstruction: treatment with expandable metallic stents: follow-up of 50 consecutive patients. Radiology 1992; 182:697-701.

Harbin WP, Mueller PR, Ferrucci JT. Complications and use patterns of fine-needle transhepatic cholangiography: a multi-institutional study. Radiology 1980; 135:15-22.

Hausegger KA, Kleinert R, Lammer J, Klein GE, Fluckiger F. Malignant biliary obstruction: histologic findings after treatment with self-expandable stents. Radiology 1992; 185:461-464.

Hausegger KA, Mischinger HJ, Karaic R, et al. Percutaneous cholangioscopy in obstructed biliary metal stents. Cardiovasc Intervent Radiol 1997; 20:191-196.

Hausegger KA, Thurnher S, Bodendorfer G, et al. Treatment of malignant biliary obstruction with polyurethane-covered Wallstents. AJR 1998; 170:403-408.

Hayashi N, Sakai T, Kitagawa M, et al. US-guided left-sided biliary drainage: nine-year experience. Radiology 1997;119-122.

Irving JD, Adam A, Dick R, Dondelinger RF, Lunderquist A, Roche A. Gianturco expandable metallic biliary stents: results of a European clinical trial. Radiology 1989; 172:321-6.

Kaufman SL. Percutaneous palliation of unresectable pancreatic cancer. Surg Clin NA 1995; 75:989-999.

Lammer J, Hausegger KA, Fluckiger F, et al. Common bile duct obstruction due to malignancy: treatment with plastic versus metal stents. Radiology 1996; 201:167-172.

Libby ED, Leung JW. Prevention of biliary stent clogging: a clinical review. Am J Gastroenterol 1996: 91:1301-1308.

Lo LD, Vogelzang RL, Braun MA, Memcek AA Jr. Percutaneous cholecystostomy for the diagnosis and treatment of acute calculous and acalculous cholecystitis. J Vasc Interv Radiol 1995; 6:629-634.

Maccioni F, Rossi M, Salvatori FM, Ricci P, Bezzi M, Rossi P. Metallic stents in benign biliary strictures: three-year follow-up. Cardiovasc Intervent Radiol 1992; 15:360-366.

Malt RA, Burhenne HJ. Biliary stricture dilatation: multicenter review of clinical management in 73 patients. Radiology 1986; 160:17-22.

Maroney TP, Ring EJ. Percutaneous transjejunal catheterization of Roux-en-Y biliary jejunal anastomoses. Radiology 1987; 164;151-153.

Moore AV Jr., Illescas FF, Mills SR, et al. Percutaneous dilation of benign biliary strictures. Radiology 1987; 163:625-628.

Mueller PR, Harbin WP, Ferrucci JT, Wittenberg J, vanSonnenberg E. Fine-needle transhepatic cholangiography: reflections after 450 cases. AJR 1981; 136:85-90.

Mueller PR, VanSonnenberg E, Ferrucci JT, et al. Biliary stricture dilatation: multicenter review of clinical management in 73 patients. Radiology 1986; 160:17-22.

Pellegrini CA, Thomas MJ, Way LW. Recurrent biliary stricture: Patterns of recurrence and outcome of surgical therapy. Am J Surg 1984; 147:175-180.

Pitt HA, Kaufman SL, Coleman J, White RI, Cameron JL. Benign postoperative biliary strictures. Ann Surg 1989; 210:417-425.

Rossi P, Bezzl M, Rossl M, et al. Metallic stents in malignant biliary obstruction: results of a multicenter European study of 240 patients. J Vasc Interv Radiol 1994; 5:279-285.

Way LW, Bernhoft RA, Thomas MJ. Biliary stricture. Surg Clin North Am 1981; 61(4):963-969.

Williams HJ, Bender CE, May GR. Benign postoperative biliary strictures: dilation with fluoroscopic guidance. Radiology 1987; 163:629-634.

CHAPTER 21 PREVIEW ANSWERS

Case 1

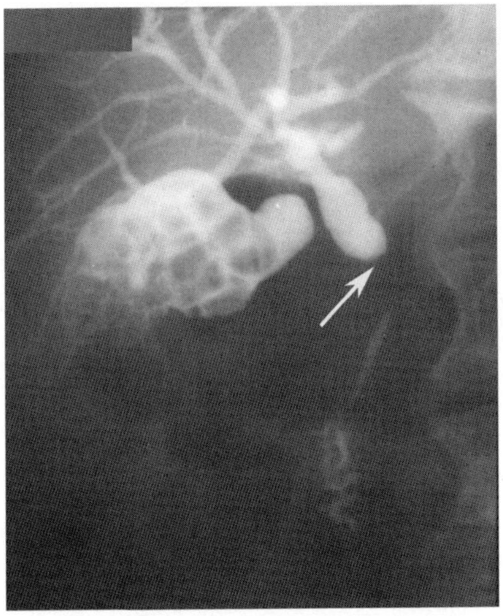

Findings: The PTC shows a tight narrowing in the middle section of the extrahepatic bile duct (arrow). The location is typical for gallbladder carcinoma. There are multiple gall-stones which appear as faceted filling defects in the gallbladder.

Case 2

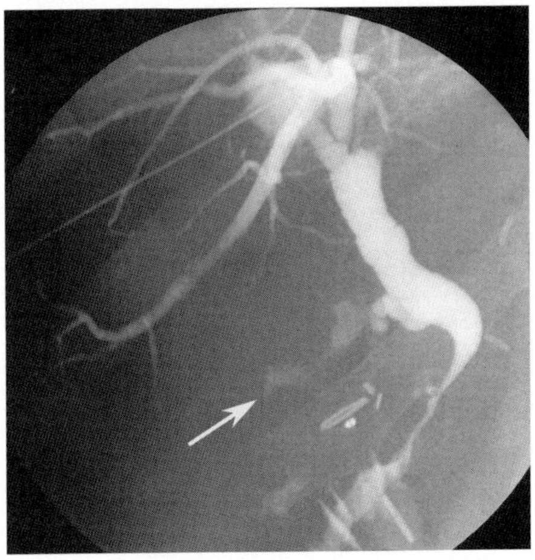

Findings: There is contrast material seen just adjacent to the cystic duct stump (arrow). This may be misinterpreted as a bile leak, but it is actually contrast in the calyces of the underlying kidney. The opacified renal calyces, pelvis and ureter often overlie the biliary tree and should not be confused with it.

Case 3

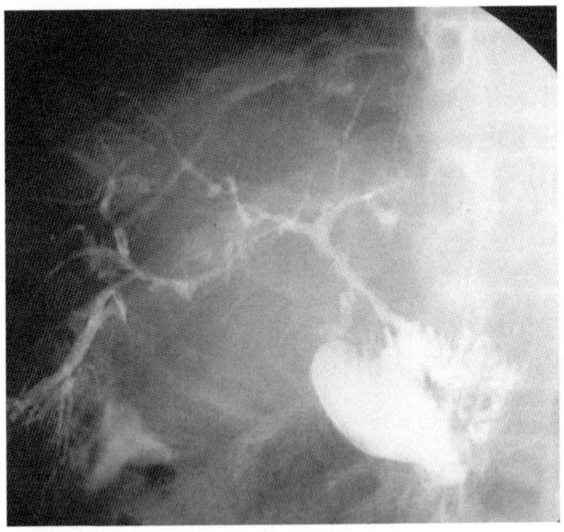

Findings: The PTC shows irregular intrahepatic ducts typical of sclerosing cholangitis. Recurrent strictures are found in a significantly higher percentage of patients who underwent transplantation for primary sclerosing cholangitis than in other liver recipients.

Case 4

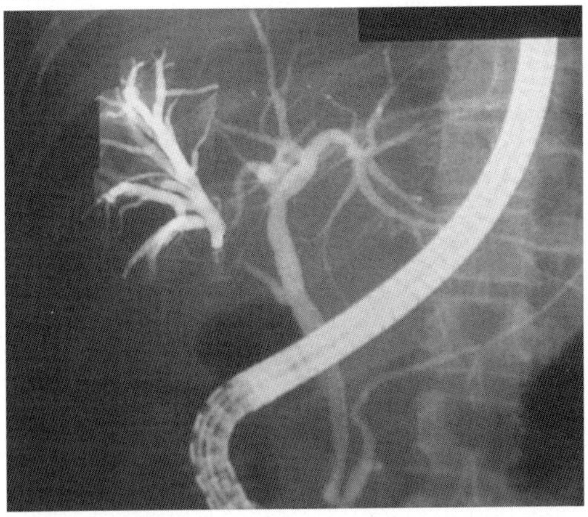

Findings: There is a tubular stump filling off the common duct which could be a long cystic duct stump, but in fact is the stump of a low inserting right posterior duct that has been divided in error. A PTC opacified the isolated right posterior ducts which have been superimposed on the original cholangiogram. The best projection in which to appreciate the absence of right ducts in the original cholangiogram would be the LPO projection.

CHAPTER 22 PREVIEW

Case 1

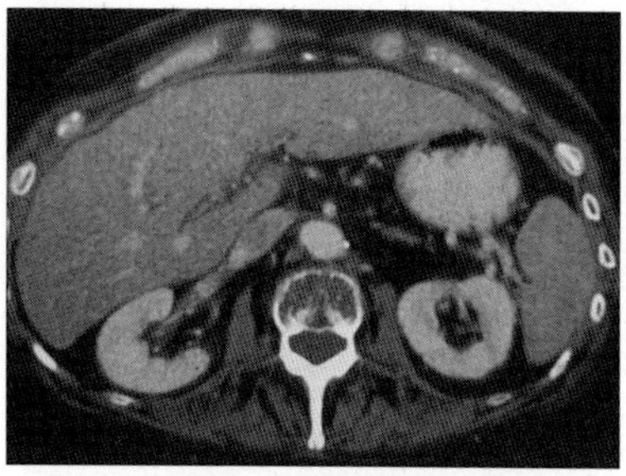

Figure 1. *Contrast-enhanced abdominal CT at the level of the stomach body.*

History: A 45-year-old man with a nasopharyngeal carcinoma was evaluated for gastrostomy placement. He had previously undergone a contrast-enhanced CT scan for evaluation of metastatic disease and a selected image from this study is shown **(Fig. 1)**.

Questions: Can this patient safely undergo percutaneous gastrostomy? What are the potential complications of percutaneous gastrostomy in this case? What are the alternative methods for gastrostomy placement?

Case 2

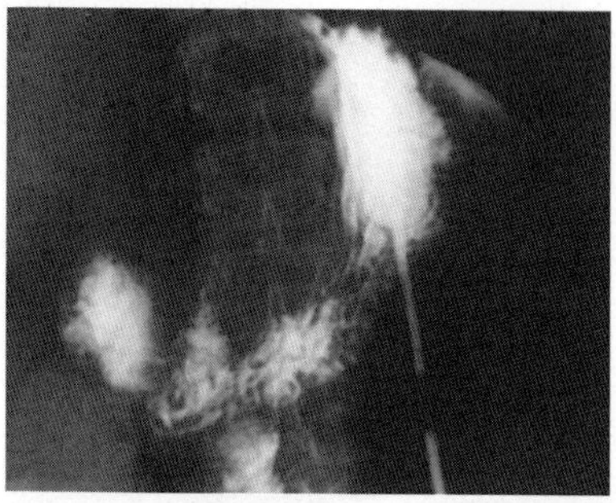

Figure 2. *Gastrostomy tube check, spot film, AP projection.*

History: A 65-year-old woman underwent endoscopic gastrostomy placement. The next day she complained of diffuse abdominal pain and a gastrostomy tube check was performed **(Fig. 2)**.

Question: Can you identify a problem?

Case 3

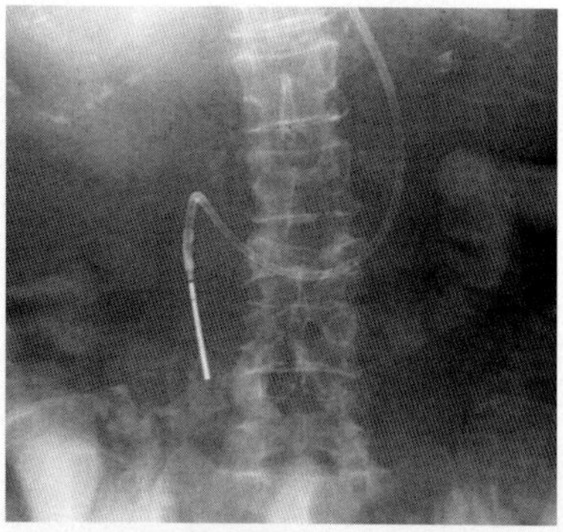

Figure 3. *KUB following bedside feeding tube placement.*

History: The ICU team has had difficulty advancing a feeding tube. A KUB was obtained **(Fig. 3)**.

Questions: What is the desired position of this tube? Is there anything that you can do to help the ICU with this problem?

Case 4

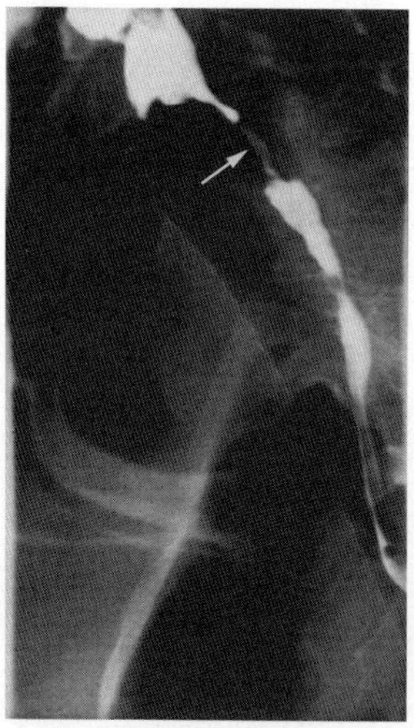

Figure 4. *Hypaque swallow, spot-film, AP projection.*

History: A 30-year-old man with epidermolysis bullosum complained of difficulty swallowing. Hypaque swallow was performed **(Fig. 4)** and an abnormality was observed in the proximal cervical esophagus (arrow).

Question: How could this patient be treated?

Chapter 22
Gastrointestinal Tract Interventions

Jeanne M. LaBerge, MD

CHAPTER OUTLINE

- Nasojejunal intubation
- Gastrostomy, gastrojejunostomy
- Direct jejunostomy
- Esophageal stricture dilatation
- GI tract stent placement
- Percutaneous decompression by external drainage

GOALS OF GI TRACT INTERVENTION

- Provide access for enteral nutrition
- Treat obstruction by dilatation, stenting or percutaneous decompression

METHODS OF ENTERIC HYPERALIMENTATION

- Nasojejunal tube placement
- Gastrostomy or gastrojejunostomy
- Jejunostomy

INTRODUCTION

Gastrointestinal (GI) tract interventions are usually performed to provide access for nutritional support or to treat GI tract strictures. Access to the GI tract can be accomplished via nasojejunal tube placement, gastrostomy, gastrojejunostomy or direct jejunostomy. GI tract strictures can be treated by dilatation or stenting. These interventions can be performed by a variety of medical specialists—endoscopists, radiologists and/or surgeons. Radiologic fluoroscopically guided approaches to these interventions are discussed in this chapter.

NASOJEJUNAL INTUBATION: BEYOND THE LIGAMENT OF TREITZ

Enteral feeding is a desirable means of providing nutritional support for hospitalized patients who cannot eat normally. Infusion of nutrition into the small bowel rather than the stomach is advisable in order to minimize gastric residuals and decrease the possibility of reflux and consequent aspiration. Therefore, nasal feeding tubes are optimally positioned in the small bowel, either in the duodenum beyond the pylorus or preferably in the jejunum beyond the ligament of Treitz **(Fig. 1)**.

Nasojejunal feeding tubes are commonly placed using one of three methods:

1) bedside placement with follow-up KUB to determine position;
2) fluoroscopic placement;
3) endoscopic placement.

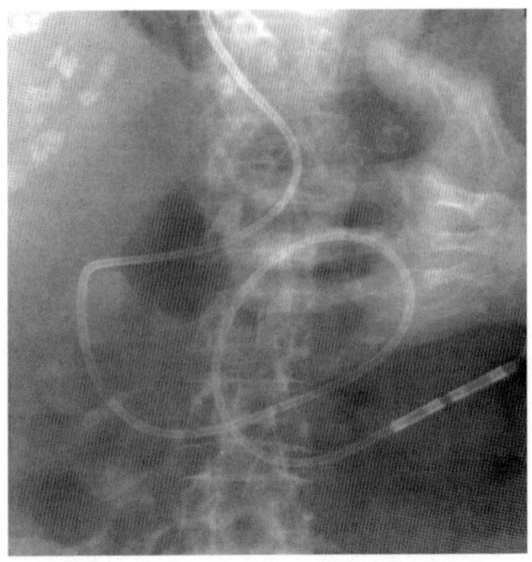

Figure 1. *Spot film obtained following fluoroscopic placement of a Ring-McLean nasojejunal feeding tube. The tip of this "over-the-guide wire" tube is in the jejunum, well beyond the ligament of Treitz.*

Fluoroscopic placement is a fast and reliable means of accomplishing nasoenteric cannulation. The radiologist can maneuver the feeding tube across the pylorus using fluoroscopy aided when necessary by the injection of a small amount of contrast material into the stomach to identify the gastric outlet **(Fig. 2)**. In more difficult cases, a shaped angiographic catheter and a torquable wire can be used access the duodenum. Forward movement of the catheter can be facilitated by the intravenous administration of metoclopramide (Reglan).

Most nasoenteric feeding tubes designed for bedside placement have a weight at the end to facilitate forward progression of the tube from the stomach into the small bowel. These tubes do not have an end-hole. Specially designed nasojejunal tubes with end-holes are used for catheter-directed fluoroscopic nasojejunal tube placement. These tubes are typically 10-12-F in size.

Fluoroscopic placement can be successfully accomplished in over 90% of patients. Failure is more common in patients who have undergone a surgical gastric diversion procedure such as Billroth II gastrojejunostomy. In such cases, endoscopic placement may be necessary. The chief complications of nasojejunal feeding tube placement are aspiration occurring during catheter placement and cardiac arrhythmia.

While nasojejunal intubation is a convenient short-term method of providing enteral nutrition, its use over the long term poses several problems. Prolonged use may result in gastroesophageal reflux, esophagitis, stricture formation and aspiration. The tubes may clog easily because of their relatively small size, and finally they are uncomfortable for patients.

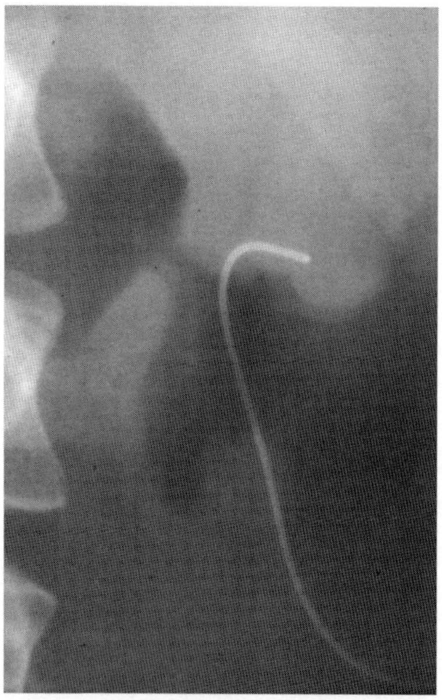

a.

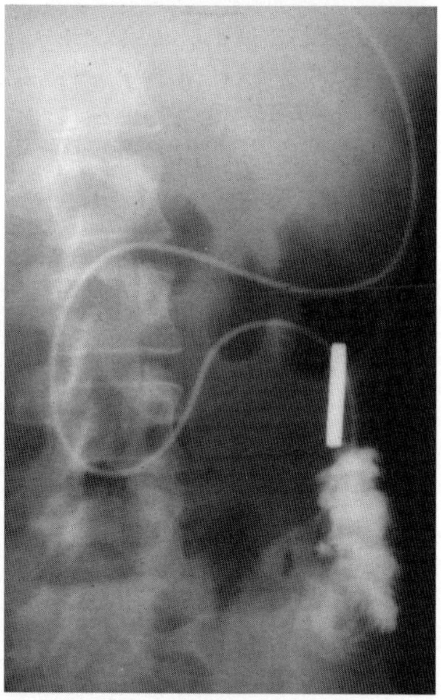

b.

Figure 2. *Fluoroscopic technique of nasojejunal tube placement. a) Fluorosopic visualization of the pylorus facilitates negotiation of a wire and catheter into the duodenum. Here the pylorus is visualized with air contrast. A shaped angiographic wire is then used to cross the pylorus under fluoroscopic guidance. b) A feeding tube is placed over a wire that had been advanced into the jejunum beyond the ligament of Treitz.*

PERCUTANEOUS GASTROSTOMY

Long-term enteral nutrition is most effectively and conveniently achieved by means of direct intubation of the stomach **(Fig. 3)** or small bowel **(Fig. 4)**. This approach allows placement of larger bore catheters than can be placed using a nasojejunal approach and tubes entering from an abdominal site are better tolerated by patients than long-term nasal tubes. Surgical gastrostomy techniques are well-established, having been first reported in 1876. Endoscopic techniques were developed in the 1970's and the first percutaneous fluoroscopic techniques were described in the early 1980's.

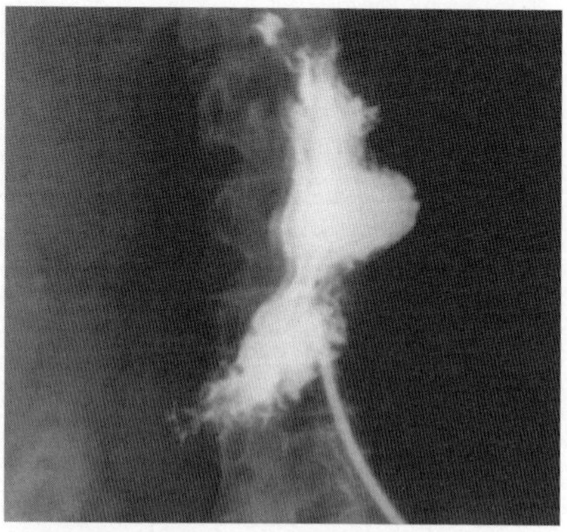

Figure 3. *Percutaneous gastrostomy. GI tract nutrition is delivered directly into the stomach.*

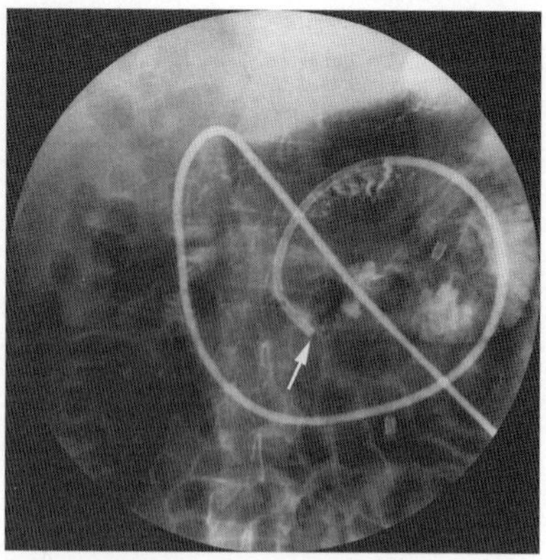

Figure 4. *Percutaneous gastrojejunostomy. Enteral nutrition is delivered to the jejunum (arrow).*

Indications

The purpose of gastrostomy is usually to provide access for enteral feeding. The most common indications are neurologic disorder or head and neck tumors that impair the patient's ability to swallow effectively. Occasionally, intubation of the stomach in this manner is performed to decompress the stomach in cases of gastric outlet obstruction.

Relative contraindications include unfavorable anatomy (inability to gain access to the stomach without transgressing other structures) or coagulopathy. The presence of a large amount of ascites represents a relative contraindication but recent reports suggest that with technical modifications, such as the use of gastric retention devices, gastrostomy may be safely and reliably performed even in patients with significant ascites.

INDICATIONS FOR PERCUTANEOUS GASTROSTOMY

- Enteral feeding
  - Neurologic disorders
  - Head and neck tumors
- Gastric decompression
  - Gastric outlet obstruction

ANATOMIC CONSIDERATIONS FOR PERCUTANEOUS GASTROSTOMY

- Avoid the left lobe of the liver
  - Use ultrasound
- Avoid small bowel and colon
  - Use CT and/or fluoroscopy

Preprocedural Preparation

Preprocedural imaging is performed at the discretion of the interventionalists. Its purpose is to insure that adjacent organs are not transgressed during the procedure. Some investigators believe that fluoroscopy alone is sufficient to avoid inadvertent injury to adjacent structures. Others feel that CT is useful in detecting the location of the colon and small bowel relative to the stomach **(Fig. 5)**. Sonography is often used at the time of the procedure to localize the left lobe of the liver. Administration of barium or water-soluble contrast the night before the procedure has been used to opacify the colon. Preprocedural imaging is particularly valuable in patients who have undergone prior gastric or abdominal surgery and may have altered anatomic relationships.

Patients should be fasting for 12 hours prior to the procedure. For convenience, a nasogastric tube may be inserted prior to the procedure.

Technique

The procedure is performed under local anesthesia with intravenous conscious sedation. The use of preprocedural antibiotics is controversial and at most institutions prophylactics are not administered.

KEY STEPS OF PERCUTANEOUS GASTROSTOMY

1) Patient preparation: NPO, NG tube
2) Preprocedural imaging: CT or ultrasound
3) Sedation
4) Stomach insufflation
5) Gastropexy

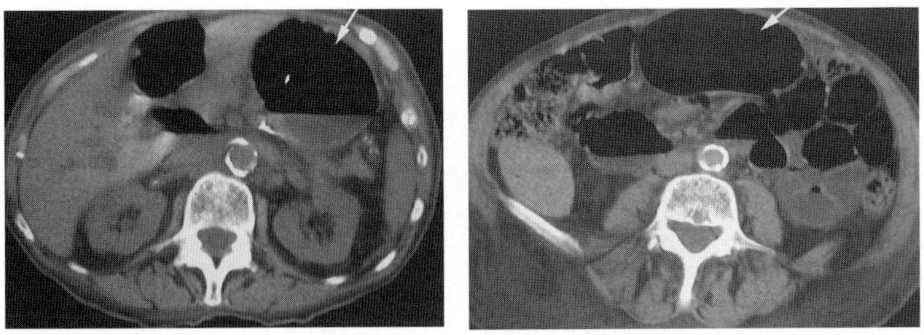

a. **b.**

Figure 5. *Preprocedural abdominal CT. Two consecutive images through the inflated stomach show the relationship of the stomach (arrow) to the small bowel and colon.*

The visceral structures to be avoided during gastric catheterization are the liver, colon and small bowel. As well, punctures should avoid major arteries in the stomach and anterior abdominal wall. A puncture site in the left upper quadrant is selected to optimize entry into the stomach and to avoid puncture of these visceral structures. A site lateral to the rectus muscle usually prevents injury to the epigastric artery. And a target site in the mid-body of the stomach usually avoids the major arteries that run along the lesser and greater curvature of the stomach.

The major steps of this procedure **(Fig. 6)** are listed below:

1) Inflation of the stomach with air via a nasogastric tube
2) Selection of puncture site and instillation of local anesthesia
3) Placement of 2-4 gastric retention devices (T-fasteners) in the stomach **(Fig. 7)**. Recent literature supports this step (gastropexy) as an effective means of securing the stomach to the anterior abdominal wall and preventing leakage of gastric contents into the peritoneal cavity
4) Puncture of the stomach, advancement of a guidewire into the stomach and dilatation of the tract
5) Insertion of a gastrostomy tube over the wire with secure attachment to the abdominal wall via skin sutures. Nasogastric suction is maintained for 12-24 hrs and then feeding may be started at increasing doses.

Catheters

Percutaneous catheters may be placed into the stomach or advanced into the jejunum (gastrojejunostomy). Some investigators advocate gastrojejunostomy as a means of providing more effective nutrition and preventing gastroesophageal reflux and aspiration. A variety of catheters have been used for these purposes **(Fig. 8)**. Percutaneous catheters are typically 12-16-F in size. Gastric tubes usually have a Cope-loop retention mechanism. Gastrojejunostomy tubes may have a locking loop or Malecot in the stomach.

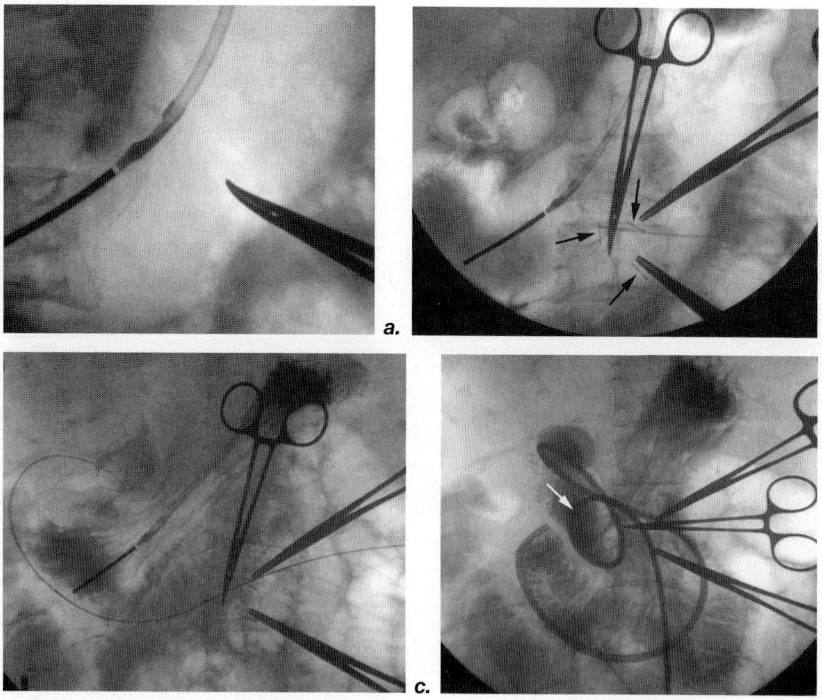

a. b. c. d.

Figure 6. *Percutaneous gastrojejunostomy technique. a) The stomach is inflated with air and an appropriate puncture site is selected. The clamp shown here marks the puncture site which lies over the mid-body of the stomach. b) Three gastric retention T-fasteners (arrows) are placed into the stomach and affixed to the abdominal wall. c) The stomach is punctured with a 19-gauge single wall puncture needle and a stiff 0.035-inch guidewire is advanced into the stomach toward the pylorus. d) The tract is dilated and then a 14-F gastrojejunostomy catheter is advanced over the wire so that the tip lies in the jejunum well beyond the ligament of Treitz. The retaining locking loop of the catheter (arrow) is coiled in the stomach.*

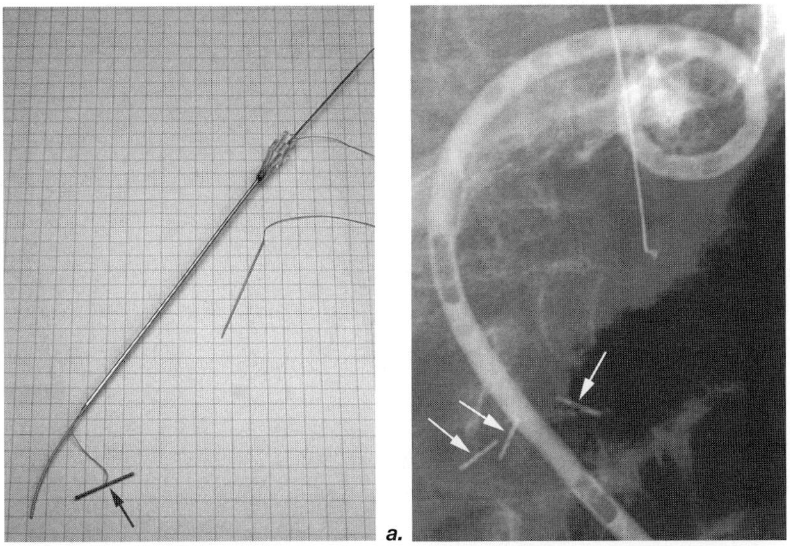

a. b.

Figure 7. Gastric retention device. a) Several types of T-fasteners are available. An example is shown here. The T-bar (arrow) is attached to a suture. When the T-bar is placed in the stomach, the suture can be pulled taut to appose the anterior gastric wall to the anterior gastric wall. b) Close-up of the gastrostomy site in another case shows three T-fasteners in place (arrows). The T-fastener skin sutures are removed after 1-2 weeks.

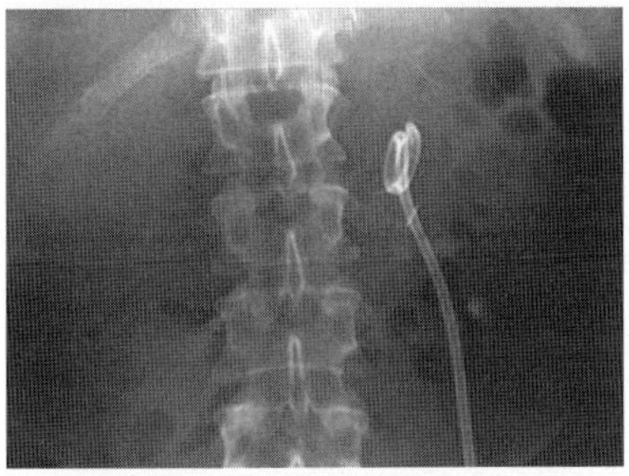

a.

Figure 8. Gastrostomy/gastrojejunostomy catheters. a) Cope loop locking pigtail catheter. This is our standard gastrostomy tube. b) Gastrojejunostomy tube designed with locking Malecot built into the catheter which acts as a restraining point to prevent inadvertent removal from the stomach.

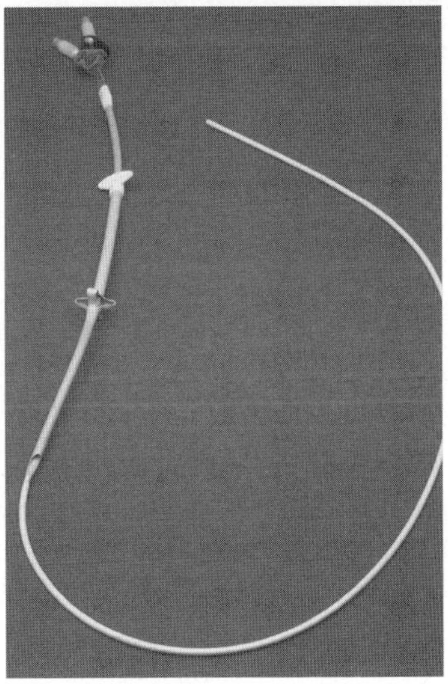

b.

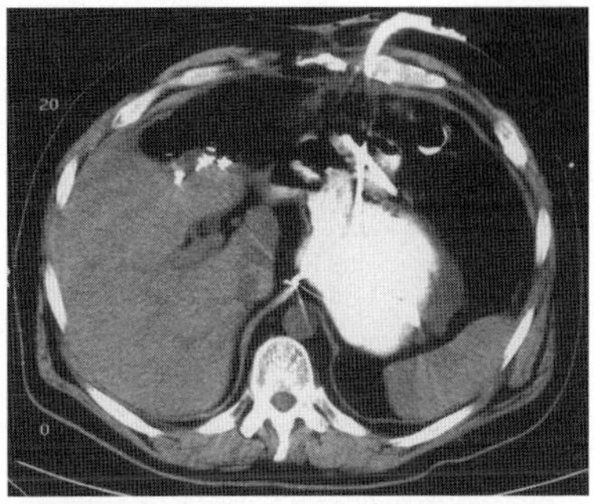

Figure 9. CT obtained following percutaneous gastrostomy. Note the posterior position of the stomach. The colon is located ventral to the stomach. Unfortunately, the gastrostomy tube has traversed the colon in transit to the stomach.

Results

Percutaneous gastrostomy or gastrojejunostomy can be performed successfully in almost all patients (> 95%). Major procedural complications are uncommon (2-3%) but include aspiration and hemorrhage, inadvertent puncture of other organs **(Fig. 9)** and peritonitis from gastric leakage. Long-term complications are chiefly due to skin entry site infection or tube-related problems such as occlusion or dislodgment.

Several recent studies have attempted to compare the results of gastrostomy placed by radiologic, endoscopic and surgical techniques. Most of these studies are limited by their retrospective design but all three approaches seem to yield similar results. It should be noted that an advantage of the endoscopic and surgical techniques is the ability to insert much larger catheters (20-22-F) which may have improved patency due to their larger size.

POTENTIAL COMPLICATIONS OF PERCUTANEOUS GASTROSTOMY

- Gastric leakage, peritonitis
- Aspiration
- Bleeding
- Bowel perforation, peritonitis
- Catheter dislodgment
- Skin entry site infection

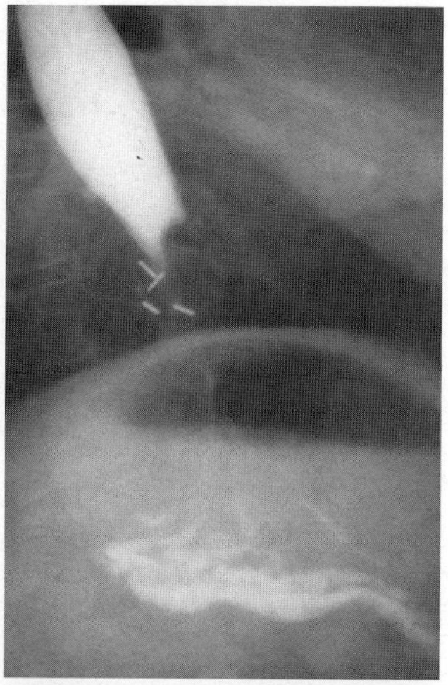

a.

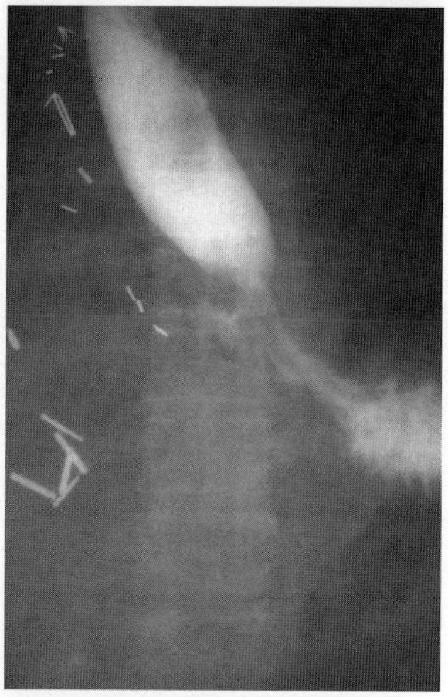

b.

Figure 10. Postoperative stricture treated by balloon dilatation. a) A pre-procedural esophagram shows a tight distal esophageal stricture at the gastroesophageal junction. A 12-mm balloon was advanced across this stricture into the stomach and the stricture was dilated. b) Postdilatation esophagram shows improvement.

PERCUTANEOUS JEJUNOSTOMY

Direct access to the jejunum may be preferable to gastrostomy in selected patients such as those with prior gastric surgery or those with obstruction of the duodenum and proximal jejunum. Surgical jejunostomy can be performed laparoscopically and is frequently performed at the time of other major surgery in patients who require enteral nutrition.

Recently a percutaneous method of jejunostomy has been reported by Cope and his collegues from the University of Pennsylvania. In his technique, the jejunum is distended by advancing a long nasoenteric balloon catheter into the proximal jejunum. Alternatively, the jejunum can be insufflated with air. The balloon is inflated and then it is percutaneously punctured. T-fasteners are used to affix the bowel loop to the abdominal wall and then a self-retaining pigtail catheter is advanced over a wire into the jejunum. Cope reported promising results in 41 patients treated by this technique. Overall success was 85%. Two patients underwent laparotomy to evaluate jejunal leakage.

Direct puncture of a biliary Roux loop is also a well-described percutaneous technique for accessing the biliary tree in patients with chronic biliary disease. *This technique is discussed in the Biliary Interventions Chapter.*

PERCUTANEOUS MANAGEMENT OF GI STRICTURE

Surgical resection or bypass is the defintive treatment of GI tract stricture. In selected patients, surgery may not be advisable and endoscopic or radiologic management may be warranted. Patients with benign disease may benefit from serial dilatation but repeated procedures are necessary. Patients with malignant disease may undergo effective palliation by dilatation and stent placement.

Benign strictures of the esophagus can be managed with serial dilatation **(Fig. 10)**. This procedure may be performed by the endoscopists or radiologist. Strictures following surgery or lye ingestion and those associated with achalasia or Barrett's esophagus may be treated in this manner. The radiologic technique involves oral passage of a balloon and fluoroscopic positioning across the stricture. Balloon sizes of 10-20 mm are typically required. Complications are uncommon but the chief concern is rupture or perforation leading to sepsis, mediastinitis or peritonitis.

Malignant strictures of the upper and lower GI tracts respond poorly to dilatation alone. Recent investigators have reported promising results with deployment of large diameter stents to maintain luminal patency. The largest experience has been reported for esophageal covered stents to treat esophageal carcinoma. These large bore stents are inserted through delivery systems ranging from 24-F to 30-F. A variety of stent designs have been used for this application. Stents are postioned across the esophageal stricture into the stomach **(Fig. 11)**. Published results suggest that these devices provide effective relief of dysphagia and are also beneficial in selected patients with tracheal-esophageal fistulas. However, life expectancy is very short in patients with esophageal cancer who usually die of their disease within 3 months.

Malignant obstruction elsewhere in the GI tract can also be treated by stenting. An example of duodenal stenting with an 18-mm uncovered Wallstent is shown in **Figure 12**. The application of stents in the lower GI tract is limited by the problem of stent occlusion and clogging from colonic contents. This procedure seems to yield greatest benefit when used to decompress the colon in patients with acute obstruction.

Stent migration and obstruction are the main long-term complications of stent placement. Perforation and bleeding have also been reported.

METALLIC STENTS FOR ESOPHAGEAL MALIGNANCY
• Wallstent 18-20 mm diameter, 18-22-F delivery • Strecker 18 mm diameter, 24-F delivery • Gianturco-Rosch 18 mm diameter, 24-30-F delivery

PERCUTANEOUS DRAINAGE OF OBSTRUCTED BOWEL

Direct percutaneous puncture of an obstructed or dilated bowel loop may be advisable in selected patients. As mentioned above, gastrostomy is occasionally performed for this indication. Percutaneous cecostomy has also been described in the management of Ogilvie syndrome. As with gastrostomy and jejunostomy, initial insertion of retention T-fasteners is advisable. Large tubes are necessary to facilitate drainage for cecostomy.

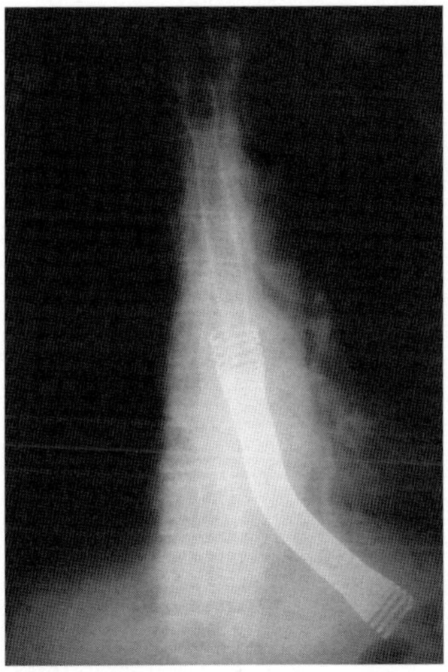

Figure 11. *A large-bore covered Wallstent has been placed across a malignant distal esophageal stricure in this patient with esophageal carcinoma.*

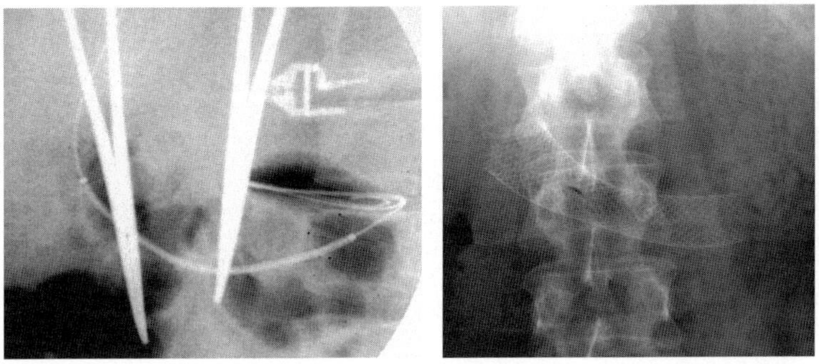

a. **b.**

Figure 12. *An uncovered Wallstent was inserted to relieve duodenal obstruction in this patient with metastatic disease and gastric outlet obstruction. a) A guidewire was first negotiated across the obstruction in the distal duodenum and then an 18-mm Wallstent was advanced over the wire and positioned across the stricture. b) The deployed Wallstent was dilated with marked improvement in luminal diameter. The patient's gastric outlet obstruction was relieved immediately after the procedure.*

REFERENCES

Cope C, Davis AG, Baum RA, Haskal ZJ, Soulen MC, Shlansky-Goldberg RD. Direct percutaneous jejunostomy: techniques and applications—ten years experience. Radiology 1998; 209:747-754.

Hillard AE, Waddell JJ, Metzler MH. Fluoroscopically guided nasoenteric feeding tube placement versus bedside placement. South Med J 1995; 88:425-428.

Hoffer EK, Cosgrove JM, Levin DQ, Herskowitz MM, Sclafani SJ. Radiologic gastrojejunostomy and percutaneous endoscopic gastrostomy: a prospective randomized comparison. J Vasc Interv Radiol 1999; 10:413-420.

Park HS, Do YS, Suh SW, Choo SW, Lim HK, et al. Upper gastrointestinal tract malignant obstruction: initial results of palliation with a flexible covered stent. Radiology 1999; 210:865-870.

Ryan MJ, Hahn PF, Boland GW, McDowell RK, Saini S, Mueller PR. Percutaneous gastrostomy with t-fastener gastropexy: results of 316 consecutive procedures. Radiology 1997; 203:496-500.

vanSonnenberg E, Varney RR, Casola G, et al. Percutaneous cecostomy for Ogilvie syndrome: laboratory observations and clinical experience. Radiology 1990; 175:679-682.

Wollman B, D'Augustino HB, Walus-Wigle JR, Easter DW, Beale A. Radiologic, endoscopic, and surgical gastrostomy: an institutional evaluation and metaanalysis of the literature. Radiology 1995; 197:699-704.

CHAPTER 22 PREVIEW ANSWERS

Case 1

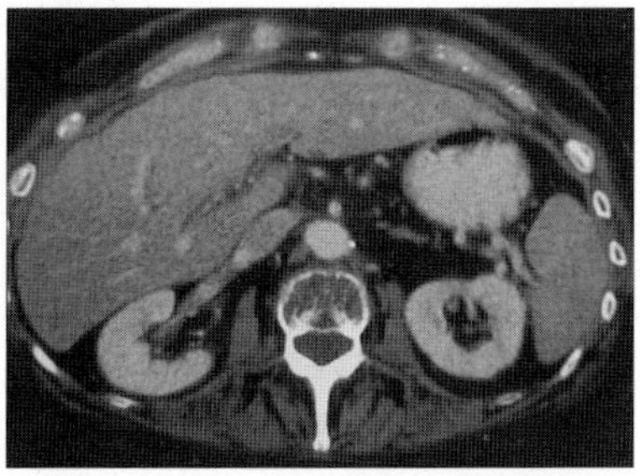

Figure 1. *Contrast-enhanced abdominal CT at the level of the stomach body.*

Findings: The left lobe of the liver is prominent, extending well beyond the midline **(Fig. 1)**. At the level shown, the stomach is behind the left lobe of the liver.

Answers: This anatomy may prevent the safe performance of percutaneous gastrostomy. Potential complications of percutaneous gastrostomy include liver laceration and bleeding. Alternatively, a surgical gastrostomy could be performed to safely avoid injury to the liver during tube placement.

Case 2

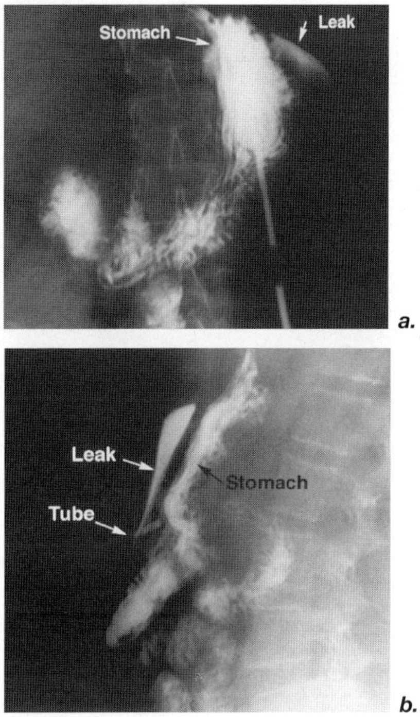

Figure 2. *Gastrostomy tube check, spot-film. a) AP projection. b) Lateral projection.*

Findings: On the AP spot-film, a small rim of contrast material is seen lateral to the stomach fundus **(Fig. 2a)**. This is extravasated material indicating a leak. The anatomy is better depicted on the lateral view **(Fig. 2b)**.

Diagnosis: Leak at the gastrostomy insertion site.

Case 3

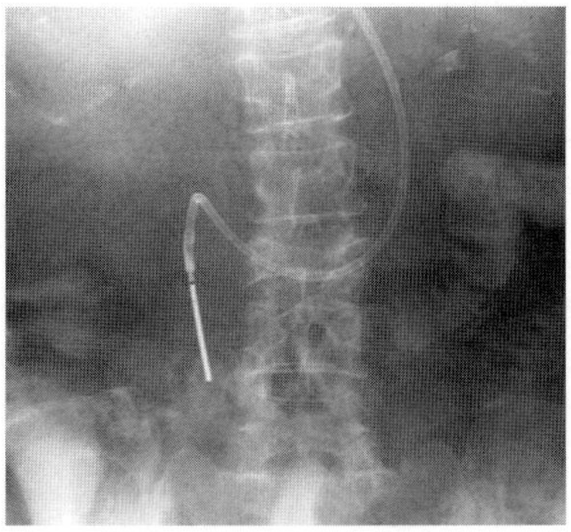

Figure 3. *KUB following bedside feeding tube placement.*

Findings: The tip of this weighted feeding tube appears to be in the duodenum.

Answer: The optimal position for the feeding tube is in the jejunum, beyond the ligament of Treitz.

Case 4

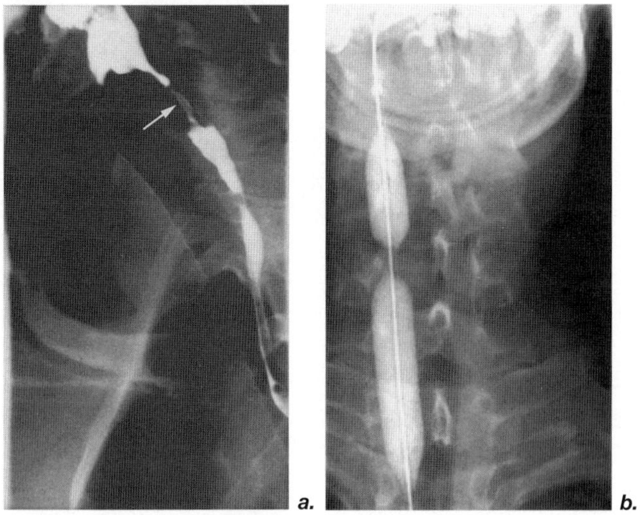

Figure 4. *Hypaque swallow, spot-film, AP projection.*

Findings: There is a tight focal stricture of the proximal esophagus **(Fig. 4a)**. This is a known complication of epidermolysis bullosum.

Treatment: Options include endoscopic serial dilatation or fluoroscopic dilatation. In this case, a 10-mm balloon was used to dilate the stricture **(Fig. 4b)** with a good clinical result.

CHAPTER 23 PREVIEW

Case 1

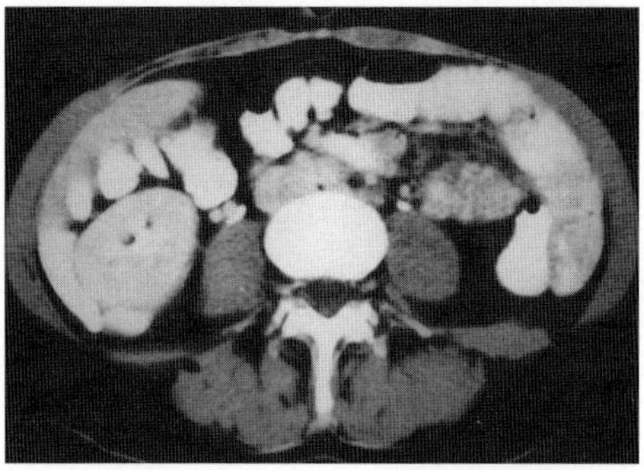

History: CT scan of a patient scheduled to undergo a right percutaneous nephrostomy.
Question: Is this scan helpful?

Case 2

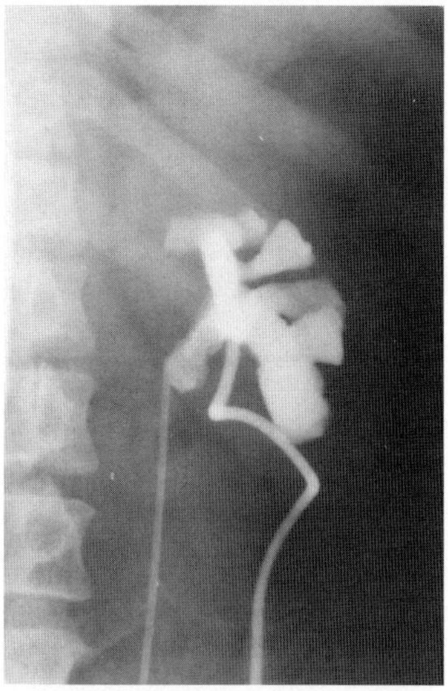

History: Status postnephrostomy.
Question: What does this radiograph demonstrate?

Case 3

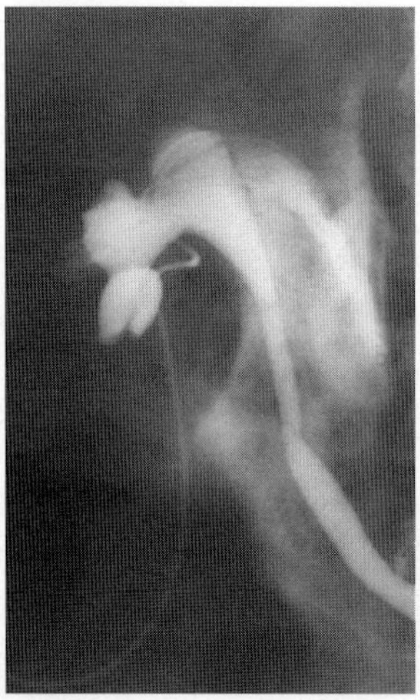

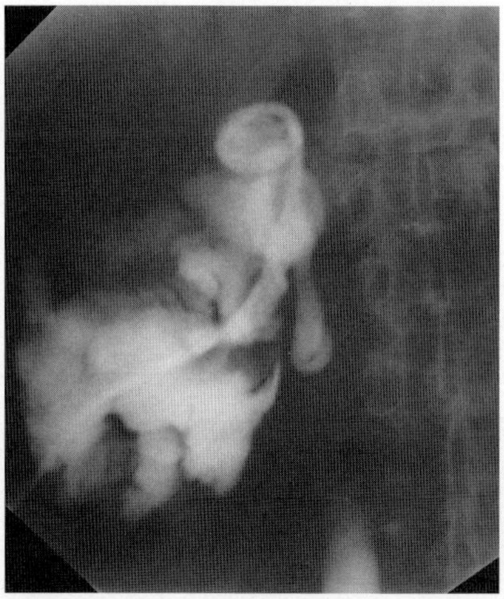

History: Two radiographs taken while placing a left-sided nephrostomy in the prone position.
Question: What is your interpretation?

Chapter 23
Percutaneous Nephrostomy and Ureteral Stenting

Roy L. Gordon, MD

NEPHROSTOMY

- Indications
- When not to place PCN
- Contraindications
- Preprocedure steps
- Guidance
- Technique
- Results
- Complications
- Tube maintenance

INTRODUCTION

This chapter will focus on the two interventions most frequently performed in the renal tract. Part 1 deals with the insertion of a tube connecting the surface of the body, via the renal parenchyma, to the collecting system of the kidney—percutaneous nephrostomy (PCN); and Part 2 deals with the insertion of catheters, also called stents, into the ureter.

PART 1. NEPHROSTOMY: PCN INDICATIONS

PCN is in widespread clinical use. The indications for PCN fall into four general categories:

a) Relief of obstruction
b) Diversion of urine
c) Assessment of renal function
d) Provision of initial access to the renal tract prior to other procedures.

a) Obstruction may result from a variety of causes, such as stones, strictures, tumors, retroperitoneal fibrosis and others. Three important consequences that may result from obstruction are: 1) sepsis with septic shock, representing a life-threatening situation demanding urgent drainage; 2) derangement of normal metabolic functions as a result of impaired renal function, requiring early but less urgent drainage; and 3) obstructive renal damage that is usually reversible with early drainage.

PCN INDICATIONS

- Relieve obstruction
- Divert urine
- Access system
- Assess function

b) Diversion is important and often successful in healing cases of renal, ureteral or bladder leaks, or fistulous tracks between the urinary tract and the bowel. Healing is promoted both by means of catheter diversion of urine out of the renal tract and also by adequate drainage of urine collections that may have accumulated in extrarenal locations.

c) Assessment of renal function by PCN is useful in cases of chronic obstruction. Whether renal function will recover with drainage can be decided by physiologic measurements of the concentration of urine draining from the PCN over time, allowing appropriate choice of surgical therapy.

d) Access to the renal tract is the starting point for the rapidly expanding array of endourologic techniques such as stent placement, calculous therapy, endoscopy, stricture dilatation, biopsy, and other intrarenal or ureteral therapies.

WHEN SHOULD A NEPHROSTOMY NOT BE PLACED?

Because PCN is such an easy and accepted therapy, this important question is too often not addressed. The following paragraph written by Andrew LeRoy in *Pollack's Clinical Urography* should always be remembered before PCN placement:

> "In addition to the standard medical review of a patient prior to PCN for relief of renal obstruction, serious ethical and moral consideration of the patient's prognosis must be undertaken if a terminal malignancy is involved. The mere presence of reversible renal obstruction does not per se mandate PCN placement. It is often a disservice to a uremic patient with untreatable malignant disease to relieve the ureteral obstruction and permit an extension of life for several months when that additional time is filled with unrelieved pain and burdened by cumbersome nephrostomy tubes, not to mention incalculable financial hardship. Frank discussions between physicians and patients are necessary to address these unfortunate situations."

Once a nephrostomy is placed in an irreversibly obstructed kidney, it cannot be removed without resulting in a continuous urine leak. In some cancer patients, PCN is required to improve renal function sufficiently to allow chemotherapy to be safely administered.

CONTRAINDICATIONS

These are few and must be viewed relative to the patient's overall condition and the urgency for PCN. For example, in a semielective PCN for stone extraction or U-P-J stricture therapy, any bleeding diathesis or urinary infection can be treated before placing the PCN. In the acute situation of urosepsis, urgent PCN may be key to controlling the infection. FFP and platelet infusions may be needed urgently before PCN when a bleeding diathesis and infection are both present. Anatomic problems such as severe scoliosis, extreme obesity, large spleen, extensive renal cysts, colon interposition, or renal transplant may be defined by using suitable preliminary imaging with CT or ultrasound (*see Case 1 in Chapter Preview*).

PCN: POTENTIAL CONTRAINDICATIONS

- Bleeding diathesis
- Untreated infection
- Neoplasm
- Anatomic
- Moral/Ethical

PREPROCEDURE STEPS

1) Review indications; Is the PCN in the patient's best interest? Would retrograde drainage be better?
2) Informed consent
3) Stop anticoagulants, aspirin, etc., in nonurgent cases; otherwise, consider fresh frozen plasma and/or platelet infusions
4) Review prior imaging studies to choose best kidney and to avoid anatomic pitfalls
5) Review laboratory results
6) Decide on guidance system
7) If infected urine is suspected, start intravenous antibiotics, e.g. Ampicillin and Gentamicin against organisms such as E. *coli*, P. *mirabilis pseudomonas*, and enterococcus. If infection is not suspected, some interventionalists will not use prophylactic antibiotic coverage.

PREPROCEDURE

- Review: prior studies
- Pause: to consider
  - antegrade vs. retrograde
  - which kidney
- Bleed: coagulation status
- Infection: antibiotics

PCN: GUIDANCE

- IV contrast
- Bony landmarks
- Ultrasound
- CT
- Retrograde

GUIDANCE

The choice of guidance system depends not only on physician preference and equipment availability, but also on whether the calyceal system is dilated or not. In a patient with good renal function, the calyceal system can be opacified by intravenous contrast agents and puncture achieved under fluoroscopic control. In the majority of patients, renal function is not good enough for this, and the two most common guidance techniques are:

a) Using bony landmarks, the collecting system is punctured with a 21- or 22-gauge Chiba needle. When urine is aspirated, contrast material is injected, delineating the collecting system. The PCN is then performed under fluoroscopy. In most patients, the kid neys occupy a fairly constant position. In the prone position, a vertical puncture 4 fingers from the midline just below the 11th rib will enter the kidney, allowing aspiration of urine and injection of contrast for opacification. The angle of puncture should be altered slightly on each additional puncture that is required.
b) The kidney can be punctured using real-time ultrasound (US) guidance in the fluoroscopy suite.

In cases with marked calyceal dilatation, the whole PCN can be performed under US guidance. In difficult cases, it may be necessary to utilize CT. In all cases fluoroscopy is of added value.

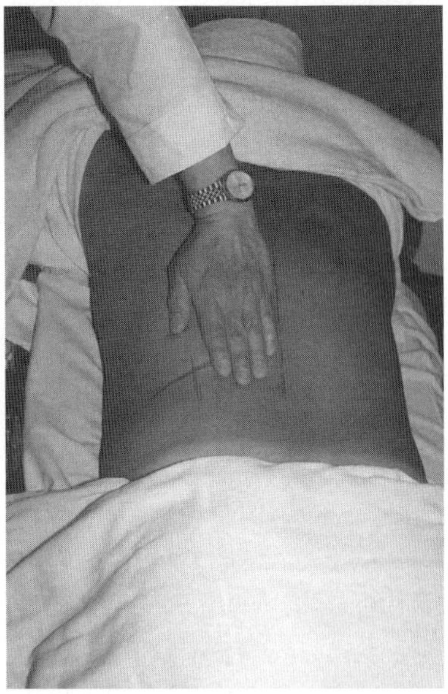

Figure 1. *Using bony landmarks the puncture site is four finger breadths from the midline just below the margin of the lowest rib that can be palpated in this location. The costal margin is the curved line marked on the patient's back in this figure.*

TECHNIQUE

Our preference is first to opacify the collecting system using a Chiba needle, and then to choose the best entry site and perform the puncture under fluoroscopy. In general, the puncture site should be as lateral as possible for convenience of nursing, avoiding a direct dorsal approach as most patients are nursed lying on their backs. A flank entry also avoids the large lumbar muscle groups. However, the entry site cannot be too lateral because of the danger of puncturing the colon. The puncture should be caudal enough to avoid the pleura and should yield the shortest, straightest track consistent with the prior considerations. The inferior parietal pleura on either side crosses the seventh rib anteriorly, the tenth rib in the midaxillary line and the twelfth rib posteriorly to join the dorsal spine of the twelfth thoracic vertebra in the midline **(Fig. 1)**.

PCN: INTERCOSTAL APPROACH

- Serious complications: 12.5%
- Try: positioning patient; control respiration

For calculous therapy, the stone-bearing calyx, or one that yields convenient access to it, must be entered. The collecting system is ideally accessed through a lower or mid calyx of the posterior group. Intercostal routes are usually associated with high complication rates and should be avoided. Careful positioning of the patient and taking advantage of the changes in renal position at different times in the respiratory cycle can be of value in avoiding an intercostal approach.

Puncture of the renal pelvis directly, without passing through renal parenchyma, can cause bleeding and/or urine leakage as well as problems with subsequent tube exchanges (*see Case 2 in Chapter Preview*). The site of entry into the collecting system should be through the calyx **(Figs. 2, 3)**. A posterolateral entry is chosen to avoid the renal vein and artery which are located anteromedially.

Entry into the kidney can be achieved by Seldinger technique, using an 18-gauge needle or needle sheath system which then accepts an .035-inch or .038-inch guidewire. Alternatively, the puncture can be made using a 21- or 22-gauge Chiba needle and an .018-inch guidewire, and then employing a conversion system such as the Cope or Neff system, an .035-inch or .038-inch wire is then inserted. This is followed by standard dilation and catheter placement. The choice of equipment is one of personal preference and experience. Less commonly used is a trocar technique, which is best reserved for greatly dilated systems.

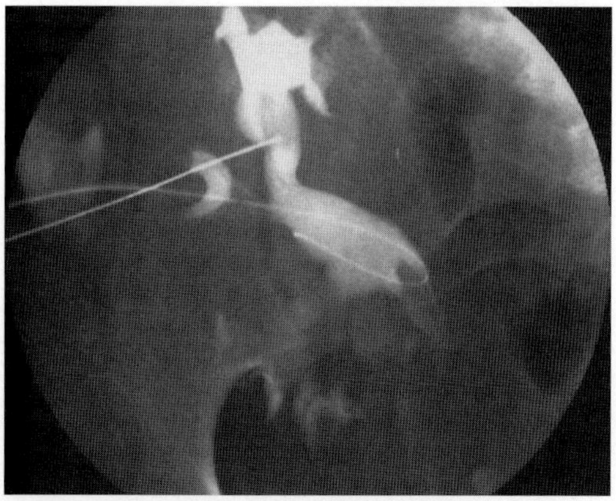

Figure 2. *PCN in transplanted kidney. A 22-gauge needle was placed under ultrasound guidance. It entered an upper pole infundibulum and was used to opacify the collecting system by the injection of contrast agent. The midpole calyx was then punctured under fluoroscopic control. The guidewire shows an ideal entry site in the calyx.*

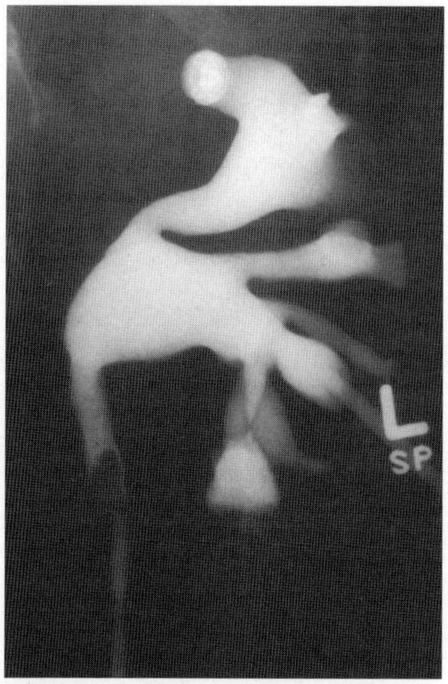

Figure 3. *To show an excellent entry site of a PCN catheter through a mid- to lower-pole calyx. Note the filling defect in the upper ureter which is a urinary stone causing partial obstruction.*

In all cases, it is important not to overdistend the collecting system, as this is a potent cause of sepsis. Access to the system is confirmed by fluoroscopy and aspiration of urine. Confirmation of needle position in the collecting system by injecting contrast material through the needle is commonly performed. If the needle is not well placed, a bothersome extravasation of contrast material may obscure further manipulations. A good alternative is a short, brisk puff of saline down the needle, which is easily seen fluoroscopically as a negative defect in the contrast-opacified collecting system. Extravasation of saline does not obscure further visualization.

A variety of guidewires is available to advance from the calyceal puncture site into the pelvis and down the ureter. It is strongly recommended that the guidewire be manipulated down the ureter to aid in subsequent nephrostomy placement. The wire should not merely be coiled in the renal pelvis. Our preference is to have a stiff Amplatz wire down the ureter before commencing track dilatation and nephrostomy placement. This avoids possible buckling of the wire in the pararenal space **(Figs. 4, 5;** *see Case 3 in Chapter Preview***)**.

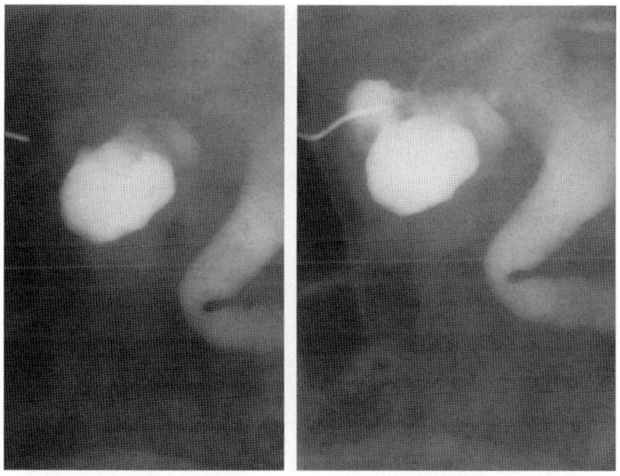

Figure 4. *Calyceal puncture under fluoroscopic guidance after prior opacification of collecting system using a 22-gauge needle.*

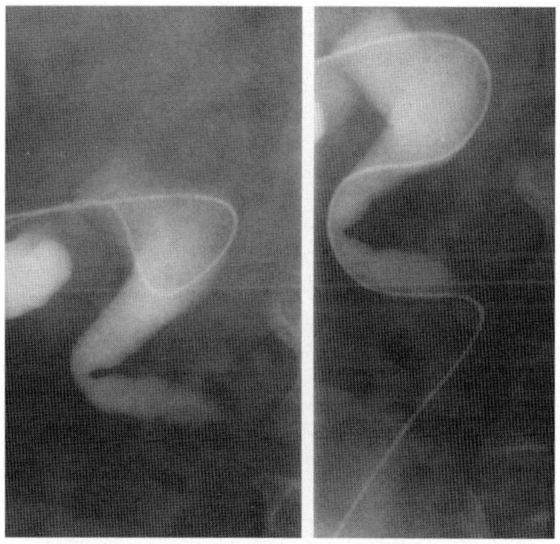

Figure 5. *To show stages in manipulation of a torquable guidewire from the calyx to the renal pelvis and then down the tortuous ureter.*

Many nephrostomy tubes are available, and we usually choose a self-retaining Cope Loop nephrostomy of 8- or 10-F size. A larger tube is required if thick pus must be drained. The Cope loop is positioned in the renal pelvis and the catheter anchored at the skin. A diagnostic nephrostogram should not be performed at the time of PCN placement in order to minimize the likelihood of causing sepsis. A small amount of contrast can be injected to confirm satisfactory placement and drainage. All diagnostic studies should be postponed until the patient's condition has stabilized.

RESULTS

PCN is safe and effective, and successful in 98% of patients. In infected cases, marked defervescence occurs in 24 to 48 hours. Relief of obstruction results in an early and brisk diuresis except in cases of long-standing obstruction where there is irreversible renal drainage.

PCN: COMPLICATIONS (%)	
• Bleeding	
— Transfusion	0-3.6
— Embolization or operation	1-2
• Pneumo/hydrothorax	0-1
• Infection	1.4-21
• Tube problems	2.5-18

COMPLICATIONS

The published complication rates for percutaneous nephrostomy vary widely from series to series. Complication rates are generally higher when PCN is used for stone extraction. Representative figures are a 4% to 6% overall rate of complications including massive bleeding necessitating treatment by surgery or embolization (1% to 2%), pneumothorax or hydrothorax (1%), peritonitis or life-threatening sepsis (1%-2%). Bleeding requiring transfusion has been reported at rates between 0% and 3.6%. Overall minor complication rates are 10% to 28% and include tube problems such as dislodgments, kinks, blockages and leaks (2.5% to 18%) and infection (1.4% to 21%). Serious arterial injuries may occur in 1% to 2% of cases, and can be effectively treated by selective arterial embolization **(Fig. 6)**. Almost all patients have bloodstained urine after PCN placement. The urine usually clears in 12 to 36 hours. This should not be confused with massive arterial bleeding. Clots will usually lyse spontaneously, and irrigation of the nephrostomy tube is usually not necessary. Punctures made too medially are more likely to cause serious bleeding by injury to large vessels in the hilum of the kidney. Puncture of adjacent organs, such as the colon, spleen or gallbladder, are fortunately rare, as is pneumothorax. Stables reported a 19% incidence of exacerbation of pyonephrosis/pyelonephritis. In an obstructed system, if the initial attempts at PCN are unsuccessful, great efforts should be made to establish adequate drainage as soon as possible. A delay of 1/2 hour or so will allow extravasated contrast to disperse, and a fresh attempt at PCN undertaken. A failed PCN can lead to urine leakage and urinoma formation. Pain is common in the first 12 to 24 hours after PCN placement, and adequate analgesia should be provided. Pain usually subsides spontaneously with time.

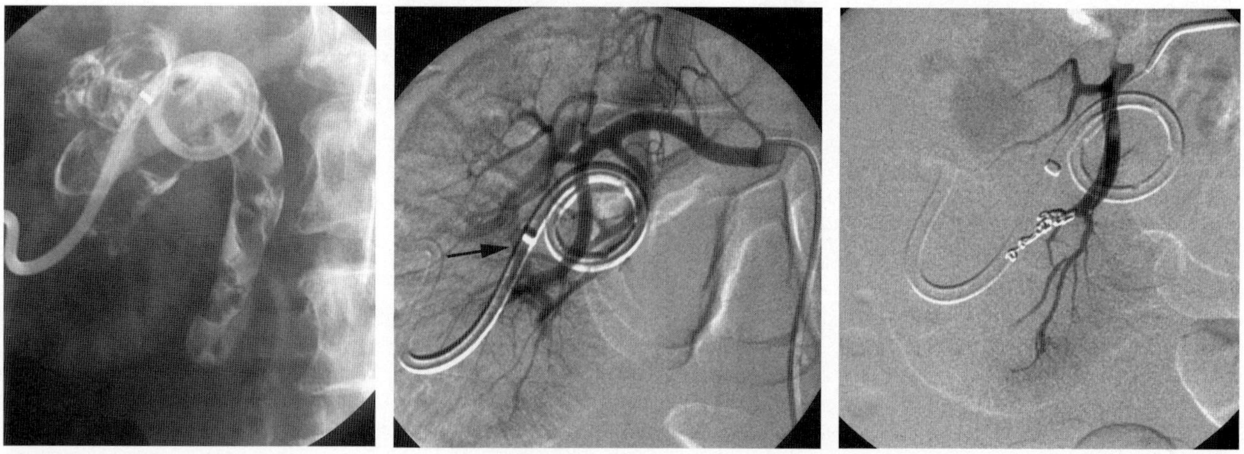

a. b. c.

***Figure 6.** Patient requiring embolization after PCN, because of confirmed bleeding. a) Right nephrostogram shows collecting system filled with clots. b) Selective right renal angiogram showing iatrogenic arteriovenous fistula. The vein (arrow) fills early. c) After selective microcoil embolization of the injured intrarenal arterial branch. There is now no longer filling of the vein. Bleeding stopped immediately.*

NEPHROSTOMY TUBE MAINTENANCE

For long-term drainage, soft, self-retaining catheters are used. We favor the Cope loop-type, but types with Malecot wings or Foley balloon catheters are also used. Catheters should be changed routinely every 2 to 3 months. It is better to change catheters electively on a fixed schedule rather than to wait for catheter blockage, which will likely result in infection. The time period between exchanges needs to be reduced in patients who are on long-term drainage, because the tendency for clogging and blockage increases with chronic drainage. Not infrequently, problems occur with encrustation of the retention sutures inside the Cope loop catheters.

The Cope loop must be exchanged over a wire, under fluoroscopic control. Nephrostomy tubes are changed routinely without antibiotic cover, unless the patient has fever or has presented with an obstructed catheter. Antibiotics are then given systemically about 30 minutes before the exchange.

STENTING

- Indications
- Contraindications
- Technique
- Internal stents
- Internal-external stents
- Results and complications
- Expandable metallic stents
- Dilatation of ureteral strictures

PART 2. URETERAL STENTING

Ureteral stents are in common use for a variety of indications as summarized below:

a) Relief of obstruction, e.g., stone, neoplasm, retroperitoneal fibrosis, radiation-induced, inflammation, iatrogenic or traumatic obstruction (see **Table 1** following). The stent may be needed permanently, or only temporarily
b) Treatment of urinary leaks from kidney or ureter
c) Stone therapy, before and/or after ESWL
d) During surgery for localization purposes, as an aid during anastomotic procedures, and postoperatively as a temporary scaffold and drain during the time that edema subsides and healing takes place.

INDICATIONS

- Obstruction
- Leaks
- ESWL
- Surgery

The contraindications to stent placement are often relative and involve evaluation of the whole clinical picture of each patient. Stent placement is never urgent in patients who have a PCN in place. However, retrograde stent placement as the primary mode to relieve obstruction may be urgently required.

a) Untreated urinary tract infection. Drainage via PCN and antibiotics are a first step, with placement of the stent being delayed
b) Untreated bleeding diathesis
c) Bladder fistula or leak. In this case, diversion of urine away from the bladder is required, but not ureteral stenting, which drains urine into the leaking bladder
d) Bladder irritability. With conditions such as acute radiation cystitis, the mere presence of urine in the bladder may cause great discomfort. In other cases, the bladder end of the stent can provoke intolerable irritation, especially if it rubs against the trigone.

CONTRAINDICATIONS

Ureteric Stents
- Untreated infection
- Bleeding diathesis
- Bladder fistula or leak
- Bladder irritability

TABLE 1: CAUSES OF URETERAL OBSTRUCTION

Intrinsic or Intraluminal
- Calculi, blood clots
- Congenital UPJ or UVJ
- TB or schistosomiasis
- Anastomotic strictures

Extrinsic
- Lymphadenopathy
- Prostate cancer
- Pelvic malignancies
- Pelvic inflammatory disease
- Pelvic abscess
- Endometriosis
- Urinoma
- Lymphocele
- Retroperitoneal fibrosis
- Iliac artery aneurysm
- Pregnancy
- Trauma
- Iatrogenic trauma

TECHNIQUE

Stents can be placed initially by the retrograde approach using a cystoscope in the bladder. Recently, some authors have reported success using a curved angiographic catheter passed up the urethra into the bladder. Using fluoroscopic control, they have succeeded in finding the ureteric orifice, and then catheterizing the ureter in the retrograde direction.

Alternatively, the ureter can be stented in an antegrade direction, using a PCN as the access route.

A stent may be completely internal, with its upper end in the renal pelvis and the lower end in the bladder, typically termed a double-J stent.

Stents may also be of the internal-external type, with one end exiting through the nephrostomy or through the urethra, or through the skin stoma that is part of some internal urinary reservoir.

INTERNAL STENTS

Internal stents have the great advantage that the patient is not encumbered with external tubes and drainage bags. The disadvantage is that internal stents, like all stents, require routine replacement at approximately 3-month intervals. The distal end of the stent has to be captured in the bladder, by cystoscope or snare, and drawn out through the urethra to allow stent exchange with guidewire technique. The longer the duration during which internal stents are used, the more prone to encrustation they become, necessitating more frequent stent exchanges.

We usually use double-J type internal stents of 8- or 10-F caliber, and of a length tailored to the patient's size (range 20-24 cm between the bases of the pigtails). For transplanted kidneys, we use special-made short stents of 8-12 cm in length. Stents are best placed over a stiff Amplatz wire, with balloon predilatation of any tight ureteral strictures. If more flexible stents with high friction coefficients, such as silastic, are chosen, these must often be inserted through a peel-away sheath. Care must be taken that the distal pigtail is formed completely within the bladder, so that retrieval from the bladder is possible. The pigtail can deform the distal ureter and give a mistaken impression of being fully within the bladder. Oblique films will eliminate this possibility, as the ureter enters the bladder posteriorly. At the upper end, the pigtail must be contained entirely within the collecting system, and not protrude into the renal parenchyma or, worse still, outside the renal capsule. Side-holes are best positioned on the inner curves of each pigtail, and should not be located on the shaft of the stent. The holes should be of a similar size as the stent lumen. Larger holes weaken the stent, making insertion more difficult without providing improved drainage. Non-infected urine will drain effectively even through small catheters of 5- and 6-F caliber **(Figs. 7, 8)**.

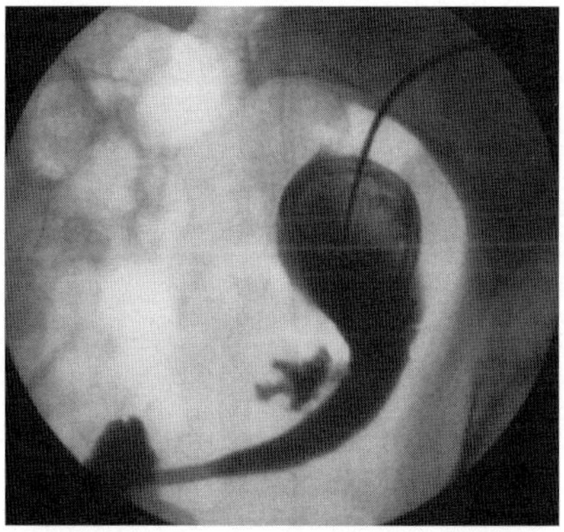

Figure 7. *A patient with a renal transplant on the left side, and a clearly seen ureteral leak. There is a double-J stent in place inserted via an antegrade percutaneous approach.*

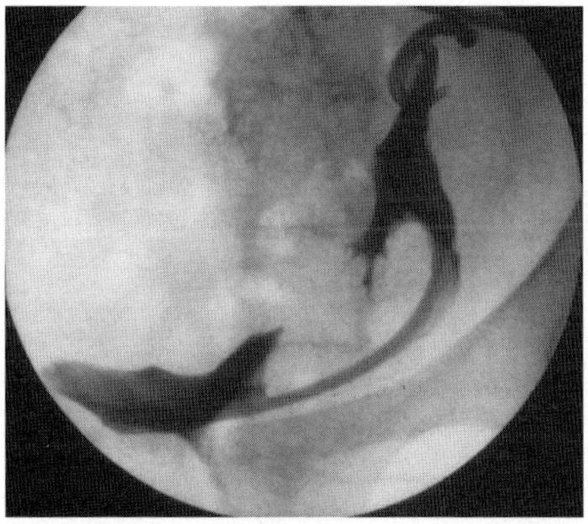

Figure 8. *One week later the leak is no longer seen. The double-J stent and percutaneous nephrostomy are in position. The renal pelvis and renal calyces are not dilated.*

Double-J stents should not be used in patients with mucus-producing reservoirs such as ileal conduits. The mucus secreted by the conduit wall rapidly plugs up these stents. These patients are best managed with stents that have only side-holes in the pigtail in the renal pelvis, no side-holes in the stent shaft, and holes in the distal stent once it has emerged from the skin stoma.

Advantages of internal stenting from the antegrade approach are that immediate drainage is obtained via the PCN. The stent placement can be postponed while infection is treated, bleeding disorders are corrected, and the nephrostomy track matures. Once the stent is placed, the nephrostomy is retained. A later nephrostogram will show that the stent is functional. The stent can be tested by capping the nephrostomy before removal of the PCN.

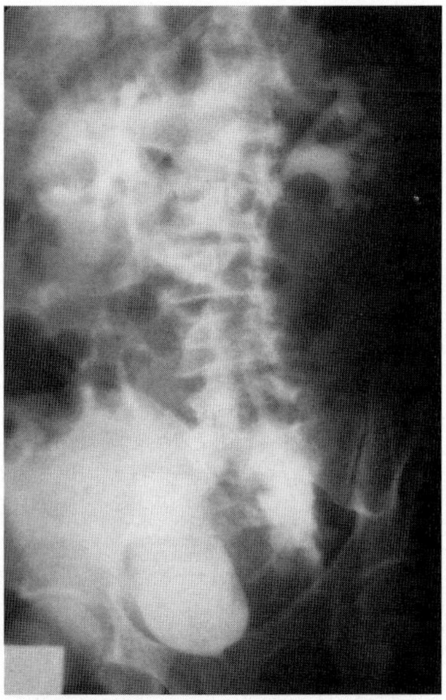

Figure 9. The left ureter was perforated during the passage of a ureteroscope. This intravenous pyelogram (IVP) shows extravasation of contrast from the ureter at the pelvic brim with compression of the left side of the bladder.

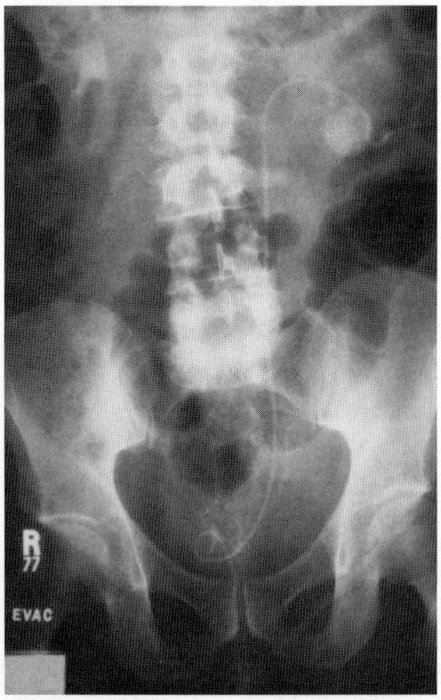

Figure 10. An antegrade internal-external ureteral stent was passed percutaneously into the left kidney and down into the bladder. This IVP 1 week later shows that the previously demonstrated leak is no longer visible.

INTERNAL-EXTERNAL STENTS

Internal-external stents have the great advantage that they can be replaced simply over a guidewire for both antegrade and retrograde stents.

Antegrade stents can have a pigtail in the bladder and exit via the kidney, with side-holes in the renal pelvis segment of the stent and in the bladder pigtail segment. Alternatively, a Cope loop may be formed in the renal pelvis, for better self-retention, with a bladder pigtail. Such catheters are good for temporary stenting in cases of urine leaks or temporary ureteral edema. They can be used to test whether a patient can tolerate a catheter in his or her bladder. This is of particular value in men with prostate cancer and ureteral obstruction. A stent should never be placed with a free end in the ureter itself, as this causes stricturing of the ureter.

Internal-external stents must be used in patients with internal urinary reservoirs, as double-J stents become blocked frequently by mucus produced in the reservoir **(Figs. 9, 10)**.

RESULTS AND COMPLICATIONS

Stents have proved most effective in curing urine leaks, and are excellent for providing relief of obstruction. Their use in surgery to provide a scaffold is most controversial. Does the stent allow good healing without stricture formation, or does the stent itself incite inflammation and stricture formation? The main problem with stents is that all stents become occluded over time. The longer stents are used, the shorter is the required interval between exchanges.

EXPANDABLE METALLIC STENTS

Preliminary clinical experience using Wallstents in strictured ureters has been reported in a few series with short follow-up. Flueckiger et al placed Wallstents in 13 ureters. Urothelial reaction blocked four ureters in 1 to 2 weeks, and blocked one ureter after four weeks, each requiring the insertion of double-J internal plastic stents. One ureter became occluded by tumor overgrowth of the stent 8 months after stenting. Follow-up was from 3 to 14 months (average 5.8 months). Lugmayr et al implanted Wallstents in 30 ureters, but, in addition, maintained a double-J internal stent through the Wallstent for the first 4 weeks to overcome obstruction by reversible hyperplastic reaction of the urothelium. These reports suggest that Wallstents may possibly be better than plastic stents, but this is yet to be convincingly demonstrated. In contrast, Wright et al suggested using Gianturco stents as a method of occluding the ureter!

DILATATION OF URETERAL STRICTURES

Early papers reported good success with balloon dilatation of ureteral strictures. These good results have clearly been shown to be maintained only in the short term, with recurrence of the stricture recurring over time **(Table 2)**. Occasional successes are likely to have been achieved in patients with early edema, that would in any event have resolved spontaneously.

TABLE 2: DILATATION OF URETEROENTERIC STRICTURES: LONG-TERM RESULTS

	No. Strictures	% Success
Initial	37	100
6 months	11	30
1 year	6	16
< 2 years	3	8

(All patients patent at greater than 6 months had repeated dilatations.)

(Shapiro MJ, et al 1988.)

CHAPTER 24 PREVIEW

Case 1

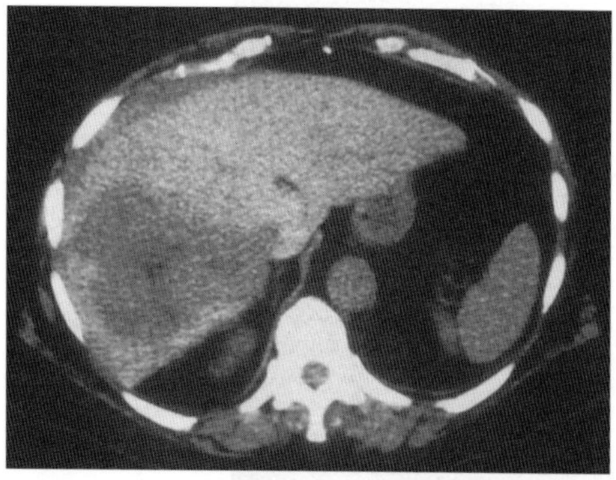

History: A 54-year-old woman presented with vague right upper quadrant pain and history of severe contrast media allergy.
Question: Should this lesion be biopsied or should other diagnostic tests be performed?

Case 2

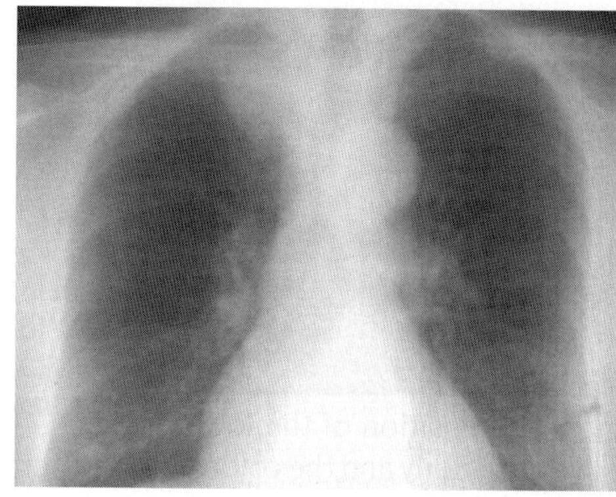

History: A 63-year-old man presented with a right apical mass discovered on a routine chest radiograph.
Question: What type of guidance could be used for percutaneous biopsy?

Case 3

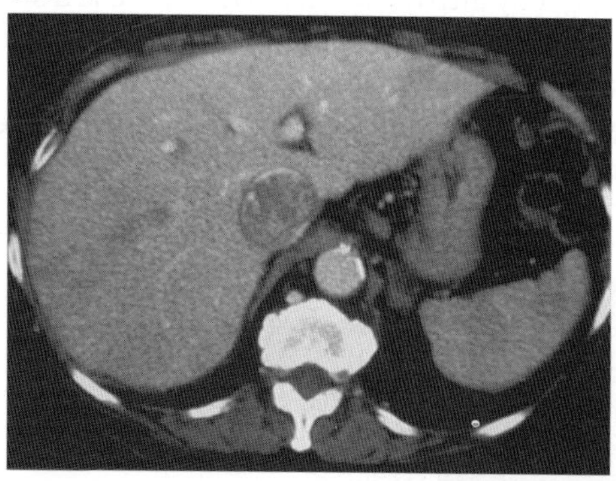

History: A 55-year-old man has a history of renal cell carcinoma and lower extremity edema. The referring surgeon would like to know if tumor or bland thrombus extends into the inferior vena cava.
Question: What is the abnormality and how could the diagnosis be established?

Case 4

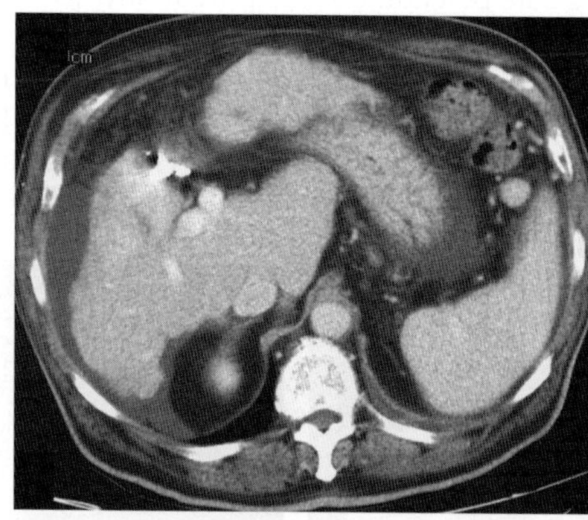

History: A 44-year-old with abdominal distension, abnormal liver function tests, and abnormal coagulation parameters.
Question: What are the potential techniques by which a liver biopsy could be obtained in this patient?

Chapter 24
Percutaneous Biopsy

Robert K. Kerlan, Jr., MD

CHAPTER OUTLINE

- Indications
- Contraindications
- Technique
    - Preprocedural evaluation
    - Guidance
    - Needle selection
- Clinical applications
    - Lung
    - Liver
    - Pancreas
    - Adrenal
    - Retroperitoneal lymph nodes
    - Bone
- Results
- Complications

INTRODUCTION

Percutaneous biopsy is among the most common interventional radiologic procedures performed in clinical practice. No other procedure has nearly the clinical or economic impact of fine-needle aspiration biopsy. Before guided-needle biopsy was developed, patients would commonly undergo thorocotomy or exploratory laparotomy to ascertain the etiology of abnormalities detected on radiographs or physical examination. Now, the same information can be obtained with minimal risk and discomfort to the patient in an outpatient setting.

INDICATIONS

In general, percutaneous biopsy is performed to establish the diagnosis of primary or metastatic malignancy, and to differentiate malignant lesions from inflammatory lesions. The most common indications for imaging guided biopsy are:

1) Solitary or multiple lung nodules
2) Multiple liver nodules
3) Intra- or retroperitoneal mass.

CONTRAINDICATIONS

The only absolute contraindication to the performance of a fine-needle aspiration biopsy is lack of a safe percutaneous access route. As most normal structures can be safely transgressed with a 22-gauge needle, the only real limitation is presented by the heart which precludes puncture of lesions located deep within the mediastinum.

Relative contraindications include:

1) Marked coagulopathy (INR >2.0, platelet count <50,000/ml)
2) Highly vascular liver tumors such as suspected hepatoma or hemangioma with no intervening normal hepatic tissue.

TECHNIQUE

Preprocedural Evaluation

All patients are given informed consent prior to the procedure. The majority of patients have the procedure solely with local anesthetic agent. Anxious patients may require a mild sedative or formal intravenous conscious sedation. The specific procedural technique is dependent upon the location of the pathology and the type of specimen required to yield the diagnosis. It should be noted that there are many acceptable techniques for performing biopsies which depend on the individual experience of the radiologist.

No preprocedural laboratory evaluation is recommended prior to superficial, small-bore biopsy. Prothrombin time, partial thromboplastin time, CBC, and platelet counts are recommended prior to deep, large-bore biopsies (Murphy, et al).

Guidance

Three methods of guidance for percutaneous biopsy are commonly used: computed tomography (CT), ultrasound, and fluoroscopy. Magnetic resonance imaging (MRI) has also been used to guide percutaneous biopsies, but requires specialized needles. The selection of guidance is dependent upon the visibility of the specific lesion for a given modality, as well as the experience and preference of the operating physician.

Needle Selection

The needle selected to perform the biopsy depends on the site of the lesion, the tissue requirement (histology versus cytology) for diagnosis, and the preference of the operator. Though larger needles produce more bleeding, the differences are statistically significant only when comparing 14-gauge to 16-gauge needles to the group of 18-, 20-, and 22-gauge needles. Dahnert et al compared 19.5-gauge cut needles, 20-gauge Franseen trephine needles, 20-gauge spinal needles, and 20-gauge Wescott slotted needles. Though small differences were detected in obtaining material for cell blocks, no clear advantages in diagnostic yield were detected between the needles that were evaluated.

A number of spring-mechanism (automated) biopsy devices (guns) have been developed over the past 5 years for obtaining histologic (core) specimens. Hopper et al evaluated a number of these devices and concluded that most 18-gauge devices with an excursion (throw) of at least 2.0 cm provided high-quality tissue.

CLINICAL APPLICATIONS

Lung Biopsy

Percutaneous lung biopsy is usually performed on an outpatient basis. Either CT **(Fig. 1)** or fluoroscopic guidance can be used. CT guidance is preferable for lesions which abut or are adjacent to the pleural surface **(Fig. 2)**. It is also very useful for lesions in the upper lobes where the effect of respiratory motion is minimal. Fluoroscopic guidance is preferred when the lesion is small but visible by fluoroscopy after determining the precise depth and safe percutaneous course by CT. As pulmonary parenchymal lesions move with respiration, the real-time guidance afforded by fluoroscopic observation is extremely beneficial in directing the needle pass. Moreover, using this technique, it is usually possible to obtain a specimen in under 30 seconds, allowing the biopsy to be performed during suspended respiration. In so doing, the chance of lacerating the pleura and creating a pneumothorax is reduced.

LUNG BIOPSY: TECHNIQUE

Guidance

- CT
- Fluoroscopy

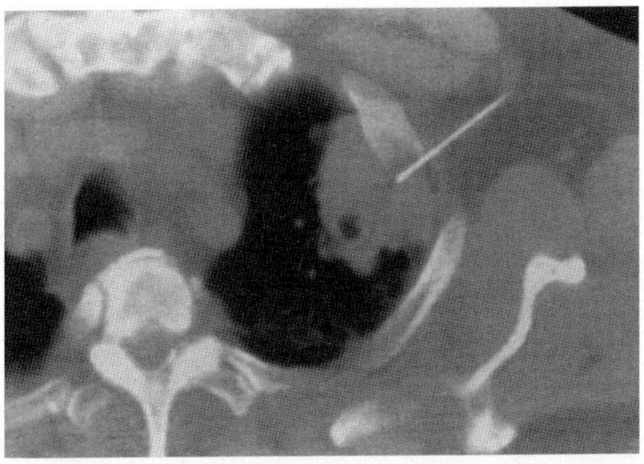

Figure 1. *A 22-gauge needle is advanced into a nodular lesion in the left upper lobe under CT guidance. The specimen was positive for adenocarcinoma.*

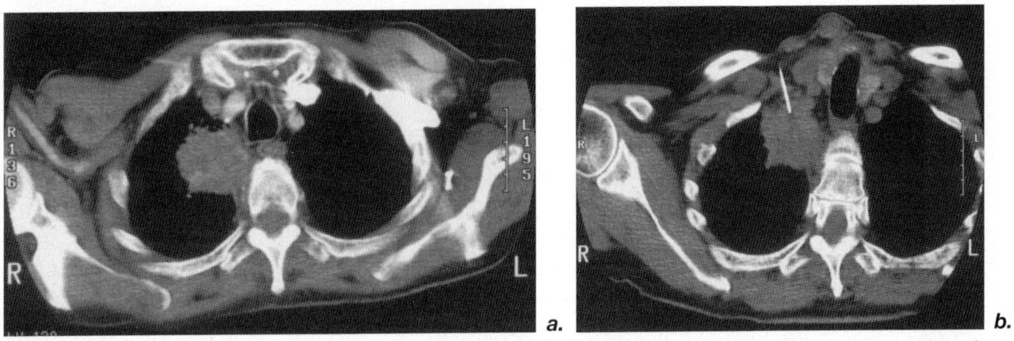

Figure 2. *58-year-old man with right apical lung nodule. a) Conventional axial images reveal interposition of aerated lung from the anterior approach. b) By angling the gantry, a percutaneous route devoid of aerated lung is identified.*

The vast majority of lung biopsies are performed only with local anesthesia. However, intravenous access is established in the event that it becomes necessary to administer sedative or other medications.

A variety of different types of needles are used to perform the lung biopsies. Selection of the precise type of needle depends upon the personal preference of the radiologist. We prefer using 22-gauge Westcott, Chiba or Menghini needles in most cases. Other investigators prefer coaxial systems with initial placement of an 18- or 19-gauge, through which multiple passes with 22- or 23-gauge biopsy needles may be made. An on-site cytopathologist is always present so that the adequacy of the specimen can be immediately assessed, avoiding the necessity of making additional, unnecessary needle passes. In most cases, the diagnosis is established on the initial pass.

To detect a pneumothorax, the patient receives an upright chest radiograph immediately following the procedure and again 90 minutes later. If no pneumothorax is present on either film, the patient is discharged home. If a pneumothorax is detected, serial radiographs are obtained at intervals dictated by the size of the pneumothorax and the condition of the patient. If a pneumothorax is less than 20% of the overall lung volume and determined to be stable on serial radiographs, the patient is discharged home.

LIVER BIOPSY

Guidance

- CT
- Ultrasound
- None

Liver Biopsy

Other than regional lymph nodes, the liver is the most common site of metastases from malignancies of the gastrointestinal tract and pancreas. Patients who have had colon cancer resected are evaluated at periodic intervals to detect the presence of liver lesions. Moreover, patients who have a primary tumor discovered at almost any site will have an evaluation for liver metastases. If a metastasis is discovered in the liver, the operation to remove the primary tumor may be limited or not performed. Therefore, an accurate diagnosis of liver metastases is extremely important to determine the appropriate management of the patient.

Percutaneous liver biopsy is usually performed on an outpatient basis using only local anesthetic agents. In patients who are unusually anxious, intravenous conscious sedation is performed. Needle selection is generally a matter of personal preference. We prefer 22-gauge Chiba, Westcott or Menghini needles.

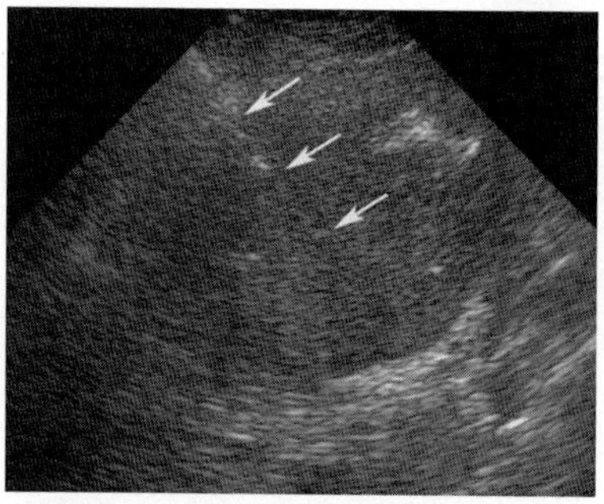

Figure 3. A 22-gauge needle (arrows) is advanced into an echogenic liver mass under sonographic guidance. Biopsy was positive for hepatocellular carcinoma.

Percutaneous liver biopsy can be performed blindly if disease is extensive. When lesions are fewer in number, ultrasound **(Fig. 3)** or CT guidance **(Fig. 4)** is needed. The guidance modality is most often dependent upon the experience and preference of the radiologist.

A percutaneous path is selected which avoids traversing the lung, gallbladder or bowel. The skin site is then anesthetized and the needle is advanced into the lesion using imaging guidance. An on-site pathologist is always present to determine the adequacy of the biopsy. If an adequate specimen has been obtained, the patient is observed for approximately 2 hours. If no pain or evidence of bleeding is observed, the patient is discharged home. The patient is counseled to return for any abdominal pain, lightheadedness, or any problem which deviates from their usual status.

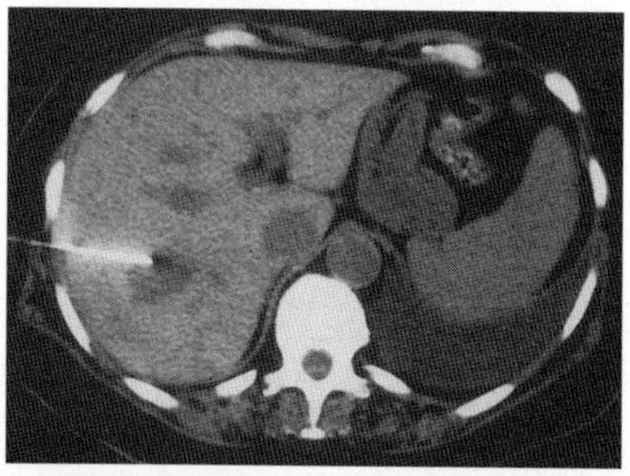

Figure 4. CT-guided percutaneous biopsy of a colonic carcinoma metastatic to the liver.

Most liver biopsies are performed with the patient in a supine position. Small, posterior lesions located toward the hepatic dome may be approached posteriorly with the patient in an ipsilateral decubitus position. In this position, overlying aerated lung may be displaced superiorly, diminishing the chance of a pneumothorax following the biopsy **(Fig. 5)**.

Large-bore histologic biopsies performed with 14-18-gauge cutting needles are also performed by some interventional radiologists. Automated biopsy devices (spring-loaded gun) are beneficial for obtaining adequate specimens. Patients obtaining large bore hepatic biopsies should undergo short-term (6-8 hour) observation for hemorrhagic complications.

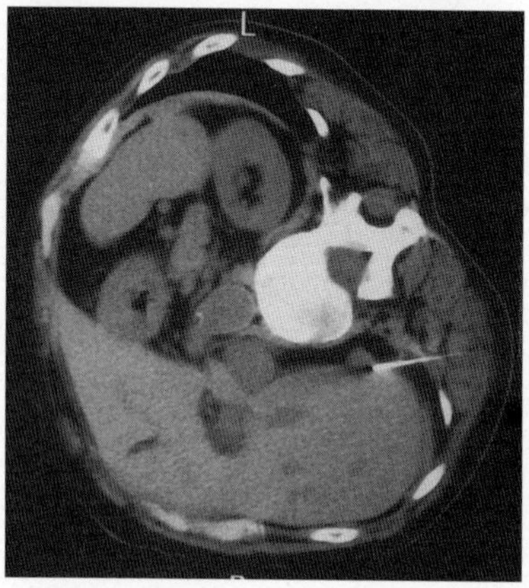

Figure 5. 56-year-old woman with ovarian carcinoma. Biopsy of the 15-mm hepatic implant is performed with the patient in an ipsilateral decubitus position displacing aerated lung superiorly.

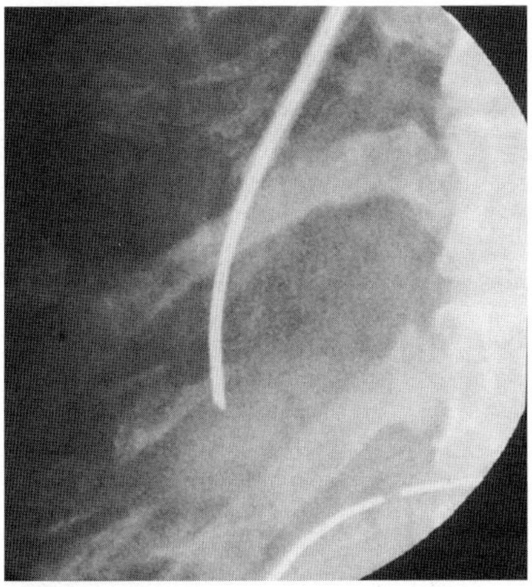

Figure 6. A transjugular biopsy needle is thrust through the wall of an hepatic vein into the hepatic parenchyma.

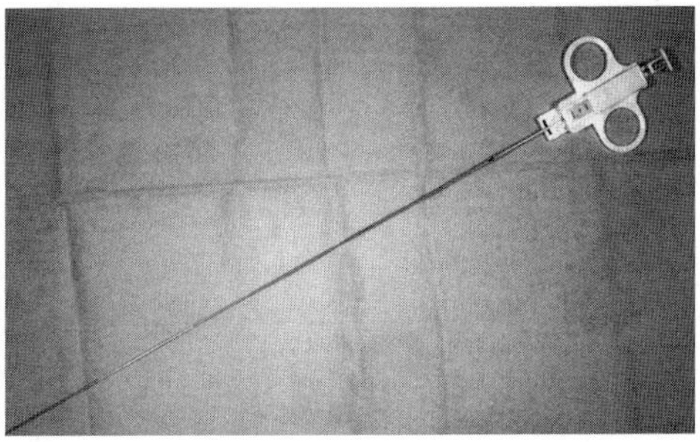

Figure 7. 20-gauge spring-loaded transjugular hepatic biopsy device.

If ascites, small liver size or coagulopathy precludes percutaneous liver biopsy, transjugular biopsy may be safer than percutaneous biopsy **(Fig. 6)**. Transjugular biopsies are performed in the angiographic suite under fluoroscopic guidance. After sterile preparation, the right internal jugular vein is punctured and a catheter is manipulated through the superior vena cava and right atrium into a right hepatic vein. The catheter is replaced for a specially designed biopsy sheath through which a biopsy needle is inserted. The biopsy needle is then thrust through the wall of the hepatic vein to obtain a histologic specimen of the hepatic parenchyma. Transjugular liver biopsy was performed for many years with a 17-gauge Colapinto needle. However, smaller gauge automated biopsy devices specially designed for the transjugular approach **(Fig. 7)** are now available which yield suitable specimens on a more consistent basis.

In patients with biliary drainage catheters, bile duct biopsy may be performed through the transhepatic route. Needles, brushes, forceps and atherectomy catheters have been used for this purpose with inconsistent results.

A common lesion of the liver, cavernous hemangioma, should not be biopsied when typical imaging features are present **(Fig. 8)**. Though these lesions can be safely biopsied with 20-22-gauge needles through normal hepatic parenchyma, intraperitoneal hemorrhage may occur if the lesions are biopsied close to the surface of the liver.

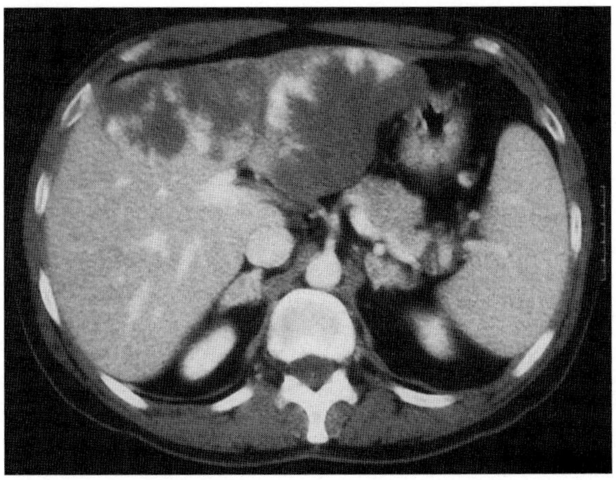

Figure 8. Cavernous hemangioma of the liver. CT scan shows the typical peripheral enhancement which is isodense with the aorta.

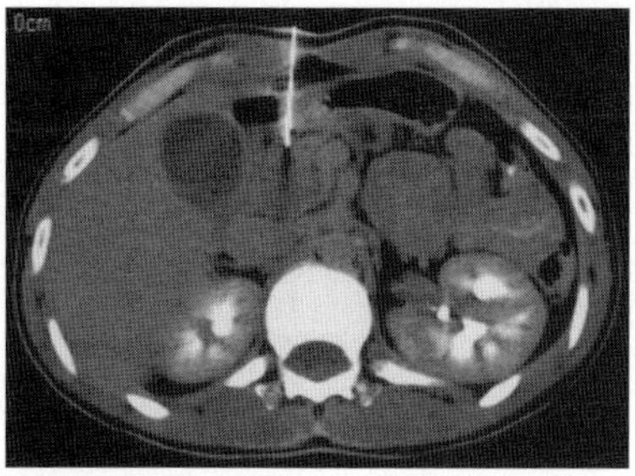

Figure 9. Biopsy of pancreatic head mass. The position of the pancreas may require the 21- or 22-gauge needle to traverse bowel loops or, as in this case, the stomach.

Pancreas Biopsy

Pancreatic biopsies are performed with an identical technique as described for fine needle aspiration liver biopsy. Due to the position of the pancreas, it may be necessary to traverse bowel loops to obtain the specimen **(Fig. 9)**. As an alternative, a posterior retroperitoneal approach may be used. A transcaval biopsy may be performed with a 22-gauge needle **(Fig. 10)**. The spleen should not be traversed due to the possibility of creating a laceration.

Adrenal Biopsy

Adrenal biopsies are also performed with a similar technique. Lesions which are suspected of harboring pheochromocytoma should not be biopsied. Because of the usual upper retroperitoneal position of the adrenal glands, using the ipsilateral lateral decubitus position may be useful to obtain an in-plane biopsy and avoid aerated lung **(Figs. 11, 12)**. Massive hemorrhage will occasionally complicate adrenal FNA biopsy.

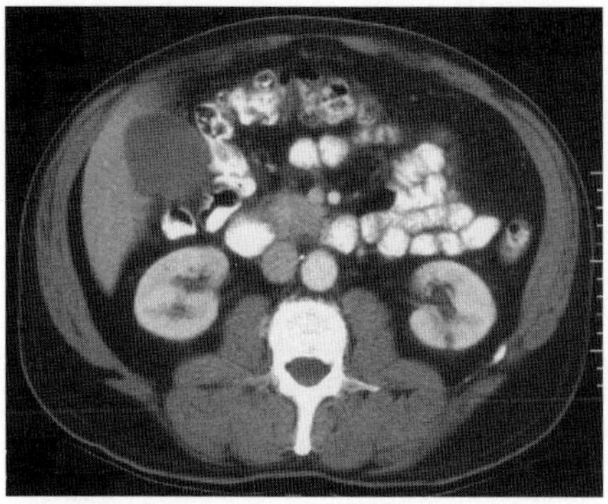

a.

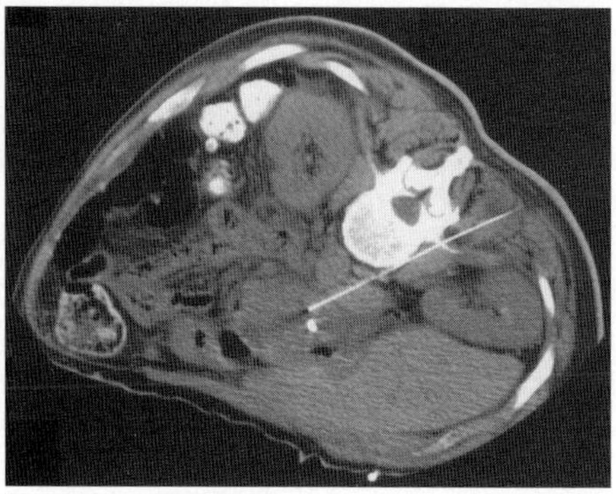

b.

Figure 10. *Biopsy of the pancreatic head from a posterior transcaval approach. a) Axial CT image with the patient supine shows colon overlying the possible mass in the pancreatic head. b) At a slightly higher level, with the patient in a semiprone position, the mass is biopsied with a 22-gauge needle from a posterior transcaval approach. The biopsy was positive for adenocarcinoma.*

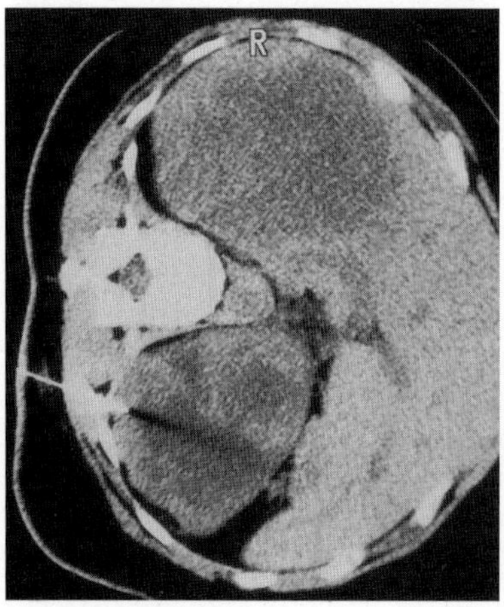

Figure 11. *Large left adrenal mass biopsied from a posterior approach with the patient in the ipsilateral lateral decubitus position.*

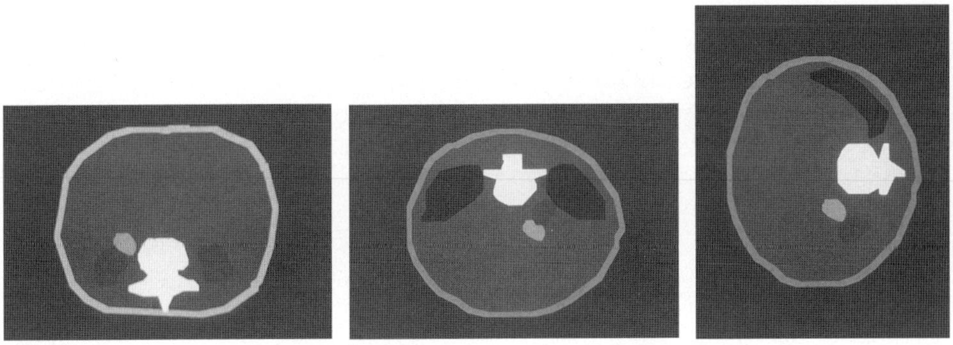

a. b. c.

Figure 12. Percutaneous adrenal biospy technique. a) On supine CT scans, the posterior approach for percutaneous biopsy of the adrenal gland may appear straightforward. b) When the patient is placed in the prone position, aerated lung will commonly descend, and interpose itself between the adrenal gland and the skin surface. c) With the patient in the ipsilateral lateral decubitus position, the aerated lung will often be displaced superiorly. In this position, the adrenal can be safely biopsied.

Retroperitoneal Lymph Node Biopsy

Minimally enlarged lymph nodes in the paracaval and paraaortic lymph node chain can usually be biopsied with computed tomographic guidance. The biopsy can be performed from either an anterior or posterior approach **(Fig. 13)**. For the posterior approach, the patient is placed in a prone position with the approach being made posteriorly through the psoas muscle. Coaxial needle systems are particularly useful for performing this type of biopsy as the larger outer trocar can be precisely adjusted, directing the smaller aspiration biopsy needle into the desired lymph node. Though these lymph nodes are adjacent to major vascular structures (such as the aorta and inferior vena cava), inadvertent puncture of these structures with a 21- or 22-gauge needle is generally of no clinical consequence.

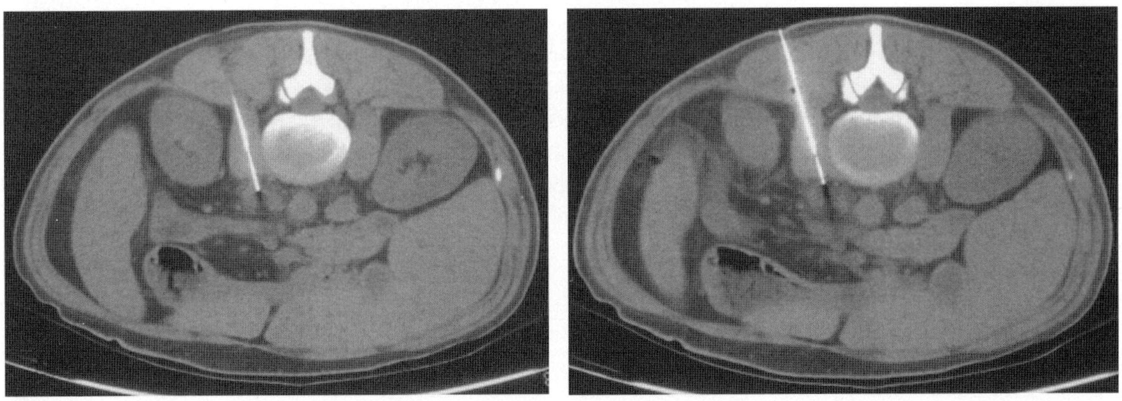

a. b.

***Figure 13.** Biopsy of retroperitoneal nodes from a posterior approach. a) A 22-gauge needle is placed coaxially through a thin-walled 19-gauge needle which has been direct toward the enlarged paraaortic lymph nodes via a transpsoas approach. b) Contiguous image at a slightly more caudal level confirms the tip of the biopsy needle within the nodal mass.*

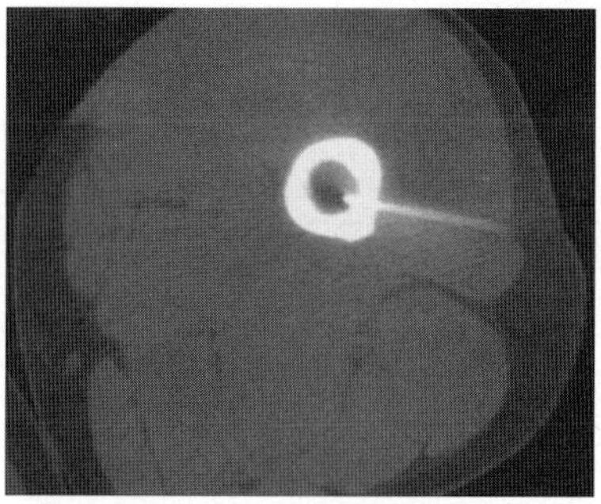

Figure 14. *A bone biopsy needle is advanced through intact femoral cortex into a medullary lytic lesion.*

Bone Biopsy

Percutaneous bone biopsy is a straightforward procedure when the abnormality destroys the cortical bone. Fine-needle aspiration biopsy directed by either fluoroscopy or CT can be performed in an identical fashion to soft-tissue biopsies elsewhere in the body.

When the overlying cortical bone is intact, the procedure is more difficult **(Fig. 14)**. The intact bone must be penetrated by a sturdy coring needle to provide access to the lesion, and often requires considerable force for successful placement. Several commercially available systems have been designed with serrated tips to penetrate the overlying bony cortex. Often, 22-gauge needles can be inserted through these trocars to establish the diagnosis. Core specimens of the cortical and underlying cancellous bone can also be obtained.

PERCUTANEOUS BIOPSY: RESULTS

1,000 biopsy procedures
- 92% sensitive
- 99% specific
- 8% false negative
- Positive predictive value 99.7%
- Negative predictive value 73.3%

(Welch TJ, et al 1989.)

RESULTS

A prospective analysis of 1,000 biopsy procedures was reported by Welch et al. In this analysis, imaging-directed biopsy was 92% sensitive and 99% specific. Approximately 7% of biopsies were falsely negative. The positive predictive value was 99.7% and negative predictive value was 73.3%.

COMPLICATIONS: NONTHORACIC BIOPSY

Overall rate 1.1%

- Hemorrhage (especially adrenal biopsy)
- Pancreatitis (especially normal pancreas)
- Pleural complications

(Welch TJ, et al 1989.)

COMPLICATIONS

The overall complication rate of abdominal biopsies in Welch's series was 1.1%, which is comparable to other series of abdominal and pelvic biopsies. The most common complication was hemorrhage related to biopsy of the adrenal gland.

Though the risk is low, pancreatic biopsies can cause pancreatitis which can be serious and lead to death. Pancreatitis appears to be more common when normal (rather than cancerous) pancreas is sampled.

There are two common and important complications of lung biopsy: hemoptysis and pneumothorax. Hemoptysis, to some degree, is noted in up to 20% of patients and is usually benign and self-limited. Severe hemoptysis requiring intervention is unusual. If significant hemoptysis is encountered, the patient should be placed in the decubitus position with the biopsy side down. This minimizes aspiration of blood into the normal lung. Emergency bronchoscopy may be useful to direct intubation into the nonhemorrhaging main stem bronchus in cases of massive bleeding.

HEMOPTYSIS: TREATMENT

- Oxygen
- Biopsy side placed down
- Selective main-stem bronchial intubation or Y-tube placement

Pneumothorax is observed in 5% to 30% of patients. Risk factors include chronic obstructive pulmonary disease, biopsies at the lung apicies, deep lesions requiring the needle to traverse significant amounts of aerated lung, and transgression of pleural surfaces located along fissures. Most biopsy-induced pneumothoraces do not require tube thorocostomy. Tube thorocostomy is indicated for pneumothoraces greater than 25%, pneumothoraces which are definitely enlarging, or pneumothoraces associated with significant symptoms. Some investigators have used aspiration of pneumothoraces as opposed to tube placement with considerable success.

Tube thorocostomy to evacuate a pneumothorax is usually performed from an anterior approach in the second or third intercostal interspace. Conventional drainage tubes ranging in size from 8-12-F as well as silastic tubes affixed to self-contained evacuation devices have been used for this purpose. Often, patients are admitted following tube thorocostomy, though in many centers, these patients are treated as outpatients with the tube attached to a Heimlich flutter valve.

PNEUMOTHORAX

- Incidence: 5%-30%
- Risk factors
  - COPD
  - Apical lesions
  - Depth
  - Number of pleural surfaces violated

PNEUMOTHORAX

Chest tube required if:

- Getting larger on serial chest films
- Patient symptomatic with pain or shortness of breath
- Larger than 25%

REFERENCES

Bernardino ME. Percutaneous biopsy. AJR 1984; 142.

Charboneau JW, Reading CC, Welch TJ. CT and sonographically guided needle biopsy: current techniques and new innovations. AJR 1990; 154:1.

Dahnert WF, Hoagland MH, Hamper UM, Erozan YS, Peirce JC. Fine-needle aspiration biopsy of abdominal lesions: diagnostic yield for different needle tip configurations. Radiology 1992; 185:263.

Gardner D, vanSonnenberg E, D'Agostino HB et al. CT-guided transthoracic needle biopsy. Cardiovasc Intervent Radiol 1991; 14:17.

Gazelle GS, Haaga JR. Biopsy needle characteristics. Cardiovasc Intervent Radiol 1991; 14:13.

Gazelle GS, Haaga JR, Rowland DY. Effect of needle gauge, level of anticoagulation, and target organ on bleeding associated with aspiration biopsy. Radiology 1992; 183:509.

Hathout GM, Lufkin RB, Jabout B, Andrews J, Castro D. MR-guided aspiration cytology in the head and neck at high field strength. J Mag Res Imag 1992; 2:93.

Hopper KD, Abendroth CS, Sturtz KW, Matthews YL, Stevens LA, Shirk SJ. Automated biopsy devices: a blinded evaluation. Radiology 1993; 187:653.

Matalon TAS, Silver B. US guidance of interventional procedures. Radiology 1990; 174:43.

Moore EH, Shepard JO, McLoud TC, Templeton PA, Kosiuk JP. Positional precautions in needle aspiration lung biopsy. Radiology 1990; 175:733.

Murphy TP, Dorfman GS, Becker J. Use of preprocedural tests by interventional radiologists. Radiology 1993; 186:213.

Smith EH. Complications of percutaneous abdominal fine-needle biopsy: review. Radiology 1991; 178:253.

Welch TJ, Sheedy PF, Johnson CD, Johnson CM, Stephens DH. CT-guided biopsy: prospective analysis of 1000 procedures. Radiology 1989; 171:493.

CHAPTER 24 PREVIEW ANSWERS

Case 1

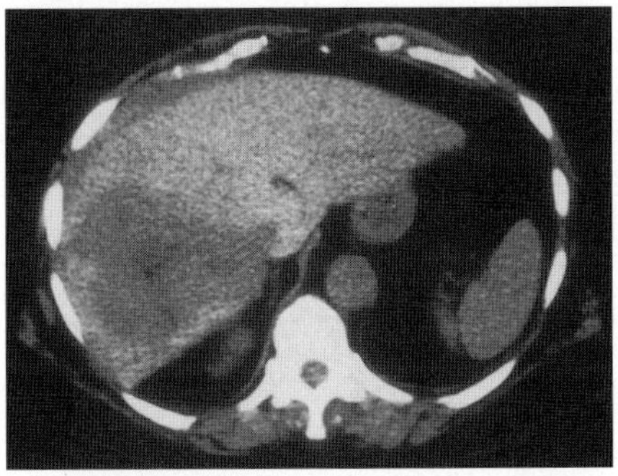

Findings: The central scarring raises possibility of cavernous hemangioma. A nuclear medicine blood pool study or MRI would be useful to confirm and biopsy could be avoided.
Diagnosis: Cavernous hemangioma.

Case 2

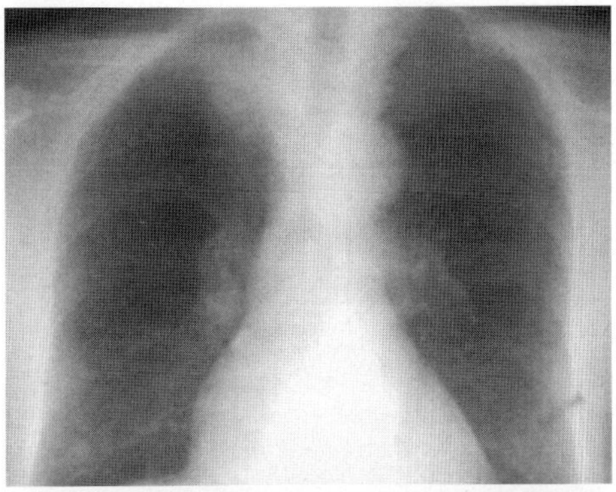

a.

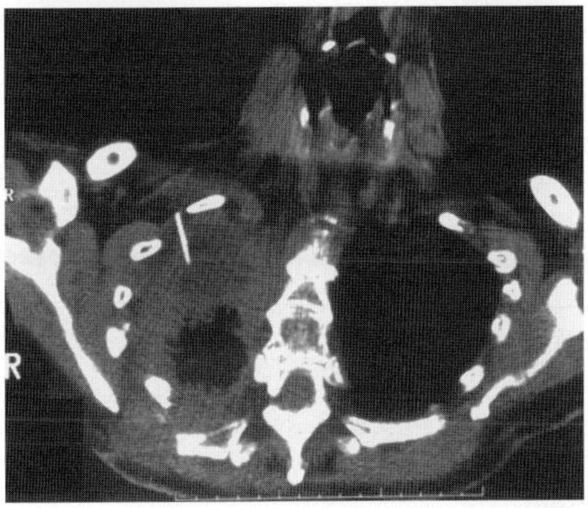

b.

Findings: a) A right apical mass is present on the radiograph. b) CT is used to guide the biopsy. The gantry is angled to avoid traversing aerated lung.
Diagnosis: Adenocarcinoma.

Case 3

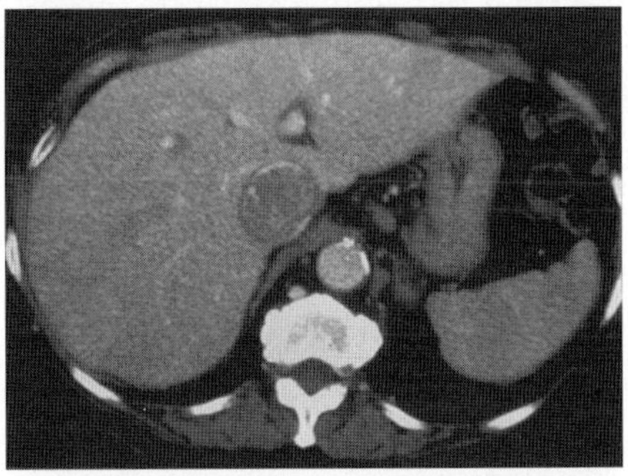

a.

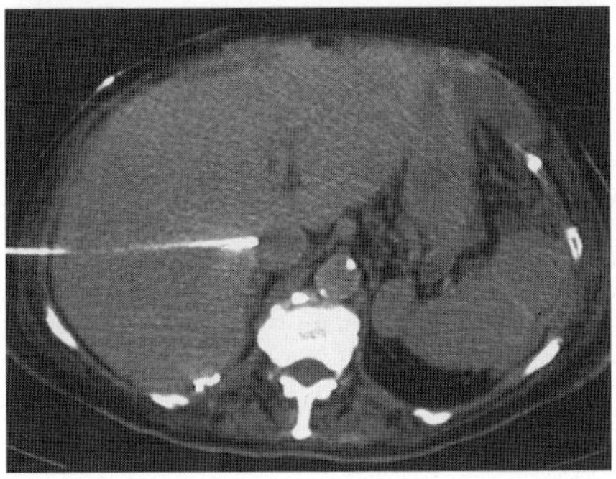

b.

Findings: a) A heterogeneous mass is present within the inferior vena cava. b) Though the mass could have been biopsied from a transvenous approach, it was elected to perform a CT-guided biopsy of the mass through the hepatic parenchyma.
Diagnosis: Adenocarcinoma extending from the primary renal tumor.

Case 4

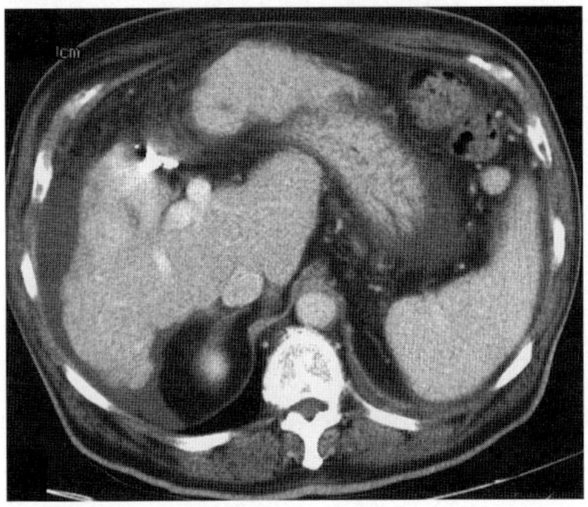

Findings: Moderate ascites surrounds a small, shrunken liver. Transjugular biopsy of the liver would be an excellent option in this patient.
Diagnosis: Cirrhosis secondary to alcohol abuse.

CHAPTER 25 PREVIEW

Case 1

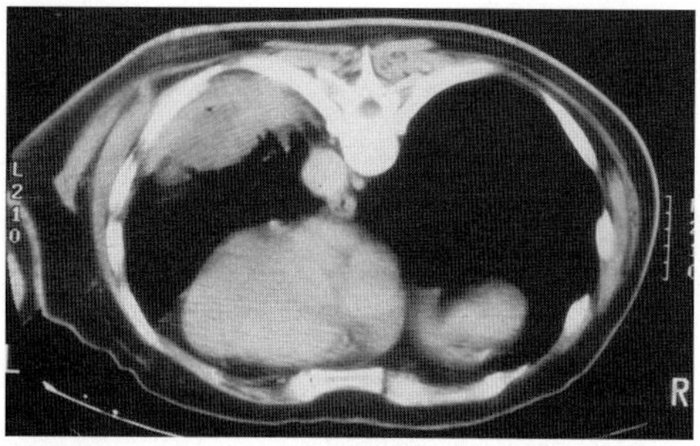

Figure 1. *Single axial image from a contrast-enhanced thoracic CT obtained with the patient in a prone position.*

History: An elderly man with pneumonia complained of persistent fever and dyspnea. Chest radiography suggested a loculated pleural effusion. A chest CT was performed **(Fig. 1)**.

Questions: What is your diagnosis? How could this patient be treated?

Case 2

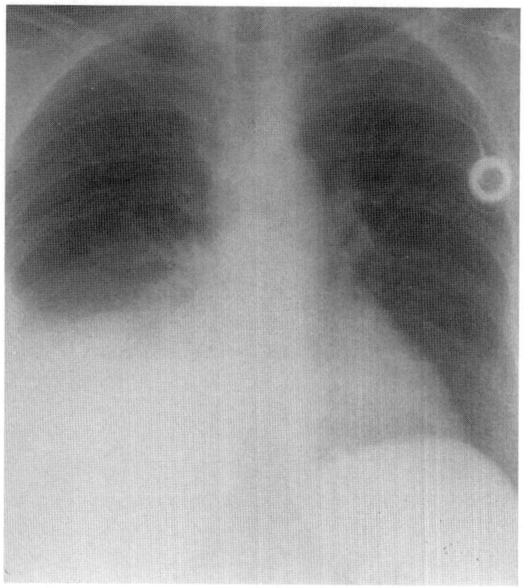

Figure 2. *Chest radiograph, frontal projection.*

History: A 50-year-old woman with metastatic ovarian cancer complained of shortness of breath on exertion. There was no history of fever or cough. A chest radiograph was obtained **(Fig. 2)**.

Questions: What is your diagnosis? How could this patient be treated?

Case 3

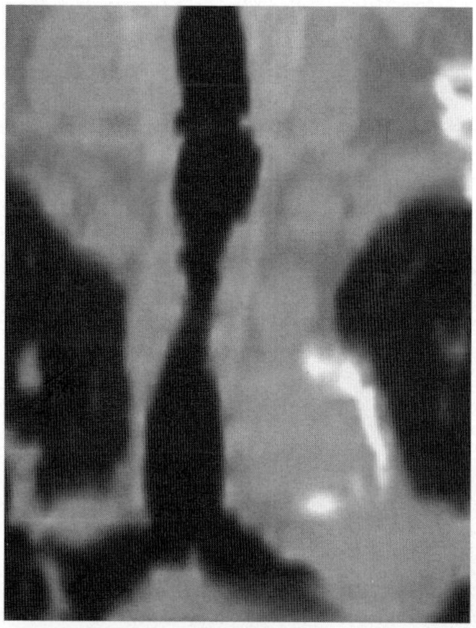

Figure 3. *Thoracic CT, coronal reconstruction.*

History: A 76-year-old woman required prolonged intubation for pneumonia and ARDS. Following extubation, she developed progressive shortness of breath. Thoracic CT was obtained and then coronal reconstruction was performed **(Fig. 3)**.

Questions: What is your diagnosis? How could this patient be treated?

Chapter 25
Thoracic Interventions

Jeanne M. LaBerge, MD,
and Roy L. Gordon, MD

CHAPTER OUTLINE

- Chest drainage
  - Air (pneumothorax)
  - Fluid (empyema)
- Pleurodesis
- Airway stents

INTRODUCTION

The involvement of interventional radiology (IR) in chest interventions has grown considerably in the past few years. In fact, procedures formerly viewed as the purview of thoracic surgeons or pulmonologists are now frequently performed by interventional radiologists. Active involvement by IR makes sense because image guidance often enhances the safety of these procedures.

In this chapter, we review three chest interventions: chest tube drainage, pleurodesis and airway stent placement. Other chest interventions including lung biopsy, bronchial artery embolization, and pulmonary angiography are discussed elsewhere in this book.

CHEST DRAINAGE

Chest tubes are inserted primarily for two reasons: to evacuate air from the pleural space (pneumothorax), or to drain fluid (pleural effusion). Since the drainage techniques differ in these two situations, each will be addressed separately.

Evaluation and Treatment of Pneumothorax

Pneumothorax presents in several settings. It may occur spontaneously, particularly in young men with no underlying lung disease **(Fig. 1)**. Spontaneous pneumothorax may also occur in patients with chronic underlying lung disease such as chronic obstructive pulmonary disease (COPD), asthma or pulmonary fibrosis. Iatrogenic pneumothorax can result from line placement, intubation or positive pressure ventilation. Pneumothorax is observed in 5%-30% of patients following lung biopsy.

PNEUMOTHORAX: CAUSES

- Spontaneous
- Secondary to underlying pulmonary disease
- Iatrogenic
  - central venous access
  - lung biopsy
  - intubation

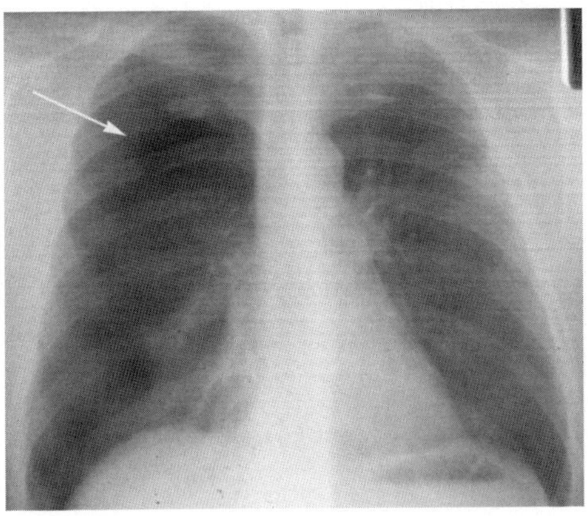

Figure 1. *PA chest radiograph in a young man with dyspnea and right shoulder pain demonstrates a spontaneous right pneumothorax (arrow). The patient was treated by insertion of a 10-F chest tube that was placed to suction for 48 hours and then removed.*

PNEUMOTHORAX: INITIAL TREATMENT

- Oxygen
- Analgesia
- Evacuation
  - aspiration
  - tube thorocostomy

PNEUMOTHORAX: INDICATIONS FOR CHEST TUBE

- Patient symptomatic with pain or shortness or breath
- Larger than 25%
- Getting larger on serial chest films

Immediate treatment of pneumothorax includes oxygen, analgesia, and evacuation of the air by aspiration or tube thorocostomy. The requirement for chest tube placement to treat pneumothorax is determined by the patient's condition. Treatment is indicated if the patient complains of significant dyspnea, if the pneumothorax is large (> 25%-30%), or if it increases in size on serial radiographs.

The appropriate anatomic locations for chest tube insertion are shown in **Figure 2**. For ambulatory patients, the chest tube is placed in the apex of the chest, usually through a second or third midline intercostal anterior space **(Fig. 3)**. For recumbent patients, the drain may be placed using a low anterior intercostal approach.

We place all chest tubes using Seldinger technique and a single-wall 19-gauge needle for entry into the pleural space. Once air has been aspirated, a 15-J floppy guidewire is coiled into the pleural space and the course of the wire is observed under fluoroscopy. The tract is dilated and then either an 8-F or a 10-F drain is inserted and connected to suction through a Pleurevac (Deknatel, Inc., Fall River, Ma.) chamber. The tube is placed to suction for 24 hours. If there is full reexpansion of the lung on chest radiograph and there is no air leak, the drain is placed to water seal for a trial period before removal.

High success rates have been reported for the treatment of iatrogenic pneuomothorax (> 95%) with an average duration of chest tube therapy for 2 days. Success is lower and a longer duration of therapy is required for patients with underlying lung disease.

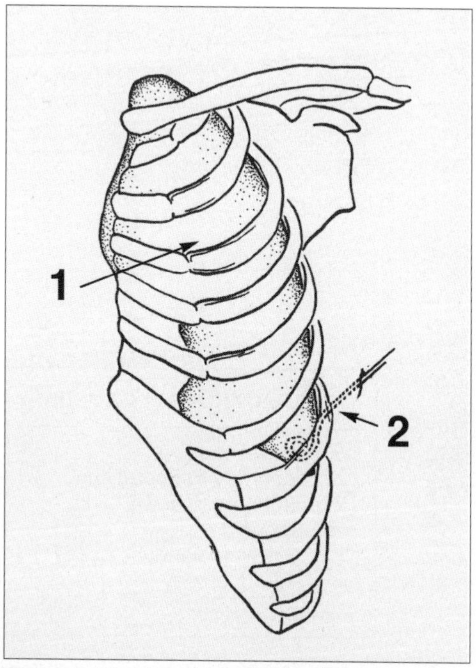

Figure 2. Diagram of the chest wall depicting the preferred location for chest tube placement. Thorocostomy for pneumothorax is placed through the second or third intercostal space anteriorly (1). For drainage of pleural effusion, the tube is placed via a lower posterior lateral intercostal approach (2).

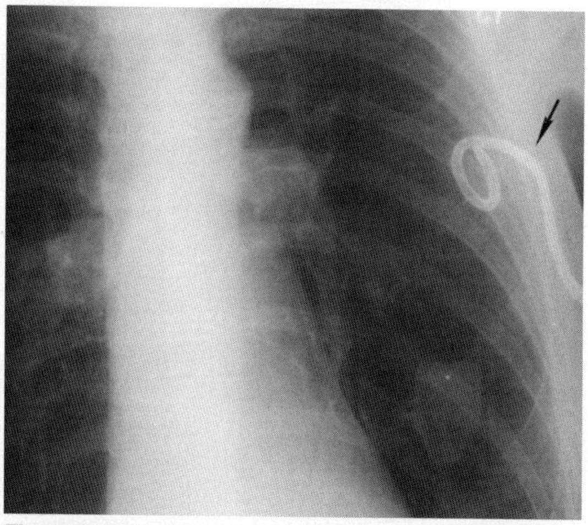

Figure 3. This patient developed a pneumothorax following lung biopsy. Symptoms of pain and dyspnea necessitated insertion of a 10-F pigtail catheter (arrow) in the upper left chest.

Drainage of Pleural Fluid

The two most common indications for drainage of intrathoracic fluid are empyema and sterile parapneumonic fluid causing dyspnea.

As with drainage for pneumothorax, we use Seldinger technique. The procedure is performed with the patient sitting in an upright position. Ultrasound is used for localization **(Fig. 4)** and the drain is inserted through a lower posterior intercostal space **(Fig. 2)**. We typically use a 12-F or 14-F tube to drain empyema. CT guidance is employed if the collection is loculated or difficult to access **(Fig. 5)**.

Empyemas are frequently loculated and may require catheter manipulation for complete evacuation. Several techniques may be used. Simply changing or upsizing the drain may empty the collection. Alternatively, the drain can be manipulated under fluoroscopic guidance and redirected into an undrained area of the empyema. Or, in some cases, instillation of urokinase (100,000 U) or streptokinase (250,000 U) may facilitate resolution of locules and drainage of the empyema.

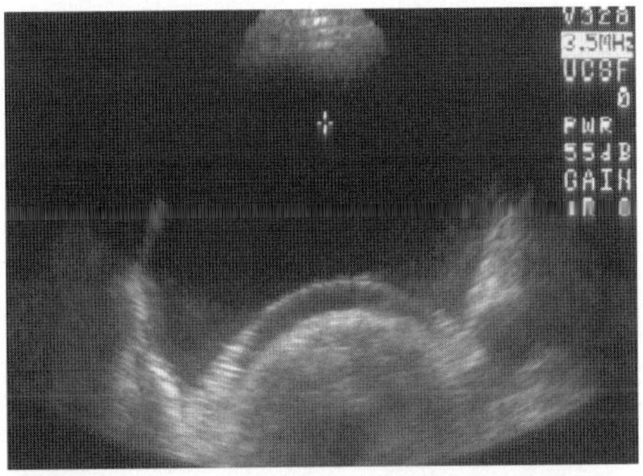

Figure 4. *Sonogram of the right lower posterior chest demonstrates a large pleural effusion. This study was used to select an appropriate puncture site for thorocostomy.*

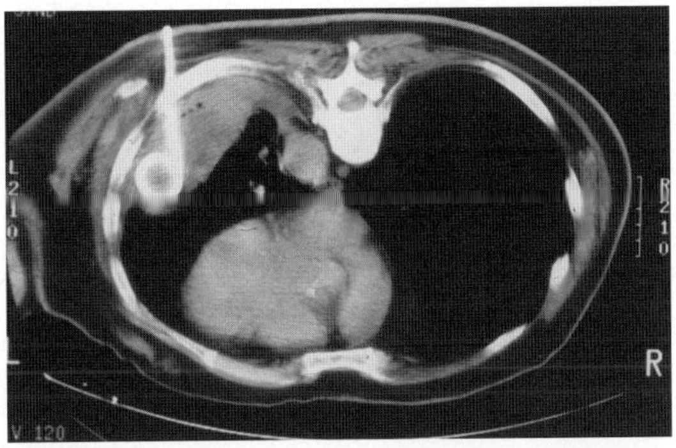

Figure 5. *An elderly man with pneumonia and persistent fevers underwent chest CT that disclosed a loculated pleural effusion containing air. Aspiration of purulent fluid confirmed the diagnosis of empyema and a 12-F chest tube was inserted using Seldinger technique and CT guidance.*

PLEURODESIS

Malignant pleural effusion develops in almost 50% of patients with disseminated cancers and commonly leads to severe shortness of breath. The mean survival time in patients with malignant pleural effusion is only 3 to 6 months. The ideal method for the palliation of patients with malignant pleural effusion has not been established. Though thoracocentesis may provide acute, symptomatic relief, rapid recurrence is common. The 30-day recurrence rate following thoracocentesis alone is almost 100%.

PLEURODESIS AGENTS

- Antibiotics
  - Tetracycline
  - Minocycline
  - Doxycycline
- Antineoplastic agents
  - Bleomycin
  - Mitoxantrone
- Biologic response modifiers
  - *Corynebacterium parvum*
  - Interleukin 2
- Miscellaneous agents
  - Talc
  - Mepacrine

In order to prevent recurrence of malignant pleural effusion, pleurodesis is often attempted. The goal of pleurodesis is to generate adhesions between the visceral and parietal pleura, obliterating the potential space and preventing fluid accumulation by the injection of sclerosing agents. Initially, a chest tube is inserted to drain the effusion. When the output decreases to less than 100 ml/day, pleurodesis is performed **(Fig. 6)**.

A number of sclerosing agents have been used for pleurodesis including antibiotics (doxycycline, minocycline, and tetracycline), antineoplastic agents (bleomycin, cisplatin, doxorubicin, and mitomycin-c), biologic response modifiers (*Corynebacterium parvum* and interleukin-2) and other miscellaneous agents such as talc and mepacrine. As evidenced by the number of agents that have been used for pleurodesis, no agent is ideal.

Numerous investigators have reported results using sterile talc as the sclerosing agent, and this is currently our agent of choice for pleurodesis. In the early reports, a talc poudrage was applied thoracoscopically. More recently, instillation of a talc slurry through large bore chest tubes has been performed with similar results. However, performing thoracoscopy or inserting large-bore chest tubes may be a source of considerable morbidity and discomfort in these dyspneic, frail and often terminally ill patients. In an effort to minimize these problems, we have attempted to perform talc pleurodesis through small-bore, percutaneously placed thoracostomy tubes.

Complications related to pleurodesis include pain, fever, infection, hemoptysis, cardiac arrhythmia, transient hypotension and subcutaneous emphysema. In some series, pain associated with the procedure occurred in 100% of patients and required narcotic analgesia. The mean duration of chest tube placement following pleurodesis has varied form 2.7-5.3 days (range 1-11 days). Tumor seeding along the chest tube tract is a recognized late complication. Uncommon but potentially life-threatening complications of the procedure include reexpansion pulmonary edema, chemical pleuritis, bronchopleural fistula, respiratory failure and death.

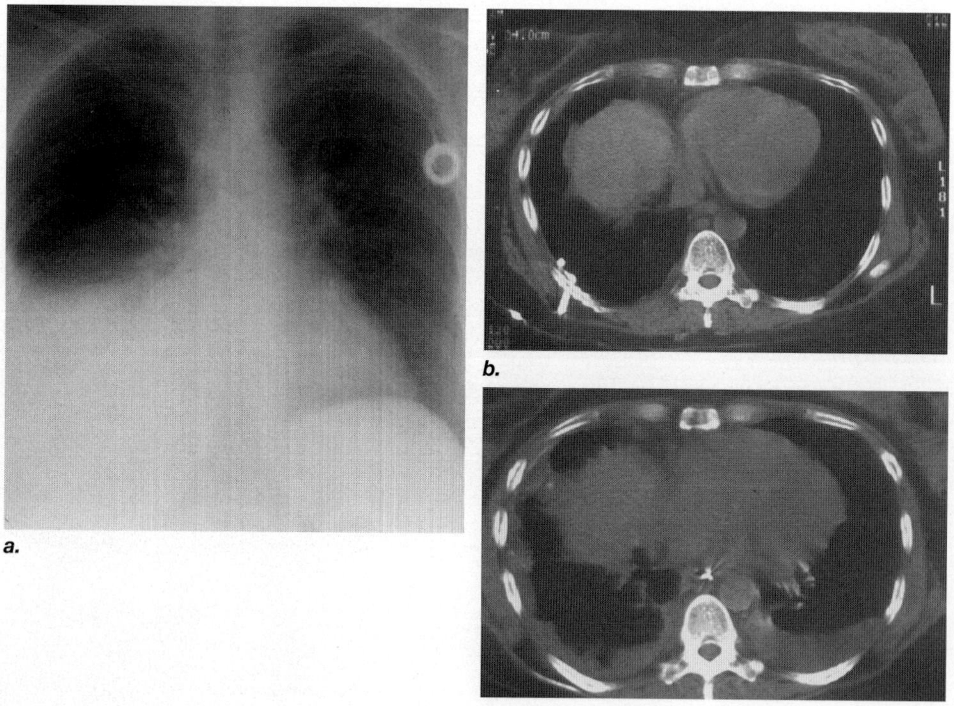

a. b. c.

__Figure 6.__ A 50-year-old woman with ovarian cancer presented with shortness of breath. a) Chest radiograph reveals a large right pleural effusion. Cytologic evaluation of a sample of aspirated fluid revealed malignant cells. A 10-F chest tube was inserted using ultrasound guidance. b) Two days later, the effusion had resolved and the tube output was 75 ml/day. CT reveals resolution of pleural fluid. Talc pleurodesis was performed. Five grams of sterile talc dissoved in 100 ml of saline was injected through the tube. Four hours later, the tube was again placed to suction and the next day the tube was removed. c) Chest CT 1 week following talc pleurodesis demonstates pleural thickening without significant effusion. The patient was asymptomatic at 6 months but later died of her underlying disease.

AIRWAY STENTS

Narrowing of the tracheobronchial tree may result in dyspnea, stridor or postobstructive pneumonia. Patients with severe or recurrent symptoms can be treated in a variety of ways including surgical resection, laser ablation or placement of a silicone stent. Expandable metallic stenting is a recent treatment alternative that offers some potential benefits over existing therapies. However, experience with metallic stents in the bronchial tree is limited and only short-term results are available.

Airway Anatomy

The normal anatomy of the central airways is shown in **Figure 7**. Structures that are amenable to stenting include the trachea and mainstem bronchi. More peripheral airways are generally too small and have too many branches to accommodate stenting.

The diameter of the trachea varies considerably but generally ranges from 14 mm-20 mm and the diameter of the mainstem bronchi varies from 8 mm-12 mm. When possible, stents are positioned so that the wire mesh of the stent does not cross any branch bronchi. While aeration of a stent-covered bronchus is still possible, the metal wires may inhibit normal clearing of airway secretions.

Diseases Affecting the Airways

Narrowing of the airway can be caused by a process intrinsic to the airway wall or by extrinsic compression of the airway from a mediastinal or cervical process **(Table 1)**. It is also useful to categorize the underlying pathology as benign or malignant since this distinction will help determine the mode of treatment **(Table 2)**.

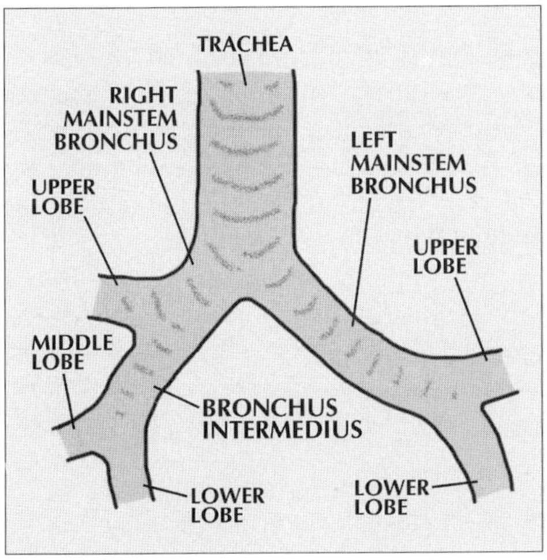

Figure 7. *The normal anatomy of the central airway. Stents can be placed in the trachea or mainstem bronchi.*

TABLE 1: AIRWAY NARROWING— EXTRINSIC VERSUS INTRINSIC

Extrinsic
- Nodes, vascular, thyroid, esophageal

Intrinsic
- Infectious: TB, fungal, AIDS
- Granulomatous: Wegener's, sarcoid
- Neoplasm: lung, mets, lymphoma
- Trauma: intubation, transplant
- Immune: amyloid, relapsing polychondritis
- Cryptogenic: tracheopathica osteopathica

TABLE 2: AIRWAY NARROWING— MALIGNANT VERSUS BENIGN

Etiology

- Malignant
  - Lung, thyroid, lymphoma
- Benign
  - Following lung transplantation
  - Traumatic
  - Inflammatory

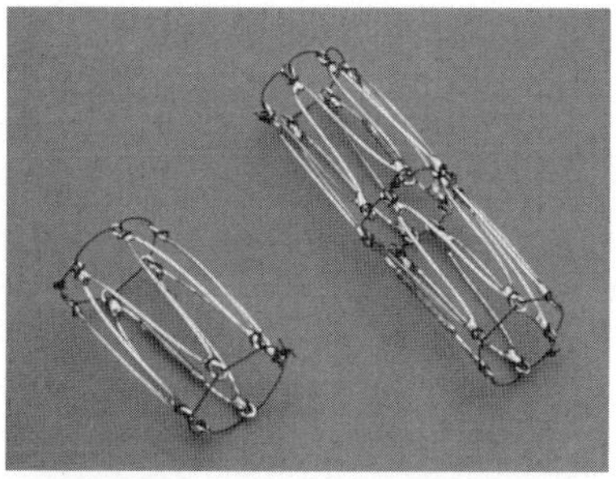

Figure 8. *Gianturco stent. This is a self-expanding stent with a ziz-zag stainless steel monofilament design.*

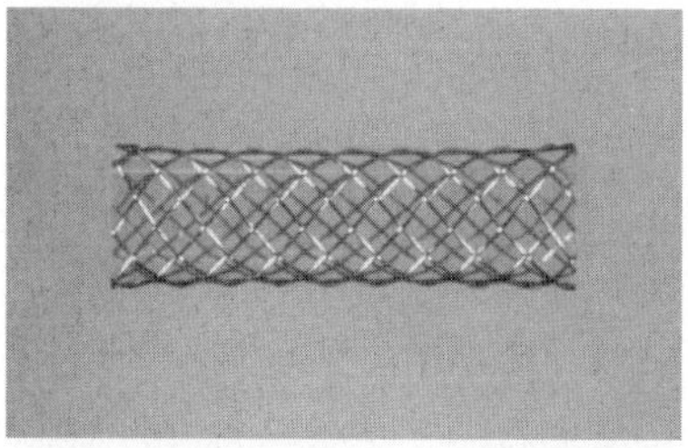

Figure 9. *Wallstent. This is a flexible self-expanding stent which shortens as it is deployed.*

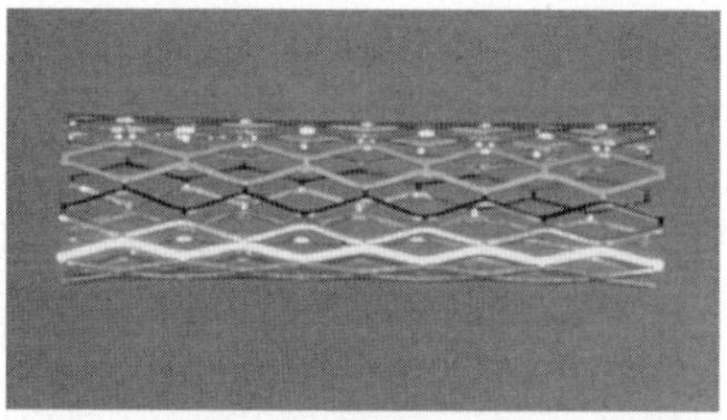

Figure 10. *Palmaz stent. This is a non-elastic balloon-expandable stent.*

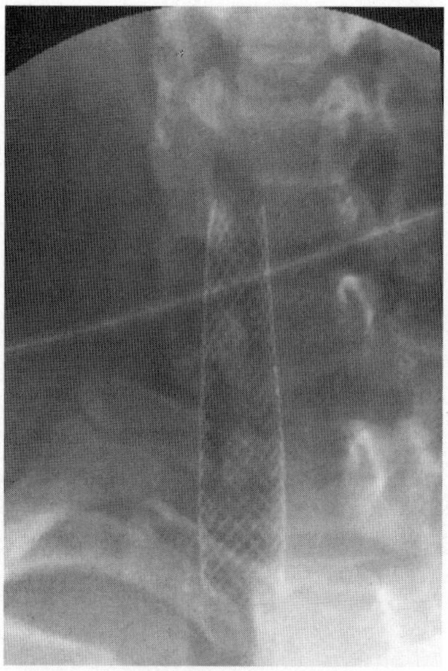

Figure 11. *Spot film obtained following tracheal stent placement. This young woman developed subglottic tracheal stenosis following extended tracheostomy. She experienced severe stridor following removal of the tracheostomy tube. An 18-mm by 40-mm Wallstent was placed across the stenosis.*

Airway Stents

A variety of expandable stents have been used to treat tracheobronchial lesions. The three most commonly used stents are described below.

Gianturco Stent **(Fig. 8)**: One of the first stents to be used in the bronchial tree was the Gianturco-Z stent. It is made of .018-inch stainless steel monofilament, bent into a zigzag configuration and formed into a cylinder. Several cylinders can be linked together in series, yielding devices of increasing length. The stents are typically available with diameters of 6 mm to 35 mm. The stent is deployed by pushing it through a delivery sheath of 8.5-F to 16.5-F. When the stent is in a suitable position, the sheath is retracted and the self-expanding stent springs open within the airway. The stent is self-expanding, exerts good radial force, is well seen on fluoroscopy, is nonshortening, is easy to deploy and has large spaces between the stent wires.

Wallstent **(Fig. 9)**: The second commonly used self-expanding stent is the Wallstent, which is woven in crisscross fashion from thin alloy wires to form a flexible cylinder. The stent is concentrically constrained onto the shaft of a flexible delivery catheter by a membrane. When this membrane is retracted, the stent opens up. Wallstents are available in a range of diameters from 6 mm to 30 mm and in lengths from 20 mm to 100 mm. The woven structure of the stent mandates that the stents of diameters of 12 mm and larger are at least 40 mm in length. Delivery systems are 7-F or 9-F in diameter.

Wallstents are self-expanding, exert moderate radial force, have good flexibility, are well seen on fluoroscopy, and show good tissue epithelialization, but have significant shortening upon deployment. Wallstents are also available with a silicone covering designed to prevent ingrowth of tissue through the interstices of the stent.

Palmaz Stent **(Fig. 10)**: The Palmaz stent differs from the previously described stents, because it is a nonelastic type that is deployed on a balloon delivery catheter. It is expanded from its collapsed state into position by inflating the carrier balloon. Delivery is via a 7-F introducer sheath. The stent is available in lengths ranging from 10 mm to 40 mm and its diameter depends on the diameter of the deployment balloon, typically in the range of 8 mm to 12 mm.

The Palmaz stent is a balloon-expandable stent that exerts strong radial force. But it is nonelastic and may collapse under pressure. It is well seen on fluoroscopy, exhibits little shortening on deployment, and has little flexibility. It is available in short lengths from 10 mm to 40 mm, and in small diameters, and is therefore good for exact placements in small airways.

Choice of stent

None of the three stents described above is ideal. In the trachea, we favor the Wallstent because it is flexible and available in sufficiently large diameters to match the size of the adult trachea **(Fig. 11)**. We prefer it to the Gianturco stent because of our concern that the Gianturco stent may exert radial force strong enough to cause it to erode into surrounding structures. In addition there are reports of Gianturco failure due to strut breakage and breakage of the restraining nylon sutures. Stent migration has occurred as a result of these mechanical failures. Palmaz stents are usually not of large enough diameter for use in the adult trachea.

In the bronchi, Palmaz stents were previously our first choice. They are available in variable lengths from 10 mm to 40 mm, to facilitate accurate stenting of a short-segment stenosis in the branching airways without covering the origins of non-involved bronchi. The stent diameter is determined by the degree of balloon expansion and so it can be accurately matched to the caliber of the airway by using increasingly large balloons. Unfortunately because the stents are not self-expanding, they may become loose if the airway caliber expands after placement. We have also experienced cases of collapse and migration of the Palmaz stents in three patients **(Fig. 12)** and we now use these stents sparingly.

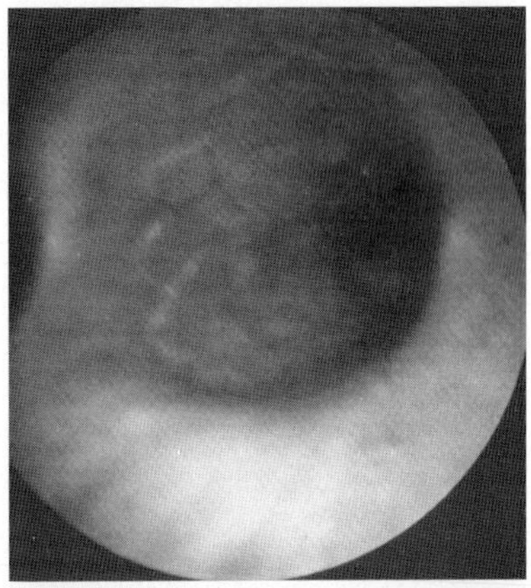

a.

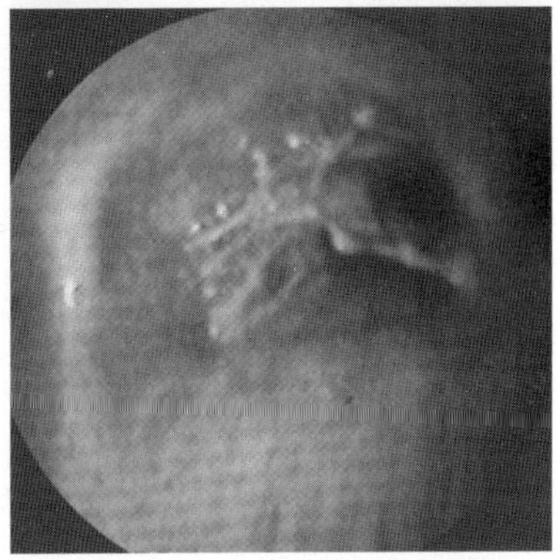

b.

Figure 12. *Endoscopic images of a Palmaz stent in the right mainstem bronchus. a) This image was obtained immediately following placement. b) This image was obtained three months later after the patient had developed recurrent right lower lobe pneumonia. The stent has collapsed and is obstructing the airway.*

Indications for Stent Placement

Metallic stents have been used to treat benign airway obstruction caused by post-lung transplantation anastomotic narrowing, infection, congenital lesions, tracheobronchial malacia, inflammatory conditions including relapsing polychondritis, Wegener's granulomatosis, AIDS, external compression from benign mediastinal masses or fibrosis, and airway collapse related to chronic obstructive pulmonary disease. Stents may also be of short-term benefit in the palliation of patients with malignant tracheobronchial obstruction. The indications for stenting are evolving as more experience is gained and follow-up data is accrued. Most reports emphasize the importance of avoiding stent placement in the presence of active inflammation.

Preprocedure Imaging

CT is essential in imaging patients that are being considered for stent placement **(Fig. 13)**. Cross-sectional imaging prior to stent placement can be useful in a number of ways. It provides accurate representation of airway anatomy in three dimensions, measurement of airway diameter, evaluation of airway anatomy distal to a narrowed segment and invisible to bronchoscopy, demonstration of dynamic changes in airway morphology during forced exhalation in patients with airway malacia, and demonstration of focal or diffuse air trapping in lung peripheral to the abnormal airway. Furthermore, in patients who have already undergone stent placement, CT is a valuable means of assessing airway morphology and dynamics, and can be useful in assessing stent dysfunction.

Figure 13. *A 60-year-old man with primary right upper lobe bronchogenic carcinoma developed severe stridor. Preliminary bronchoscopy revealed stenosis of the right mainstem bronchus. a) Spiral CT with coronal reconstruction demonstrates compression of the right mainstem bronchus (arrow). b) Spot film obtained following stent placement demonstrates a Wallstent in the trachea and a Wallstent extending into the right mainstem bronchus. Subsequently, a hole was made through the wire stent mesh of the Wallstent to allow placement of a Palmaz stent into the left bronchus.* **a.**

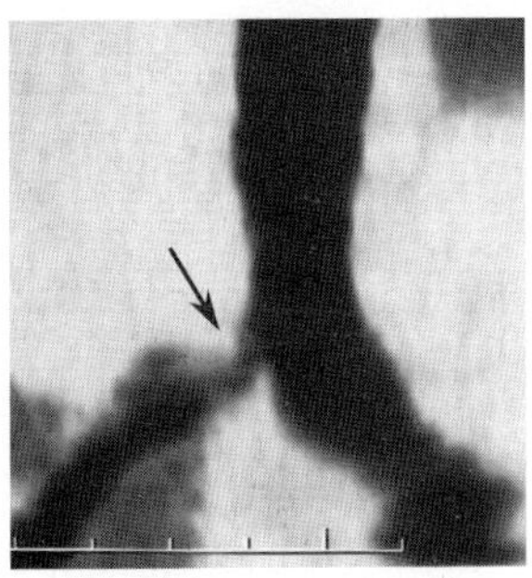

b.

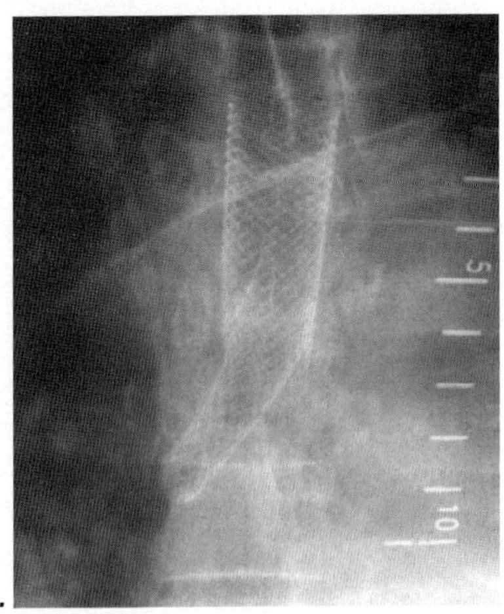

Spiral CT is of particular value in the evaluation of stent patients. It allows the acquisition of volumetric data and enables two-dimensional or three-dimensional reconstructions that can be instrumental in understanding airway anatomy at a site of airway stenosis. Further, dynamic CT can also be used to obtain expiratory images at the site of airway narrowing visible on inspiratory images **(Fig. 14)**, or to demonstrate airway collapse and air trapping in some patients with relatively normal inspiratory images.

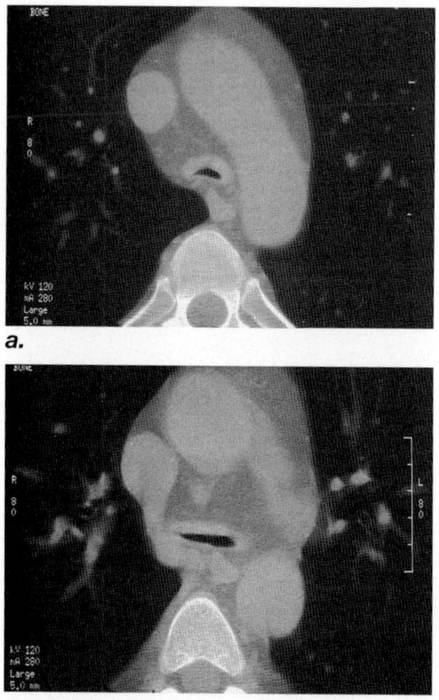

Figure 14. A patient with relapsing polychondritis complained of dyspnea. Expiratory chest CT was obtained prior to stent placement. a) Axial image of the trachea at the level of the thoracic arch reveals a thick-walled trachea that collapses on forced expiration. b) Tracheal collapse is also seen at the level of the carina (See Figs. 15, 16).

Stent Placement Technique

At UCSF, tracheobronchial stenting is undertaken by a multidisciplinary team of physicians including an anesthesiologist, pulmonologist and interventional radiologist. Stent placement is performed under general anesthesia. A specially designed endotracheal tube is used to allow ventilation during bronchoscopy and stent deployment. Initial bronchosopy is performed to confirm the presence and location of the tracheobronchial stenosis. The location of the stenosis is then identified with fluoroscopy. The stent is delivered and deployed under fluoroscopic guidance **(Fig. 15)**. Follow-up pulmonary function tests and CT are usually performed the next day **(Fig. 16)** and then at 3 or 6 months.

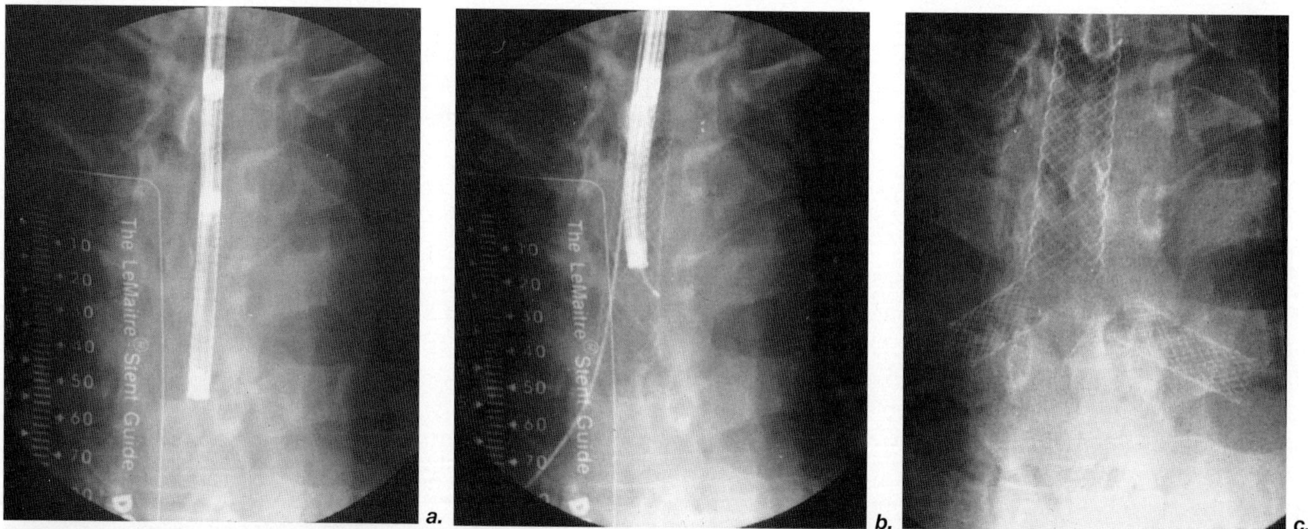

Figure 15. Stent placement for relapsing polychondritis. a) The procedure was performed using fluoroscopic and endoscopic guidance. A radiopaque marker is used to estimate airway diameter and location. b) Following initial deployment of a 20-mm Wallstent in the trachea, the bronchoscope was reinserted and the mainstem bronchi were inspected. c) Spot film was obtained following insertion of stents in the right and left mainstem bronchi.

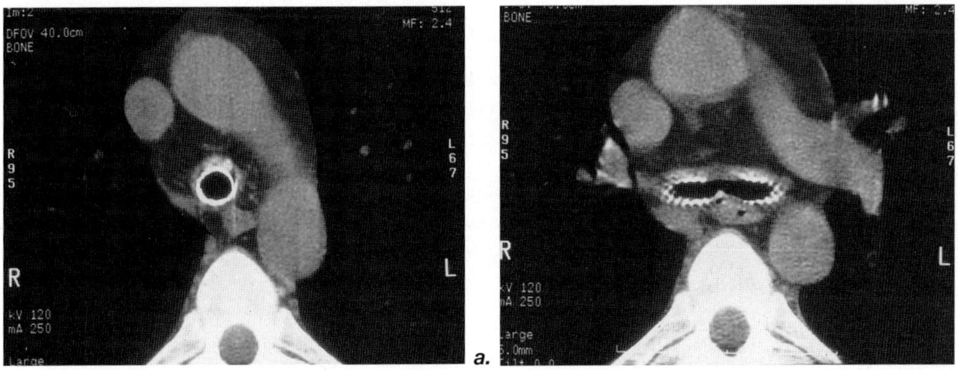

Figure 16. *Follow-up CT was obtained 1 month after stent placement. a) Axial image of the trachea at the level of the arch shows marked improvement in airway diameter. b) Axial image at the carina shows the appearance of the stents in the mainstem bronchi.*

Currently, our stent of choice for use in the bronchi and trachea is the Wallstent. Assessment of the diameter and length of the bronchial segment to be stented determines the selection of an appropriately sized Wallstent. This assessment is made from the preliminary CT scan and from combined bronchoscopic and fluoroscopic estimates. If the diameter of the chosen Wallstent is too large for the airway being treated, it will not expand completely and consequently will be of longer length than needed. For this reason we tend to choose shorter length stents with the option of placing a second coaxial stent of the same diameter that can be overlapped to achieve accurate coverage. We make every attempt not to cover the origin of noninvolved bronchi in order to avoid interference with ventilation or the clearance of secretions. Our experience to date suggests that noncovered Wallstents epithelialize well, whereas silicone covered Wallstents tend to become blocked by secretions. We would only choose the covered version in benign cases with a possible airway fistula. We have not used Gianturco stents in the bronchi because of the mechanical problems reported by other workers detailed above.

Results

Tracheobronchial stents have been assessed in small groups of patients treated primarily for benign tracheobronchial diseases. Although the patient populations and techniques differ, stent placement has been associated with symptomatic and functional improvement in the majority of patients. Significant complications can occur, including airway inflammation, stent migration, airway erosion, and stent fracture and collapse. More serious complications are uncommon.

REFERENCES

Chest Tubes:

Klein JS, Schultz S. Percutaneous drainage of intrathoracic fluid collections. In: Sandhu J, Meglin AJ, Trerotola SO eds. SCVIR Syllabus: Thoracic and Visceral Nonvascular Interventions. San Francisco: Society of Cardiovascular & Interventional Radiology, 1997; 181-196.

Parker LA, Levin DL. Evaluation and treatment of pneumothorax. In: Sandhu J, Meglin AJ, Trerotola SO eds. SCVIR Syllabus: Thoracic and Visceral Nonvascular Interventions. San Francisco: Society of Cardiovascular & Interventional Radiology, 1997; 171-174.

Pleurodesis:

Bloom AI, Wilson MW, Kerlan RK Jr, Gordon RL, LaBerge JM. Talc pleurodesis through small-bore percutaneous tubes. Cardiovasc Intervent Radiol 1999; 22:433-435.

Grossi F, Pennucci MC, Tixi L, et al. Management of malignant pleural effusions. Drugs 1998; 55:47-58.

Hartman DL, Gaither JM, Kesler KA, Mylet DM, Brown JW, Mathur PN. Comparison of insufflated talc under thoracoscopic guidance with standard tetracycline and bleomycin pleurodesis for control of malignant pleural effusions. J Thor Cardiovasc Surg 1993; 105(4):743-747.

Walker-Renard PB, Vaughan LM, Sahn S. Chemical pleurodesis for malignant pleural effusions. Annals of Internal Medicine 1994; 120(1):56-64.

Airway stents: Gianturco Stents

Higgins R, McNeil K, Dennis C. Airway stenoses after lung transplantation: Management with expanding metal stents. J Ht Lung Transpl 1994; 13:774-778.

Nasef SA, Dromer C, Velley F. Expanding wire stents in benign tracheobronchial disease. Ann Throacic Surg 1992; 54:937-940.

Petersen BD, Uchida BT, Barton RE, Keller FS, Rösch J. Gianturco-Rösch Z stents in tracheobronchial stenoses. J Vasc Interv Radiol 1995; 6(6):925-931.

Rousseau H, et al. Self-expandable prosthesis in the tracheobronchial tree. Radiology 1993; 188:199-203.

Wilson GE, Walshaw MJ, Hind CR. Treatment of large airway obstruction in lung cancer using expandable metal stents inserted under direct vision via the fibreoptic bronchoscope. Thorax 1996; 51:248-252.

Airway stents: Palmaz Stents

Filler RM, Forte V, Chait P. Tracheobronchial stenting for the treatment of airway obstruction. J Ped Surg 1998; 33(2):304-311.

Hautmann H, Huber RM. Stent-flexibility and essential feature in the treatment of dynamic airway collapse. Eur Respir J 1996; 9:609-611.

Slonim SM, Razavi M, Kee S, Semba CP, Dake MD. Transbronchial Palmaz stent placement for tracheo-bronchial stenosis. J Vasc Interv Radiol 1998; 9(1 Pt 1):153-160.

Airway stents: Wallstents

Carré P, Rousseau H, Lombart L, et al. Balloon dilatation and self-expanding metal wallstent insertion for management of bronchostenosis following lung transplantation. Chest 1994; 105:343-48.

Lehman JD, Gordon RL, Kerlan RK Jr, Laberge JM, Wilson MW, Golden JA, Webb WR. Expandable metallic stents in benign tracheobronchial obstruction. J Thor Imag 1998; 13(2):105-15.

Rousseau H, Dahan M, Lauque D, et al. Self-expandable prosthesis in the tracheobronchial tree. Radiology 1993; 188:199-203.

CHAPTER 25 PREVIEW ANSWERS

Case 1

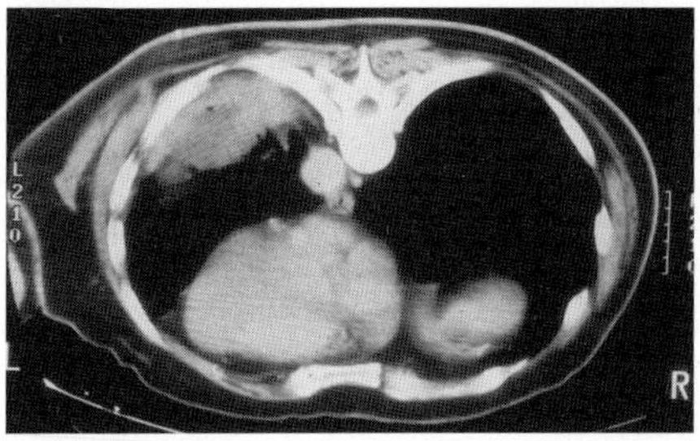

Figure 1a. *Contrast-enhanced chest CT reveals a loculated posterior pleural effusion with rim enhancement and a small amount of air within the fluid collection.*

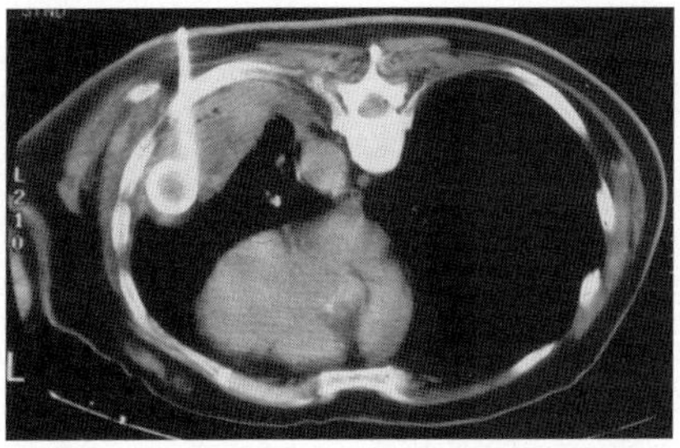

Figure 1b. *A 12-F pigtail chest tube was inserted using CT guidance. The contents were aspirated and the drain was connected to suction through a Pleuravac.*

Findings: Chest CT discloses a loculated pleural effusion **(Fig. 1a)**. The effusion is most likely infected given the history of persistent fever and the presence of air within the collection. The patient was scanned in the prone position in anticipation of a percutaneous aspiration and drainage procedure. Needle aspiration with a 19-gauge single-wall puncture needle yielded purulent fluid. A 12-F pigtail chest tube was inserted using CT guidance **(Fig. 1b)**. The patient was treated with appropriate antibiotics and the chest tube was removed 5 days later.
Diagnosis: Loculated empyema.
Treatment: Percutaneous CT-guided drainage and appropriate antibiotics.

Case 2

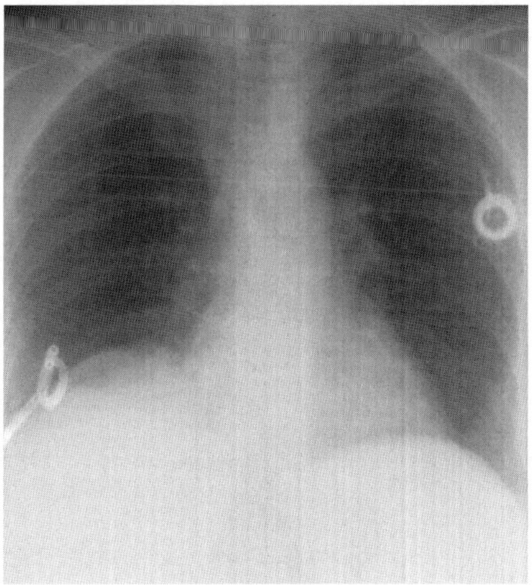

Findings: Frontal chest radiograph reveals a large right pleural effusion. Given the patient's history of metastatic ovarian cancer, this effusion is most likely malignant. She underwent diagnostic thoracentesis and cytologic evaluation revealed malignant cells. A 10-F thoracostomy tube was inserted percutaneously using ultrasound guidance **(Fig. 2)**. Three days later, the chest tube output was less than 100 ml/day. She underwent talc pleurodesis and then the chest tube was removed and the patient was discharged.
Diagnosis: Malignant pleural effusion.
Treatment: Percutaneous thorocostomy and talc pleurodesis.

Figure 2. *Frontal chest radiograph following placement of a 10-F chest tube (arrow). The tube was inserted through a lower lateral posterior intercostal approach using ultrasound guidance.*

Case 3

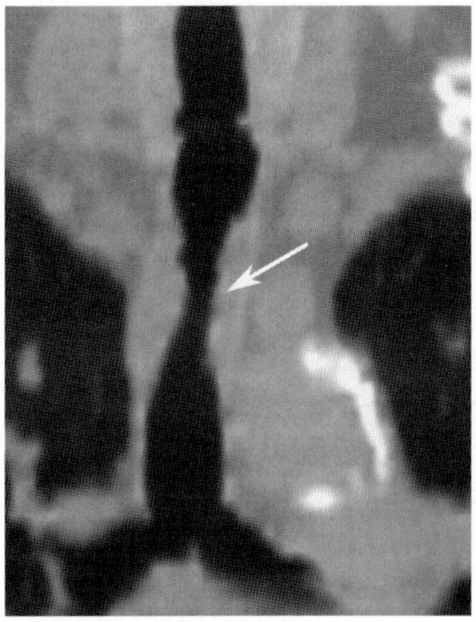

Figure 3a. *Coronal CT reconstruction.*

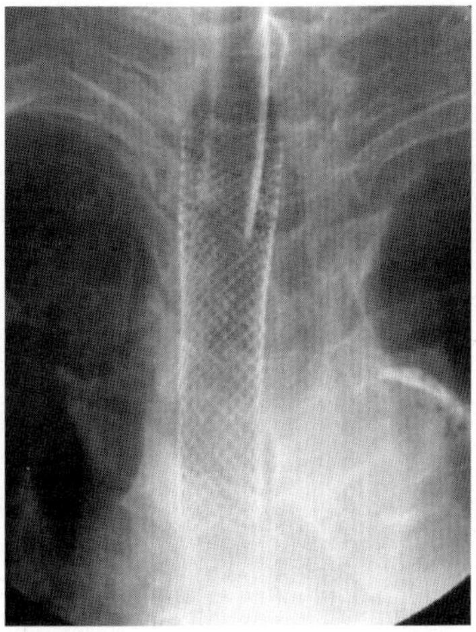

Figure 3b. *Fluoroscopic spot film obtained following stent deployment and balloon dilation.*

Findings: Chest CT with coronal reconstruction shows a focal narrowing of the midthoracic trachea **(Fig. 3a)** (arrow). This is a fibrotic reaction to prolonged intubation. The patient was initially treated by laser ablation using rigid bronchoscopy. However, her symptoms persisted and she was referred for tracheal stent placement. An 18-mm by 50-mm Wallstent was deployed across the stenosis and the stent was dilated with an 18-mm balloon **(Fig. 3b)**. The patient's symptoms resolved following stent placement and she was discharged.
Diagnosis: Postintubation tracheal stenosis.
Treatment: Stent placement (18-mm x 50-mm Wallstent).

CHAPTER 26 PREVIEW

Case 1

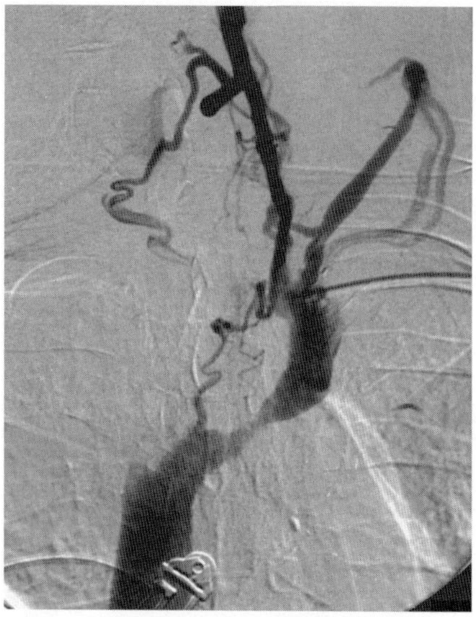

History: Dialysis fistulagram outflow in a hemodialysis patient with elevated venous pressures at dialysis.
Questions: What is the diagnosis? What is the suggested treatment?

Case 2

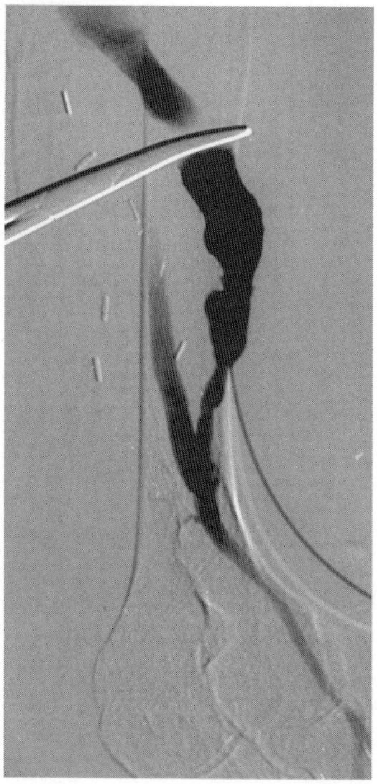

History: Dialysis fistulogram in a patient with abnormal urea recirculation at dialysis. The graft is externally compressed to allow reflux of contrast across the arterial anastomosis.
Questions: What is the diagnosis? What is the suggested treatment?

Case 3

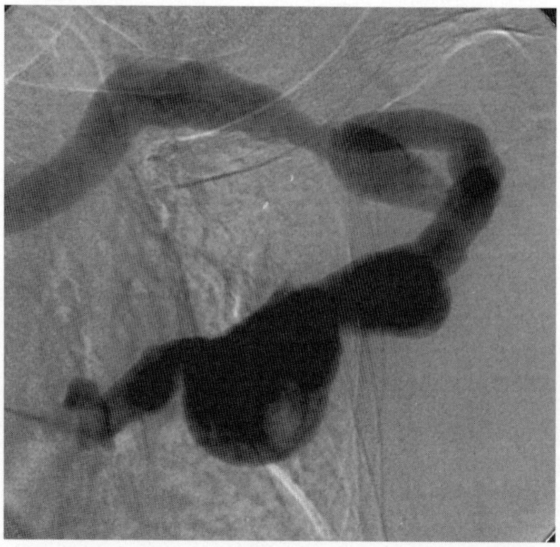

History: Dialysis fistulogram in a hemodialysis patient with elevated venous pressures at dialysis and difficulty accessing the graft.
Questions: What type of dialysis A-V graft is this? What are the causes of elevated pressures and difficulty with grafting access? What is the suggested treatment?

Chapter 26
Dialysis Access Interventions

Mark W. Wilson, MD

CHAPTER OUTLINE

Types of hemodialysis access
- Direct arteriovenous fistulas
- Synthetic interposition grafts
- Nonsynthetic interposition grafts

Locations of hemodialysis access fistulas
- Forearm
- Upper arm
- Groin
- Chest (axillary artery to axillary vein)

Complications of hemodialysis access fistulas
- Stenosis
- Thrombosis
- Graft pseudoaneurysms
- Venous aneurysms and varicosities

Treatment of hemodialysis fistula complications
- Balloon dilatation
- Vascular stenting
- Transcatheter thrombolysis
- Percutaneous thrombectomy
- Percutaneous versus surgical recanalization

INTRODUCTION

More than 110,000 patients currently undergo chronic hemodialysis for end-stage renal disease (ESRD) in the United States. Prior to the 1960's, ESRD was uniformly fatal early in its course. Technological advances during the 1960's, such as improved hemodialysis equipment and the development of reliable vascular access, made the widespread use of hemodialysis a feasible treatment for patients with ESRD. The most significant improvement in vascular access came with the development of the Brescia-Cimino radiocephalic arteriovenous fistula (AVF). Shortly thereafter, synthetic AV grafts were developed.

Preservation of patent, well functioning vascular access is a key component in maintaining an acceptable quality of life in patients with chronic renal failure. Tunnelled and nontunnelled dialysis catheter systems have helped greatly, but they carry with them their own set of problems such as infection risk and insufficient flow rates to provide adequate dialysis in a reasonable amount of time. Therefore, maintenance of hemodynamic access fistula (HDAF) patency and function has evolved into a multidisciplinary effort on the part of nephrologists, surgeons, and interventional radiologists. The National Kidney Foundation recently published its Dialysis Outcomes Quality Initiative (DOQI) guidelines which, among other things, describe in detail the preferred methods of obtaining and preserving vascular access for ESRD patients. This document reinforces the importance of the role of interventional radiologists in maintaining HDAF function and patency.

NATIONAL KIDNEY FOUNDATION DIALYSIS OUTCOMES QUALITY INITIATIVE (DOQI)

- Describes the preferred methods of vascular access
- Describes methods for maintaining vascular acccess
- Reinforces the importance of interventional radiology in maintaining hemodialysis vascular access

TYPES OF HEMODIALYSIS ACCESS FISTULAS (HDAF's)

- Direct arteriovenous fistulas (25%)
- Interposition grafts (75%)

TYPES OF HEMODIALYSIS INTERPOSITION GRAFTS

- Prosthetic (PTFE)
- Nonprosthetic
  - Bovine carotid artery heterograft
  - Cadaveric vein
  - Homologous saphenous vein

LOCATIONS OF SYNTHETIC INTERPOSITION GRAFTS

- Brachial artery to antecubital vein
- Radial artery to cephalic vein
- Brachial artery to axillary vein
- Femoral artery to femoral vein
- Femoral artery to saphenous vein

TYPES OF HEMODYNAMIC ACCESS FISTULAS

The direct arteriovenous access fistula has remained the most reliable means of hemodialysis access since its inception more than a quarter of a century ago. In the United States, however, most HDAF's (almost 75%) are formed by interposition grafts. There are two types of interposition grafts: prosthetic and nonprosthetic. The standard prosthetic graft material in use today is polytetrafluoroethylene (PTFE). While lengths may vary, the standard diameter of the synthetic interposition grafts is 6 mm. Nonprosthetic interposition grafts are rarely used today; however, bovine carotid artery heterografts, cadaveric vein, and homologous saphenous vein grafts have been utilized.

LOCATIONS OF HEMODYNAMIC ACCESS FISTULAS

HDAF's formed from direct AVF's can most often be found in the forearm and are usually formed by anastomosing the cephalic vein with the radial artery at wrist level or just above **(Fig. 1)**. Synthetic HDAF's are usually inserted in the upper extremity in the following configurations:

- Loop graft from brachial artery to antecubital vein **(Fig. 2)**
- Straight graft from radial artery to cephalic vein
- Curved graft from brachial artery to axillary vein **(Fig. 3)**

Synthetic groin HDAF's, while less common, have the following configurations:

- Femoral artery to femoral vein **(Fig. 4)**
- Femoral artery to saphenous vein

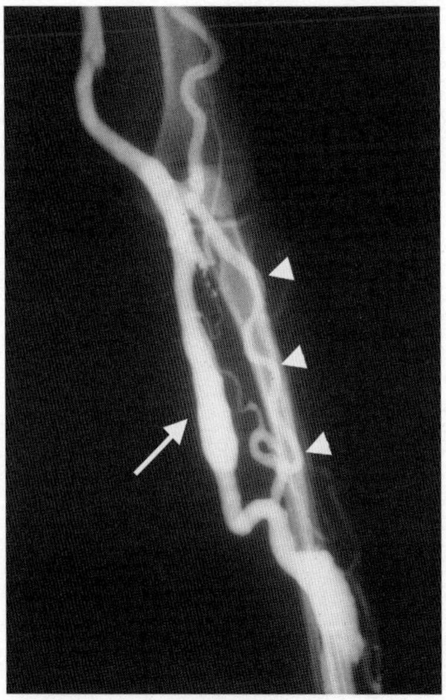

Figure 1. *Dialysis fistulogram demonstrating direct arteriovenous fistula between the cephalic vein (arrow) and radial artery (arrowheads). Note the dilated and tortuous nature of both the artery and vein, resulting from the high rate of blood flow across the fistula.*

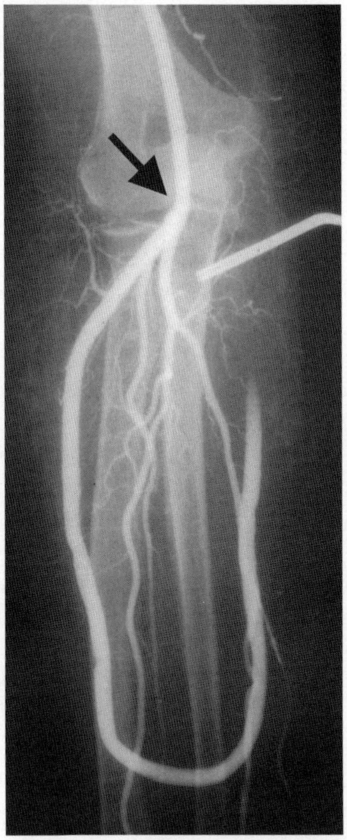

Figure 2. *Fistulogram demonstrating synthetic loop graft connecting brachial artery to the median antecubital vein. Compression of the venous end of the graft causes reflux of contrast across the patent arterial anastomosis (arrow).*

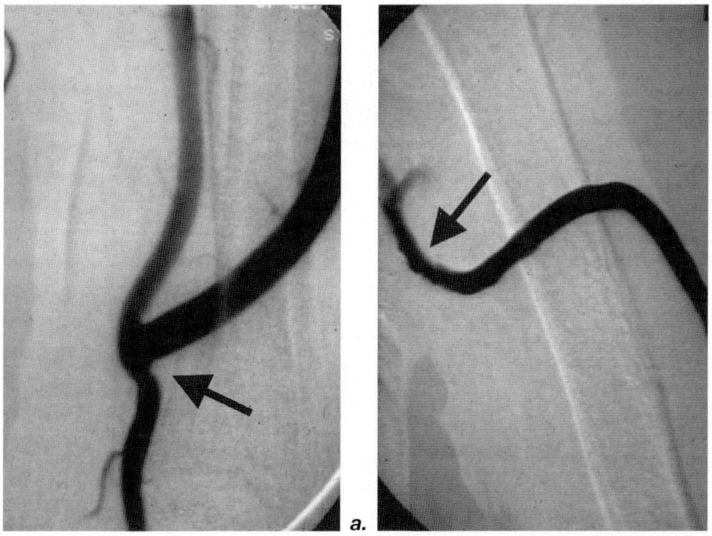

Figure 3. a) Fistulogram of a synthetic curved graft connecting the brachial artery to (b) the axillary vein. Both anastomoses are widely patent (arrows).

On occasion, when the more conventional sites of access are expended, creative means such as a transthoracic axillary artery to axillary vein interposition graft must be employed **(Fig. 5)**.

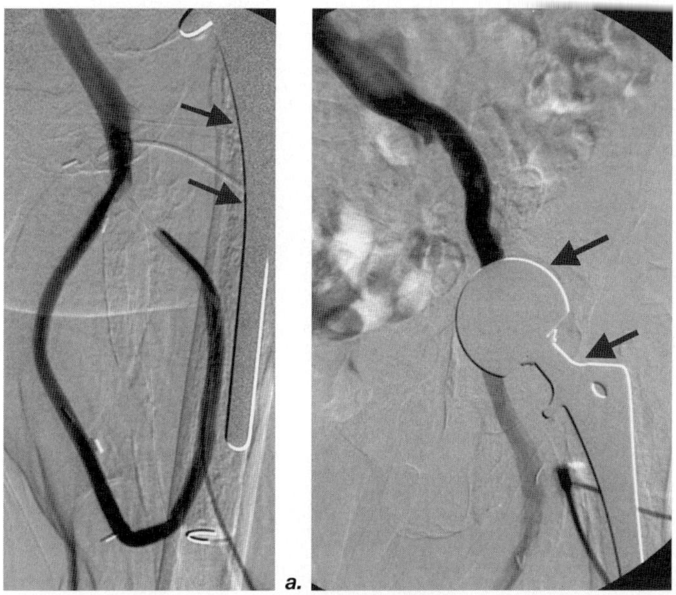

Figure 4. a) Fistulogram of a synthetic left groin loop graft and (b) the venous outflow in the pelvis. Note the subtraction artifact from the left hip prosthesis (arrows).

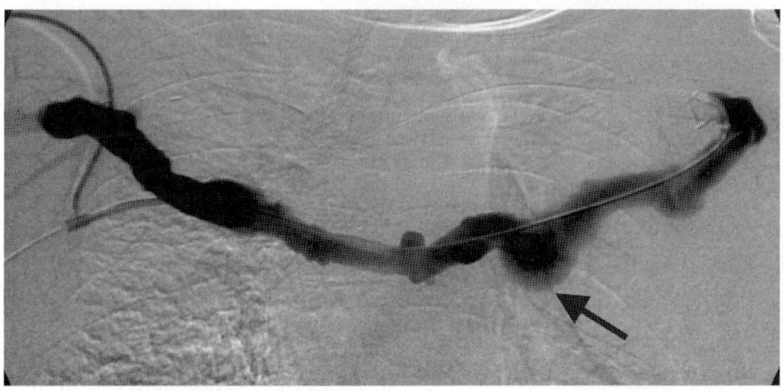

Figure 5. Fistulogram of a patent transthoracic HDAF connecting the right axillary artery to the left axillary vein. Cadaveric vein was used for the conduit, making it susceptible to venous aneurysm formation (arrow).

HDAF NONHEMODYNAMIC COMPLICATIONS

- Infection
- Pulmonary embolism
- Congestive heart failure (systemic hemodynamic complication)

HDAF HEMODYNAMIC COMPLICATIONS

- Thrombosis
- Stenosis
- Venous aneurysms and varicosities
- Graft pseudoaneurysms
- Arterial and venous steal syndromes

COMPLICATIONS OF HEMODYNAMIC ACCESS FISTULAS

HDAF complications may be of a hemodynamic or nonhemodynamic nature. The most common nonhemodynamic complication is infection, which usually requires removal of the infected graft material. Congestive heart failure (really a systemic hemodynamic complication) and clinically significant pulmonary embolism are rare complications.

HDAF hemodynamic complications include the following:

- Thrombosis **(Fig. 6)**
- Stenosis **(Fig. 7)**
- Graft pseudoaneurysms **(Fig. 8)**
- Venous aneurysms and varicosities **(Fig. 9)**
- Arterial or venous steal

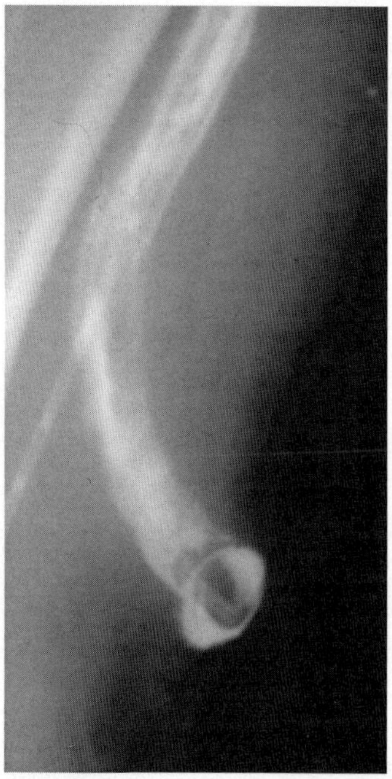

Figure 6. *Fistulogram demonstrating thrombosis of an upper extremity synthetic HDAF. Injected contrast material is insinuated in the interstices of and around the thrombus.*

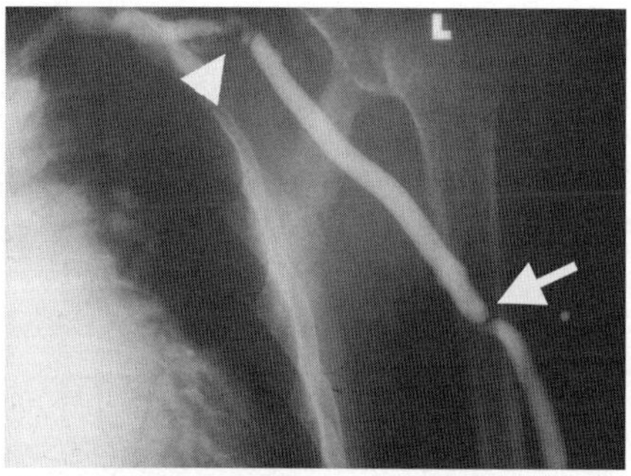

Figure 7. *Dialysis fistulogram demonstrating venous outflow stenoses in the mid (arrow) and central (arrowhead) portions of the cephalic vein.*

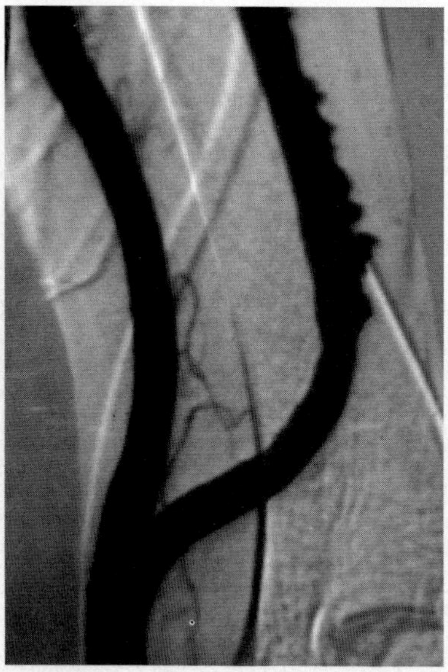

Figure 8. *Dialysis fistulogram demonstarting several small puncture site pseudoaneuryms along the synthetic graft conduit.*

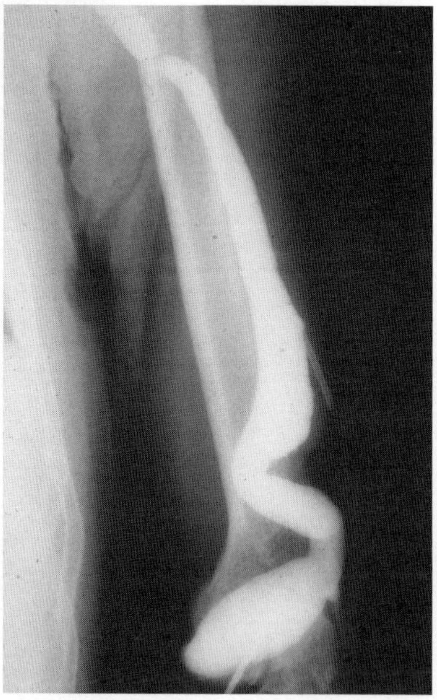

Figure 9. *Fistulogram demonstrating large aneurysmal dilatation of the outflow vein. This may be at risk for clot formation and subsequent HDAF thrombosis.*

There are multiple etiologies of HDAF failure, as summarized in **Table 1**. Regardless of the primary etiology, one of two final common pathways appear to be operative in nearly all HDAF failures: 1) intimal hyperplasia leading to stenosis; and 2) slow or stagnant blood flow leading to thrombosis.

FINAL COMMON PATHWAYS OF HDAF FAILURES

- Intimal hyperplasia causing stenosis
- Slow or stagnant blood flow causing thrombosis

TABLE 1: ETIOLOGIES OF HDAF HEMODYNAMIC COMPLICATIONS

- Excessive flow rates and turbulence (which accelerates intimal hyperplasia and causes platelet damage)
- Central venous obstruction
- Hypercoagulable states
- Graft kinking
- Repeated cannulations
- Improper puncture site compression following dialysis
- Central venous hypertension
- Dialysis-induced platelet damage (increases adherence to endothelium)

Signs of HDAF stenosis include a diminished thrill in the graft, conversion of a thrill to a pulse, venous congestion in the upper extremity, elevated venous pressures, and/or abnormal urea recirculation detected at dialysis. The vast majority of stenoses occur at the venous anastomosis or along the venous outflow, with relatively few occurring in the graft itself or at the arterial anastomosis **(Table 2)**.

Thrombosis accounts for 87% of all graft failures. It is detected simply by the lack of a palpable pulse or by the lack of blood return when the graft is punctured. While graft thrombosis is most often the result of a venous outflow stenosis, stagnant flow in a venous aneurysm and graft pseudoaneurysm may form a nidus for thrombus formation throughtout the graft as well.

Graft monitoring with early intervention has been demonstrated to result in a threefold reduction in graft thrombosis, a fourfold reduction in graft replacement, and improvement in dialysis efficiency.

BENEFITS OF HDAF MONITORING AND EARLY INTERVENTION

- Threefold reduction in graft thrombosis
- Fourfold reduction in graft replacement
- Improved efficiency of dialysis

TABLE 2: DISTRIBUTION OF HDAF STENOSES

Location	%
Venous anastomosis	47%*
Peripheral basilic vein	19%*
Within 1 cm of venous anastomosis	11%*
Central basilic vein	8%
Subclavian vein	6%
Arterial anastomosis	4%
Midgraft	2%
Axillary vein	2%

(77% of stenoses occur at or near the venous anastomosis)*

(Kanterman, 1995.)

STENOSIS TREATMENT OPTIONS

- Balloon dilatation
- Endovascular stenting
- Surgical revision

TREATMENT OF HEMODIALYSIS FISTULA COMPLICATIONS

Balloon Dilatation and Vascular Stenting

Treatment options for stenosis include surgical revision, balloon dilatation, and endovascular stenting.

The lesion is first identified on the diagnostic fistulogram and carefully traversed with a guidewire. Pressure measurements across the stenotic region may be obtained if there is any doubt of its hemodynamic significance. Balloon size is selected based on the fistulogram. Balloon venoplasty is usually performed with high-pressure balloons typically ranging in size from 6 mm to 12 mm in diameter, with a shaft size of 5-F to 7-F. One or more dilatations are performed with the goal of obliterating, or at least relieving, any hemodynamically significant stenosis **(Fig. 10)**. Balloon insertion though a vascular sheath is helpful in avoiding excessive bleeding at the puncture site. Heparinization is optional, based on the morphology of the abnormality.

Endovascular stenting is indicated in the treatment of stenoses resistant to angioplasty, treatment of venous dissection or rupture, and exclusion of pseudoaneurysms. The two types of stents in common use today are the Palmaz Stent (Johnson & Johnson/Cordis, New Brunswick, N.J.) and the Wallstent (Boston Scientific, Minneapolis, Minn.). A host of other stainless steel and nitinol stents are being investigated for intravacular use, as well as covered stents.

While there have been reports of stent placement across venous anastomotic stenoses which are resistant to balloon dilatation alone, stents are most frequently employed more centrally in the outflow circuit of the graft. The subclavian and innominate veins can be sites of stenoses in the dialysis patient population, with prior catheter placement and flow-related intimal hyperplasia being among the etiologies for stenoses at these locations. Ideally, balloon dilatation should be the treatment mode of first choice to avoid the later formation of stent-related intimal hyperplasia that could occur with primary stenting **(Figs. 11, 12)**. However, stent placement may be necessary for stenoses resistant to balloon venoplasty. In such cases, it is important to avoid stenting across the internal jugular vein orifice whenever possible, so as not to eliminate the option for future hemodialysis catheter placement through this vein.

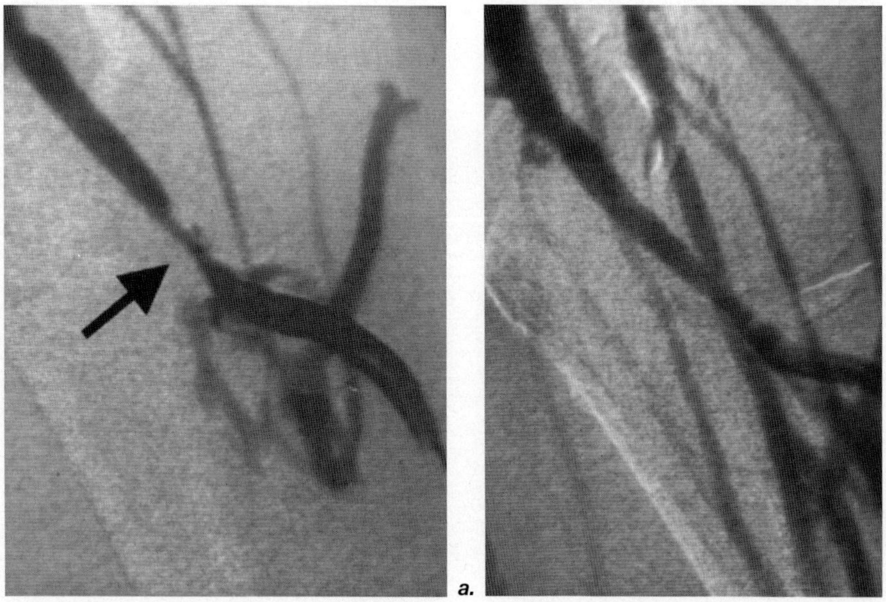

a. b.

Figure 10. *a) Dialysis fistulogram demonstrating venous anastomotic stenosis at the median antecubital vein (arrow). b) Balloon dilatation resulted in improved caliber at the anastomosis.*

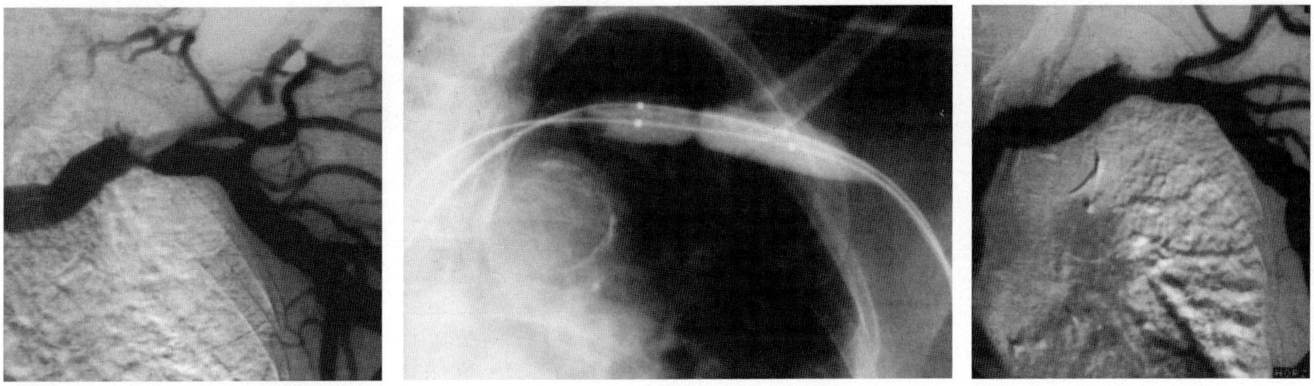

a. b. c.

Figure 11. a) Dialysis fistulogram demonstrating left subclavian vein stenosis. b) Given the large diameter of the native vein, parallel 10-mm angioplasty balloons were inserted from the left arm and the groin to facilitate venoplasty of the lesion. c) The parallel balloon dilatation resulted in improved caliber of the subclavian vein.

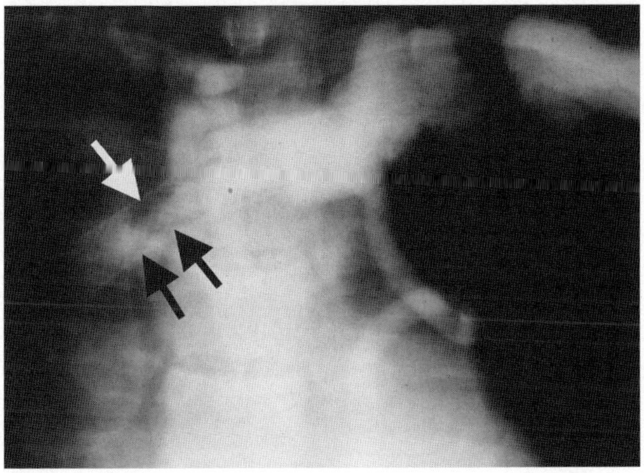

Figure 12. Dialysis fistulogram demostrating intimal hyperplasia formation in the previously stented left innominate vein (arrows).

To deploy the stent, the lesion must first be identified and completely defined on the diagnostic fistulogram. After carefully traversing the stenosis with a guidewire, the stent is inserted through a vascular sheath and deployed primarily or following initial predilatation of the lesion **(Fig. 13)**. The stent may be further enlarged with a balloon. Antibiotic prophylaxis, heparinization, postprocedural aspirin administration are additional therapeutic adjuncts that are at the discretion of the operator.

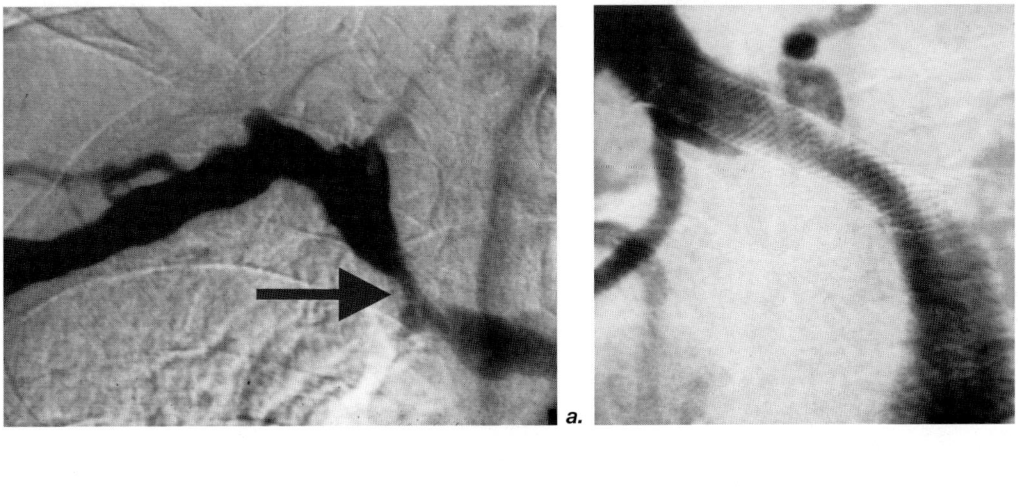

a. b.

Figure 13. a) Right innominate vein stenosis refractory to balloon dilatation (arrow). b) Placement of a 12-mm diameter Wallstent across the lesion resulted in markedly improved caliber.

HDAF THROMBOSIS TREATMENT OPTIONS

- Pharmacologic transcatheter thrombolysis
- Nonpharmacologic percutaneous thrombectomy

Transcatheter Thrombolysis

When HDAF thrombosis is suspected, the patient may either go directly to surgery for thrombectomy and graft revision or to interventional radiology for a diagnostic fistulogram. Once the presence and extent of stenosis and occlusion are determined, the patient may at that time still undergo either surgical revision or percutaneous transcatheter therapy. Transcatheter interventions include pharmacologic thrombolysis or pharmacomechanical thrombolysis along with balloon angioplasty of underlying stenoses and maceration of any residual clot. Premedication for thrombolysis varies among interventional services, but common medications include IV antibiotics (given at start of procedure), heparin (given during the procedure, systemically or mixed with the thrombolytic agent), and aspirin (postprocedure).

Historically, urokinase (UK) (Abbokinase, Abbott Laboratories, Abbott Park, Ill.) was the dominant agent used for transcatheter thrombolysis. More recently, recombinant tissue plasminogen activator (rt-PA) (Activase, Alteplase, recombinant, Genentech, Inc., South San Francisco, Calif.) has been employed for HDAF and peripheral vascular thrombolysis. Contraindications to thrombolysis, regardless of the agent, include recent cerebrovascular accident, known CNS malignancy (primary or metastatic), recent surgery, and active internal bleeding. Graft infection is also a contraindication to thrombolysis due to the risk of precipitating sepsis by the release of infected thrombus.

PHARMACOLOGIC TRANSCATHETER THROMBOLYSIS TECHNIQUES

- Pulse-spray pharmacomechanical thrombolysis
- Lyse-and-wait technique

There are two general approaches to thrombolysis of occluded HDAF's: 1) pulse-spray pharmacomechanical thrombolysis (PSPMT) and 2) the lyse-and-wait technique. PSPMT is performed through a criss-crossed multiside-hole infusion catheter placement, which allows the entire graft to be subjected to the agent. The catheters are pulsed with the thrombolytic agent with or without heparin mixed in. Typical pulse-spray protocols utilize two syringes (one for each catheter), each containing 0.5–1.0 mg of rt-PA or 125,000 units of UK, dissolved in sterile water or normal saline, for a total volume of 5 ml. This volume would include any heparin added to the mixture, which is usually between 2,500–5,000 units. Each catheter is pulsed with 0.2 to 0.5 ml boluses of the mixture until it is completely consumed. At the conclusion of the pulse-spray, the infusion catheters are exchanged for balloon catheters to sweep out the loosened thrombus and to dilate any remaining stenoses **(Fig. 14)**. A fibrinous plug is often present adjacent to the arterial anastomosis. Removal of this plug by balloon sweep or other methods is required to restore brisk graft flow **(Fig. 15)**.

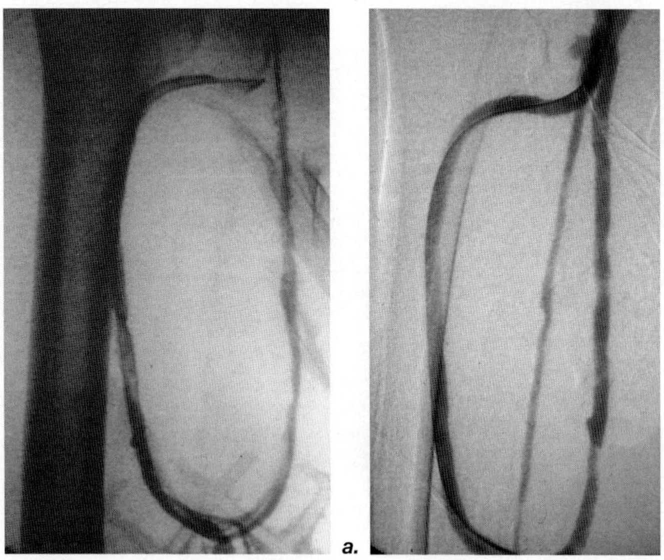

a. b.

Figure 14. *a) Dialysis fistulogram demonstrating thrombosed right groin HDAF. b) Pulse-spray pharmacomechanical thrombolysis and balloon sweeping resulted in restoration of HDAF patency.*

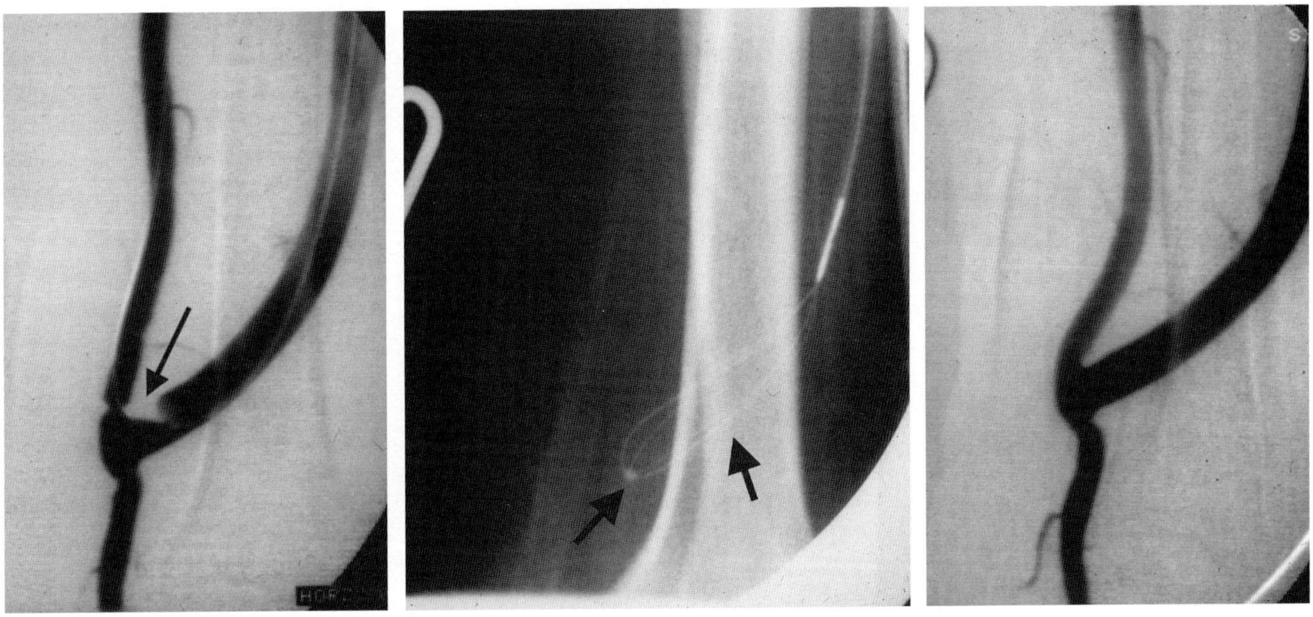

a. b. c.

Figure 15. a) Fibrinous plug (arrow) at the arterial anastomosis, refractory to balloon sweeping, in a newly recanalized HDAF. b) Careful sweeping of the plug with a stone basket catheter (arrows) resulted in (c) a widely patent arterial anastomosis.

The lyse-and-wait method involves the slow injection of a thrombolytic agent and heparin (up to 250,000 IU of UK or 1-2 mg of rt-PA, and up to 5000 U of heparin) into the graft with the arterial and venous ends compressed manually, and then waiting for 30 minutes to 2 hours for lysis to occur. After the waiting period, any residual thrombus or stenosis are again dealt with by balloon sweeping and balloon dilatation, respectively. Both the PSPMT and lyse-and-wait methods have comparable technical success rates, ranging from 75% to 98%. Patency rates invariably decline over time, with many patients requiring repeated percutaneous recanalization and eventual surgical revision or selection of a completely new implantation site.

Complications of HDAF thrombolysis include local bleeding **(Fig. 16)**, internal hemorrhage (e.g. CNS bleed), and graft infection.

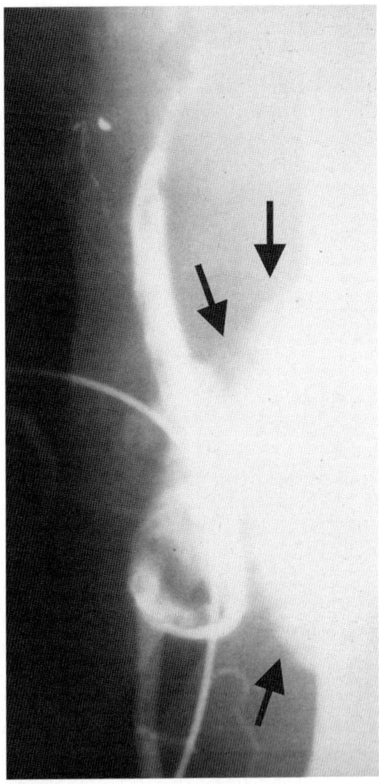

Figure 16. Large focus of extravasation from newly recanalized HDAF (arrows). The bleeding is occurring at previous puncture sites. Hemostasis was achieved with careful manual compression.

PERCUTANEOUS NONPHARMACOLOGIC THROMBECTOMY TECHNIQUES

- Balloon thrombectomy
- Pulse-spray with normal saline
- Manual thrombo-aspiration
- Specialized percutaneous thrombectomy device techniques

Percutaneous Thrombectomy

In recent years, interest has grown in the use of nonpharmacologic techniques of HDAF recanalization. These include percutaneous balloon thrombectomy, pulse-spray with normal saline, manual thrombo-aspiration, and the use of specialized percutaneous thrombectomy devices. We have the greatest experience with the Trerotola Device and the Amplatz Clot Buster.

The Trerotola Percutaneous Thrombectomy Device (Arrow International, Inc., Reading, Pa.) consists of a rotating basket catheter attached to a self-contained hand-held motor **(Fig. 17)**. The catheter is introduced through a vascular sheath and then macerates the thrombus in the graft **(Fig. 18)**, after which balloon sweeping and dilatation are performed to complete the recanalization.

The Amplatz Clot Buster (Microvena Corp., White Bear Lake, Minn.) consists of an air-driven motor which forces radially directed jets of saline through the tip of an attached flexible catheter **(Fig. 19)**. This catheter is also introduced through a vascular sheath to macerate the thrombus in the graft **(Fig. 20)**, after which balloon sweeping and dilatation are performed to complete the recanalization.

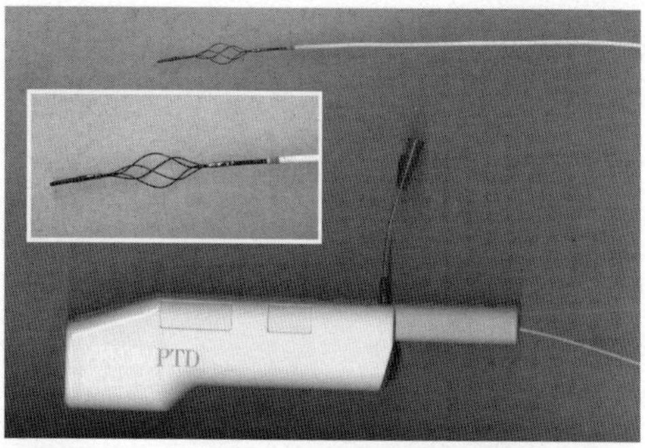

Figure 17. *The Trerotola Percutaneous Thrombectomy Device. The inset demonstrates a magnified view of the rotating basket catheter which macerates thrombus in dialysis grafts.*

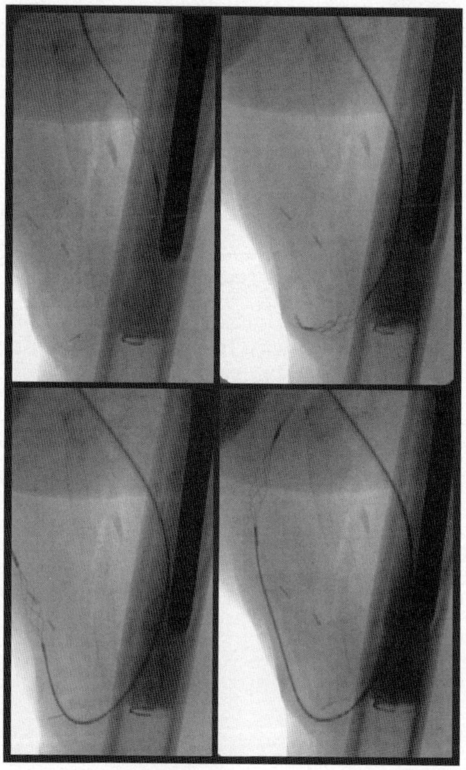

Figure 18. *The Trerotola Percutaneous Thrombectomy Device being advanced through an occluded left groin HDAF.*

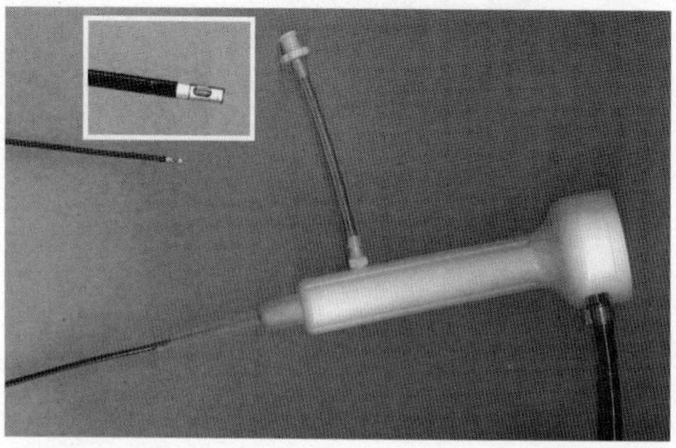

Figure 19. *The Amplatz Clot Buster. The inset demonstrates a magnified view of the catheter tip which shoots out radial streams of saline to macerate thrombus in dialysis grafts.*

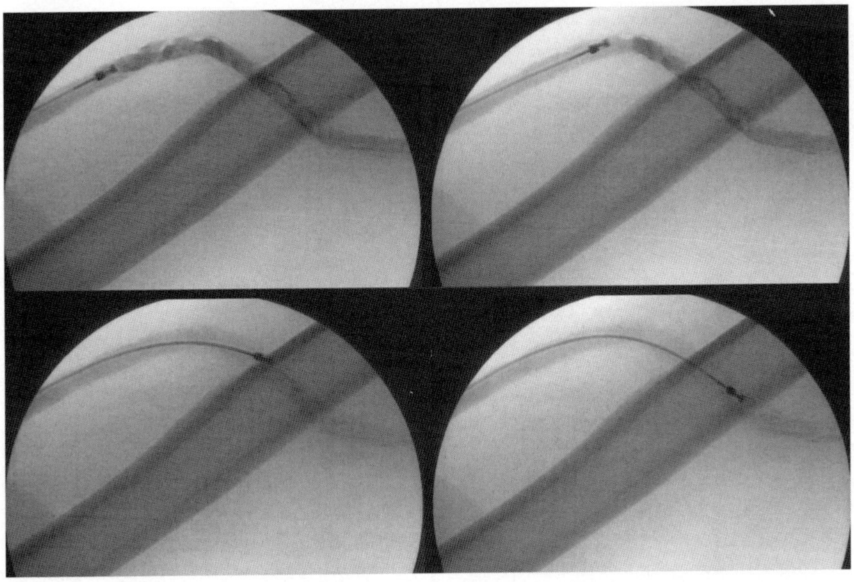

Figure 20. *The Amplatz Clot Buster being advanced through an occluded right upper extremity HDAF.*

Results of the various percutaneous thrombectomy techniques are similar and comparable to pharmacologic transcatheter thrombolysis.

Percutaneous Versus Surgical Recanalization

Percutaneous HDAF recanalization offers several advantages over surgical revision and it is the procedure of choice at UCSF. Patency can quickly be achieved on an outpatient basis and dialysis can be resumed immediately. Transcatheter therapy has also been shown to be at least as efficacious as surgical recanalization. However, regardless of the approach employed, regular monitoring and early intervention are key to maintaining HDAF longevity.

REFERENCES

Beathard GA. Mechanical versus pharmacomechanical thrombolysis for the treatment of thrombosed dialysis access grafts. Kidney Int 1994; 45:1401-1406.

Cynamon J, Lakritz PS, Wahl SI, Bakal CW, Sprayregen S. Hemodialysis graft declotting: description of the "lyse and wait" technique. J Vasc Interv Radiol 1997; 8(5):825-829.

Duszak R Jr, Sacks D. Dialysis graft declotting with very low dose urokinase: is it feasible to use "less and wait?" J Vasc Interv Radiol 1999; 10(2 Pt 1):123-128.

Gaylord G. Rapid thrombolysis of dialysis access grafts (abstr). J Vasc Interv Radiol 1991; 2:44.

Kanterman RY, Vesely TM, Pilgram TK, et al. Dialysis access grafts: anatomic location of venous stenosis and results of angioplasty. Radiology 1995; 195:135-139.

Poulain F, Raynaud A, Bourquelot P, et al. Local thrombolysis and thromboaspiration in the treatment of acutely thrombosed arteriovenous hemodialysis access fistulas. Cardiovasc Intervent Radiol 1991; 14:98-101.

Sandhu JS, Wilson MW. Use of a stone basket to treat lysis-resistant clot after pulse-spray thrombolysis of an occluded hemodialysis graft. AJR 1994; 163:957-959.

Schwartz CI, McBrayer CV, Sloan JH, et al. Thrombosed dialysis grafts: comparison of treatment with transluminal angioplasty and surgical revision. Radiology 1995; 194:337-341.

Sofocleous CT, Cooper SG, Schur I, Patel RI, Iqbal A, Walker S. Retrospective comparison of the Amplatz thrombectomy device with modified pulse-spray pharmacomechanical thrombolysis in the treatment of thrombosed hemodialysis access grafts. Radiology 1999; 213(2):561-567.

Trerotola SO, Vesely TM, Lund GB, Soulen MC, Ehrman KO, Cardella JF. Treatment of thrombosed hemodialysis access grafts: Arrow-Trerotola percutaneous thrombolytic device versus pulse-spray thrombolysis. Arrow-Trerotola Percutaneous Thrombolytic Device Clinical Trial. Radiology 1998; 206(2):403-414.

Valji K. Transcatheter treatment of thrombosed hemodialysis access grafts. AJR 1995; 164:823-829.

Valji K, Bookstein JJ, Roberts AC, et al. Pharmacomechanical thrombolysis and angioplasty in the management of clotted hemodialysis grafts: early and late clinical results. Radiology 1991; 178:243-247.

Vorwerk D, Guenther RW, Mann H, et al. Venous stenosis and occlusion in hemodialysis shunts: follow-up results of stent placement in 65 patients. Radiology 1995; 195:140-146.

CHAPTER 26 PREVIEW ANSWERS

Case 1

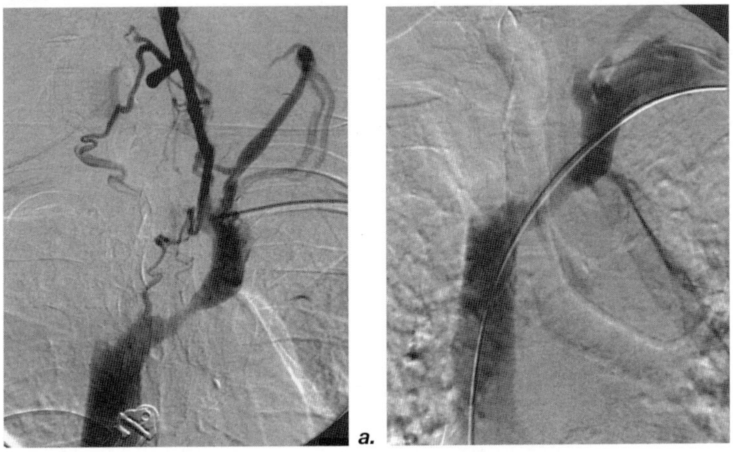

a. *b.*

Findings: a) A left innominate vein stenosis is identified. (Venous outflow stenosis, anastomotic, peripheral, or central, can all cause elevated venous pressures during hemodialysis.) If allowed to continue, HDAF flow may be compromised to the point of thrombosis. b) This central stenosis responded well to balloon venoplasty alone.

Case 2

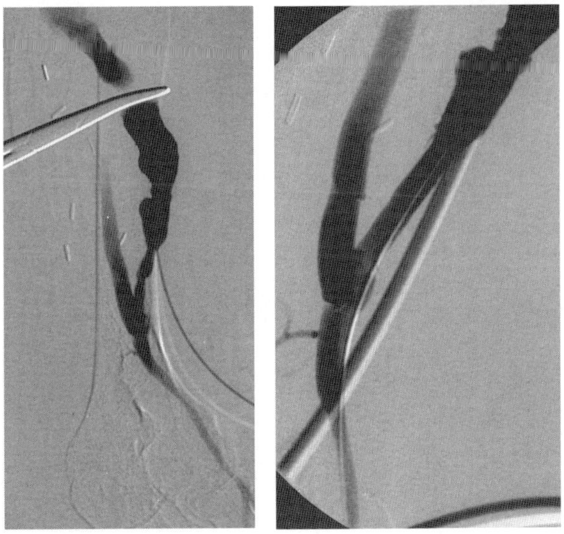

a. *b.*

Findings: Urea recirculation results from impaired flow through a hemodialysis access system. In the case of the HDAF in case figure (a), an arterial anastomotic stenosis is present. This causes impaired flow through the graft, resulting in recirculation of dialysed blood. b) Balloon dilatation at the arterial anastomosis improved the HDAF caliber and flow, correcting the recirculation problem.

Case 3

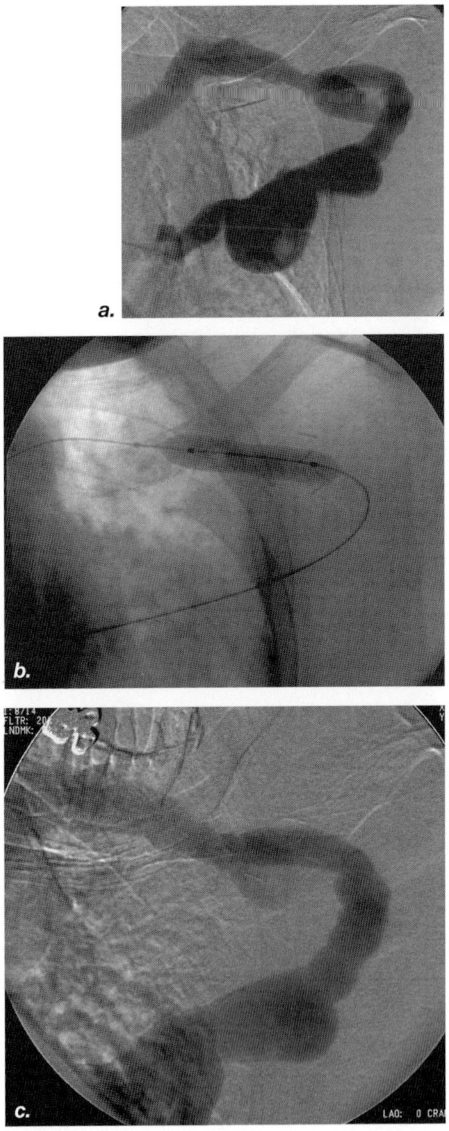

a. *b.* *c.*

Findings: a) This is a right axillary artery to left axillary vein fistula that traverses the anterior thoracic wall subcutaneously. The bulges seen in the graft are aneurysms of the cadaveric vein that was used as the graft material (aneurysm formation of this degree is rare in synthetic grafts). b) The elevated dialysis pressures can be attributed to the venous anastomotic stenosis, which was successfully balloon dilated. c) Postvenoplasty image shows improved caliber.

Subject Index

Note: Page numbers followed by f indicate figures; page numbers followed by t indicate tables.

A

B

D

E